DESIGN of CONCRETE STRUCTURES

Sixteenth Edition

David Darwin

Ph.D., P.E., Distinguished Member of ASCE
Honorary Member of ACI, Fellow of SEI
Deane E. Ackers Distinguished Professor and Chair
of Civil, Environmental & Architectural Engineering
University of Kansas

Charles W. Dolan

Ph.D., P.E., Honorary Member of ACI
Fellow of PCI
H. T. Person Professor of Engineering, Emeritus
University of Wyoming

McGraw Hill

DESIGN OF CONCRETE STRUCTURES

Published by McGraw-Hill Education, 2 Penn Plaza, New York, NY 10121. Copyright ©2021 by McGraw-Hill Education. All rights reserved. Printed in the United States of America. No part of this publication may be reproduced or distributed in any form or by any means, or stored in a database or retrieval system, without the prior written consent of McGraw-Hill Education, including, but not limited to, in any network or other electronic storage or transmission, or broadcast for distance learning.

Some ancillaries, including electronic and print components, may not be available to customers outside the United States.

This book is printed on acid-free paper.

1 2 3 4 5 6 7 8 9 LWI 24 23 22 21 20

ISBN 978-1-260-57511-8
MHID 1-260-57511-X

Cover Image: *Courtesy of Michael G. Walmsley*

About the Authors

David Darwin has been a member of the faculty at the University of Kansas since 1974, where he has served as director of the Structural Engineering and Materials Laboratory since 1982 and currently chairs the Department of Civil, Environmental, and Architectural Engineering. He was appointed the Deane E. Ackers Distinguished Professor of Civil Engineering in 1990. Dr. Darwin served as President of the American Concrete Institute (ACI) in 2007–2008 and is a member and past chair of ACI Committees 224 on Cracking and 408 on Bond and Development of Reinforcement. He is also a member of ACI Committee 318 Building Code for Concrete Structures and ACI-ASCE Committee 445 on Shear and Torsion. Dr. Darwin is an acknowledged expert on concrete crack control and bond between steel reinforcement and concrete. He received the ACI Arthur R. Anderson Award for his research efforts in plain and reinforced concrete, the ACI Structural Research Award, the ACI Joe W. Kelly Award for his contributions to teaching and design, the ACI Foundation—Concrete Research Council Arthur J. Boase Award for his research on reinforcing steel and concrete cracking, and the Concrete Research Council Robert E. Philleo Award for concrete material research and bridge construction practices. He has also received a number of awards from the American Society of Civil Engineers, including the Walter L. Huber Civil Engineering Research Prize, the Moisseiff Award, and the State-of-the-Art of Civil Engineering Award twice, the Richard R. Torrens Award, and the Dennis L. Tewksbury Award, and has been honored for his teaching by both undergraduate and graduate students at the University of Kansas. He is past editor of the ASCE *Journal of Structural Engineering*. Professor Darwin is a Distinguished Member of ASCE, an Honorary Member of ACI, and a Fellow of the Structural Engineering Institute of ASCE. He is a licensed professional engineer and serves as a consultant in the fields of concrete materials and structures. He has been honored with the Distinguished Alumnus Award from the University of Illinois Civil and Environmental Engineering Alumni Association. Between his M.S. and Ph.D. degrees, he served four years with the U.S. Army Corps of Engineers. He received the B.S. and M.S. degrees from Cornell University in 1967 and 1968 and the Ph.D. from the University of Illinois at Urbana-Champaign in 1974.

Charles W. Dolan is a consulting engineer and emeritus faculty member of the University of Wyoming. At the University of Wyoming from 1991 to 2012, he served as Department Head from 1998 to 2001 and as the first H. T. Person Chair of Engineering from 2002 to 2012, for which he received the University of Wyoming's John P. Ellbogen lifetime teaching award. A member of American Concrete Institute (ACI) Committee 318 Building Code for Concrete Structures for 17 years, he has chaired the Building Code Subcommittees on Prestressed Concrete and Code Reorganization. He has served as chair of the ACI Technical Activities Committee, ACI Committee 358 on Transit Guideways, and ACI-ASCE Committee 423 on Prestressed Concrete. A practicing engineer for over 40 years, including 20 years at Berger/ABAM, he was the project engineer on the Walt Disney World Monorail, the Detroit Downtown Peoplemover guideway, and the original Dallas–Fort Worth Airport transit system guideway. He developed the conceptual design of the Vancouver BC SkyTrain structure and the Dubai Palm Island monorail. He received the ASCE T. Y. Lin Award for outstanding contributions to the field of prestressed concrete, the ACI Arthur R. Anderson award for advancements in the design of reinforced and prestressed concrete structures, and the Prestress/Precast Concrete Institute's (PCI) Martin P. Korn award for advances in design and research in prestressed concrete. An Honorary Member of ACI and a Fellow of PCI, he is internationally recognized as a leader in the design of specialty transit structures and development of fiber-reinforced polymers for concrete reinforcement. Dr. Dolan is a registered professional engineer and lectures widely on the design and behavior of structural concrete. He received his B.S. from the University of Massachusetts in 1965 and his M.S. and Ph.D. from Cornell University in 1967 and 1989.

Contents

Preface

The sixteenth edition of *Design of Concrete Structures* continues the dual objectives of establishing a firm understanding of the behavior of structural concrete and of developing proficiency in the methods of design practice. It is generally recognized that mere training in special design skills and codified procedures is inadequate for a successful career in professional practice. As new research becomes available and new design methods are introduced, these procedures are subject to frequent changes. To understand and keep abreast of these rapid developments and to engage safely in innovative design, the engineer needs a thorough grounding in the fundamental performance of concrete and steel as structural materials and in the behavior of reinforced concrete members and structures. At the same time, the main business of the structural engineer is to design structures safely, economically, and efficiently. Consequently, with this basic understanding as a firm foundation, familiarity with current design procedures is essential. This edition, like the preceding ones, addresses both needs.

The text presents the basic mechanics of structural concrete and methods for the design of individual members subjected to bending, shear, torsion, and axial forces. It additionally addresses in detail applications of the various types of structural members and systems, including an extensive presentation of slabs, beams, columns, walls, footings, retaining walls, and the integration of building systems.

The 2019 ACI Building Code, which governs design practice in most of the United States and serves as a model code in many other countries, underwent a number of significant changes, many due to increases in the specified strengths of reinforcing steels that can used for building construction.

Changes of note include the addition of Grade 100 steel for use as principal reinforcement for gravity and lateral loads and the recognition that changes were needed in the Code, even for Grade 80 reinforcement. The use of steels with grades above 60, long the standard in U.S. practice, has led to changes in the approaches to both strength and serviceability, including the limits on both maximum and minimum reinforcement; development lengths of straight, hooked, and headed reinforcement; and requirements for the effective moment of inertia when calculating deflections. Shear design has changed through the addition of a size effect term that recognizes that shear stress at failure decreases as member depth increases. Inclusion of the size effect affects foundation walls, as well as beams and slabs—a point that is highlighted in this edition. The techniques used for two-way slab design were deleted from the 2019 ACI Building Code with the understanding that those techniques would be covered by textbooks. That information has been retained in Chapters 13, 22, and 23. Finally, the requirements for the strut-and-tie method have been updated.

In addition to changes in the ACI Code, the text also includes the modified compression field theory method of shear design presented in the 2017 edition of the American Association of State Highway and Transportation Officials (AASHTO) *LRFD Bridge Design Specifications*. Chapters on yield line and strip methods, on the McGraw-Hill Education website in the previous edition, have been returned to the printed version of the text.

A strength of the text is the analysis chapter, which includes load combinations for use in design, a description of envelope curves for moment and shear, guidelines for proportioning members under both gravity and lateral loads, and procedures for developing preliminary designs of reinforced concrete structures. The chapter also includes the ACI moment and shear coefficients.

Present-day design is performed using computer programs, either general-purpose commercially available software or individual programs written for special needs. Procedures given throughout the book guide the student and engineer through the increasingly complex methodology of design, with the emphasis on understanding the design process. Once mastered, these procedures are easily converted into flow charts to aid in preparing design aids or to validate commercial computer program output.

The text is suitable for either a one- or two-semester course in the design of concrete structures. If the curriculum permits only a single course, probably taught in the fourth undergraduate year, the following will provide a good basis: the introduction and treatment of materials found in Chapters 1 through 3; the material on flexure, shear, and anchorage in Chapters 4, 5, and 6; Chapter 7 on serviceability; Chapter 9 on short columns; the introduction to one-way slabs found in Chapter 12; and footings, Chapter 15. Time may or may not permit classroom coverage of frame analysis or building systems, Chapters 11 and 19, but these could well be assigned as independent reading, concurrent with the earlier work of the course. In the authors' experience, such complementary outside reading tends to enhance student motivation.

The text is more than adequate for a second course, most likely taught in the senior year or first year of graduate study. The authors have found that this is an excellent opportunity to provide students with a more general understanding of reinforced concrete structural design, often beginning with analysis and building systems, Chapters 11 and 19, followed by the increasingly important behavioral topics of torsion, Chapter 8; slender columns, Chapter 10; the strut-and-tie method of Chapter 17; and the design and detailing of joints, Chapter 18. It should also offer an opportunity for a much-expanded study of slabs, including Chapter 13, plus the methods for slab analysis and design based on plasticity theory found in Chapters 23 and 24, yield line analysis, and the strip method of design. Other topics appropriate to a second course include retaining walls, Chapter 16, and the introduction to earthquake-resistant design in the expanded Chapter 20. Prestressed concrete in Chapter 22 is sufficiently important to justify a separate course in conjunction with anchoring to concrete, Chapter 21, and strut-and-tie methods, Chapter 17. If time constraints do not permit this, Chapter 22 provides an introduction and can be used as the text for a one-credit-hour course.

At the end of each chapter, the user will find extensive reference lists, which provide an entry into the literature for those wishing to increase their knowledge through individual study. For professors, the Instructor's Solution Manual is available online at the McGraw-Hill Education website.

A word must be said about units. In the United States, customary inch-pound units remain prominent. Accordingly, inch-pound units are used throughout the text, although some graphs and basic data in Chapter 2 are given in dual units. Appendix B gives the SI equivalents of inch-pound units.

A brief historical note may be of interest. This book is the sixteenth edition of a textbook originated in 1923 by Leonard C. Urquhart and Charles E. O'Rourke, both professors of structural engineering at Cornell University. Over its remarkable 97-year history, new editions have kept pace with research, improved materials, and new methods of analysis and design. The second, third, and fourth editions firmly established the work as a leading text for elementary courses in the subject area. Professor George Winter, also of Cornell, collaborated with Urquhart in preparing the fifth and sixth editions. Winter and Professor Arthur Nilson were responsible for the seventh, eighth, and ninth editions, which substantially expanded both the scope and the depth of the presentation. The tenth, eleventh, and twelfth editions were prepared by Professor Nilson subsequent to Professor Winter's passing in 1982.

Professor Nilson was joined by Professor David Darwin of the University of Kansas and by Professor Charles Dolan of the University of Wyoming for the thirteenth, fourteenth, and fifteenth editions, although Professor Nilson passed away prior to completion of the fifteenth. Like Professors Winter and Nilson, the current authors have been deeply involved in research and teaching in the fields of reinforced and prestressed concrete, as well as professional Code-writing committees, and have spent significant time in professional practice, invaluable in developing the perspective and structural judgment that sets this book apart.

Special thanks are due to the McGraw-Hill Education project team, notably, Sarah Paratore, Sue Nodine, Carey Lange, and Jane Mohr.

We gladly acknowledge our indebtedness to the original authors. Although it is safe to say that neither Urquhart or O'Rourke would recognize much of the detail and that Winter would be impressed by the many changes, the approach to the subject and the educational philosophy that did so much to account for the success of the early editions would be familiar. The imprint of Arthur Nilson—our longstanding mentor, colleague, and friend—remains clear in the organization and approach taken to the material in this text.

David Darwin
Charles W. Dolan

1 *Introduction*

1.1 CONCRETE, REINFORCED CONCRETE, AND PRESTRESSED CONCRETE

Concrete is a stonelike material obtained by permitting a carefully proportioned mixture of cement, sand and gravel or other coarse aggregate, and water to harden in forms of the shape and dimensions of the desired structure. The bulk of the material consists of fine and coarse aggregate. Cement and water interact chemically to bind the aggregate particles into a solid mass. Additional water, over and above that needed for this chemical reaction, is necessary to give the mixture the workability that enables it to fill the forms and surround the embedded reinforcing steel prior to hardening. Concretes with a wide range of properties can be obtained by appropriate adjustment of the proportions of the constituent materials. Special cements (such as high early strength cements), special aggregates (such as various lightweight or heavyweight aggregates), admixtures (such as plasticizers, air-entraining agents, silica fume, and fly ash), and special curing methods (such as steam-curing) permit an even wider variety of properties to be obtained.

These properties depend to a very substantial degree on the proportions of the mixture, on the thoroughness with which the various constituents are intermixed, and on the conditions of humidity and temperature in which the mixture is maintained from the moment it is placed in the forms until it is fully hardened. The process of controlling conditions after placement is known as *curing*. To protect against the unintentional production of substandard concrete, a high degree of skillful control and supervision is necessary throughout the process, from the proportioning by weight of the individual components, through mixing and placing, until the completion of curing.

The factors that make concrete a universal building material are so pronounced that it has been used, in more primitive kinds and ways than at present, for thousands of years, starting with lime mortars from 12,000 to 6000 BCE in Crete, Cyprus, Greece, and the Middle East. The facility with which, while plastic, it can be deposited and made to fill forms or molds of almost any practical shape is one of these factors. Its high fire and weather resistance is an evident advantage. Most of the constituent materials, with the exception of cement and additives, are usually available at low cost locally or at small distances from the construction site. Its compressive strength, like that of natural stones, is high, which makes it suitable for members primarily subject to compression, such as columns and arches. On the other hand, again as in natural stones, it is a relatively brittle material whose tensile strength is low compared with its compressive strength. This prevents its economical use as the sole building material in structural members that are subject to tension either entirely (such as in tie-rods) or over part of their cross sections (such as in beams or other flexural members).

1

To offset this limitation, it was found possible, in the second half of the nineteenth century, to use steel with its high tensile strength to reinforce concrete, chiefly in those places where its low tensile strength would limit the carrying capacity of the member. The reinforcement, usually round steel rods with appropriate surface deformations to provide interlocking, is placed in the forms in advance of the concrete. When completely surrounded by the hardened concrete mass, it forms an integral part of the member. The resulting combination of two materials, known as *reinforced concrete*, combines many of the advantages of each: the relatively low cost, good weather and fire resistance, good compressive strength, and excellent formability of concrete and the high tensile strength and much greater ductility and toughness of steel. It is this combination that allows the almost unlimited range of uses and possibilities of reinforced concrete in the construction of buildings, bridges, dams, tanks, reservoirs, and a host of other structures.

It is possible to produce steels, at relatively low cost, whose yield strength is 3 to 4 times and more that of ordinary reinforcing steels. Likewise, it is possible to produce concrete 4 to 5 times as strong in compression as the more ordinary concretes. These high-strength materials offer many advantages, including smaller member cross sections, reduced dead load, and longer spans. However, there are limits to the strengths of the constituent materials beyond which certain problems arise. To be sure, the strength of such a member would increase roughly in proportion to those of the materials. However, the high strains that result from the high stresses that would otherwise be permissible would lead to large deformations and consequently large deflections of such members under ordinary loading conditions. Equally important, the large strains in such high-strength reinforcing steel would induce large cracks in the surrounding low tensile strength concrete, cracks that not only would be unsightly but also could significantly reduce the durability of the structure. This limits the useful yield strength of high-strength reinforcing steel to 100 ksi[†] according to many codes and specifications; 60 and 80 ksi steel is most commonly used.

Construction known as *prestressed concrete*, however, does use steels and concretes of very high strength in combination. The steel, in the form of wires, strands, or bars, is embedded in the concrete under high tension that is held in equilibrium by compressive stresses in the concrete after hardening. Because of this precompression, the concrete in a flexural member will crack on the tension side at a much larger load than when not so precompressed. Prestressing greatly reduces both the deflections and the tensile cracks at ordinary loads in such structures and thereby enables these high-strength materials to be used effectively. Prestressed concrete has extended, to a very significant extent, the range of spans of structural concrete and the types of structures for which it is suited.

1.2 STRUCTURAL FORMS

The figures that follow show some of the principal structural forms of reinforced concrete. Pertinent design methods for many of them are discussed later in this volume.

Floor support systems for buildings include the monolithic slab-and-beam floor shown in Fig. 1.1, the one-way joist system of Fig. 1.2, and the flat plate floor, without beams or girders, shown in Fig. 1.3. The flat slab floor of Fig. 1.4, frequently used for more heavily loaded buildings, is similar to the flat plate floor, but makes use of increased slab thickness in the vicinity of the columns, as well as flared

---

[†]Abbreviation for kips per square inch, or thousands of pounds per square inch.

FIGURE 1.1
One-way reinforced concrete floor slab with monolithic supporting beams. (*Courtesy of Portland Cement Association*)

FIGURE 1.2
One-way joist floor system, with closely spaced ribs supported by monolithic concrete beams; transverse ribs provide for lateral distribution of localized loads. (*Courtesy of Portland Cement Association*)

column tops, to reduce stresses and increase strength in the support region. The choice among these and other systems for floors and roofs depends upon functional requirements, loads, spans, and permissible member depths, as well as on cost and esthetic factors.

Where long clear spans are required for roofs, concrete shells permit use of extremely thin surfaces, often thinner, relatively, than an eggshell. The folded plate roof of Fig. 1.5 is simple to form because it is composed of flat surfaces; such roofs have been employed for spans of 200 ft and more. The cylindrical shell of Fig. 1.6 is also relatively easy to form because it has only a single curvature; it is similar to the folded plate in its structural behavior and range of spans and loads. Shells of this type were once quite popular in the United States and remain popular in other parts of the world.

Doubly curved shell surfaces may be generated by simple mathematical curves such as circular arcs, parabolas, and hyperbolas, or they may be composed of complex combinations of shapes. Hemispherical concrete domes are commonly used for storage of bulk materials. The dome shown in Fig. 1.7 is for storage of dry cement,

FIGURE 1.3
Flat plate floor slab, carried directly by columns without beams or girders. (*Courtesy of Portland Cement Association*)

FIGURE 1.4
Flat slab floor, without beams but with slab thickness increased at the columns and with flared column tops to provide for local concentration of forces. (*Courtesy of Portland Cement Association*)

and the piping around the perimeter is for the pneumatic movement of the cement. Domed structures are commonly constructed using shotcrete, a form of concrete that is sprayed onto a liner and requires formwork or backing on only one side. The dome in Fig. 1.7 was constructed by inflating a membrane, spraying insulation on the membrane, placing the reinforcement on the insulation, then spraying the concrete on

FIGURE 1.5
Folded plate roof of 125 ft span that, in addition to carrying ordinary roof loads, carries the second floor as well using a system of cable hangers; the ground floor is kept free of columns.
(*Photograph by Arthur H. Nilson*)

FIGURE 1.6
Cylindrical shell roof providing column-free interior space. (*Photograph by Arthur H. Nilson*)

both sides of the insulation to the prescribed thickness using the insulation as a backing form, as shown in Fig. 1.8. Piers and wharf facilities (shown in Fig. 1.7), silos, waters tanks, reservoirs, and other industrial facilities are commonly constructed of reinforced or prestressed concrete.

FIGURE 1.7
Hemispherical cement storage dome in New Zealand. (*Photograph courtesy of Michael Hunter, Domtec, Inc.*)

FIGURE 1.8
Shotcrete being applied to the interior of a dome structure. (*Photograph courtesy of Michael Hunter, Domtec, Inc.*)

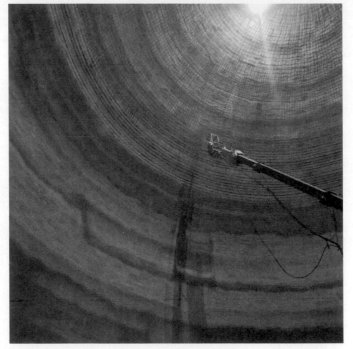

Bridge design has provided the opportunity for some of the most challenging and creative applications of structural engineering. The award-winning Napoleon Bonaparte Broward Bridge, shown in Fig. 1.9, is a six-lane, cable-stayed structure that spans St. John's River at Dame Point, Jacksonville, Florida. It has a 1300 ft center span. Figure 1.10 shows the Bennett Bay Centennial Bridge, a four-span continuous, segmentally cast-in-place box girder structure. Special attention was given to esthetics in this award-winning design. The spectacular Natchez Trace Parkway Bridge in Fig. 1.11, a two-span arch structure using hollow precast concrete elements, carries a two-lane highway 155 ft above the valley floor.

FIGURE 1.9
Napoleon Bonaparte Broward
Bridge, with a 1300 ft center
span at Dame Point,
Jacksonville, Florida.
(*HNTB Corporation, Kansas
City, Missouri*)

FIGURE 1.10
Bennett Bay Centennial
Bridge, Coeur d'Alene,
Idaho, a four-span continuous
concrete box girder structure
of length 1730 ft. (*HNTB
Corporation, Kansas City,
Missouri*)

FIGURE 1.11
Natchez Trace Parkway
Bridge near Franklin,
Tennessee, an award-winning
two-span concrete arch
structure rising 155 ft above
the valley floor. (*Designed by
Figg Bridge Group*)

FIGURE 1.12
Premier on Pine under
construction. The cover photo
is the competed building.
(*Photograph provided by Cary
Kopczynski and Company,
Structural Engineers*)

Buildings clad in glass or other fascia materials do not immediately indicate the underlying structural framing. The *Premiere on Pine* building in downtown Seattle is a case in point. The 42-story building contains condominiums, underground parking, and a hotel-type sky lounge. The 450,000 square foot flat slab cast-in-place concrete construction uses 15,000 psi concrete for columns to increase available floor space and to resist gravity and earthquake loads; see Fig. 1.12.

The structural forms shown in Figs. 1.1 to 1.12 hardly constitute a complete inventory but are illustrative of the shapes appropriate to the properties of reinforced or prestressed concrete. They illustrate the adaptability of the material to a great variety of one-dimensional (beams, girders, columns), two-dimensional (slabs, arches, rigid frames), and three-dimensional (shells, tanks) structures and structural components. This variability allows the shape of the structure to be adapted to its function in an economical manner, and furnishes the architect and design engineer with a wide variety of possibilities for esthetically satisfying structural solutions.

1.3 LOADS

Loads that act on structures can be divided into three broad categories: dead loads, live loads, and environmental loads.

Dead loads are those that are constant in magnitude and fixed in location throughout the lifetime of the structure. Usually the major part of the dead load is the weight of the structure itself. This can be calculated with good accuracy from the design configuration, dimensions of the structure, and density of the material. For buildings, floor fill, finish floors, and plastered ceilings are usually included as dead loads, and an allowance is made for suspended loads such as piping and lighting fixtures. For bridges, dead loads may include wearing surfaces, sidewalks, and curbs, and an allowance is made for piping and other suspended loads.

Live loads consist chiefly of occupancy loads in buildings and traffic loads on bridges. They may be either fully or partially in place or not present at all, and may also change in location. Their magnitude and distribution at any given time are uncertain, and even their maximum intensities throughout the lifetime of the structure are not known with precision. The minimum live loads for which the floors and roof of a building should be designed are usually specified in the building code that governs at the site of construction. Representative values of minimum live loads to be used in a wide variety of buildings are found in *Minimum Design Loads and Other Associated Criteria for Buildings and Other Structures* (Ref. 1.1), a portion of which is reprinted in Table 1.1. The table gives uniformly distributed live loads for various types of occupancies; these include impact and concentrated load provisions where necessary. These loads are expected maxima and considerably exceed average values.

In addition to these uniformly distributed loads, it is recommended that, as an alternative to the uniform load, floors be designed to support safely certain concentrated loads if these produce a greater stress. For example, according to Ref. 1.1, office floors are to be designed to carry a load of 2000 lb distributed over an area 2.5 ft square (6.25 ft^2), to allow for heavy equipment, and stair treads must safely support a 300 lb load applied on the center of the tread. Certain reductions are often permitted in live loads for members supporting large areas with the understanding that it is unlikely that the entire area would be fully loaded at one time (Refs. 1.1 and 1.2).

TABLE 1.1
Minimum uniformly distributed live loads in pounds per square foot (psf)

Occupancy or Use	Live Load, psf	Occupancy or Use	Live Load, psf
Apartments (see Residential)		Hospitals	
Access floor systems		Operating rooms, laboratories	60
Office use	50	Patient rooms	40
Computer use	100	Corridors above first floor	80
Armories and drill rooms[a]	150	Hotels (see Residential)	
Assembly areas and theaters		Libraries	
Fixed seats (fastened to floor)[a]	60	Reading rooms	60
Lobbies[a]	100	Stack rooms[a,e]	150
Movable seats[a]	100	Corridors above first floor	80
Platforms (assembly)[a]	100	Manufacturing	
Stage floors[a]	150	Light[a]	125
Balconies and decks[b]		Heavy[a]	250
Catwalks for maintenance access	40	Office buildings	
Corridors		File and computer rooms shall be designed for heavier loads based on anticipated occupancy	
First floor	100	Lobbies and first floor corridors	100
Other floors, same as occupancy served except as indicated		Offices	50
Dining rooms and restaurants[a]	100	Corridors above first floor	80
Dwellings (see Residential)		Penal institutions	
Fire escapes	100	Cell blocks	40
On single-family dwellings only	40	Corridors	100
Garages (passenger vehicles only)[a,c,d]	40	Recreational uses	
Trucks and buses[c]		Bowling alleys[a]	75

(continued)

Tabulated live loads cannot always be used. The type of occupancy should be considered and the probable loads computed as accurately as possible. Warehouses for heavy storage may be designed for loads as high as 500 psf or more; unusually heavy operations in manufacturing buildings may require an increase in the 250 psf value specified in Table 1.1; special provisions must be made for all definitely located heavy concentrated loads.

Live loads for highway bridges are specified by the American Association of State Highway and Transportation Officials (AASHTO) in its *LRFD Bridge Design Specifications* (Ref. 1.3). For railway bridges, the American Railway Engineering and Maintenance-of-Way Association (AREMA) has published the *Manual of Railway Engineering* (Ref. 1.4), which specifies traffic loads.

Environmental loads consist mainly of snow loads, wind pressure and suction, earthquake load effects (that is, inertia forces caused by earthquake motions), soil and hydraulic pressures on subsurface portions of structures, loads from possible ponding of rainwater on flat surfaces, and forces caused by temperature differentials. Like live loads, environmental loads at any given time are uncertain in both magnitude and distribution. Reference 1.1 contains much information on environmental loads, which is often modified locally depending, for instance, on local climatic or seismic conditions.

TABLE 1.1
(Continued)

Occupancy or Use	Live Load, psf	Occupancy or Use	Live Load, psf
Dances halls	100	Schools	
Gymnasiums[a]	100	Classrooms	40
Residential		Corridors above first floor	80
One- and two-family dwellings		First floor corridors	100
Uninhabitable attics without storage[f]	10	Sidewalks, vehicular driveways, and yards, subject to trucking[a,j]	250
Uninhabitable attics with storage[g]	20	Stairs and exit-ways	100
Habitable attics and sleeping areas	30	One- and two-family residences only	40
All other areas except stairs	40	Storage areas above ceilings	20
All other residential occupancies, hotels, and multifamily houses		Storage warehouses (shall be designed for heavier loads if required for anticipated storage)	
Private rooms and corridors serving them	40	Light[a]	125
Public rooms and corridors serving them	100	Heavy[a]	250
Roofs		Stores	
Ordinary flat, pitched, and curved roofs[h]	20	Retail	
Roofs used for roof gardens	100	First floor	100
Roofs used for assembly purposes		Upper floors	73
Roofs used for other occupancies[i]		Wholesale, all floors[a]	125
Awnings and canopies		Walkways and elevated platforms (other than exit-ways)	60
Fabric construction supported by a lightweight rigid skeleton structure	5	Yards and terraces, pedestrians	100
All other construction	20		

[a] Live load reduction for this use is not permitted unless specific exceptions apply.

[b] 1.5 times live load for area served. Not required to exceed 100 psf.

[c] Floors of garages or portions of a building used for storage of motor vehicles shall be designed for the uniformly distributed live loads of this table or for concentrated loads specified in Ref. 1.1.

[d] Design for trucks and buses shall be in accordance with Ref. 1.3; however, provisions for fatigue and dynamic load are not required.

[e] The loading applies to stack room floors that support nonmobile, double-faced library book stacks subject to the following limitations: (1) The nominal book stack unit height shall not exceed 90 in.; (2) the nominal shelf depth shall not exceed 12 in. for each face; and (3) parallel rows of double-faced book stacks shall be separated by aisles not less than 36 in. wide.

[f] See Ref. 1.1 for description of uninhabitable attic areas without storage. This live load need not be assumed to act concurrently with any other live load requirement.

[g] See Ref. 1.1 for description of uninhabitable attic areas with storage and where this provision applies.

[h] Where uniform roof live loads are reduced to less than 20 psf in accordance with Section 4.8.2 of Ref. 1.1 and are applied to the design of structural members arranged so as to create continuity, the reduced roof live load shall be applied to adjacent spans or to alternate spans, whichever produces the greatest unfavorable load effect.

[i] Roofs used for other special purposes shall be designed for appropriate loads as approved by the authority having jurisdiction.

[j] Other uniform loads in accordance with an approved method that contains provisions for truck loadings shall also be considered where appropriate.

Data Source: *Minimum Design Loads and Other Associated Criteria for Buildings and Other Structures* (ASCE/SEI 7-16). American Society of Civil Engineers, Reston, VA, 2010.

Figure 1.13, from the 1972 edition of Ref. 1.1, gives snow loads for the continental United States and is included here for illustration only. The 2016 edition of Ref. 1.1 gives much more detailed information. In either case, specified values represent not average values, but expected upper limits. A minimum roof load of 20 psf is often specified to provide for construction and repair loads and to ensure reasonable stiffness.

FIGURE 1.13

Snow load in pounds per square foot (psf) on the ground, 50-year mean recurrence interval. (*Minimum Design Loads for Buildings and Other Structures, ANSI A58.1–1972, American National Standards Institute, New York, 1972.*)

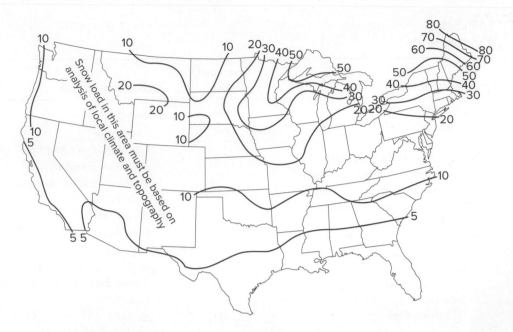

Much progress has been made in developing rational methods for predicting horizontal forces on structures due to wind and seismic action. Reference 1.1 summarizes current thinking regarding wind forces and earthquake loads. Reference 1.5 presents detailed recommendations for lateral forces from earthquakes.

Reference 1.1 specifies design wind pressures per square foot of vertical wall surface. Depending upon locality, these equivalent static forces vary from about 10 to 50 psf. Factors include basic wind speed, exposure (urban vs. open terrain, for example), height of the structure, the importance of the structure (that is, consequences of failure), and gust effect factors to account for the fluctuating nature of the wind and its interaction with the structure.

Seismic forces may be found for a particular structure by elastic or inelastic dynamic analysis, considering expected ground accelerations and the mass, stiffness, and damping characteristics of the construction. In less seismically active areas, the design is often based on equivalent static forces calculated from provisions such as those of Refs. 1.1 and 1.5. The base shear is found by considering such factors as location, type of structure and its occupancy, total dead load, and the particular soil condition. The total lateral force is distributed to floors over the entire height of the structure in such a way as to approximate the distribution of forces obtained from a dynamic analysis.

1.4 SERVICEABILITY, STRENGTH, AND STRUCTURAL SAFETY

To serve its purpose, a structure must be safe against collapse and serviceable in use. Serviceability requires that deflections be adequately small; that cracks, if any, be kept to tolerable limits; and that vibrations be minimized. Safety requires that the strength of the structure be adequate for all loads that may foreseeably act on it. If the strength of a structure, built as designed, could be predicted accurately, and if the loads and their internal effects (bending moments, shears, axial forces, and torsional moments) were known accurately, safety could be ensured by providing a carrying capacity just barely in excess of the known loads. However, there are a number of sources of uncertainty in the analysis, design, and construction of reinforced concrete structures. These sources of uncertainty, which require a definite margin of safety, may be listed as follows:

1. Actual loads may differ from those assumed.
2. Actual loads may be distributed in a manner different from that assumed.
3. The assumptions and simplifications inherent in any analysis may result in calculated load effects—moments, shears, etc.—different from those that, in fact, act in the structure.
4. The actual structural behavior may differ from that assumed, owing to imperfect knowledge.
5. Actual member dimensions may differ from those specified.
6. Reinforcement may not be in its proper position.
7. Actual material strength may be different from that specified.

In the establishment of safety requirements, consideration must be given to the consequences of failure. In some cases, a failure would be merely an inconvenience. In other cases, loss of life and significant loss of property may be involved. A further consideration should be the nature of the failure, should it occur. A gradual failure with ample warning permitting remedial measures is preferable to a sudden, unexpected collapse.

It is evident that the selection of an appropriate margin of safety is not a simple matter. However, progress has been made toward rational safety provisions in design codes (Refs. 1.6 to 1.11).

a. Variability of Loads

Since the maximum load that occurs during the life of a structure is uncertain, it can be considered a random variable. In spite of this uncertainty, the engineer must provide an adequate structure. A probability model for the maximum load can be devised by means of a probability density function for loads (Ref. 1.8), as represented by the frequency curve of Fig. 1.14a. The exact form of this distribution curve, for a particular type of loading such as office loads, can be determined only on the basis of statistical data obtained from large-scale load surveys. A number of such surveys have been completed. For types of loads for which such data are scarce, fairly reliable information can be obtained from experience, observation, and judgment.

For such a frequency curve (Fig. 1.14a), the area under the curve between two abscissas, such as loads Q_1 and Q_2, represents the probability of occurrence of loads Q of magnitude $Q_1 < Q < Q_2$. A specified service load Q_d for design is selected conservatively in the upper region of Q in the distribution curve, as shown. The probability of occurrence of loads larger than Q_d is then given by the shaded area to the right of Q_d. It is seen that this specified service load is considerably larger than the mean load Q acting on the structure. This mean load is much more typical of average load conditions than the design load Q_d.

b. Strength

The strength of a structure depends on the strength of the materials from which it is made. For this purpose, minimum material strengths are specified in standardized ways. Actual material strengths cannot be known precisely and therefore also constitute random variables (see Section 2.6). Structural strength depends, furthermore, on the care with which a structure is built, which in turn reflects the quality of supervision

FIGURE 1.14
Frequency curves for
(*a*) loads *Q*, (*b*) strengths *S*,
and (*c*) safety margin *M*.

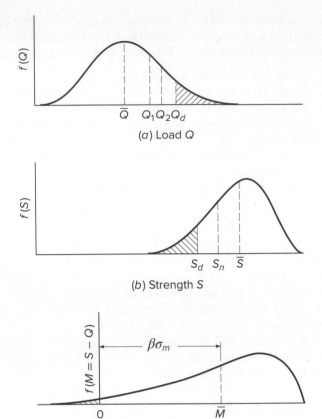

(*a*) Load *Q*

(*b*) Strength *S*

(*c*) Safety margin *M = S − Q*

and inspection. Member sizes may differ from specified dimensions, reinforcement may be out of position, poorly placed concrete may show voids, etc.

The strength of the entire structure or of a population of repetitive structures, such as highway overpasses, can also be considered a random variable with a probability density function of the type shown in Fig. 1.14*b*. As in the case of loads, the exact form of this function cannot be known but can be approximated from known data, such as statistics of actual, measured materials and member strengths and similar information. Considerable information of this type has been, or is being, developed and used.

c.   Structural Safety

A given structure has a *safety margin M* if

$$M = S - Q > 0 \tag{1.1}$$

that is, if the strength of the structure is larger than the load acting on it. Since *S* and *Q* are random variables, the safety margin $M = S - Q$ is also a random variable. A plot of the probability function of *M* may appear as in Fig. 1.14*c*. Failure occurs when *M* is less than zero. Thus, the probability of failure is represented by the shaded area in the figure.

Even though the precise form of the probability density functions for *S* and *Q*, and therefore for *M*, is not known, much can be achieved in the way of a rational approach to structural safety. One such approach is to require that the mean safety margin M be a specified number β of standard deviations σ_m above zero. It can be demonstrated that this results in the requirement that

$$\psi_s \overline{S} \geq \psi_L \overline{Q} \qquad (1.2)$$

where ψ_s is a partial safety coefficient smaller than 1.0 applied to the mean strength $\overline{S}$ and ψ_L is a partial safety coefficient larger than 1.0 applied to the mean load $\overline{Q}$. The magnitude of each partial safety coefficient depends on the variance of the quantity to which it applies, S or Q, and on the chosen value of β, the reliability index of the structure. As a general guide, a value of the safety index β between 3 and 4 corresponds to a probability of failure of the order of 1:100,000 (Ref. 1.9). The value of β is often established by calibration against well-proved and established designs.

In practice, it is more convenient to introduce partial safety coefficients with respect to code-specified loads that considerably exceed average values, rather than with respect to mean loads as in Eq. (1.2); similarly, the partial safety coefficient for strength is applied to *nominal strength*[†] generally computed somewhat conservatively, rather than to mean strengths as in Eq. (1.2). A restatement of the safety requirement in these terms is

$$\phi S_n \geq \gamma Q_d \qquad (1.3a)$$

in which ϕ is a *strength reduction factor* applied to nominal strength S_n and γ is a *load factor* applied to calculated or code-specified design loads Q_d. Furthermore, recognizing the differences in variability between, say, dead loads D and live loads L, it is both reasonable and easy to introduce different load factors for different types of loads. The preceding equation can thus be written

$$\phi S_n \geq \gamma_d D + \gamma_l L \qquad (1.3b)$$

in which γ_d is a load factor somewhat greater than 1.0 applied to the calculated dead load D and γ_l is a larger load factor applied to the code-specified live load L. When additional loads, such as the wind load W, are to be considered, the reduced probability that maximum dead, live, and wind or other loads will act simultaneously can be incorporated by using modified load factors such that

$$\phi S_n \geq \gamma_{d_i} D + \gamma_{l_i} L + \gamma_{w_i} W + \cdots \qquad (1.3c)$$

Present U.S. design codes follow the format of Eqs. (1.3b) and (1.3c).

1.5 DESIGN BASIS

The single most important characteristic of any structural member is its actual strength, which must be large enough to resist, with some margin to spare, all foreseeable loads that may act on it during the life of the structure, without failure or other distress. It is logical, therefore, to proportion members, that is, to select concrete dimensions and reinforcement, so that member strengths are adequate to resist forces resulting from certain hypothetical overload stages, significantly above loads expected actually to occur in service. This design concept is known as *strength design*.

[†] Throughout this book quantities that refer to the strength of members, calculated by accepted analysis methods, are furnished with the subscript n, which stands for "nominal." This notation is in agreement with the ACI Code. It is intended to convey that the actual strength of any member is bound to deviate to some extent from its calculated, nominal value because of inevitable variations of dimensions, materials properties, and other parameters. Design in all cases is based on this nominal strength, which represents the best available estimate of the actual member strength.

For reinforced concrete structures at loads close to and at failure, one or both of the materials, concrete and steel, are invariably in their nonlinear inelastic range. That is, concrete in a structural member reaches its maximum strength and subsequent fracture at stresses and strains far beyond the initial elastic range in which stresses and strains are fairly proportional. Similarly, steel close to and at failure of the member is usually stressed beyond its elastic domain into and even beyond the yield region. Consequently, the nominal strength of a member must be calculated on the basis of this inelastic behavior of the materials.

A member designed by the strength method must also perform in a satisfactory way under normal service loading. For example, beam deflections must be limited to acceptable values, and the number and width of flexural cracks at service loads must be controlled. Serviceability limit conditions are an important part of the total design, although attention is focused initially on strength.

Historically, members were proportioned so that stresses in the steel and concrete resulting from normal service loads were within specified limits. These limits, known as *allowable stresses*, were only fractions of the failure stresses of the materials. For members proportioned on such a service load basis, the margin of safety was provided by stipulating allowable stresses under service loads that were appropriately small fractions of the compressive concrete strength and the steel yield stress. We now refer to this basis for design as *service load design*. Allowable stresses, in practice, were set at about one-half the concrete compressive strength and one-half the yield stress of the steel.

Because of the difference in realism and reliability, the strength design method has displaced the older service load design method. However, the older method provides the basis for some serviceability checks and is the design basis for many older structures. Throughout this text, strength design is presented almost exclusively.

1.6   DESIGN CODES AND SPECIFICATIONS

The design of concrete structures such as those of Figs. 1.1 to 1.12 is generally done within the framework of codes giving specific requirements for materials, structural analysis, member proportioning, etc. The International Building Code (Ref. 1.2) is an example of a consensus code governing structural design and is often adopted by local municipalities. The responsibility of preparing material-specific portions of the codes rests with various professional groups, trade associations, and technical institutes. In contrast with many other industrialized nations, the United States does not have an official, government-sanctioned, national code.

The American Concrete Institute (ACI) has long been a leader in such efforts. As one part of its activity, the American Concrete Institute has published the widely recognized *Building Code Requirements for Structural Concrete and Commentary* (Ref. 1.12), which serves as a guide in the design and construction of reinforced concrete buildings. The ACI Code has no official status in itself. However, it is generally regarded as an authoritative statement of current good practice in the field of reinforced concrete. As a result, it has been incorporated by reference into the International Building Code and similar codes that are, in turn, adopted by law into municipal and regional building codes that do have legal status. Its provisions thereby attain, in effect, legal standing. Most reinforced concrete buildings and related construction in the United States are designed in accordance with the current ACI Code. It has also served as a model document for many other countries. The commentary incorporated in Ref. 1.12 provides background material and rationale for the Code provisions. The

American Concrete Institute also publishes important journals and standards, as well as recommendations for the analysis and design of special types of concrete structures such as shown in Fig. 1.7.

Most highway bridges in the United States are designed according to the requirements of the AASHTO bridge specifications (Ref. 1.3), which not only contain the provisions relating to loads and load distributions mentioned earlier but also include detailed provisions for the design and construction of concrete bridges. Some of the provisions follow ACI Code provisions closely, although a number of significant differences will be found.

The design of railway bridges is done according to the specifications of the AREMA *Manual of Railway Engineering* (Ref. 1.4). It, too, is patterned after the ACI Code in most respects, but it contains much additional material pertaining to railway structures of all types.

No code or design specification can be construed as a substitute for sound engineering judgment in the design of concrete structures. In structural practice, circumstances are frequently encountered where code provisions can serve only as a guide, and the engineer must rely upon a firm understanding of the basic principles of structural mechanics applied to reinforced or prestressed concrete, and an intimate knowledge of the nature of the materials.

1.7   SAFETY PROVISIONS OF THE ACI CODE

The safety provisions of the ACI Code are given in the form of Eqs. (1.3b) and (1.3c) using strength reduction factors and load factors. These factors are based on statistical information, experience, engineering judgment, and compromise. In words, the design strength ϕS_n of a structure or member must be at least equal to the required strength U calculated from the factored loads, that is,

Design strength ≥ Required strength

or

$$\phi S_n \geq U \tag{1.4}$$

The nominal strength S_n is computed (usually somewhat conservatively) by accepted methods. The required strength U is calculated by applying appropriate load factors to the respective service loads: dead load D; live load L; wind load W; earthquake load E; snow load S; rain load R; cumulative effects T due to differential settlement and restrained volume change due to creep, shrinkage, and temperature change; fluid pressure F; and earth pressure H. Loads are defined in a general sense, to include either loads or the related internal effects such as moments, shears, and thrusts. Thus, in specific terms for a member subjected, say, to moment, shear, axial load, and torsional moment

$$\phi M_n \geq M_u \tag{1.5a}$$

$$\phi V_n \geq V_u \tag{1.5b}$$

$$\phi P_n \geq P_u \tag{1.5c}$$

$$\phi T_n \geq T_u \tag{1.5d}$$

where the subscripts n denote the nominal strengths in flexure, shear, and axial load, respectively, and the subscripts u denote the factored load moment, shear, axial

TABLE 1.2
Factored load combinations for determining required strength U in the ACI Code

Primary Load[a]	Factored Load or Load Effect U
Basic[b]	$U = 1.2D + 1.6L$
Dead	$U = 1.4D$
Live	$U = 1.2D + 1.6L + 0.5(L_r \text{ or } S \text{ or } R)$
Roof, snow, rain[c]	$U = 1.2D + 1.6(L_r \text{ or } S \text{ or } R) + 0.5(1.0L \text{ or } 0.5W)$
Wind[c,d]	$U = 1.2D + 1.0W + 1.0L + 0.5(L_r \text{ or } S \text{ or } R)$
	$U = 0.9D + 1.0W$
Earthquake[c,e]	$U = 1.2D + 1.0E + 1.0L + 0.2S$
	$U = 0.9D + 1.0E$

[a] Where the following represent the loads or related internal moments or forces resulting from the listed factors: D = dead load; E = earthquake; L = live load; L_r = roof live load; R = rain; S = snow; and W = wind. In addition to the loads shown in this table, the ACI Code also requires consideration of loads due to F = fluids; H = earth pressure; and T = cumulative effects of differential settlement and restraint of volume change (creep, shrinkage, temperature change).

[b] The "basic" load condition of $U = 1.2D + 1.6L$ reflects the fact that interior members in buildings generally are not subjected to L_r or S or R and that $1.4D$ rarely governs design.

[c] The load factor on live load L in these load combinations may be reduced up to 0.5, except for garages, areas occupied as places of public assembly, and areas where L is greater than 100 psf.

[d] Versions of ASCE/SEI 7 before 2010 provided wind speeds based on service-level design. If service-level winds are used, $1.6W$ should be used for strength design.

[e] The vertical effects of earthquake are additive to the the dead load effects.

load and torsion. In computing the factored load effects on the right, load factors may be applied either to the service loads themselves or to the internal load effects calculated from the service loads.

The load factors specified in the ACI Code, to be applied to calculated dead loads and those live and environmental loads specified in the appropriate codes or standards, are summarized in Table 1.2. A maximum load factor of 1.0 is used for wind load W and earthquake load E because these loads are expressed at strength level in Ref. 1.1. In addition to the load combinations shown in Table 1.2, Chapter 5 of the ACI Code addresses how load effects due to differential settlement, creep, shrinkage, temperature change, fluid pressure, and earth pressure should be handled depending on the load combination and whether they add or counteract the effects of the primary load. The load combinations in Table 1.2 are consistent with the concepts introduced in Section 1.4 and with ASCE/SEI 7, *Minimum Design Loads and Other Associated Criteria for Buildings and Other Structures* (Ref. 1.1). For individual loads, lower factors are used for loads known with greater certainty, such as dead load, compared with loads of greater variability, such as live loads. Further, for load combinations such as dead plus live loads plus wind forces, reductions are applied to one load or the other that reflect the improbability that an excessively large live load coincides with an unusually high windstorm. The factors also reflect, in a general way, uncertainties with which internal load effects are calculated from external loads in systems as complex as highly indeterminate, inelastic reinforced concrete structures which, in addition, consist of variable-section members (because of tension cracking, discontinuous reinforcement, etc.). Finally, the load factors also distinguish between two situations, particularly when horizontal forces are present

TABLE 1.3
Strength reduction factors in the ACI Code

Strength Condition	Strength Reduction Factor ϕ
Tension-controlled sections[a]	0.90
Compression-controlled sections[b]	
Members with spiral reinforcement	0.75
Other reinforced members	0.65
Shear and torsion	0.75
Bearing on concrete	0.65
Post-tensioned anchorage zones	0.85
Strut-and-tie models[c]	0.75

[a] Chapter 22 discusses reductions in ϕ for pretensioned members where strand embedment is less than the development length.

[b] Chapter 4 contains a discussion of the linear variation of ϕ between tension and compression-controlled sections. Chapter 9 discusses the conditions that allow an increase in ϕ for spirally reinforced columns.

[c] Chapter 17 describes strut-and-tie models.

in addition to gravity, that is, the situation where the effects of all simultaneous loads are additive, as distinct from that in which various load effects counteract one another. For example, wind load produces an overturning moment, and the gravity forces produce a counteracting stabilizing moment.

In all cases in Table 1.2, the controlling equation is the one that gives the largest factored load effect U.

The strength reduction factors ϕ in the ACI Code are given different values depending on the state of knowledge, that is, the accuracy with which various strengths can be calculated. Thus, the value for bending is higher than that for shear or bearing. Also, ϕ values reflect the probable importance, for the survival of the structure, of the particular member and of the probable quality control achievable. For both these reasons, a lower value is used for columns than for beams. Table 1.3 gives some of the ϕ values specified in Chapter 21 of the ACI Code.

The joint application of strength reduction factors (Table 1.3) and load factors (Table 1.2) is aimed at producing approximate probabilities of understrength of the order of 1/100 and of overloads of 1/1000. This results in a probability of structural failure on the order of 1/100,000.

1.8   DEVELOPING FACTORED GRAVITY LOADS

To be of use in design, the live loads in Table 1.1 and information on the self-weight of the structural members and other dead loads must be converted into forces acting on the structure. By way of several examples, this section describes the conventions by which this is done for gravity loads.

Figure 1.15 shows a hospital building with a reinforced concrete frame. The masonry fascia and steel entrance give little indication of the underlying structure. Figure 1.16 shows the same building under construction. The slabs, beams, columns, and stairwells are identified. Temporary formwork and shoring for the cast-in-place

FIGURE 1.15
Hospital building
(*Photograph by Charles W. Dolan*)

Area detailed in Figure 1.16

FIGURE 1.16
Details of framing system
(*Photograph by Charles W. Dolan*)

Formwork and temporary construction loads

Slab and beams

Columns

Walls and stairwells (for lateral support)

Suspended ceiling and utilities

framing system are visible in the upper stories. This structure is designed for wind and gravity loads because wind loads exceed the earthquake effects at this location. The stairwell walls provide lateral stability and resistance to wind load. The remaining structural elements are designed for gravity loads.

Figure 1.17 shows a schematic floor plan of the building and a photograph of the one-way joist floor system. The slab is 5 in. thick, and the joists (narrow beams not shown in the floor plan) are 6 in. wide, 24 in. deep, spaced at 5 ft, and run in the East–West direction between supporting girders that run North–South between columns. The bays adjacent to building line C are selected to illustrate the development of factored gravity loads to be used in design. Operating rooms are located in this portion of the building. Preliminary sizing of a typical floor indicates that the girder cross section will be 24 in. deep by 16 in. wide. In addition to the live load, the floor supports a suspended ceiling and duct work below weighing 6.5 psf. Normalweight concrete, producing reinforced concrete with a unit weight of 150 pcf, is used for construction.

FIGURE 1.17
(*a*) Building floor plan and
(*b*) joist floor system
(*Photograph by Charles
W. Dolan*)

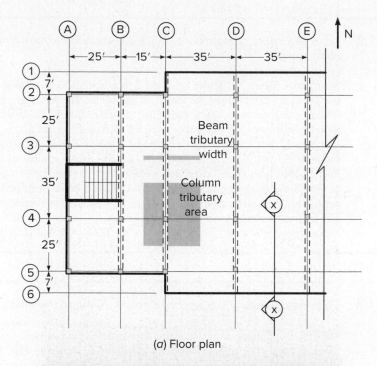

(*a*) Floor plan

(*b*) Section X-X joist floor system

EXAMPLE 1.1   **Loads on slab/joist system.**   Determine the service and factored loads acting on the slab/joist system between lines C and D.

Solution.   Slab loads are typically defined as surface loads in pounds per square foot (psf). From Table 1.1, the live load for hospital operating rooms is 60 psf. The slab is 5 in. thick. The joists are 6 in. wide by 24 in. deep, extending 19 in. below the bottom of the slab and spaced 5 ft on center. The joists can be considered as adding to the thickness of the slab for the purpose of calculating the dead load of the system. The equivalent increase in slab thickness equals the cross-sectional area of the joist below the bottom of the slab divided by the spacing of the joists in in. or $(6 \times 19)/(5 \times 12) = 1.9$ in., giving an equivalent total slab thickness for calculation of dead load of 6.9 in. The dead load of the slab/joist system is then the equivalent total slab thickness in feet times the concrete density, 150 pcf, resulting in a slab dead weight of $(6.9/12) \times 150$ pcf $= 86.3$ psf. The service load q_s on the slab is then 86.3 psf $+ 6.5$ psf superimposed dead load $+ 60$ psf live load, giving $q_s = 152.8$ psf. The factored load for this example is determined using the basic load factor condition from Table 1.2. Thus, the factored load on the slab q_u is $1.2 \times (86.3$ psf $+ 6.5$ psf$) + 1.6 \times 60$ psf $= 207.4$ psf, which is rounded to $q_u = 207$ psf.

EXAMPLE 1.2   **Load on girder.**   Determine the factored load applied to the interior girder on line C between lines 3 and 4.

SOLUTION.   Beam and girder loads are typically defined in pounds per linear foot (plf) or kips per linear foot (klf) along the length of the beam. The loads are developed using a 1 ft wide tributary strip perpendicular to the girder, shown in Fig. 1.17a. In this example, the length of the tributary strip goes halfway across the slabs loading the girder and is, thus, 7.5 ft long on the B-C side and 17.5 ft long on the C-D side of building line C for a total length of 25 feet. From Example 1.1, the factored load on the slab, 207 psf, is applied to the 1 ft wide strip, giving a load on the girder of 207 psf × 1 ft wide × 25 ft total tributary width = 5175 plf. To this must be added the factored girder self-weight. Only the 19 in. deep portion below the slab need be added, thus the load must be increased by 16 in. wide × 19 in. deep × 150 pcf/144 in^2/ft^2 = 317 plf. The uniform factored design load on the girder w_u is then 5175 plf + 1.2 × 317 plf = 5555.4 plf. Using three significant figures, the load to be used in design is w_u = 5.56 kip/ft.

EXAMPLE 1.3   **Load on column.**   Determine the factored axial load transferred to column C4.

SOLUTION.   Column axial loads are expressed in pounds or kips and are established using a tributary area. The tributary area for column C4, shown in Fig. 1.17a, is a rectangular area measure halfway between column lines (that is, half the distance to the adjacent columns) equal to (15 ft/2 + 35 ft/2) × (25 ft/2 + 35 ft/2) = 25 × 30 = 750 ft^2. From Example 1.1, the factored load of the slab is q_u = 207 psf to which must be added the factored weight of the beam from Example 2, 1.2 × 317 plf, within the tributary area. Thus, the load transferred to the column is P_u = 207 psf × 750 ft^2 + 1.2 × 317 plf × (35 ft/2 + 25 ft/2) = 166,650 lb or 166.7 kips. Using three significant figures, the factored axial load would be P_u = 167 kips. The determination of axial loads for use in design is discussed further in Section 11.3.

1.9   CONTRACT DOCUMENTS AND INSPECTION

Design information is transmitted from the *licensed design professional*, sometimes called the engineer of record, to the contractor through the *contract documents*. These documents typically consist of *plans*, *specifications*, and *estimates*. A typical set of plans includes the graphical information describing the architectural, structural, mechanical, and electrical components of the building. The structural plans contain the concrete sections, reinforcement, reinforcement details and placement, and other technical information based on the engineer's calculations. The licensed design professional commonly includes sketches or drawings of specific details in the calculations to allow the information to be accurately incorporated into the plans.

Specifications consist of two parts: the contract terms and conditions and the technical specifications. Contract terms and conditions include what is to be constructed, the time and cost of the construction, bonding requirements, and other specific issues between the owner and the contractor. Technical specifications contain the detailed information the contractor needs to complete a project and are provided by the licensed design professional. They include the ASTM specifications for concrete and steel, requirements for concrete strength and placement, grade of reinforcement to be used, considerations for hot and cold weather concreting, and other project specific information. During design, the licensed design professional makes decisions about the materials and details needed to comply with the design intent and the building code requirements. A contractor is not required to be familiar with the ACI Building Code nor is the contractor responsible for assuring that code requirements are satisfied. Therefore, the licensed design professional must include all relevant project and code requirements in the plans and specifications. Chapter 26 of the ACI Building Code (Ref. 1.12) contains a comprehensive description of information that the engineer must include in the plans and specifications. Pointers to this chapter are included throughout the ACI Code to assist the licensed design professional in finding and recording the correct information.

Inclusion of information in the project specifications does not, by itself, assure that construction will be executed according to the design intent. Rather, compliance requirements and inspection provide the licensed design professional and the owner with confirmation that the design intent is being met. Compliance requirements associated with project specifications are provided in Chapter 26 of the ACI Code. These compliance requirements complement the technical specifications by providing direct feedback to the licensed design professional. For example, if the specifications require the concrete strength to be 6000 psi, the compliance requirement would be that the strength test results, based on ASTM specifications for testing concrete, demonstrate the concrete meets or exceeds the specified strength. Actual testing of materials is done by testing agencies and technicians certified by the American Concrete Institute or other qualification agencies.

Inspection further advances compliance with the design intent. Inspection can range from onsite observations to detailed investigation of reinforcement placement. Onsite observations are conducted intermittently as a general overview of the construction with the intent of confirming overall design intent. Such observations may lead to discussions with the contractor regarding the way the work is done but do not direct the contractor's work. In areas prone to earthquakes, *special inspection* may be required. Special inspection requires the licensed design professional or a certified designee to conduct the inspections. These inspections specifically examine those elements of the design required to resist earthquake load effects and to certify that the construction meets the design details. The General Building Code and the ACI Building Code specify situations where special inspection is required.

The licensed design professional is sometimes required to provide an estimate of the cost of construction. This estimate addresses several issues. The estimate initially provides the owner with information to indicate that the available project funding is adequate. It can also provide a basis for estimating the degree of completion during construction. Cost estimates solely by the engineer are most often associated with engineered projects, such as bridges, piers, and industrial facilities. Cost estimates for building construction are typically provided by the architect with input from the engineer.

REFERENCES

1.1. *Minimum Design Loads and Other Associated Criteria for Buildings and Other Structures*, ASCE/SEI 7-16, American Society of Civil Engineers, Reston, VA, 2016.

1.2. *International Building Code*, International Code Council, Washington, DC, 2019.

1.3. *AASHTO LRFD Bridge Design Specifications*, 8th ed., American Association of State Highway and Transportation Officials (AASHTO), Washington, DC, 2017.

1.4. *Manual of Railway Engineering*, American Railway Engineering and Maintenance-of-Way Association (AREMA), Landover, MD, 2018.

1.5. *Building Seismic Safety Council NEHRP Recommended Provisions for Seismic Regulations for New Buildings and Other Structures*, 2000 edition, Part 1, "Provisions," FEMA 450, Part 2, "Commentary," FEMA 450, Federal Emergency Management Agency, Washington, DC, March 2003.

1.6. J. G. MacGregor, S. A. Mirza, and B. Ellingwood, "Statistical Analysis of Resistance of Reinforced and Prestressed Concrete Members," *J. ACI*, vol. 80, no. 3, 1983, pp. 167–176.

1.7. J. G. MacGregor, "Load and Resistance Factors for Concrete Design," *J. ACI*, vol. 80, no. 4, 1983, pp. 279–287.

1.8. B. Ellingwood, *Development of a Probability Based Load Criterion for American National Standard A58: Building Code Requirements for Minimum Design Loads in Buildings and Other Structures*, vol. 13, National Bureau of Standards, U.S. Department of Commerce, 1980.

1.9. J. G. MacGregor, "Safety and Limit States Design for Reinforced Concrete," *Can. J. Civ. Eng.*, vol. 3, no. 4, 1976, pp. 484–513.

1.10. A. S. Nowak and M. M. Szerszen, "Calibration of Design Code for Buildings (ACI 318): Part 1—Statistical Models for Resistance," *ACI Struct. J.*, vol. 100, no. 3, 2003, pp. 377–382.

1.11. M. M. Szerszen and A. S. Nowak, "Calibration of Design Code for Buildings (ACI 318): Part 2—Reliability Analysis and Resistance Factors," *ACI Struct. J.*, vol. 100, no. 3, 2003, pp. 383–391.

1.12. *Building Code Requirements for Structural Concrete and Commentary* (ACI 318-19), American Concrete Institute, Farmington Hills, MI, 2019.

Problems

All problems refer to Table 1.1 for live loads and Fig. P1.1 for the building layout. No live load reduction factors are considered. Figure P1.1 provides a plan and elevation of a reinforced concrete building 7 bays long by 4 bays wide. The building has beams along building lines A through E and one-way slabs spanning between building lines A through E. A central stairwell/elevator shaft between building lines 5 and 6 provides the lateral support, so only gravity loads need to be calculated. The bay dimensions, beam dimensions, and occupancy uses are given in the individual problem statements below. Construction is with normalweight concrete with a density of 150 pcf for the purposes of calculating dead load.

1.1. The building in Fig. P1.1 is used for general office space. The slab is 8 in. thick. The beams are 12 in. wide and have a total depth of 18 in., the bay dimensions are 18.5 ft in the X direction and 21 ft in the Y direction, and the superimposed service dead load is 25 psf. Calculate the slab service load in psf and the interior beam service load in klf. (**Solution:** $q_s = 175$ psf, $w_s = 3.36$ klf.)

1.2. The building in Fig. P1.1 is used for general office space. The slab is 8 in. thick. The beams are 12 in. wide and have a total depth of 18 in., the bay dimensions are 18.5 ft in the X direction and 21 ft in the Y direction, and the superimposed service dead load is 25 psf. Calculate the factored axial column load transferred to column C3 on the third floor. (**Solution:** $P_u = 92.5$ kips.)

1.3. The building in Fig. P1.1 is used for general office space. The slab is 8 in. thick. The beams are 12 in. wide and have a total depth of 18 in., the bay dimensions are 18.5 ft in the X direction and 21 ft in the Y direction, and the superimposed service dead load is 25 psf. Calculate the slab factored load in psf and the beam factored load in klf. Comment on your solution in comparison with Problem 1.1.

1.4. A slab in Fig. P1.1 is used for lobby space. The slab is 10 in. thick. The beams are 14 in. wide and have a total depth of 24 in., the bay dimensions are 21 ft in the X direction and 26 ft in the Y direction, and the superimposed service dead load is 15 psf. Calculate the slab factored load in psf and the beam factored load in klf.

1.5. The building in Fig. P1.1 is used for light storage space. The slab is 10 in. thick. The beams are 16 in. wide and have a total depth of 20 in., the bay dimensions are 20 ft in the X direction and 25 ft in the Y direction, and the superimposed sprinkler dead load is 4 psf. Calculate the slab factored load in psf and the beam factored load in klf.

FIGURE P1.1
Building plan and elevation

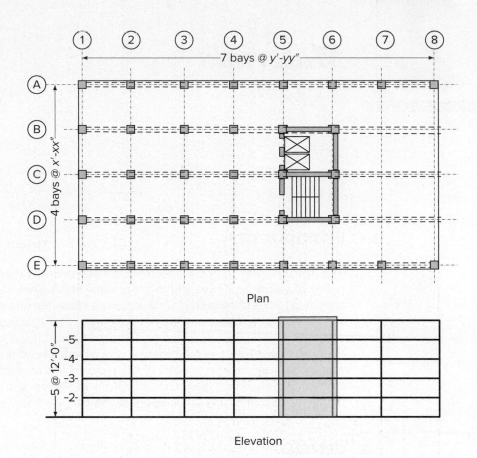

Plan

Elevation

1.6. The roof on the building in Fig. P1.1 has a slab that is 7 in. thick. The beams are 12 in. wide and have a total depth of 16 in., the bay dimensions are 19 ft in the X direction and 21 ft in the Y direction, and the superimposed service dead load is 6 psf. Calculate the slab factored load in psf and the beam factored load in klf given the roof snow load is 30 psf.

2 *Materials*

2.1 INTRODUCTION

The structures and component members treated in this text are composed of concrete reinforced with steel bars, and in some cases prestressed with steel wire, strand, or alloy bars. An understanding of the materials characteristics and behavior under load is fundamental to understanding the performance of structural concrete, and to safe, economical, and serviceable design of concrete structures. Although prior exposure to the fundamentals of material behavior is assumed, a brief review is presented in this chapter, as well as a description of the types of bar reinforcement and prestressing steels in common use. Numerous references are given as a guide for those seeking more information on any of the topics discussed.

2.2 CEMENT

A cementitious material is one that has the adhesive and cohesive properties necessary to bond inert aggregates into a solid mass of adequate strength and durability. This technologically important category of materials includes not only cements proper but also limes, asphalts, and tars as they are used in road building, and others. For making structural concrete, *hydraulic cements* are used exclusively. Water is needed for the chemical process (hydration) in which the cement powder sets and hardens into one solid mass. Of the various hydraulic cements that have been developed, *portland cement*, which was first patented in England in 1824, is by far the most common.

Portland cement is a finely powdered, grayish material that consists chiefly of calcium and aluminum silicates.[†] The common raw materials from which it is made are limestones, which provide CaO, and clays or shales, which furnish SiO_2 and Al_2O_3. These are ground, blended, fused to clinkers in a kiln, and cooled. Gypsum and additional unreacted limestone are added and the mixture is ground to the required fineness. The material is shipped in bulk or in bags containing 94 lb of cement.

Over the years, five standard types of portland cement have been developed. Type I, *normal* portland cement, is used for over 90 percent of construction in the United States. Concretes made with Type I portland cement generally need one to two weeks to reach sufficient strength so that forms of beams and slabs can be

---

[†] See ASTM C150, "Standard Specification for Portland Cement." This and other ASTM references are published and periodically updated by ASTM International (formerly the American Society for Testing and Materials), West Conshohoken, PA.

removed and reasonable loads applied; they reach their design strength after 28 days and continue to gain strength thereafter at a decreasing rate. To speed construction when needed, *high early strength cements* such as Type III have been developed. They are costlier than ordinary portland cement, but within 7 to 14 days they reach the strength achieved using Type I at 28 days. Type III portland cement contains the same basic compounds as Type I, but the relative proportions differ and it is ground more finely.

When cement is mixed with water to form a soft paste, it gradually stiffens until it becomes a solid. This process is known as *setting* and *hardening*. The cement is said to have set when it has gained sufficient rigidity to support an arbitrarily defined pressure, after which it continues for a long time to harden, that is, to gain further strength. The water in the paste dissolves material at the surfaces of the cement grains and forms a gel that gradually increases in volume and stiffness. This leads to a rapid stiffening of the paste 2 to 4 hours after water has been added to the cement. *Hydration* continues to proceed deeper into the cement grains, at decreasing speed, with continued stiffening and hardening of the mass. The principal products of hydration are calcium silicate hydrate, which is insoluble, and calcium hydroxide, which is soluble.

In ordinary concrete, the cement is probably never completely hydrated. The gel structure of the hardened paste seems to be the chief reason for the volume changes that are caused in concrete by variations in moisture, such as the shrinkage of concrete as it dries.

For complete hydration of a given amount of cement, an amount of water equal to about 25 percent of that of cement, by weight—that is, a *water-cement ratio* of 0.25—is needed chemically. An additional amount must be present, however, to provide mobility for the water in the cement paste during the hydration process so that it can reach the cement particles and to provide the necessary workability of the concrete mix. For normal concretes, the water-cement ratio is generally in the range of about 0.40 to 0.60, although for high-strength concretes, ratios as low as 0.21 have been used. In this case, the needed workability is obtained through the use of admixtures.

Any amount of water above that consumed in the chemical reaction produces pores in the cement paste. The strength of the hardened paste decreases in inverse proportion to the fraction of the total volume occupied by pores. Put differently, since only the solids, and not the voids, resist stress, strength increases directly as the fraction of the total volume occupied by the solids. That is why the strength of the cement paste depends primarily on, and decreases directly with, an increasing water-cement ratio.

The chemical process involved in the setting and hardening liberates heat, known as *heat of hydration*. In large concrete masses, such as dams, this heat is dissipated very slowly and results in a temperature rise and volume expansion of the concrete during hydration, with subsequent cooling and contraction. To avoid the serious cracking and weakening that may result from this process, special measures must be taken for its control.

2.3 AGGREGATES

In ordinary structural concretes the aggregates occupy 65 to 75 percent of the volume of the hardened mass. The remainder consists of hardened cement paste, uncombined water (that is, water not involved in the hydration of the cement), and air voids. The latter two do not contribute to the strength of the concrete. In general, the more densely the aggregate can be packed, the better the durability and economy of the concrete. For this reason the gradation of the particle sizes in the aggregate, to produce close packing, is important. It is also important that the aggregate have good strength, durability, and weather resistance; that its surface be free from impurities such as loam, clay, silt, and organic matter that may weaken the bond with cement paste; and that no unfavorable chemical reaction take place between it and the cement.

Natural aggregates are generally classified as fine and coarse. *Fine aggregate* (typically natural sand) is any material that will pass a No. 4 sieve, that is, a sieve with four openings per linear inch. Material coarser than this is classified as *coarse aggregate*. When favorable gradation is desired, aggregates are separated by screening into two or three size groups of sand and several size groups of coarse aggregate. These can then be combined according to grading criteria to provide a densely packed aggregate. The *maximum size of coarse aggregate* in reinforced concrete is governed by the requirement that it must easily fit into the forms and between the reinforcing bars. For this purpose it should not be larger than one-fifth of the narrowest dimension of the forms or one-third of the depth of slabs, nor three-quarters of the minimum distance between reinforcing bars. Requirements for satisfactory aggregates are found in ASTM C33, "Standard Specification for Concrete Aggregates," and authoritative information on aggregate properties and their influence on concrete properties, as well as guidance in selection, preparation, and handling of aggregate, is found in Refs. 2.1 and 2.2.

The unit weight of *normalweight concrete*, that is, concrete with natural aggregates, varies from about 140 to 152 pounds per cubic foot (pcf) and can generally be assumed to be 145 pcf. For special purposes, lightweight concretes, on one hand, and heavy concretes, on the other, are used.

A variety of *lightweight* aggregates are available. Some unprocessed aggregates, such as pumice or cinders, are suitable for insulating concretes, but for structural lightweight concrete, *processed aggregates* are used because of better control. These consist of expanded shales, clays, slates, slags, or pelletized fly ash. They are light in weight because of the porous, cellular structure of the individual aggregate particles, which is achieved by gas or steam formation in processing the aggregates in rotary kilns at high temperatures (generally in excess of 2000°F). Requirements for satisfactory lightweight aggregates are found in ASTM C330, "Standard Specification for Lightweight Aggregates for Structural Concrete."

Structural lightweight concretes have unit weights between 70 and 120 pcf, with most in the range of 105 to 120 pcf. Lower density lightweight concretes typically have compressive strengths of 1000 to 2500 psi and are chiefly used as fill, such as over light-gage steel floor panels. Lightweight concretes with unit weights between 90 and 120 pcf have compressive strengths comparable to those of normalweight concretes. Similarities and differences in structural characteristics of lightweight and normalweight concretes are discussed in Sections 2.8 and 2.9.

Heavyweight concrete is sometimes required for shielding against gamma and X-radiation in nuclear reactors and similar installations, for protective structures, and for special purposes, such as counterweights of lift bridges. Heavy aggregates are used for such concretes. These consist of heavy iron ores or barite (barium sulfate) rock crushed to suitable sizes. Steel in the form of scrap, punchings, or shot (as fines) is also used. Unit weights of heavyweight concretes with natural heavy rock aggregates range from about 200 to 230 pcf; if iron punchings are added to high-density ores, weights as high as 270 pcf are achieved. The weight may be as high as 330 pcf if ores are used for the fines only and steel for the coarse aggregate.

2.4 PROPORTIONING AND MIXING CONCRETE

The various components of a mix are proportioned so that the resulting concrete has adequate strength, proper workability for placing, and low cost. The third calls for use of the minimum amount of cement (the most costly of the components) that will achieve adequate properties. The better the gradation of aggregates, that is, the smaller the volume of voids, the less cement paste is needed to fill these voids. In addition to the water required for hydration, water is needed for wetting the surface of the aggregate. As water is added, the plasticity and fluidity of the mix increase (that is, its workability improves), but the strength decreases because of the larger volume of voids created by the free water. To reduce the free water while retaining the workability, cement must be added. Therefore, as for the cement paste, the *water-cement ratio* is the chief factor that controls the strength of the concrete. For a given water-cement ratio, one selects the minimum amount of cement that will secure the desired workability.

Figure 2.1 shows the decisive influence of the water-cement ratio on the compressive strength of concrete. Its influence on tensile strength, as measured by the nominal flexural strength or modulus of rupture, is also seen to be pronounced but much less than its effect on compressive strength. This seems to be so because, in addition to the

FIGURE 2.1

Effect of water-cement ratio on 28 day compressive and flexural tensile strength. (*Adapted from Ref. 2.3.*)

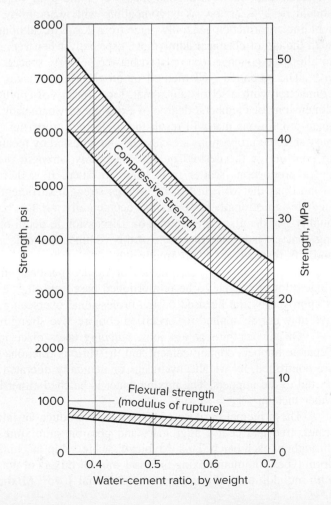

void ratio, the tensile strength depends on the strength of the bond between coarse aggregate and mortar (that is, cement paste plus fine aggregate). Tests show that this bond strength is only slightly affected by the water-cement ratio (Ref. 2.4).

It is customary to define the *proportions* of a concrete mix in terms of the total weight of each component needed to make up 1 yd^3 of concrete, such as 517 lb of cement, 300 lb of water, 1270 lb of sand, and 1940 lb of coarse aggregate, plus the total volume of air, in percent. Air content is typically 4 to 7 percent when air is deliberately *entrained* in the mix and 1 to 2 percent when it is not. The weights of the fine and coarse aggregates are based on material in the *saturated surface dry condition*, in which, as the description implies, the aggregates are fully saturated but have no water on the exterior of the particles.

Various methods of proportioning are used to obtain mixes of the desired properties from the cements and aggregates at hand. One is the *trial-batch method*. Selecting a water-cement ratio from information such as that in Fig. 2.1, one produces several small trial batches with varying amounts of aggregate to obtain the required strength, consistency, and other properties with a minimum amount of paste. Concrete *consistency* is most frequently measured by the *slump test*. A metal mold in the shape of a truncated cone 12 in. high is filled with fresh concrete in a carefully specified manner. Immediately upon being filled, the mold is lifted off, and the slump of the concrete is measured as the difference in height between the mold and the pile of concrete. The slump is a good measure of the total water content in the mix and should be kept as low as is compatible with workability. Slumps for concretes in building construction generally range from 2 to 5 in., although higher slumps are used with the aid of chemical admixtures, especially when very fluid mixtures are needed to allow the concrete to be placed between closely spaced reinforcing bars.

The so-called ACI method of proportioning makes use of the slump test in connection with a set of tables that, for a variety of conditions (types of structures, dimensions of members, degree of exposure to weathering, etc.), permit one to estimate proportions that will result in the desired properties (Ref. 2.5). These preliminary selected proportions are checked and adjusted by means of trial batches to result in concrete of the desired quality. Inevitably, strength properties of a concrete of given proportions scatter from batch to batch. It is therefore necessary to select proportions that will furnish an average strength sufficiently greater than the specified design strength for even the accidentally weaker batches to be of adequate quality (for details, see Section 2.6). Discussion in detail of practices for proportioning concrete is beyond the scope of this volume; this topic is treated fully in Refs. 2.5 and 2.6, respectively, for normalweight and lightweight concrete.

If the results of trial batches or field experience are not available, the ACI Code allows concrete to be proportioned based on other experience or information, if approved by the licensed design professional overseeing the project. This alternative may not be applied for specified compressive strengths greater than 5000 psi.

On all but the smallest jobs, *batching* takes place in special batching plants. Separate hoppers contain cement and the various fractions of aggregate. Proportions are controlled, by weight, by means of manually operated or automatic scales connected to the hoppers. The mixing water is batched either by measuring tanks or by water meters.

The principal purpose of *mixing* is to produce an intimate mixture of cement, water, fine and coarse aggregate, and possible admixtures of uniform consistency throughout each batch. This is typically achieved in machine mixers of the revolving-drum type. Minimum mixing time is 1 min for mixers of not more than 1 yd^3 capacity, with an additional 15 sec for each additional 1 yd^3. Mixing can be continued for a considerable time without adverse effect. This fact is particularly important in connection with ready mixed concrete.

On large projects, particularly in the open country where ample space is available, movable mixing plants are installed and operated at the site. On the other hand, in construction under congested city conditions, on smaller jobs, and frequently in highway construction, *ready mixed concrete* is used. Such concrete is batched in a stationary plant and then hauled to the site in trucks in one of three ways: (1) mixed completely at the stationary plant and hauled in a truck agitator, (2) transit-mixed, that is, batched at the plant but mixed in a truck mixer, or (3) partially mixed at the plant with mixing completed in a truck mixer. Concrete should be discharged from the mixer or agitator within a limited time after the water is added to the batch. Although specifications often provide a single value for all conditions, the maximum mixing time should be based on the concrete temperature because higher temperatures lead to increased rates of *slump loss* and rapid setting. Conversely, lower temperatures increase the period during which the concrete remains workable. A good guide for maximum mixing time is to allow 1 hour at a temperature of 70°F, plus (or minus) 15 min for each 5°F drop (or rise) in concrete temperature for concrete temperatures between 40 and 90°F. Ten minutes may be used at 95°F, the practical upper limit for normal mixing and placing.

Much information on proportioning and other aspects of design and control of concrete mixtures will be found in Refs. 2.7 and 2.8.

2.5 CONVEYING, PLACING, CONSOLIDATING, AND CURING

Conveying of most building concrete from the mixer or truck to the forms is done in bottom-dump buckets or by pumping through steel pipelines. The chief danger during conveying is that of *segregation*, the separation of the individual components of concrete because of their dissimilarity. In overly wet concrete standing in containers or forms, the heavier coarse aggregate particles tend to settle, and the lighter materials, particularly water, tend to rise. Lateral movement, such as flow within the forms, tends to separate the coarse aggregate particles from the finer components of the mix.

Placing is the process of transferring the fresh concrete from the conveying device to its final place in the forms. Prior to placing, loose rust must be removed from reinforcement, forms must be cleaned, and hardened surfaces of previous concrete lifts must be cleaned and treated appropriately. Placing and consolidating are critical in their effect on the final quality of the concrete. Proper placement must avoid segregation, displacement of forms or of reinforcement in the forms, and poor bond between successive layers of concrete. Immediately upon placing, the concrete should be *consolidated*, usually by means of vibrators. Consolidation prevents honeycombing, ensures close contact with forms and reinforcement, and serves as a partial remedy to possible prior segregation. Consolidation is achieved by high-frequency, power-driven *vibrators*. These are of the *internal* type, immersed in the concrete, or of the *external* type, attached to the forms. The former are preferable but must be supplemented by the latter where narrow forms or other obstacles make immersion impossible (Ref. 2.9). Vibration is not needed for *self-consolidating concrete*, a fluid concrete that consolidates under its own weight, discussed in more detail in Section 2.7.

Fresh concrete gains strength most rapidly during the first few days and weeks. Structural design is generally based on the *28-day strength*, about 70 percent of which is reached at the end of the first week after placing. The final concrete strength depends greatly on the conditions of moisture and temperature during this initial period. The maintenance of proper conditions during this time is known as *curing*. Thirty percent of the strength or more can be lost by premature drying out of the concrete; similar amounts may be lost by permitting the concrete temperature to drop to 40°F or lower during the first few days unless the concrete is kept continuously moist for a long time thereafter. Freezing of fresh concrete may reduce its strength by 50 percent or more.

To prevent such damage, concrete should be protected from loss of moisture for at least 7 days and, in more sensitive work, up to 14 days. When high early strength cements are used, curing periods can be cut in half. Curing can be achieved by keeping exposed surfaces continually wet through sprinkling, ponding, or covering with plastic film or by the use of sealing compounds, which, when properly used, form evaporation-retarding membranes. In addition to improving strength, proper moist-curing provides better shrinkage control. To protect concrete against low temperatures during cold weather, the mixing water, and occasionally the aggregates, is heated; thermal insulation is used where possible; and special admixtures are employed. When air temperatures are very low, external heat may have to be supplied in addition to insulation (Refs. 2.7, 2.8, 2.10, and 2.11).

2.6   QUALITY CONTROL

The quality of mill-produced materials, such as structural or reinforcing steel, is ensured by the producer, who must exercise systematic quality controls, usually specified by pertinent ASTM standards. Concrete, in contrast, is produced at or close to the site, and its final qualities are affected by a number of factors, which have been discussed briefly. Thus, systematic quality control must be instituted at the construction site.

The main measure of the structural quality of concrete is its *compressive strength*. Tests for this property are made on cylindrical specimens of height equal to twice the diameter, usually 6×12 in. or 4×8 in. Impervious molds of this shape are filled with concrete during construction as specified by ASTM C172, "Standard Method of Sampling Freshly Mixed Concrete," and ASTM C31, "Standard Practice for Making and Curing Concrete Test Specimens in the Field." The cylinders are moist-cured at about 70°F, generally for 28 days, and then tested in the laboratory at a specified rate of loading. The compressive strength obtained from such tests is known as the *cylinder strength,* which is compared to the *specified compressive strength* f_c', the main property specified for design.

To provide structural safety, continuous control is necessary to ensure that the strength of the concrete as furnished is in satisfactory agreement with the value called for by the designer. The ACI Code specifies that at least two 6×12 in. or three 4×8 in. cylinders must be tested for each 150 yd^3 of concrete or for each 5000 ft^2 of surface area actually placed, but not less than once a day. As mentioned in Section 2.4, the results of strength tests of different batches mixed to identical proportions show inevitable scatter. The scatter can be reduced by closer control, but occasional tests below the cylinder strength specified in the design cannot be avoided.

To ensure adequate concrete strength in spite of such scatter, the ACI Code stipulates that concrete quality is satisfactory if

1. Every average of any three consecutive strength tests equals or exceeds f_c', and
2. No strength test (the average of two or three cylinder tests depending on cylinder size) falls below the required f_c' by more than 500 psi if f_c' is 5000 psi or less or by more than $0.10 f_c'$ if f_c' exceeds 5000 psi.

It is evident that if concrete were proportioned so that its mean strength were just equal to the required strength f_c', it would not pass these quality requirements, because about one-half of the strength test results would fall below the required f_c'.

FIGURE 2.2
Frequency curves and
average strengths for various
degrees of control of
concretes with specified
design strength f_c'.
(*Adapted from Ref. 2.12.*)

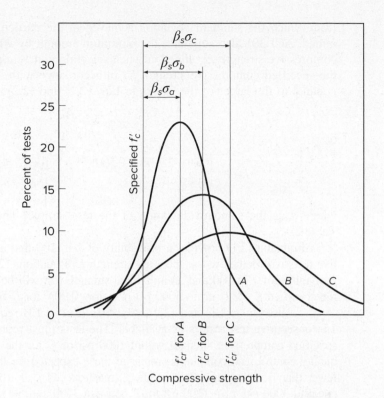

It is therefore necessary to proportion the concrete so that its mean strength f_{cr}', used as the basis for selection of suitable proportions, exceeds the specified design strength f_c' by an amount sufficient to ensure that the two quoted requirements are met. The minimum amount by which the required mean strength must exceed f_c' can be determined only by statistical methods because of the random nature of test scatter. Requirements have been derived, based on statistical analysis, to be used as a guide to proper proportioning of the concrete at the plant so that the probability of strength deficiency at the construction site is acceptably low.

The basis for these requirements is illustrated in Fig. 2.2, which shows three normal frequency distribution curves giving the distribution of strength test results. The specified design strength is f_c'. The curves correspond to three different degrees of quality control, curve *A* representing the best control, that is, the least scatter, and curve *C* the worst control, with the most scatter. The degree of control is measured statistically by the standard deviation σ (σ_a for curve *A*, σ_b for curve *B*, and σ_c for curve *C*), which is relatively small for producer A and relatively large for producer C. All three distributions have the same probability of strength less than the specified value f_c'; that is, each has the same fractional part of the total area under the curve to the left of f_c'. For any normal distribution, that fractional part is defined by the index β_s, a multiplier applied to the standard deviation σ; β_s is the same for all three distributions of Fig. 2.2. As demonstrated in the figure, to satisfy the requirement that, say, 1 test in 100 will fall below f_c' (with the value of β_s thus determined), for producer A with the best quality control the mean strength f_{cr}' can be much closer to the specified f_c' than for producer C with the most poorly controlled operation.

On the basis of such studies, ACI Code 26.4.3.1 requires that mixture proportions be established in accordance with ACI 301, "Specifications for Structural Concrete" (Ref. 2.13). ACI 301 requires concrete production facilities to maintain records from which the standard deviation achieved in the particular facility can be determined. ACI 301 also stipulates the minimum amount by which the required average compressive strength f'_{cr}, aimed at when selecting concrete proportions, must exceed the specified compressive strength f'_c. In accordance with ACI 301, the value of f'_{cr} is equal to the larger of the values in Eqs. (2.1) and (2.2).

$$f'_{cr} = f'_c + 1.34ks_s \tag{2.1}$$

or

$$f'_{cr} = \begin{cases} f'_c + 2.33ks_s - 500 & \text{for } f'_c \leq 5000 \text{ psi} \tag{2.2a} \\ 0.9f'_c + 2.33ks_s & \text{for } f'_c > 5000 \text{ psi} \tag{2.2b} \end{cases}$$

where s_s is the standard deviation of the test sample. The value of k is given in Table 2.1.

Equation (2.1) provides a probability of 1 in 100 that averages of three consecutive tests will be below the specified strength f'_c. Equations (2.2a) and (2.2b) provide a probability of 1 in 100 that an individual strength test will be more than 500 psi below the specified f'_c for f'_c up to 5000 psi or below $0.90f'_c$ for f'_c over 5000 psi.

To use Eqs. (2.1) and (2.2), ACI 301 (Ref. 2.13) requires that a minimum of 15 consecutive test results be available. The tests must represent concrete with (1) a specified compressive strength within 1000 psi of f'_c for the project and (2) materials, quality control, and conditions similar to those expected for the building in question. If fewer than 15 tests have been made, f'_{cr} must exceed $f'_c + 1000$ psi for f'_c less than or equal to 3000 psi, $f'_c + 1200$ psi for f'_c between 3000 and 5000 psi, and $0.1 f'_c + 700$ psi for f'_c over 5000 psi.

It is seen that this method of control recognizes the fact that occasional deficient batches are inevitable. The requirements for f'_{cr} ensure (1) a small probability that such strength deficiencies, as are bound to occur, will be large enough to represent a serious danger and (2) an equally small probability that a sizable portion of the structure, as represented by three consecutive strength tests, will be made of below-par concrete.

Both the requirements described earlier in this section for determining if concrete, as produced, is of satisfactory quality and the process just described of selecting f'_{cr} are based on the same basic considerations but are applied independently, as demonstrated in Examples 2.1 and 2.2.

TABLE 2.1

Modification factor k for sample standard deviation s_s when less than 30 tests are available

No. of Tests[†]	Modification Factor k for Sample Standard Deviation
Less than 15	See paragraph following Eqs. (2.1) and (2.2)
15	1.16
20	1.08
25	1.03
30 or more	1.00

[†] Interpolate for intermediate values.

EXAMPLE 2.1   **Average required strength.**   A building design calls for specified concrete strength f_c' of 4000 psi. Calculate the average required strength f_{cr}' if (*a*) 30 consecutive tests for concrete with similar strength and materials produce a sample standard deviation s_s of 535 psi, (*b*) 15 consecutive tests for concrete with similar strength and materials produce a sample standard deviation s_s of 510 psi, and (*c*) less than 15 tests are available.

SOLUTION.

(*a*) 30 tests available. Using $s_s = 535$ psi and $k = 1.0$ (from Table 2.1), Eq. (2.1) gives

$$f_{cr}' = f_c' + 1.34 k s_s = 4000 + 1.34 \times 1.0 \times 535 = 4720 \text{ psi}^\dagger$$

Because the specified strength f_c' is less than 5000 psi, Eq. (2.2*a*) must be used.

$$f_{cr}' = f_c' + 2.33 k s_s - 500 = 4000 + 2.33 \times 1.0 \times 535 - 500 = 4750 \text{ psi}$$

The required average strength f_{cr}' is equal to the larger value, 4750 psi.

(*b*) 15 tests available. Because only 15 tests are available, s_s the factor $k = 1.16$ from Table 2.1.

$$1.16 \times s_s = 1.16 \times 510 = 590 \text{ psi}$$

Using $s_s = 510$ and $k = 1.16$, Eqs. (2.1) and (2.2*a*) give, respectively,

$$f_{cr}' = 4000 + 1.34 \times 1.16 \times 510 = 4790 \text{ psi}$$

$$f_{cr}' = 4000 + 2.33 \times 1.16 \times 510 - 500 = 4880 \text{ psi}$$

The larger value, 4880 psi, is selected as the required average strength f_{cr}'.

(*c*) Less than 15 tests available. Because f_c' is between 3000 and 5000 psi, the required average strength is

$$f_{cr}' = f_c' + 1200 = 4000 + 1200 = 5200 \text{ psi}$$

This example demonstrates that in cases where test data are available, good quality control, represented by a low sample standard deviation s_s, can be used to reduce the required average strength f_{cr}'. The example also demonstrates that a lack of certainty in the value of the standard deviation due to the limited availability of data results in higher values for f_{cr}', as shown in parts (*b*) and (*c*). As additional test results become available, the higher safety margins can be reduced.

EXAMPLE 2.2   **Satisfactory test results.**   The first eight compressive strength test results for the building described in Example 2.1*c* are 4730, 4280, 3940, 4370, 5180, 4870, 4930, and 4850 psi.

(*a*) Are the test results satisfactory, and (*b*) in what fashion, if any, should the mixture proportions of the concrete be altered?

SOLUTION.

(*a*) For concrete to be considered satisfactory, every arithmetic mean of any three consecutive tests must equal or exceed f_c', and no individual test may fall below $f_c' - 500$ psi. The eight tests meet these criteria. The average of all sets of three consecutive tests exceeds f_c' [for example, $(4730 + 4280 + 3940)/3 = 4320$, $(4280 + 3940 + 4370)/3 = 4200$, etc.], and no test is less than $f_c' - 500$ psi $= 4000 - 500 = 3500$ psi.

(*b*) To determine if the mixture proportions must be altered, we note that the solution to Example 2.1*c* requires that f_{cr}' equal or exceed 5200 psi. The average of the first eight tests is

---

† ASTM International specifies that concrete cylinder strengths be recorded to the nearest 10 psi. Hence the values used for test results and f_{cr}' are rounded accordingly.

4640 psi, well below the value of f'_{cr}. Thus, the mixture proportions should be modified by decreasing the water-cement ratio to increase the concrete strength. Once at least 15 tests are available, the value of f'_{cr} can be recalculated using Eqs. (2.1) and (2.2) with the appropriate factor k from Table 2.1. The mixture proportions can then be adjusted based on the new value of f'_{cr}, the strength of the concrete being produced, and the level of quality control, as represented by the sample standard deviation s_s.

In spite of advances, building in general and concrete making in particular retain some elements of an art; they depend on many skills and imponderables. It is the task of systematic *inspection* to ensure close correspondence between plans and specifications and the finished structure. Inspection during construction should be performed by a competent engineer, preferably the one who produced the design or one who is responsible to the design engineer. The inspector's main functions in regard to materials quality control are sampling, examination, and field testing of materials; control of concrete proportioning; inspection of batching, mixing, conveying, placing, compacting, and curing; and supervision of the preparation of specimens for laboratory tests. In addition, the inspector must inspect foundations, formwork, placing of reinforcing steel, and other pertinent features of the general progress of work; keep records of all the inspected items; and prepare periodic reports. The importance of thorough inspection to the correctness and adequate quality of the finished structure cannot be emphasized too strongly.

This brief account of concrete technology represents the merest outline of an important subject. Anyone in practice who is actually responsible for any of the phases of producing and placing concrete must be familiar with the details in much greater depth.

2.7   ADMIXTURES

In addition to the main components of concretes, *admixtures* are often used to improve concrete performance. There are admixtures to accelerate or retard setting and hardening, improve workability, increase strength, improve durability, decrease permeability, and impart other properties (Ref. 2.14). The beneficial effects of particular admixtures are well established. Chemical admixtures should meet the requirements of ASTM C494, "Standard Specification for Chemical Admixtures for Concrete."

Air-entraining agents are widely used. They cause the formation of small dispersed air bubbles in the concrete. These improve workability and durability (chiefly resistance to freezing and thawing) and reduce segregation during placing. They decrease concrete density because of the increased void ratio and thereby decrease strength; however, this decrease can be largely offset by a reduction of mixing water without loss of workability. The chief use of air-entrained concretes is in pavements and structures exposed to the elements (Ref. 2.7).

Accelerating admixtures are used to reduce setting time and accelerate early strength development. Calcium chloride is the most widely used accelerator because of its cost effectiveness, but it should not be used in prestressed concrete and should be used with caution in reinforced concrete in a moist environment, because of its tendency to promote corrosion of steel, or in architectural concrete, because of its tendency to discolor concrete. Nonchloride, noncorrosive accelerating admixtures are available, the principal one being calcium nitrite (Ref. 2.14).

Set-retarding admixtures are used primarily to offset the accelerating effect of high ambient temperature and to keep the concrete workable during the entire placing period. This helps to eliminate cracking due to form deflection and also keeps concrete workable long enough that succeeding lifts can be placed without the development of "cold" joints.

Certain organic compounds are used to reduce the water requirement of a concrete mix for a given slump. Such compounds are termed *water-reducing admixtures* or *plasticizers*. Reduction in water demand may result in either a reduction in the water-cement ratio for a given slump and cement content or an increase in slump for the same water-cement ratio and cement content. Plasticizers work by reducing the interparticle forces that exist between cement grains in the fresh paste, thereby increasing the paste fluidity. *High-range water-reducing admixtures*, often termed as *superplasticizers*, are used to produce high-strength concrete (see Section 2.12) with a very low water-cement ratio while maintaining the higher slumps needed for proper placement and compaction of the concrete. They are also used to produce flowable concrete at conventional water-cement ratios. Superplasticizers differ from conventional water-reducing admixtures in that they do not act as retarders at high dosages; therefore, they can be used at higher dosage rates without severely slowing hydration (Ref. 2.14). The specific effects of water-reducing admixtures vary with different cements, changes in water-cement ratio, mixing temperature, ambient temperature, and other job conditions, and trial batches are generally required.

When superplasticizers are combined with *viscosity-modifying admixtures*, they can be used to produce *self-consolidating concrete* (SCC) (Ref. 2.15). Self-consolidating concrete is highly fluid and does not require vibration to remove entrapped air. The viscosity-modifying agents allow the concrete to remain cohesive even with a very high degree of fluidity. As a result, SCC can be used for members with congested reinforcement, such as beam-column joints in earthquake-resistant structures, and is widely used for precast concrete, especially precast prestressed concrete, a manufactured product (prestressed concrete is discussed in Chapter 22). The high fluidity of the mix, however, has been shown to have a negative impact on the bond strength between the concrete and prestressing steel located in the upper portions of a member, a shortcoming that should be considered in design (Ref. 2.16) but is not currently addressed in the ACI Code, and the composition of SCC mixtures may result in moduli of elasticity, creep, and shrinkage properties that differ from those of more traditional mixtures.

Fly ash and *silica fume* are pozzolans, highly active silicas, that combine with calcium hydroxide, the soluble product of cement hydration (Section 2.2), to form more calcium silicate hydrate, the insoluble product of cement hydration (Refs. 2.17 and 2.18). Pozzolans qualify as *supplementary cementitious materials*, also referred to as *mineral admixtures*, which are used to replace a part of the portland cement in concrete mixes. Fly ash, which is specified under ASTM C618, "Standard Specification for Coal Fly Ash and Raw or Calcified Natural Pozzolan for Use in Concrete," is precipitated electrostatically as a by-product of the exhaust fumes of coal-fired power stations. It is very finely divided and reacts with calcium hydroxide in the presence of moisture to form a cementitious material. It tends to increase the strength of concrete at ages over 28 days. Silica fume, which is specified under ASTM C1240, "Standard Specification for Silica Fume Used in Cementitious Mixtures," is a by-product resulting from the manufacture, in electric-arc furnaces, of ferro-silicon alloys and silicon metal. It is extremely finely divided and is highly cementitious when combined with portland cement. In contrast to fly ash, silica fume contributes mainly to strength gain at early ages, from 3 to 28 days. Both fly ash and silica fume, particularly the latter, have been important in the production of high-strength concrete (see Section 2.12).

Slag cement, which is specified under ASTM C989, "Standard Specification for Slag Cement for Use in Concrete and Mortars," is another supplementary cementitious material. It is produced by water quenching and grinding slag from the production of pig iron, the key ingredient used to make steel (Ref. 2.19). Slag cement consists primarily of calcium silicates, making it very similar to portland cement. As a result of the similarity, slag cement can be used in higher quantities than fly ash or silica fume, and the resulting material generally has similar or improved properties to those exhibited by concrete made with 100 percent portland cement.

When slag cement, silica fume, fly ash, or a combination is used, it is customary to refer to the *water-cementitious material ratio* rather than the water-cement ratio. This typically may be as low as 0.25 for high-strength concrete, and ratios as low as 0.21 have been used (Refs. 2.20 and 2.21).

Historically, the high durability and high thermal mass of concrete structures have played a key role in *sustainable development*, that is, development that minimizes both its impact on the environment and the resources used both during and after construction. In sustainable development, the "cost" of concrete lies primarily in the manufacture of portland cement. The production of a ton of portland cement requires roughly the energy needed to operate a typical U.S. household for two weeks and generates approximately 0.9 ton of CO_2 (a greenhouse gas). The latter translates to about 250 lb of CO_2 for every cubic yard of concrete that is placed. The energy and greenhouse gases involved in the production of concrete, however, can be viewed as investments because properly designed reinforced concrete structures that take advantage of concrete's thermal mass provide significant reductions in the energy and CO_2 needed for heating and cooling, and concrete's inherent durability results in structures with long service lives. Because by-products, such as the mineral admixtures fly ash and blast furnace slag, involve minimal energy usage or greenhouse gas production, they have the potential to further improve the sustainability of concrete construction when used as a partial replacement for portland cement.

2.8   PROPERTIES IN COMPRESSION

a.   Short-Term Loading

Performance of a structure under load depends to a large degree on the stress-strain relationship of the material from which it is made, under the type of stress to which the material is subjected in the structure. Since concrete is used mostly in compression, its compressive stress-strain curve is of primary interest. Such a curve is obtained by appropriate strain measurements in cylinder tests (Section 2.6) or on the compression side in beams. Figure 2.3 shows a typical set of such curves for normalweight concrete, obtained from uniaxial compressive tests performed at normal, moderate testing speeds on concretes that are 28 days old. Figure 2.4 shows corresponding curves for lightweight concretes having a density of 100 pcf.

All of the curves have somewhat similar character. They consist of an initial relatively straight elastic portion in which stress and strain are closely proportional, then begin to curve to the horizontal, reaching the maximum stress, that is, the compressive strength, at a strain that ranges from about 0.002 to 0.003 for normalweight concretes, and from about 0.003 to 0.0035 for lightweight concretes (Refs. 2.22 and 2.23), the larger values in each case corresponding to the higher strengths. All curves show a descending branch after the peak stress is reached; however, the characteristics of the curves after peak stress are highly dependent upon the method of testing. If

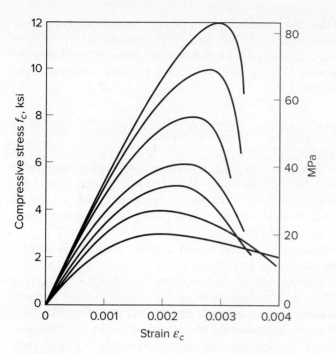

FIGURE 2.3
Typical compressive stress-strain curves for normal-density concrete with $w_c = 145$ pcf. (*Adapted from Refs. 2.22 and 2.23.*)

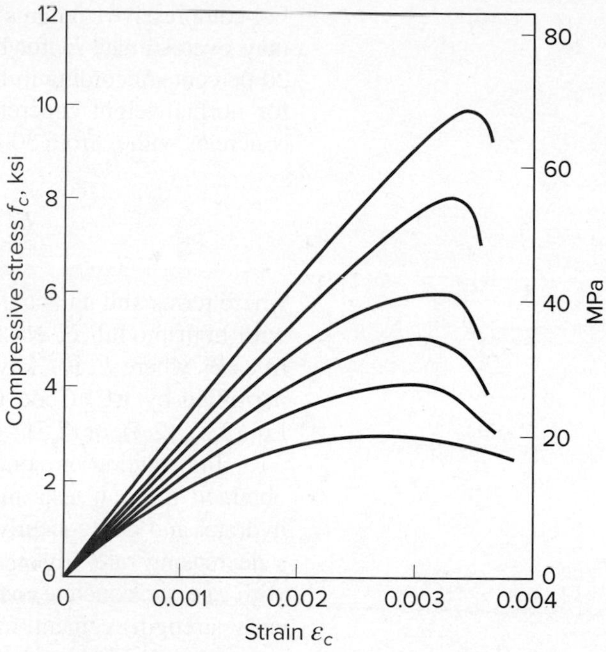

FIGURE 2.4
Typical compressive stress-strain curves for lightweight concrete with $w_c = 100$ pcf. (*Adapted from Refs. 2.22 and 2.23.*)

special procedures are followed in testing to ensure a constant strain rate while cylinder resistance is decreasing, long stable descending branches can be obtained (Ref. 2.24). In the absence of special devices, unloading past the point of peak stress may be rapid, particularly for the higher-strength concretes, which are generally more brittle than low-strength concrete.

In present practice, the specified compressive strength f_c' is commonly in the range from 3000 to 6000 psi for normalweight cast-in-place concrete, and up to about 10,000 psi for precast prestressed concrete members. Lightweight concrete strengths are typically below these values. High-strength concretes, with f_c' to 15,000 psi or more, are used with increasing frequency, particularly for heavily loaded columns in high-rise concrete buildings and for long-span bridges (mostly prestressed) where a significant reduction in dead load may be realized by minimizing member cross section dimensions. (See Section 2.12.)

The *modulus of elasticity* E_c (in psi units), that is, the slope of the initial straight portion of the stress-strain curve, is seen to be larger as the strength of the concrete increases. For concretes in the strength range to about 6000 psi, it can be computed with reasonable accuracy from the empirical equation found in ACI Code 19.2.2

$$E_c = 33w_c^{1.5}\sqrt{f_c'} \tag{2.3}$$

where w_c is the unit weight of the hardened concrete in pcf and f_c' is its strength in psi. Equation (2.3) was obtained by testing structural concretes with values of w_c from 90 to 155 pcf. For normalweight concrete, with $w_c = 145$ pcf, E_c may be taken as

$$E_c = 57{,}000\sqrt{f_c'} \tag{2.4}$$

For compressive strengths in the range from 6000 to 12,000 psi, the ACI Code equation may overestimate E_c for both normalweight and lightweight material by as much as 20 percent. According to Refs. 2.22 and 2.23, the following equation is recommended for normalweight concretes with f_c' from 3000 to 12,000 psi, and for lightweight concretes with f_c' from 3000 to 9000 psi:

$$E_c = (40,000\sqrt{f_c'} + 1,000,000)\left(\frac{w_c}{145}\right)^{1.5} \tag{2.5}$$

where terms and units are as defined in Eqs. (2.3) and (2.4). When coarse aggregates with high moduli of elasticity are used, however, Eq. (2.4) may *underestimate* E_c. In cases where E_c is a key design criterion, or where it is specified based on tests, as permitted by ACI Code 19.2.2, it should be measured, rather than estimated, using Eq. (2.3), (2.4), or (2.5).

Information on concrete strength properties such as those discussed is usually obtained through tests made 28 days after placing. However, cement continues to hydrate, and consequently concrete continues to gain strength, long after this age, at a decreasing rate. Figure 2.5 shows a typical curve of the gain of concrete strength with age for concrete made using Type I (normal) cement and also Type III (high early strength) cement, each curve normalized with respect to the 28-day compressive strength. High early strength cements produce more rapid strength gain at early ages, although the rate of strength gain at later ages is generally less. Concretes made with Type III cement are often used in precasting plants, and often the strength f_c' is specified at 7 days, rather than 28 days.

Note that the shape of the stress-strain curve for various concretes of the same cylinder strength, and even for the same concrete under various conditions of loading, varies considerably. An example of this is shown in Fig. 2.6, where different specimens of the same concrete are loaded at different rates of strain, from one corresponding to a relatively fast loading (0.001 per min) to one corresponding to an extremely slow application of load (0.001 per 100 days). It is seen that the descending branch of the curve, indicative of internal disintegration of the material, is much more pronounced at fast than at slow rates of loading. It is also seen that the peaks of the curves, that is, the maximum strengths reached, are somewhat lower at slower rates of strain.

When compressed in one direction, concrete, like other materials, expands in the direction transverse to that of the applied stress. The ratio of the transverse to the longitudinal strain is known as *Poisson's ratio* and depends somewhat on strength,

FIGURE 2.5

Effect of age on compressive strength f_c' for moist-cured concrete. (*Adapted from Ref. 2.25.*)

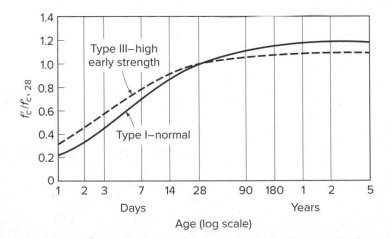

FIGURE 2.6
Stress-strain curves at
various strain rates,
concentric compression.
(*Adapted from Ref. 2.26.*)

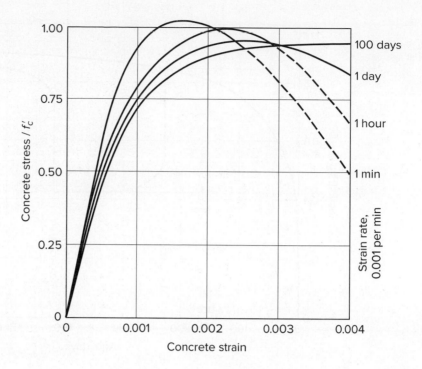

b.   Long-Term Loading

In some engineering materials, such as steel, strength and the stress-strain relationships
are independent of rate and duration of loading, at least within the usual ranges of rate of
stress, temperature, and other variables. In contrast, Fig. 2.6 illustrates the fact that the
influence of time, in this case of rate of loading, on the behavior of concrete under load
is pronounced. The main reason is that concrete creeps under load, while steel does not
exhibit creep under conditions prevailing in buildings, bridges, and similar structures.

Creep is the slow deformation of a material over considerable lengths of time
at constant stress or load. The nature of the creep process is shown schematically in
Fig. 2.7. This particular concrete was loaded after 28 days, resulting in instantaneous
strain $\varepsilon_{\text{inst}}$. The load was then maintained for 230 days, during which time creep
increased the total deformation to almost 3 times its instantaneous value. If the load
were maintained, the deformation would follow the solid curve. If the load is
removed, as shown by the dashed curve, most of the elastic instantaneous strain $\varepsilon_{\text{inst}}$
is recovered, and some creep recovery is seen to occur. If the concrete is reloaded
at some later date, instantaneous and creep deformations develop again, as shown.

Creep deformations for a given concrete are practically proportional to the
magnitude of the applied stress; at any given stress, and even at the same ratio of
stress to compressive strength, high-strength concretes show less creep than lower-strength concretes (Ref. 2.27). As shown in Fig. 2.7, with elapsing time, creep
proceeds at a decreasing rate and ceases after 2 to 5 years at a final value which,
depending on concrete strength and other factors, is 1.2 to 3 times the magnitude of
the instantaneous strain. If, instead of being applied quickly and thereafter kept
constant, the load is increased slowly and gradually, as is the case in many structures

composition, and other factors. At stresses lower than about $0.7f'_c$, Poisson's ratio for
concrete is between 0.15 and 0.20.

FIGURE 2.7
Typical creep curve (concrete
loaded to 600 psi at age
28 days).

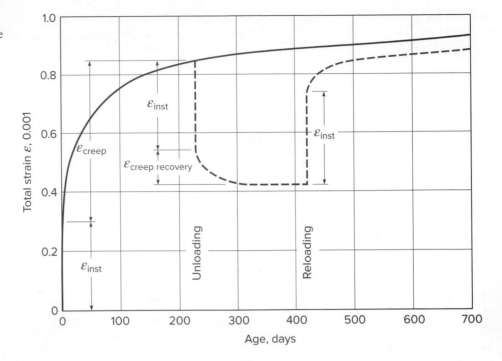

FIGURE 2.7
Typical creep curve (concrete
loaded to 600 psi at age
28 days).

during and after construction, then instantaneous and creep deformations proceed simultaneously. The effect is shown in Fig. 2.6; that is, the previously discussed difference in the shape of the stress-strain curve for various rates of loading is chiefly the result of the creep deformation of concrete.

For stresses not exceeding about one-half the cylinder strength, creep strains are approximately proportional to stress. Because initial elastic strains are also proportional to stress in this range, this permits definition of the *creep coefficient*

$$C_{cu} = \frac{\varepsilon_{cu}}{\varepsilon_{ci}} \tag{2.6}$$

where ε_{cu} is the final asymptotic value of the additional creep strain and ε_{ci} is the initial, instantaneous strain when the load is first applied. Creep may also be expressed in terms of the *specific creep* δ_{cu}, defined as the additional time-dependent strain per psi stress. For a given stress, f_c, $\varepsilon_i = f_c/E_c$ and $\varepsilon_{cu} = \delta_{cu}f_c$. Thus, based on Eq. (2.6),

$$C_{cu} = E_c\delta_{cu} \tag{2.7}$$

In addition to the stress level, creep depends on the average ambient relative humidity, being more than twice as large for 50 percent as for 100 percent humidity (Ref. 2.8). This is so because part of the reduction in volume under sustained load is caused by outward migration of free pore water, which evaporates into the surrounding atmosphere. Other factors of importance include the type of cement and aggregate, age of the concrete when first loaded, and concrete strength (Ref. 2.8). The creep coefficient for high-strength concrete is much less than that for low-strength concrete. However, sustained load stresses are apt to be higher so that the creep *deformation* may be as great for high-strength concrete, even though the creep coefficient is less.

The values of specific creep and creep coefficient in Table 2.2, quoted from Ref. 2.28 and extended for high-strength concrete, are for average humidity conditions, for concretes loaded at the age of 7 days.

TABLE 2.2
Typical creep parameters

Compressive Strength		Specific Creep δ_{cu}		Creep Coefficient C_{cu}
psi	MPa	10^{-6} per psi	10^{-6} per MPa	
3,000	21	1.00	145	3.1
4,000	28	0.80	116	2.9
6,000	41	0.55	80	2.4
8,000	55	0.40	58	2.0
10,000	69	0.28	41	1.6
12,000	83	0.22	33	1.4

To illustrate, if the concrete in a column with $f_c' = 4000$ psi is subject to a long-term load that causes sustained stress f_c of 1200 psi, then after several years under load, the final value of the creep strain will be about $\varepsilon_{cu} = \delta_{cu} f_c = 1200 \times 0.80 \times 10^{-6} = 0.00096$. Thus, if the column were 20 ft long, creep would shorten it by $\varepsilon_{cu} \times \ell = 0.00096 \times 20$ ft $\times$ 12 in./ft $= 0.23$ in. or about $\frac{1}{4}$ in.

The creep coefficient at any time C_{ct} can be related to the ultimate creep coefficient C_{cu}. In Ref. 2.25, Branson suggests the equation

$$C_{ct} = \frac{t^{0.60}}{10 + t^{0.60}} C_{cu} \tag{2.8}$$

where t = time in days after loading.

In many special situations, for example, slender members or frames, or in prestressed construction, the designer must take account of the combined effects of creep and shrinkage (Section 2.11). In such cases, rather than rely on the sample values of Table 2.2, more accurate information on creep parameters should be obtained, such as from Ref. 2.25 or 2.28.

Sustained loads affect not only the deformation but also the strength of concrete. The cylinder strength f_c' is determined at normal rates of test loading (about 35 psi/sec). For concentrically loaded unreinforced concrete prisms and cylinders, the *strength under sustained load* is significantly lower than f_c', on the order of 75 percent of f_c' for loads maintained for a year or more (Refs. 2.26, 2.29, and 2.30). Thus, a member subjected to a sustained overload causing compressive stress of over 75 percent of f_c' may fail after a period of time, even though the load is not increased.

c. **Fatigue**

When concrete is subject to fluctuating rather than sustained loading, its *fatigue strength*, as for all other materials, is considerably lower than its static strength. When plain concrete in compression is stressed cyclically from zero to maximum stress, its fatigue limit is from 50 to 60 percent of the static compressive strength, for 2,000,000 cycles. For other types of applied stress, such as flexural compressive stress in reinforced concrete beams or flexural tension in unreinforced beams or on the tension side of reinforced beams, the fatigue limit likewise appears to be about 55 percent of the corresponding static strength. These figures, however, are for general guidance only. It is known that the fatigue strength of concrete depends not only on its static strength but also on moisture condition, age, and rate of loading (see Ref. 2.31).

2.9 PROPERTIES IN TENSION

While concrete is best employed in a manner that uses its favorable compressive strength, its behavior in tension is also important. The conditions under which cracks form and propagate on the tension side of reinforced concrete flexural members depend strongly on both the tensile strength and the fracture properties of the concrete, the latter dealing with the ease with which a crack progresses once it has formed. Concrete tensile stresses also occur as a result of shear, torsion, and other actions, and in most cases member behavior changes upon cracking. Thus, it is important to be able to predict, with reasonable accuracy, the tensile strength of concrete and to understand the factors that control crack propagation.

a. Tensile Strength

There are considerable experimental difficulties in determining the true tensile strength of concrete. In *direct tension* tests, minor misalignments and stress concentrations in the gripping devices are apt to mar the results. For many years, tensile strength has been measured in terms of the *modulus of rupture f_r*, the computed flexural tensile stress at which a test beam of plain concrete (shown in Fig. 2.8) fractures. Because this nominal stress is computed on the assumption that concrete is an elastic material, and because this bending stress is localized at the outermost surface, it is apt to be larger than the strength of concrete in uniform axial tension. It is thus a measure of, but not identical with, the real axial tensile strength.

The *splitting tensile strength test* also provides a measure of the tensile strength of concrete. A concrete cylinder, the same as is used for compressive tests, is inserted in a compression testing machine in the horizontal position, so that compression is applied uniformly along two opposite generators, as shown in Fig. 2.9. Pads are inserted between the compression platens of the machine and the cylinder to equalize and distribute the pressure. When an elastic cylinder so loaded, a nearly uniform

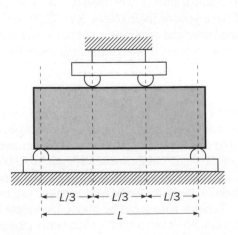

FIGURE 2.8
Schematic of flexure test to determine the modulus of rupture.

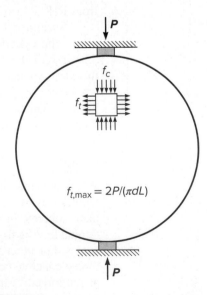

FIGURE 2.9
Schematic of splitting tensile strength test. Cylinder diameter $= d$ and length $= L$.

tensile stress of magnitude $2P/(\pi dL)$ exists at right angles to the plane of load application. Correspondingly, such cylinders, when tested, split into two halves along that plane, at a stress f_{ct} that can be computed from the above expression. P is the applied compressive load at failure, and d and L are the diameter and length of the cylinder, respectively. Because of local stress conditions at the load lines and the presence of stresses at right angles to the aforementioned tension stresses, the results of the split-cylinder tests likewise are not identical with (but are believed to be a good measure of) the true axial tensile strength. The results of all types of tensile tests show considerably more scatter than those of compression tests.

Tensile strength, however determined, does not correlate well with the compressive strength f_c'. It appears that for normalweight concrete, the tensile strength depends primarily on the strength of bond between hardened cement paste and aggregate, whereas for lightweight concretes it depends largely on the tensile strength of the porous aggregate. The compressive strength, on the other hand, is much less determined by these particular characteristics.

Better correlation is found between the various measures of tensile strength and the square root of the compressive strength. Typical ranges of values for direct tensile strength, split-cylinder strength, and modulus of rupture are summarized in Table 2.3. The direct tensile strength, for example, ranges from $3\sqrt{f_c'}$ to $5\sqrt{f_c'}$ for normalweight concrete and from $2\sqrt{f_c'}$ to $3\sqrt{f_c'}$ for all-lightweight concrete, while the modulus of rupture ranges from $8\sqrt{f_c'}$ to $12\sqrt{f_c'}$ for normalweight concrete and from $6\sqrt{f_c'}$ to $8\sqrt{f_c'}$ for all-lightweight concrete. In these expressions, f_c' is expressed in psi units, and the resulting tensile strengths are obtained in psi. The relationship between the modulus of rupture f_r and $8\sqrt{f_c'}$ and $12\sqrt{f_c'}$ is illustrated in Fig. 2.10.

These approximate expressions show that tensile and compressive strengths are by no means proportional and that any increase in compressive strength, such as that achieved by lowering the water-cement ratio, is accompanied by a much smaller percentage increase in tensile strength.

According to ACI Code 19.2.4, the modulus of rupture f_r is equal to $7.5\lambda\sqrt{f_c'}$, where λ can be based on either the equilibrium density of the concrete w_c (density after drying at 50 percent relative humidity of 73.4 ± 3.5 °F) or the composition of the aggregate. When based on w_c, λ equals 1.00 for $w_c \geq 135$ lb/ft^3 and 0.75 for $w_c \leq 100$ lb/ft^3. λ varies linearly between 1.00 and 0.75 for w_c between 135 and 100 lb/ft^3. When based on aggregate composition, λ equals 1.00 for normalweight concrete, 0.85 for "sand-lightweight" concrete, and 0.75 for "all-lightweight" concrete, giving values of $7.5\sqrt{f_c'}$, $6.4\sqrt{f_c'}$, and $5.6\sqrt{f_c'}$, respectively, for the three concrete types. In this case, values of λ between 0.75 and 1.00 are possible depending on the particular blend of lightweight and normalweight aggregates used in the concrete. Based on the test results shown in Fig. 2.10, $f_r = 7.5\sqrt{f_c'}$ is a safe lower bound for normalweight concretes with compressive strengths above 2000 psi.

TABLE 2.3
Approximate range of tensile strengths of concrete

	Normalweight Concrete, psi	Lightweight Concrete, psi
Direct tensile strength f_t'	3 to $5\sqrt{f_c'}$	2 to $3\sqrt{f_c'}$
Split-cylinder strength f_{ct}	6 to $8\sqrt{f_c'}$	4 to $6\sqrt{f_c'}$
Modulus of rupture f_r	8 to $12\sqrt{f_c'}$	6 to $8\sqrt{f_c'}$

FIGURE 2.10
Modulus of rupture versus
compressive strength for
normalweight concrete.
(*Based on data summarized
in Ref. 2.32*)

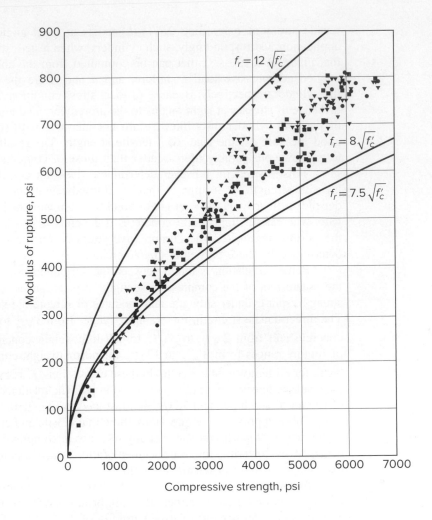

b.   Tensile Fracture

The failure of concrete in tension involves both the formation and the propagation of cracks. The field of fracture mechanics deals with the latter. While reinforced concrete structures have been successfully designed and built for over 150 years without the use of fracture mechanics, the brittle response of high-strength concretes (Section 2.12), in tension as well as compression, increases the importance of the fracture properties of the material as distinct from tensile strength. Research dealing with the shear strength of high-strength concrete beams and the bond between reinforcing steel and high-strength concrete indicates relatively low increases in these structural properties with increases in concrete compressive strength (Refs. 2.33 and 2.34). While shear and bond strength are associated with the $\sqrt{f_c'}$ for normal-strength concrete, tests of high-strength concrete indicate that increases in shear and bond strengths are well below values predicted using $\sqrt{f_c'}$, indicating that concrete tensile strength alone is not the governing factor. An explanation for this behavior is provided by research at the University of Kansas and elsewhere (Refs. 2.35 and 2.36) that demonstrates that the fracture energy, the energy required to fully open a crack (that is, after the crack has started to grow), is largely independent of compressive strength, water-cement ratio, and age, but is a function of the coarse aggregate. These points are illustrated in Fig. 2.11 for concretes with two types of coarse aggregate

FIGURE 2.11
Fracture energy versus
compressive strength for
concretes with basalt and
limestone coarse aggregates.
(*Based on data presented
by Ref. 2.35*)

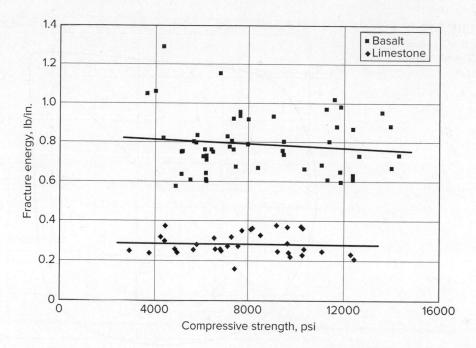

and compressive strengths between 3,000 and 15,000 psi. Design expressions reflecting this research are not yet available. The behavior is, however, recognized in the ACI Code by limitations on the maximum value of $\sqrt{f_c'}$ that may be used to calculate shear and bond strength, as will be discussed in Chapters 5 and 6.

2.10 STRENGTH UNDER COMBINED STRESS

In many structural situations, concrete is subjected simultaneously to various stresses acting in various directions. For instance, in beams much of the concrete is subject simultaneously to compression and shear stresses, and in slabs and footings to compression in two perpendicular directions plus shear. By methods well known from the study of engineering mechanics, any state of combined stress, no matter how complex, can be reduced to three principal stresses acting at right angles to one another on an appropriately oriented elementary cube in the material. Any or all of the principal stresses can be either tension or compression. If any one of them is zero, a state of *biaxial* stress is said to exist; if two of them are zero, the state of stress is *uniaxial*, either simple compression or simple tension. In most cases, only the uniaxial strength properties of a material are known from simple tests, such as the cylinder strength f_c' and the tensile strength f_t'. For predicting the strengths of structures in which concrete is subject to biaxial or triaxial stress, it would be desirable to be able to calculate the strength of concrete in such states of stress, knowing from tests only either f_c' or f_c' and f_t'.

In spite of extensive and continuing research, no general theory of the strength of concrete under combined stress has yet emerged. Modifications of various strength theories, such as maximum stress, maximum strain, the Mohr-Coulomb, and the octahedral shear stress theories, all of which are discussed in structural mechanics texts, have been adapted with varying partial success to concrete. At present, none of these theories has been generally accepted, and many have obvious internal contradictions. The main difficulty in developing an adequate general strength theory lies in the highly nonhomogeneous nature of concrete, and in the degree to which

FIGURE 2.12
Strength of concrete in
biaxial stress. (*Adapted from
Ref. 2.39.*)

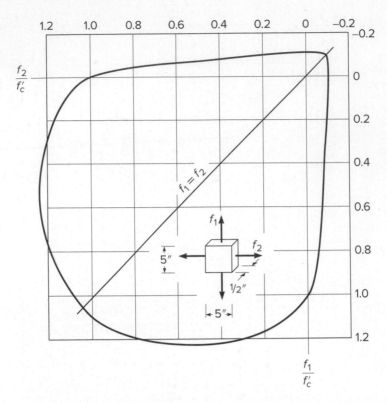

its behavior at high stresses and at fracture is influenced by microcracking and other discontinuity phenomena (Refs. 2.8 and 2.37).

The strength of concrete, however, has been well established by tests, at least for the biaxial stress state (Refs. 2.38 and 2.39). Results may be presented in the form of an interaction diagram such as Fig. 2.12, which shows the strength in direction 1 as a function of the stress applied in direction 2. All stresses are normalized in terms of the uniaxial compressive strength f_c'. It is seen that in the quadrant representing biaxial compression a strength increase as great as about 20 percent over the uniaxial compressive strength is attained, the amount of increase depending upon the ratio of f_2 to f_1. In the biaxial tension quadrant, the strength in direction 1 is almost independent of stress in direction 2. When tension in direction 2 is combined with compression in direction 1, the compressive strength is reduced almost linearly, and vice versa. For example, lateral compression of about one-half the uniaxial compressive strength will reduce the tensile strength by almost one-half compared with its uniaxial value. This fact is of great importance in predicting diagonal tension cracking in deep beams or shear walls, for example.

Experimental investigations into the triaxial strength of concrete have been few, due mainly to the practical difficulty of applying load in three directions simultaneously without introducing significant restraint from the loading equipment (Ref. 2.40). From information now available, the following conclusions can be drawn relative to the triaxial strength of concrete: (1) in a state of equal triaxial compression, concrete strength may be an order of magnitude larger than the uniaxial compressive strength; (2) for equal biaxial compression combined with a smaller value of compression in the third direction, a strength increase greater than 20 percent can be expected; and (3) for stress states including compression combined with tension in at least one other direction, the intermediate principal stress is of little consequence, and the compressive strength can be predicted safely based on Fig. 2.12.

In fact, the strength of concrete under combined stress cannot yet be calculated rationally, and, equally important, in many situations in concrete structures it is nearly impossible to calculate all of the acting stresses and their directions; these are two of the main reasons for continued reliance on tests. Because of this, the design of reinforced concrete structures continues to be based more on extensive experimental information than on consistent analytical theory, particularly in the many situations where combined stresses occur.

2.11 SHRINKAGE AND TEMPERATURE EFFECTS

The deformations discussed in Section 2.8 were induced by stresses caused by external loads. Influences of a different nature cause concrete, even when free of any external loading, to undergo deformations and volume changes. The most important of these are shrinkage and the effects of temperature variations.

a. Shrinkage

As discussed in Sections 2.2 and 2.4, any workable concrete mix contains more water than is needed for hydration. If the concrete is exposed to air, the larger part of this free water evaporates in time, the rate and completeness of drying depending on ambient temperature and humidity conditions. As the concrete dries, it shrinks in volume, due initially to the capillary tension that develops in the water remaining in the concrete (Ref. 2.7). Conversely, if dry concrete is immersed in water, it expands, regaining much of the volume loss from prior shrinkage. Shrinkage, which continues at a decreasing rate for years, depending on the configuration of the member, is a detrimental property of concrete in several respects. When not adequately controlled, it will cause unsightly and often deleterious cracks, as in slabs and walls. In structures that are statically indeterminate (and most concrete structures are), it can cause large and harmful stresses. In prestressed concrete it leads to partial loss of initial prestress. For these reasons it is essential that shrinkage be minimized and controlled.

As is clear from the nature of the process, a key factor in determining the amount of final shrinkage is the unit water content of the fresh concrete. This is illustrated in Fig. 2.13, which shows the amount of shrinkage for varying amounts of mixing water. The same aggregates were used for all tests, but in addition to and independently of the water content, the amount of cement was also varied from 376 to 1034 lb/yd^3 of concrete. This very large variation of cement content causes a 20 to 30 percent variation in shrinkage strain for water contents between 250 to 350 lb/yd^3, the range used for most structural concretes. Increasing the cement content increases the cement paste constituent of the concrete, where the shrinkage actually takes place, while reducing the aggregate content. Since most aggregates do not contribute to shrinkage, an increase in aggregate content can significantly decrease shrinkage. This is shown in Fig. 2.14, which compares the shrinkage of concretes with various aggregate contents with the shrinkage obtained for neat cement paste (cement and water alone). For example, increasing the aggregate content from 71 to 74 percent (at the same water-cement ratio) results in a 20 percent reduction in shrinkage (Ref. 2.28). Increased aggregate content may be obtained through the use of (1) a larger maximum size coarse aggregate (which also reduces the water content required for a given workability), (2) a concrete with lower workability, and (3) chemical admixtures to increase workability at lower water contents. It is evident that an effective means of reducing shrinkage involves both a reduction in water content and an

FIGURE 2.13
Effect of water content on
drying shrinkage.
(*Adapted from Ref. 2.3.*)

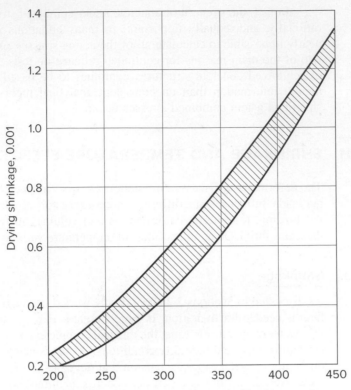

Pounds of water per cubic yard of fresh concrete

FIGURE 2.14
Influence of aggregate
content in concrete (by
volume) on the ratio of the
shrinkage of concrete to the
shrinkage of neat cement
paste. (*Adapted from Ref. 2.28,
based on data in Ref. 2.41.*)

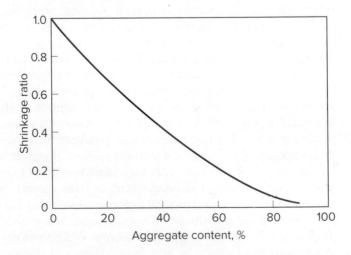

Aggregate content, %

increase in aggregate content. In addition, prolonged and careful curing is beneficial
for shrinkage control.

Values of final shrinkage for ordinary concretes are generally on the order of
400×10^{-6} to 800×10^{-6}, depending on the initial water content, ambient temperature and humidity conditions, and the nature of the aggregate. Highly absorptive
aggregates with low moduli of elasticity, such as some sandstones and slates, result
in shrinkage values 2 or more times those obtained with less absorptive materials,
such as granites and some limestones. Some lightweight aggregates, in view of their

great porosity, can result in much larger shrinkage values than ordinary concretes. On the other hand, pre-wetted fine lightweight aggregate can be used to provide *internal curing* and reduce early-age shrinkage.

For some purposes, such as predicting the time-dependent loss of force in pre-stressed concrete beams, it is important to estimate the amount of shrinkage as a function of time. Long-term studies (Ref. 2.25) show that, for moist-cured concrete at any time t after the initial 7 days, shrinkage can be predicted satisfactorily by the equation

$$\varepsilon_{sh,t} = \frac{t}{35 + t}\, \varepsilon_{sh,u} \tag{2.9}$$

where $\varepsilon_{sh,t}$ is the unit shrinkage strain at time t in days and $\varepsilon_{sh,u}$ is the ultimate value after a long period of time. Equation (2.9) pertains to "standard" conditions, defined in Ref. 2.25 to exist for humidity not in excess of 40 percent and for an average thickness of member of 6 in., and it applies both for normalweight and lightweight concretes. Modification factors are applied for nonstandard conditions, and separate equations are given for steam-cured members. Other, more detailed models for shrinkage are available that incorporate the ratio of volume to surface are for the member and environment factors, such as relative humidity (Ref. 2.42).

For structures in which a reduction in cracking is of particular importance, such as bridge decks, pavement slabs, and liquid storage tanks, *expansive cement concrete* or concretes containing *shrinkage compensating admixtures* or *shrinkage reducing admixtures* may be appropriate. Expansive cement concrete is made with shrinkage-compensating cement, which is constituted and proportioned so that the concrete will increase in volume after setting and during hardening. When the concrete is restrained by reinforcement or other means, the tendency to expand will result in compression. With subsequent drying, the shrinkage so produced, instead of causing a tension stress in the concrete that would result in cracking, merely reduces or relieves the expansive strains caused by the initial expansion (Ref. 2.43). Expansive cement is produced by adding a source of reactive aluminate to ordinary portland cement; approximately 90 percent of shrinkage-compensating cement is made up of the constituents of conventional portland cement. Of the three main types of expansive cements produced, only type K is commercially available in the United States; it is about 20 percent more expensive than ordinary portland cement (Ref. 2.44). Requirements for expansive cement are given in ASTM C845, "Standard Specification for Expansive Hydraulic Cement." Shrinkage-compensating admixtures, consisting of calcium oxide or magnesium oxide, also provide early expansion, and they are converted, respectively, to calcium hydroxide or magnesium hydroxide. Concretes containing shrinkage-compensating admixtures behave much like concrete containing an expansive cement. Shrinkage-reducing admixtures do not cause appreciable expansion but reduce shrinkage by reducing the surface tension of water and, thus, capillary tension. The usual admixtures can be used in concretes containing expansive or shrinkage-reducing agents, but trial mixes are necessary because some admixtures, particularly air-entraining agents, are not compatible with all expansive or shrinkage-reducing agents.

b. Effect of Temperature Change

Like most other materials, concrete expands with increasing temperature and contracts with decreasing temperature. The effects of such volume changes are similar to those caused by shrinkage; that is, temperature contraction can lead to objectionable cracking, particularly when superimposed on shrinkage. In indeterminate structures, deformations due to temperature changes can cause large and occasionally harmful stresses.

The coefficient of thermal expansion and contraction varies somewhat, depending upon the type of aggregate and richness of the mix. It is generally within the range of 4×10^{-6} to 7×10^{-6} per °F. A value of 5.5×10^{-6} per °F is generally accepted as satisfactory for calculating stresses and deformations caused by temperature changes (Ref. 2.7).

2.12   HIGH-STRENGTH CONCRETE

There are a number of applications in which *high-strength concrete* will provide improved structural performance. Although the exact definition is arbitrary, the term generally refers to concrete having uniaxial compressive strength in the range of about 8000 to 20,000 psi or higher. Such concretes can be made using carefully selected but widely available cements, sands, and coarse aggregates; certain admixtures including high-range water-reducing superplasticizers, fly ash, and silica fume; and very careful quality control during production (Refs. 2.45 and 2.46). In addition to higher strength in compression, most other engineering properties are improved, leading to use of the alternative term *high-performance concrete*.

The most common application of high-strength concretes has been in the columns of tall concrete buildings, where normal concrete would result in unacceptably large cross sections, with loss of valuable floor space. It has been shown that the use of the more expensive high-strength concrete mixes in columns not only saves floor area but also is more economical than increasing the amount of steel reinforcement. Concrete of up to 12,000 psi was specified for the lower-story columns of 311 South Wacker Drive in Chicago (see Fig. 2.15), a pioneering structure with a total height of 946 ft. Once holding the height record, it has been superseded by taller buildings; the present record is held by the tallest building and the tallest structure of any type in the world, the Burj Khalifa in Dubai, United Arab Emirates, shown in Fig. 19.2, which has a total height of 2717 ft.

For bridges, too, smaller cross sections bring significant advantages, and the resulting reduction in dead load permits longer spans. The higher elastic modulus and lower creep coefficient result in reduced initial and long-term deflections, and in the case of prestressed concrete bridges, initial and time-dependent losses of prestress force are less. Other applications of high-strength concrete include offshore oil structures, parking garages, bridge deck overlays, dam spillways, warehouses, and heavy industrial slabs.

An essential requirement for high-strength concrete is a low water–cementitious material ratio. For normal concretes, this usually falls in the range from about 0.40 to 0.60 by weight, but for high-strength mixes it may be 0.25 or even lower. To permit proper placement of what would otherwise be a zero slump mix, high-range water-reducing admixtures, or superplasticizers, are essential and may increase slumps to as much as 6 or 8 in. and even higher when viscosity-modifying admixtures are used to produce self-consolidating concrete. Other additives include fly ash and, most notably, silica fume (see Section 2.7).

Much research has been devoted to establishing the fundamental and engineering properties of high-strength concretes, as well as the engineering characteristics of structural members made with the material (Refs. 2.29, 2.30, and 2.46 to 2.50). A large body of information is available, permitting the engineer

FIGURE 2.15
311 South Wacker Drive, Chicago, which is among the
world's tallest buildings. High-strength concrete with $f_c' =$
12,000 psi was used in the lower stories. (*Courtesy of Portland
Cement Association*)

FIGURE 2.16
High-strength concrete test cylinder after uniaxial loading to
failure; note the typically smooth fracture surface, with little
aggregate interlock. (*Photograph by Arthur H. Nilson*)

to use high-strength concrete with confidence when its advantages justify the
higher cost. The compressive strength curves in Figs. 2.3 and 2.4 illustrate important differences compared with normal concrete, including a higher elastic modulus and an extended range of linear elastic response. Creep coefficients are
reduced, as indicated in Table 2.2. Disadvantages include brittle behavior in compression (see Fig. 2.16), somewhat reduced ultimate strain capacity, and an
increased tendency to crack when drying shrinkage is restrained (Ref. 2.51), the
latter resulting from the lower creep exhibited by the material. Strength under
sustained load is a higher fraction of standard cylinder strength (Refs. 2.20 and
2.30), and high-strength concrete exhibits improved durability and abrasion resistance (Refs. 2.52 and 2.53). As broader experience has been gained in practical
applications, and as design codes have been gradually updated to recognize the
special properties of higher-strength concrete, it is now recognized as the material
of choice where minimum member sizes are desired for compressive loading and
where maximum member stiffness is needed.

2.13 REINFORCING STEELS FOR CONCRETE

The useful strength of ordinary reinforcing steels in tension as well as compression, that is, the yield strength, is about 15 times the compressive strength of common structural concrete and well over 100 times its tensile strength. On the other hand, steel is a high-cost material compared with concrete. It follows that the two materials are best used in combination if the concrete is made to resist the compressive stresses and the steel the tensile stresses. Thus, in reinforced concrete beams, the concrete resists the compressive force, longitudinal steel reinforcing bars are located close to the tension face to resist the tension force, and usually additional steel bars are so placed to resist the inclined tension stresses that are caused by the shear force in the beams. Reinforcement, however, is also used for resisting compressive forces primarily where it is desired to reduce the cross-sectional dimensions of compression members, as in the lower-floor columns of multistory buildings. Even if no such necessity exists, a minimum amount of reinforcement is placed in all compression members to safeguard them against the effects of small accidental bending moments that might crack and even fail an unreinforced member.

For most effective reinforcing action, it is essential that steel and concrete deform together, that is, that there be a sufficiently strong *bond* between the two materials to ensure that no relative movements of the steel bars and the surrounding concrete occur. This bond is provided primarily by the *natural roughness* of the mill scale on the surface of hot-rolled reinforcing bars and by the closely spaced rib-shaped *surface deformations* that provide a high degree of interlock between the bars and the surrounding concrete.

Additional features that make for the satisfactory joint performance of steel and concrete are the following:

1. The *thermal expansion coefficients* of the two materials, about 6.5×10^{-6} per °F for steel vs. an average of 5.5×10^{-6} per °F for concrete, are sufficiently close to forestall cracking and other undesirable effects of differential thermal deformations.
2. While the *corrosion resistance* of bare steel is poor, the concrete that surrounds the steel reinforcement provides excellent corrosion protection, minimizing corrosion problems and corresponding maintenance costs.
3. The *fire resistance* of unprotected steel is impaired by its high thermal conductivity and by the fact that its strength decreases sizably at high temperatures. Conversely, the thermal conductivity of concrete is relatively low. Thus, damage caused by even prolonged fire exposure, if any, is generally limited to the outer layer of concrete, and a moderate amount of concrete cover provides sufficient thermal insulation for the embedded reinforcement.

Steel is used in two different ways in concrete structures: as reinforcing steel and as prestressing steel. Reinforcing steel is placed in the forms prior to casting of the concrete. Stresses in the steel, as in the hardened concrete, are caused only by the loads on the structure, except for possible parasitic stresses from shrinkage or similar causes. In contrast, in prestressed concrete structures, large tension forces are applied to the reinforcement prior to letting it act jointly with the concrete in resisting external loads. The steels for these two uses are very different and will be discussed separately.

2.14   REINFORCING BARS

The most common type of reinforcing steel (as distinct from prestressing steel) is in the form of round bars, often called *rebars*, available in a large range of diameters from about $\frac{3}{8}$ to $1\frac{3}{8}$ in. for ordinary applications and in two heavy bar sizes of about $1\frac{3}{4}$ and $2\frac{1}{4}$ in. These bars are furnished with surface deformations for the purpose of increasing resistance to slip between steel and concrete. Minimum requirements for these deformations (spacing, projection, etc.) have been developed in experimental research. Different bar producers use different patterns, all of which satisfy these requirements. Figure 2.17 shows a variety of current types of deformations.

For many years, bar sizes have been designated by numbers, Nos. 3 to 11 being commonly used and Nos. 14 and 18 representing the two special large-sized bars previously mentioned. Designation by number, instead of by diameter, was introduced because the surface deformations make it impossible to define a single easily measured value of the diameter. The numbers are so arranged that the unit in the number designation corresponds closely to the number of $\frac{1}{8}$ in. of diameter size. A No. 5 bar, for example, has a nominal diameter of $\frac{5}{8}$ in. Bar sizes are rolled into the surface of the bars for easy identification.

In addition to the usual numbering, bars can also be designated in accordance with the International System of Units (SI), with the size being identified using the nominal diameter in millimeters. Thus, Nos. 3 to 11 bars can be marked with Nos. 10 to 36, and Nos. 14 and 18 bars with Nos. 43 and 57. Both systems are used in the ASTM standards, but the customary system is used in the ACI Code. To recognize the dual system of identifying and marking the bars, the customary bar designation system is retained throughout this text, followed by the SI bar designations in parentheses, such as No. 6 (No. 19). Table A.1 of Appendix A gives areas and weights of standard bars. Tables A.2 and A.3 give similar information for groups of bars.

FIGURE 2.17
Types of deformed reinforcing bars. (*Photograph by Arthur H. Nilson*)

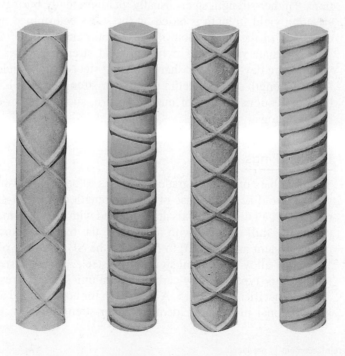

a. **Grades and Strengths**

In reinforced concrete, a long-term trend is evident toward the use of higher-strength materials, both steel and concrete. Reinforcing bars with 40 ksi yield strength, once standard, have largely been replaced by bars with 60 ksi yield strength, both because they are more economical and because their use tends to reduce steel congestion in the forms. Bars with yield strengths of 80 and 100 ksi are often used in columns and walls and as confining reinforcement. Bars with a yield strength of 120 ksi are also available but not yet recognized by the ACI Code. Table 2.4 lists all presently available reinforcing steels, their grade designations, the ASTM specifications that define their properties (including deformations) in detail, and their two main minimum specified strength values. Grade 40 bars are no longer available in sizes larger than No. 6 (No. 19) and Grade 50 bars are available in sizes up to No. 8 (No. 25).[†]

The conversion to SI units described above also applies to the strength grades. Thus, Grade 40 is also designated as Grade 280 (for a yield strength of 280 MPa), Grades 60 and 80 are designated Grades 420 and 550, and Grades 100 and 120 are designated Grades 690 and 830. The values 280, 420, 550, 690, and 830 result in minimum yield strengths of 40.6, 60.9, 79.8, 100.1, and 120.4 ksi. Grades based on inch-pound units are used in this text.

Most reinforced concrete in the U.S. is constructed using ASTM A615 carbon-steel bars. ASTM A706 low-alloy steel bars are usually specified, however, for structures designed for seismic loading because they are more ductile than A615 bars. ASTM A1035 is often used when high-strength steel is needed.

Welding of reinforcing bars in making splices, or for convenience in fabricating reinforcing cages for placement in the forms, may result in metallurgical changes that reduce both strength and ductility, and special restrictions must be placed both on the type of steel used and the welding procedures. The provisions of ASTM A706 relate to welding, as well as ductility.

The ACI Code permits reinforcing steels up to $f_y = 100$ ksi for most applications. Higher-strength steels usually yield gradually but have no yield plateau. In this situation yield strength is based on the stress determined by the 0.2% offset method, explained in the description of stress-strain curves that follows (see Section 2.13c). This alternate method of defining yield strength allows current design methods, which were developed for sharp-yielding steels with a yield plateau, to be used with higher-strength steels. Steel in this higher-strength range is often used in cases where high deflections are not of major concern, such as in lower-story columns of high-rise buildings.

b. **Bar Markings**

To allow bars of various grades and sizes to be easily distinguished, which is necessary to avoid accidental use of lower-strength or smaller-size bars than called for in the design, all deformed bars are furnished with rolled-in markings. These identify the producing mill (usually with an initial), the bar size (Nos. 3 to 18 under the inch-pound system and Nos. 10 to 57 under the SI), the type of steel (*S* for carbon steel, *W* for low-alloy steel, a rail sign for rail steel, *A* for axle steel, and *CL*, *CM*, or *CS* for the various types of low-carbon chromium steel, corresponding, respectively, to ASTM Specifications A615, A706, A996 for both rail and axle steel, and A1035), and an additional marking to identify higher-strength steels. Grade 60 (420) bars have

---

[†] In practice, very little Grade 50 reinforcement is produced.

TABLE 2.4
Summary of minimum ASTM strength requirements

Product	ASTM Specification	Designation	Minimum Yield Strength, psi (MPa)	Minimum Tensile Strength, psi (MPa)
Reinforcing bars	A615	Grade 40	40,000 (280)	60,000 (420)
		Grade 60	60,000 (420)	90,000 (620)
		Grade 80	80,000 (550)	105,000 (725)
		Grade 100	100,000 (690)	115,000 (790)
	A706	Grade 60	60,000 (420) [78,000 (540) maximum]	80,000 (550)[a]
		Grade 80	80,000 (550) [98,000 (540) maximum]	100,000 (690)[a]
	A996	Grade 40	40,000 (280)	60,000 (420)
		Grade 50	50,000 (350)	80,000 (550)
		Grade 60	60,000 (420)	90,000 (620)
	A1035	Grade 100	100,000 (690)	150,000 (1030)
		Grade 120	120,000 (830)	150,000 (1030)
Deformed bar mats	A184	Same as Grades 40 and 60 A615 and A706 reinforcing bars		
Zinc-coated bars	A767, A1094	Same as reinforcing bars		
Epoxy-coated bars	A775, A934, A1055	Same as reinforcing bars		
Stainless steel bars	A955	Available in Grades 60, 75, and 80		
Wire	A1064	Grades 70 to 80[b]	70,000 (485)	80,000 (550)
Plain			80,000 (550)	90,000 (620)
Deformed		Grades 75 to 80[b]	75,000 (515)	85,000 (585)
			80,000 (550)	90,000 (620)
Welded wire reinforcement	A1064			
Plain		Grades 65 to 80[b]	65,000 (450)	75,000 (515)
W1.2 and larger			80,000 (550)	90,000 (620)
Smaller than W1.2		Grade 56	56,000 (385)	70,000 (485)
Deformed		Grades 70 to 80[b]	70,000 (485)	80,000 (550)
			80,000 (550)	90,000 (620)
Prestressed reinforcement	A416			
Low-relaxation, seven-wire strand		Grade 250	225,000 (1555)	250,000 (1725)
		Grade 270	243,000 (1675)	270,000 (1860)
Wire	A421	BA wire	199,750 (1377) or 204,000 (1407)[c]	235,000 (1620) or 240,000 (1620)[c]
		WA wire	199,750 (1377) to 212,500 (1465)[c]	235,000 (1620) to 250,000 (1725)[c]
Bars	A722	Type I (plain)	127,500 (800)	150,000 (1035)
		Type II (deformed)	120,000 (825)	150,000 (1035)
Compacted strand	A779	Grades 245 to 270 (normal-relaxation)	214,800 (1481) to 235,000 (1620)[c]	247,000 (1700) to 270,000 (1860)[c]
		Grades 245 to 270 (low-relaxation)	222,300 (1533) to 243,000 (1675)[c]	247,000 (1700) to 270,000 (1860)[c]

[a] But not less than 1.25 times the actual yield strength.
[b] Intermediate grades above 70 increase by 2.5 (72.5, 75, etc.).
[c] Minimum strengths depend on wire or strand size.

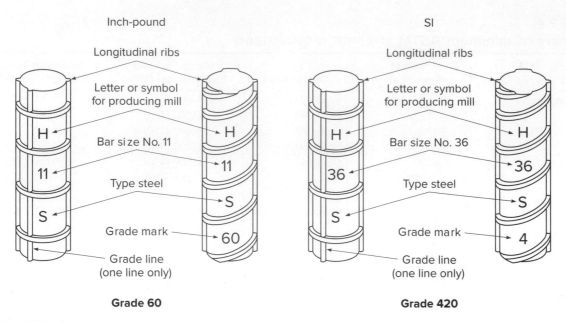

FIGURE 2.18
Marking system for reinforcing bars meeting ASTM Specifications A615, A706, and A996. (*Adapted from Ref. 2.54.*)

either one longitudinal line or the number 60 (4); Grade 80 (550) bars have either three longitudinal lines or the number 80 (6); Grade 100 (690) bars have either three or four longitudinal lines or the number 100 (6 or 7 in SI)*; and Grade 120 (830) bars have either four longitudinal lines or the number 120 (8). The identification marks are shown in Fig. 2.18 for Grade 60 (420) bars.

c. Stress-Strain Curves

The two chief numerical characteristics that determine the character of bar reinforcement are its *yield point* (generally identical in tension and compression) and its *modulus of elasticity E_s*. The latter is practically the same for all reinforcing steels (but not for prestressing steels) and is taken as $E_s = 29,000,000$ psi.

In addition, however, the shape of the stress-strain curve, and particularly of its initial portion, has significant influence on the performance of reinforced concrete members. Typical stress-strain curves for U.S. reinforcing steels are shown in Fig. 2.19. The complete stress-strain curves are shown in the left part of the figure; the right part gives the initial portions of the curves magnified 10 times.

Low-carbon steels, typified by the Grade 40 curve, show an elastic portion followed by a *yield plateau*, that is, a horizontal portion of the curve where strain continues to increase at constant stress. For such steels, the yield point is that stress at which the yield plateau establishes itself. With further strains, the stress begins to increase again, though at a slower rate, a process that is known as *strain-hardening*. The curve flattens out when the *tensile strength* is reached; it then turns down until fracture occurs. Higher-strength carbon steels, for example, those with 60 ksi yield

---

* Grade 100 (690) A615 bars use four longitudinal lines and, in SI, the number 7, while Grade 100 A1035 bars use three longitudinal lines and, in SI, the number 6 to designate grade. They can, however, be distinguished by the marking indicating the type of steel.

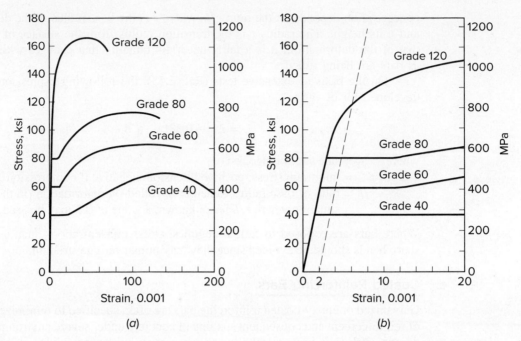

FIGURE 2.19
Typical stress-strain curves for reinforcing bars.

stress or higher, either have a yield plateau of much shorter length or enter strain-hardening immediately without any continued yielding at constant stress. In the latter case, the yield stress f_y is determined using the 0.2 percent offset method. Using this method, a line with a strain intercept of 0.2 percent (or 0.002) is drawn parallel to the initial elastic portion of the stress-strain curve. The yield stress f_y is defined by the point at which this line intercepts the stress-strain curve, as shown in Fig. 2.19*b*. Low-alloy, high-strength steels rarely show any yield plateau and usually enter strain-hardening immediately upon beginning to yield.

d. Fatigue Strength

In highway bridges and some other situations, both steel and concrete are subject to large numbers of stress fluctuations. Under such conditions, steel, just like concrete (Section 2.8c), is subject to *fatigue*. In metal fatigue, one or more microscopic cracks form after cyclic stress has been applied a significant number of times. These fatigue cracks occur at points of stress concentrations or other discontinuities and gradually increase with increasing numbers of stress fluctuations. This reduces the remaining uncracked cross-sectional area of the bar until it becomes too small to resist the applied force. At this point the bar fails in a sudden, brittle manner.

For reinforcing bars it has been found (Refs. 2.31 and 2.55) that the fatigue strength, that is, the stress at which a given stress fluctuation between f_{max} and f_{min} can be applied 2 million times or more without causing failure, is practically independent of the grade of steel. It has also been found that the stress range, that is, the algebraic difference between maximum and minimum stress, $f_1 = f_{max} - f_{min}$, that can be sustained without fatigue failure depends on f_{min}. Further, in deformed bars, the degree of stress concentration at the location where the deformation joins the main cylindrical body of the bar tends to reduce the safe stress range. This stress concentration depends on the ratio r/h, where r is the base radius of the deformation and h its height. The radius r is the transition radius from the surface of the bar to that of the deformation; it is a fairly uncertain quantity that changes with roll wear as bars are being rolled.

On the basis of extensive tests (Ref. 2.55), the following expression has been developed for design:

$$f_r = 21 - 0.33f_{min} + 8\,\frac{r}{h} \tag{2.10}$$

where f_r = safe stress range, ksi
         f_{min} = minimum stress; positive if tension, negative if compression
         r/h = ratio of base radius to height of rolled-on deformation (in the common situation where r/h is not known, a value of 0.3 may be used)

Where bars are exposed to fatigue regimes, stress concentrations such as welds or sharp bends should be avoided since they may impair fatigue strength.

e.   Coated Reinforcing Bars

Galvanized or epoxy-coated reinforcing bars are often specified to minimize corrosion of reinforcement and consequent spalling of concrete under severe environmental conditions, such as in bridge decks or parking garages subject to deicing chemicals, port and marine structures, and wastewater treatment plants.

Epoxy-coated bars, presently more widely used than galvanized bars, are governed by ASTM A775, "Standard Specification for Epoxy-Coated Reinforcing Steel Bars," ASTM A934, "Standard Specification for Epoxy-Coated Prefabricated Steel Reinforcing Bars," and ASTM A1055, "Standard Specification for Zinc-Epoxy Dual-Coated Steel Reinforcing Bars," which includes requirements for the coating material, surface preparation prior to coating, method of application, and limits on coating thickness. Under ASTM A775, the coating is applied to straight bars in a production-line operation, and the bars are cut and bent after coating. Under ASTM A934, bars are bent to final shape prior to coating. ASTM A1055 covers bars that are sprayed with zinc followed by a conventional epoxy coating. Cut ends and small spots of damaged coating are suitably repaired after fabrication. Extra care is required in the field to ensure that the coating is not damaged during shipment and placing and that repairs are made if necessary.

ASTM A767, "Standard Specification for Zinc-Coated (Galvanized) Steel Bars for Concrete Reinforcement," and ASTM A1094, "Standard Specification for Continuous Hot-Dip Galvanized Steel Bars for Concrete Reinforcement," include requirements for the zinc-coating material, the galvanizing process, the class or weight of coating, finish and adherence of the coating, and the method of fabrication. Supplementary requirements pertain to coating of sheared ends and repair of damaged coating when bars are fabricated after galvanizing.

2.15   WELDED WIRE REINFORCEMENT

Apart from single reinforcing bars, *welded wire reinforcement* (also described as *welded wire fabric*) is often used for reinforcing slabs and other surfaces, such as shells, and for shear reinforcement in thin beam webs, particularly in prestressed beams. Welded wire reinforcement consists of sets of longitudinal and transverse cold-drawn steel wires at right angles to each other and welded together at all points of intersection. The size and spacing of wires may be the same in both directions or may be different, depending on the requirements of the design.

The notation used to describe the type and size of welded wire fabric involves a letter-number combination. ASTM uses the letter "W" to designate plain wire and letter "D" to describe deformed wire. In most cases, deformed wire is produced by indenting the wire during the cold-drawing process. The number following the letter gives the cross-sectional area of the wire in hundredths of a square inch. For example, a W5.0 wire is a smooth wire with a cross-sectional area of 0.05 in^2. A W5.5 wire has a cross-sectional area of 0.055 in^2. D6.0 indicates a deformed wire with a cross-sectional area of 0.06 in^2. Welded wire fabric having a designation $4 \times 4 - W5.0 \times W5.0$ has wire spacings 4 in. in each way with smooth wire of cross-sectional area 0.05 in^2 in each direction. Sizes and spacings for common types of welded wire fabric and cross-sectional areas of steel per foot, as well as weight per 100 ft^2, are shown in Table A.12 of Appendix A.

ASTM Specification A1064 covers both smooth and deformed welded wire reinforcement, as shown in Table 2.4. Deformed wire larger than D31 must be treated as plain wire because these larger size wires exhibit reduced bond strength compared to deformed bars.

2.16 PRESTRESSING STEELS

Prestressing steel is used in three forms: round wires, strands, and alloy steel bars. Prestressing wire ranges in diameter from 0.192 to 0.276 in. It is made by cold-drawing high-carbon steel after which the wire is stress-relieved by heat treatment to produce the prescribed mechanical properties. Wires are normally bundled in groups of up to about 50 individual wires to produce prestressing tendons of the required strength. Strands, more common than wire in U.S. practice, are fabricated with six wires wound around a seventh of slightly larger diameter. The pitch of the spiral winding is between 12 and 16 times the nominal diameter of the strand. Strand diameters range from 0.250 to 0.700 in. Alloy steel bars for prestressing are available in diameters from 0.750 to 1.375 in. as plain round bars and from 0.625 to 3.00 in. as deformed bars, with the largest size deformed bars serving as ground anchors. Specific requirements for prestressing steels are found in ASTM A421, "Standard Specification for Uncoated Stress-Relieved Steel Wire for Prestressed Concrete"; ASTM A416, "Standard Specification for Low-Relaxation, Seven-Wire Strand for Prestressed Concrete"; and ASTM A722, "Standard Specification for High-Strength Steel Bars for Prestressed Concrete." Table A.15 of Appendix A provides design information for U.S. prestressing steels.

a. Grades and Strengths

The tensile strengths of prestressing steels range from about 2.5 to 6 times the yield strengths of commonly used reinforcing bars. The grade designations correspond to the minimum specified tensile strength in ksi. For the widely used seven-wire strand, three grades are available: Grade 250 ($f_{pu} = 250$ ksi), Grade 270, and Grade 300, although the last is not yet recognized in ASTM A416. Grade 270 strand is used most often. For alloy steel bars, two grades are used: the regular Grade 150 is most common, but special Grade 160 bars may be ordered. Round wires may be obtained in Grades 235, 240, and 250, depending on diameter.

b. Stress-Strain Curves

Figure 2.20 shows stress-strain curves for prestressing wires, strand, and alloy bars of various grades. For comparison, the stress-strain curve for a Grade 60 reinforcing bar is also shown. It is seen that, in contrast to reinforcing bars, prestressing steels do not show a sharp yield point or yield plateau; that is, they do not yield at constant or nearly constant stress. Yielding develops gradually, and in the inelastic range the curve continues to rise smoothly until the tensile strength is reached. Because well-defined yielding is not observed in these steels, the yield strength is somewhat arbitrarily defined as the stress at a total elongation of 1 percent for strand and wire and at 0.7 percent for alloy steel bars. Figure 2.20 shows that the yield strengths so defined represent a good limit below which stress and strain are fairly proportional and above which strain increases much more rapidly with increasing stress. It is also seen that the spread between tensile strength and yield strength is smaller in prestressing steels than in reinforcing steels. It may further be noted that prestressing steels have significantly less ductility.

While the modulus of elasticity E_s for deformed bars is taken as 29,000,000 psi, the effective modulus of prestressing steel varies, depending on the type of steel (for example, strand vs. wire or bars) and type of use, and is best determined by test or supplied by the manufacturer. The modulus of elasticity of prestressing steel E_p has been shown to have values of 26,000,000 psi for unbonded strand (that is, strand not embedded in concrete), 27,000,000 psi for bonded strand and alloy steel bars, and about 29,000,000 psi for smooth wires, the same as for deformed reinforcing bars. ACI Commentary 20.3.2.1, however, indicates that values between 28,500,000 and 29,000,000 psi are often used in design but states that values based on tests may be needed when checking elongation during stressing operations.

FIGURE 2.20
Typical stress-strain curves for prestressing steels.

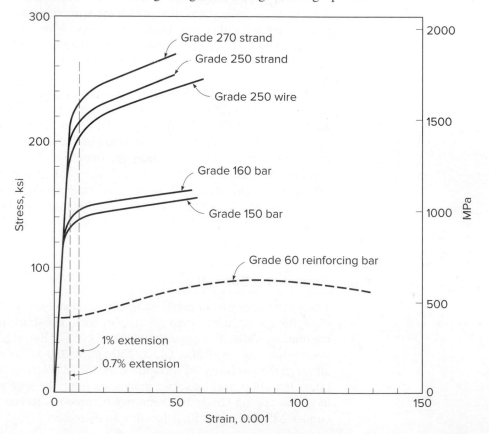

c. Relaxation

When prestressing steel is stressed to the levels that are customary during initial tensioning and at service loads, it exhibits a property known as *relaxation*. Relaxation is defined as the loss of stress in stressed material held at constant length. (The same basic phenomenon is known as creep when defined in terms of change in strain of a material under constant stress.) To be specific, if a length of prestressing steel is stressed to a sizable fraction of its yield strength f_{py} (say, 80 to 90 percent) and held at a constant strain between fixed points such as the ends of a beam, the steel stress f_p will gradually decrease from its initial value f_{pi}. In prestressed concrete members this stress relaxation is important because it modifies the internal stresses in the concrete and changes the deflections of the beam some time after initial prestress was applied.

The amount of relaxation varies, depending on the type and grade of steel, the time under load, and the initial stress level. A satisfactory estimate for stress-relieved compacted strand and wires can be obtained from Eq. (2.11), which was derived from more than 400 relaxation tests of up to 9 years' duration:

$$\frac{f_p}{f_{pi}} = 1 - \frac{\log t}{10}\left(\frac{f_{pi}}{f_{py}} - 0.55\right) \qquad (2.11)$$

where f_p is the final stress after t hours, f_{pi} is the initial stress, and f_{py} is the nominal yield stress (Ref. 2.56). In Eq. (2.11), log t is to the base 10, and f_{pi}/f_{py} not less than 0.55; below that value essentially no relaxation occurs.

The tests on which Eq. (2.11) is based were carried out on round, stress-relieved wires and are equally applicable to stress-relieved strand. In the absence of other information, results may be used for alloy steel bars as well.

Low-relaxation strand, now the industry standard, is specified under ASTM A416. Such steel must exhibit relaxation after 1000 hours of not more than 2.5 percent when initially stressed to 70 percent of specified tensile strength and not more than 3.5 percent when loaded to 80 percent of tensile strength. For low-relaxation strand, Eq. (2.11) is replaced by

$$\frac{f_p}{f_{pi}} = 1 - \frac{\log t}{45}\left(\frac{f_{pi}}{f_{py}} - 0.55\right) \qquad (2.12)$$

2.17 FIBER REINFORCEMENT

In addition to using bars, wire, welded wire reinforcement, and prestressing steel, the ACI Code recognizes that discrete steel fibers can, on a limited basis, be used to improve the tensile properties of concrete. The application—to serve as a design alternative to minimum shear reinforcement—is discussed in Chapter 5.

Steel fibers may provide a small increase in concrete tensile strength, but their main contribution is to increase the toughness of the concrete, that is, allowing the concrete to continue to carry a tensile load once the peak tensile strength has been reached, as shown for mortar in Fig 2.21.

To qualify as an alternative to minimum shear reinforcement, fiber-reinforced concrete must exhibit minimum values of residual strength when tested in flexure in accordance with ASTM C1609. Figure 2.22 shows an example load-deflection curve for a fiber-reinforced concrete beam with length L loaded in flexure.

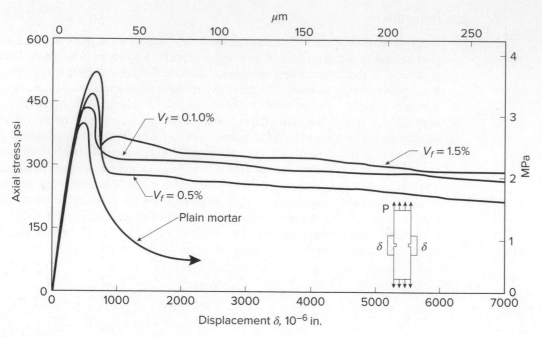

FIGURE 2.21
Stress-displacement curves for mortars with volume fractions V_f of steel fibers ranging from 0 to 1.5%. (*Adapted from Ref. 2.57.*)

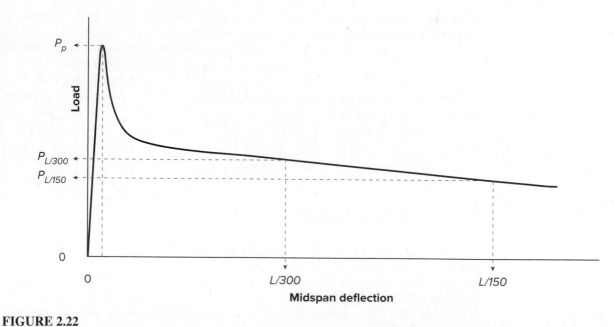

FIGURE 2.22
Load-deflection curve for fiber-reinforced beam loaded in flexure. (*Adapted from ASTM C1609 / C1609M - 19. Standard Test Method for Flexural Performance of Fiber-Reinforced Concrete (Using Beam With Third-Point Loading). ASTM International. https://www.astm.org/Standards/ C1609.htm*)

ACI Code 26.12.7.1 requires that (1) the residual strength $P_{L/300}$ obtained at a midspan deflection of $L/300$ be at least equal to the greater of 90 percent of the measured peak strength P_p obtained from the test and 90 percent of the strength corresponding to a stress of $7.5\sqrt{f_c'}$ and (2) the residual strength $P_{L/150}$ obtained at a midspan deflection of $L/150$ be at least equal to the greater of 75 percent of the measured peak strength P_p obtained from the test and 75 percent of the strength corresponding to $7.5\sqrt{f_c'}$.

REFERENCES

2.1. ACI Committee 221, "Guide for Use of Normal Weight and Heavyweight Aggregate in Concrete," American Concrete Institute, Farmington Hills, MI, 1996.

2.2. ACI Committee 213, "Guide for Structural Lightweight Aggregate Concrete," American Concrete Institute, Farmington Hills, MI, 2014.

2.3. G. E. Troxell, H. E. Davis, and J. W. Kelly, *Composition and Properties of Concrete*, 2nd ed., McGraw-Hill, New York, 1968.

2.4. T. T. C. Hsu and F. O. Slate, "Tensile Bond Strength between Aggregate and Cement Paste or Mortar," *J. ACI*, vol. 60, no. 4, 1963, pp. 465–486.

2.5. ACI Committee 211, "Standard Practice for Selecting Proportions for Normal, Heavyweight, and Mass Concrete," American Concrete Institute, Farmington Hills, MI, 1991.

2.6. ACI Committee 211, "Standard Practice for Selecting Proportions for Structural Lightweight Concrete," American Concrete Institute, Farmington Hills, MI, 1998.

2.7. S. Mindess, J. F. Young, and D. Darwin, *Concrete*, 2nd ed., Prentice-Hall, Upper Saddle River, NJ, 2003.

2.8. M. L. Wilson and S. A. Kosmatka, *Design and Control of Concrete Mixtures*, 16th ed., Portland Cement Association, Skokie, IL, 2016.

2.9. ACI Committee 309, "Guide for Consolidation of Concrete," American Concrete Institute, Farmington Hills, MI, Part 2, 2005.

2.10. ACI Committee 304, "Guide for Measuring, Transporting, and Placing Concrete," American Concrete Institute, Farmington Hills, MI, Part 2, 2000.

2.11. ACI Committee 306, "Guide to Cold Weather Concreting," American Concrete Institute, Farmington Hills, MI, 2016.

2.12. ACI Committee 214, "Guide to Evaluation of Strength Test Results of Concrete," American Concrete Institute, Farmington Hills, MI, 2011.

2.13. ACI Committee 301, "Specifications for Structural Concrete," American Concrete Institute, Farmington Hills, MI, 2016.

2.14. ACI Committee 212, "Report on Chemical Admixtures for Concrete," American Concrete Institute, Farmington Hills, MI, 2016.

2.15. ACI Committee 237, "Self-Consolidating Structural Concrete," American Concrete Institute, Farmington Hills, MI, Part 2, 2007.

2.16. R. J. Peterman, "The Effects of As-Cast Depth and Concrete Fluidity on Strand Bond," *PCI J.*, vol. 52, no. 3, 2007, pp. 72–101.

2.17. ACI Committee 232, "Report in the Use of Fly Ash in Concrete," American Concrete Institute, Farmington Hills, MI, 2018.

2.18. ACI Committee 234, "Guide for the Use of Silica Fume in Concrete," American Concrete Institute, Farmington Hills, MI, 2006.

2.19. ACI Committee 233, "Guide to Use of Slag Cement in Concrete and Mortar," American Concrete Institute, Farmington Hills, MI, 2017.

2.20. V. M. Malhotra, "Fly Ash, Silica Fume, and Rice-Husk Ash in Concrete: A Review," *Concr. Intl.*, vol. 15, no. 4, 1993, pp. 23–28.

2.21. G. Detwiler, "High-Strength Silica Fume Concrete—Chicago Style," *Concr. Intl.*, vol. 14, no. 10, 1992, pp. 32–36.

2.22. R. L. Carrasquillo, A. H. Nilson, and F. O. Slate, "Properties of High Strength Concrete Subject to Short Term Loads," *J. ACI*, vol. 78, no. 3, 1981, pp. 171–178.

2.23. F. O. Slate, A. H. Nilson, and S. Martinez, "Mechanical Properties of High-Strength Lightweight Concrete," *J. ACI*, vol. 83, no. 4, 1986, pp. 606–613.

2.24. S. H. Ahmad and S. P. Shah, "Behavior of Hoop Confined Concrete under High Strain Rates," *J. ACI*, vol. 82, no. 5, 1985, 634–647.

2.25. D. E. Branson, *Deformation of Concrete Structures*, McGraw-Hill, New York, 1977.

2.26. H. Rüsch, "Researches toward a General Flexural Theory for Structural Concrete," *J. ACI*, vol. 32, no. 1, 1960, pp. 1–28.

2.27. A. S. Ngab, A. H. Nilson, and F. O. Slate, "Shrinkage and Creep of High-Strength Concrete," *J. ACI*, vol. 78, no. 4, 1981, pp. 255–261.

2.28. A. M. Neville, *Properties of Concrete*, 4th ed., John Wiley & Sons, New York, 1996.

2.29. M. M. Smadi, F. O. Slate, and A. H. Nilson, "High, Medium, and Low-Strength Concretes Subject to Sustained Overloads," *J. ACI*, vol. 82, no. 5, 1985, pp. 657–664.

2.30. M. M. Smadi, F. O. Slate, and A. H. Nilson, "Shrinkage and Creep of High, Medium, and Low-Strength Concretes, Including Overloads," *ACI Mater. J.*, vol. 84, no. 3, 1987, pp. 224–234.

2.31. "Fatigue of Concrete Structures," *Special Publication* SP-75, American Concrete Institute, Detroit, MI, 1982.

2.32. J. M. Raphael, "Tensile Strength of Concrete," *J. ACI*, vol. 81, no. 2, 1984, pp. 158–165.

2.33. M. P. Collins and D. Kuchma, "How Safe Are Our Large, Lightly Reinforced Concrete Beams, Slabs, and Footings?" *ACI Struct. J.*, vol. 96, no. 4, 1999, pp. 482–490.

2.34. J. Zuo and D. Darwin, "Splice Strength of Conventional and High Relative Rib Area Bars in Normal and High Strength Concrete," *ACI Struct. J.*, vol. 97, no. 4, 2000, pp. 630–641.

2.35. D. Darwin, S. Barham, R. Kozul, and S. Luan, "Fracture Energy of High-Strength Concrete," *ACI Mater. J.*, vol. 98, no. 5, 2001, pp. 410–417.

2.36. E. A. Jensen and W. Hansen, "Fracture Energy Test for Highway Concrete—Determining the Effect of Coarse Aggregate on Crack Propagation Resistance," *Transp. Res. Rec.*, vol. 1730, 2001, pp. 10–16.

2.37. T. T. C. Hsu, F. O. Slate, G. M. Sturman, and G. Winter, "Microcracking of Plain Concrete and the Shape of the Stress-Strain Curve," *J. ACI*, vol. 60, no. 2, 1963, pp. 209–224.

2.38. H. Kupfer, H. K. Hilsdorf, and H. Rüsch, "Behavior of Concrete under Biaxial Stresses," *J. ACI*, vol. 66, no. 8, 1969, pp. 656–666.

2.39. M. E. Tasuji, F. O. Slate, and A. H. Nilson, "Stress-Strain Response and Fracture of Concrete in Biaxial Loading," *J. ACI*, vol. 75, no. 7, 1978, pp. 306–312.

2.40. K. H. Gerstle et al., "Strength of Concrete under Multiaxial Stress States," *Proc. Douglas McHenry International Symposium on Concrete and Concrete Structures*, ACI Special Publication SP-55, American Concrete Institute, Farmington Hills, MI, 1978, pp. 103–131.

2.41. G. Pickett, "Effect of Aggregate on Shrinkage of Concrete and Hypothesis Concerning Shrinkage," *J. ACI*, vol. 52, no. 6, 1956, pp. 581–589.

2.42. ACI Committee 209, "Guide for Modeling and Calculating Shrinkage and Creep in Hardened Concrete," American Concrete Institute, Farmington Hills, MI, 2008.

2.43. ACI Committee 223, "Guide for the Use of Shrinkage-Compensating Concrete," American Concrete Institute, Farmington Hills, MI, 2010.

2.44. A. Neville, "Whither Expansive Cement," *Concr. Intl.*, vol. 16, no. 9, 1994, pp. 34–35.

2.45. ACI Committee 363, "Report on High-Strength Concrete," American Concrete Institute, Farmington Hills, MI, 2010.

2.46. ACI Committee 363, "Guide to Quality Control and Testing of High-Strength Concrete," ACI Committee 363, American Concrete Institute, Farmington Hills, MI, 2011.

2.47. A. H. Nilson, "Properties and Performance of High-Strength Concrete," *Proc. of IABSE Symposium on Concrete Structures for the Future*, Paris-Versailles, 1987, pp. 389–394.

2.48. K. A. Paulson, A. H. Nilson, and K. C. Hover, "Long-Term Deflection of High-Strength Concrete Beams," *ACI Mater. J.*, vol. 88, no. 2, 1991, pp. 197–206.

2.49. N. J. Carino and J. R. Clifton, "High-Performance Concrete: Research Needs to Enhance Its Use," *Concr. Intl.*, vol. 13, no. 9, 1991, pp. 70–76.

2.50. A. Azizinamini, R. Pavel, E. Hatfield, and S. K. Ghosh, "Behavior of Spliced Reinforcing Bars Embedded in High-Strength Concrete," *ACI Struct. J.*, vol. 96, no. 5, 1999, pp. 826–835.

2.51. D. Darwin, J. Browning, and W. D. Lindquist, "Control of Cracking in Bridge Decks: Observations from the Field," *Cement, Concrete and Aggregates*, ASTM International, vol. 26, no. 2, 2004, pp. 148–154.

2.52. A. E. Fiorato, "PCA Research on High-Strength Concrete," *Concr. Intl.*, vol. 11, no. 4, 1989, pp. 44–50.

2.53. D. Whiting, "Durability of High-Strength Concrete," *Proc. of Katharine and Bryant Mather International Conference*, Special Publication SP-100, American Concrete Institute, Detroit, MI, 1987, pp. 169–186.

2.54. *Manual of Standard Practice*, 28th ed., Concrete Reinforcing Steel Institute, Schaumburg, IL, 2009.

2.55. W. G. Corley, J. M. Hanson, and T. Helgason, "Design of Reinforced Concrete for Fatigue," *J. Struct. Div.*, ASCE, vol. 104, no. ST6, 1978, pp. 921–932.

2.56. W. G. Corley, M. A. Sozen, and C. P. Siess, "Time-Dependent Deflections of Prestressed Concrete Beams," *Highway Res. Board Bull.* No. 307, 1961, pp. 1–25.

2.57. ACI Committee 544, "Report on Fiber-Reinforced Concrete," American Concrete Institute, Farmington Hills, MI, 1996.

PROBLEMS

2.1. The specified concrete strength f_c' for a new building is 5000 psi. Calculate the required average f_{cr}' for the concrete (a) if there are no prior test results for concrete with a compressive strength within 1000 psi of f_c' made with similar materials, (b) if 20 test results for concrete with $f_c' = 5500$ psi made with similar materials produce a sample standard deviation s_s of 560 psi, and (c) if 30 tests with $f_c' = 4500$ psi made with similar materials produce a sample standard deviation s_s of 540 psi.

2.2. Ten consecutive strength tests are available for a new concrete mixture with $f_c' = 4000$ psi: 4830, 4980, 3840, 4370, 4410, 4890, 4450, 3970, 4780, and 4040 psi.

    (a) Do the strength results represent concrete of satisfactory quality? Explain your reasoning.

    (b) If f_{cr}' has been selected based on 30 consecutive test results from an earlier project with a sample standard deviation s_s of 570 psi, must the mixture proportions be adjusted? Explain.

2.3. The specified concrete strength f_c' for the columns in a high-rise building is 12,000 psi. Calculate the required average f_{cr}' for the concrete (a) if there are no prior test results for concrete with a compressive strength within 1000 psi of f_c' made with similar materials, (b) if 15 test results for concrete with $f_c' = 11,000$ psi made with similar materials produce a sample standard deviation s_s of 930 psi, and (c) if 30 tests with $f_c' = 12,000$ made with similar materials produce a sample standard deviation s_s of 950 psi.

3
Design of Concrete Structures and Fundamental Assumptions

3.1 INTRODUCTION

Design is the determination of the general shape and specific dimensions so that a structure will perform the function for which it was created and will safely withstand the influences that will act on it throughout its useful life. These influences are primarily the loads and other forces to which it will be subjected, as well as other detrimental agents, such as temperature fluctuations and foundation settlements.

The basic form of the structure is defined by its intended use. In the case of a building, an architect may present an overall concept and with the engineer develop a structural system. For bridges and industrial facilities, the engineer is often directly involved in selecting both the concept and the structural system. Regardless of the application, the design of concrete structures follows the same general sequence. First, an initial structural system is defined, the initial member sizes are selected, and a mathematical model of the structure is generated. Second, gravity and lateral loads are determined based on the selected system, member sizes, and external loads. Building loads typically are defined in ASCE/SEI 7 (Ref 3.1), as discussed in Chapter 1. Third, the loads are applied to the structural model and the load effects calculated for each member. This step may be done on a preliminary basis or by using computer modeling software. This step is more complex for buildings in Seismic Design Categories D though F where the seismic analysis requires close coordination of the structural framing system and the earthquake loads (discussed in Chapter 20). Fourth, maximum load effects at critical member sections are identified and each critical section is designed for moment, axial load, shear, and torsion as needed. At this step, the process may become iterative. For example, if the member initially selected is too small, its size must be increased, load effects recalculated for the larger member, and the members redesigned. If the initial member is too large, a smaller section is selected. Loads, however, are usually not recalculated for small changes in member size as gravity effects are often conservative. Fifth, each member is checked for serviceability. Sixth, the reinforcement for each member is detailed, that is, the number and size of reinforcing bars are selected for the critical sections to provide the required strength. Seventh, connections are designed to ensure that the building performs as intended. Finally, the design information is incorporated in the construction documents. This process is illustrated in Fig. 3.1. In addition to the design methodology, Fig. 3.1 indicates the chapters in this book and in the ACI Code (Ref. 3.2) where the topics are covered. The ACI Code is written based on the assumption that the user understands concrete structural behavior and the design process, whereas this text builds that understanding. The text is organized so that the fundamental theory is presented first, followed by the Code interpretation of the theory. Thus, the text remains relevant even as Code provisions are updated.

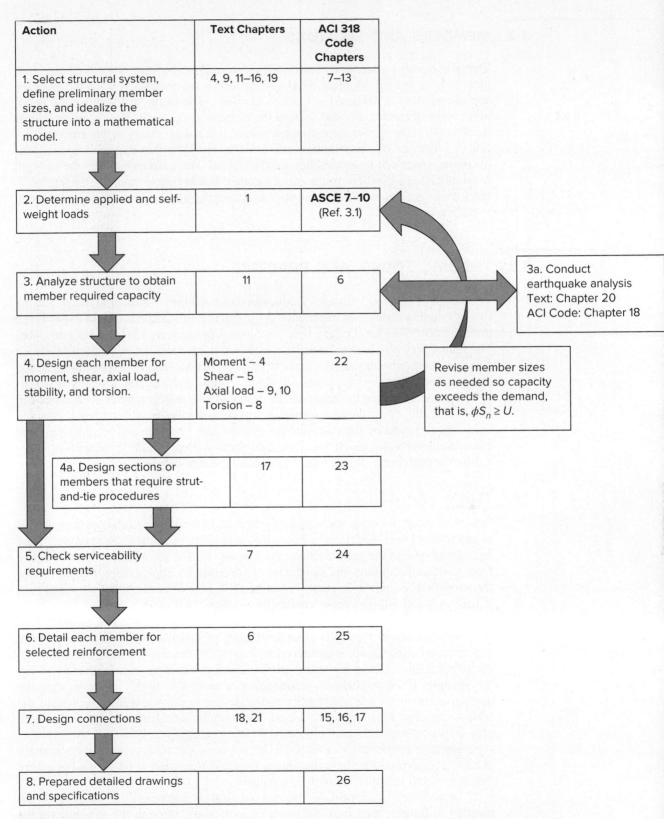

Action	Text Chapters	ACI 318 Code Chapters
1. Select structural system, define preliminary member sizes, and idealize the structure into a mathematical model.	4, 9, 11–16, 19	7–13
2. Determine applied and self-weight loads	1	**ASCE 7–10** (Ref. 3.1)
3. Analyze structure to obtain member required capacity	11	6
4. Design each member for moment, shear, axial load, stability, and torsion.	Moment – 4 Shear – 5 Axial load – 9, 10 Torsion – 8	22
4a. Design sections or members that require strut-and-tie procedures	17	23
5. Check serviceability requirements	7	24
6. Detail each member for selected reinforcement	6	25
7. Design connections	18, 21	15, 16, 17
8. Prepared detailed drawings and specifications		26

3a. Conduct earthquake analysis
Text: Chapter 20
ACI Code: Chapter 18

Revise member sizes as needed so capacity exceeds the demand, that is, $\phi S_n \geq U$.

FIGURE 3.1
Design development sequence.

3.2   MEMBERS AND SECTIONS

The term *member* refers to an individual portion of the structure, such as a beam, column, slab, or footing. Moment, axial load, and shear are distributed along the member, and the member is designed at discrete locations. The engineer identifies the maximum value of these loads and designs the member at these discrete locations so that the strength at the *section* exceeds these values. It is not necessary to design every section of a member. The requirement $\phi S_n \geq U$ [Eq. (1.4)] implies that reinforcement for maximum loads can be carried beyond the critical section to ensure that the strength requirements are satisfied for the entire member. In addition to strength, the reinforcement is designed to provide overall structural integrity and to ensure that it is anchored to the concrete.

3.3   THEORY, CODES, AND PRACTICE

The design of concrete structures requires an understanding of structural theory and the role of building codes, and experience in the practice of structural design itself. These three elements interact. For example, a structural failure may lead to a code revision. The failure may also lead to research, which in turn provides a new theoretical model. Changes in practice may also be made to preclude similar failures, even without a code change. The following discussion of theory, codes, and practice provides a framework for understanding the behavior and design of concrete structures. As described in Section 3.1, this text follows a format of providing the theory of behavior of concrete structures followed by the code interpretation of that behavior and includes practical considerations for the design. Insight to the interplay of each of these elements is essential for the engineer to design safe, serviceable, and economical structures.

a.   Theory

Structural theory includes mathematical, physical, or empirical models of the behavior of structures. These models have evolved over decades of research and practice. They are used to predict the nominal strength of members. The most robust theories derive from statics, equilibrium, and mechanics of materials. Examples include equations for the strength of a concrete section for bending (Chapter 4) and bending plus axial load (Chapters 9 and 10). For these conditions, mathematical models provide representations of actual behavior that agree within a few percent of experimental results.

In other cases, an empirical understanding of structural behavior, derived from experimental observation, is combined with theory to develop the prediction of member strength. In this case, equations are then fitted to the experimental data to predict the strength. If the experimental strength of a section is highly variable, then the predictive equations are adjusted for use in design to provide a lower bound of the section capacity. This approach is used, for example, to calculate the shear strength of a section (Chapter 5) and anchorage capacity (Chapter 21).

Because theoretical and empirical expressions are used to predict the strength at a *specific section*, the strength at many locations may need to be verified to ensure that the overall member strength is adequate.

In addition to supporting the applied load at each section, members work together to transfer load from the point of application, through the structure, to the point where the force exits the structure, such as the foundation or other support location. For example, as described in Section 1.8, live load is applied to a slab, the live load and slab gravity loads are carried to a girder, the girder load is carried to the column, and the column carries the load to the foundation. Similarly, when wind applies a force to an external wall, the wind load is transferred by the wall to the floors, which act as diaphragms, and in turn transfer the load to the lateral load resisting system for the building, such as a moment frame or shear wall. The *load path* is the sequence of members and connections that transfers the factored loads through the structure.

The engineer's responsibility is to provide at least one load path, and preferably multiple paths, for any force applied to the structure. When multiple load paths are present, the loads will follow the stiffest path, that is, the sequence of members and connections that tend to deform the least. In *"The Wisdom of the Structure"* (Ref. 3.3), Halvard Birkeland points out that a structure will exhaust every possible load path before collapsing. Load paths are dependent on equilibrium and the ability to deform and redistribute loads.

When applied to design, structural theory is typically presented in a deterministic format. That is, an equation results in a single nominal strength for a given section. Thus, the moment capacity of a beam can be calculated. The theory, however, provides no guidance as to approach the selection of member size or reinforcement to attain the desired capacity. Equally valid solutions for the given nominal strength range from large cross sections containing small amounts of reinforcement to small sections with large quantities of reinforcement. Building codes provide some guidance for these decisions.

b. Codes

Building codes provide minimum requirements for the life safety and serviceability for structures. In their simplest application, codes present the theory needed to ensure that sectional and member strengths are provided and define the limits on that theory. For example, a structure could be constructed using a large unreinforced concrete beam that relies solely on the tensile strength of the concrete. Such a structure would be brittle, and an unanticipated load would lead to sudden collapse. Codes prohibit such designs. In a similar manner, codes prescribe the maximum and minimum amount of reinforcement allowed in a member. Codes also address serviceability considerations, such as deflection and crack control. In addition to providing the theoretical or empirical basis for design, codes may also contain restrictions resulting from failures in practice that were not predicted by the theory upon which the code is based.

Structural integrity provisions in concrete building codes require reinforcement to limit *progressive* or *disproportional collapse*. Disproportional collapse occurs when the failure of a single member leads to the failure of multiple adjacent members. The failure of a single apartment wall in the Ronan Point apartment complex in 1968 led to the failure of several other units (Ref. 3.4). In response to this collapse, codes added requirements for *integrity reinforcement* based on a rational assessment of the failure. This integrity reinforcement is a *prescriptive provision*, that is, the requirements are detailed in the code and must be incorporated in the structure without associated detailed calculations.

Codes are written in terse language, based on the assumption that the user is a competent engineer, and typically adopt lower-bound approaches to structural safety. (See Section 23.2 for a description of *upper-* and *lower-bound theorems*.) Because codes provide the minimum requirements for safety and serviceability, the engineer is allowed to exceed these requirements—providing less than code requirements is imprudent. A *commentary* accompanies most codes and assists in understanding, provides references or background, and offers rationale for the provisions.

c. Practice

Structural engineering practice encompasses both the art and the technical practice of structural design. Throughout history, many extraordinary structures, such as the gothic cathedrals, have been designed and constructed without the benefit of modern theory and codes. While theory and codes provide the mechanics for establishing the strength and serviceability of structures, neither provides the aesthetic, economic, or functional guidance needed for member selection. Questions such as "Should a beam be slender or stout within the code limits?" or "How should the concrete mixture be adjusted for corrosive environments?" need to be answered by the engineer. To respond, the engineer relies on judgment, personal experience, and the broader experience of the profession to adapt the design to meet the overall project requirements. Inclusion of long-standing design guidelines for the selection of member sizes is an example of how that broader experience of the profession is used.

The following sections introduce the fundamental assumptions needed to develop the equations for member design that are presented in this text starting in Chapter 4.

3.4 FUNDAMENTAL ASSUMPTIONS FOR REINFORCED CONCRETE BEHAVIOR

Structural mechanics is one of the main tools in the process of design. As here understood, it is the body of knowledge that permits one to predict, with a good degree of certainty, how a structure of given shape and dimensions will behave when acted upon by known forces or other mechanical influences. The chief items of behavior that are of practical interest are (1) the strength of the structure, that is, the magnitude of loads of a given distribution that will cause the structure to fail, and (2) the deformations, such as deflections and extent of cracking, that the structure will undergo when loaded under service conditions.

The fundamental propositions on which the mechanics of reinforced concrete is based are as follows:

1. The internal forces, such as bending moments, shear forces, normal and shear stresses, and torsional moments, at any section of a member are in equilibrium with the effects of the external loads at that section. This proposition is not an assumption but a fact, because any body or any portion thereof can be at rest only if all forces acting on it are in equilibrium.

2. The strain in an embedded reinforcing bar (unit extension or compression) is the same as that of the surrounding concrete. Expressed differently, it is assumed that perfect bonding exists between concrete and steel at the interface, so that no slip occurs between the two materials. Hence, as the one deforms, so must the other. With modern deformed bars (see Section 2.14), a high degree of mechanical interlocking is provided in addition to the natural surface adhesion, so this assumption is very close to correct.

3. Cross sections that were plane prior to loading continue to be plane in the member under load. Accurate measurements have shown that when a reinforced concrete member is loaded close to failure, this assumption is not absolutely accurate. However, the deviations are usually minor, and the results of theory based on this assumption check well with extensive test information.

4. In view of the fact that the tensile strength of concrete is only a small fraction of its compressive strength (see Section 2.9), the concrete in that part of a member which is in tension is usually cracked. While these cracks, in well-designed

members, are generally so narrow as to be hardly visible (they are known as *hairline* cracks), they effectively render the cracked concrete incapable of resisting tension stress. Correspondingly, it is assumed that concrete is not capable of resisting any tension stress whatever. This assumption is evidently a simplification of the actual situation because, in fact, concrete prior to cracking, as well as the concrete located between cracks, does resist tension stresses of small magnitude. Later in discussions of the resistance of reinforced concrete beams to shear, it will become apparent that under certain conditions this particular assumption is dispensed with and advantage is taken of the modest tensile strength that concrete can develop.

5. The theory is based on the actual stress-strain relationships and strength properties of the two constituent materials (see Sections 2.8 and 2.14) or some reasonable equivalent simplifications thereof. The fact that nonelastic behavior is reflected in modern theory, that concrete is assumed to be ineffective in tension, and that the joint action of the two materials is taken into consideration results in analytical methods that are considerably more complex, and also more challenging, than those that are adequate for members made of a single, substantially elastic material.

These five propositions permit one to predict by calculation the performance of reinforced concrete members in a number of important cases. Because, however the joint action of two materials as dissimilar and complicated as concrete and steel is complex, it cannot be fully represented using a purely analytical treatment. For this reason, methods of design and analysis, while using these assumptions, are very largely based on the results of extensive and continuing experimental and analytical research. They are modified and improved as the results of additional research become available.

3.5 BEHAVIOR OF MEMBERS SUBJECT TO AXIAL LOADS

Many of the fundamentals of the behavior of reinforced concrete, through the full range of loading from zero to ultimate, can be illustrated clearly in the context of members subject to simple axial compression or tension. The basic concepts illustrated here will be recognized in later chapters in the analysis and design of beams, slabs, eccentrically loaded columns, and other members subject to more complex loadings.

a. Axial Compression

In members that sustain chiefly or exclusively axial compression loads, such as building columns, it is economical to make the concrete carry most of the load. Still, some steel reinforcement is always provided for various reasons. For one, very few members are subjected to truly axial load; steel is essential for resisting any bending that may exist. For another, if part of the total load is carried by steel with its much greater strength, the cross-sectional dimensions of the member can be reduced—the more so, the larger the amount of reinforcement.

The two chief forms of reinforced concrete columns are shown in Fig. 3.2. In the square column, the four longitudinal bars serve as main reinforcement. They are held in place by transverse small-diameter steel *ties* that prevent displacement of the main bars during construction operations and counteract any tendency of the compression-loaded bars to buckle out of the concrete by bursting the thin outer cover. A round column is shown with eight main reinforcing bars. These are surrounded by a closely

FIGURE 3.2
Reinforced concrete columns.

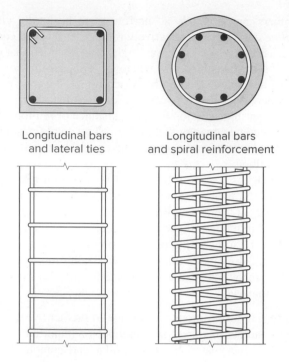

Longitudinal bars
and lateral ties

Longitudinal bars
and spiral reinforcement

spaced *spiral* that serves the same purpose as the more widely spaced ties but also acts to confine the concrete within it, thereby increasing its resistance to axial compression. The discussion that follows applies to tied columns.

When axial load is applied, the compression strain is the same over the entire cross section and, in view of the bonding between concrete and steel, is the same in the two materials (see propositions 2 and 3 in Section 3.4). To illustrate the action of such a member as load is applied, Fig. 3.3 shows two representative stress-strain curves, one for a concrete with compressive strength $f_c' = 4000$ psi and the other for a steel with yield stress $f_y = 60,000$ psi. The curves for the two materials are drawn on the same graph using different vertical stress scales. Curve *b* has the shape that would be obtained in a concrete cylinder test. The rate of loading in most structures is considerably slower than that in a cylinder test, and this affects the shape of the curve. Curve *c*, therefore, is drawn as being characteristic of the performance of concrete under slow loading. Under these conditions, tests have shown that the maximum reliable compressive strength of reinforced concrete is about $0.85f_c'$, as shown.

ELASTIC BEHAVIOR   At low stresses, up to about $f_c'/2$, the concrete is seen to behave nearly elastically, that is, stresses and strains are quite closely proportional; the straight line *d* represents this range of behavior with little error for both rates of loading. For the given concrete, the range extends to a strain of about 0.0005. The steel, on the other hand, is seen to be elastic nearly to its yield point of 60 ksi, or to the much greater strain of about 0.002.

Within this elastic range, the compression strain in the concrete, at any given load, is equal to the compression strain in the steel,

$$\varepsilon_c = \frac{f_c}{E_c} = \varepsilon_s = \frac{f_s}{E_s}$$

FIGURE 3.3
Concrete and steel stress-strain curves.

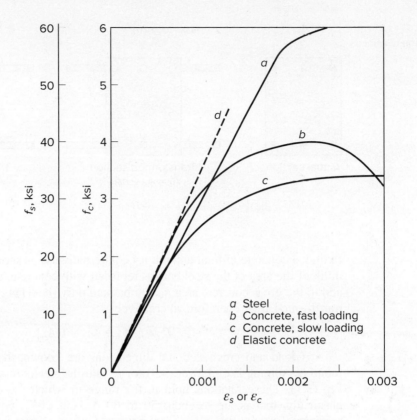

a Steel
b Concrete, fast loading
c Concrete, slow loading
d Elastic concrete

from which the relation between the steel stress f_s and the concrete stress f_c is obtained as

$$f_s = \frac{E_s}{E_c} f_c = nf_c \qquad (3.1)$$

where $n = E_s/E_c$ is known as the *modular ratio*.

Let

    A_c = net area of concrete, that is, gross area minus area occupied by reinforcing bars
    A_g = gross area
    A_{st} = total area of reinforcing bars
    P = axial load

Then

$$P = f_c A_c + f_s A_{st} = f_c A_c + nf_c A_{st}$$

or

$$P = f_c(A_c + nA_{st}) \qquad (3.2)$$

The term $A_c + nA_{st}$ can be interpreted as the area of a fictitious concrete cross section, the *transformed area*, which when subjected to the particular concrete stress f_c results in the same axial load P as the actual section composed of both steel and concrete. This transformed concrete area is seen to consist of the actual concrete area plus n times the area of the reinforcement. It can be visualized as shown in Fig. 3.4. That is, in Fig. 3.4b the three bars along each of the two faces are thought of as being removed and replaced, at the same distance from the axis of the section, with added areas of

FIGURE 3.4
Transformed section in axial
compression.

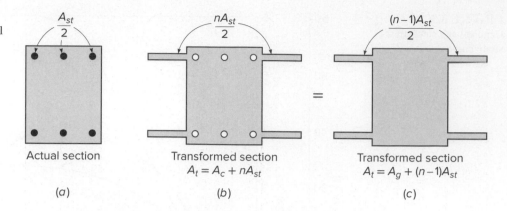

 Actual section
(a)

Transformed section
$A_t = A_c + nA_{st}$
(b)

Transformed section
$A_t = A_g + (n-1)A_{st}$
(c)

fictitious concrete of total amount nA_{st}. Alternatively, as shown in Fig. 3.4c, one can think of the area of the steel bars as replaced with concrete, in which case one has to add to the gross concrete area A_g so obtained only $(n-1)A_{st}$ to obtain the same total transformed area. Therefore, alternatively,

$$P = f_c[A_g + (n-1)A_{st}] \tag{3.3}$$

If load and cross-sectional dimensions are known, the concrete stress can be found by solving Eq. (3.2) or (3.3) for f_c, and the steel stress can be calculated from Eq. (3.1). These relations hold in the range in which the concrete behaves nearly elastically, that is, up to about 50 to 60 percent of f_c'. For reasons of safety and serviceability, concrete stresses in structures under normal conditions are kept within this range. Therefore, these relations permit one to calculate *service load stresses*.

EXAMPLE 3.1   **Axial load to produce given stress.**   A column made of the materials defined in Fig. 3.3 has a cross section of 16 × 20 in. and is reinforced by six No. 9 (No. 29) bars, placed as shown in Fig. 3.4. (See Tables A.1 and A.2 of Appendix A for bar diameters and areas and Section 2.14 for a description of bar size designations.) Determine the axial load that will stress the concrete to 1200 psi. The modular ratio n may be assumed equal to 8. (In view of the scatter inherent in E_c, it is customary and satisfactory to round off the value of n to the nearest integer and never justified to use more than two significant figures.)

SOLUTION.   One finds $A_g = 16 \times 20 = 320$ in², and from Appendix A, Table A.2, six No. 9 (No. 29) bars provide steel area $A_{st} = 6.00$ in² or 1.88 percent of the gross area. The load on the column, from Eq. (3.3), is $P = 1200[320 + (8 - 1)6.00] = 434,000$ lb. Of this total load, the concrete is seen to carry $P_c = f_c A_c = f_c(A_g - A_{st}) = 1200(320 - 6) = 377,000$ lb, and the steel $P_s = f_s A_{st} = (nf_c)A_{st} = 9600 \times 6 = 57,600$ lb, which is 13.3 percent of the total axial load.

INELASTIC RANGE   Inspection of Fig. 3.3 shows that the elastic relationships that have been used so far cannot be applied beyond a strain of about 0.0005 for the given concrete. To obtain information on the behavior of the member at larger strains and, correspondingly, at larger loads, it is therefore necessary to make direct use of the information in Fig. 3.3.

EXAMPLE 3.2   **Axial load to produce given strain.** Calculate the magnitude of the axial load that will produce a strain or unit shortening $\varepsilon_c = \varepsilon_s = 0.0010$ in the column of Example 3.1.

SOLUTION. At this strain the steel is seen to be still elastic, so that the steel stress $f_s = \varepsilon_s E_s = 0.001 \times 29{,}000{,}000 = 29{,}000$ psi. The concrete is in the inelastic range, so that its stress cannot be directly calculated, but it can be read from the stress-strain curve for the given value of strain.

1. If the member has been loaded at a fast rate, curve *b* holds at the instant when the entire load is applied. The stress for $\varepsilon = 0.001$ can be read from Fig. 3.3 as $f_c = 3200$ psi. Consequently, the total load can be obtained from

$$P = f_c A_c + f_s A_{st} \qquad (3.4)$$

   which applies in the inelastic as well as in the elastic range. Hence, $P = 3200(320 - 6) + 29{,}000 \times 6 = 1{,}005{,}000 + 174{,}000 = 1{,}179{,}000$ lb. Of this total load, the steel is seen to carry 174,000 lb, or 14.7 percent.

2. For slowly applied or sustained loading, curve *c* represents the behavior of the concrete. Its stress at a strain of 0.001 can be read as $f_c = 2400$ psi. Then $P = 2400 \times 314 + 29{,}000 \times 6 = 754{,}000 + 174{,}000 = 928{,}000$ lb. Of this total load, the steel is seen to carry 18.8 percent.

Comparison of the results for fast and slow loading shows the following. Owing to creep of concrete, a given shortening of the column is produced by a smaller load when slowly applied or sustained over some length of time than when quickly applied. More importantly, the farther the stress is beyond the proportional limit of the concrete, and the more slowly the load is applied or the longer it is sustained, the smaller the share of the total load carried by the concrete and the larger the share carried by the steel. In the sample column, the steel was seen to carry 13.3 percent of the load in the elastic range, 14.7 percent for a strain of 0.001 under fast loading, and 18.8 percent at the same strain under slow or sustained loading.

STRENGTH   The quantity of chief interest to the structural designer is *strength*, that is, the maximum load that the structure or member will carry. Information on stresses, strains, and similar quantities serves chiefly as a tool for determining carrying capacity. The performance of the column discussed so far emphasizes two points: (1) in the range of large stresses and strains that precede attainment of the maximum load and subsequent failure, elastic relationships cannot be used; (2) the member behaves differently under fast and under slow or sustained loading and shows less resistance to the latter than to the former. In usual construction, many types of loads, such as the weight of the structure and any permanent equipment housed therein, are sustained, and others are applied at slow rates. For this reason, to calculate a reliable magnitude of compressive strength, curve *c* of Fig. 3.3 must be used as far as the concrete is concerned.

The steel reaches its tensile strength (peak of the curve) at strains on the order of 0.08 (see Fig. 2.19). Concrete, on the other hand, fails by crushing at the much smaller strain of about 0.003 and, as seen from Fig. 3.3 (curve *c*), reaches its maximum stress in the strain range of 0.002 to 0.003. Because the strains in steel and concrete are equal in axial compression, the load at which the steel begins to yield can be calculated from the information in Fig. 3.3.

If the small knee prior to yielding of the steel is disregarded, that is, if the steel is assumed to be sharp-yielding, the strain at which it yields is

$$\varepsilon_y = \frac{f_y}{E_s} \tag{3.5}$$

or for Grade 60 reinforcement

$$\varepsilon_y = \frac{60,000}{29,000,000} = 0.00207$$

At this strain, curve c of Fig. 3.3 indicates a stress of 3200 psi in the concrete; therefore, by Eq. (3.4), the load in the member when the steel starts yielding is $P_y = 3200 \times 314 + 60,000 \times 6 = 1,365,000$ lb. At this load the concrete has not yet reached its full strength, which, as mentioned before, can be assumed as $0.85f_c' = 3400$ psi for slow or sustained loading, and therefore the load on the member can be further increased. During this stage of loading, the steel keeps yielding at constant stress. Finally, the nominal capacity of the member is reached when the concrete crushes while the steel yields, that is,

$$P_n = 0.85f_c'\, A_c + f_y\, A_{st} \tag{3.6}$$

Numerous careful tests have shown the reliability of Eq. (3.6) in predicting the ultimate strength of a concentrically loaded reinforced concrete column, provided its slenderness ratio is small so that buckling will not reduce its strength.

For the particular numerical example, $P_n = 3400 \times 314 + 60,000 \times 6 = 1,068,000 + 360,000 = 1,428,000$ lb. At this stage the steel carries 25.2 percent of the load.

SUMMARY   In the elastic range, the steel carries a relatively small portion of the total load of an axially compressed member. As member strength is approached, there occurs a redistribution of the relative shares of the load resisted by concrete and steel, the latter taking an increasing amount. The nominal capacity, at which the member is on the point of failure, consists of the contribution of the steel when it is stressed to the yield point plus that of the concrete when its stress has attained a value of $0.85f_c'$, as reflected in Eq. (3.6).

b.   Axial Tension

The tension strength of concrete is only a small fraction of its compressive strength. It follows that reinforced concrete is not well suited for use in tension members because the concrete will contribute little, if anything, to their strength. Still, there are situations in which reinforced concrete is stressed in tension, chiefly in tie-rods in structures such as arches. Such members consist of one or more bars embedded in concrete in a symmetric arrangement similar to compression members (see Figs. 3.2 and 3.4).

When the tension force in the member is small enough for the stress in the concrete to be considerably below its tensile strength, both steel and concrete behave elastically. In this situation, all the expressions derived for elastic behavior in compression in Section 3.5a are identically valid for tension. In particular, Eq. (3.2) becomes

$$P = f_{ct}(A_c + nA_{st}) \tag{3.7}$$

where f_{ct} is the tensile stress in the concrete.

When the load is further increased, however, the concrete reaches its tensile strength at a stress and strain on the order of one-tenth of what it could sustain in compression. At this stage, the concrete cracks across the entire cross section. When this happens, it ceases to resist any part of the applied tension force, since, evidently, no force can be transmitted across the air gap in the crack. At any load larger than that which caused the concrete to crack, the steel is called upon to resist the entire tension force. Correspondingly, at this stage,

$$P = f_s A_{st} \tag{3.8}$$

With further increased load, the tensile stress f_s in the steel reaches the yield point f_y. When this occurs, the tension members cease to exhibit small, elastic deformations but instead stretch a sizable and permanent amount at substantially constant load. This does not impair the strength of the member. Its elongation, however, becomes so large (on the order of 1 percent or more of its length) as to render it unserviceable. Therefore, the maximum useful strength P_{nt} of a tension member is the force that will just cause the steel stress to reach the yield point. That is,

$$P_{nt} = f_y A_{st} \tag{3.9}$$

To provide adequate safety, the force permitted in a tension member under normal service loads should be limited to about $\frac{1}{2} P_{nt}$. Because the concrete has cracked at loads considerably smaller than this, concrete does not contribute to the carrying capacity of the member in service. It does serve, however, as fire and corrosion protection and often improves the appearance of the structure.

There are situations, though, in which reinforced concrete is used in axial tension under conditions in which the occurrence of tension cracks must be prevented. A case in point is a circular tank. To provide watertightness, the hoop tension caused by the fluid pressure must be prevented from causing the concrete to crack. In this case, Eq. (3.7) can be used to determine a safe value for the axial tension force P by using, for the concrete tension stress f_{ct}, an appropriate fraction of the tensile strength of the concrete, that is, of the stress that would cause the concrete to crack.

3.6 BENDING OF HOMOGENEOUS BEAMS

Reinforced concrete beams are nonhomogeneous in that they are made of two entirely different materials. The methods used in the analysis of reinforced concrete beams are therefore different from those used in the design or investigation of beams composed entirely of steel, wood, or any other structural material. The fundamental principles involved are, however, essentially the same. Briefly, these principles are as follows.

At any cross section there exist internal forces that can be resolved into components normal and tangential to the section. Those components that are normal to the section are the *bending* stresses (tension on one side of the neutral axis and compression on the other, Fig. 3.5). Their function is to resist the bending moment at the section. The tangential components are known as the *shear* stresses, and they resist the transverse or shear forces.

Fundamental assumptions relating to flexure and flexural shear are as follows:

1. A cross section that was plane before loading remains plane under load. This means that the unit strains in a beam above and below the neutral axis are proportional to the distance from that axis.
2. The bending stress f at any point depends on the strain at that point in a manner given by the stress-strain diagram of the material. If the beam is made of a homogeneous material whose stress-strain diagram in tension and compression is that of Fig. 3.5a, the following holds. If the maximum strain at the outer fibers is smaller than the strain ε_p up to which stress and strain are proportional for the given material, then the compression and tension stresses on either side of the axis are proportional to the distance from the axis, as shown in Fig. 3.5b. However, if

FIGURE 3.5
Elastic and inelastic stress
distributions in homogeneous
beams.

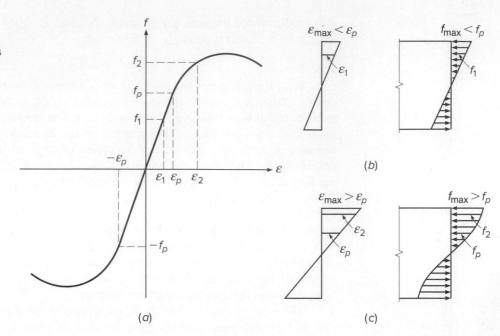

the maximum strain at the outer fibers is larger than ε_p, this is no longer true. The situation that then occurs is shown in Fig. 3.5c; that is, in the outer portions of the beam, where $\varepsilon > \varepsilon_p$, stresses and strains are no longer proportional. In these regions, the magnitude of stress at any level, such as f_2 in Fig. 3.5c, depends on the strain ε_2 at that level in the manner given by the stress-strain diagram of the material. In other words, for a given strain in the beam, the stress at a point is the same as that given by the stress-strain diagram for the same strain.

3. The distribution of the shear stresses v over the depth of the section depends on the shape of the cross section and of the stress-strain diagram. These shear stresses are largest at the neutral axis and equal to zero at the outer fibers. The shear stresses on horizontal and vertical planes through any point are equal.

4. Owing to the combined action of shear stresses (horizontal and vertical) and flexure stresses, at any point in a beam there are inclined stresses of tension and compression, the largest of which form an angle of 90° with each other. The intensity of the inclined maximum or principal stress t at any point is given by

$$t = \frac{f}{2} \pm \sqrt{\frac{f^2}{4} + v^2} \qquad (3.10)$$

where f = intensity of normal fiber stress
v = intensity of tangential shearing stress

The inclined stress makes an angle α with the horizontal such that $\tan 2\alpha = 2v/f$.

5. Since horizontal and vertical shear stresses are equal and the flexural stresses are zero along the neutral axis, the inclined tensile and compressive stresses at any point on the neutral axis form an angle of 45° with the horizontal, the intensity of each being equal to the unit shear at the point.

6. When the stresses in the outer fibers are smaller than the proportional limit f_p, the beam behaves *elastically*, as shown in Fig. 3.5b. In this case the following pertains:

    (a) The neutral axis passes through the center of gravity of the cross section.

(b) The intensity of the bending stress normal to the section increases directly with the distance from the neutral axis and is a maximum at the extreme fibers. The stress at any given point in the cross section is represented by the equation

$$f = \frac{My}{I} \tag{3.11}$$

where f = bending stress at a distance y from neutral axis
M = external bending moment at section
I = moment of inertia of cross section about neutral axis

The maximum bending stress occurs at the outer fibers and is equal to

$$f_{max} = \frac{Mc}{I} = \frac{M}{S} \tag{3.12}$$

where c = distance from neutral axis to outer fiber
$S = I/c$ = section modulus of cross section

(c) The shear stress (horizontal equals vertical) v at any point in the cross section is given by

$$v = \frac{VQ}{Ib} \tag{3.13}$$

where V = total shear at section
Q = statical moment about neutral axis of that portion of cross section lying between a line through point in question parallel to neutral axis and nearest face (upper or lower) of beam
I = moment of inertia of cross section about neutral axis
b = width of beam at a given point

(d) The intensity of shear along a vertical cross section in a rectangular beam varies as the ordinates of a parabola, the intensity being zero at the outer fibers of the beam and a maximum at the neutral axis. For a total depth h, the maximum shear stress is $\frac{3}{2}V/bh$, since at the neutral axis $Q = bh^2/8$ and $I = bh^3/12$ in Eq. (3.13).

With the fundamentals now established for homogeneous beams, the next step is to address the behavior and design of reinforced concrete beams. This is done in Chapter 4 for bending and in Chapter 5 for shear.

REFERENCES

3.1 *Minimum Design Loads and Other Associated Criteria for Buildings and Other Structures*, ASCE/SEI 7-16, American Society of Civil Engineers, Reston, VA, 2016.

3.2 *Building Code Requirements for Structural Concrete* (ACI 318–19), American Concrete Institute, Farmington Hills, MI, 2019.

3.3 H. Birkeland, "The Wisdom of the Structure," *ACI J.*, vol. 75, no. 4, 1978, pp. 105–111.

3.4 C. Pearson and N. Delatte, "Ronan Point Apartment Tower Collapse and Its Effect on Building Codes," *J. Perform. Constr. Facil.*, vol. 19, no. 2, 2005, pp. 172–177.

PROBLEMS

Problems 3.1 through 3.5 reinforce the understanding of elastic and inelastic behavior of a member under axial load.

3.1. A 16 × 20 in. column is made of the same concrete and reinforced with the same six No. 9 (No. 29) bars as the column in Examples 3.1 and 3.2, except that a steel with yield strength $f_y = 40$ ksi is used. The stress-strain diagram of this reinforcing steel is shown in Fig. 2.19 for $f_y = 40$ ksi. For this column determine (a) the axial load that will stress the concrete to 1200 psi; (b) the load at which the steel starts yielding; (c) the maximum load; and (d) the share of the total load carried by the reinforcement at these three stages of loading. Compare results with those calculated in the examples for $f_y = 60$ ksi, keeping in mind, in regard to relative economy, that the price per pound for reinforcing steels with 40 and 60 ksi yield points is about the same.

3.2. The area of steel, expressed as a percentage of gross concrete area, for the column of Problem 3.1 is lower than would often be used in practice. Recalculate the comparisons of Problem 3.1, using f_y of 40 ksi and 60 ksi as before, but for a 16×20 in. column reinforced with eight No. 11 (No. 36) bars. Compare your results with those of Problem 3.1.

3.3. A square concrete column with dimensions 22×22 in. is reinforced with a total of eight No. 10 (No. 32) bars arranged uniformly around the column perimeter. Material strengths are $f_y = 60$ ksi and $f_c' = 4000$ psi, with stress-strain curves as given by curves a and c of Fig. 3.3. Calculate the percentages of total load carried by the concrete and by the steel as load is gradually increased from 0 to failure, which is assumed to occur when the concrete strain reaches a limit value of 0.0030. Determine the loads at strain increments of 0.0005 up to the failure strain, and graph your results, plotting load percentages vs. strain. The modular ratio may be assumed at $n = 8$ for these materials.

3.4. A 20×24 in. column is made of the same concrete as used in Examples 3.1 and 3.2. It is reinforced with six No. 11 (No. 36) bars with $f_y = 60$ ksi. For this column section, determine (a) the axial load that the section will carry at a concrete stress of 1400 psi; (b) the load on the section when the steel begins to yield; (c) the maximum load if the section is loaded slowly; and (d) the maximum load if the section is loaded rapidly. The area of one No. 11 (No. 36) bar is 1.56 in^2. Determine the percent of the load carried by the steel and the concrete for each combination.

3.5. A 24 in. diameter column is made of the same concrete as used in Examples 3.1 and 3.2. The area of reinforcement equals 2.1 percent of the gross cross section (that is, $A_s = 0.021A_g$) and $f_y = 60$ ksi. For this column section, determine (a) the axial load the section will carry at a concrete stress of 1200 psi; (b) the load on the section when the steel begins to yield; (c) the maximum load if the section is loaded slowly; (d) the maximum load if the section is loaded rapidly; and (e) the maximum load if the reinforcement in the column is raised to 6.5 percent of the gross cross section and the column is loaded slowly. Comment on your answer, especially the percent of the load carried by the steel and the concrete for each combination.

4

Flexural Analysis and Design of Beams

4.1 INTRODUCTION

The fundamental assumptions upon which the analysis and design of reinforced concrete members are based were introduced in Section 3.4, and the application of those assumptions to the simple case of axial loading was developed in Section 3.5. Bending of homogeneous beams was covered in Section 3.6. The student should review Sections 3.4, 3.5, and 3.6 at this time. In developing methods for the analysis and design of beams in this chapter, the same assumptions apply, and identical concepts will be used. This chapter includes analysis and design for flexure, including the dimensioning of the concrete cross section and the selection and placement of reinforcing steel. Other important aspects of beam design, including shear reinforcement, bond, and anchorage of reinforcing bars, and the important questions of serviceability (for example, limiting deflections and controlling concrete cracking) will be treated in Chapters 5, 6, and 7.

4.2 REINFORCED CONCRETE BEAM BEHAVIOR

Plain concrete beams are inefficient as flexural members because the tensile strength in bending (modulus of rupture, see Section 2.9) is a small fraction of the compressive strength. As a consequence, such beams fail on the tension side at low loads long before the strength of the concrete on the compression side has been fully utilized. For this reason, steel reinforcing bars are placed on the tension side as close to the extreme tension fiber as is compatible with proper fire and corrosion protection of the steel. In such a reinforced concrete beam, the tension caused by the bending moments is chiefly resisted by the steel reinforcement, while the concrete alone is usually capable of resisting the corresponding compression. Such joint action of the two materials is ensured if relative slip is prevented. This is achieved by using deformed bars with their high bond strength at the steel-concrete interface (see Section 2.14) and, if necessary, by special anchorage of the ends of the bars. A simple example of such a beam, with the customary designations for the cross-sectional dimensions, is shown in Fig. 4.1. For simplicity, the discussion that follows deals with beams of rectangular cross section, even though members of other shapes are very common in most concrete structures. A beam elevation is shown in Fig. 4.1*a*. Figure 4.1*b* shows the beam cross section, followed by the strain distribution and the corresponding stresses acting on the cross section in Fig. 4.1*c*. This representation of a beam cross section, followed by a strain and stress distribution is used throughout this text.

When the load on such a beam is gradually increased from zero to the magnitude that will cause the beam to fail, several different stages of behavior can be clearly

FIGURE 4.1
Behavior of reinforced
concrete beam under
increasing load.

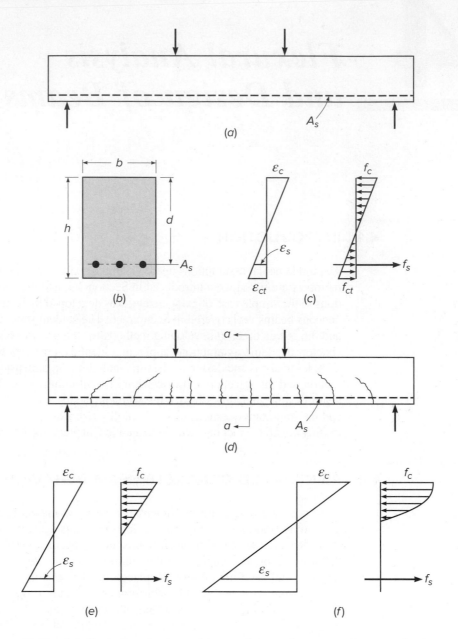

distinguished. At low loads, as long as the maximum tensile stress in the concrete is smaller than the modulus of rupture, the entire concrete section is effective in resisting stress, in compression on one side and in tension on the other side of the neutral axis. In addition, the reinforcement, deforming the same amount as the adjacent concrete, is also subject to tensile stresses. At this stage, all stresses in the concrete are of small magnitude and are proportional to strains. The distribution of strains and stresses in concrete and steel over the depth of the section is shown in Fig. 4.1c.

When the load is further increased, the tensile strength of the concrete is soon reached, and at this stage tension cracks develop. These propagate quickly upward to or close to the level of the neutral axis, which in turn shifts upward with progressive cracking. The general shape and distribution of these tension cracks is shown in Fig. 4.1d. In well-designed beams, the width of these cracks is so small (hairline cracks) that they are not objectionable from the viewpoint of either corrosion protection or appcarance. Their presence, however, profoundly affects the behavior of the beam under load. At a cracked section, that is, at a cross section located at a crack such as *a-a* in Fig. 4.1*d*, it is appropriate to treat the concrete as transmitting no tensile stresses. Hence, just as in tension members (Section 3.5b), the steel is called upon to resist the entire tension. At moderate loads, if the concrete stresses do not exceed approximately $f'_c/2$, stresses and strains continue to be closely proportional (see Fig. 3.3). The distribution of strains and stresses at or near a cracked section is then that shown in Fig. 4.1*e*. When the load is still further increased, stresses and strains rise correspondingly and are no longer proportional. The ensuing nonlinear relation between stresses and strains is that given by the concrete stress-strain curve. Therefore, just as in homogeneous beams (see Fig. 3.5), the distribution of concrete stresses on the compression side of the beam is of the same shape as the stress-strain curve. Figure 4.1*f* shows the distribution of strains and stresses close to the ultimate load.

Eventually, the carrying capacity of the beam is reached. Failure can be caused in one of two ways. When relatively moderate amounts of reinforcement are employed, at some value of the load the steel will reach its yield point. At that stress, the reinforcement yields suddenly and stretches a large amount (see Fig. 2.19), and the tension cracks in the concrete widen visibly and propagate upward, with simultaneous significant deflection of the beam. When this happens, the strains in the remaining compression zone of the concrete increase to such a degree that crushing of the concrete, the *secondary compression failure*, ensues at a load only slightly larger than that which caused the steel to yield. Effectively, therefore, attainment of the yield point in the steel determines the carrying capacity of moderately reinforced beams. Such yield failure is gradual and is preceded by visible signs of distress, such as the widening and lengthening of cracks and the marked increase in deflection.

On the other hand, if large amounts of reinforcement or normal amounts of steel of very high strength are employed, the compressive strength of the concrete may be exhausted before the steel starts yielding. Concrete fails by crushing when strains become so large that they disrupt the integrity of the concrete. Exact criteria for this occurrence have yet to be established, but it has been observed that rectangular beams fail in compression when the concrete strains reach values of 0.003 to 0.004. Compression failure through crushing of the concrete is sudden, of an almost explosive nature, and occurs without warning. For this reason it is good practice to dimension beams in such a manner that, should they be overloaded, failure would be initiated by yielding of the steel rather than by crushing of the concrete.

The analysis of stresses and strength in the different stages just described are discussed in the next several sections.

a. Stresses Elastic and Section Uncracked

As long as the tensile stress in the concrete is smaller than the modulus of rupture, so that no tension cracks develop, the strain and stress distribution as shown in Fig. 4.1*c* is essentially the same as in an elastic, homogeneous beam (Fig. 3.5*b*). The only difference is the presence of another material, the steel reinforcement. As shown in Section 3.5a, in the elastic range, for any given value of strain, the stress in the steel is *n* times that of the concrete [Eq. (3.1)]. In the same section, it was shown that one can take account of this fact in calculations by replacing the actual steel-and-concrete cross section with a fictitious section thought of as consisting of concrete only. In this "transformed section," the actual area of the reinforcement is replaced with an equivalent concrete area equal to nA_s located at the level of the steel. The transformed, uncracked section pertaining to the beam of Fig. 4.1*b* is shown in Fig. 4.2. The open

FIGURE 4.2
Uncracked transformed beam
section.

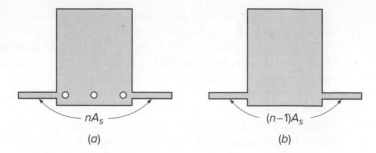

circles in Fig. 4.2*a* represent the reinforcement that has been removed and replaced by
the equivalent area of concrete. It is often convenient to use the representation shown
in Fig. 4.2*b*, treating the member as monolithic within its original boundaries while
adding an equivalent area of concrete equal to $(n - 1)A_s$.

Once the transformed section has been obtained, the usual methods of analysis
of elastic homogeneous beams apply. That is, the section properties (location of
neutral axis, moment of inertia, section modulus, etc.) are calculated in the usual
manner, and, in particular, stresses are calculated with Eqs. (3.11) to (3.13).

EXAMPLE 4.1   A rectangular beam has the dimensions (see Fig. 4.3) $b = 10$ in., $h = 25$ in., and $d = 23$ in.
and is reinforced with three No. 8 (No. 25) bars so that $A_s = 2.37$ in.^{2}. The concrete compressive strength f_c' is 4000 psi, and the tensile strength in bending (modulus of rupture) is
475 psi. The yield point of the steel f_y is 60,000 psi, the stress-strain curves of the materials
being those of Fig. 3.3. Determine the stresses caused by a bending moment $M = 45$ ft-kips.

Solution.   With a value $n = E_s/E_c = 29,000,000/3,600,000 = 8$, one has to add to the
rectangular outline an area $(n - 1)A_s = 7 \times 2.37 = 16.59$ in.^{2}, rounded slightly and distributed
as shown in Fig. 4.3, to obtain the uncracked, transformed section. Conventional calculations
show that the location of the neutral axis of this section is given by $\bar{y} = 13.2$ in. from the top
of the section, and its moment of inertia about this axis is 14,740 in.^{4}. For $M = 45$ ft-kips =
540,000 in-lb, the concrete compression stress at the top fiber is, from Eq. (3.11),

$$f_c = \frac{M\bar{y}}{I} = \frac{540,000 \times 13.2}{14,740} = 484 \text{ psi}$$

FIGURE 4.3
Transformed beam section of
Example 4.1.

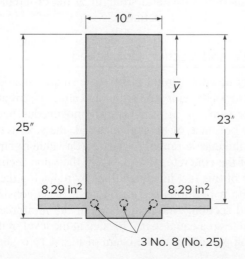

3 No. 8 (No. 25)

and, similarly, the concrete tension stress at the bottom fiber, 11.8 in. from the neutral axis, is

$$f_{ct} = \frac{540,000 \times 11.8}{14,740} = 432 \text{ psi}$$

Since this value is below the given tensile bending strength of the concrete, 475 psi, no tension cracks will form, and calculation by the uncracked, transformed section is justified. The stress in the steel, from Eqs. (3.1) and (3.11), is

$$f_s = n\frac{My}{I} = 8\left(\frac{540,000 \times 9.8}{14,740}\right) = 2870 \text{ psi}$$

By comparing f_c and f_s with the concrete cylinder strength and the yield point, respectively, it is seen that at this stage the actual stresses are quite small compared with the available strengths of the two materials.

---

b.   Stresses Elastic and Section Cracked

When the tensile stress f_{ct} exceeds the modulus of rupture, cracks form, as shown in Fig. 4.1*d*. If the concrete compressive stress is less than approximately $\frac{1}{2}f_c'$ and the steel stress has not reached the yield point, both materials continue to behave elastically, or very nearly so. This situation generally occurs in structures under normal service conditions and loads, since at these loads the stresses are generally of the order of magnitude just discussed. At this stage, for simplicity and with little if any error, it is assumed that tension cracks have progressed all the way to the neutral axis and that sections plane before bending are plane in the deformed member. The situation with regard to strain and stress distribution is that shown in Fig. 4.1*e*.

To calculate stresses, and strains if desired, the device of the transformed section can still be used. One need only take account of the fact that all of the concrete that is stressed in tension is assumed cracked, and therefore effectively absent. As shown in Fig. 4.4*a*, the transformed section then consists of the concrete in compression on one side of the axis and n times the steel area on the other. The distance to the neutral axis, in this stage, is conventionally expressed as a fraction kd of the effective depth d. (Once the concrete is cracked, any material located below the steel

FIGURE 4.4
Cracked transformed section
and stresses on the section.

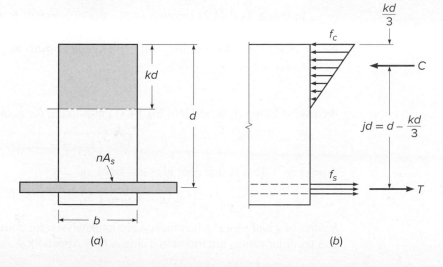

(a)   (b)

is ineffective, which is why d is the effective depth of the beam.) To determine the location of the neutral axis, the moment of the tension area about the axis is set equal to the moment of the compression area, which gives

$$b\frac{(kd)^2}{2} - nA_s(d - kd) = 0 \tag{4.1}$$

Having obtained kd by solving this quadratic equation, one can determine the moment of inertia and other properties of the transformed section as in the preceding case. Alternatively, one can proceed from basic principles by accounting directly for the forces that act on the cross section. These are shown in Fig. 4.4b. The concrete stress, with maximum value f_c at the outer edge, is distributed linearly as shown. The entire steel area A_s is subject to the stress f_s. Correspondingly, the total compression force C and the total tension force T are

$$C = \frac{f_c}{2} bkd \qquad \text{and} \qquad T = A_s f_s \tag{4.2}$$

The requirement that these two forces be equal numerically has been taken care of by the manner in which the location of the neutral axis has been determined.

Equilibrium requires that the couple constituted by the two forces C and T be equal numerically to the external bending moment M. Hence, taking moments about compression resultant C gives

$$M = Tjd = A_s f_s jd \tag{4.3}$$

where jd is the internal lever arm between C and T. From Eq. (4.3), the steel stress is

$$f_s = \frac{M}{A_s jd} \tag{4.4}$$

Conversely, taking moments about the tension force T gives

$$M = Cjd = \frac{f_c}{2} bkdjd = \frac{f_c}{2} kjbd^2 \tag{4.5}$$

from which the concrete stress is

$$f_c = \frac{2M}{kjbd^2} \tag{4.6}$$

In using Eqs. (4.2) through (4.6), it is convenient to have equations by which k and j may be found directly, to establish the neutral axis distance kd and the internal lever arm jd. First defining the *reinforcement ratio* as

$$\rho = \frac{A_s}{bd} \tag{4.7}$$

then substituting $A_s = \rho bd$ into Eq. (4.1) and solving for k, one obtains

$$k = \sqrt{(\rho n)^2 + 2\rho n} - \rho n \tag{4.8}$$

From Fig. 4.4b it is seen that $jd = d - kd/3$, or

$$j = 1 - \frac{k}{3} \tag{4.9}$$

Values of k and j for elastic cracked section analysis, for common reinforcement ratios and modular ratios, are found in Table A.6 of Appendix A.

EXAMPLE 4.2   The beam of Example 4.1 is subject to a bending moment $M = 90$ ft-kips (rather than 45 ft-kips as previously). Calculate the relevant properties and stresses.

SOLUTION.   If the section were to remain uncracked, the tensile stress in the concrete would now be twice its previous value, that is, 864 psi. Since this exceeds by far the modulus of rupture of the given concrete (475 psi), cracks will have formed and the analysis must be adapted consistent with Fig. 4.4. Equation (4.1), with the known quantities b, n, and A_s inserted, gives the distance to the neutral axis $kd = 7.6$ in., or $k = 7.6/23 = 0.33$. From Eq. (4.9), $j = 1 - 0.33/3 = 0.89$. With these values the steel stress is obtained from Eq. (4.4) as $f_s = 22,300$ psi, and the maximum concrete stress from Eq. (4.6) as $f_c = 1390$ psi.

Comparing the results with the pertinent values for the same beam when subject to one-half the moment, as previously calculated, one notices that (1) the neutral axis has migrated upward so that its distance from the top fiber has changed from 13.2 to 7.6 in.; (2) even though the bending moment has only been doubled, the steel stress has increased from 2870 to 22,300 psi, or about 7.8 times, and the concrete compression stress has increased from 484 to 1390 psi, or 2.9 times; (3) the moment of inertia of the cracked transformed section is easily calculated to be 5910 in⁴, compared with 14,740 in⁴ for the uncracked section. This affects the magnitude of the deflection, as discussed in Chapter 7. Thus, it is seen how radical is the influence of the formation of tension cracks on the behavior of reinforced concrete beams.

c. Flexural Strength

It is of interest in structural practice to calculate those stresses and deformations that occur in a structure in service under the design load. For reinforced concrete beams, this can be done by the methods just presented, which assume elastic behavior of both materials. It is equally, if not more, important that the structural engineer be able to predict with satisfactory accuracy the strength of a structure or structural member. By making this strength larger by an appropriate amount than the largest loads that can be expected during the lifetime of the structure, an adequate margin of safety is ensured. Historically, methods based on elastic analysis, like those just presented or variations thereof, have been used for this purpose. It is clear, however, that at or near the ultimate load, stresses are no longer proportional to strains. In regard to axial compression, this has been discussed in detail in Section 3.5, and in regard to bending, it has been pointed out that at high loads, close to failure, the distribution of stresses and strains is that of Fig. 4.1*f* rather than the elastic distribution of Fig. 4.1*e*. More realistic methods of analysis, based on actual inelastic rather than assumed elastic behavior of the materials and on results of extremely extensive experimental research, have been developed to predict the member strength. They are now used almost exclusively in structural design practice.

If the distribution of concrete compressive stresses at or near ultimate load (Fig. 4.1*f*) had a well-defined and invariable shape—parabolic, trapezoidal, or otherwise—it would be possible to derive a completely rational theory of bending strength, just as the theory of elastic bending with its known triangular shape of stress distribution (Figs. 3.5*b* and 4.1*c* and *e*) is straightforward and rational. Actually, inspection of Figs. 2.3, 2.4, and 2.6, and of many more concrete stress-strain curves that have been published, shows that the geometric shape of the stress distribution is quite varied and depends on a number of factors, such as the cylinder strength and the rate and duration of loading. For this and other reasons, a wholly rational flexural theory for reinforced concrete has not yet been developed

FIGURE 4.5
Stress and strain distributions
at ultimate load.

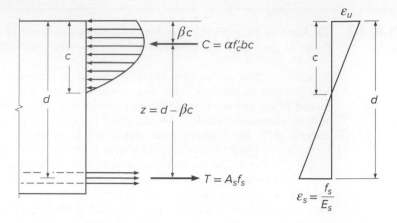

(Refs. 4.1 to 4.3). Present methods of analysis, therefore, are based in part on known laws of mechanics and are supplemented, where needed, by extensive test information.

Let Fig. 4.5 represent the distribution of internal stresses and strains when the beam is about to fail. One desires a method to calculate that moment M_n (nominal moment) at which the beam will fail either by tension yielding of the steel or by crushing of the concrete in the outer compression fiber. For the first mode of failure, the criterion is that the steel stress equal the yield point, $f_s = f_y$. It has been mentioned before that an exact criterion for concrete compression failure is not yet known, but that for rectangular beams, strains of 0.003 to 0.004 have been measured immediately preceding failure. If one assumes, usually slightly conservatively, that the concrete is about to crush when the maximum strain reaches $\varepsilon_u = 0.003$, comparison with a great many tests of beams and columns of a considerable variety of shapes and conditions of loading shows that a satisfactorily accurate and safe strength prediction can be made (Ref. 4.4). In addition to these two criteria (yielding of the steel at a stress of f_y and crushing of the concrete at a strain of 0.003), it is not really necessary to know the exact shape of the concrete stress distribution in Fig. 4.5. What is necessary is to know, for a given distance c of the neutral axis, (1) the total resultant compression force C in the concrete and (2) its vertical location, that is, its distance from the outer compression fiber.

In a rectangular beam, the area that is in compression is bc, and the total compression force on this area can be expressed as $C = f_{av}bc$, where f_{av} is the average compression stress on the area bc. Evidently, the average compressive stress that can be developed before failure occurs becomes larger, the higher the cylinder strength f_c' of the particular concrete. Let

$$\alpha = \frac{f_{av}}{f_c'} \tag{4.10}$$

Then

$$C = \alpha f_c'bc \tag{4.11}$$

For a given distance c to the neutral axis, the location of C can be defined as some fraction β of this distance. Thus, as indicated in Fig. 4.5, for a concrete of given strength it is necessary to know only α and β to completely define the effect of the concrete compressive stresses.

Extensive direct measurements, as well as indirect evaluations of numerous beam tests, have shown that the following values for α and β are satisfactorily accurate (see Ref. 4.5, where α is designated as k_1k_3 and β as k_2):

α equals 0.72 for $f'_c \leq 4000$ psi and decreases by 0.04 for every 1000 psi above 4000 up to 8000 psi. For $f'_c > 8000$ psi, $\alpha = 0.56$.

β equals 0.425 for $f'_c \leq 4000$ psi and decreases by 0.025 for every 1000 psi above 4000 up to 8000 psi. For $f'_c > 8000$ psi, $\beta = 0.325$.

The decrease in α and β for high-strength concretes is related to the fact that such concretes are more brittle; that is, they show a more sharply curved stress-strain plot with a smaller near-horizontal portion (see Figs. 2.3 and 2.4). Figure 4.6 shows these simple relations.

If this experimental information is accepted, the maximum moment can be calculated from the laws of equilibrium and from the assumption that plane cross sections remain plane. Equilibrium requires that

$$C = T \quad \text{or} \quad \alpha f'_c bc = A_s f_s \tag{4.12}$$

Also, the bending moment, being the couple of the forces C and T, can be written as either

$$M = Tz = A_s f_s (d - \beta c) \tag{4.13}$$

or

$$M = Cz = \alpha f'_c bc (d - \beta c) \tag{4.14}$$

For failure initiated by yielding of the tension steel, $f_s = f_y$. Substituting this value in Eq. (4.12), one obtains the distance to the neutral axis

$$c = \frac{A_s f_y}{\alpha f'_c b} \tag{4.15a}$$

Alternatively, using $A_s = \rho bd$, the neutral axis distance is

$$c = \frac{\rho f_y d}{\alpha f'_c} \tag{4.15b}$$

FIGURE 4.6

Variation of α and β with concrete strength f'_c.

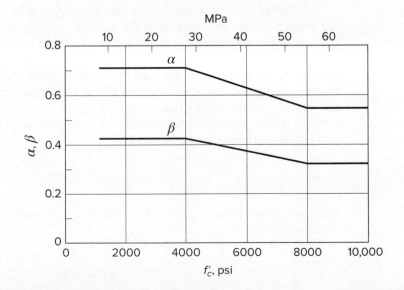

giving the distance to the neutral axis when tension failure occurs. The nominal moment M_n is then obtained from Eq. (4.13) with the value for c just determined, and $f_s = f_y$; that is,

$$M_n = \rho f_y b d^2 \left(1 - \frac{\beta f_y \rho}{\alpha f_c'} \right) \qquad (4.16a)$$

With the specific, experimentally obtained values for α and β given previously, this becomes

$$M_n = \rho f_y b d^2 \left(1 - 0.59 \frac{\rho f_y}{f_c'} \right) \qquad (4.16b)$$

If, for larger reinforcement ratios, the steel does not reach yield at failure, then the strain in the concrete becomes $\varepsilon_u = 0.003$, as previously discussed. The steel stress f_s, not having reached the yield point, is proportional to the steel strain ε_s; that is, according to Hooke's law,

$$f_s = \varepsilon_s E_s$$

From the strain distribution shown in Fig. 4.5, the steel strain ε_s can be expressed in terms of the distance c by evaluating similar triangles, after which it is seen that

$$f_s = \varepsilon_u E_s \frac{d - c}{c} \qquad (4.17)$$

Then, from Eq. (4.12),

$$\alpha f_c' b c = A_s \varepsilon_u E_s \frac{d - c}{c} \qquad (4.18)$$

and this quadratic may be solved for c, the only unknown for the given beam. With both c and f_s known, the nominal moment of the beam, so heavily reinforced that failure occurs by crushing of the concrete, may be found from either Eq. (4.13) or Eq. (4.14).

Whether or not the steel has yielded at failure can be determined by comparing the actual reinforcement ratio with the *balanced reinforcement ratio* ρ_b, representing that amount of reinforcement necessary for the beam to fail by crushing of the concrete at the same load that causes the steel to yield. This means that the neutral axis must be so located that at the load at which the steel starts yielding, the concrete reaches its compressive strain limit ε_u. Correspondingly, setting $f_s = f_y$ in Eq. (4.17) and substituting the yield strain ε_y for f_y / E_s gives the value of c defining the unique position of the neutral axis corresponding to simultaneous crushing of the concrete and initiation of yielding in the steel

$$c = \frac{\varepsilon_u}{\varepsilon_u + \varepsilon_y} d \qquad (4.19)$$

Substituting that value of c into Eq. (4.12), with $A_s f_s = \rho b d f_y$, gives the balanced reinforcement ratio

$$\rho_b = \frac{\alpha f_c'}{f_y} \frac{\varepsilon_u}{\varepsilon_u + \varepsilon_y} \qquad (4.20)$$

EXAMPLE 4.3   Determine the nominal moment M_n at which the beam of Examples 4.1 and 4.2 will fail.

SOLUTION.   For this beam the reinforcement ratio $\rho = A_s/(bd) = 2.37/(10 \times 23) = 0.0103$. The balanced reinforcement ratio is found from Eq. (4.20) to be 0.0284. Since the amount of steel in the beam is less than that which would cause failure by crushing of the concrete, the beam will fail in tension by yielding of the steel. Its nominal moment, from Eq. (4.16b), is

$$M_n = \rho f_y b d^2 \left(1 - 0.59 \frac{\rho f_y}{f_c'}\right) = 0.0103 \times 60,000 \times 10 \times 23^2 \left(1 - 0.59 \frac{0.0103 \times 60,000}{4000}\right)$$

$$= 2,970,000 \text{ in-lb} = 248 \text{ ft-kips}$$

When the beam reaches M_n, the distance to its neutral axis, from Eq. (4.15b), is

$$c = \frac{\rho f_y d}{\alpha f_c'} = \frac{0.0103 \times 60,000 \times 23}{0.72 \times 4000} = 4.94$$

It is informative to compare this result with those of Examples 4.1 and 4.2. In the previous calculations, it was found that at low loads, when the concrete had not yet cracked in tension, the neutral axis was located at a distance of 13.2 in. from the compression edge; at higher loads, when the tension concrete was cracked but stresses were still sufficiently small to be elastic, this distance was 7.6 in. Immediately before the beam fails, as has just been shown, this distance has further decreased to 4.9 in. For these same stages of loading, the stress in the steel increased from 2870 psi in the uncracked section to 22,300 psi in the cracked elastic section and to 60,000 psi at the nominal moment capacity. This migration of the neutral axis toward the compression edge and the increase in steel stress as load is increased is a graphic illustration of the differences between the various stages of behavior through which a reinforced concrete beam passes as its load is increased from zero to the value that causes it to fail. The examples also illustrate the fact that nominal moments cannot be determined accurately by elastic calculations.

4.3   DESIGN OF TENSION-REINFORCED RECTANGULAR BEAMS

For reasons that were explained in Chapter 1, the present design of reinforced concrete structures is based on the concept of providing sufficient strength to resist hypothetical overloads. The *nominal strength* of a proposed member is calculated based on the best current knowledge of member and material behavior. That nominal strength is modified by a *strength reduction factor* ϕ, less than unity, to obtain the *design strength*. The *required strength*, should the hypothetical overload stage actually be realized, is found by applying *load factors* γ, greater than unity, to the loads actually expected. These expected *service loads* include the calculated dead load, the calculated or legally specified live load, and environmental loads such as those due to wind, earthquake action, or temperature. Thus reinforced concrete members are proportioned so that, as shown in Eq. (1.5),

$$\phi M_n \geq M_u$$

$$\phi P_n \geq P_u$$

$$\phi V_n \geq V_u$$

$$\phi T_n \geq T_u$$

where the subscripts n denote the nominal strengths in flexure, thrust, shear, and torsion respectively, and the subscripts u denote the factored load moment, thrust, shear, and torsion. The strength reduction factors ϕ differ, depending upon the type of strength to be calculated, the importance of the member in the structure, and other considerations discussed in detail in Chapter 1.

A member proportioned on the basis of adequate strength at a specified overload stage must also perform in a satisfactory way under normal service load conditions. In specific terms, the deflection must be limited to an acceptable value, and concrete tensile cracks, which inevitably occur, must be of narrow width and well distributed throughout the tensile zone. Therefore, after proportioning for adequate strength, deflections are calculated and compared against limiting values (or otherwise controlled), and crack widths limited by specific means. This approach to design, referred to in Europe, and to some extent in U.S. practice, as *limit states design*, is the basis of the ACI Code, and it is the approach followed in this and later chapters.

a.   Equivalent Rectangular Stress Distribution

The method presented in Section 4.2c for calculating the flexural strength of reinforced concrete beams, derived from basic concepts of structural mechanics and pertinent experimental research information, also applies to situations other than the case of rectangular beams reinforced on the tension side. It can be used and gives valid answers for beams of other cross-sectional shapes, reinforced in other manners, and for members subject not only to simple bending but also to the simultaneous action of bending and axial force (compression or tension). However, the pertinent equations for these more complex cases become increasingly cumbersome and lengthy. What is more important, it becomes increasingly difficult for the designer to visualize the physical basis for the design methods and formulas; this could lead to a blind reliance on formulas, with a resulting lack of actual understanding. This is not only undesirable on general grounds but also, practically, more likely to lead to numerical errors in design work than when the designer at all times has a clear picture of the physical situation in the member being dimensioned or analyzed. Fortunately, it is possible, using a conceptual model, to formulate the strength analysis of reinforced concrete members in a different manner, which gives the same answers as the general analysis just developed but which is much more easily visualized and much more easily applied to cases of greater complexity than that of the simple rectangular beam. Its consistency is shown, and its application to more complex cases has been checked against the results of a vast number of tests on a great variety of types of members and conditions of loading (Ref. 4.4).

It was noted in the preceding section that the actual geometric shape of the concrete compressive stress distribution varies considerably and that, in fact, one need not know this shape exactly, provided one does know two things: (1) the magnitude C of the resultant of the concrete compressive stresses and (2) the location of this resultant. Information on these two quantities was obtained from the results of experimental research and expressed in the two parameters α and β.

Evidently, then, one can think of the actual complex stress distribution as replaced by a fictitious one of some simple geometric shape, provided that this fictitious distribution results in the same total compression force C applied at the same location as in the actual member when it is on the point of failure. Historically, a number of simplified, fictitious equivalent stress distributions have been proposed by investigators in various countries. The one generally accepted worldwide, and in the ACI Code, was first proposed by C. S. Whitney (Ref. 4.4) and was subsequently elaborated and checked experimentally by others (see, for example,

FIGURE 4.7

Actual and equivalent rectangular stress distributions at ultimate load.

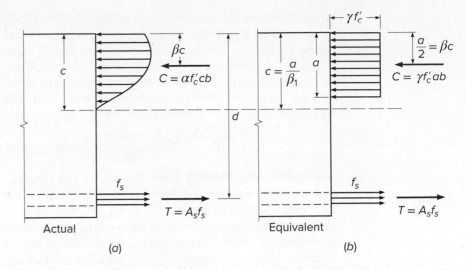

Refs. 4.5 and 4.6). The actual stress distribution immediately before failure and the fictitious equivalent distribution are shown in Fig. 4.7.

It is seen that the actual stress distribution is replaced by an equivalent one of simple rectangular outline. The intensity $\gamma f_c'$ of this equivalent constant stress and its depth $a = \beta_1 c$ are easily calculated from the two conditions that (1) the total compression force C and (2) its location, that is, distance from the top fiber, must be the same in the equivalent rectangular as in the actual stress distribution. From Fig. 4.7a and b the first condition gives

$$C = \alpha f_c' \, cb = \gamma f_c' \, ab \qquad \text{from which} \qquad \gamma = \alpha \frac{c}{a}$$

With $a = \beta_1 c$, this gives $\gamma = \alpha/\beta_1$. The second condition simply requires that in the equivalent rectangular stress block, the force C be located at the same distance βc from the top fiber as in the actual distribution. It follows that $\beta_1 = 2\beta$.

To supply the details, the upper two lines of Table 4.1 present the experimental evidence of Fig. 4.6 in tabular form. The lower two lines give the just-derived parameters β_1 and γ for the rectangular stress block. It is seen that the intensity factor for compressive stress γ is essentially independent of f_c' and can be taken as 0.85 throughout. Hence, regardless of f_c', the concrete compression force at failure in a rectangular beam of width b is

$$C = 0.85 f_c' ab \tag{4.21}$$

TABLE 4.1

Concrete stress block parameters

	f_c' psi				
	≤ 4000	5000	6000	7000	≥ 8000
α	0.72	0.68	0.64	0.60	0.56
β	0.425	0.400	0.375	0.350	0.325
$\beta_1 = 2\beta$	0.85	0.80	0.75	0.70	0.65
$\gamma = \alpha/\beta_1$	0.85	0.85	0.85	0.86	0.86

Also, for commonly used concretes with $f_c' \leq 4000$ psi, the depth of the rectangular stress block is $a = 0.85c$, with c being the distance to the neutral axis. For higher-strength concretes, this distance is $a = \beta_1 c$, with the β_1 values shown in Table 4.1. This is expressed as follows: For f_c' between 2500 and 4000 psi, β_1 shall be taken as 0.85; for f_c' above 4000 psi, β_1 shall be reduced linearly at a rate of 0.05 for each 1000 psi of strength in excess of 4000 psi, but β_1 shall not be taken as less than 0.65. In mathematical terms, the relationship between β_1 and f_c' can be expressed as

$$\beta_1 = 0.85 - 0.05 \frac{f_c' - 4000}{1000} \quad \text{and} \quad 0.65 \leq \beta_1 \leq 0.85 \quad (4.22)$$

The equivalent rectangular stress distribution can be used for deriving the equations that have been developed in Section 4.2c. The failure criteria, of course, are the same as before: yielding of the steel at $f_s = f_y$ or crushing of the concrete at $\varepsilon_u = 0.003$. Because the rectangular stress block is easily visualized and its geometric properties are extremely simple, many calculations are carried out directly without reference to formally derived equations, as will be seen in the following sections.

b. Balanced Strain Condition

A reinforcement ratio ρ_b producing balanced strain conditions can be established based on the condition that, at balanced failure, the steel strain is exactly equal to ε_y when the strain in the concrete simultaneously reaches the crushing strain of $\varepsilon_u = 0.003$. Referring to Fig. 4.5,

$$c = \frac{\varepsilon_u}{\varepsilon_u + \varepsilon_y} d \quad (4.23)$$

which is seen to be identical to Eq. (4.19). Then from the equilibrium requirement that $C = T$

$$\rho_b f_y bd = 0.85 f_c' ab = 0.85 \beta_1 f_c' bc$$

from which

$$\rho_b = 0.85 \, \beta_1 \frac{f_c'}{f_y} \frac{c}{d} = 0.85 \beta_1 \frac{f_c'}{f_y} \frac{\varepsilon_u}{\varepsilon_u + \varepsilon_y} \quad (4.24)$$

This is easily shown to be equivalent to Eq. (4.20).

c. Underreinforced Beams

A compression failure in flexure, should it occur, gives little if any warning of distress, while a tension failure, initiated by yielding of the steel, typically is gradual. Distress is obvious from observing the large deflections and widening of concrete cracks associated with yielding of the steel reinforcement, and measures can be taken to avoid total collapse. In addition, most beams for which failure initiates by yielding possess substantial strength based on strain-hardening of the reinforcing steel, which is not accounted for in the calculations of M_n.

Because of these differences in behavior, it is prudent to require that beams be designed such that failure, if it occurs, will be by yielding of the steel, not by crushing of the concrete. This can be done, theoretically, by requiring that the reinforcement ratio ρ be less than the balance ratio ρ_b given by Eq. (4.24). Such a beam is described as being *underreinforced*.

In actual practice, the upper limit on ρ should be below ρ_b for the following reasons: (1) for a beam with ρ exactly equal to ρ_b, the compressive strain limit of the concrete would be reached, theoretically, at precisely the same moment that the steel reaches its yield stress, without significant yielding before failure; (2) material properties are never known precisely; (3) strain-hardening of the reinforcing steel, not accounted for in design, may lead to a brittle concrete compression failure even though ρ may be somewhat less than ρ_b; (4) the actual steel area provided, considering standard reinforcing bar sizes, will always be equal to or larger than required, based on selected reinforcement ratio ρ, tending toward overreinforcement; and (5) the extra ductility provided by beams with lower values of ρ increases the deflection capability substantially and thus provides warning prior to failure.

d. ACI Code Provisions for Underreinforced Beams

While the nominal strength of a member may be calculated based on principles of mechanics, the mechanics alone cannot establish safe limits for maximum reinforcement ratios, as discussed in Chapter 3. These limits are defined by the ACI Code. The limitations take two forms. First, the Code addresses the minimum tensile reinforcement strain allowed at nominal strength in the design of beams. Second, the Code defines strength reduction factors that may depend on the tensile strain at nominal strength. Both limitations are based on the *net tensile strain* ε_t of the reinforcement farthest from the compression face of the concrete at the depth d_t. The net tensile strain is exclusive of prestress, temperature, and shrinkage effects. For beams with a single layer of reinforcement, the depth to the centroid of the steel d is the same as d_t. For beams with multiple layers of reinforcement, d_t is greater than the depth to the centroid of the reinforcement d, as shown in Fig. 4.8. Substituting d_t for d and ε_t for ε_y in Eq. (4.23), the net tensile strain may be represented as

$$\varepsilon_t = \varepsilon_u \frac{d_t - c}{c} \tag{4.25}$$

Then based on Eq. (4.24), the reinforcement ratio to produce a selected value of net tensile strain is

$$\rho = 0.85\beta_1 \frac{f_c'}{f_y} \frac{d_t}{d} \frac{\varepsilon_u}{\varepsilon_u + \varepsilon_t} \tag{4.26a}$$

or somewhat conservatively

$$\rho = 0.85\beta_1 \frac{f_c'}{f_y} \frac{\varepsilon_u}{\varepsilon_u + \varepsilon_t} \tag{4.26b}$$

To ensure truly underreinforced behavior, ACI Code Table 21.2.2 establishes a minimum net tensile strain $\varepsilon_{t,\min}$ at nominal member strength for beams subjected to axial loads less than $0.10 f_c' A_g$, where A_g is the gross area of the cross section.

$$\varepsilon_{t,\min} = \varepsilon_{ty} + 0.003 \tag{4.26c}$$

where $\varepsilon_{ty} = f_y/E_s$.

FIGURE 4.8

Beam with two layers of reinforcement showing difference between the effective depth d and the distance to the reinforcement farthest from the compressive face of the concrete d_t.

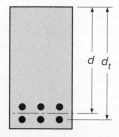

FIGURE 4.9
Variation of strength reduction factor with net tensile strain in the steel.

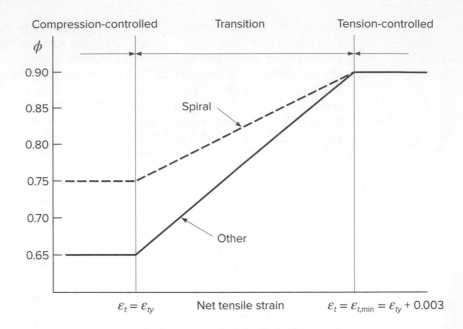

The ACI Code defines members that meet this requirement, that is, $\varepsilon_t \geq \varepsilon_{t,\min}$, as *tension-controlled*. The corresponding strength reduction factor is $\phi = 0.9$. The requirement that $\varepsilon_t \geq \varepsilon_{t,\min}$ applies to all grades of reinforcing steel, including prestressing steel. The ACI Code additionally defines *compression-controlled* members as those having a net tensile strain $\varepsilon_t \leq \varepsilon_y$. The strength reduction factor for compression-controlled members is 0.65; a value of $\phi = 0.75$ may be used if the members are spirally reinforced. Members with net tensile strains between ε_y and $\varepsilon_{t,\min}$ are classified as *transition*, and the ACI Code allows a linear interpolation of ϕ based on ε_t, as shown in Fig. 4.9 for members with axial loads greater than $0.10f'_c A_g$. For the purposes of defining compression-controlled members and calculating ϕ, ACI Code 21.2.2.1 permits a value of $\varepsilon_y = 0.002$ to be used for Grade 60 reinforcement, in place of the calculated value 0.00207. A value of $\varepsilon_{ty} = 0.200$ is *required* for all prestressed reinforcement.

Based on Eq. (4.26b), the maximum reinforcement ratio for a tension-controlled beam is

$$\rho_{\max} = 0.85\beta_1 \frac{f'_c}{f_y} \frac{\varepsilon_u}{\varepsilon_u + \varepsilon_{t,\min}} \qquad (4.26d)$$

Calculation of the nominal moment capacity frequently involves determination of the depth of the equivalent rectangular stress block a. Since $c = a/\beta_1$, it is sometimes more convenient to calculate c/d_t ratios than either ρ or the net tensile strain. The assumption that plane sections remain plane ensures a direct correlation between net tensile strain and the c/d_t ratio. Values of c/d_t corresponding to tension-controlled sections with $\varepsilon_t = \varepsilon_{t,\min}$ for Grades 60, 80, and 100 reinforcement are shown in Fig. 4.10. A strength reduction factor of 0.90 is permitted for sections with c/d_t values less than or equal to the values shown.

Comparing Eqs. (4.26a) and (4.26b), it can be seen that the maximum reinforcement ratio in Eq. (4.26d) is exact for beams with a single layer of reinforcement and slightly conservative for beams with multiple layers of reinforcement, where d_t is greater than d. Because $\varepsilon_t \geq \varepsilon_{t,\min}$ ensures that steel is yielding in tension, $f_s = f_y$

FIGURE 4.10
Minimum net tensile strain $\varepsilon_{t,min}$ and maximum c/d_t for tension-controlled sections for Grades 60, 80, and 100 reinforcement.

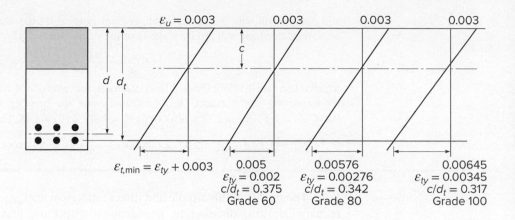

$\varepsilon_{t,min} = \varepsilon_{ty} + 0.003$

0.005	0.00576	0.00645
$\varepsilon_{ty} = 0.002$	$\varepsilon_{ty} = 0.00276$	$\varepsilon_{ty} = 0.00345$
$c/d_t = 0.375$	$c/d_t = 0.342$	$c/d_t = 0.317$
Grade 60	Grade 80	Grade 100

FIGURE 4.11
Singly reinforced rectangular beam.

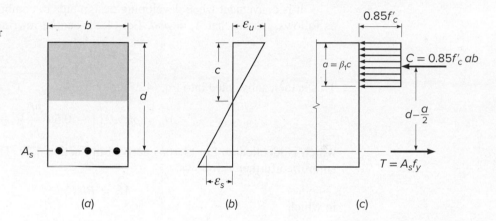

at failure. Referring to Fig. 4.11, the nominal flexural strength M_n is obtained by summing moments about the centroid of the compression force C.

$$M_n = A_s f_y \left(d - \frac{a}{2} \right) \tag{4.27}$$

The depth of the equivalent stress block a can be found based on equilibrium, $C = T$. Hence, $0.85 f_c' ab = A_s f_y$, giving

$$a = \frac{A_s f_y}{0.85 f_c' b} \tag{4.28}$$

EXAMPLE 4.4   Using the equivalent rectangular stress distribution, directly calculate the nominal strength of the beam previously analyzed in Example 4.3. Recall that $b = 10$ in., $d = 23$ in., $A_s = 2.37$ in^2., $f_c' = 4000$ psi, $f_y = 60{,}000$ psi, $\varepsilon_y = 0.002$, and $\beta_1 = 0.85$.

SOLUTION.   The distribution of stresses, internal forces, and strains is shown in Fig. 4.11. With $\varepsilon_u = 0.003$ and $\varepsilon_{t,min} = 0.002 + 0.003 = 0.005$ for Grade 60 reinforcement, the maximum reinforcement ratio is calculated from Eq. (4.26d) as

$$\rho_{max} = 0.85 \times 0.85 \, \frac{4000}{60{,}000} \, \frac{0.003}{0.003 + 0.005} = 0.0181$$

and comparison with the actual reinforcement ratio of 0.0103 confirms that the member is underreinforced and will fail by yielding of the steel. Alternatively, recalling that $c = 4.94$ in.,

$$\frac{c}{d_t} = \frac{c}{d} = \frac{4.94}{23} = 0.215$$

which is less than 0.375 for Grade 60 reinforcement, the value of c/d_t corresponding to $\varepsilon_{t,min} = 0.005$ for Grade 60 reinforcement, also confirming that the member is underreinforced. Hence, $0.85f_c'\, ab = A_s f_y$, or $a = 2.37 \times 60{,}000/(0.85 \times 4000 \times 10) = 4.18$. The nominal moment is

$$M_n = A_s f_y\left(d - \frac{a}{2}\right) = 2.37 \times 60{,}000(23 - 2.09) = 2{,}970{,}000 \text{ in-lb} = 248 \text{ ft-kips}$$

The results of this simple and direct numerical analysis, based on the equivalent rectangular stress distribution, are identical with those previously determined from the general strength analysis described in Section 4.2c.

It is convenient when developing design aids to combine Eqs. (4.27) and (4.28) as follows. Noting that $A_s = \rho bd$, Eq. (4.28) can be rewritten as

$$a = \frac{\rho f_y d}{0.85 f_c'} \tag{4.29}$$

This is then substituted into Eq. (4.27) to obtain

$$M_n = \rho f_y bd^2\left(1 - 0.59\,\frac{\rho f_y}{f_c'}\right) \tag{4.30}$$

which is identical to Eq. (4.16b) derived in Section 4.2c. This basic equation can be simplified further as follows:

$$M_n = Rbd^2 \tag{4.31}$$

in which

$$R = \rho f_y\left(1 - 0.59\,\frac{\rho f_y}{f_c'}\right) \tag{4.32}$$

The *flexural resistance factor R* depends only on the reinforcement ratio and the strengths of the materials and is easily tabulated. Tables A.5a and A.5b and Graphs A.1a and A.1b of Appendix A give R values for ordinary combinations of steel and concrete and the full practical range of reinforcement ratios.

In accordance with the safety provisions of the ACI Code, the nominal flexural strength M_n is reduced by imposing the strength reduction factor ϕ to obtain the *design strength* ϕM_n

$$\phi M_n = \phi A_s f_y\left(d - \frac{a}{2}\right) \tag{4.33}$$

or, alternatively,

$$\phi M_n = \phi\rho f_y bd^2\left(1 - 0.59\,\frac{\rho f_y}{f_c'}\right) \tag{4.34}$$

or

$$\phi M_n = \phi Rbd^2 \tag{4.35}$$

EXAMPLE 4.4
(*continued*)

Calculate the design moment capacity ϕM_n for the beam analyzed earlier in Example 4.4.

SOLUTION. Comparing ρ with ρ_{max} or c/d_t for the beam with the value of c/d_t corresponding to $\varepsilon_{t,min} = 0.005$ demonstrates that $\varepsilon_t > 0.005$. Therefore, $\phi = 0.90$ and the design capacity is

$$\phi M_n = 0.9 \times 248 = 223 \text{ ft-kips}$$

e. Minimum Reinforcement Ratio

Another mode of failure may occur in very lightly reinforced beams. If the flexural strength of the cracked section is less than the moment that produced cracking of the previously uncracked section, the beam will fail immediately and without warning of distress upon formation of the first flexural crack. To ensure against this type of failure, a *lower limit* can be established for the reinforcement ratio by equating the cracking moment, calculated from the concrete modulus of rupture (Section 2.9), to the strength of the cracked section.

For a rectangular section having width b, total depth h, and effective depth d (see Fig. 4.1b), the section modulus with respect to the tension fiber is $bh^2/6$. For typical cross sections, it is satisfactory to assume that $h/d = 1.1$ and that the internal lever arm at flexural failure is $0.95d$. If the modulus of rupture is taken as $f_r = 7.5\sqrt{f_c'}$, as usual, then an analysis equating the cracking moment to the flexural strength results in

$$A_{s,\min} = \frac{1.6\sqrt{f_c'}}{f_y}bd \qquad (4.36a)$$

This development can be generalized to apply to beams having a T cross section (see Section 4.7 and Fig. 4.17). The corresponding equations depend on the proportions of the cross section and on whether the beam is bent with the flange (slab) in tension or in compression. For T beams of typical proportions that are bent with the flange in compression, analysis confirms that the minimum steel area should be

$$A_{s,\min} = \frac{2.7\sqrt{f_c'}}{f_y}b_w d \qquad (4.36b)$$

where b_w is the width of the web, or stem, projecting below the slab. For T beams that are bent with the flange in tension, from a similar analysis, the minimum steel area is

$$A_{s,\min} = \frac{6.2\sqrt{f_c'}}{f_y}b_w d \qquad (4.36c)$$

The ACI Code requirements for minimum steel area are based on the results just discussed, but there are some differences. According to ACI Code 9.6.1, at any section where tensile reinforcement is required by analysis, with some exceptions as noted below, the area A_s provided must not be less than

$$A_{s,\min} = \frac{3\sqrt{f_c'}}{f_y}b_w d \geq \frac{200b_w d}{f_y} \qquad (4.37a)$$

This applies to both positive and negative bending sections. The inclusion of the additional limit of $200b_w d/f_y$ is merely for historical reasons; it happens to give the same minimum reinforcement ratio that was imposed in earlier codes for then-common material strengths. Note that in Eq. (4.37a) the section width b_w is used; it is understood that for rectangular sections $b_w = b$. Note further that the ACI coefficient of 3 is a conservatively rounded value compared with 2.7 in Eq. (4.36b) for T beams with the flange in compression, and is very conservative when applied to rectangular beam sections, for which a rational analysis gives 1.6 in Eq. (4.36a). This probably reflects the view that the minimum steel for the negative bending sections of a continuous T beam (which are, in effect, rectangular sections, as discussed in Section 4.7c) should be no less than that for the positive bending sections, where the moment is generally smaller.

ACI Code 9.6.1 treats *statically determinate* T beams with the flange in *tension* as a special case, for which the minimum steel area is equal to or greater than the value given by Eq. (4.37a) with b_w replaced by either $2b_w$ or the width of the *flange*, whichever is smaller.

ACI Code Eq. (4.37a) is conveniently expressed in terms of a *minimum tensile reinforcement ratio* ρ_{min} by dividing both sides by $b_w d$.

$$\rho_{min} = \frac{3\sqrt{f_c'}}{f_y} \geq \frac{200}{f_y} \tag{4.37b}$$

According to ACI Code 9.6.1, the requirements of Eq. (4.37a) need not be imposed if, at every section, the area of tensile reinforcement provided is at least one-third greater than that required by analysis. This provides sufficient reinforcement for large members such as grade beams, where the usual equations would require excessive amounts of steel.

For structural slabs and footings of uniform thickness, the minimum area of tensile reinforcement in the direction of the span is that required for shrinkage and temperature steel (see Section 12.3 and Table 12.2), and the above minimums need not be imposed. The maximum spacing of such steel is the smaller of 3 times the total slab thickness or 18 in.

f. Examples of Rectangular Beam Analysis and Design

Flexural problems can be classified broadly as *analysis problems* or *design problems*. In analysis problems, the section dimensions, reinforcement, and material strengths are known, and the moment capacity is required. In the case of design problems, the required moment capacity is given, as are the material strengths, and it is required to find the section dimensions and reinforcement. Examples 4.5 and 4.6 illustrate analysis and design, respectively.

EXAMPLE 4.5   **Flexural strength of a given member.**   A rectangular beam has width 12 in. and effective depth 17.5 in. It is reinforced with four No. 8 (No. 25) bars in one row. If $f_y = 60,000$ psi and $f_c' = 4000$ psi, what is the nominal flexural strength, and what is the maximum moment that can be utilized in design, according to the ACI Code?

SOLUTION.   From Table A.2 of Appendix A, the area of four No. 8 (No. 25) bars is 3.16 in.^{2}. Assuming that the beam is underreinforced and using Eq. (4.28),

$$a = \frac{A_s f_y}{0.85\, f_c'\, b} = \frac{3.16 \times 60}{0.85 \times 4 \times 12} = 4.65 \text{ in.}$$

The depth of the neutral axis is $c = a/\beta_1 = 4.65/0.85 = 5.47$, giving

$$\frac{c}{d_t} = \frac{5.47}{17.5} = 0.313$$

which is less than 0.375, the value corresponding to $\varepsilon_t = 0.005$, as shown in Fig. 4.10. Thus, the beam is, as assumed, underreinforced, and from Eq. (4.27)

$$M_n = A_s f_y \left(d - \frac{a}{2} \right) = 3.16 \times 60 \left(17.5 - \frac{4.65}{2} \right) = 2880 \text{ in-kips}$$

The fact that the beam is underreinforced could also have been established by calculating $\rho = 3.16/(12 \times 17.5) = 0.0151$, which is less than $\rho_{\max}$, which is calculated using Eq. (4.26d).

$$\rho_{\max} = 0.85\beta_1 \frac{f_c'}{f_y} \frac{\varepsilon_u}{\varepsilon_u + \varepsilon_y + 0.003} = 0.85 \times 0.85 \left(\frac{4}{60}\right)\left(\frac{0.003}{0.003 + 0.005}\right) = 0.0181$$

Thus, $\phi = 0.90$, and the design strength is taken as

$$\phi M_n = 0.90 \times 2880 = 2590 \text{ in-kips}$$

The ACI Code limit for the minimum reinforcement ratio

$$\rho_{\min} = \frac{3\sqrt{f_c'}}{f_y} \geq \frac{200}{f_y} = \frac{3\sqrt{4000}}{60,000} \geq \frac{200}{60,000} = 0.0033$$

is satisfied for this beam.

EXAMPLE 4.6   **Concrete dimensions and steel area to resist a given moment.**   Find the concrete cross section and the steel area required for a simply supported rectangular beam with a span of 15 ft that is to carry a calculated dead load of 1.27 kips/ft and a service live load of 2.15 kips/ft, as shown in Fig. 4.12. Material strengths are $f_c' = 4000$ psi and $f_y = 60,000$ psi.

SOLUTION.   Load factors are first applied to the given service loads to obtain the factored load for which the beam is to be designed, and the corresponding moment:

$$w_u = 1.2 \times 1.27 + 1.6 \times 2.15 = 4.96 \text{ kips/ft}$$

$$M_u = \frac{1}{8} \times 4.96 \times 15^2 \times 12 = 1670 \text{ in-kips}$$

The concrete dimensions depend on the designer's choice of reinforcement ratio. To minimize the concrete section, it is desirable to select the maximum permissible reinforcement ratio. To maintain $\phi = 0.9$, the maximum reinforcement ratio corresponding to a net tensile strain of 0.005 for Grade 60 reinforcement is selected (see Fig. 4.9). Then, from Eq. (4.26d)

$$\rho_{\max} = 0.85\beta_1 \frac{f_c'}{f_y} \frac{\varepsilon_u}{\varepsilon_u + \varepsilon_{t,\min}} = 0.85 \times 0.85 \left(\frac{4}{60}\right)\left(\frac{0.003}{0.003 + 0.005}\right) = 0.0181$$

Setting the required flexural strength equal to the design strength from Eq. (4.34), and substituting the selected values for ρ and material strengths,

$$M_u = \phi M_n$$

$$1670 = 0.90 \times 0.0181 \times 60bd^2\left(1 - 0.59 \frac{0.0181 \times 60}{4}\right)$$

from which

$$bd^2 = 2040 \text{ in}^3$$

A beam with width $b = 10$ in. and $d = 14.3$ in. satisfies this requirement. The required steel area is found by applying the chosen reinforcement ratio to the required concrete dimensions:

$$A_s = 0.0181 \times 10 \times 14.3 = 2.59 \text{ in}^2$$

Two No. 10 (No. 32) bars provide 2.54 in^2, which is very close to the required area.

FIGURE 4.12
Structural loads for
Example 4.6.

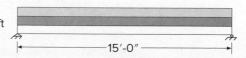

Service live load = 2.15 kips/ft
Calculated dead load = 1.27 kips/ft
(including beam self-weight)

15'-0"

Assuming 2.5 in. concrete cover from the centroid of the bars, the required total depth is $h = 16.8$ in. In actual practice, however, the concrete dimensions b and h are always rounded up to the nearest inch, and often to the nearest multiple of 2 in. (see Section 4.4). The actual d is then found by subtracting the required concrete cover dimension from h. For the present example, $b = 10$ in. and $h = 18$ in. are selected, resulting in effective depth $d = 15.5$ in. Improved economy then may be possible, refining the steel area based on the actual, larger, effective depth. One can obtain the revised steel requirement directly by solving Eq. (4.34) for ρ, with $\phi M_n = M_u$. A quicker solution can be obtained by iteration. First a reasonable value of a is assumed, and A_s is found from Eq. (4.33). From Eq. (4.28) a revised estimate of a is obtained, and A_s is revised. This method converges very rapidly. For example, assume $a = 5$ in. Then

$$A_s = \frac{M_u}{\phi f_y\left(d - \dfrac{a}{2}\right)} = \frac{1670}{0.90 \times 60(15.5 - 5/2)} = 2.38 \text{ in}^2$$

Checking the assumed a gives

$$a = \frac{A_s f_y}{0.85\, f_c'\, b} = \frac{2.38 \times 60}{0.85 \times 4 \times 10} = 4.20 \text{ in.}$$

This is close enough to the assumed value that no further calculation is required. The required steel area of 2.38 in^2 could be provided using three No. 8 (No. 25) bars, but for simplicity of construction, two No. 10 (No. 32) bars will be used as before.

A somewhat larger beam cross section using less steel may be more economical and will tend to reduce deflections. As an alternative solution, the beam will be redesigned with a lower reinforcement ratio of $\rho = 0.60\rho_{max} = 0.60 \times 0.0181 = 0.0109$. Setting the required strength equal to the design strength [Eq. (4.34)] as before,

$$1670 = 0.90 \times 0.0109 \times 60bd^2\left(1 - 0.59\frac{0.0109 \times 60}{4}\right)$$

and

$$bd^2 = 3140 \text{ in}^3$$

A beam with $b = 10$ in. and $d = 17.7$ in. meets the requirement, for which

$$A_s = 0.0109 \times 10 \times 17.7 = 1.93 \text{ in}^2$$

Two No. 9 (No. 29) bars are sufficient, providing an area of 2.00 in^2. If the total concrete height is rounded to 20 in., a 17.5 in. effective depth results, increasing the required steel area to 1.96 in^2. Two No. 9 (No. 29) bars remain the best choice.

It is apparent that an infinite number of solutions to the stated problem are possible, depending upon the reinforcement ratio selected. That ratio may vary from an upper limit of ρ_{max} to a lower limit of $3\sqrt{f_c'}/f_y \geq 200/f_y$ for beams, according to the ACI Code. To compare the two solutions (using the theoretical dimensions, unrounded for the comparison, and assuming h is 2.5 in. greater than d in each case), increasing the concrete section area by 14 percent achieves a steel saving of 20 percent. The second solution would likely be more economical and would be preferred, unless beam dimensions must be minimized for architectural or functional reasons. Economical designs typically have reinforcement ratios between $0.50\rho_{max}$ and $0.75\rho_{max}$.

There is a type of problem, occurring frequently, that does not fall strictly into either the analysis or the design category. The concrete dimensions are given and are known to be adequate to carry the required moment, and it is necessary only to find the steel area. Typically, this is the situation at critical design sections of continuous beams, in which the concrete dimensions are often kept constant, although the steel reinforcement varies along the span according to the required flexural resistance. Dimensions b, d, and h are determined at the maximum moment section, usually at one of the supports. At other supports, and at midspan locations, where moments are usually smaller, the concrete dimensions are known to be adequate and only the tensile steel remains to be found. An identical situation was encountered in the design problem of Example 4.6, in which concrete dimensions were rounded up from the minimum required values, and the required steel area was to be found. In either case, the iterative approach demonstrated in Example 4.6 is convenient.

EXAMPLE 4.7   **Determination of steel area.** Using the same concrete dimensions as were used for the second solution of Example 4.6 ($b = 10$ in., $d = 17.5$ in., and $h = 20$ in.) and the same material strengths, find the steel area required to resist a moment M_u of 1300 in-kips.

SOLUTION.   Assume $a = 4.0$ in. Then

$$A_s = \frac{1300}{0.90 \times 60(17.5 - 4.0/2)} = 1.55 \text{ in}^2$$

Checking the assumed a gives

$$a = \frac{1.55 \times 60}{0.85 \times 4 \times 10} = 2.74 \text{ in.}$$

Next assume $a = 2.6$ in. and recalculate A_s:

$$A_s = \frac{1300}{0.90 \times 60(17.5 - 1.3)} = 1.49 \text{ in}^2$$

No further iteration is required. Use $A_s = 1.49$ in^2. Two No. 8 (No. 25) bars, $A_s = 1.58$ in^2, will be used. A check of the reinforcement ratio shows $\rho < \rho_{max}$ and $\phi = 0.9$.

EXAMPLE 4.8   **Determination of steel area for a fixed concrete section.** Architectural considerations limit the height of a 20 ft long simple span beam to 16 in. and the width to 12 in. The following loads and material properties are given: $w_d = 0.79$ kips/ft, $w_l = 1.65$ kips/ft, $f'_c = 5000$ psi, and $f_y = 60,000$ psi. Determine the reinforcement for the beam.

SOLUTION.   Calculating the factored loads gives

$$w_u = 1.2 \times 0.79 + 1.6 \times 1.65 = 3.59 \text{ kips/ft}$$

$$M_u = 3.59 \times \frac{20^2}{8} = 179 \text{ ft-kips} = 2150 \text{ in-kips}$$

Assume $a = 4.0$ in. and $\phi = 0.90$. The effective depth is $(16 - 2.5)$ in. $= 13.5$ in. Calculating A_s gives

$$A_s = \frac{M_u/\phi}{f_y(d - a/2)} = \frac{2150/0.90}{60(13.5 - 2.0)} = 3.46 \text{ in}^2$$

Try two No. 10 (No. 32) and one No. 9 (No. 29) bar, $A_s = 3.54$ in^2.

Check $a = 3.54 \times 60/(0.85 \times 5 \times 12) = 4.16$ in. from Eq. (4.28). This is more than assumed; therefore, continue to check the moment capacity.

$$M_n = 3.54 \times 60(13.5 - 4.16/2) = 2426 \text{ in-kips}$$

Using a ϕ of 0.90 gives $\phi M_n = 2183$ in-kips, which is adequate; however, the net tensile strain must be checked to validate the selection of $\phi = 0.9$. In this case $c = a/\beta_1 = 4.16/0.80 = 5.20$ in. The c/d ratio is $0.385 > 0.375$, indicating that the criterion for a tension-controlled section for Grade 60 reinforcement of $\varepsilon_t > 0.005$ is not satisfied. In this case, the net tensile strain is

$$\varepsilon_t = \varepsilon_u \frac{d-c}{c} = 0.003 \frac{13.5 - 5.2}{5.2} = 0.00479$$

To maintain the architectural depth limitations, two design options are possible: increase the width of the beam or increase the strength of the concrete.

Option 1: Increase the beam width by 2 inches.

$$a = \frac{3.81 \times 60}{0.85 \times 5 \times 14} = 3.84 \text{ in.}$$

$$c = \frac{3.84}{0.80} = 4.80 \text{ in.}$$

$$\frac{c}{d_t} = \frac{4.80}{13.5} = 0.356 < 0.375 \text{ and } \phi = 0.90$$

$$M_n = 3.81 \times 60 \left(13.5 - \frac{3.84}{2}\right) = 265 \text{ in-kips}$$

$$M_u = \phi M_n = 0.90 \times 2650 = 2380 \text{ in-kips}$$

The nominal and design moments increase slightly and $\phi = 0.90$, allowing the section to meet the design requirements.

Option 2: Increase the concrete strength to 6000 psi.

$$a = \frac{3.81 \times 60}{0.85 \times 6 \times 12} = 3.74 \text{ in.}$$

$$c = \frac{3.74}{0.75} = 4.98 \text{ in.}$$

$$\frac{c}{d_t} = \frac{4.98}{13.5} = 0.369 < 0.375 \text{ and } \phi = 0.90$$

$$M_n = 3.81 \times 60 \left(13.5 - \frac{3.74}{2}\right) = 2660 \text{ in-kips}$$

$$M_u = \phi M_n = 0.90 \times 2660 = 2390 \text{ in-kips}$$

Again, the nominal and design moments increase slightly and $\phi = 0.90$, allowing the section to meet the design requirements.

In actuality, the first solution deviates less than 1 percent from the desired value and would likely be acceptable. The remaining portion of the example demonstrates the design implications of requiring a tension-controlled section. Option 1 is preferable to maintain a common concrete strength in the structure. Option 2 may require different concrete strengths in the structure and would be selected only if the width dimension of the section could not be revised.

In solving these examples, the basic equations have been used to develop familiarity with them. In actual practice, however, design aids such as Table A.4 of Appendix A, giving values of maximum and minimum reinforcement ratios, and Table A.5 and Graph A.1, providing values of flexural resistance factor *R*, are more convenient. The example problems are repeated in Section 4.4 to demonstrate the use of these aids.

g. Overreinforced Beams

According to the ACI Code, all beams are to be designed for yielding of the tension steel with $\varepsilon_t \geq \varepsilon_{t,\min}$ or $\rho \leq \rho_{\max}$. Occasionally, however, such as when analyzing the capacity of existing construction or members built under earlier Codes, it may be necessary to calculate the flexural strength of an overreinforced compression-controlled member, for which f_s is less than f_y at flexural failure.

In this case, the steel strain, in Fig. 4.11b, will be less than the yield strain, but can be expressed in terms of the concrete strain ε_u and the still-unknown distance c to the neutral axis:

$$\varepsilon_s = \varepsilon_u \frac{d - c}{c} \tag{4.38}$$

From the equilibrium requirement that $C = T$, one can write

$$0.85\beta_1 f_c' bc = \rho \varepsilon_s E_s bd$$

Substituting the steel strain from Eq. (4.38) in the last equation, and defining $k_u = c/d$, one obtains a quadratic equation in k_u as follows:

$$k_u^2 + m\rho k_u - m\rho = 0$$

Here, $\rho = A_s/bd$ as usual, and m is a material parameter given by

$$m = \frac{E_s \varepsilon_u}{0.85\beta_1 f_c'} \tag{4.39}$$

Solving the quadratic equation for k_u,

$$k_u = \sqrt{m\rho + \left(\frac{m\rho}{2}\right)^2} - \frac{m\rho}{2} \tag{4.40}$$

The neutral axis depth for the overreinforced beam can then easily be found from $c = k_u d$, after which the stress-block depth $a = \beta_1 c$. With steel strain ε_s then calculated from Eq. (4.38), and with $f_s = E_s \varepsilon_s$, the nominal flexural strength is

$$M_n = A_s f_s \left(d - \frac{a}{2}\right) \tag{4.41}$$

The strength reduction factor ϕ will equal 0.65 for beams in this range or slightly higher if the net tensile strain is in the transition zone shown in Fig. 4.9.

4.4 DESIGN AIDS

Basic equations were developed in Section 4.3 for the analysis and design of reinforced concrete beams, and these were used directly in the examples. In practice, the design of beams and other reinforced concrete members is greatly facilitated by the use of aids such as computer software and those in Appendix A of this text and in Refs. 4.7 through 4.9. Tables A.1, A.2, A.4 through A.7, and Graph A.1 of Appendix A relate directly to this chapter, and the student can scan this material to become familiar with the coverage. Other aids will be discussed, and their use demonstrated, in later chapters.

Equation (4.35) gives the flexural design strength ϕM_n of an underreinforced rectangular beam with a reinforcement ratio at or below ρ_b. The flexural resistance factor R, from Eq. (4.32), is given in Table A.5a for lower reinforcement ratios or Table A.5b for higher reinforcement ratios. Alternatively, R can be obtained from Graph A.1. For *analysis* of the capacity of a section with known concrete dimensions b and d, having known reinforcement ratio ρ, and with known materials strengths, the design strength ϕM_n can be obtained directly by Eq. (4.35).

For *design* purposes, where concrete dimensions and reinforcement are to be found and the factored load moment M_u is to be resisted, there are two possible approaches. One approach starts with selecting the optimum reinforcement ratio and then calculating concrete dimensions, as follows:

1. Set the required strength M_u equal to the design strength ϕM_n from Eq. (4.35):

$$M_u = \phi R b d^2$$

2. With the aid of Table A.4, select an appropriate reinforcement ratio between $\rho_{\max}$ and $\rho_{\min}$. Often a ratio of about $0.60\rho_{\max}$ will be an economical and practical choice. Selection of $\rho \le \rho_{\max}$ ($\varepsilon_t \ge \varepsilon_{t,\min}$) ensures that ϕ remains equal to 0.90.
3. From Table A.5, for the specified material strengths and selected reinforcement ratio, find the flexural resistance factor R. Then

$$bd^2 = \frac{M_u}{\phi R}$$

4. Choose b and d to meet that requirement. Unless construction depth must be limited or other constraints exist (see Section 11.6), an effective depth about 2 to 3 times the width is often appropriate.
5. Calculate the required steel area

$$A_s = \rho b d$$

Then, referring to Table A.2, choose the size and number of bars, giving preference to the larger bar sizes to minimize placement costs.
6. Refer to Table A.7 to ensure that the selected beam width provides room for the bars chosen, with adequate concrete cover and spacing. (These points are discussed further in Section 4.5.)

The alternative approach starts with selecting concrete dimensions (see Section 11.6 for practical guidelines), after which the required reinforcement is found, as follows:

1. Select beam width b and effective depth d. Then calculate the required R:

$$R = \frac{M_u}{\phi b d^2}$$

2. Using Table A.5 for specified material strengths, find the reinforcement ratio $\rho < \rho_{\max}$ that provides the required value of R.
3. Calculate the required steel area

$$A_s = \rho b d$$

and from Table A.2 select the size and number of bars.
4. Using Table A.7, confirm that the beam width is sufficient to contain the selected reinforcement.

Use of design aids to solve the example problems of Section 4.3 is illustrated as follows.

EXAMPLE 4.9   **Flexural strength of a given member.**   Find the nominal flexural strength and design strength of the beam in Example 4.5, which has $b = 12$ in. and $d = 17.5$ in. and is reinforced with four No. 8 (No. 25) bars. Make use of the design aids of Appendix A. Material strengths are $f_c' = 4000$ psi and $f_y = 60,000$ psi.

SOLUTION.   From Table A.2, four No. 8 (No. 25) bars provide $A_s = 3.16$ in^2, and with $b = 12$ in. and $d = 17.5$ in., the reinforcement ratio is $\rho = 3.16/(12 \times 17.5) = 0.0150$. According to Table A.4, this is less than $\rho_{max} = 0.0181$ and above $\rho_{min} = 0.0033$. Then from Table A.5b, with $f_c' = 4000$ psi, $f_y = 60,000$ psi, and $\rho = 0.015$, the value $R = 781$ psi is found. The nominal and design strengths with $\phi = 0.90$ from Example 4.5 are, respectively,

$$M_n = Rbd^2 = 781 \times 12 \times \frac{17.5^2}{1000} = 2870 \text{ in-kips}$$

$$\phi M_n = 0.90 \times 2870 = 2580 \text{ in-kips}$$

If R had been interpolated based on $\rho = 0.0151$, as used in Example 4.5, the solution would have been as before.

EXAMPLE 4.10   **Concrete dimensions and steel area to resist a given moment.**   Find the cross section of concrete and the area of steel required for the beam in Example 4.6, making use of the design aids of Appendix A. $M_u = 1670$ in-kips, $f_c' = 4000$ psi, and $f_y = 60,000$ psi. Use a reinforcement ratio of $0.60\rho_{max}$.

SOLUTION.   From Table A.4, the maximum reinforcement ratio is $\rho_{max} = 0.0181$. For economy, a value of $\rho = 0.60\rho_{max} = 0.0109$ will be used. For that value, by interpolation from Table A.5a, the required value of R is 596. Then

$$bd^2 = \frac{M_u}{\phi R} = \frac{1670 \times 1000}{0.90 \times 596} = 3113 \text{ in}^3$$

Concrete dimensions $b = 10$ in. and $d = 17.6$ in. satisfy this, but the depth will be rounded to 17.5 in. to provide a total beam depth of 20.0 in. It follows that

$$R = \frac{M_u}{\phi bd^2} = \frac{1670 \times 1000}{0.90 \times 10 \times 17.5^2} = 606 \text{ psi}$$

and from Table A.5a, by interpolation, $\rho = 0.0112$. This leads to a steel requirement of $A_s = 0.0112 \times 10 \times 17.5 = 1.96$ in^2 as before.

EXAMPLE 4.11   **Determination of steel area.**   Find the steel area required for the beam in Example 4.7, with concrete dimensions $b = 10$ in. and $d = 17.5$ in. known to be adequate to carry the factored load moment of 1300 in-lb. Material strengths are $f_c' = 4000$ psi and $f_y = 60,000$ psi.

SOLUTION. Note that in cases in which the concrete dimensions are known to be adequate and only the reinforcement must be found, the iterative method used earlier is not required. The necessary flexural resistance factor is

$$R = \frac{M_u}{\phi b d^2} = \frac{1300 \times 1000}{0.90 \times 10 \times 17.5^2} = 472 \text{ psi}$$

According to Table A.5a, with the specified material strengths, this corresponds to a reinforcement ratio of $\rho = 0.0085$, giving a steel area of

$$A_s = 0.0085 \times 10 \times 17.5 = 1.49 \text{ in}^2$$

as before. Two No. 8 (No. 25) bars will be used, providing $A_s = 1.58 \text{ in}^2$.

The tables and graphs of Appendix A give basic information and are used extensively throughout this text for illustrative purposes. The reader should be aware, however, of the greatly expanded versions of these tables, plus many other useful aids, that are found in Refs. 4.7 through 4.9 and in commercial design software.

4.5 PRACTICAL CONSIDERATIONS IN THE DESIGN OF BEAMS

To focus attention initially on the basic aspects of flexural design, the preceding examples were carried out with only minimum regard for certain practical considerations that always influence the actual design of beams. These relate to optimal concrete proportions for beams, rounding of dimensions, standardization of dimensions, required cover for main and auxiliary reinforcement, and selection of bar combinations. Good judgment on the part of the design engineer is particularly important in translating from theoretical requirements to practical design. Several of the more important aspects are discussed here; much additional guidance is provided by the publications of ACI (Refs. 4.7 and 4.8) and CRSI (Refs. 4.9 to 4.11).

a. Concrete Protection for Reinforcement

To provide the steel with adequate concrete protection against fire and corrosion, the designer must maintain a certain minimum thickness of concrete cover outside of the outermost steel. The thickness required will vary, depending upon the type of member and conditions of exposure. According to ACI Code 20.5.1, for cast-in-place concrete, concrete protection at surfaces not exposed directly to the ground or weather should be not less than $\frac{3}{4}$ in. for slabs and walls and $1\frac{1}{2}$ in. for beams and columns. If the concrete surface is to be exposed to the weather or in contact with the ground, a protective covering of at least 2 in. is required [$1\frac{1}{2}$ in. for No. 5 (No. 16) and smaller bars], except that if the concrete is cast in direct contact with the ground without the use of forms, a cover of at least 3 in. must be furnished.

In general, the centers of main flexural bars in beams should be placed $2\frac{1}{2}$ to 3 in. from the top or bottom surface of the beam to furnish at least $1\frac{1}{2}$ in. of clear cover for the bars and the stirrups (see Fig. 4.13). In slabs, 1 in. to the center of the bar is ordinarily sufficient to give the required $\frac{3}{4}$ in. cover.

To simplify construction and thereby to reduce costs, the overall concrete dimensions of beams, b and h, are almost always rounded up to the nearest inch, and often to the next multiple of 2 in. As a result, the actual effective depth d, found

FIGURE 4.13

Requirements for concrete cover in beams and slabs not exposed to weather or in contact with ground.

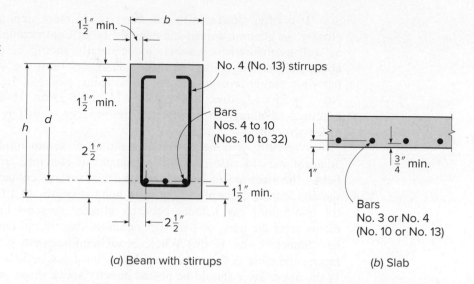

(*a*) Beam with stirrups   (*b*) Slab

by subtracting the sum of cover distance, stirrup diameter, and one-half the main reinforcing bar diameter from the total depth h, is seldom an even dimension. For slabs, the total depth is generally rounded up to the nearest $\frac{1}{2}$ in. The differences between h and d shown in Fig. 4.13 are not exact, but are satisfactory for design purposes for beams with No. 4 (No. 13) stirrups and No. 10 (No. 32) longitudinal bars or smaller, and for slabs using No. 4 (No. 13) or smaller bars. If larger bars are used for the main flexural reinforcement or for the stirrups, as is frequently the case, the corresponding dimensions are easily calculated.

Recognizing the closer tolerances that can be maintained under plant-control conditions, ACI Code 20.5.1 permits some reduction in concrete protection for reinforcement in precast concrete.

b. Concrete Section Proportions

Reinforced concrete beams may be wide and shallow, or relatively narrow and deep. Consideration of maximum material economy often leads to proportions with effective depth d in the range from about 2 to 3 times the width b (or web width b_w for T beams). However, constraints may dictate other choices, and as will be discussed in Section 11.6, maximum material economy may not translate to maximum structural economy. For example, with one-way concrete joists supported by monolithic beams (see Fig. 1.2 and Chapter 19), use of beams and joists with the same total depth will permit the use of a single flat-bottom form, resulting in fast, economical construction and permitting level ceilings. The beams will generally be wide and shallow, with heavier reinforcement than otherwise, but the result will be an overall saving in construction cost. In other cases, it may be necessary to limit the total depth of floor or roof construction for architectural or other reasons. An advantage of reinforced concrete is its adaptability to such special needs.

c. Selection of Bars and Bar Spacing

As noted in Section 2.14, common reinforcing bar sizes range from No. 3 to No. 11 (No. 10 to No. 36), the bar number corresponding closely to the number of eighth-inches (millimeters) of bar diameter. The two larger sizes, No. 14 (No. 43) [$1\frac{3}{4}$ in. (43 mm) diameter] and No. 18 (No. 57) [$2\frac{1}{4}$ in. (57 mm) diameter] are used mainly in columns.

It is often desirable to mix bar sizes to meet steel area requirements more closely. In general, mixed bars should be of comparable diameter, for practical as well as theoretical reasons, and generally should be arranged symmetrically about the vertical centerline. Many designers limit the variation in diameter of bars in a single layer to two bar sizes, using, say, No. 10 and No. 8 (No. 32 and No. 25) bars together, but not Nos. 11 and 6 (Nos. 36 and 19). There is some practical advantage to minimizing the number of different bar sizes used for a given structure.

Normally, it is necessary to maintain a certain minimum distance between adjacent bars to ensure proper placement of concrete around them. Air pockets below the steel are to be avoided, and full surface contact between the bars and the concrete is desirable to optimize bond strength. ACI Code 25.2 specifies that the minimum clear distance between adjacent bars not be less than the nominal diameter of the bars, or 1 in. (For columns, these requirements are increased to $1\frac{1}{2}$ bar diameters and $1\frac{1}{2}$ in.) Where beam reinforcement is placed in two or more layers, the clear distance between layers must not be less than 1 in., and the bars in the upper layer should be placed directly above those in the bottom layer. In no case should the clear spacing of reinforcement be less than $\frac{4}{3}$ of the maximum aggregate size, a requirement that good practice suggests should be applied to the clear cover of reinforcement as well.

The maximum number of bars that can be placed in a beam of given width is limited by bar diameter and spacing requirements and is also influenced by stirrup diameter, concrete cover requirement, and the maximum size of concrete aggregate specified. Table A.7 of Appendix A gives the maximum number of bars that can be placed in a single layer in beams, assuming $1\frac{1}{2}$ in. concrete cover and the use of No. 4 (No. 13) stirrups. When using the minimum bar spacing in conjunction with a large number of bars in a single plane of reinforcement, the designer should be aware that problems may arise in the placement and consolidation of concrete, especially when multiple layers of bars are used or when the bar spacing is smaller than the size of the vibrator head.

There are also restrictions on the *minimum* number of bars that can be placed in a single layer, based on requirements for the distribution of reinforcement to control the width of flexural cracks (see Section 7.3). Table A.8 gives the minimum number of bars that satisfy ACI Code requirements, which will be discussed in Chapter 7.

In large girders and columns, it is sometimes advantageous to "bundle" tensile or compressive reinforcement with two, three, or four bars in contact to provide for better deposition of concrete around and between adjacent bundles. These bars may be assumed to act as a unit, with not more than four bars in any bundle, provided that stirrups or ties enclose the bundle. No more than two bars should be bundled in one plane; typical bundle shapes are triangular, square, or L-shaped patterns. Individual bars in a bundle, cut off within the span of flexural members, should terminate at different points. ACI Code 25.6.1 requires at least 40 bar diameters stagger between points of cutoff. Where spacing limitations and minimum concrete cover requirements are based on bar diameter, a unit of bundled bars is treated as a single bar with a diameter that provides the same total area.

ACI Code 25.6.1 states that bars larger than No. 11 (No. 36) shall not be bundled in beams, although the AASHTO Specifications permit bundling of No. 14 and No. 18 (No. 43 and No. 57) bars in highway bridges.

4.6 RECTANGULAR BEAMS WITH TENSION AND COMPRESSION REINFORCEMENT

If a beam cross section is limited because of architectural or other considerations, it may happen that the concrete cannot develop the compression force required to resist the given bending moment. In this case, reinforcement is added in the compression zone, resulting in a *doubly reinforced* beam, that is, one with compression as well as tension reinforcement (see Fig. 4.14). The use of compression reinforcement has decreased markedly with the use of strength design methods, which account for the full-strength potential of the concrete on the compressive side of the neutral axis. However, there are situations in which compressive reinforcement is used for reasons other than strength. It has been found that the inclusion of some compression steel will reduce the long-term deflections of members (see Section 7.5). In addition, in some cases, bars will be placed in the compression zone for minimum-moment loading (see Section 11.2) or as stirrup support bars continuous throughout the beam span (see Chapter 5). It may be desirable to account for the presence of such reinforcement in flexural design, although in many cases it is neglected in flexural calculations.

a. Tension and Compression Steel Both at Yield Stress

If, in a doubly reinforced beam, the tensile reinforcement ratio ρ is less than or equal to ρ_b, the strength of the beam may be approximated within acceptable limits by disregarding the compression bars. The strength of such a beam will be controlled by tensile yielding, and the lever arm of the resisting moment will ordinarily be little affected by the presence of the compression bars.

If the tensile reinforcement ratio is larger than ρ_b, a somewhat more elaborate analysis is required. In Fig. 4.14a, a rectangular beam cross section is shown with compression steel A'_s placed a distance d' from the compression face and with tensile steel A_s at effective depth d. It is assumed initially that both A'_s and A_s are stressed to f_y at failure. The total resisting moment can be thought of as the sum of two parts. The first part, M_{n1}, is provided by the couple consisting of the force in the compression steel A'_s and the force in an equal area of tension steel

$$M_{n1} = A'_s f_y (d - d') \tag{4.42a}$$

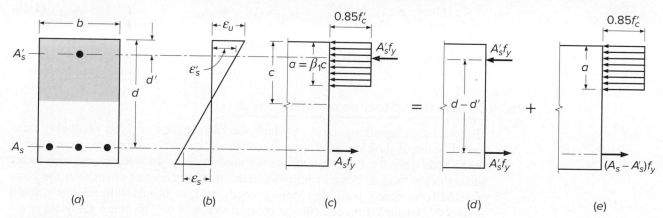

FIGURE 4.14
Doubly reinforced rectangular beam.

as shown in Fig. 4.14d. The second part, M_{n2}, is the contribution of the remaining tension steel $A_s - A_s'$ acting with the compression concrete

$$M_{n2} = (A_s - A_s')f_y\left(d - \frac{a}{2}\right) \tag{4.42b}$$

as shown in Fig. 4.14e, where the depth of the stress block is

$$a = \frac{(A_s - A_s')f_y}{0.85f_c'\, b} \tag{4.43a}$$

With the definitions $\rho = A_s/bd$ and $\rho' = A_s'/bd$, this can be written

$$a = \frac{(\rho - \rho')f_y d}{0.85f_c'} \tag{4.43b}$$

The total nominal moment is then

$$M_n = M_{n1} + M_{n2} = A_s'\, f_y\, (d - d') + (A_s - A_s')f_y\left(d - \frac{a}{2}\right) \tag{4.44}$$

In accordance with the safety provisions of the ACI Code, the net tensile strain is checked; and if $\varepsilon_t \geq \varepsilon_{t,\min}$, this nominal capacity is reduced by the factor $\phi = 0.90$ to obtain the design strength.

It is highly desirable, for reasons given earlier, that failure, should it occur, be precipitated by tensile yielding rather than crushing of the concrete. This can be ensured by setting an *upper limit* on the tensile reinforcement ratio. By setting the tensile steel strain in Fig. 4.14b equal to ε_y to establish the location of the neutral axis for the failure condition and then summing horizontal forces shown in Fig. 4.14c (still assuming the compressive steel to be at the yield stress at failure), it is easily shown that the balanced reinforcement ratio $\bar{\rho}_b$ for a doubly reinforced beam is

$$\bar{\rho}_b = \rho_b + \rho' \tag{4.45}$$

where ρ_b is the balanced reinforcement ratio for the corresponding singly reinforced beam and is calculated from Eq. (4.24). The ACI Code establishes the strength reduction factor ϕ based on the net tensile strain, not the reinforcement ratio. The maximum reinforcement ratio for $\phi = 0.90$ is

$$\bar{\rho}_{\max} = \rho_{\max} + \rho' \tag{4.46}$$

Since $\bar{\rho}_{\max}$ corresponds to $\varepsilon_t = \varepsilon_{t,\min}$, no check of ε_t is required to determine the strength reduction factor ϕ if $\bar{\rho} \leq \rho_{\max}$.

b.   Compression Steel below Yield Stress

The preceding equations, through which the fundamental analysis of doubly reinforced beams is developed clearly and concisely, are valid *only* if the compression steel yields when the beam reaches its nominal capacity. In many cases, such as for wide, shallow beams, beams with more than the usual concrete cover over the compression bars, beams with high yield strength steel, or beams with relatively small amounts of tensile reinforcement, the compression bars will be below the yield stress at failure. It is necessary, therefore, to develop more generally applicable equations to account for the possibility that the compression reinforcement has not yielded when the doubly reinforced beam fails in flexure.

Whether or not the compression steel yields at failure can be determined as follows. Referring to Fig. 4.14b, and taking as the limiting case $\varepsilon'_s = \varepsilon_y$, one obtains, from geometry,

$$\frac{c}{d'} = \frac{\varepsilon_u}{\varepsilon_u - \varepsilon_y} \quad \text{or} \quad c = \frac{\varepsilon_u}{\varepsilon_u - \varepsilon_y} d'$$

Summing forces in the horizontal direction (Fig. 4.14c) gives the *minimum* tensile reinforcement ratio $\bar{\rho}_{cy}$ that will ensure yielding of the compression steel at failure:

$$\bar{\rho}_{cy} = 0.85\beta_1 \frac{f'_c}{f_y} \frac{d'}{d} \frac{\varepsilon_u}{\varepsilon_u - \varepsilon_y} + \rho' \tag{4.47}$$

If the *tensile reinforcement ratio* is *less than* this limiting value, the neutral axis is sufficiently high that the compression steel stress at failure will be less than the yield stress. In this case, it can easily be shown on the basis of Fig. 4.14b and c that the balanced reinforcement ratio is

$$\bar{\rho}_b = \rho_b + \rho' \frac{f'_s}{f_y} \tag{4.48}$$

where

$$f'_s = E_s \varepsilon'_s = E_s \left[\varepsilon_u - \frac{d'}{d}(\varepsilon_u + \varepsilon_y) \right] \leq f_y \tag{4.49a}$$

To determine $\rho_{\max}$, $\varepsilon_t = \varepsilon_{t,\min}$ is substituted for ε_y in Eq. (4.49a), giving

$$f'_s = E_s \left[\varepsilon_u - \frac{d'}{d}(\varepsilon_u + \varepsilon_{t,\min}) \right] \leq f_y \tag{4.49b}$$

Hence, the maximum reinforcement ratio permitted for $\phi = 0.90$ is

$$\bar{\rho}_{\max} = \rho_{\max} + \rho' \frac{f'_s}{f_y} \tag{4.50}$$

where f'_s is given by Eq. (4.49b). A simple comparison shows that Eqs. (4.48) and (4.50), with f'_s given by Eqs. (4.49a) and (4.49b), respectively, are the generalized forms of Eqs. (4.45) and (4.46).

It should be emphasized that Eqs. (4.49a) and (4.49b) for compression steel stress apply *only for beams with exact strain values in the extreme tensile reinforcement of ε_y and $\varepsilon_t = \varepsilon_{t,\min}$*, respectively.

If the tensile reinforcement ratio is less than $\bar{\rho}_b$, as given by Eq. (4.48), and less than $\bar{\rho}_{cy}$, as given by Eq. (4.47), then the tensile steel is at the yield stress at failure but the compression steel is not, and new equations must be developed for compression steel stress and flexural strength. The compression steel stress can be expressed in terms of the still-unknown neutral axis depth as

$$f'_s = \varepsilon_u E_s \frac{c - d'}{c} \tag{4.51}$$

Consideration of horizontal force equilibrium (Fig. 4.14c with compression steel stress equal to f'_s) then gives

$$A_s f_y = 0.85\beta_1 f'_c bc + A'_s \varepsilon_u E_s \frac{c - d'}{c} \tag{4.52}$$

This is a quadratic equation in c, the only unknown, and is easily solved for c. The nominal flexural strength is found using the value of f'_s from Eq. (4.41), and $a = \beta_1 c$ in the expression

$$M_n = 0.85f'_c\, ab\left(d - \frac{a}{2}\right) + A'_s f'_s (d - d') \tag{4.53}$$

This nominal capacity is reduced by the strength reduction factor ϕ to obtain the design strength. As the yield strength of the reinforcement is increased, yielding of the compression reinforcement becomes more unlikely. For example, the yield strain of Grade 100 reinforcement is 0.0034, yet the maximum strain in the concrete is 0.003; thus Grade 100 reinforcement cannot yield in compression.

If compression bars are used in a flexural member, precautions must be taken to ensure that these bars will not buckle outward under load, spalling off the outer concrete. ACI Code 9.7.6.4 imposes the requirement that such bars be enclosed by closed stirrups or hoops much in the same way that compression bars in columns are enclosed by transverse ties (Section 9.2). Such transverse reinforcement must be used throughout the distance where the compression reinforcement is required.

c. Examples of Analysis and Design of Beams with Tension and Compression Steel

As was the case for beams with only tension reinforcement, doubly reinforced beam problems can be placed in one of two categories: analysis problems or design problems. For *analysis*, in which the concrete dimensions, reinforcement, and material strengths are given, one can find the flexural strength directly from the equations in Section 4.6a or 4.6b. First, it must be confirmed that the tensile reinforcement ratio is less than $\bar{\rho}_b$ given by Eq. (4.48), with compression steel stress from Eq. (4.49a). Once it is established that the tensile steel has yielded, the tensile reinforcement ratio defining compression steel yielding is calculated from Eq. (4.47), and compared to the actual tensile reinforcement ratio. If it is greater than $\bar{\rho}_{cy}$, then $f'_s = f_y$, and M_n is found from Eq. (4.44). If it is less than $\bar{\rho}_{cy}$, then $f'_s < f_y$. In this case, c is calculated by solving Eq. (4.52), f'_s comes from Eq. (4.53), and M_n is found from Eq. (4.53).

For *design*, in which case the factored moment M_u to be resisted is known and the section dimensions and reinforcement are to be found, a direct solution is impossible. The steel areas to be provided depend on the steel stresses, which are not known before the section is proportioned. It can be assumed that the compression steel stress is equal to the yield stress, but this must be confirmed; if it has not yielded, the design must be adjusted. The design procedure can be outlined as follows:

1. Calculate the maximum moment that can be resisted by the underreinforced section with $\rho = \rho_{max}$ to ensure that $\phi = 0.90$. The corresponding tensile steel area is $A_s = \rho bd$, and, as usual,

$$M_n = A_s f_y\left(d - \frac{a}{2}\right)$$

with

$$a = \frac{A_s f_y}{0.85f'_c\, b}$$

2. Find the excess moment, if any, that must be resisted, and set $M_2 = M_n$, as calculated in step 1.

$$M_1 = \frac{M_u}{\phi} - M_2$$

Now A_s from step 1 is defined as A_{s2}, that is, that part of the tension steel area in the doubly reinforced beam that works with the compression force in the concrete. In Fig. 4.14e, $A_s - A_s' = A_{s2}$.

3. Tentatively assume that $f_s' = f_y$. Then

$$A_s' = \frac{M_1}{f_y(d - d')}$$

4. Add an additional amount of tensile steel $A_{s1} = A_s'$. Thus, the total tensile steel area A_s is A_{s2} from step 2 plus A_{s1}.

5. Analyze the doubly reinforced beam to see if $f_s' = f_y$; that is, check the tensile reinforcement ratio against $\bar\rho_{cy}$.

6. If $\rho < \bar\rho_{cy}$, then the compression steel stress is less than f_y and the compression steel area must be increased to provide the needed force. This can be done as follows. The stress block depth is found from the requirement of horizontal equilibrium (Fig. 4.14e),

$$a = \frac{(A_s - A_s')f_y}{0.85f_c'b} \quad \text{or} \quad a = \frac{[A_s - A_s'(f_s'/f_y)]f_y}{0.85f_c'b}$$

and the neutral axis depth is $c = a/\beta_1$. From Eq. (4.51),

$$f_s' = \varepsilon_u E_s \frac{c - d'}{c}$$

The revised compression steel area, acting at f_s', must provide the same force as the trial steel area that was assumed to act at f_y. Therefore,

$$A_{s,\text{revised}}' = A_{s,\text{trial}}' \frac{f_y}{f_s'}$$

The tensile steel area need not be revised, because it acts at f_y as assumed. Using a spreadsheet or MathCAD, steps 3 through 6 are easily solved by varying f_s'.

EXAMPLE 4.12 **Flexural strength of a given member.** A rectangular beam, shown in Fig. 4.15, has a width of 12 in. and an effective depth to the centroid of the tension reinforcement of 24 in. The tension reinforcement consists of six No. 10 (No. 32) bars in two rows. For simplicity in calculating ε_t, d_t will be taken as d. Compression reinforcement consisting of two No. 8 (No. 25) bars is placed 2.5 in. from the compression face of the beam. If $f_y = 60,000$ psi and $f_c' = 5000$ psi, what is the design moment capacity of the beam?

SOLUTION. The steel areas and ratios are

$$A_s = 7.62 \text{ in}^2 \qquad \rho = \frac{7.62}{12 \times 24} = 0.0265$$

$$A_s' = 1.58 \text{ in}^2 \qquad \rho' = \frac{1.58}{12 \times 24} = 0.0055$$

Check the beam first as a singly reinforced beam to see if the compression bars can be disregarded,

$$\rho_{\max} = 0.0243 \quad \text{from Table A.4 or Eq. (4.26d)}$$

FIGURE 4.15
Doubly reinforced beam of
Example 4.12.

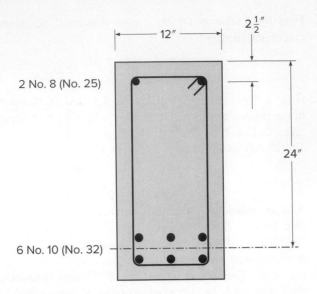

2 No. 8 (No. 25)

6 No. 10 (No. 32)

The actual $\rho = 0.0265$ is larger than ρ_{max}, so the beam must be analyzed as doubly reinforced. From Eq. (4.47), with $\beta_1 = 0.80$,

$$\overline{\rho}_{cy} = 0.85 \times 0.80 \times \frac{5}{60} \times \frac{2.5}{24} \times \frac{0.003}{0.003 - 0.00207} + 0.0055 = 0.0245$$

The tensile reinforcement ratio is greater than this, so the compression bars yield when the beam fails. The maximum reinforcement ratio thus can be found from Eq. (4.46),

$$\overline{\rho}_{max} = 0.0213 + 0.0055 = 0.0268$$

The actual tensile reinforcement ratio is below the maximum value, as required. Then, from Eq. (4.43a),

$$a = \frac{(A_s - A'_s)f_y}{0.85\,f'_c b} = \frac{(7.62 - 1.58)60}{0.85 \times 5 \times 12} = 7.11 \text{ in.}$$

$$c = a/\beta_1 = \frac{7.11}{0.80} = 8.89 \text{ in.}$$

$$\varepsilon_t = \varepsilon_u \left(\frac{d_t - c}{c} \right) = 0.003 \left(\frac{24 - 8.89}{8.89} \right) = 0.0051 > 0.005$$

and thus,

$$\phi = 0.90$$

and from Eq. (4.44),

$$M_n = 1.58 \times 60\,(24 - 2.5) + 6.04 \times 60 \left(24 - \frac{7.11}{2} \right) = 9450 \text{ in-kips}$$

The design strength is

$$\phi M_n = 0.90 \times 9450 = 8500 \text{ in-kips}$$

EXAMPLE 4.13   **Design of a doubly reinforced beam.**   A rectangular beam that must carry a service live load of 2.47 kips/ft and a calculated dead load of 1.05 kips/ft on an 18 ft simple span is limited in cross section for architectural reasons to 10 in. width and 20 in. total depth. If $f_y = 60,000$ psi and $f'_c = 4000$ psi, what steel area(s) must be provided?

Solution. The service loads are first increased by load factors to obtain the factored load of $1.2 \times 1.05 + 1.6 \times 2.47 = 5.21$ kips/ft. Then $M_u = 5.21 \times 18^2/8 = 211$ ft-kips $= 2530$ in-kips. To satisfy spacing and cover requirements (see Section 4.5), assume that the tension steel centroid will be 4 in. above the bottom face of the beam and that compression steel, if required, will be placed 2.5 in. below the beam's top surface. Then $d = 16$ in. and $d' = 2.5$ in.

First, check the capacity of the section if singly reinforced. Table A.4 shows that ρ_{max}, the maximum value of ρ for $\phi = 0.90$, to be 0.0181. So $A_s = 10 \times 16 \times 0.0181 = 2.90$ in^2. Then with

$$a = \frac{2.90 \times 60}{0.85 \times 4 \times 10} = 5.12 \text{ in.}$$

$c = a/\beta_1 = 5.12/0.85 = 6.02$ in., and the maximum nominal moment that can be developed is

$$M_n = 2.90 \times 60(16 - 5.12/2) = 2340 \text{ in-kips}$$

Alternatively, using $R = 913$ from Table A.5b, the nominal moment is $M_n = 913 \times 10 \times 16^2/1000 = 2340$ in-kips. Because the corresponding design moment $\phi M_n = 2100$ in-kips is less than the required capacity 2530 in-kips, compression steel is needed as well as additional tension steel.

The remaining moment to be carried by the compression steel couple is

$$M_1 = \frac{2530}{0.90} - 2340 = 470 \text{ in-kips}$$

Assume d is less than the value required to develop the compression reinforcement yield stress, in which case a reduced stress in the compression reinforcement must be used. Using the strain distribution in Fig. 4.11b, ε'_s and f'_c can be calculated as

$$\varepsilon'_s = 0.003 \frac{6.02 - 2.5}{6.02} = 0.00175 \quad \text{and} \quad f'_s = 0.00175 \times 29{,}000 = 50.9 \text{ ksi}$$

Try $f'_s = 50$ ksi for the compression reinforcement to obtain the required area of compression steel.

$$A'_s = \frac{470}{50(16 - 2.5)} = 0.70 \text{ in}^2$$

The total area of tensile reinforcement at 60 ksi is

$$A_s = 2.90 + 0.70\left(\frac{50}{60}\right) = 3.48 \text{ in}^2$$

Two No. 7 (No. 22) bars will be used for the compression reinforcement, and four No. 9 (No. 29) bars will be used for the tension reinforcement, as shown in Fig. 4.16. To place the tension bars in a 10 in. wide beam, two rows of reinforcement of two bars each will be used.

A final check is made to ensure that the design selections meet the problem requirements.

$$A_s - A'_s = 4.0 - 1.20\left(\frac{50}{60}\right) = 3.00 \text{ in}^2$$

$$a = \frac{(A_s - A'_s)f_y}{0.85 f'_c b} = \frac{3.00 \times 60}{0.85 \times 4 \times 10} = 5.29 \text{ in.}$$

$$c = \frac{a}{\beta_1} = \frac{5.29}{0.85} = 6.23 \text{ in.}$$

$$f'_s = E_s \varepsilon_c\left(\frac{c - d'}{c}\right) = 29{,}000 \times 0.003\left(\frac{6.23 - 2.5}{6.23}\right) = 52.1 \text{ ksi}$$

Which is close enough to the assumed value to continue.

$$\varepsilon_t = \varepsilon_c\left(\frac{d_t - c}{c}\right) = 0.003\left(\frac{17.5 - 6.23}{6.23}\right) = 0.0054$$

FIGURE 4.16
Doubly reinforced beam of
Example 4.13.

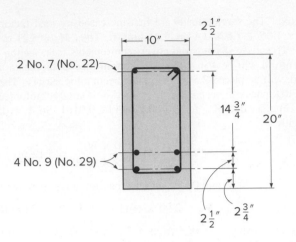

ε_t is greater than the 0.005 limit allowing $\phi = 0.90$. Then

$$\phi M_n = \phi\left[(A_s - A_s')f_y\left(d - \frac{a}{2}\right) + A_s'\,f_s'\,(d - d')\right]$$

$$= 0.90\left[3.00 \times 60\left(16 - \frac{5.29}{2}\right) + 1.20 \times 50(16 - 2.5)\right] = 2890 \text{ in-kips}$$

This is greater than M_u, so no further refinement is necessary.

d.   Tensile Steel below the Yield Stress

All doubly reinforced beams designed according to the ACI Code must be underreinforced, in the sense that the tensile reinforcement ratio is limited to ensure yielding at beam failure. Two cases were considered in Sections 4.6a and 4.6b, respectively: (a) both tension steel and compression steel yield and (b) tension steel yields but compression steel does not. Two other combinations may be encountered in analyzing the capacity of existing beams: (c) tension steel does not yield, but compression steel does, and (d) neither tension steel nor compression steel yields. The last two cases are unusual, and in fact, it would be difficult to place sufficient tension reinforcement to create such conditions, but it is possible. The solution in such cases is obtained as a simple extension of the treatment of Section 4.6b. An equation for horizontal equilibrium is written, in which both tension and compression steel stress are expressed in terms of the unknown neutral axis depth c. The resulting quadratic equation is solved for c, after which steel stresses can be calculated and the nominal flexural strength determined.

4.7   T BEAMS

With the exception of precast systems, reinforced concrete floors, roofs, decks, and beams are almost always monolithic. Forms are built for beam soffits and sides and for the underside of slabs, and the entire construction is cast at once, from the bottom of the deepest beam to the top of the slab. Beam stirrups and bent bars extend up into the slab. It is evident, therefore, that a part of the slab will act with the upper part of the beam to resist longitudinal compression. The resulting beam cross section is T-shaped rather than rectangular. The slab forms the beam flange, while the part of the beam projecting below the slab forms what is called the *web* or *stem*. The upper part of such a T beam is stressed laterally due to slab action in that direction. Although transverse compression at the level of the bottom of the slab may increase the longitudinal compressive strength by as much as 25 percent, transverse tension at the top surface reduces the longitudinal compressive strength (see Section 2.10). Neither effect is usually taken into account in design.

a.   Effective Flange Width

The next issue to be resolved is that of the effective width of flange. In Fig. 4.17a, it is evident that if the flange is but little wider than the stem width, the entire flange can be considered effective in resisting compression. For the floor system shown in Fig. 4.17b, however, it may be equally obvious that elements of the flange midway between the beam stems are less highly stressed in longitudinal compression than those elements directly over the stem. This is so because of shearing deformation of the flange, which relieves the more remote elements of some compressive stress.

The effect of this *shear-lag* on the stresses in the flange of a beam under positive bending is illustrated in Fig. 4.18, with higher stress near the web and lower stresses farther out in the flange. Rather than attempting to work with the variable stresses, it is convenient to make use of an *effective flange width b_f*, which may be smaller than the actual flange width but is considered to be uniformly stressed. This effective flange width has been found to depend on the span length and the relative thickness of the flange.

FIGURE 4.17
Effective flange width of T beams.

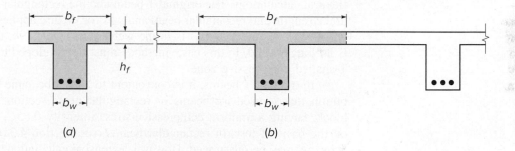

(a)   (b)

FIGURE 4.18
Stress distribution in T beam.

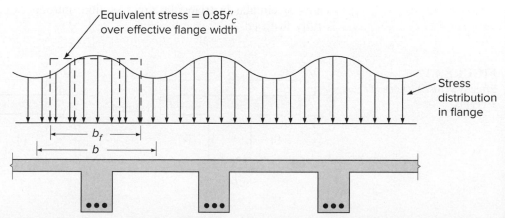

The criteria for effective width b_f given in ACI Code 6.3.2 can be summarized as follows:

1. For T beams with flanges on both sides of the web, the overhanging slab width on either side of the beam web shall not exceed one-eighth of the beam clear span ℓ_n, 8 times the thickness of the slab h, or go beyond one-half the clear distance to the next beam s_w.
2. For beams having a slab on one side only, the effective overhanging slab width shall not exceed one-twelfth the beam clear span ℓ_n, 6 times the thickness of the slab h, or go beyond one-half the clear distance to the next beam s_w.
3. For isolated beams in which the flange is used only for the purpose of providing additional compressive area, the flange thickness shall not be less than one-half the width of the web b_w, and the total flange width shall not be more than 4 times the web width b_w.

b.   Strength Analysis

The neutral axis of a T beam may be either in the flange or in the web, depending upon the proportions of the cross section, the amount of tensile steel, and the strengths of the materials. If the calculated depth to the neutral axis is less than or equal to the flange thickness h_f, the beam can be analyzed as if it were a rectangular beam of width equal to b_f, the effective flange width. The reason is illustrated in Fig. 4.19a, which shows a T beam with the neutral axis in the flange. The compressive area is indicated by the shaded portion of the figure. If the additional concrete indicated by areas 1 and 2 had been added when the beam was cast, the physical cross section would have been rectangular with a width b_f. No bending strength would have been added because areas 1 and 2 are entirely in the tension zone, and tension concrete is disregarded in flexural calculations. The original T beam and the rectangular beam are equal in flexural strength, and rectangular beam analysis for flexure applies.

When the neutral axis is in the web, as in Fig. 4.19b, the preceding argument is no longer valid. In this case, methods must be developed to account for the actual T-shaped compressive zone.

In treating T beams, it is convenient to adopt the same equivalent stress distribution that is used for beams of rectangular cross section. The rectangular stress block, having a uniform compressive-stress intensity $0.85f_c'$, was devised originally on the basis of tests of rectangular beams (see Section 4.3a), and its suitability for T beams may be questioned. However, extensive calculations based on actual stress-strain curves (reported in Ref. 4.12) indicate that its use for T beams, as well as for beams of circular or triangular cross section, introduces only minor deviations and is fully justified.

FIGURE 4.19
Effective cross sections of T beams.

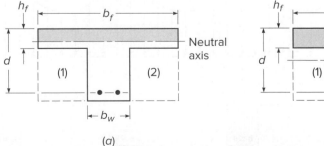
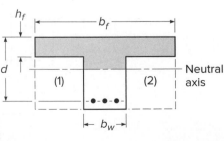

(a)   (b)

FIGURE 4.20
Strain and equivalent stress
distributions for T beams.

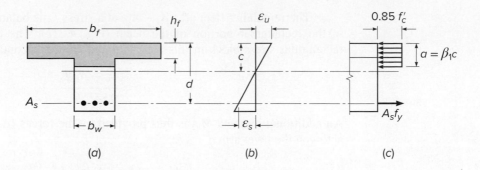

(a)   (b)   (c)

FIGURE 4.21
Computational model for
design and analysis of
T beams.

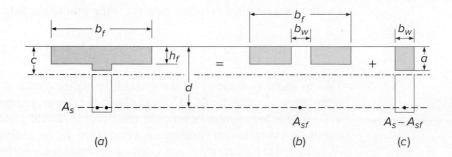

(a)   (b)   (c)

Accordingly, a T beam may be treated as a rectangular beam if the depth of the equivalent stress block is less than or equal to the flange thickness. Figure 4.20 shows a tensile-reinforced T beam with effective flange width b_f, web width b_w, effective depth to the steel centroid d, and flange thickness h_f. If for trial purposes the stress block is assumed to be completely within the flange,

$$a = \frac{A_s f_y}{0.85 f'_c \, b_f} = \frac{\rho f_y d}{0.85 f'_c} \qquad (4.54)$$

where $\rho = A_s / b_f d$. If a is less than or equal to the flange thickness h_f, the member may be treated as a rectangular beam of width b_f and depth d. If a is greater than h_f, a T beam analysis is required as follows.

It is assumed that the strength of the T beam is controlled by yielding of the tensile steel. This is nearly always the case because of the large compressive concrete area provided by the flange. In addition, an upper limit can be established for the reinforcement ratio to ensure that this is so, as will be shown.

As a computational device, it is convenient to divide the total tensile steel into two parts, as shown in Fig. 4.21. The first part, A_{sf}, represents the steel area that, when stressed to f_y, is required to balance the longitudinal compressive force in the overhanging portions of the flange that are stressed uniformly at $0.85 f'_c$ (Fig. 4.21b). Thus,

$$A_{sf} = \frac{0.85 f'_c (b_f - b_w) h_f}{f_y} \qquad (4.55)$$

The force $A_{sf} f_y$ and the equal and opposite force $0.85 f'_c (b_f - b_w) h_f$ act with a lever arm $d - h_f/2$ to provide the nominal resisting moment

$$M_{n1} = A_{sf} f_y \left(d - \frac{h_f}{2} \right) \qquad (4.56)$$

The remaining steel area $A_s - A_{sf}$, at a stress f_y, is balanced by the compression in the rectangular portion of the beam (Fig. 4.21c). The depth of the equivalent rectangular stress block in this zone is found from horizontal equilibrium.

$$a = \frac{(A_s - A_{sf})f_y}{0.85f_c' b_w} \qquad (4.57)$$

An additional moment M_{n2} is thus provided by the forces $(A_s - A_{sf})f_y$ and $0.85f_c' ab_w$ acting at the lever arm $d - a/2$.

$$M_{n2} = (A_s - A_{sf})f_y \left(d - \frac{a}{2}\right) \qquad (4.58)$$

and the total nominal resisting moment is the sum of the parts:

$$M_n = M_{n1} + M_{n2} = A_{sf} f_y \left(d - \frac{h_f}{2}\right) + (A_s - A_{sf}) f_y \left(d - \frac{a}{2}\right) \qquad (4.59)$$

This moment is reduced by the strength reduction factor ϕ in accordance with the safety provisions of the ACI Code to obtain the design strength.

As for rectangular beams, the tensile steel should yield prior to sudden crushing of the compression concrete, as assumed in the preceding development. Yielding of the tensile reinforcement and Code compliance are ensured if the net tensile strain ε_t is greater than $\varepsilon_{t,\min}$, in which case a strength reduction factor $\phi = 0.90$ may be used. From the geometry of the section,

$$\frac{c}{d_t} \leq \frac{\varepsilon_u}{\varepsilon_u + \varepsilon_t} \qquad (4.60)$$

Setting $\varepsilon_u = 0.003$ and $\varepsilon_t = 0.005$ for Grade 60 reinforcement provides a maximum c/d_t ratio of 0.375, as shown in Fig. 4.10. Thus, as long as the depth to the neutral axis is less than $0.375d_t$, the net tensile strain requirements are satisfied, as they are for rectangular beam sections. This will occur if $\rho_w = A_s/b_w d$ is less than

$$\rho_{w,\max} = \rho_{\max} + \rho_f \qquad (4.61)$$

where $\rho_f = A_{sf}/b_w d$ and $\rho_{\max}$ is as previously defined for a rectangular cross section [Eq. (4.26d)].

The practical result of applying Eq. (4.61) is that the stress block of T beams will almost always be within the flange, except for unusual geometry or combinations of material strength. Consequently, rectangular beam equations may be applied in most cases.

The ACI Code restriction that the tensile reinforcement ratio for beams not be less than $\rho_{\min} = 3\sqrt{f_c'}/f_y$ and $\geq 200/f_y$ (see Section 4.3d) applies to T beams as well as rectangular beams. For T beams, the ratio ρ should be calculated for this purpose based on the web width b_w.

c.   **Proportions of Cross Section**

When designing T beams, in contrast to analyzing the capacity of a given section, normally the slab dimensions and beam spacing will have been established by transverse flexural requirements. Consequently, the only additional section dimensions that must be determined from flexural considerations are the width and depth of the web and the area of the tensile steel.

If the stem dimensions were selected on the basis of concrete stress capacity in compression, they would be very small because of the large compression flange width furnished by the presence of the slab. Such a design would not represent the optimum solution because of the large tensile steel requirement resulting from the small effective depth, because of the excessive web reinforcement that would be required for shear, and because of large deflections associated with such a shallow member. It is better practice to select the proportions of the web (1) so as to keep an arbitrarily low web reinforcement ratio ρ_w or (2) so as to keep web-shear stress at desirably low limits (Chapter 5) or (3) for continuous T beams, on the basis of the flexural requirements at the supports, where the effective cross section is rectangular and of width b_w.

In addition to the main reinforcement calculated according to the preceding requirements, it is necessary to ensure the integrity of the compressive flange of T beams by providing steel in the flange in the direction transverse to the main span. In typical construction, the slab steel serves this purpose. In other cases, additional bars must be added to permit the effective overhanging flanges to carry, as cantilever beams, the loads directly applied. According to ACI Code 7.7.2, the spacing of such bars must not exceed 5 times the thickness of the flange or in any case exceed 18 in.

d. Examples of Analysis and Design of T Beams

For *analyzing* the capacity of a T beam with known concrete dimensions and tensile steel area, it is reasonable to start with the assumption that the stress block depth a does not exceed the flange thickness h_f. In that case, all ordinary rectangular beam equations (see Section 4.3) apply, with beam width taken equal to the effective width of the flange. If, upon checking that assumption, a proves to exceed h_f, then T beam analysis must be applied. Equations (4.55) through (4.59) can be used, in sequence, to obtain the nominal flexural strength, after which the design strength is easily calculated.

For *design*, the following sequence of calculations may be followed:

1. Establish flange thickness h_f based on flexural requirements of the slab, which normally spans transversely between parallel T beams.
2. Determine the effective flange width b_f according to ACI limits.
3. Choose web dimensions b_w and d based on either of the following:
   - (a) Negative bending requirements at the supports, if a continuous T beam
   - (b) Shear requirements, setting a reasonable upper limit on the nominal unit shear stress v_u in the beam web (see Chapter 5)
4. With all concrete dimensions thus established, calculate a trial value of A_s, assuming that a does not exceed h_f, with beam width equal to flange width b_f. Use ordinary rectangular beam design methods.
5. For the trial A_s, check the depth of stress block a to confirm that it does not exceed h_f. If it should exceed that value, revise A_s, using the T beam equations.
6. Check to ensure that $\varepsilon_t \geq \varepsilon_{t,\min}$ or c/d is less than the appropriate limit in Fig. 4.9 to ensure that $\phi = 0.90$. (This will almost invariably be the case.)
7. Check to ensure that $\rho_w \geq \rho_{w,\min}$.

EXAMPLE 4.14   **Moment capacity of a given section.**   The isolated T beam shown in Fig. 4.22 is composed of a flange 28 in. wide and 6 in. deep cast monolithically with a web of 10 in. width that extends 24 in. below the bottom surface of the flange to produce a beam of 30 in. total depth. Tensile reinforcement consists of six No. 10 (No. 32) bars placed in two horizontal rows separated by 1 in. clear spacing. The centroid of the bar group is 26 in. from the top of the beam. The concrete has a strength of 3000 psi, and the yield strength of the steel is 60,000 psi. What is the design moment capacity of the beam?

FIGURE 4.22
T beam of Example 4.14.

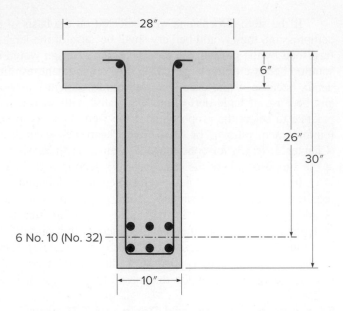

SOLUTION.   It is easily confirmed that the flange dimensions are satisfactory according to the ACI Code for an isolated beam. The entire flange can be considered effective. For six No. 10 (No. 32) bars, $A_s = 7.62$ in^2. First check the location of the neutral axis, on the assumption that rectangular beam equations may be applied. Using Eq. (4.28) with $b_f = b$

$$a = \frac{A_s f_y}{0.85\, f_c'\, b_f} = \frac{7.62 \times 60}{0.85 \times 3 \times 28} = 6.40 \text{ in.}$$

This exceeds the flange thickness, and so a T beam analysis is required. From Eq. (4.55) and Fig. 4.20b,

$$A_{sf} = 0.85 \frac{f_c'}{f_y}(b_f - b_w)h_f = 0.85 \times \frac{3}{60}(28 - 10) \times 6 = 4.59 \text{ in}^2$$

Then, from Eq. (4.56),

$$M_{n1} = A_{sf} f_y \left(d - \frac{h_f}{2}\right) = 4.59 \times 60(26 - 3) = 6330 \text{ in-kips}$$

Then, from Fig. 4.20c,

$$A_s - A_{sf} = 7.62 - 4.59 = 3.03 \text{ in}^2$$

and from Eqs. (4.54) and (4.55)

$$a = \frac{A_s f_y}{0.85 f_c' b_w} = \frac{3.03 \times 60}{0.85 \times 3 \times 10} = 7.13 \text{ in.}$$

$$M_{n2} = (A_s - A_{sf}) f_y \left(d - \frac{a}{2}\right) = 3.03 \times 60(26 - 3.56) = 4080 \text{ in-kips}$$

The depth to the neutral axis is $c = a/\beta_1 = 7.13/0.85 = 8.39$ and $d_t = 27.5$ in. to the lowest bar. The c/d_t ratio is $8.39/27.5 = 0.305 < 0.375$, so the $\varepsilon_t > 0.005$ for Grade 60 reinforcement requirement is met and $\phi = 0.90$. When the ACI strength reduction factor is incorporated, the design strength is

$$\phi M_n = \phi(M_{n1} + M_{n2}) = 0.90(6330 + 4080) = 9370 \text{ in-kips}$$

EXAMPLE 4.15   **Determination of steel area for a given moment.**   A floor system, shown in Fig. 4.23, consists of a 3 in. concrete slab supported by continuous T beams with a clear span $\ell_n = 24$ ft, 47 in. on centers. Web dimensions, as determined by negative-moment requirements at the supports, are $b_w = 11$ in. and $d = 20$ in. What tensile steel area is required at midspan to resist a factored moment of 6400 in-kips if $f_y = 60,000$ psi and $f_c' = 3000$ psi?

SOLUTION.   First determining the effective flange width b_f,

$$16h_f + b_w = 16 \times 3 + 11 = 59 \text{ in.}$$

$$2\frac{\ell_n}{8} + b_w = 2 \times \frac{24 \times 12}{8} + 11 = 83 \text{ in.}$$

$$\text{Centerline beam spacing} = 47 \text{ in.}$$

The centerline T beam spacing controls in this case, and $b_f = 47$ in. The concrete dimensions b_w and d are known to be adequate in this case, since they have been selected for the larger negative support moment applied to the effective rectangular section $b_w d$. The tensile steel at midspan is most conveniently found by trial. Assuming the stress-block depth a is equal to the flange thickness of $h_f = 3$ in., one gets

$$d - \frac{a}{2} = 20 - \frac{3}{2} = 18.50 \text{ in.}$$

Trial:

$$A_s = \frac{M_u}{\phi f_y (d - a/2)} = \frac{6400}{0.90 \times 60 \times 18.50} = 6.41 \text{ in}^2$$

Checking the assumed value for a,

$$a = \frac{A_s f_y}{0.85 f_c' b_f} = \frac{6.41 \times 60}{0.85 \times 3 \times 47} = 3.21 \text{ in.}$$

Since a is greater than h_f, a T beam design is required and $\phi = 0.90$ is assumed.

$$A_{sf} = \frac{0.85 f_c' (b_f - b_w) h_f}{f_y} = \frac{0.85 \times 3 \times (47 - 11) \times 3}{60} = 4.59 \text{ in}^2$$

$$\phi M_{n1} = \phi A_{sf} f_y \left(d - \frac{h_f}{2} \right) = 0.90 \times 4.59 \times 60 \times \left(20 - \frac{3}{2} \right) = 4590 \text{ in-kips}$$

$$\phi M_{n2} = M_u - \phi M_{n1} = 6400 - 4590 = 1810 \text{ in-kips}$$

Assume $a = 4.0$ in.:

$$A_s - A_{sf} = \frac{\phi M_{n2}}{\phi f_y (d - a/2)} = \frac{1810}{0.90 \times 60 \times (20 - 4/2)} = 1.86 \text{ in}^2$$

Check:

$$a = \frac{(A_s - A_{sf}) f_y}{0.85 f_c' b_w} = \frac{1.86 \times 60}{0.85 \times 3 \times 11} = 3.98 \text{ in.}$$

FIGURE 4.23
T beam of Example 4.15.

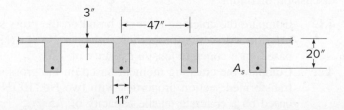

This is satisfactorily close to the assumed value of 4 in. Then

$$A_s = A_{sf} + A_s - A_{sf} = 4.59 + 1.86 = 6.45 \text{ in}^2$$

Checking to ensure that the net tensile strain requirement is met to allow $\phi = 0.90$,

$$c = \frac{a}{\beta_1} = \frac{3.98}{0.85} = 4.68$$

$$\frac{c}{d_t} = \frac{4.68}{20} = 0.23 < 0.325$$

indicating that the design is satisfactory.

The close agreement should be noted between the approximate tensile steel area of 6.41 in^2 found by assuming the stress-block depth equal to the flange thickness and the more exact value of 6.45 in^2 found by T beam analysis. The approximate solution would be satisfactory in most cases.

REFERENCES

4.1.   H. Rusch, "Researches toward a General Flexural Theory of Structural Concrete," *J. ACI*, vol. 32, no. 1, 1960, pp. 1–28.

4.2.   L. B. Kriz, "Ultimate Strength Criteria for Reinforced Concrete," *J. Eng. Mech. Div. ASCE*, vol. 85, no. EM3, 1959, pp. 95–110.

4.3.   L. B. Kriz and S. L. Lee, "Ultimate Strength of Overreinforced Beams," *Proc. ASCE*, vol. 86, no. EM3, 1960, pp. 95–106.

4.4.   C. S. Whitney, "Design of Reinforced Concrete Members under Flexure or Combined Flexure and Direct Compression," *J. ACI*, vol. 33, Mar–Apr 1937, pp. 483–498.

4.5.   A. H. Mattock, L. B. Kriz, and E. Hogenstad, "Rectangular Concrete Stress Distribution in Ultimate Strength Design," *J. ACI*, vol. 32, no. 8, 1961, pp. 875–928.

4.6.   P. H. Kaar, N. W. Hanson, and H. T. Capell, "Stress-Strain Curves and Stress Block Coefficients for High-Strength Concrete," *Proc. Douglas McHenry Symposium*, ACI Special Publication SP-55, 1978.

4.7.   *The Reinforced Concrete Design Manual in Accordance with ACI 318-11*, SP-17(11), American Concrete Institute, Farmington Hills, MI, 2012.

4.8.   *ACI Detailing Manual*, ACI Special Publication SP-66, American Concrete Institute, Farmington Hills, MI, 2004.

4.9.   *CRSI Design Guide Suite—Buildings*, 9th ed., Concrete Reinforcing Steel Institute, Schaumburg, IL.

4.10. *CRSI Design Guide for Economical Concrete Construction*, Concrete Reinforcing Steel Institute, Schaumburg, IL.

4.11. *Manual of Standard Practice*, 29th ed., Concrete Reinforcing Steel Institute, Schaumburg, IL, 2009.

4.12. C. W. Dolan, *Ultimate Capacity of Reinforced Concrete Sections Using a Continuous Stress-Strain Function*, M.S. Thesis, Cornell University, Ithaca, NY, June 1967.

PROBLEMS

Problems 4.1 through 4.8 address service-level behavior, Problems 4.9 through 4.16 are sectional analysis problems, Problems 4.17 through 4.22 are sectional design problems, and Problems 4.23 through 4.30 are comprehensive problems requiring discussion of findings.

4.1.   Compare the cracking moment based on the gross section properties and the transformed section properties with four No. 11 (No. 36) bars in Fig. P4.1*a* based on a concrete tensile capacity of $7.5\sqrt{f_c'}$.

4.2.   Compare the cracking moment based on the gross section properties and the transformed section properties with two No. 10 (No. 32) bars in Fig. P4.1*b* based on a concrete tensile capacity of $7.5\sqrt{f_c'}$.

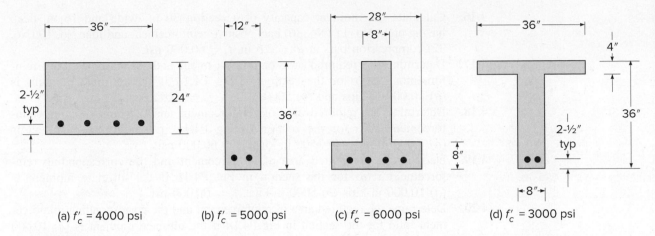

(a) f'_c = 4000 psi   (b) f'_c = 5000 psi   (c) f'_c = 6000 psi   (d) f'_c = 3000 psi

FIGURE P4.1

4.3. Compare the cracking moment based on the gross section properties and the transformed section properties with four No. 9 (No. 29) bars in Fig. P4.1c based on a concrete tensile capacity of $7.5\sqrt{f'_c}$.

4.4. Compare the cracking moment based on the gross section properties and the transformed section properties with two No. 11 (No. 36) bars in Fig. P4.1d based on a concrete tensile capacity of $7.5\sqrt{f'_c}$.

4.5. Determine the cracking moment based on the gross section properties in Fig. P4.1d if the section is prestressed such that there is a 300 psi compression stress in the extreme tension zone and the concrete tensile capacity is $7.5\sqrt{f'_c}$.

4.6. Determine the service level moment capacity of the section in Fig. P4.1a if the allowable stress for concrete is $0.45f'_c$ and the allowable stress for the reinforcement is 30,000 psi. Use the areas of reinforcement from Problem 4.1.

4.7. Determine the service level moment capacity of the section in Fig. P4.1b if the allowable stress for concrete is $0.45f'_c$ and the allowable stress for the reinforcement is 30,000 psi. Use the areas of reinforcement from Problem 4.2.

4.8. Determine the service level moment capacity of the section in Fig. P4.1c if the allowable stress for concrete is $0.45f'_c$ and the allowable stress for the reinforcement is 30,000 psi. Use the areas of reinforcement from Problem 4.3.

4.9. Determine the nominal moment capacity of the section in Fig. 4.1a using the reinforcement areas from Problem 4.1. f_y = 60,000 psi.

4.10. Determine the nominal moment capacity of the section in Fig. 4.1b using the reinforcement areas from Problem 4.2. f_y = 60,000 psi.

4.11. Determine the nominal moment capacity of the section in Fig. 4.1c using the reinforcement areas from Problem 4.3. f_y = 60,000 psi.

4.12. Determine the nominal moment capacity of the section in Fig. 4.1d using the reinforcement areas from Problem 4.4. f_y = 60,000 psi.

4.13. Determine the nominal and design moment capacity of the section in Fig. 4.1d using the reinforcement areas from Problem 4.4. f_y = 80,000 psi.

4.14. Determine the nominal and design moment capacity of the section in Fig. 4.1a using the reinforcement areas from Problem 4.1 and two No. 10 (No. 32) compression bars at d' = 2.5 in. f_y = 60,000 psi.

4.15. Calculate the moment capacity of the section in Fig. 4.1a using eight No. 11 (No. 36) bars positive reinforcement and four No. 10 (No. 32) compression bars at d' = 2.5 in. f_y = 60,000 psi.

4.16. Calculate the moment capacity of a section 30 in. wide and 16 in. deep having eight No. 11 (No. 36) bars positive reinforcement and four No. 10 (No. 32) compression bars at $d' = 2.5$ in. $f_y = 60,000$ psi.

4.17. Determine the required area of reinforcement and the corresponding reinforcement ratio for the section in Fig P4.1a if the ultimate moment is (a) 10,000 in-kips and (b) 5000 in-kips. $f_y = 60,000$ psi.

4.18. Determine the required area of reinforcement and the corresponding reinforcement ratio for the section in Fig P4.1b if the ultimate moment is (a) 7,000 in-kips (b) 3500 in-kips. $f_y = 60,000$ psi.

4.19. Determine the required area of reinforcement and the corresponding reinforcement ratio for the section in Fig P4.1c if the ultimate moment is (a) 10,000 in-kips (b) 5000 in-kips. $f_y = 60,000$ psi.

4.20. Determine the required area of reinforcement and the corresponding reinforcement ratio for the section in Fig P4.1d if the ultimate moment is (a) 10,000 in-kips and (b) 5000 in-kips. $f_y = 60,000$ psi. Comment on your solutions.

4.21. Determine the required area of reinforcement and the corresponding reinforcement ratio for the section in Fig P4.1a if the ultimate moment is (a) 10,000 in-kips and (b) 5000 in-kips. $f_y = 80,000$ psi.

4.22. Determine the required area of reinforcement and the corresponding reinforcement ratio for the section in Fig P4.1b if the ultimate moment is (a) 7,000 in-kips (b) 3500 in-kips. $f_y = 80,000$ psi.

4.23. A rectangular beam made using concrete with $f_c' = 6000$ psi and steel with $f_y = 60,000$ psi has a width $b = 20$ in., an effective depth of $d = 17.5$ in., and a total depth of $h = 20$ in. The concrete modulus of rupture $f_r = 530$ psi. The elastic moduli of the concrete and steel are, respectively, $E_c = 4,030,000$ psi and $E_s = 29,000,000$ psi. The tensile steel consists of four No. 11 (No. 36) bars.
  (a) Find the maximum service load moment that can be resisted without stressing the concrete above $0.45f_c'$ or the steel above $0.40f_y$.
  (b) Determine whether the beam cracks before reaching the service load.
  (c) Calculate the nominal flexural strength of the beam.
  (d) Calculate the ratio of the nominal flexural strength of the beam to the maximum service load moment, and compare your findings to the ACI load factors and strength reduction factor.

4.24. A rectangular reinforced concrete beam with dimensions $b = 14$ in., $d = 25$ in., and $h = 28$ in. is reinforced with three No. 10 (No. 32) bars. Material strengths are $f_y = 60,000$ psi and $f_c' = 5000$ psi.
  (a) Find the moment that produces the first cracking at the bottom surface of the beam, basing your calculation on I_g, the moment of inertia of the gross concrete section.
  (b) Repeat the calculation, using I_{ut}, the moment of inertia of the uncracked transformed section.
  (c) Determine the maximum moment that can be carried without stressing the concrete beyond $0.45f_c'$ or the steel beyond $0.60f_y$.
  (d) Find the nominal flexural strength of this beam.
  (e) Calculate the ratio of the flexural strength from part (d) to the service capacity from part (c).
  (f) Comment on your results, paying particular attention to comparing parts (a) and (b) and comparing the result in part (e) with the load factors in the ACI Code.

4.25. A rectangular, tension-reinforced beam is to be designed for dead load of 500 lb/ft plus self-weight and service live load of 1200 lb/ft, with a 22 ft simple span. Material strengths are $f_y = 60$ ksi and $f_c' = 3$ ksi for steel and concrete, respectively. The total beam depth must not exceed 16 in. Calculate the required beam width and tensile steel requirement, using a reinforcement ratio of approximately $0.60\rho_{max}$. Use ACI load factors and strength reduction factors. The effective depth may be assumed to be 2.5 in. less than the total depth.

4.26. A four-span continuous beam of constant rectangular cross section is supported at A, B, C, D, and E. The factored moments resulting from analysis are as follows:

At Supports, ft-kips	At Midspan, ft-kips
$M_a = 138$	$M_{ab} = 158$
$M_b = 220$	$M_{bc} = 138$
$M_c = 200$	$M_{cd} = 138$
$M_d = 220$	$M_{de} = 158$
$M_e = 138$	

Determine the required final concrete dimensions for this beam, using $d = 1.75b$, and determine the reinforcement requirements at each critical moment section. Your final reinforcement ratio should not exceed $= 0.6\rho_{max}$. Use $f_y = 60,000$ psi and $f_c' = 6000$ psi.

4.27. A two-span continuous concrete beam is to be supported by three concrete walls spaced 30 ft on centers. A service live load of 1.5 kips/ft is to be carried in addition to the self-weight of the beam. Use pattern loading; that is, consider two loading conditions: (1) live load on both spans and (2) live load on a single span. A constant rectangular cross section is to be used with $d = 2b$, but reinforcement is to be varied according to requirements. Find the required concrete dimensions and reinforcement at all critical sections. Allow for No. 3 (No. 10) stirrups. Use a span-to-depth ratio of 15 as the first estimate of the depth. Adjust the depth if the reinforcement ratio is too high. Include sketches, drawn to scale, of the critical cross sections. Use $f_y = 60,000$ psi and $f_c' = 6000$ psi.

4.28. A rectangular concrete beam of width $b = 24$ in. is limited by architectural considerations to a maximum total depth $h = 16$ in. It must carry a total factored load moment $M_u = 400$ ft-kips. Design the flexural reinforcement for this member, using compression steel if necessary. Allow 3 in. to the center of the bars from the compression or tension face of the beam. Material strengths are $f_y = 60,000$ psi and $f_c' = 4000$ psi. Select reinforcement to provide the needed areas, and show a sketch of your final design, including provision for No. 4 (No. 13) stirrups.

4.29. A precast T beam is to be used as a bridge over a small roadway. Concrete dimensions are $b = 48$ in., $b_w = 16$ in., $h_f = 5$ in., and $h = 25$ in. The effective depth $d = 20$ in. Concrete and steel strengths are 6000 psi and 60,000 psi, respectively. Using approximately one-half the maximum tensile reinforcement permitted by the ACI Code (select the actual size of bar and number to be used), determine the design moment capacity of the girder. If the beam is used on a 30 ft simple span, and if in addition to its own weight it must support railings, curbs, and suspended loads totaling 0.475 kip/ft, what uniform service live load limit should be posted?

4.30 Using Eq. (4.27) and assuming that $d = 0.9h$, show that A_s is approximately equal to $M_u/4h$ for Grade 60 reinforcement and where M_u is in kip-ft.

5 Shear and Diagonal Tension in Beams

5.1 INTRODUCTION

Chapter 4 dealt with the flexural behavior and flexural strength of beams. Beams must also have an adequate safety margin against other types of failure, some of which may be more dangerous than flexural failure. This may be so because of greater uncertainty in predicting certain other modes of collapse, or because of the catastrophic nature of some other types of failure, should they occur.

Shear failure of reinforced concrete, more properly called *diagonal tension failure*, is one example. Shear failure is difficult to predict accurately. In spite of many decades of experimental research (Refs. 5.1 to 5.6) and the use of highly sophisticated analytical tools (Refs. 5.7 and 5.8), it is not fully understood. Furthermore, if a beam without properly designed shear reinforcement is overloaded to failure, shear collapse is likely to occur suddenly, with no advance warning of distress. This is in strong contrast with the nature of flexural failure. For typically underreinforced beams, flexural failure is initiated by gradual yielding of the tension steel, accompanied by obvious cracking of the concrete and large deflections, giving ample warning and providing the opportunity to take corrective measures. Because of these differences in behavior, reinforced concrete beams are generally provided with special *shear reinforcement* to ensure that flexural failure would occur before shear failure if the member were severely overloaded.

Figure 5.1 shows a shear-critical beam tested under third point loading. With no shear reinforcement provided, the member failed immediately upon formation of the critical crack in the high-shear region near the right support.

It is important to realize that shear analysis and design are not really concerned with shear as such. The shear stresses in most beams are far below the direct shear strength of the concrete. The real concern is with *diagonal tension stress*, resulting from the combination of shear stress and longitudinal flexural stress. Most of this chapter deals with analysis and design for diagonal tension, and it provides background for understanding and using the shear provisions of the ACI Code. Members without web reinforcement are studied first to establish the location and orientation of cracks and the diagonal cracking load. Methods are then developed for the design of shear reinforcement according to the present ACI Code, both in ordinary beams and in special types of members, such as deep beams.

Over the years, alternative methods of shear design have been proposed, based on variable angle truss models and diagonal compression field theory (Refs. 5.9 and 5.10). These approaches will be reviewed briefly later in this chapter, with one such approach, the modified compression field theory, presented in detail.

FIGURE 5.1
Shear failure of reinforced concrete beam: (*a*) overall view and (*b*) detail near right support. (*Photograph by Arthur H. Nilson.*)

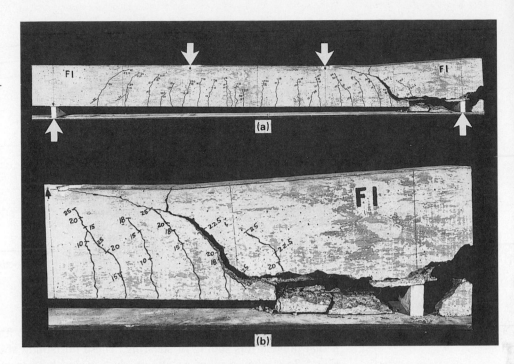

Finally, there are some circumstances in which consideration of direct shear is appropriate. One example is in the design of composite members combining precast beams with a cast-in-place top slab. Horizontal shear stresses on the interface between components are important. The shear-friction theory, useful in this and other cases, will be presented following development of methods for the analysis and design of beams for diagonal tension.

5.2 DIAGONAL TENSION IN HOMOGENEOUS ELASTIC BEAMS

The stresses acting in homogeneous beams were briefly reviewed in Section 3.6. It was pointed out that when the material is elastic (stresses proportional to strains), shear stresses

$$v = \frac{VQ}{Ib} \qquad (3.13)$$

act at any section in addition to the bending stresses

$$f = \frac{My}{I} \qquad (3.11)$$

except for those locations at which the shear force V happens to be zero.

The role of shear stresses is easily visualized by the performance under load of the laminated beam of Fig. 5.2; it consists of two rectangular pieces bonded together along the contact surface. If the adhesive is strong enough, the member will deform as one single beam, as shown in Fig. 5.2*a*. On the other hand, if the adhesive is weak, the two pieces will separate and slide relative to each other, as shown in Fig. 5.2*b*.

FIGURE 5.2
Shear in homogeneous
rectangular beams.

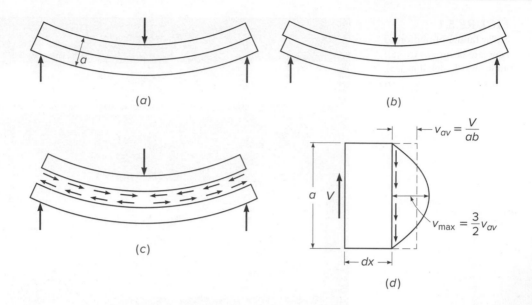

Evidently, then, when the adhesive is effective, there are forces or stresses acting in it that prevent this sliding or shearing. These horizontal shear stresses are shown in Fig. 5.2c as they act, separately, on the top and bottom pieces. The same stresses occur in horizontal planes in single-piece beams; they are different in intensity at different distances from the neutral axis.

Figure 5.2d shows a differential length of a single-piece rectangular beam acted upon by a shear force of magnitude V. Upward translation is prevented; that is, vertical equilibrium is provided by the vertical shear stresses v. Their average value is equal to the shear force divided by the cross-sectional area $v_{av} = V/ab$, but their intensity varies over the depth of the section. As is easily computed from Eq. (3.13), the shear stress is zero at the outer fibers and has a maximum of $1.5v_{av}$ at the neutral axis, the variation being parabolic as shown. Other values and distributions are found for other shapes of the cross section, the shear stress always being zero at the outer fibers and of maximum value at the neutral axis.

Figure 5.3 shows a simply supported beam under uniform load. If a small square element located at the neutral axis of such a beam is isolated, as shown in Fig. 5.3b, the vertical shear stresses on it, equal and opposite on the two faces for reasons of equilibrium, act as shown. However, if these were the only stresses present, the element would not be in equilibrium; it would spin. Therefore, on the two horizontal faces there exist equilibrating horizontal shear stresses of the same magnitude. That is, at any point within the beam, the horizontal shear stresses of Fig. 5.3b are equal in magnitude to the vertical shear stresses of Fig. 5.2d.

As demonstrated in any strength-of-materials text for an element cut at 45° these shear stresses combine in such a manner that their effect is as shown in Fig. 5.3c. That is, the action of the two pairs of shear stresses on the vertical and horizontal faces is the same as that of two pairs of normal stresses, one tensile and one compressive, acting on the 45° faces and of numerical value equal to that of the shear stresses. If an element of the beam is considered that is located neither at the neutral axis nor at the outer edges, its vertical faces are subject not only to the shear stresses but also to the familiar bending stresses whose magnitude is given by Eq. (3.11) (Fig. 5.3d). The six stresses that now act on the element can again be combined into a pair of inclined

FIGURE 5.3

Stress trajectories in homogeneous rectangular beam.

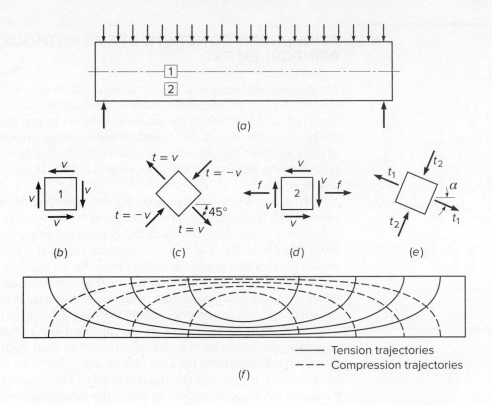

(a)

(b)   (c)   (d)   (e)

(f)

——— Tension trajectories

– – – Compression trajectories

compressive stresses and a pair of inclined tensile stresses that act at right angles to each other. They are known as *principal stresses* (Fig. 5.3e). Their value, as mentioned in Section 3.6, is given by

$$t = \frac{f}{2} \pm \sqrt{\frac{f^2}{4} + v^2} \tag{3.10}$$

and their inclination α by $\tan 2\alpha = 2v/f$.

Since the magnitudes of the shear stresses v and the bending stresses f change both along the beam and vertically with distance from the neutral axis, the inclinations as well as the magnitudes of the resulting principal stresses t also vary from one place to another. Figure 5.3f shows the inclinations of these principal stresses for a rectangular beam uniformly loaded. That is, these stress trajectories are lines which, at any point, are drawn in that direction in which the particular principal stress, tension or compression, acts at that point. It is seen that at the neutral axis the principal stresses in a beam are always inclined at 45° to the axis. In the vicinity of the outer fibers they are horizontal near midspan.

An important point follows from this discussion. Tensile stresses, which are of particular concern in view of the low tensile strength of the concrete, are not confined to the horizontal bending stresses f that are caused by bending alone. Tensile stresses of various inclinations and magnitudes, resulting from shear alone (at the neutral axis) or from the combined action of shear and bending, exist in all parts of a beam and can impair its integrity if not adequately provided for. It is for this reason that the inclined tensile stresses, known as *diagonal tension*, must be carefully considered in reinforced concrete design.

5.3   REINFORCED CONCRETE BEAMS WITHOUT SHEAR REINFORCEMENT

The discussion of shear in a homogeneous elastic beam applies very closely to a plain concrete beam *without* reinforcement. As the load is increased in such a beam, a tension crack will form where the tensile stresses are largest and will immediately cause the beam to fail. Except for beams of very unusual proportions, the largest tensile stresses are those caused at the outer fiber by bending alone, at the section of maximum bending moment. In this case, shear has little, if any, influence on the strength of a beam.

However, when tension reinforcement is provided, the situation is quite different. Even though tension cracks form in the concrete, the required flexural tension strength is furnished by the steel, and much higher loads can be carried. Shear stresses increase proportionally to the loads. In consequence, diagonal tension stresses of significant intensity are created in regions of high shear forces, chiefly close to the supports. The longitudinal tension reinforcement has been so calculated and placed that it is chiefly effective in resisting longitudinal tension near the tension face. It does not reinforce the tensionally weak concrete against the diagonal tension stresses that occur elsewhere, caused by shear alone or by the combined effect of shear and flexure. Eventually, these stresses attain magnitudes sufficient to open additional tension cracks in a direction perpendicular to the local tension stress. These are known as *diagonal* cracks, in distinction to the vertical flexural cracks. The latter occur in regions of large moments, the former in regions in which the shear forces are high. In beams in which no reinforcement is provided to counteract the formation of large diagonal tension cracks, their appearance has far-reaching and detrimental effects. For this reason, methods of predicting the loads at which these cracks will form are desired.

a.   Criteria for Formation of Diagonal Cracks

It is seen from Eq. (3.10) that the diagonal tension stresses t represent the combined effect of the shear stresses v and the bending stresses f. These in turn are, respectively, proportional to the shear force V and the bending moment M at the particular location in the beam [Eqs. (3.11) and (3.13)]. Depending on configuration, support conditions, and load distribution, a given location in a beam may have a large moment combined with a small shear force, or the reverse, or large or small values for both shear and moment. Evidently, the relative values of M and V will affect the magnitude as well as the direction of the diagonal tension stresses. Figure 5.4 shows a few typical beams and their moment and shear diagrams and draws attention to locations at which various combinations of high or low V and M occur.

At a location of large shear force V and small bending moment M, there will be little flexural cracking, if any, prior to the development of a diagonal tension crack. Consequently, the average shear stress prior to crack formation is

$$v = \frac{V}{bd} \tag{5.1}$$

The exact distribution of these shear stresses over the depth of the cross section is not known. It cannot be computed from Eq. (3.13) because this equation does not account for the influence of the reinforcement and because concrete is not an elastic homogeneous material. The value computed from Eq. (5.1) must therefore be regarded merely as a measure of the average intensity of shear stresses in the section. The maximum

FIGURE 5.4
Typical locations of critical combinations of shear and moment.

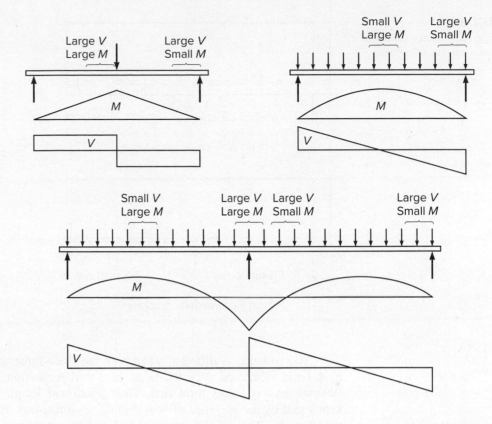

value, which occurs at the neutral axis, will exceed this average by an unknown but moderate amount.

If flexural stresses are negligibly small at the particular location, the diagonal tensile stresses, as in Fig. 5.3*b* and *c*, are inclined at about 45° and are numerically equal to the shear stresses, with a maximum at the neutral axis. Consequently, diagonal cracks form mostly at or near the neutral axis and propagate from that location, as shown in Fig. 5.5*a*. These *web-shear* cracks can be expected to form when the diagonal tension stress in the vicinity of the neutral axis becomes equal to the tensile strength of the concrete. The former, as was indicated, is of the order of, and somewhat larger than, $v = V/bd$; the latter, as discussed in Section 2.9, varies from about $3\sqrt{f_c'}$ to about $5\sqrt{f_c'}$. An evaluation of a very large number of beam tests is in fair agreement with this reasoning (Ref. 5.1). It was found that in regions with large shear and small moment, diagonal tension cracks form at an average or nominal shear stress v_{cr} of about $3.5\sqrt{f_c'}$, that is,

$$v_{cr} = \frac{V_{cr}}{bd} = 3.5\sqrt{f_c'} \qquad (5.2a)$$

where V_{cr} is that shear force at which the formation of the crack was observed.[†] Web-shear cracking is relatively rare and occurs chiefly near supports of deep, thin-webbed beams or at inflection points of continuous beams.

[†]Actually, diagonal tension cracks form at places where a compressive stress acts in addition to and perpendicular to the diagonal tension stress, as shown in Fig. 5.3*d* and *e*. The crack, therefore, occurs at a location of biaxial stress rather than uniaxial tension. However, the effect of this simultaneous compressive stress on the cracking strength appears to be small, in agreement with the information in Fig. 2.12.

FIGURE 5.5
Diagonal tension cracking in
reinforced concrete beams.

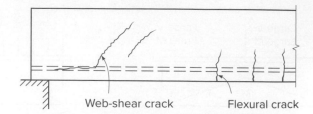

(*a*) Web-shear cracking

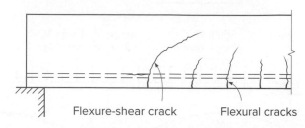

(*b*) Flexure-shear cracking

The situation is different when both the shear force and the bending moment have large values. At such locations, in a well-proportioned and reinforced beam, flexural tension cracks form first. Their width and length are well controlled and kept small by the presence of longitudinal reinforcement. However, when the diagonal tension stress at the upper end of one or more of these cracks exceeds the tensile strength of the concrete, the crack bends in a diagonal direction and continues to grow in length and width (see Fig. 5.5*b*). These cracks are known as *flexure-shear* cracks and are more common than web-shear cracks.

It is evident that at the instant at which a diagonal tension crack of this type develops, the average shear stress is larger than that given by Eq. (5.1). This is so because the preexisting tension crack has reduced the area of uncracked concrete that is available to resist shear to a value smaller than that of the uncracked area *bd* used in Eq. (5.1). The amount of this reduction will vary, depending on the unpredictable length of the preexisting flexural tension crack. Furthermore, the simultaneous bending stress *f* combines with the shear stress *v* to increase the diagonal tension stress *t* further [see Eq. (3.10)]. No way has been found to calculate reliable values of the diagonal tension stress under these conditions, and recourse must be made to test results.

A large number of beam tests have been evaluated for this purpose (Ref. 5.1). They show that in the presence of large moments (for which adequate longitudinal reinforcement has been provided) the nominal shear stress at which diagonal tension cracks form and propagate is, in most cases, conservatively given by

$$v_{cr} = \frac{V_{cr}}{bd} = 1.9\sqrt{f_c'} \qquad (5.2b)$$

Comparison with Eq. (5.2*a*) shows that large bending moments can reduce the shear force at which diagonal cracks form to roughly one-half the value at which they would form if the moment were zero or nearly so. This is in qualitative agreement with the discussion just given.

It is evident, then, that the shear at which diagonal cracks develop depends on the ratio of shear force to bending moment, or, more precisely, on the ratio of shear stress v to bending stress f near the top of the flexural crack. Neither of these can be accurately calculated. It is clear, though, that $v = K_1(V/bd)$, where, by comparison with Eq. (5.1), constant K_1 depends chiefly on the depth of penetration of the flexural crack. On the other hand [see Eq. (4.6)], $f = K_2(M/bd^2)$, where K_2 also depends on crack configuration. Hence, the ratio

$$\frac{v}{f} = \frac{K_1}{K_2} \frac{Vd}{M}$$

must be expected to affect that load at which flexural cracks develop into flexure-shear cracks, the unknown quantity K_1/K_2 to be explored by tests. Equation (5.2a) gives the cracking shear for very large values of Vd/M, and Eq. (5.2b) for very small values. Moderate values of Vd/M result in magnitudes of v_{cr} intermediate between these extremes.

In addition to the effect of bending moment, there are two other important factors dealing with the shear strength of reinforced concrete members without shear reinforcement that must be considered. The first is tied to the experimental observation that increasing the quantity of flexural reinforcement, which, in turn, limits the width of flexural cracks, results in an increase in the average shear stress at which flexure-shear cracks form. For typical laboratory test specimens with depths between 8 and 13 in., the combined relationship can be conservatively expressed by the empirical relationship (Ref. 5.11)

$$v_{cr} = \frac{V_{cr}}{bd} = 59\left(f_c' \rho \frac{Vd}{M}\right)^{1/3} \tag{5.3a}$$

where $\rho = A_s/bd$, as before, and 59 is an empirical constant. The effect of concrete compressive strength in Eq. (5.3a) is represented by $f_c'^{1/3}$ because the developers of the equation found that it gave a better match with test results than $\sqrt{f_c'}$. A graph of this relation and comparison with test data are given in Fig. 5.6.

FIGURE 5.6
Correlation of Eq. (5.3a) with test results.

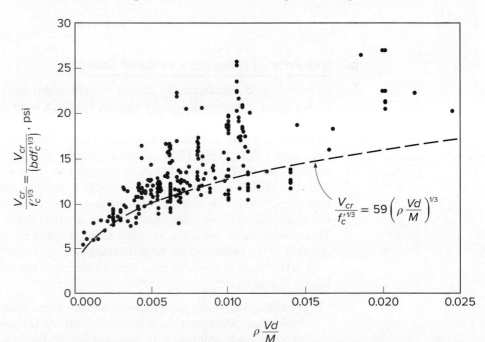

FIGURE 5.7

Comparison of ratio of shear strength predicted by Eq. (5.3a) to test results with the effective depth of the members d.

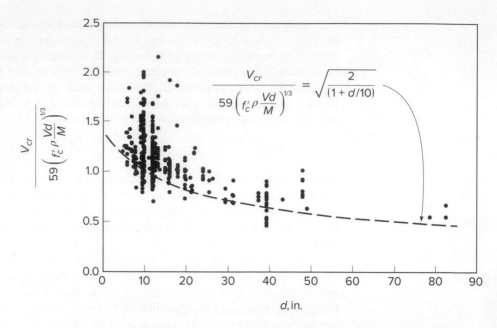

$$\frac{V_{cr}}{59 \left(f_c' \rho \dfrac{Vd}{M} \right)^{1/3}} = \sqrt{\frac{2}{(1 + d/10)}}$$

The second factor becomes apparent when specimens are tested that match the size of larger beams used in practice. As the depth of a member increases above that typical of laboratory specimens, Eq. (5.3a) becomes progressively unconservative (that is, progressively overestimates v_{cr}). This is illustrated in Fig. 5.7 for members with effective depths d of up to 84 in. This lower relative strength, known as the *size effect*, is directly tied to the fact that shear failure results from both the formation *and* the propagation of a critical shear crack at failure. As demonstrated in Fig. 5.7, the size effect can be quite adequately described for members that fail in shear using the concepts of fracture mechanics as

$$\sqrt{\frac{2}{(1 + d/10)}} \tag{5.3b}$$

b.   Behavior of Diagonally Cracked Beams

In regard to flexural cracks, as distinct from diagonal tension cracks, it was explained in Section 4.2 that cracks on the tension side of a beam are permitted to occur and are in no way detrimental to the strength of the member. One might expect a similar situation in regard to diagonal cracking caused chiefly by shear. The analogy, however, is not that simple. Flexural tension cracks are harmless only because adequate longitudinal reinforcement has been provided to resist the flexural tension stresses that the cracked concrete is no longer able to transmit. In contrast, the beams now being discussed, although furnished with the usual longitudinal reinforcement, are not equipped with any other reinforcement to offset the effects of diagonal cracking. This makes the diagonal cracks much more decisive in subsequent performance and strength of the beam than the flexural cracks.

Two types of behavior have been observed in the many tests on which present knowledge is based:

1. The diagonal crack, once formed, spreads either immediately or at only slightly higher load, traversing the entire beam from the tension reinforcement to the compression face, splitting it in two and failing the beam. This process is sudden

and without warning and occurs chiefly in the shallower beams, that is, beams with span-depth ratios of about 8 or more. Beams in this range of dimensions are very common. Complete absence of shear reinforcement would make them very vulnerable to accidental large overloads, which would result in catastrophic failures without warning. For this reason, it is good practice to provide a minimum amount of shear reinforcement even if calculation does not require it, because such reinforcement restrains growth of diagonal cracks, thereby increasing ductility and providing warning in advance of actual failure. Only in situations where an unusually large safety factor against inclined cracking is provided, that is, where actual shear stresses are very small compared with v_{cr}, as in some slabs and most footings, is it permissible to omit shear reinforcement.

2. Alternatively, the diagonal crack, once formed, spreads toward and partially into the compression zone but stops short of penetrating to the compression face. In this case no sudden collapse occurs, and the failure load may be significantly higher than that at which the diagonal crack first formed. This behavior is chiefly observed in the deeper beams with smaller span to depth ratios and will be analyzed now.

Figure 5.8*a* shows a portion of a beam, arbitrarily loaded, in which a diagonal tension crack has formed. Consider the part of the beam to the left of the crack, shown in solid lines. There is an external upward shear force $V_{ext} = R_l - P_1$ acting on this portion.

Once a crack is formed, no tension force perpendicular to the crack can be transmitted across it. However, as long as the crack is narrow, it can still transmit forces in its own plane through interlocking of the surface roughnesses. Sizable interlock forces V_i of this kind have in fact been measured, amounting to one-third and more of the total shear force. The components V_{ix} and V_{iy} of V_i are shown in Fig. 5.8*a*.

FIGURE 5.8

Forces at a diagonal crack in a beam without web reinforcement.

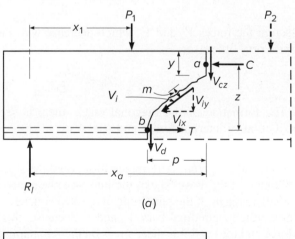

(a)

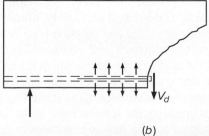

(b)

The other internal vertical forces are those in the uncracked portion of the concrete V_{cz} and across the longitudinal steel, acting as a dowel, V_d. Thus, the internal shear force is

$$V_{\text{int}} = V_{cz} + V_d + V_{iy}$$

Equilibrium requires that $V_{\text{int}} = V_{\text{ext}}$ so that the part of the shear resisted by the uncracked concrete is

$$V_{cz} = V_{\text{ext}} - V_d - V_{iy} \tag{5.4}$$

In a beam provided with longitudinal reinforcement only, the portion of the shear force resisted by the steel in dowel action is usually quite small. In fact, the reinforcing bars on which the dowel force V_d acts are supported against vertical displacement chiefly by the thin concrete layer below. The bearing pressure caused by V_d creates, in this concrete, vertical tension stresses as shown in Fig. 5.8b. Because of these stresses, diagonal cracks often result in splitting of the concrete along the tension reinforcement, as shown. (See also Fig. 5.1.) This reduces the dowel force V_d and also permits the diagonal crack to widen. This, in turn, reduces the interface force V_i and frequently leads to immediate failure.

Next consider moments about point a at the intersection of V_{cz} and C; the external moment $M_{\text{ext},a}$ acts at a and happens to be $R_l x_a - P_1(x_a - x_1)$ for the loading shown. The internal moment is

$$M_{\text{int},a} = T_b z + V_d p - V_i m$$

Here p is the horizontal projection of the diagonal crack and m is the moment arm of the force V_i with respect to point a. The designation T_b for T is meant to emphasize that this force in the steel acts at point b rather than vertically below point a. Equilibrium requires that $M_{\text{int},a} = M_{\text{ext},a}$ so that the longitudinal tension in the steel at b is

$$T_b = \frac{M_{\text{ext},a} - V_d p + V_i m}{z} \tag{5.5}$$

Neglecting the forces V_d and V_i, which decrease with increasing crack opening, one has, with very little error,

$$T_b = \frac{M_{\text{ext},a}}{z} \tag{5.6}$$

The formation of the diagonal crack, then, is seen to produce the following redistribution of internal forces and stresses:

1. At the vertical section through point a, the average shear stress before crack formation was V_{ext}/bd. After crack formation, the shear force is resisted by a combination of the dowel shear, the interface shear, and the shear force on the much smaller area by of the remaining uncracked concrete. As tension splitting develops along the longitudinal bars, V_d and V_i decrease; this, in turn, increases the shear force and the resulting shear stress on the remaining uncracked concrete area.

2. The diagonal crack, as described previously, usually rises above the neutral axis and traverses some part of the compression zone before it is arrested by the compression stresses. Consequently, the compression force C also acts on an area equal to y times the width of the beam, which is smaller than the area on which it acted before the crack was formed. Correspondingly, formation of the crack has increased the compression stresses in the remaining uncracked concrete.

3. Prior to diagonal cracking, the tension force in the steel at point b was caused by, and was proportional to, the bending moment in a vertical section through the

same point b. As a consequence of the diagonal crack, however, Eq. (5.6) shows that the tension in the steel at b is now caused by, and is proportional to, the bending moment at a. Since the moment at a is evidently larger than that at b, formation of the crack has caused a sudden increase in the steel stress at b.

If the two materials are capable of resisting these increased stresses, equilibrium will establish itself after internal redistribution and further load can be applied before failure occurs. Such failure can then develop in various ways. For one, if only enough steel has been provided at b to resist the moment at that section, the increase of the steel force, described in item 3, will cause the steel to yield because of the larger moment at a, thus failing the beam. If the beam is properly designed to prevent this occurrence, it is usually the concrete at the head of the crack that will eventually crush. This concrete is subject simultaneously to large compression and shear stresses, and this biaxial stress combination is conducive to earlier failure than would take place if either of these stresses were acting alone. Finally, if there is splitting along the reinforcement, it will cause the bond between steel and concrete to weaken to such a degree that the reinforcement may pull loose. This either may be the cause of failure of the beam or may occur simultaneously with crushing of the remaining uncracked concrete.

It was noted earlier that relatively deep beams will usually show continued and increasing resistance after formation of a critical diagonal tension crack, but relatively shallow beams will fail almost immediately upon formation of the crack. The amount of reserve strength, if any, was found to be erratic. In fact, in several test series in which two specimens as identical as one can make them were tested, one failed immediately upon formation of a diagonal crack, while the other reached equilibrium under the described redistribution and failed at a higher load.

For this reason, this reserve strength is discounted in modern design procedures. As previously mentioned, most beams are furnished with at least a minimum of web reinforcement. For those flexural members that are not, such as slabs, footings, and others as described in Section 5.5b, design is based on that shear force V_{cr} or shear stress v_{cr} at which formation of inclined cracks must be expected.

5.4 REINFORCED CONCRETE BEAMS WITH WEB REINFORCEMENT

Economy of design demands, in most cases, that a flexural member be capable of developing its full moment capacity rather than having its strength limited by premature shear failure. This is also desirable because structures, if overloaded, should not fail in the sudden and explosive manner characteristic of many shear failures, but should show adequate ductility and warning of impending distress. The latter, as pointed out earlier, is typical of flexural failure caused by yielding of the longitudinal bars, which is preceded by gradual excessively large deflections and noticeable widening of cracks. Therefore, if a fairly large safety margin relative to the available shear strength of the concrete alone does not exist, special shear reinforcement, known as *web reinforcement*, is used to increase this strength.

a. Types of Web Reinforcement

Typically, web reinforcement is provided in the form of vertical *stirrups*, spaced at varying intervals along the axis of the beam depending on requirements, as shown in Fig. 5.9a. Relatively small bars are used, generally Nos. 3 to 5 (Nos. 10 to 16).

FIGURE 5.9
Types of web reinforcement.

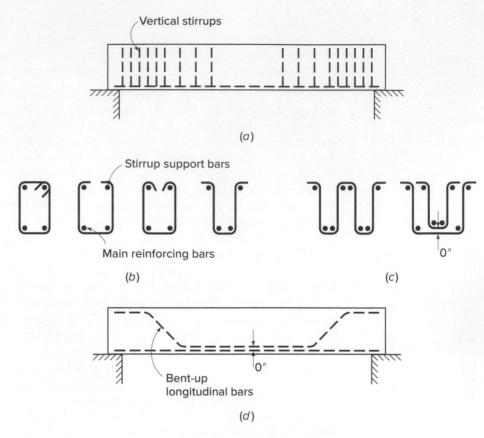

(a)

(b)

(c)

(d)

Simple U-shaped bars similar to Fig. 5.9*b* are most common. Multiple-leg stirrups, such as shown in Fig. 5.9*c*, are required in beams with widths larger than the depth, where U-shaped stirrups at the boundary of the members may not provide adequate shear capacity across the width of the members (Refs. 5.13 to 5.15). Stirrups are formed to fit around the main longitudinal bars at the bottom and hooked or bent around longitudinal bars at the top of the member to improve anchorage and provide support during construction. Detailed requirements for anchorage of stirrups will be discussed in Chapter 6.

Alternatively, shear reinforcement may be provided by bending up a part of the longitudinal steel where it is no longer needed to resist flexural tension, as suggested by Fig. 5.9*d*. In continuous beams, these bent-up bars may also provide all or part of the necessary reinforcement for negative moments. The requirements for longitudinal flexural reinforcement often conflict with those for diagonal tension, and because the savings in steel resulting from use of the capacity of bent bars as shear resistance is small, most designers prefer to include vertical stirrups to provide for all the shear requirement, counting on the bent part of the longitudinal bars, if bent bars are used, only to increase the overall safety against diagonal tension failure.

Welded wire reinforcement is also used for shear reinforcement, particularly for small, lightly loaded members with thin webs, and for certain types of precast, prestressed beams.

b. **Behavior of Web-Reinforced Concrete Beams**

Web reinforcement has no noticeable effect prior to the formation of diagonal cracks. In fact, measurements show that the web steel is practically free of stress prior to crack formation. After diagonal cracks have developed, web reinforcement augments the shear resistance of a beam in four separate ways:

1. Part of the shear force is resisted by the bars that traverse a particular crack. The mechanism of this added resistance is discussed below.
2. The presence of these same bars restricts the growth of diagonal cracks and reduces their penetration into the compression zone. This leaves more uncracked concrete available at the head of the crack for resisting the combined action of shear and compression, already discussed.
3. The stirrups also counteract the widening of the cracks, so that the two crack faces stay in close contact. This makes for a significant and reliable interface force V_i (see Fig. 5.8).
4. As shown in Fig. 5.9, the stirrups are arranged so that they tie the longitudinal reinforcement into the main bulk of the concrete. This provides some measure of restraint against the splitting of concrete along the longitudinal reinforcement, shown in Figs. 5.1 and 5.8b, and increases the share of the shear force resisted by dowel action.

From this it is clear that failure will be imminent when the stirrups start yielding. This not only exhausts their own resistance but also permits a wider crack opening with consequent reduction of the beneficial restraining effects, points 2 to 4, above.

It becomes clear from this description that member behavior, once a crack is formed, is quite complex and dependent in its details on the particulars of crack configuration (length, inclination, and location of the main or critical crack). The latter, in turn, is quite erratic and has so far defied purely analytical prediction. For this reason, the concepts that underlie present design practice are not wholly rational. They are based partly on rational analysis, partly on test evidence, and partly on successful long-time experience with structures in which certain procedures for designing web reinforcement have resulted in satisfactory performance.

BEAMS WITH VERTICAL STIRRUPS. Since web reinforcement is ineffective in the uncracked beam, the magnitude of the shear force or stress that causes cracking to occur is the same as in a beam without web reinforcement and is approximated by Eq. (5.3a). Most frequently, web reinforcement consists of *vertical stirrups*; the forces acting on the portion of such a beam between the crack and the nearby support are shown in Fig. 5.10. They are the same as those of Fig. 5.8, except that each stirrup traversing the crack exerts a force $A_v f_v$ on the given portion of the beam. Here A_v is the cross-sectional area of the stirrup (in the case of the U-shaped stirrup

FIGURE 5.10

Forces at a diagonal crack in a beam with vertical stirrups.

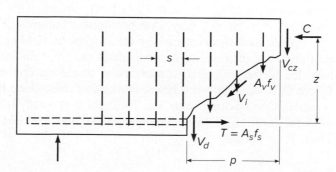

FIGURE 5.11

Redistribution of internal
shear forces in a beam with
stirrups. (*Adapted from Ref. 5.3.*)

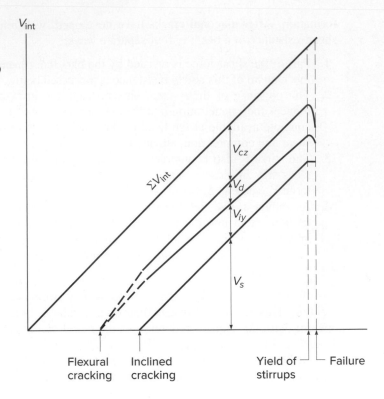

of Fig. 5.9*b* it is twice the area of one bar), and f_v is the tensile stress in the stirrup.
Equilibrium in the vertical direction requires

$$V_{ext} = V_{cz} + V_d + V_{iy} + V_s \tag{a}$$

where $V_s = nA_v f_v$ is the vertical force in the stirrups, n being the number of stirrups
traversing the crack. If s is the stirrup spacing and p the horizontal projection of the
crack, as shown, then $n = p/s$.

The approximate distribution of the four components of the internal shear force
with increasing external shear V_{ext} is shown schematically in Fig. 5.11. It is seen that
after inclined cracking, the portion of the shear $V_s = nA_v f_v$ carried by the stirrups
increases linearly, while the sum of the three other components, $V_{cz} + V_d + V_{iy}$, stays
nearly constant. When the stirrups yield, their contribution remains constant at the
yield value $V_s = nA_v f_{yt}$, where f_{yt} represents the yield strength of the stirrup (or
transverse) reinforcement. However, because of widening of the inclined cracks and
longitudinal splitting, V_{iy} and V_d fall off rapidly. This overloads the remaining
uncracked concrete and very soon precipitates failure.

While total shear carried by the stirrups at yielding is known, the individual
magnitudes of the three other components are not. Limited amounts of test evidence
have led to the conservative assumption in present-day methods that just prior to
failure of a web-reinforced beam, the sum of these three internal shear components
is equal to the cracking shear V_{cr}. This sum is generally (somewhat loosely) referred
to as the *contribution of the concrete* to the total shear resistance and is denoted V_c.
Thus, $V_c = V_{cr}$ and

$$V_c = V_{cz} + V_d + V_{iy} \tag{b}$$

The number of stirrups n spaced a distance s apart was seen to depend on the length p of the horizontal projection of the diagonal crack. This length is conservatively assumed to be equal to the effective depth of the beam; thus $n = d/s$, implying a crack somewhat flatter than 45°. Then, at failure, when $V_{\text{ext}} = V_n$, Eqs. (a) and (b) yield for the nominal shear strength

$$V_n = V_c + \frac{A_v f_{yt} d}{s} \tag{5.7a}$$

Dividing both sides of Eq. (5.7a) by bd, the same relation is expressed in terms of the nominal shear stress:

$$v_n = \frac{V_n}{bd} = v_c + \frac{A_v f_{yt}}{bs} \tag{5.7b}$$

BEAMS WITH INCLINED BARS. The function of *inclined web reinforcement* (Fig. 5.9d) can be discussed in very similar terms. Figure 5.12 again indicates the forces that act on the portion of the beam to one side of the diagonal crack that results in eventual failure. The crack with horizontal projection p and inclined length $i = p/\cos\theta$ is crossed by inclined bars horizontally spaced a distance s apart. The inclination of the bars is α and that of the crack θ, as shown. The distance between bars measured parallel to the direction of the crack is seen from the irregular triangle to be

$$a = \frac{s}{\sin\theta\,(\cot\theta + \cot\alpha)} \tag{a}$$

The number of bars crossing the crack $n = i/a$, after some transformation, is

$$n = \frac{p}{s}\,(1 + \cot\alpha\,\tan\theta) \tag{b}$$

The vertical component of the force in one bar or stirrup is $A_v f_v \sin\alpha$, so that the total vertical component of the forces in all bars that cross the crack is

$$V_s = nA_v f_v \sin\alpha = A_v f_v \frac{p}{s}\,(\sin\alpha + \cos\alpha\,\tan\theta) \tag{5.8}$$

As in the case of vertical stirrups, shear failure occurs when the stress in the web reinforcement reaches the yield point. Also, the same assumptions are made as in the case of stirrups, namely, that the horizontal projection of the diagonal crack is equal to the effective depth d, and that $V_{cz} + V_d + V_{iy}$ is equal to V_c. Lastly, the

FIGURE 5.12

Forces at a diagonal crack in a beam with inclined web reinforcement.

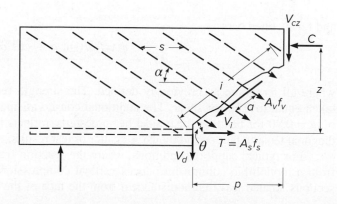

inclination θ of the diagonal crack, which varies somewhat depending on various influences, is generally assumed to be 45°. On this basis, when failure is caused by shear, the nominal strength is

$$V_n = V_c + \frac{A_v f_{yt} \, d(\sin \alpha + \cos \alpha)}{s} \tag{5.9}$$

It is seen that Eq. (5.7a), developed for vertical stirrups, is only a special case, for $\alpha = 90°$, of the more general expression in Eq. (5.9).

Note that Eqs. (5.7) and (5.9) apply only if web reinforcement is so spaced that any conceivable diagonal crack is traversed by at least one stirrup or inclined bar. Otherwise web reinforcement would not contribute to the shear strength of the beam, because diagonal cracks that could form between widely spaced web reinforcement would fail the beam at the load at which it would fail if no web reinforcement were present. This imposes upper limits on the permissible spacing s to ensure that the web reinforcement is actually effective as calculated.

To summarize, at this time the nature and mechanism of diagonal tension failure are clearly understood qualitatively, but some of the quantitative assumptions that have been made in the preceding development cannot be proved by rational analysis. However, the calculated results are in acceptable and generally conservative agreement with a very large body of empirical data, and structures designed on this basis have proved satisfactory. Newer methods, introduced in Section 5.8, provide alternatives that are slowly being incorporated into the ACI Code and the AASHTO Bridge Specifications (Ref. 5.12). Chapter 17 presents a detailed description of one such alternative, the so-called strut-and-tie model, which appears in Chapter 23 of the ACI Code.

5.5 ACI CODE PROVISIONS FOR SHEAR DESIGN

According to ACI Code 9.5.1.1, the design of beams for shear is to be based on the relation

$$V_u \le \phi V_n \tag{5.10}$$

where V_u is the total shear force applied at a given section of the beam due to factored loads and $V_n = V_c + V_s$ is the nominal shear strength, equal to the sum of the contributions of the concrete and the web steel if present. Thus for vertical stirrups

$$V_u \le \phi V_c + \frac{\phi A_v f_{yt} d}{s} \tag{5.11a}$$

and for inclined bars

$$V_u \le \phi V_c + \frac{\phi A_v f_{yt} d(\sin \alpha + \cos \alpha)}{s} \tag{5.11b}$$

where all terms are as previously defined. The strength reduction factor ϕ is to be taken equal to 0.75 for shear. The additional conservatism, compared with the value of $\phi = 0.90$ for bending for typical beam designs, reflects both the sudden nature of diagonal tension failure and the large scatter of test results.

For typical support conditions, where the reaction from the support surface or from a monolithic column introduces vertical compression at the end of the beam, sections located less than a distance d from the face of the support may be designed

FIGURE 5.13
Location of critical section for shear design: (*a*) end-supported beam; (*b*) beam supported by columns; (*c*) concentrated load within *d* of the face of the support; (*d*) member loaded near the bottom; (*e*) beam supported by girder of similar depth; and (*f*) beam supported by monolithic vertical element.

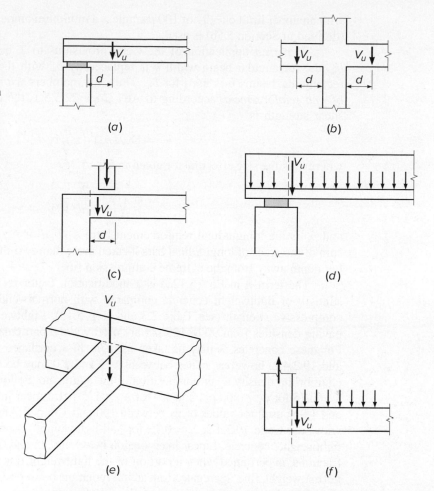

for the same shear V_u as that computed at a distance *d*, as shown in Fig. 5.13*a* and *b*. However, the critical design section should be taken at the face of the support if concentrated loads act within that distance (Fig. 5.13*c*), if the beam is loaded near its bottom edge (as may occur for an inverted T beam, as shown in Fig. 5.13*d*), or if the reaction causes vertical tension rather than compression [for example, if the beam is supported by a girder of similar depth (Fig. 5.13*e*) or at the end of a monolithic vertical element (Fig. 5.13*f*)].

a. Shear Strength Provided by the Concrete

To establish the nominal shear strength contribution of the concrete (including the contributions from aggregate interlock, dowel action of the main reinforcing bars, and that of the uncracked concrete), ACI Code 22.5.5.1 uses an approach that simplifies that shown in Eqs. (5.3*a*) and (5.3*b*). It does so by (1) neglecting the effect of bending moment on the shear stress at which diagonal tension cracks form $v_{cr}^{\dagger}$ and (2) representing the contribution of concrete compressive strength on v_{cr} using $\sqrt{f_c'}$ in place of $f_c'^{1/3}$,

---

[†]The effect of bending moment on the shear stress at which diagonal tension cracks form is considered for prestressed concrete members, as discussed in Chapter 22.

with an upper limit on $\sqrt{f_c'}$ of 100 psi unless a minimum amount of web reinforcement (defined in Section 5.5b) is used.

To permit application of the Code provisions to T beams having web width b_w, the rectangular beam width b is replaced by b_w with the understanding that for rectangular beams b is used for b_w. Thus, for members *without a minimum amount of web reinforcement*, according to ACI Code 22.5.5.1, the concrete contribution to shear strength is

$$V_c = 8\lambda_s\lambda(\rho_w)^{1/3}\sqrt{f_c'}b_w d \qquad (5.12a)$$

where λ_s is the size effect factor given in Eq. (5.3b)

$$\lambda_s = \sqrt{\frac{2}{(1 + d/10)}} \qquad (5.12b)$$

and ρ_w is the longitudinal reinforcement ratio $A_s/b_w d$ or A_s/bd, with A_s taken as the sum of the areas of longitudinal bars located more than two-thirds of the overall member depth away from the extreme compression fiber.

The term λ in Eq. (5.12a) is a modification factor reflecting the lower tensile strength of lightweight concrete compared with normalweight concrete of the same compressive strength (see Table 2.3 and Ref. 5.13). Lightweight aggregate concretes having densities from 90 to 135 pcf are used widely, particularly for precast elements. For these concretes, λ may be taken as 0.75. In accordance with ACI Code 19.2.4.1 and 19.2.4.2, however, values between 0.75 and 1.0 may be used based on either the equilibrium density w_c or composition of the aggregate. Following ACI Code 19.2.4.1, $\lambda = 0.75$ for $w_c \le 100$ pcf and 1.0 for $w_c \ge 135$ pcf. Linear interpolation between 0.75 and 1.0 is used for values of w_c between 100 and 135 pcf. Alternatively, in accordance with ACI Code 19.2.4.2, $\lambda = 0.75$ for "all-lightweight" concrete and 0.85 for "sand-lightweight" concrete. Linear interpolation between 0.75 and 0.85, based on volumetric fractions, is permitted when a portion of the lightweight fine aggregate is replaced by normalweight fine aggregate. Linear interpolation between 0.85 and 1.0 is also permitted for concretes containing normalweight fine aggregate and a blend of lightweight and normalweight coarse aggregate. For normalweight concrete, $\lambda = 1.0$.

For members *with a minimum amount of web reinforcement*, the ACI Code recognizes that because web reinforcement limits crack width, it both increases the effective contribution of concrete to the shear strength of the member V_c and reduces the size effect, as represented by λ_s. Based on these observations, ACI Code 22.5.5.1 permits the concrete contribution to shear strength to be calculated for members *with a minimum amount of web reinforcement* as either

$$V_c = 2\lambda\sqrt{f_c'}b_w d \qquad (5.12c)$$

or

$$V_c = 8\lambda(\rho_w)^{1/3}\sqrt{f_c'}b_w d \qquad (5.12d)$$

Equation (5.12c) is close to the conservative value shown in Eq. (5.2b). It has been used successfully for many years, and because of its simplicity, is often used in practice. Equation (5.12d) is Eq. (5.12a) with λ_s set to 1.0. The value of V_c calculated using Eq. (5.12d) will exceed V_c calculated using (5.12c) for ρ_w greater than 0.0156.

For members with a circular cross section, ACI Code 22.5.2.2 provides that V_c in Eqs. (5.12a), (5.12c), and (5.12d) be calculated using b_w equal to the diameter of solid sections and twice the wall thickness for hollow sections and d equal to 0.8 times the diameter of the member.

The upper limit on $\sqrt{f_c'}$ of 100 psi is based on experimental results (Refs. 5.14 to 5.17) for beams constructed using concrete with f_c' above 6000 psi (see Section 2.12) showing that the concrete contribution to shear strength V_c increases more slowly than $\sqrt{f_c'}$ as f_c' increases. This effect, however, is greatly reduced in the presence of web reinforcement. Thus, values of $\sqrt{f_c'}$ greater than 100 psi may be used in computing V_c if a minimum amount of web reinforcement is used (see Section 5.5b).

b.   Minimum Web Reinforcement

If V_u, the shear force at factored loads, is no larger than ϕV_c, calculated by Eq. (5.12a), then theoretically no web reinforcement is required. Even in such a case, however, ACI Code 9.6.3 requires provision of at least a minimum area of web reinforcement equal to

$$A_{v,\min} = 0.75 \sqrt{f_c'} \, \frac{b_w s}{f_{yt}} \geq 50 \frac{b_w s}{f_{yt}} \tag{5.13a}$$

where s = longitudinal spacing of web reinforcement, in.

       f_{yt} = yield strength of web steel, psi, and

   $A_{v,\min}$ = total cross-sectional area of web steel within distance s, in.^{2}.

This provision holds unless V_u is one-half or less of the design strength provided by the concrete based on Eq. (5.12c), that is, unless

$$V_u \leq \phi \lambda \sqrt{f_c'} b_w d \tag{5.13b}$$

corresponding to

$$V_c = \lambda \sqrt{f_c'} b_w d \tag{5.13c}$$

Specific exceptions to this requirement for minimum web steel are made for concrete joist floor construction; for beams with total depth h not greater than 10 in.; and for beams integral with slabs with h not greater than 24 in. and not greater than the larger of 2.5 times the thickness of the flange and 0.5 times the thickness of the web. These members are excluded because of their capacity to redistribute internal forces before diagonal tension failure, as confirmed by both tests and successful design experience. In addition, beams constructed of steel fiber reinforced, normalweight concrete with f_c' not exceeding 6000 psi, total depth h not greater than 24 in., and V_u not greater than $\phi 2 \sqrt{f_c'} b_w d$ are not required to meet the requirements for minimum web reinforcement because beams meeting these requirements have been shown to have shear strength in excess of $3.5 \sqrt{f_c'} b_w d$ (Ref. 5.16).[†]

For beams without web reinforcement, Eq. (5.12a) will govern in place of Eqs. (5.13b) and (5.13c) in cases where the product $8\lambda_s(\rho_w)^{1/3}$ is less than 1.0. This will occur, for example, for members with reinforcement ratios ρ_w of 0.008, 0.010, and 0.012 with effective depths d that exceed, respectively, 41, 49, and 57 in. For shallower members, Eqs. (5.13b) and (5.13c) will govern.

For high-strength concrete beams, the limitation of 100 psi imposed on the value of $\sqrt{f_c'}$ used in calculating V_c is waived by ACI Code 22.5.3.2 if such beams are designed with minimum web reinforcement equal to the amount required by Eq. (5.13a). In this case, the concrete contribution to shear strength may be calculated based on the full concrete compressive strength. Tests described in Refs. 5.17 and 5.18 indicate that for beams with concrete strength above about 6000 psi, the concrete contribution V_c was significantly less than predicted by the ACI Code equations, although the steel contribution V_s was higher. The total nominal shear strength V_n

---

[†]To qualify, the fiber-reinforced concrete must conform to requirements in ACI Code 26.4.1.5, 26.4.2.2(i), and 26.12.7.1 that specify a minimum deformed steel fiber content of 100 lb/yd^3 and minimum residual flexural strength values when the concrete is tested in accordance with ASTM C1609, "*Standard Test Method for Flexural Performance of Fiber-Reinforced Concrete (Using Beam with Third-Point Loading)*."

was greater than predicted by ACI Code methods in all cases. The use of minimum web steel for high-strength concrete beams is intended to enhance the post-cracking capacity, thus resulting in safe designs even though the concrete contribution to shear strength is overestimated.[‡]

EXAMPLE 5.1

Beam without web reinforcement.   A rectangular beam, with ρ_w estimated to be 0.01, is designed to carry a shear force V_u of 27 kips. No web reinforcement is used, and f_c' is 4000 psi. What is the minimum cross section if controlled by shear?

SOLUTION.   If no web reinforcement is used, the cross-sectional dimensions must be selected so that the applied shear V_u is no larger than design strength given in Eq. (5.13b).

$$V_u = \phi \lambda \sqrt{f_c'} b_w d$$

$$b_w d = \frac{27,000}{0.75 \times 1.0 \sqrt{4000}} = 569 \text{ in}^2$$

A beam with $b_w = 18$ in. and $d = 32$ in. is required. As described earlier, because the beam is relatively shallow, Eq. (5.13b) will govern this design because $8\lambda_s(\rho_w)^{1/3}$ in Eq. (5.12a) exceeds 1.0. For comparison, however, the cross section will also be checked based on Eq. (5.12a) with

$$\lambda_s = \sqrt{\frac{2}{(1 + d/10)}} = \sqrt{\frac{2}{(1 + 32/10)}} = 0.69$$

and $\rho_w = 0.01$, giving $8\lambda_s(\rho_w)^{1/3} = 8 \times 0.69 \times (0.01)^{1/3} = 1.19$.

$$V_u = \phi 8\lambda_s \lambda (\rho_w)^{1/3} \sqrt{f_c'} b_w d$$

$$b_w d = \frac{27,000}{0.75 \times 8 \times 0.69 \times 1.0 \times (0.01)^{1/3} \sqrt{4000}} = 479 \text{ in}^2$$

As expected, the section calculated using Eq. (5.12a) is too small, and the cross section with $b_w = 18$ in. and $d = 32$ in. will be used. Alternately, if the minimum amount of web reinforcement given by Eq. (5.13a) is used, the concrete shear resistance may be taken at its full value ϕV_c, with V_c given by Eq. (5.12c), and it is easily confirmed that a beam with $b_w = 12$ in. and $d = 24$ in. will be sufficient.

---

c.   Region in which Web Reinforcement is Required

If the required shear strength V_u is greater than the design shear strength ϕV_c provided by the concrete in any portion of a beam, with V_c based on Eq. (5.12a), there is a theoretical requirement for web reinforcement. Following ACI Code 22.5.5.1, the quantity of web reinforcement would be based on V_c given in Eq. (5.12c) or Eq. (5.12d). In addition, web reinforcement at least equal to the amount given by Eq. (5.13a) must be provided, unless the factored shear force is low enough to satisfy Eq. (5.13b).

The portion of any span through which web reinforcement is theoretically necessary can be found from the shear diagram for the span, superimposing a plot of the shear strength of the concrete. Where the shear force V_u exceeds ϕV_c, shear reinforcement must provide for the excess. The additional length through which at least the minimum web steel is needed can be found by superimposing a plot of $\phi \lambda \sqrt{f_c'} b_w d$, as given in Eq. (5.13$b$).

[‡] The shortcomings of the ACI Code "$V_c + V_s$" approach to shear design, particularly the provisions relating to the concrete contribution V_c, have provided motivation for the development of more rational procedures, as will be discussed in Section 5.8.

EXAMPLE 5.2   **Limits of web reinforcement.**   A simply supported rectangular beam 16 in. wide having an effective depth of 22 in. carries a total factored load of 9.4 kips/ft on a 20 ft clear span. It is reinforced with 7.62 in^2 of tensile steel, which continues uninterrupted into the supports. If $f_c' = 4000$ psi, throughout what part of the beam is web reinforcement required?

SOLUTION.   The variation of V_u along the beam is shown in Fig. 5.14a. The maximum external shear force occurs at the ends of the span, where $V_u = 9.4 \times 20/2 = 94$ kips. At the critical section for shear, a distance d from the support, $V_u = 94 - 9.4 \times 1.83 = 76.8$ kips. The shear force varies linearly to zero at midspan.

The size effect factor $\lambda_s = \sqrt{2/(1 + d/10)} = \sqrt{2/(1 + 22/10)} = 0.79$, and the reinforcement ratio $\rho_w = A_s/b_w d = 7.62/(16 \times 22) = 0.0216$. Using these values to calculate V_c for a member without web reinforcement gives

$$V_c = 8\lambda_s \lambda (\rho_w)^{1/3} \sqrt{f_c'} b_w d = 8 \times 0.79 \times 1.0 \times (0.0216)^{1/3} \sqrt{4000} \times 16 \times 22 = 39,200 \text{ lb}$$

By inspection, web reinforcement is needed.

Adopting Eq. (5.12c) for V_c where web reinforcement is used gives

$$V_c = 2\lambda \sqrt{f_c'} b_w d = 2 \times 1.0\sqrt{4000} \times 16 \times 22 = 44,500 \text{ lb}$$

Hence $\phi V_c = 0.75 \times 44.5 = 33.4$ kips. This value is superimposed on the shear diagram in Fig. 5.14a, and from geometry, the point at which web reinforcement theoretically is no longer required is

$$10\left(\frac{94.0 - 33.4}{94.0}\right) = 6.45 \text{ ft}$$

FIGURE 5.14
Shear design example.

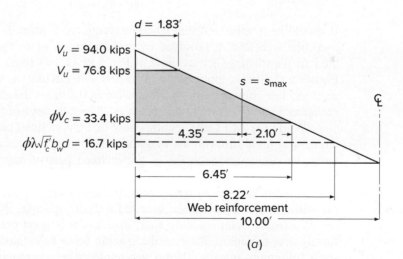

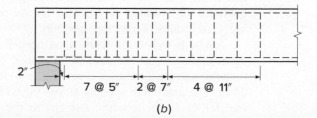

from the support face. According to the ACI Code, however, at least a minimum amount of web reinforcement is required wherever the shear force exceeds $\phi\lambda\sqrt{f_c'}b_w d$, or 16.7 kips in this case. As shown in Fig. 5.14a, this applies to a distance

$$10\left(\frac{94.0 - 16.7}{94.0}\right) = 8.22 \text{ ft}$$

from the support face. To summarize, at least the minimum web steel must be provided within a distance of 8.22 ft from the supports, and within 6.45 ft the web steel must provide for the shear force corresponding to the shaded area.

d. Design of Web Reinforcement

The design of web reinforcement, under the provisions of the ACI Code, is based on Eq. (5.11a) for vertical stirrups and Eq. (5.11b) for inclined stirrups or bent bars. In design, it is usually convenient to select a trial web-steel area A_v based on standard stirrup sizes [usually in the range from No. 3 to 5 (No. 10 to 16) for stirrups, and according to the longitudinal bar size for bent-up bars], for which the required spacing s can be found. Equating the design strength ϕV_n to the required strength V_u and transposing Eqs. (5.11a) and (5.11b) accordingly, one finds that the required spacing of web reinforcement is, for vertical stirrups,

$$s = \frac{\phi A_v f_{yt} d}{V_u - \phi V_c} \tag{5.14a}$$

and for bent bars

$$s = \frac{\phi A_v f_{yt} d(\sin\alpha + \cos\alpha)}{V_u - \phi V_c} \tag{5.14b}$$

It should be emphasized that when conventional U stirrups such as in Fig. 5.9b are used, the web area A_v provided by each stirrup is *twice* the cross-sectional area of the bar; for stirrups such as those of Fig. 5.9c, A_v is 4 times the area of the bar used. Equation (5.14a) is applicable to members with circular, as well as rectangular, cross sections. For circular members, d is taken as 0.8 times the effective depth, as defined earlier in Section 5.5a, and A_v is taken as 2 times the area of the bar, hoop, or spiral.

While the ACI Code requires only that the inclined part of a bent bar make an angle of at least 30° with the longitudinal part, bars are usually bent at a 45° angle. Only the center three-fourths of the inclined part of any bar is to be considered effective as web reinforcement.

It is undesirable to space vertical stirrups closer than about 4 in.; the size of the stirrups should be chosen to avoid a closer spacing. When vertical stirrups are required over a comparatively short distance, it is good practice to space them uniformly over the entire distance, the spacing being calculated for the point of greatest shear (minimum spacing). If the web reinforcement is required over a long distance, and if the shear varies materially throughout this distance, it is more economical to compute the spacings required at several sections and to place the stirrups accordingly, in groups of varying spacing.

Where web reinforcement is needed, ACI Code 9.7.6.2 requires the legs of vertical shear reinforcement to be spaced no more than $d/2$ along the length of the member and no more than d across the width of the member, with neither exceeding 24 in. The maximum spacing across the width of the member is to provide a uniform transfer of force from the stirrups to the concrete across the beam web of wide

FIGURE 5.15
Maximum spacing of
web reinforcement for
$V_s \le 4\sqrt{f'_c}\,b_w d$ (a) along the
length of the member as
governed by diagonal crack
interception and (b) across
the width of the member.

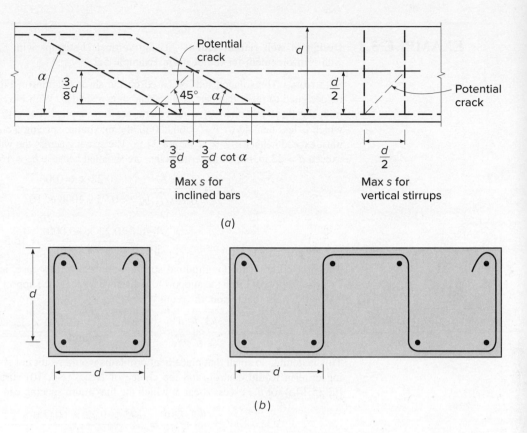

beams. When V_s exceeds $4\sqrt{f'_c}b_w d$, these maximum spacings are halved. Inclined stirrups or bars must be spaced so that every 45° line, representing a potential diagonal crack and extending from the mid-depth $d/2$ of the member to the longitudinal tension bars, is crossed by at least one line of web reinforcement. These limitations are shown in Fig. 5.15 for both vertical stirrups and inclined bars, for situations in which the excess shear does not exceed the stated limit.

For design purposes, Eq. (5.13) giving the minimum web-steel area A_v is more conveniently inverted to permit calculation of maximum spacing s for the selected A_v. Thus, for the usual case of vertical stirrups, with $V_s \le 4\sqrt{f'_c}\,b_w d$, the maximum spacing of stirrups is the smallest of

$$s_{\max} = \frac{A_v f_{yt}}{0.75\sqrt{f'_c}\,b_w} \le \frac{A_v f_{yt}}{50 b_w} \tag{5.15a}$$

$$s_{\max} = \frac{d}{2} \tag{5.15b}$$

$$s_{\max} = 24 \text{ in.} \tag{5.15c}$$

For longitudinal bars bent at 45°, Eq. (5.15b) is replaced by $s_{\max} = 3d/4$, as confirmed by Fig. 5.15.

To avoid excessive crack width in beam webs, ACI Code 20.2.2.4 limits the yield strength of the reinforcement to $f_{yt} = 60,000$ psi or less for reinforcing bars and 80,000 psi or less for welded wire reinforcement. In no case, according to ACI Code 22.5.1, is V_s to exceed $8\sqrt{f'_c}b_w d$, regardless of the amount of web steel used.

EXAMPLE 5.3

Design of web reinforcement. Using vertical U stirrups with $f_{yt} = 60,000$ psi, design the web reinforcement for the beam in Example 5.2.

SOLUTION. The solution will be based on the shear diagram in Fig. 5.14a. The stirrups must be designed to resist that part of the shear shown shaded. With No. 3 (No. 10) stirrups used for trial, the three maximum spacing criteria are first applied. For $\phi V_s = V_u - \phi V_c = 43,400$ lb, which is less than $4\phi\sqrt{f_c'}b_w d = 66,800$ lb, the maximum spacing along the length of the member must exceed neither $d/2 = 11$ in. nor 24 in. The spacing across the width of the member may not exceed $d = 22$ in. nor 24 in., criteria that are satisfied because $b_w = 16$ in. Also, from Eq. (5.15a),

$$s_{max} = \frac{A_v f_{yt}}{0.75\sqrt{f_c'}\,b_w} = \frac{0.22 \times 60,000}{0.75\sqrt{4000} \times 16} = 17.4 \text{ in.}$$

$$\leq \frac{A_v f_{yt}}{50 b_w} = \frac{0.22 \times 60,000}{50 \times 16} = 16.5 \text{ in.}$$

The first criterion for longitudinal spacing controls in this case, and a maximum spacing of 11 in. is imposed. From the support to a distance d from the support, the excess shear $V_u - \phi V_c$ is 43,400 lb. In this region, the required spacing is

$$s = \frac{\phi A_v f_{yt}\,d}{V_u - \phi V_c} = \frac{0.75 \times 0.22 \times 60,000 \times 22}{43,400} = 5.0 \text{ in.}$$

This is neither so small that placement problems would result nor so large that maximum spacing criteria would control, and the choice of No. 3 (No. 10) stirrups is confirmed. Solving Eq. (5.14a) for the excess shear at which the maximum spacing can be used gives

$$V_u - \phi V_c = \frac{\phi A_v f_{yt}\,d}{s} = \frac{0.75 \times 0.22 \times 60,000 \times 22}{11} = 19,800 \text{ lb}$$

With reference to Fig. 5.14a, this is attained at a distance x_1 from the point of zero excess shear, where $x_1 = 6.45 \times 19,800/60,600 = 2.10$ ft. This is 4.35 ft from the support face. With this information, a satisfactory spacing pattern can be selected. The first stirrup is usually placed at a distance $s/2$ from the support. The following spacing pattern is satisfactory:

$$1 \text{ space at 2 in.} = 2 \text{ in.}$$
$$7 \text{ spaces at 5 in.} = 35 \text{ in.}$$
$$2 \text{ spaces at 7 in.} = 14 \text{ in.}$$
$$4 \text{ spaces at 11 in.} = \underline{44 \text{ in.}}$$
$$\text{Total} = 95 \text{ in.} = 7 \text{ ft 11 in.}$$

The resulting stirrup pattern is shown in Fig. 5.14b. As an alternative solution, it is possible to plot a curve showing required spacing as a function of distance from the support. Once the required spacing at some reference section, say at the support, is determined,

$$s_0 = \frac{\phi A_v f_{yt}\,d}{V_u - \phi V_c} = \frac{0.75 \times 0.22 \times 60,000 \times 22}{94,000 - 33,400} = 3.59 \text{ in.}$$

it is easy to obtain the required spacings elsewhere. In Eq. (5.14a), only $V_u - \phi V_c$ changes with distance from the support. For uniform load, this quantity is a linear function of distance from the point of zero excess shear, 6.45 ft from the support face. Hence, at 1 ft intervals,

$$s_1 = 3.59 \times 6.45/5.45 = 4.25 \text{ in.}$$
$$s_2 = 3.59 \times 6.45/4.45 = 5.20 \text{ in.}$$
$$s_3 = 3.59 \times 6.45/3.45 = 6.70 \text{ in.}$$
$$s_4 = 3.59 \times 6.45/2.45 = 9.45 \text{ in.}$$
$$s_5 = 3.59 \times 6.45/1.45 = 15.97 \text{ in.}$$

FIGURE 5.16
Required stirrup spacings for
Example 5.3.

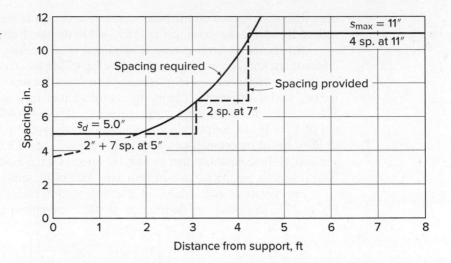

This is plotted in Fig. 5.16 together with the maximum spacing of 11 in., and a practical spacing pattern is selected. The spacing at a distance d from the support face is selected as the minimum requirement, in accordance with the ACI Code. The pattern of No. 3 (No. 10) U-shaped stirrups selected (shown on the graph) is identical with the previous solution. In most cases, the experienced designer would find it unnecessary actually to plot the spacing diagram of Fig. 5.16 and would select a spacing pattern directly after calculating the required spacing at intervals along the beam.

Although not required by the ACI Code, it is good design practice to continue the stirrups (at maximum spacing) through the middle region of the beam, even though the calculated shear is low. Doing so satisfies the dual purposes of providing continuing support for the top longitudinal reinforcement that is required wherever stirrups are used and providing additional shear capacity in the region to handle load cases not considered in developing the shear diagram. If this were done, the number of stirrups would increase from 14 to $16\frac{1}{2}$ per half-span (that is, one stirrup at midspan), respectively.

5.6 EFFECT OF AXIAL FORCES

The beams considered in the preceding sections were subjected to shear and flexure only. Reinforced concrete beams may also be subjected to axial forces, acting simultaneously with shear and flexure, due to a variety of causes. These include external axial loads, longitudinal prestressing, and restraint forces introduced as a result of shrinkage of the concrete or temperature changes. Beams may have their strength in shear significantly modified in the presence of axial tension or compression, as is evident from a review of Sections 5.1 through 5.4.

Prestressed concrete members are treated by somewhat specialized methods, according to present practice, based largely on results of testing prestressed concrete beams. They will be considered separately in Chapter 22, and only nonprestressed reinforced concrete beams will be treated here.

The main effect of axial load is to modify the diagonal cracking load of the member. It was shown in Section 5.3 that diagonal tension cracking will occur when the principal tensile stress in the web of a beam, resulting from combined action of shear and bending, reaches the tensile strength of the concrete. It is clear that the introduction of longitudinal force, which modifies the magnitude and direction of the principal tensile stresses, may significantly alter the diagonal cracking load. Axial compression will increase the cracking load, while axial tension will decrease it. The effect of axial compression or tension on the maximum tensile stress t in Eq. (3.10) can be easily visualized by adding the stress due to axial load to the stress due to flexure f.

For members carrying only flexural and shear loading, the shear force at which diagonal cracking occurs V_{cr} is predicted by Eq. (5.3a), based on a combination of theory and experimental evidence. Furthermore, for reasons that were explained in Section 5.4b, in beams with web reinforcement, the contribution of the concrete to shear strength V_c is taken equal to the diagonal cracking load V_{cr}. Thus, according to the ACI Code, values of $V_c = V_{cr}$ for members without axial load are given by Eqs. (5.12a), (5.12c), and (5.12d). Based on comparisons with tests, however, the ACI Code has added modified versions of these equations that include the effect of axial load. Indeed, the earlier equations represent the special case of zero axial load of those that follow.

For members with axial load and web reinforcement below $A_{v,\min}$, as given in Eq. (5.13a), the concrete contribution to shear strength is given by

$$V_c = \left[8\lambda_s \lambda(\rho_w)^{1/3} \sqrt{f_c'} + \frac{N_u}{6A_g} \right] b_w d \tag{5.16a}$$

For members with axial load and web reinforcement of at least $A_{v,\min}$, the concrete contribution to shear strength is given by

$$V_c = \left[2\lambda \sqrt{f_c'} + \frac{N_u}{6A_g} \right] b_w d \tag{5.16b}$$

or

$$V_c = \left[8\lambda(\rho_w)^{1/3} \sqrt{f_c'} + \frac{N_u}{6A_g} \right] b_w d \tag{5.16c}$$

where N_u is the axial load, taken as positive for compression and negative for tension, and A_g is the gross area of the concrete cross section. A_g does not include voids if a member is hollow. According to ACI Code 22.5.5, $N_u/6A_g$ may not be taken greater than $0.05 f_c'$, and V_c may not be taken greater than $5\lambda\sqrt{f_c'}b_w d$ nor less than zero.

EXAMPLE 5.4

Effect of axial forces on V_c. A beam with dimensions $b = 12$ in., $d = 24$ in., and $h = 27$ in., with $f_c' = 4000$ psi, carries a single concentrated factored load of 100 kips at midspan. Find the maximum shear strength of the concrete V_c at the first critical section for shear at a distance d from the support (a) if no axial forces are present, (b) if axial compression of 60 kips acts, and (c) if axial tension of 60 kips acts. In each case, compute V_c by both the more complex and simplified expressions of the ACI Code. Neglect the self-weight of the beam. At the section considered, tensile reinforcement consists of three No. 10 (No. 32) bars with a total area of 3.81 in.^{2}.

SOLUTION. At the critical section, $V_u = 50$ kips, while $\lambda_s = \sqrt{2/(1 + d/10)} = \sqrt{2/(1 + 24/10)} = 0.77$ and $\rho = 3.81/(12 \times 24) = 0.013$.

(a) If $N_u = 0$, Eq. (5.16a) gives

$$V_c = \left[8\lambda_s \lambda(\rho_w)^{1/3} \sqrt{f_c'} + \frac{N_u}{6A_g} \right] b_w d$$

$$= \left[8 \times 0.77 \times 1.0(0.013)^{1/3}\sqrt{4000} + 0 \right] 12 \times \frac{24}{1000} = 26.4 \text{ kips}$$

Since ϕV_c based on Eq. (5.16b) is below V_u, web reinforcement will be needed and Eq. (5.16b) will be used to calculate V_c.

$$V_c = \left[2\lambda\sqrt{f_c'} + \frac{N_u}{6A_g}\right]b_w d = [2 \times 1.0\sqrt{4000} + 0]\, 12 \times \frac{24}{1000} = 36.4 \text{ kips}$$

(b) With a compression force of 60 kips introduced and N_u taken as positive, $N_u/6A_g = 60{,}000/(6 \times 12 \times 24) = 35$ psi, which is less than the upper limit of $0.05 f_c' = 0.05 \times 4000 = 200$ psi.

V_c based on Eq. (5.16b) gives

$$V_c = \left[2\lambda\sqrt{f_c'} + \frac{N_u}{6A_g}\right]b_w d = [2 \times 1.0\sqrt{4000} + 35]\, 12 \times \frac{24}{1000} = 46.5 \text{ kips}$$

By inspection, V_c does not exceed $5\lambda\sqrt{f_c'}b_w d$.

(c) With an axial tension of 60 kips acting (N_u is now negative), the reduced V_c is also found using Eq. (5.16b).

$$V_c = \left[2\lambda\sqrt{f_c'} + \frac{N_u}{6A_g}\right]b_w d = [2 \times 1.0\sqrt{4000} - 35]\, 12 \times \frac{24}{1000} = 26.3 \text{ kips}$$

a reduction of nearly 30 percent from the value for $N_u = 0$.

In all cases above, the strength reduction factor $\phi = 0.75$ must be applied to V_c to obtain the design strength.

5.7 BEAMS WITH VARYING DEPTH

Reinforced concrete members having varying depth are frequently used in the form of haunched beams for bridges or portal frames, as shown in Fig. 5.17a, as precast roof girders such as shown in Fig. 5.17b, or as cantilever slabs. Generally the depth increases in the direction of increasing moments. For beams with varying depth, the inclination of the internal compressive and tensile stress resultants may significantly affect the shear for which the beam should be designed. In addition, the shear resistance of such members may differ from that of prismatic beams.

Figure 5.17c shows a cantilever beam, with fixed support at the left end, carrying a single concentrated load P at the right. The depth increases linearly in the direction of increasing moment. In such cases, the internal tension in the steel and the compressive stress resultant in the concrete are inclined, and introduce components transverse to the axis of the member. With reference to Fig. 5.17d, showing a short length dx of the beam, if the slope of the top surface is θ_1 and that of the bottom is θ_2, the net shear force $\overline{V}_u$ for which the beam should be designed is very nearly equal to

$$\overline{V}_u = V_u - T\tan\theta_1 - C\tan\theta_2$$

where V_u is the external shear force equal to the load P here, and $C = T = M_u/z$. The internal lever arm $z = d - a/2$ as usual. Thus, in a case for which the beam depth increases in the direction of increasing moment, the shear for which the member should be designed is approximately

$$\overline{V}_u = V_u - \frac{M_u}{z}(\tan\theta_1 + \tan\theta_2) \tag{5.17a}$$

FIGURE 5.17
Effect of varying beam depth
on shear.

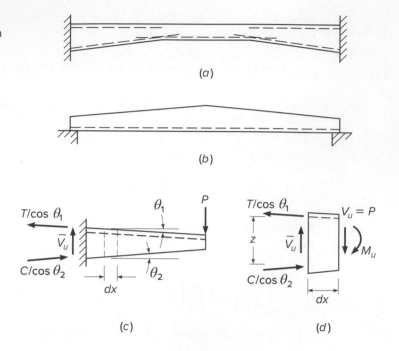

For the infrequent case in which the member depth decreases in the direction of increasing moment, it is easily confirmed that the corresponding equation is

$$\overline{V}_u = V_u + \frac{M_u}{z} (\tan \theta_1 + \tan \theta_2) \tag{5.17b}$$

These equations are approximate because the direction of the internal forces is not exactly as assumed; however, the equations may be used without significant error provided the slope angles do not exceed about 30°.

There has been very little research studying the shear strength of beams having varying depth. Tests reported in Ref. 5.19 on simple span beams with haunches at slopes up to about 15° and with depths both increasing and decreasing in the direction of increasing moments indicate no appreciable change in the cracking load V_{cr} compared with that for prismatic members. Furthermore, the strength of the haunched beams, which contained vertical stirrups as web reinforcement, was not significantly decreased or increased, regardless of the direction of decreasing depth. Based on this information, *it appears safe to design beams with varying depth for shear using equations for V_c and V_s developed for prismatic members*, provided the actual depth d at the section under consideration is used in the calculations.

5.8   ALTERNATIVE MODELS FOR SHEAR ANALYSIS AND DESIGN

The ACI Code method of design for shear and diagonal tension in beams, presented in preceding sections of this chapter, is essentially empirical. While generally leading to safe designs, the ACI Code "$V_c + V_s$" approach lacks a physical model for the behavior of beams subject to shear combined with bending, and its shortcomings are now generally recognized. The "concrete contribution" V_c is generally considered to be some combination of force transfer by dowel action of the main steel, aggregate interlock along a diagonal crack, and shear in the uncracked concrete beyond the end of the crack. The values of each contribution are not identified. Furthermore, as discussed in Section 5.4, Eqs. (5.12a), (5.12c), and (5.12d) ignore the influence of $V_u d/M_u$ (Ref. 5.3).

Ad hoc procedures are built into the ACI Code to adjust for some of these deficiencies, but it follows that it is necessary to include equations, also empirically developed for the most part, for specific classes of members (such as, deep beams vs. normal beams, beams with axial loads, prestressed vs. nonprestressed beams, high-strength concrete beams)—with restrictions on the range of applicability of such equations. And it is necessary to incorporate seemingly arbitrary provisions for the maximum nominal shear stress and for the extension of flexural reinforcement past the theoretical point of need. The end result is that the number of ACI Code equations for shear design has grown from 4 prior to 1963 to 38 in the current Code.

With this as background, attention has been given to the development of design approaches based on rational behavioral models, generally applicable, rather than on empirical evidence alone (Ref. 5.6).

The *truss model* was originally introduced by Ritter (Ref. 5.20) and Morsch (Ref. 5.21) at the turn of the last century. A simplified version has long provided the basis for the ACI Code design of shear steel. The essential features of the truss model are reviewed with reference to Fig. 5.18a, which shows one-half the span of a simply supported, uniformly loaded beam. The combined action of flexure and shear produces the pattern of cracking shown. Reinforcement consists of the main flexural steel near the tension face and vertical stirrups distributed over the span.

The structural action can be represented by the truss of Fig. 5.18b, with the main steel providing the tension chord, the concrete top flange acting as the compression chord, the stirrups providing the vertical tension web members, and the concrete between inclined cracks acting as 45° compression diagonals. The truss is formed by lumping all the stirrups cut by section a-a into one vertical member and all the diagonal concrete struts cut by section b-b into one compression diagonal. Experience shows that for typical cases, the results of the model described are quite conservative, particularly for beams with small amounts of web reinforcement. As noted above, in the ACI Code the observed excess shear capacity is taken equal to the shear at the commencement of diagonal cracking and is referred to as the *concrete contribution* V_c.

The truss concept has been greatly extended by the work of Schlaich, Marti, Collins, MacGregor, and others (Refs. 5.6, 5.22 to 5.27). It was realized that the angle of inclination of the concrete struts is generally not 45° but may range between about 25° and 65°, depending to a large extent on the arrangement of reinforcement. This led to what has become known as the *variable-angle truss model*, shown in Fig. 5.18c, which illustrates the five basic components of the improved model: (a) struts or concrete compression members uniaxially loaded; (b) ties or steel tension members; (c) joints at the intersection of truss members assumed to be pin-connected; (d) compression fans, which form at "disturbed" regions, such as at the supports or under concentrated loads, transmitting the forces into the beam; and (e) diagonal compression fields, occurring where parallel compression struts transmit force from one stirrup to another. As in the ACI Code development, stirrups are typically assumed to reach yield stress at failure. With the force in all the verticals known and equal to $A_v f_{yt}$, the truss of Fig. 5.18c becomes statically determinate. Direct design equations can be based on the variable-angle truss model for ordinary cases. The model also permits direct numerical solution for the required reinforcement for special cases. The truss model does not include components of the shear failure mechanism such as aggregate interlock and friction,

FIGURE 5.18
Truss model for beams with
web reinforcement:
(*a*) uniformly loaded beam;
(*b*) simple truss model; and
(*c*) more realistic model.

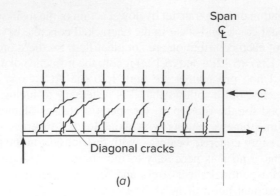

(*a*)

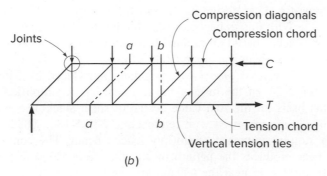

(*b*)

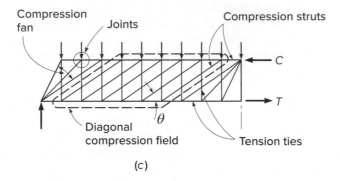

(*c*)

dowel action of the longitudinal steel, and shear carried across uncracked concrete.
Furthermore, in the format originally proposed, the truss model does not account for
compatibility requirements; that is, it is based on *plasticity theory*. One form of the truss
model is incorporated in Chapter 23 of the ACI Code; strut-and-tie models are discussed
in detail in Chapter 17.

a. Compression Field Theory

The Canadian National Standard for reinforced concrete (Ref. 5.28) includes a method
of shear design that is essentially the same as the present ACI method but also includes
an alternative "general method" based on the variable-angle truss and the *compression field theory* (Refs. 5.25 and 5.29). The latter is incorporated in *AASHTO LRFD
Bridge Design Specifications* (Ref. 5.12). In its complete form, known as the *modified compression field theory*, it accounts for requirements of compatibility as well as
equilibrium and incorporates stress-strain characteristics of both materials. Thus, it is
capable of predicting not only the failure load but also the complete load-deformation

FIGURE 5.19
Basis of compression field theory for shear: (*a*) beam with shear and longitudinal steel; (*b*) tension in horizontal bars due to shear; (*c*) diagonal compression on beam web; (*d*) vertical tension in stirrups; and (*e*) equilibrium diagram of forces due to shear. (*Adapted from Ref. 5.25.*)

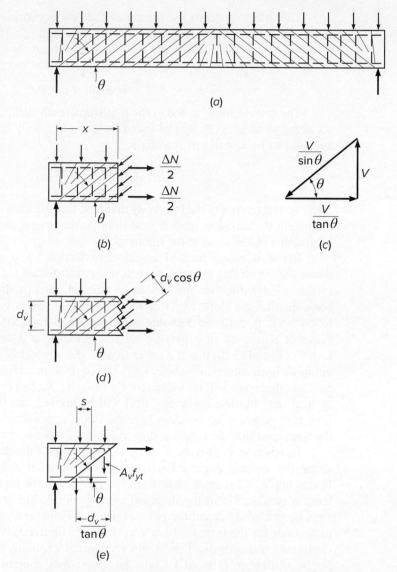

response. The most basic elements of the compression field theory, applied to members carrying combined flexure and shear, will be clear from Fig. 5.19. Figure 5.19*a* shows a simple-span concrete beam, reinforced with longitudinal bars and transverse stirrups, and carrying a uniformly distributed loading along the top face. The light diagonal lines are an idealized representation of potential tensile cracking in the concrete.

Figure 5.19*b* illustrates that the net shear V at a section a distance x from the support is resisted by the vertical component of the diagonal compression force in the concrete struts. The horizontal component of the compression in the struts must be equilibrated by the total tension force ΔN in the longitudinal steel. Thus, with reference to Fig. 5.19*b* and *c*, the magnitude of the longitudinal tension resulting from shear is

$$\Delta N = \frac{V}{\tan \theta} = V \cot \theta \tag{5.18}$$

where θ is the angle of inclination of the diagonal struts. These forces superimpose on the longitudinal forces owing to flexure, not shown in Fig. 5.19*b*.

The effective depth for shear calculations, according to this method, is taken at the distance between longitudinal force resultants d_v. Thus, from Fig. 5.19d, the diagonal compressive stress in a web having width b_v is

$$f_d = \frac{V}{b_v d_v \sin \theta \cos \theta} \qquad (5.19)$$

The tensile force in the vertical stirrups, each having area A_v and assumed to act at the yield stress f_{yt}, can be found from the free body of Fig. 5.19e. With stirrups assumed to be at uniform spacing s,

$$A_v f_{yt} = \frac{Vs \tan \theta}{d_v} \qquad (5.20)$$

Note, with reference to the free-body diagram, that the transverse reinforcement within the length $d_v / \tan \theta$ can be designed to resist the lowest shear that occurs within this length, that is, the shear at the right end.

In the ACI Code method developed in Section 5.4, it was assumed that the angle θ was 45°. With that assumption, and if d is substituted for d_v, Eq. (5.20) is identical to that used earlier for the design of vertical stirrups. It is generally recognized, however, that the slope angle of the compression struts is not necessarily 45°, and following Refs. 5.12 and 5.28 that angle can range from 20° to 75°, provided the same value of θ is used in satisfying all requirements at a section. It is evident from Eqs. (5.18) and (5.20) that if a lower slope angle is selected, less vertical reinforcement but more horizontal reinforcement will be required. In addition, the compression in the concrete diagonals will be increased. Conversely, if a higher slope angle is used, more vertical steel but less horizontal steel will be needed, and the diagonal thrust will be less. It is generally economical to use a slope angle θ somewhat less than 45°, with the limitation that the concrete diagonal struts not be overstressed in compression.

In addition to providing an improved basis for the design of reinforcement for shear, the variable angle truss model gives important insights into detailing needs. For example, it becomes clear from the above that the increase in longitudinal steel tension resulting from the diagonal compression in the struts requires that flexural steel be extended beyond the point at which it is theoretically not needed for flexure, to account for the increased horizontal tensile force resulting from the thrust in the compression diagonals. This is not recognized explicitly in the ACI Code method for beam design. (The ACI Code, however, does contain the requirement that the flexural steel be extended a distance equal to the greater of d or 12 bar diameters beyond the point indicated by flexural requirements.) Also, it is clear from the basic concept of the truss model that stirrups must be capable of developing their full tensile strength throughout the entire stirrup height. For wide beams, focus on truss action indicates that special attention should be given to lateral distribution of web reinforcement. It is often the practice to use conventional U stirrups for wide beams, with the vertical tension from the stirrups concentrated around the outermost bars. According to the discussion above, diagonal compression struts transmit forces only at the joints. Lack of stirrup joints at the interior of the wide-beam web would force joints to form only at the exterior longitudinal bars, which would concentrate the diagonal compression at the outer faces of the beam and possibly result in premature failure. It is best to form a truss joint at each of the longitudinal bars, and as in the ACI Code, multiple leg stirrups should always be used in wide beams (see Fig. 5.9c).

References 5.12 and 5.28 incorporate a refined version of the approach just described, known as the modified compression field theory (MCFT), in which the cracked concrete is treated as a new material with its own stress-strain characteristics, including the ability to carry tension following crack formation. The compressive strength and the stress-strain curve of the concrete in the diagonal compression struts decrease as the diagonal tensile strain in the concrete increases. Equilibrium, compatibility, and constitutive relationships are formulated in terms of average stresses and average strains. Variability in the angle of inclination of the compression struts and stress-strain softening effects in the response of the concrete are taken into account. Consideration is also given to local stress conditions at crack locations. The method is capable of accurately predicting the response of complex elements such as shear walls, diaphragms, and membrane elements subjected to in-plane shear and axial loads through the full range of loading, from zero load to failure (Refs. 5.26 and 5.27). The version of the method adopted in Ref. 5.12 has been simplified to allow its use for routine design.

b.   Design Provisions

The version of the MCFT adopted in the *AASHTO LRFD Bridge Design Specifications* (Ref. 5.12) is, like the shear provisions in the ACI Code, based on nominal shear capacity, with V_n equal to the lesser of

$$V_n = V_c + V_s \tag{5.21}$$

$$V_n = 0.25 f_c' b_v d_v \tag{5.22}$$

where b_v = web width (the same as b_w in the ACI Code) and d_v = effective depth in shear, taken as equal to the flexural lever arm (the distance between the centroids of the tensile and compressive forces), but not less than the greater of $0.9d$ or $0.72h$.

The values of V_c and V_s differ from those used by the ACI, with

$$V_c = \beta \sqrt{f_c'} \, b_v d_v \tag{5.23}$$

and

$$V_s = \frac{A_v f_{yt} d_v (\cot \theta + \cot \alpha) \sin \alpha}{s} \tag{5.24}$$

where A_v, f_{yt}, s, α, and θ are as defined before. β is the *concrete tensile stress factor* and is based on the ability of diagonally cracked concrete to resist tension, which also controls the angle of the diagonal tension crack θ. In Ref. 5.12, the values of β and θ are determined based on the strain in the longitudinal tension reinforcement, which can be approximated by[†]

$$\varepsilon_s = \frac{|M_u|/d_v - 0.5 N_u + |V_u|}{E_s A_s} \le 0.006 \tag{5.25}$$

The sign convention for N_u is the same as used in Section 5.6 and the ACI Code: compression is positive and tension is negative (the opposite sign convention is used in Ref. 5.12). M_u should not be taken less than $V_u d_v$; when calculating A_s, the area of bars terminated less than their development length (see Chapter 6) from the section under consideration should be reduced in proportion to the decreased development; ε_s should be taken as zero if the value calculated in Eq. (5.25) is negative; and ε_s should be doubled if N_u is high enough to cause cracking to the flexural compression face of the member. For sections closer than d_v to the face of the support, ε_s calculated at d_v from the face of the support may be used to determine β and θ.

[†] Equation (5.25) is a simplification of $\varepsilon_s = \frac{|M_u/d_v| - 0.5 N_u + 0.5 |V_u| \cot \theta}{E_s A_s}$, with $0.5|V_u| \cot \theta$ approximated by $|V_u|$. The simplification eliminates the need for an iterative solution between ε_s and θ.

FIGURE 5.20
Equilibrium diagram for calculating tensile force in reinforcement. (*Adapted from Ref. 5.12.*)

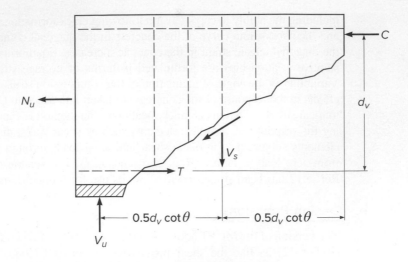

For members with at least the minimum shear reinforcement, the concrete tensile stress factor is given by

$$\beta = \frac{4.8}{1 + 750\varepsilon_s} \tag{5.26}$$

The angle θ, in degrees, is given by

$$\theta = 29 + 3500\varepsilon_s \tag{5.27}$$

As shown in Eq. (5.18), the strength of the *longitudinal* reinforcement must be adequate to carry the additional forces induced by shear. Referring to Fig. 5.20, this leads to

$$A_s f_y \geq T = \frac{|M_u|}{\phi_f} - \frac{0.5N_u}{\phi_c} + \left(\frac{|V_u|}{\phi_v} - 0.5V_s \right) \cot \theta \tag{5.28}$$

where ϕ_f, ϕ_c, and ϕ_v are, respectively, the capacity reduction factors for flexure, axial load (tension or compression), and shear. V_s need not be taken greater than V_u/ϕ. Since the inclination of the compression struts changes, tension in the longitudinal reinforcement does not exceed that required to resist the maximum moment alone.

For members with less than the minimum transverse reinforcement, the angle θ is given by Eq. (5.27), while the value of β becomes a function of ε_s and a crack spacing parameter s_{xe}.

$$\beta = \frac{4.8}{1 + 750\varepsilon_s} \frac{51}{39 + s_{xe}} \tag{5.29}$$

The crack spacing parameter is

$$s_{xe} = s_x \frac{1.38}{a_g + 0.63} \tag{5.30}$$

where 12.0 in. $\leq s_{xe} \leq 80.0$ in., s_x = lesser of the shear depth d_v or the spacing between layers of longitudinal crack control reinforcement, each layer with an area of steel of at least $0.003 b_v s_x$, and a_g = maximum size of the coarse aggregate. Note that $s_{xe} = s_x$ for $\frac{3}{4}$ in. coarse aggregate.

Since θ is not, in general, equal to 45°, the critical section might appropriately be taken as $d_v \cot \theta$ from the face of the support if all the load were applied to the upper surface of the member. For simplicity, however, the critical section is taken a distance d_v from the face of the support when the reaction introduces compression into the end region of the member, similar to the loading cases shown in Fig. 5.13a and b. For all other cases, the crucial section is taken at the face of the support, as shown in Fig. 5.13c to f.

AASHTO requires a minimum amount of transverse reinforcement $A_v = \sqrt{f_c'}\,b_v s/f_{yt}$ (compared to $0.75\sqrt{f_c'}\,b_w s/f_{yt}$ for ACI), when $V_u > 0.5\phi V_c$, and specifies maximum spacings of transverse reinforcement of $s \le 0.8d_v \le 24$ in. when $v_u < 0.125f_c'$ and $s \le 0.4d_v \le 12$ in. when $v_u \ge 0.125f_c'$. AASHTO allows f_{yt} up to 75 ksi (compared to 60 ksi for ACI). Because the predictions obtained with the MCFT are generally more accurate than those obtained with the ACI method, AASHTO allows the use of $\phi = 0.90$ for shear in normalweight concrete and $\phi = 0.80$ in lightweight concrete.

EXAMPLE 5.5   **Design by modified compression field approach.** Resolve the problem given in Examples 5.2 and 5.3 based on the MCFT. Use ACI load factors and $\phi = 0.9$ for shear, as used in *AASHTO LRFD Bridge Design Specifications* (Ref. 5.12). Assume an aggregate size a_g of $\frac{3}{4}$ in.

SOLUTION. For simplicity, the effective depth for shear d_v will be set at the minimum allowable value $= 0.9d = 0.9 \times 22 = 19.8$ in. The values of M_u and V_u are tabulated in Table 5.1.

The critical section for shear is located a distance $d_v = 19.8$ in. $= 1.65$ ft from the support where $V_u = 94 - 9.4 \times 1.65 = 78.5$ kips. Calculating $0.125f_c'b_v d_v = 0.125 \times 4000 \times 16 \times 19.8 = 158,400$ lb leads to maximum stirrup spacing equal to the smallest of $0.8d_v = 0.8 \times 19.8 = 15.8$ in., 24 in., and [for No. 3 (No. 10) stirrups]

$$s_{max} = \frac{A_v f_{yt}}{\sqrt{f_c'}\,b_v} = \frac{0.22 \times 60,000}{\sqrt{4000} \times 16} = 13.0 \text{ in.}$$

TABLE 5.1

Modified compression field design example using $\phi = 0.9$ for shear

Distance from Support, ft	M_u, ft-kips	V_u, kips	$\varepsilon_s \times 1000$	θ	ϕV_C for at Least Minimum Stirrups			ϕV_C for Less Than Minimum Stirrups		
					β	ϕV_c, kips	V_s, kips	s, in.	β	ϕV_c, kips	$\phi V_c/2$, kips
0	0	94.0	0.85	32.0	2.93	52.8	45.7	9.2	2.54	45.8	22.9
1	89	84.6	0.77	31.7	3.05	55.0	32.9	12.9	2.64	47.7	23.8
1.65[†]	144	78.5	0.75	31.6	3.07	55.4	25.6	16.5	2.66	48.1	24.0
2	169	75.2	0.80	31.8	2.99	54.0	23.6	17.9	2.60	46.8	23.4
3	240	65.8	0.96	32.3	2.80	50.4	17.1	24.2	2.43	43.7	21.9
4	301	56.4	1.08	32.8	2.65	47.8	9.5	42.6	2.30	41.5	20.7
5	353	47.0	1.18	33.1	2.55	45.9	1.2	336	2.21	39.8	19.9
6	395	37.6	1.25	33.4	2.47	44.6	—	—	2.15	38.7	19.4
7	428	28.2	1.30	33.6	2.43	43.8	—	—	2.11	38.0	19.0
8	451	18.8	1.32	33.6	2.41	43.5	—	—	2.09	37.7	18.8
9	465	9.4	1.32	33.6	2.41	43.5	—	—	2.09	37.7	18.9
10	470	0.0	1.29	33.5	2.44	44.0	—	—	2.12	38.2	19.1

[†]d_v from face of support.

Using Eq. (5.25), the strain in the longitudinal tension steel is approximated as

$$\varepsilon_s = \frac{|M_u|/19.8 + |V_u|}{29{,}000 \times 7.62}$$

with M_u and V_u in in-kips and kips, respectively.

The values of M_u, V_u, and ε_s are tabulated in Table 5.1. These values are used to calculate θ using Eq. (5.27) and β using Eqs. (5.26) and (5.29) for sections with and without minimum stirrups, respectively. Where the section meets the minimum stirrup criterion, the values of β are used to calculate the values of V_c, which are then used, along with the values of θ, to calculate V_s and the required stirrup spacing s (see Table 5.1).

For transverse reinforcement less than the minimum, the values of β are based on ε_s and s_x. The latter may be taken as the lesser of d_v or the spacing of longitudinal crack control reinforcement. In this case, $d_v = 19.8$ in. controls since crack control reinforcement is not used. The equivalent crack spacing parameter $s_{xe} = s_x$ because $a_g = 0.75$ in. These values of β are used to determine the point where $\phi V_c/2 \geq V_u$, the point at which stirrups may be terminated (Table 5.1). The values of V_u, ϕV_c with at least minimum stirrups, and $\phi V_c/2$ for less than minimum stirrups are plotted in Fig. 5.21a. The following stirrup spacings can be used for this case:

$$
\begin{aligned}
1 \text{ space at 6 in.} &= 6 \text{ in.} \\
6 \text{ spaces at 13 in.} &= \underline{78 \text{ in.}} \\
\text{Total} &= 84 \text{ in.} = 7 \text{ ft}
\end{aligned}
$$

For this example, V_s is selected based on V_u at each point, not the minimum V_u on a crack with angle θ. This simplifies the design procedure and results in a somewhat more conservative design. Even so, only 7 No. 3 (No. 10) stirrups are needed, or 9 stirrups if the stirrups are continued at the maximum spacing through the middle region of the beam. These values compare favorably with the minimum number of stirrups per half-span, 11 and 14, previously calculated (Example 5.3) using the two methods required by the ACI Code. The resulting stirrup pattern is shown in Fig. 5.21b.

FIGURE 5.21
Modified compression field design for Example 5.5.

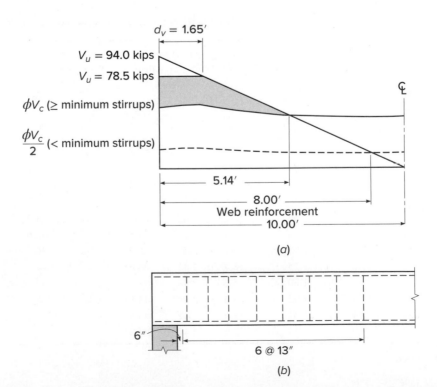

(a)

(b)

By way of comparison, had $\phi_{\text{shear}} = 0.75$ been used in this example, the stirrup spacing would have been

$$
\begin{aligned}
&\text{1 space at 4 in.} &= \quad\;\; 4 \text{ in.} \\
&\text{4 spaces at 9 in.} &= \; 36 \text{ in.} \\
&\text{4 spaces at 13 in.} &= \; \underline{52 \text{ in.}} \\
&\text{Total} &= \; 92 \text{ in.} = 7 \text{ ft } 8 \text{ in.}
\end{aligned}
$$

for a total of 9 stirrups.

The MCFT recognizes that shear increases the force in the flexural steel, although, as explained earlier, the maximum tensile force in the steel is not affected. Equation (5.28) should be used to calculate the tensile force T along the span, which will then govern the locations where tensile steel may be terminated. This will be discussed further in Chapter 6.

The MCFT is not included in the ACI Code. ACI Code 1.10.1, however, permits the use of "any system of design or construction . . . , the adequacy of which has been shown by successful use or by analysis or test," if approved by the appropriate building official. The application of the MCFT in Canada and in U.S. bridge practice provides the evidence needed to demonstrate "successful use."

5.9 SHEAR-FRICTION DESIGN METHOD

Generally, in reinforced concrete design, shear is used merely as a convenient measure of diagonal tension, which is the real concern. In contrast, there are circumstances such that direct shear may cause failure of reinforced concrete members. Such situations occur commonly in precast concrete structures, particularly in the vicinity of connections, as well as in composite construction combining cast-in-place concrete with either precast concrete or structural steel elements. Potential failure planes can be established for such cases along which direct shear stresses are high, and failure to provide adequate reinforcement across such planes may produce disastrous results.

The necessary reinforcement may be determined on the basis of the *shear-friction method* of design (Refs. 5.30 to 5.36). The basic approach is to assume that the concrete may crack in an unfavorable manner, or that slip may occur along a predetermined plane of weakness. Reinforcement must be provided crossing the potential or actual crack or shear plane to prevent direct shear failure.

The shear-friction theory is very simple, and the behavior is easily visualized. Figure 5.22a shows a cracked block of concrete, with the crack crossed by reinforcement. A shear force V_n acts parallel to the crack, and the resulting tendency for the upper block to slip relative to the lower is resisted largely by friction along the concrete interface at the crack. Since the crack surface is naturally rough and irregular, the effective coefficient of friction may be quite high. In addition, the irregular surface will cause the two blocks of concrete to separate slightly, as shown in Fig. 5.22b.

If reinforcement is present normal to the crack, then slippage and subsequent separation of the concrete will stress the steel in tension. Tests have confirmed that well-anchored steel will be stressed to its yield strength when shear failure is obtained (Ref. 5.32). The resulting tensile force sets up an equal and opposite pressure between the concrete faces on either side of the crack. It is clear from the free body of Fig. 5.22c that the maximum value of this interface pressure is $A_{vf}f_y$, where A_{vf} is the total area of steel crossing the crack and f_y is its yield strength.

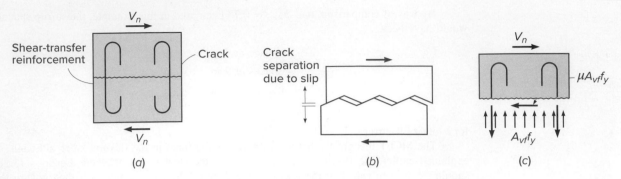

FIGURE 5.22
Basis of shear-friction design method: (*a*) applied shear; (*b*) enlarged representation of crack surface; and (*c*) free-body sketch of concrete above crack.

The concrete resistance to sliding may be expressed in terms of the normal force times a coefficient of friction μ. By setting the summation of horizontal forces equal to zero

$$V_n = \mu A_{vf} f_y \qquad (5.31)$$

Based on tests, μ may be taken as 1.4 for cracks in monolithic concrete, but V_n should not be assumed to be greater than $0.2f_c' A_c$, $(480 + 0.08f_c')A_c$, or $1600A_c$ (Refs. 5.30, 5.35, and 5.36).

The relative movement of the concrete on opposite sides of the crack also subjects the individual reinforcing bars to shearing action, and the dowel resistance of the bars to this shearing action contributes to shear resistance. However, it is customary to neglect the dowel effect for simplicity in design and to compensate for this by using an artificially high value of the friction coefficient.

The provisions of ACI Code 22.9 are based on Eq. (5.31). The design strength is equal to ϕV_n, where $\phi = 0.75$ for shear-friction design, and V_n must not exceed the smallest of $0.2f_c' A_c$, $(480 + 0.08f_c')A_c$, and $1600A_c$ for monolithic or intentionally roughened normalweight concrete or the smaller of $0.2f_c' A_c$ and $800A_c$ lb for other cases. When concretes of different strengths are cast against each another, V_n should be based on the lower value of f_c'. Recommendations for friction factor μ are as follows:

Concrete placed monolithically	1.4λ
Concrete placed against hardened concrete with surface intentionally roughened	1.0λ
Concrete placed against hardened concrete not intentionally roughened	0.6λ
Concrete anchored to as-rolled structural steel by headed studs or reinforcing bars	0.7λ

where λ is 1.0 for normalweight concrete. In other cases, λ is as described in Section 5.5a and specified in ACI Code 19.2.4, but not greater than 0.85. The yield strength of the reinforcement f_y may not exceed 60,000 psi. Direct tension across the shear plane, if present, must be carried by additional reinforcement, and permanent net compression across the shear plane may be taken as additive to the force in the shear-friction reinforcement $A_{vf} f_y$ when calculating the required A_{vf}.

FIGURE 5.23
Shear-friction reinforcement
inclined with respect to
crack face.

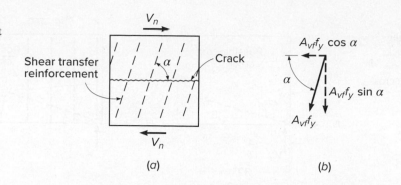

(a)   (b)

When shear is transferred between concrete newly placed against hardened concrete, the surface roughness is an important variable; an intentionally roughened surface is defined to have a full amplitude of approximately $\frac{1}{4}$ in. In any case, the old surface must be clean and free of laitance. When shear is to be transferred between as-rolled steel and concrete, the steel must be clean and without paint, according to ACI Code 25.5.6.1(d).

If V_u is the shear force to be resisted at factored loads, then with $V_u = \phi V_n$, the required steel area is found by transposition of Eq. (5.31):

$$A_{vf} = \frac{V_u}{\phi \mu f_y} \tag{5.32}$$

In some cases, the shear-friction reinforcement may not cross the shear plane at 90° as described in the preceding paragraphs. If the shear-friction reinforcement is inclined to the shear plane so that the shear force is applied in the direction to increase tension in the steel, as in Fig. 5.23a, then the component of that tension parallel to the shear plane, shown in Fig. 5.23b, contributes to the resistance to slip. Then the shear strength may be computed from

$$V_n = A_{vf} f_y (\mu \sin \alpha + \cos \alpha) \tag{5.33}$$

in lieu of Eq. (5.31). Here α is the angle between the shear-friction reinforcement and the shear plane. If α is larger than 90°, that is, if the inclination of the steel is such that the tension in the bars tends to be reduced by the shear force, then the assumption that the steel stress equals f_y is not valid, and a better arrangement of bars should be made.

Certain precautions should be observed in applying the shear-friction method of design. Reinforcement, of whatever type, should be well anchored to develop the yield strength of the steel, by the full development length or by hooks or bends, in the case of reinforcing bars, or by proper heads and welding, in the case of studs joining concrete to structural steel. The concrete should be well confined, and the liberal use of hoops has been recommended (Ref. 5.30). Care must be taken to consider all possible failure planes and to provide sufficient well-anchored steel across these planes.

EXAMPLE 5.6   **Design of beam bearing detail.**   A precast beam must be designed to resist a support reaction, at factored loads, of $V_u = 100$ kips applied to a 3×3 steel angle, as shown in Fig. 5.24. In lieu of a calculated value, a horizontal force N_{uc}, owing to restrained volume change, will be assumed to be 20 percent of the vertical reaction, or 20 kips. Determine the required auxiliary reinforcement, using steel of yield strength $f_y = 60,000$ psi. Concrete strength $f_c' = 5000$ psi.

SOLUTION.   A potential crack will be assumed at 20°, initiating at a point 4 in. from the end of the beam, as shown in Fig. 5.24a. The total required steel A_{vf} is the sum of that required to

FIGURE 5.24
Design of beam bearing
shoe: (*a*) diagonal crack;
(*b*) horizontal crack;
(*c*) reinforcement; (*d*) cross
section.

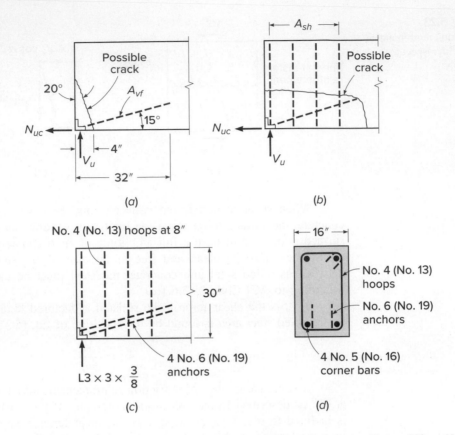

(a)

(b)

(c)

(d)

resist the resultant of V_u and N_{uc} acting parallel to the cracks = $V_u \cos 20° + N_{uc} \sin 20°$. Equation (5.32) is modified accordingly:

$$A_{vf} = \frac{V_u \cos 20° + N_{uc} \sin 20°}{\phi \mu f_y}$$

$$= \frac{100 \times 0.940 + 20 \times 0.340}{0.75 \times 1.4 \times 60} = \frac{101 \text{ kips}}{63 \text{ ksi}}$$

$$= 1.60 \text{ in}^2$$

The net compression normal to the potential crack would be no less than $V_u \sin 20° - N_{uc} \cos 20° = 15.4$ kips. This could be counted upon to reduce the required shear-friction steel, according to the ACI Code, but it will be discounted conservatively here. Four No. 6 (No. 19) bars will be used, providing an area of 1.76 in². They will be welded to the 3 × 3 angle and will extend into the beam a sufficient distance to develop the yield strength of the bars. According to the ACI Code, the development length for a No. 6 (No. 19) bar is 26 in., 32 in. without the ψ_s factor (see Chapter 6). Considering the uncertainty of the exact crack location, the bars will be extended 32 in. into the beam as shown in Fig. 5.24*a*. The bars will be placed at an angle of 15° with the bottom face of the member. For the crack oriented at an angle of 20°, as assumed, the area of the crack is

$$A_c = 16\left(\frac{4}{\sin 20°}\right) = 187 \text{ in}^2$$

Thus, according to the ACI Code, the maximum nominal shear strength of the surface is not to exceed $V_n = 0.2f_c' A_c = 187$ kips, $V_n = (480 + 0.08f_c')A_c = 165$ kips, or $V_n = 1600A_c = 299$ kips. The maximum design strength to be used is $\phi V_n = 0.75 \times 165 = 124$ kips. As calculated earlier, the applied shear on the interface at factored loads is

$$V_u = 100 \cos 20° + 20 \sin 20° = 101 \text{ kips}$$

and so the design is judged satisfactory to this point.

A second possible crack must be considered, as shown in Fig. 5.24b, resulting from the tendency of the entire anchorage weldment to pull horizontally out of the beam.

The required steel area A_{sh} and the concrete shear stress will be calculated based on the development of the full yield tension in the bars A_{vf}. (Note that the factor ϕ need not be used here because it has already been introduced in computing A_{vf}.)

$$A_{sh} = \frac{A_{vf}f_y \cos 15°}{\mu f_y}$$

$$= \frac{1.76 \times 0.966}{1.4}$$

$$= 1.21 \text{ in}^2$$

Four No. 4 (No. 13) hoops will be used, providing an area of 1.60 in^2.

The maximum shear force that can be transferred, according to the ACI Code limits, will be based conservatively on a horizontal plane 32 in. long. No strength reduction factor need be included in the calculation of this maximum value because it was already introduced in determining the steel area A_{vf} by which the shear force is applied. Accordingly,

$$V_n \leq (480 + 0.08f_c') \times 16 \times 32 = 451 \text{ kips}$$

The maximum shear force that could be applied in the given instance is the value used to calculate A_{sh},

$$V_u = 1.76 \times 60 \cos 15° = 102 \text{ kips}$$

which is well below the specified maximum.

The first hoop will be placed 2 in. from the end of the member, with the others spaced at 8 in., as shown in Fig. 5.24c. Also shown in Fig. 5.24d are four No. 5 (No. 16) bars that will provide anchorage for the hoop steel.

REFERENCES

5.1. "Shear and Diagonal Tension," pt. 2, ACI-ASCE Committee 326, *J. ACI*, vol. 59, no. 2, 1962, pp. 277–333.

5.2. B. Bresler and J. G. MacGregor, "Review of Concrete Beams Failing in Shear," *J. Struct. Div.*, ASCE, vol. 93, no. ST1, 1967, pp. 343–372.

5.3. "The Shear Strength of Reinforced Concrete Members," ASCE-ACI Committee 426, *Proc. ASCE*, vol. 99, no. ST6, 1973, pp. 1091–1187 (with extensive bibliography).

5.4. "The Shear Strength of Reinforced Concrete Members—Slabs," ASCE-ACI Task Committee 426, *Proc. ASCE*, vol. 100, no. ST8, 1974, pp. 1543–1591.

5.5. *Shear in Reinforced Concrete*, vols. 1 and 2, *Special Publication* SP-42, American Concrete Institute, Detroit, 1974.

5.6. "Recent Approaches to Shear Design of Structural Concrete," ASCE-ACI Committee 445, *J. Struct. Eng.*, ASCE, vol. 124, no. 12, 1998, pp. 1375–1417.

5.7. A. H. Nilson (ed.), *Finite Element Analysis of Reinforced Concrete*, American Society of Civil Engineers, New York, 1982.

5.8. J. Isenberg (ed.), *Finite Element Analysis of Reinforced Concrete Structures II*, American Society of Civil Engineers, New York, 1993, pp. 203–232.

5.9. M. P. Collins, "Toward a Rational Theory for RC Members in Shear," *J. Struct. Div.*, ASCE, vol. 104, no. ST4, April 1978, pp. 649–666.

5.10. T. T. C. Hsu, *Unified Theory of Reinforced Concrete*, CRC Press, Inc., Boca Raton, FL, 1993.

5.11. T. C. Zsutty, "Shear Strength Prediction for Separate Categories of Simple Beam Tests," *J. ACI*, vol. 68, no. 2, 1971, pp. 138–143.

5.12. *AASHTO LRFD Bridge Design Specifications*, 8th ed., American Association of State Highway and Transportation Officials (AASHTO), Washington, DC, 2017.

5.13. F. Leonhardt and R. Walther, "The Stuttgart Shear Tests," *C&CA Translation*, No. 111, Cement and Concrete Association, London, UK, 1964, 134 pp.

5.14. N. S. Anderson and J. A. Ramirez, "Detailing of Stirrup Reinforcement," *ACI Struct. J.*, vol. 86, no. 5, 1989, pp. 507–515.

5.15. A. S. Lubell, E. C. Bentz, and M. P. Collins, "Shear Reinforcement Spacing in Wide Members," *ACI Struct. J.*, vol. 106, no. 2, 2009, pp. 205–214.

5.16.   G. J. Parra-Montesinos, "Shear Strength of Beams with Deformed Steel Fibers," *Concr. Int.*, vol. 28, no. 11, 2006, pp. 57–66.

5.17.   A. H. Elzanaty, A. H. Nilson, and F. O. Slate, "Shear Capacity of Reinforced Concrete Beams Using High-Strength Concrete," *J. ACI*, vol. 83, no. 2, 1986, pp. 290–296.

5.18.   M. P. Collins and D. Kuchma, "How Safe Are Our Large, Lightly Reinforced Concrete Beams, Slabs, and Footings?" *ACI Struct. J.*, vol. 96, no. 4, 1999, pp. 282–290.

5.19.   S. Y. Debaiky and E. I. Elmiema, "Behavior and Strength of Reinforced Concrete Haunched Beams in Shear," *J. ACI*, vol. 79, no. 3, 1982, pp. 184–194.

5.20.   W. Ritter, "Die Bauweise Hennebique" (The Hennebique System), *Schweizerische Bauzeitung*, XXXIII, no. 7, 1899.

5.21.   E. Morsch, *Der Eisenbetonbau, seine Theorie und Anwendung* (Reinforced Concrete Theory and Application), Verlag Konrad Wittner, Stuttgart, 1912.

5.22.   J. Schlaich, K. Schafer, and M. Jennewein, "Toward a Consistent Design of Structural Concrete," *J. Prestressed Concr. Inst.*, vol. 32, no. 3, 1987, pp. 74–150.

5.23.   P. Marti, "Truss Models in Detailing," *Concr. Int.*, vol. 7, no. 12, 1985, pp. 66–73. (See also P. Marti, "Basic Tools of Reinforced Concrete Beam Design," *J. ACI*, vol. 82, no. 1, 1985, pp. 46–56.)

5.24.   J. G. MacGregor and J. K. Wight, *Reinforced Concrete*, 4th ed., Prentice-Hall, Englewood Cliffs, NJ, 2005.

5.25.   M. P. Collins and D. Mitchell, *Prestressed Concrete Structures*, Prentice-Hall, Englewood Cliffs, NJ, 1991.

5.26.   F. J. Vecchio and M. P. Collins, "Modified Compression Field Theory for Reinforced Concrete Elements Subjected to Shear," *J. ACI*, vol. 83, no. 2, 1986, pp. 219–231.

5.27.   F. J. Vecchio and M. P. Collins, "Predicting the Response of Reinforced Concrete Beams Subjected to Shear Using the Modified Compression Field Theory," *J. ACI*, vol. 85, no. 3, 1988, pp. 258–268.

5.28.   CSA Committee 23.3–14, *Design of Concrete Structures*, Canadian Standards Association, Etobicoke, Ontario, 2014, 297 pp.

5.29.   M. P. Collins, D. Mitchell, P. Adebar, and F. J. Vecchio, "A General Shear Design Method," *ACI Struct. J.*, vol. 93, no. 1, 1996, pp. 36–45.

5.30.   P. W. Birkeland and H. W. Birkeland, "Connections in Precast Concrete Construction," *J. ACI*, vol. 63, no. 3, 1966, pp. 345–368.

5.31.   R. F. Mast, "Auxiliary Reinforcement in Precast Concrete Connections," *J. Struct. Div.*, ASCE, vol. 94, no. ST6, June 1968, pp. 1485–1504.

5.32.   A. H. Mattock and N. M. Hawkins, "Shear Transfer in Reinforced Concrete—Recent Research," *J. Prestressed Concr. Inst.*, vol. 17, no. 2, 1972, pp. 55–75.

5.33.   A. H. Mattock, "Shear Transfer in Concrete Having Reinforcement at an Angle to the Shear Plane," *Special Publication* SP-42, American Concrete Institute, Detroit, 1974.

5.34.   *PCI Design Handbook*, 6th ed., Precast Prestressed Concrete Institute, Chicago, 2004.

5.35.   L. F. Kahn and A. D. Mitchell, "Shear Friction Tests with High-Strength Concrete," *ACI Struct. J.*, vol. 99, no. 1, 2002, pp. 98–103.

5.36.   A. H. Mattock, "Shear Friction and High-Strength Concrete," *ACI Struct. J.*, vol. 98, no. 1, 2001, pp. 50–59.

PROBLEMS

5.1.   A rectangular beam is 10 in. wide and has an effective depth of 13.5 in. Flexural reinforcement consists of two No. 8 (No. 25) bars. For $f_c' = 4000$ psi and no shear reinforcement, determine the nominal shear capacity in accordance with the ACI Code.

5.2.   A rectangular beam is 14 in. wide and has an effective depth of 20.5 in. Flexural reinforcement consists of three No. 9 (No. 29) bars. The beam contains No. 3 (No. 10) stirrups spaced at 9 in. For $f_c' = 3000$ psi and $f_{yt} = 60,000$ psi, calculate the nominal shear capacity of the section.

5.3.   A rectangular beam is 16 in. wide and has an effective depth of 26 in. Flexural reinforcement consists of three No. 10 (No. 32) bars. The beam contains No. 3 (No. 10) stirrups spaced at 13 in. For $f_c' = 4000$ psi and $f_{yt} = 60,000$ psi, calculate the nominal shear capacity of the section.

5.4.   A rectangular beam is 16 in. wide and has an effective depth of 26 in. Flexural reinforcement consists of six No. 10 (No. 32) bars. The beam contains No. 4 (No. 13) stirrups spaced at 13 in. For $f_c' = 4000$ psi and $f_{yt} = 60,000$ psi, calculate the nominal shear capacity of the section.

5.5. The T beam shown in Fig. P5.5 has an effective depth $d = 22$ in., a web width $b_w = 8$ in., and a flange width of $b_f = 36$ in. Flexural reinforcement consists of four No. 8 (No. 25) bars. For $f_c' = 5000$ psi, $f_{yt} = 60,000$ psi, and No. 3 (No. 10) stirrups spaced at 10 in, determine the nominal shear capacity of the section.

FIGURE P5.5

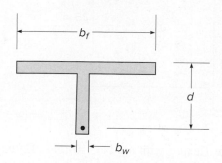

5.6. A rectangular beam is 16 in. wide and has an effective depth of 26.5 in. Flexural reinforcement consists of five No. 9 (No. 29) bars. For $f_c' = 4000$ psi and $f_{yt} = 60,000$ psi, determine the required spacing of No. 4 (No. 13) stirrups for a factored shear of 90 kips.

5.7. The T beam shown in Fig. P5.5 has an effective depth $d = 24$ in., a web width $b_w = 8$ in., and a flange width of $b_f = 36$ in. Flexural reinforcement consists of six No. 8 (No. 25) bars. For $f_c' = 5000$ psi, $f_{yt} = 60,000$ psi, and a factored shear of 50 kips, determine the spacing of No. 3 (No. 10) stirrups.

5.8. A simple span rectangular beam has an effective length of 18 ft, a width of 14 in., and an effective depth of 24 in. It is reinforced with three No. 9 (No. 29) bars longitudinally and No. 3 (No. 10) stirrups at 12 in. on center over the entire length. Determine the maximum factored load the beam can carry in plf. $f_c' = 5000$ psi, $f_y = 60,000$ psi, and $f_{yt} = 40,000$ psi.

5.9. A beam is to be designed for loads causing a maximum factored shear of 60.0 kips, using concrete with $f_c' = 5000$ psi. Proceeding on the basis that the concrete dimensions will be determined by diagonal tension, select the appropriate width and effective depth (a) for a beam in which no web reinforcement is to be used, (b) for a beam in which only the minimum web reinforcement is provided, as given by Eq. (5.13a), and (c) for a beam in which web reinforcement provides shear strength $V_s = 2V_c$. Follow the ACI Code requirements, and let $d = 2b$ in each case. Assume that $\rho_w = 0.012$.

5.10. A rectangular beam having $b = 10$ in. and $d = 17.5$ in. spans 15 ft face to face of simple supports. It is reinforced for flexure with three No. 9 (No. 29) bars that continue uninterrupted to the ends of the span. It is to carry service dead load $D = 1.27$ kips/ft (including self-weight) and service live load $L = 3.70$ kips/ft, both uniformly distributed along the span. Design the shear reinforcement, using No. 3 (No. 10) vertical U stirrups. Equation (5.12c) for V_c may be used. Material strengths are $f_c' = 4000$ psi and $f_y = 60,000$ psi.

5.11. Redesign the shear reinforcement for the beam of Problem 5.10, basing V_c on Eq. (5.12d). Comment on your results, with respect to design time and probable construction cost difference.

5.12. Design the shear reinforcement, using No. 4 (No. 13) vertical U stirrups for the independent T beam shown in Fig. P5.12. The beam spans 24 ft face to face between simple supports, has an effective depth $d = 31$ in., and is

FIGURE P5.12

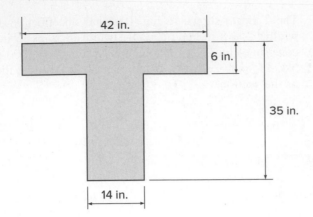

reinforced for flexure with six No. 10 (No. 32) bars in two layers that continue uninterrupted to the ends of the span. It is to carry service dead load $D = 2.67$ kips/ft (including self-weight) and service live load $L = 5.36$ kips/ft, both uniformly distributed along the span. Equation (5.12c) for V_c may be used. Material strengths are $f_c' = 5000$ psi and $f_y = 60,000$ psi.

5.13. A beam of 11 in. width and effective depth of 16 in. carries a factored uniformly distributed load of 5.3 kips/ft, including its own weight, in addition to a central, concentrated factored load of 12 kips. It spans 18 ft, and restraining end moments at full factored load are 137 ft-kips at each support. It is reinforced with three No. 9 (No. 29) bars for both positive and negative bending. If $f_c' = 4000$ psi, through what part of the beam is web reinforcement theoretically required (a) if Eq. (5.12c) is used and (b) if Eq. (5.12d) is used? Comment.

5.14. The beam of Problem 5.10 will be subjected to a factored axial compression load of 88 kips on the 10 × 20 in. gross cross section, in addition to the loads described earlier. What is the effect on concrete shear strength V_c?

5.15. The beam of Problem 5.10 will be subjected to a factored axial tension load of 44 kips on the 10 × 20 in. gross cross section, in addition to the loads described earlier. What is the effect on concrete shear strength V_c?

5.16. Redesign the shear reinforcement for the beam of Problem 5.10, using the modified compression field theory with (a) $\phi_{\text{shear}} = 0.90$ and (b) $\phi_{\text{shear}} = 0.75$.

5.17. Redesign the shear reinforcement for the beam of Problem 5.12, using the modified compression field theory with (a) $\phi_{\text{shear}} = 0.90$ and (b) $\phi_{\text{shear}} = 0.75$.

5.18. A precast concrete beam with cross-sectional dimensions $b = 10$ in. and $h = 24$ in. is designed to act in a composite sense with a cast-in-place top slab having depth $h_f = 5$ in. and width 48 in. At factored loads, the maximum compressive stress in the flange at midspan is 2400 psi; at the supports of the 28 ft simple span the flange force must be zero. Vertical stirrups provided for flexural shear will be extended into the slab and suitably anchored to provide also for transfer of the flange force by shear friction. Find the minimum number of No. 4 (No. 13) stirrups that must be provided, based on shear-friction requirements. Concrete in both precast and cast-in-place parts will have $f_c' = 4000$ psi and $f_y = 60,000$ psi. The top surface of the precast web will be intentionally roughened according to the ACI Code definition.

5.19. Redesign the beam-end reinforcement of Example 5.6, given that a roller support will be provided so that $N_{uc} = 0$.

6

Bond, Anchorage, and Development Length

6.1 FUNDAMENTALS OF FLEXURAL BOND

If the reinforced concrete beam of Fig. 6.1a were constructed using plain round reinforcing bars, and, furthermore, if those bars were greased or otherwise lubricated before the concrete were cast, the beam would be very little stronger than if it were built of plain concrete, without reinforcement. If a load were applied, as shown in Fig. 6.1b, the bars would tend to maintain their original length as the beam deflected. The bars would slip longitudinally with respect to the adjacent concrete, which would experience tensile strain due to flexure. Proposition 2 of Section 3.4, the assumption that the strain in an embedded reinforcing bar is the same as that in the surrounding concrete, would not be valid. For reinforced concrete to behave as intended, it is essential that *bond forces* be developed on the interface between concrete and steel, such as to prevent significant slip from occurring at that interface.

Figure 6.1c shows the bond forces that act on the concrete at the interface as a result of bending, while Fig. 6.1d shows the equal and opposite bond forces acting on the reinforcement. It is through the action of these interface bond forces that the slip indicated in Fig. 6.1b is prevented.

Some years ago, when plain bars without surface deformations were used, initial bond strength was provided only by the relatively weak chemical adhesion and mechanical friction between steel and concrete. Once adhesion and static friction were overcome at larger loads, small amounts of slip led to interlocking of the natural roughness of the bar with the concrete. However, this natural bond strength is so low that in beams reinforced with plain bars, the bond between steel and concrete was frequently broken. Such a beam will collapse as the bar is pulled through the concrete. To prevent this, end anchorage was provided, chiefly in the form of hooks, as in Fig. 6.2. If the anchorage is adequate, such a beam will not collapse, even if the bond is broken over the entire length between anchorages. This is so because the member acts as a tied arch, as shown in Fig. 6.2, with the uncracked concrete shown shaded representing the arch and the anchored bars the tie-rod. In this case, over the length in which the bond is broken, bond forces are zero. This means that over the entire unbonded length the force in the steel is constant and equal to $T = M_{max}/jd$. As a consequence, the total steel elongation in such beams is larger than in beams in which bond is preserved, resulting in larger deflections and greater crack widths.

To improve this situation, deformed bars are now universally used in the United States and many other countries (see Section 2.14). With such bars, the shoulders of the projecting deformations bear on the surrounding concrete and result in greatly increased bond strength. It is then possible in most cases to dispense with special anchorage devices such as hooks. In addition, crack widths as well as deflections are reduced.

FIGURE 6.1
Bond forces due to flexure:
(*a*) beam before loading;
(*b*) unrestrained slip between
concrete and steel; (*c*) bond
forces acting on concrete; and
(*d*) bond forces acting on
steel.

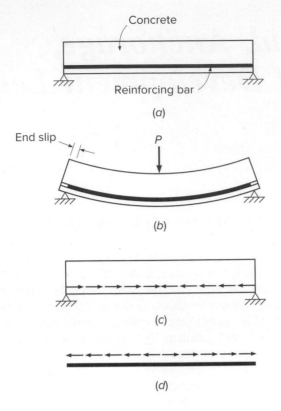

FIGURE 6.2
Tied-arch action in a beam
with little or no bond.

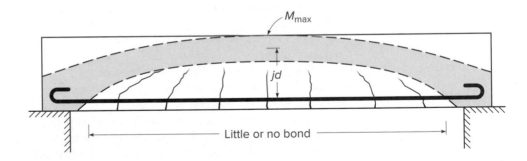

a.   **Bond Force Based on Simple Cracked Section Analysis**

In a short piece of a beam of length *dx*, such as shown in Fig. 6.3*a*, the moment at one end will generally differ from that at the other end by a small amount *dM*. If this piece is isolated, and if one assumes that, after cracking, the concrete does not resist any tension stresses, the internal forces are those shown in Fig. 6.3*a*. The change in bending moment *dM* produces a change in the bar force

$$dT = \frac{dM}{jd} \qquad (a)$$

where *jd* is the internal lever arm between tensile and compressive force resultants. Since the bar or bars must be in equilibrium, this change in bar force is resisted at the contact surface between steel and concrete by an equal and opposite force produced by bond, as indicated by Fig. 6.3*b*.

FIGURE 6.3
Forces acting on elemental length of beam: (*a*) free-body sketch of reinforced concrete element and (*b*) free-body sketch of steel element.

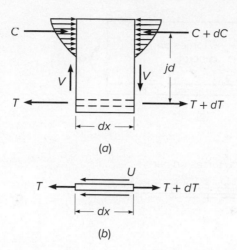

If U is the magnitude of the local bond force per unit length of bar, then, by summing horizontal forces

$$U\,dx = dT \qquad\qquad (b)$$

Thus,

$$U = \frac{dT}{dx} \qquad\qquad (6.1)$$

indicating that the local unit bond force is proportional to the rate of change of bar force along the span. Alternatively, substituting Eq. (*a*) in Eq. (6.1), the unit bond force can be written as

$$U = \frac{1}{jd}\frac{dM}{dx} \qquad\qquad (c)$$

from which

$$U = \frac{V}{jd} \qquad\qquad (6.2)$$

Equation (6.2) is the "elastic cracked section equation" for flexural bond force, and it indicates that the bond force per unit length is proportional to the shear at a particular section, that is, to the rate of change of bending moment.

Note that Eq. (6.2) applies to the *tension* bars in a concrete zone that is assumed to be fully cracked, with the concrete resisting no tension. It applies, therefore, to the tensile bars in simple spans, or, in continuous spans, either to the bottom bars in the positive bending region between inflection points or to the top bars in the negative bending region between the inflection points and the supports. It does not apply to compression reinforcement, which bears against the concrete at the end of the bar.

b. Actual Distribution of Flexural Bond Force

The actual distribution of bond force along deformed reinforcing bars is much more complex than that represented by Eq. (6.2), and Eq. (6.1) provides a better basis for understanding beam behavior. Figure 6.4 shows a beam segment subject to pure bending. The concrete fails to resist tensile stresses only where the actual crack is located; there the steel tension is maximum and has the value predicted by simple theory: $T = M/jd$. Between cracks, the concrete *does* resist moderate amounts of tension, introduced by bond forces acting along the interface in the direction shown in Fig. 6.4*a*.

FIGURE 6.4
Variation of steel and bond
forces in a reinforced
concrete member subject to
pure bending: (*a*) cracked
concrete segment; (*b*) bond
forces acting on reinforcing
bar; (*c*) variation of
tensile force in steel; and
(*d*) variation of bond force
along steel.

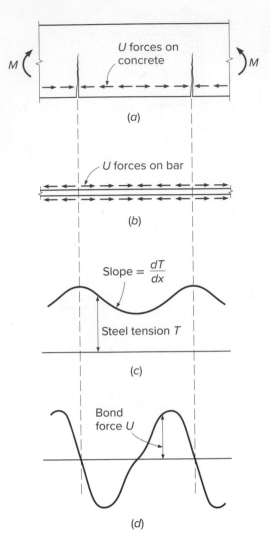

This reduces the tensile force in the steel, as illustrated by Fig. 6.4*c*. From Eq. (6.1), it is clear that U is proportional to the rate of change of bar force, and thus will vary as shown in Fig. 6.4*d*; unit bond forces are highest where the slope of the steel force curve is greatest and are zero where the slope is zero. Very high local bond forces adjacent to cracks have been measured in tests (Refs. 6.1 and 6.2). They are so high that inevitably some slip occurs between concrete and steel adjacent to each crack.

Beams are seldom subject to pure bending moment; they generally carry transverse loads producing shear and moment that vary along the span. Figure 6.5*a* shows a beam carrying a distributed load. The cracking indicated is typical. The steel force T predicted by simple cracked section analysis is proportional to the moment diagram and is as shown by the dashed line in Fig. 6.5*b*. The actual value of T, however, is less than that predicted by the simple analysis everywhere except at the actual crack locations. The actual variation of T is shown by the solid line of Fig. 6.5*b*. In Fig. 6.5*c*, the bond forces predicted by the simplified theory are shown by the dashed line, and the actual variation is shown by the solid line. Note that the value of U is equal to that given by Eq. (6.2) only at those locations where the slope of the steel force diagram equals that of the simple theory. Elsewhere, if the slope is greater than assumed, the local bond force is greater; if the slope is less, local bond force is less. Just to the left of the cracks, for the present example, U is higher than predicted by Eq. (6.2), and in

FIGURE 6.5
Effect of flexural cracks on
bond forces in beam:
(*a*) beam with flexural
cracks; (*b*) variation of tensile
force *T* in steel along span;
and (*c*) variation of bond
force per unit length *U* along
span.

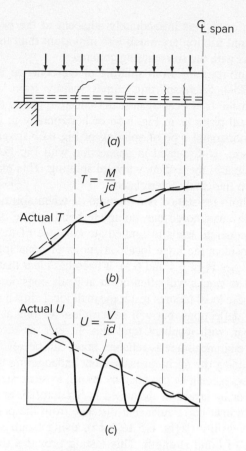

all probability will result in local bond failure. Just to the right of the cracks, U is much lower than predicted and in fact is generally negative very close to the crack; that is, the bond forces act in the reverse direction.

It is evident that actual bond forces in beams bear very little relation to those predicted by Eq. (6.2), except in the general sense that they are highest in the regions of high shear.

6.2   BOND STRENGTH AND DEVELOPMENT LENGTH

For reinforcing bars in tension, two types of bond failure have been observed. The first is *direct pullout* of the bar, which occurs when ample confinement is provided by the surrounding concrete. This could be expected when relatively small-diameter bars are used with sufficiently large concrete cover distances and bar spacing. The second type of failure is *splitting* of the concrete along the bar when cover, confinement, or bar spacing is insufficient to resist the lateral concrete tension resulting from the wedging effect of the bar deformations. Present-day design methods require that both possible failure modes be accounted for.

a.   Bond Strength

If the bar is sufficiently confined by a mass of surrounding concrete, then as the tensile force on the bar is increased, adhesive bond and friction are overcome, the concrete eventually crushes locally ahead of the bar deformations, and bar pullout results. The surrounding concrete remains intact, except for the crushing that takes place ahead of the deformations immediately adjacent to the bar interface. For deformed bars, adhesion and friction are much less important than the mechanical interlock of the deformations with the surrounding concrete.

Bond failure resulting from splitting of the concrete is more common in beams than direct pullout. Such splitting comes mainly from wedging action when the deformations on the bars bear against the concrete (Refs. 6.3 and 6.4). It may occur either in a vertical plane as in Fig. 6.6a or horizontally in the plane of the bars as in Fig. 6.6b. The horizontal type of splitting of Fig 6.6b frequently begins at a diagonal crack. In this case, as discussed in connection with Fig. 5.8b and shown in Fig. 5.1, dowel action increases the tendency toward splitting. This indicates that shear and bond failures are often intricately interrelated.

When pullout resistance is overcome or when splitting has spread all the way to the end of an unanchored bar, complete bond failure occurs. Sliding of the steel relative to the concrete leads to immediate collapse of the beam.

If one considers the large local variations of bond force caused by flexural and diagonal cracks (see Figs. 6.4 and 6.5), it becomes clear that local bond failures immediately adjacent to cracks will often occur at loads considerably below the failure load of the beam. These local failures result in small local slips and some widening of cracks and increase of deflections, but will be harmless as long as failure does not propagate all along the bar, with resultant total slip. In fact, as discussed in connection with Fig. 6.2, when end anchorage is reliable, bond can be severed along the entire length of the bar, excluding the anchorages, without endangering the carrying capacity of the beam. End anchorage can be provided by hooks, as suggested by Fig. 6.2 and discussed in Section 6.4, or by heads, as discussed in Section 6.5, or much more commonly, by extending the straight bar a sufficient distance from the point of maximum stress.

Extensive testing (Refs. 6.5 to 6.15), using beam specimens, has established limiting values of bond strength. This testing provides the basis for current design requirements.

b. Development Length

The preceding discussion suggests the concept of *development length* of a reinforcing bar. In the ACI Code, the development length is defined as that length of embedment necessary to develop the specified yield strength of the bar, controlled by either pullout or splitting. With reference to Fig. 6.7, the moment, and therefore the steel stress, is evidently maximum at point *a* (neglecting the weight of the beam) and zero at the supports. If the bar stress is f_s at *a*, then the total tension force $A_b f_s$ must be transferred

FIGURE 6.6

Splitting of concrete along reinforcement.

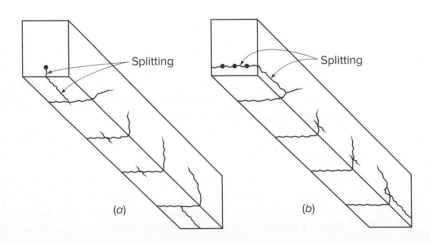

(a)   (b)

FIGURE 6.7
Development length.

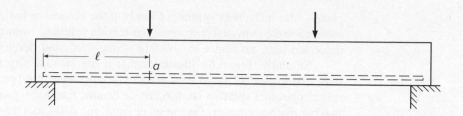

from the bar to the concrete in the distance ℓ by bond forces. To fully develop the yield strength of the bar $A_b f_y$, the distance ℓ must be at least equal to the development length of the bar, established by tests. In the beam shown in Fig. 6.7, if the actual length ℓ is equal to or greater than the development length ℓ_d, no premature bond failure will occur. That is, the beam will fail in bending or shear rather than by bond failure. This will be so even if in the vicinity of cracks local slip may have occurred over small regions along the beam.

It is seen that the main requirement for safety against bond failure is this: the length of the bar, from any point of given steel stress (f_s or at most f_y) to its nearby free end, must be at least equal to its development length. If this requirement is satisfied, the magnitude of the nominal flexural bond force along the beam, as given by Eq. (6.2), is of only secondary importance, since the integrity of the member is ensured even in the face of possible minor local bond failures. However, if the actual available length is inadequate for full development, special anchorage, such as by hooks or heads, must be provided.

c. Factors Influencing Development Length

Experimental research has identified the factors that influence bond strength and, thus, development length, and analysis of the test data has resulted in the empirical equations used in present design practice. The most basic factors will be clear from review of the preceding paragraphs and include concrete tensile strength, cover distance, spacing of the reinforcing bars, and the presence of transverse steel reinforcement.

Clearly, the *tensile strength* of the concrete is important because the most common type of bond failure in beams is the type of splitting shown in Fig. 6.6. Although tensile strength does not appear explicitly in experimentally derived equations for development length (see Section 6.3), a term representing the tensile strength of concrete, typically $\sqrt{f_c'}$, appears in the denominator of those equations and reflects the influence of concrete tensile strength.

As discussed in Section 2.9, the fracture energy of concrete plays an important role in bond failure because a splitting crack must propagate after it has formed. Since, as shown in Fig. 2.11, fracture energy is largely independent of compressive strength, bond strength increases more slowly than $\sqrt{f_c'}$, and as data for higher-strength concretes have become available, $f_c'^{1/4}$ has been shown to provide a better representation of the effect of concrete strength on bond than $\sqrt{f_c'}$ (Refs. 6.12 to 6.14). This point is recognized by ACI Committee 408, Bond and Development of Reinforcement (Ref. 6.15), in proposed design expressions based on $f_c'^{1/4}$ and within the ACI Code, which sets an upper limit on the value of $\sqrt{f_c'}$ for use in design.

For lightweight concretes, the tensile strength is usually less than for normal-density concrete having the same compressive strength; accordingly, if lightweight concrete is used, development lengths must be increased.

Cover distance—conventionally measured from the *center* of the bar to the nearest concrete face and measured either in the plane of the bars or perpendicular to that plane—also influences splitting. Clearly, if the vertical or horizontal cover is increased, more concrete is available to resist the tension resulting from the wedging effect of the deformed bars, resistance to splitting is improved, and development length is less.

Similarly, Fig. 6.6b illustrates that if the *bar spacing* is increased (such as, if only two instead of three bars are used), more concrete per bar will be available to resist horizontal splitting (Ref. 6.16). In beams, bars are typically spaced about one or two bar diameters apart. On the other hand, for slabs, footings, and certain other types of member, bar spacings are typically much greater, and the required development length is reduced, if not limited by cover.

Transverse reinforcement, such as that provided by stirrups of the types shown in Fig. 5.8, improves the resistance of tensile bars to both vertical or horizontal splitting failure because the tensile force in the transverse steel tends to prevent opening of the actual or potential crack. The effectiveness of such transverse reinforcement depends on its cross-sectional area and spacing along the development length. Its effectiveness does not depend on its yield strength f_{yt}, because transverse reinforcement rarely yields during a bond failure (Refs. 6.12 to 6.15).

Based on the results of a statistical analysis of test data available in the 1970s (Ref. 6.10), it was found that the length ℓ_d needed to develop stress f_s in a reinforcing bar could be expressed (with some modification and updating to reflect more recent test results) as

$$\ell_d = \left(\frac{\dfrac{f_s}{\sqrt{f_c'}} - 200}{12\left(\dfrac{c_b + K_{tr}}{d_b}\right)} \right) d_b \tag{6.3}$$

where d_b = bar diameter

$\quad c_b$ = smaller of minimum cover or one-half of bar spacing *measured to center of bar*

$\quad K_{tr} = 40A_{tr}/sn$, which represents effect of confining reinforcement

$\quad A_{tr}$ = area of transverse reinforcement normal to plane of splitting through the bars being developed

$\quad s$ = spacing of transverse reinforcement

$\quad n$ = number of bars developed or spliced at same location

Equation (6.3) has been simplified to

$$\ell_d = \left(\frac{3}{40} \frac{f_s}{\sqrt{f_c'}\left(\dfrac{c_b + K_{tr}}{d_b}\right)} \right) d_b \tag{6.4}$$

which serves as the basis for calculating development length in the ACI Code.

An important difference between Eqs. (6.3) and (6.4) is that Eq. (6.3) reflects the experimental result that the required development length increases disproportionately more than the bar stress f_s, while in Eq. (6.4) development length is proportional to f_s. To address this shortcoming, a special factor must be added to Eq. (6.4) to account for the extra length needed for bars with higher yield strengths, as will be shown in Section 6.3. Although both equations are written to express development length as a multiple of bar diameter, the presence of d_b in the denominator leads to development lengths that actually increase with the square of the bar diameter and, thus, the area of the bar.

Equation (6.4) captures the effects of concrete strength, concrete cover, and transverse reinforcement on ℓ_d and serves as the basis for design in the ACI Code. For full development of the bar, f_s is set equal to f_y.

In addition to the factors just discussed, other influences have been identified. The vertical location of horizontal bars relative to beam depth has been found to have an effect (Ref. 6.17). If bars are placed in the forms during construction such that a substantial *depth of concrete is placed below those bars*, there is a tendency for excess water, often used in the mix for workability, and for entrapped air to rise to the top of the concrete during consolidation. Air and water tend to accumulate on the underside of the bars. Tests have shown a loss in bond strength for bars with more than 12 in. of fresh concrete cast beneath them, and accordingly the development length must be increased. This effect increases as the slump of the concrete increases and is greatest for bars cast near the upper surface of a concrete placement (Ref. 6.18).

Epoxy-coated reinforcing bars are used regularly in projects where the structure may be subjected to corrosive environmental conditions or deicing chemicals, such as for highway bridge decks and parking garages. Zinc and epoxy dual-coated bars have also been produced. Studies have shown that bond strength is reduced because the epoxy coating reduces the friction between the concrete and the bar, and the required development length must be increased (Refs. 6.19 to 6.23). Early evidence showed that if cover and bar spacing were large, the effect of the epoxy coating would not be so pronounced, and as a result, a smaller increase was felt justified under these conditions (Ref. 6.20). Although later research (Ref. 6.12) does not support this conclusion, provisions to allow for a smaller increase remain in the ACI Code. Since the bond strength of epoxy-coated bars is already reduced because of lack of adhesion, an upper limit has been established for the product of development length factors accounting for the depth of concrete cast below horizontal bars and epoxy coating.

Not infrequently, tensile reinforcement somewhat in excess of the calculated requirement is provided, for example, as a result of upward rounding A_s when bars are selected or when minimum steel requirements govern. Logically, in this case, the required development length may be reduced by the ratio of steel area required to steel area actually provided. The modification for *excess reinforcement* should be applied only where anchorage or development for the full yield strength of the bar is not required.

Finally, based on bars with very short development lengths (most with values of $\ell_d/d_b < 15$), it was observed that *smaller-diameter bars* required lower development lengths than predicted by Eq. (6.4). As a result, the required development lengths for No. 6 (No. 19) and smaller bars were reduced below the values required by Eq. (6.4).[†]

Reference 6.15 presents a detailed discussion of the factors that control the bond and development of reinforcing bars in tension. Except as noted, these influences are accounted for in the basic equation for development length in the ACI Code. All modification factors for development length are defined explicitly in the Code, with appropriate restrictions. Details are given next.

6.3 ACI CODE PROVISIONS FOR DEVELOPMENT OF TENSION REINFORCEMENT

The approach to bond strength incorporated in the ACI Code follows from the discussion presented in Section 6.2. The fundamental requirement is that the calculated force in the reinforcement at each section of a reinforced concrete member be developed on each side of that section by adequate embedment length, hooks, mechanical

---

[†] The use of Eq. (6.4) for low values of ℓ_d/d_b greatly underestimates the actual value of bond strength and makes it appear that a lower value of ℓ_d can be used safely. An evaluation of test results for small bars with more realistic development lengths ($\ell_d/d_b \geq 16$), however, has shown that the special provision in the ACI Code for smaller bars is not justified (Refs. 6.14, 6.15, and 6.24). Because of the unconservative nature of the small bar provision, ACI Committee 408 (Ref. 6.15) recommends that it not be applied in design.

anchorage, or a combination of these, to ensure against pullout. Local high bond forces, such as are known to exist adjacent to cracks in beams, are not considered to be significant. Generally, the force to be developed is calculated based on the yield stress in the reinforcement; that is, the bar strength is to be fully developed.

In the ACI Code, the required development length for deformed bars in tension is based on Eq. (6.4). A single basic equation is given that includes *all* the influences discussed in Section 6.2 and thus appears highly complex because of its inclusiveness. It does, however, permit the designer to see the effects of all the controlling variables and allows more rigorous calculation of the required development length when it is critical. The ACI Code also includes simplified equations that can be used for most cases in ordinary design, provided that some restrictions are accepted on bar spacing, cover values, and minimum transverse reinforcement. These alternative equations can be further simplified for normalweight concrete and uncoated bars.

In the following presentation of development length, the basic ACI equation is given first and its terms are defined and discussed. After this, the alternative equations, also part of the ACI Code, are presented. Note that, in any case, development length ℓ_d must not be less than 12 in.

a. Equation for Development Length for Bars and Wires in Tension

According to ACI Code 25.4.2.3, for deformed bars or deformed wires,

$$\ell_d = \left(\frac{3}{40} \frac{f_y}{\lambda\sqrt{f_c'}} \frac{\psi_t\psi_e\psi_s\psi_g}{\left(\frac{c_b + K_{tr}}{d_b} \right)} \right) d_b \qquad (6.5)$$

in which the term $(c_b + K_{tr})/d_b$ may not be taken greater than 2.5. In Eq. (6.5), the terms are defined and values established as follows.

ψ_t = casting position factor
More than 12 in. of fresh concrete is placed below horizontal reinforcement: 1.3
Other situations: 1.0

ψ_e = epoxy coating factor
Epoxy-coated or zinc and epoxy dual-coated bars or wires with cover less than $3d_b$ or clear spacing less than $6d_b$:   1.5
All other epoxy-coated or zinc and epoxy dual-coated bars or wires: 1.2
Uncoated and zinc-coated (galvanized) reinforcement: 1.0
However, the product of $\psi_t\psi_e$ need not be taken greater than 1.7.

ψ_s = reinforcement size factor
No. 6 (No. 19) and smaller bars and deformed wires: 0.8[†]
No. 7 (No. 22) and larger bars: 1.0

ψ_g = reinforcement grade factor
Grade 40 or 60   1.0
Grade 80   1.15
Grade 100   1.3

[†] ACI Committee 408 recommends a value of 1.0 for all bar sizes based on experimental evidence. The ACI Code value of 0.8, however, is used in what follows.

λ = lightweight aggregate concrete factor

    When lightweight aggregate concrete is used: 0.75

    When normalweight concrete is used: 1.0

c_b = spacing or cover dimension, in.

    Use the smaller of either the distance from the center of the bar to the nearest concrete surface or one-half the center-to-center spacing of the bars being developed.

K_{tr} = transverse reinforcement index: $40A_{tr}/sn$

    where A_{tr} = total cross-sectional area of all transverse reinforcement that is within the spacing s and that crosses the potential plane of splitting through the reinforcement being developed, in^2

          s = maximum spacing of transverse reinforcement within ℓ_d center to center, in.

          n = number of bars or wires being developed along the plane of splitting

As a simplification, the designer is permitted to use $K_{tr} = 0$ even if transverse reinforcement is present.

According to ACI Code 25.4.2.2, for reinforcement with $f_y \geq 80,000$ psi spaced closer than 6 in. on center, transverse reinforcement must be provided such that K_{tr} is not smaller than $0.5d_b$. According to ACI Code 9.7.1.4 and 10.7.1.3, however, K_{tr} must be at least $0.5d_b$ in all beams and columns where reinforcement with $f_y \geq 80,000$ psi is developed or spliced, independent of the bar spacing.

The limit of 2.5 on $(c + K_{tr})/d_b$ is imposed to avoid pullout failure. With that term taken equal to its limit of 2.5, evaluation of Eq. (6.4) results in $\ell_d = 0.03d_b f_y/\sqrt{f'_c}$, the experimentally derived limit found in earlier ACI Codes when pullout failure controls. Note that in Eq. (6.5) and in all other ACI Code equations relating to the development length and splices of reinforcement, *values of $\sqrt{f'_c}$ are not to be taken greater than 100 psi* because, as explained in Section 6.2c and recognized in ACI Commentary 25.4.1.4, bond strength increases more slowly than $\sqrt{f'_c}$. The $\sqrt{f'_c}$, however, is reasonably accurate for values of f'_c up to 10,000 psi.

b. Simplified Equations for Development Length

Calculation of required development length (in terms of bar diameter) by Eq. (6.5) requires that the term $(c_b + K_{tr})/d_b$ be calculated for each particular combination of cover, spacing, and transverse reinforcement. Alternatively, according to the Code, a simplified form of Eq. (6.5) may be used in which $(c_b + K_{tr})/d_b$ is set equal to 1.5, provided that certain restrictions are placed on cover, spacing, and transverse reinforcement. Two cases of practical importance are:

1. Minimum clear cover of $1.0d_b$, minimum clear spacing of $1.0d_b$, and at least the Code required minimum stirrups or ties (see Section 5.5b) throughout ℓ_d
2. Minimum clear cover of $1.0d_b$ and minimum clear spacing of $2d_b$

For either of these common cases, it is easily confirmed from Eq. (6.4) that for No. 7 (No. 22) and larger bars

$$\ell_d = \left(\frac{f_y \psi_t \psi_e \psi_g}{20\lambda\sqrt{f'_c}}\right)d_b \tag{6.6a}$$

TABLE 6.1
Simplified tension development length in bar diameters according to the ACI Code

	No. 6 (No. 19) and Smaller Bars and Deformed Wires[†]	No. 7 (No. 22) and Larger Bars
Clear spacing of bars or wires being developed or spliced $\geq d_b$, clear cover $\geq d_b$, and stirrups or ties throughout ℓ_d not less than the Code minimum	$\ell_d = \left(\dfrac{f_y \psi_t \psi_e \psi_g}{25 \lambda \sqrt{f_c'}} \right) d_b$	$\ell_d = \left(\dfrac{f_y \psi_t \psi_e \psi_g}{20 \lambda \sqrt{f_c'}} \right) d_b$
Clear spacing of bars or wires being developed or spliced $\geq 2d_b$, and clear cover $\geq d_b$	Same as above	Same as above
Other cases	$\ell_d = \left(\dfrac{3f_y \, \psi_t \psi_e \psi_g}{50 \lambda \sqrt{f_c'}} \right) d_b$	$\ell_d = \left(\dfrac{3f_y \, \psi_t \psi_e \psi_g}{40 \lambda \sqrt{f_c'}} \right) d_b$

[†] For reasons discussed in Section 6.3a, ACI Committee 408 recommends that ℓ_d for No. 7 (No. 22) and larger bars be used for all bar sizes.

and for No. 6 (No. 19) bars and smaller

$$\ell_d = \left(\frac{f_y \psi_t \psi_e \psi_g}{25 \lambda \sqrt{f_c'}} \right) d_b \tag{6.6b}$$

If these restrictions on spacing are not met, then, provided that Code-imposed minimum spacing requirements are met (see Section 4.5c), the term $(c_b + K_{tr})/d_b$ will have a value not less than 1.0 (rather than 1.5 as before) whether or not transverse steel is used. The values given by Eqs. (6.6a) and (6.6b) are then multiplied by the factor 1.5/1.0.

Thus, if the designer accepts certain restrictions on bar cover, spacing, and transverse reinforcement, simplified calculation of development requirements is possible. The simplified equations are summarized in Table 6.1.

Further simplification is possible for the most common condition of normalweight concrete and uncoated reinforcement. Then λ and ψ_e in Table 6.1 take the value 1.0, and the development length, in terms of bar diameters, is simply a function of f_y, f_c', and the bar location factor ψ_t. Thus, development lengths are easily tabulated for the usual combinations of material strengths and bottom or top bars and for the restrictions on bar spacing, cover, and transverse steel defined.[†] Results are given in Table A.10 of Appendix A.

Regardless of whether development length is calculated using the basic Eq. (6.5) or the more approximate Eqs. (6.6a) and (6.6b), development length may be reduced where reinforcement in a flexural member is in excess of that required by analysis, except where anchorage or development for f_y is specifically required in seismic-force-resisting systems in structures assigned to Seismic Design Categories D, E, or F (discussed in Chapter 20). According to ACI Code 25.4.10, the reduction in ℓ_d is made according to the ratio (A_s required/A_s provided).

[†] Note that, for convenient reference, the term top bar is used for any horizontal reinforcing bar placed with more than 12 in. of fresh concrete cast below the development length or splice. This definition may require that bars relatively near the bottom of a deep member be treated as top bars.

EXAMPLE 6.1 **Development length in tension.**   Figure 6.8 shows a beam-column joint in a continuous building frame. Based on frame analysis, the negative steel required at the end of the beam is 1.80 in^2; two No. 9 (No. 29) bars are used, providing $A_s = 2.00$ in^2. Beam dimensions are $b = 10$ in., $d = 18$ in., and $h = 21$ in. The design will include No. 3 (No. 10) stirrups spaced four at 3 in., followed by a constant 5 in. spacing in the region of the support, with 1.5 in. clear cover. Normal-weight concrete is to be used, with $f'_c = 4000$ psi, and the reinforcing bars have $f_y = 60,000$ psi.

Find the minimum distance ℓ_d at which the negative bars can be cut off, based on development of the required steel area at the face of the column, (a) using the simplified equations of Table 6.1, (b) using Table A.10 of Appendix A, and (c) using the more accurate Eq. (6.5).

SOLUTION.   Checking for lateral spacing of the No. 9 (No. 29) bars determines that the clear distance between the bars is $10 - 2(1.50 + 0.38 + 1.128) = 4$ in., or 3.55 times the bar diameter d_b. The clear cover of the No. 9 (No. 29) bars to the side face of the beam is $1.50 + 0.38 = 1.88$ in., or 1.67 bar diameters, and that to the top of the beam is $3.00 - 1.128/2 = 2.44$ in., or 2.16 bar diameters. These dimensions meet the restrictions stated in the second row of Table 6.1. Then for top bars, uncoated, Grade 60 reinforcement cast in normalweight concrete, we have values of $\psi_t = 1.3$, $\psi_e = 1.0$, $\psi_g = 1.0$, and $\lambda = 1.0$. From Table 6.1,

$$\ell_d = \left(\frac{f_y \psi_t \psi_e \psi_g}{20\lambda\sqrt{f'_c}}\right) d_b = \frac{60,000 \times 1.3 \times 1.0 \times 1.0}{20 \times 1.0\sqrt{4000}} 1.128 = 62 \times 1.128 = 70 \text{ in.}$$

This can be reduced by the ratio of steel required to that provided, so that the final development length is $70 \times 1.80/2.00 = 63$ in.

Alternatively, from the lower portion of Table A.10, $\ell_d/d_b = 62$. The required length to point of cutoff is $62 \times 1.128 \times 1.80/2.00 = 63$ in., as before.

The more accurate Eq. (6.5) will now be used. The center-to-center spacing of the No. 9 (No. 29) bars is $10 - 2(1.50 + 0.38 + 1.128/2) = 5.11$ in., one-half of which is 2.56 in. The side cover to the bar centerline is $1.50 + 0.38 + 1.128/2 = 2.44$ in., and the top cover to the bar centerline is 3.00 in. The smallest of these three dimensions controls, giving $c_b = 2.44$ in. Potential splitting would be in the horizontal plane of the bars, and in calculating A_{tr}, two times the stirrup bar area is used.[†] Based on No. 3 (No. 10) stirrups at 5 in. spacing:

$$K_{tr} = \frac{40A_{tr}}{sn} = \frac{40 \times 0.11 \times 2}{5 \times 2} = 0.88 \quad \text{and} \quad \frac{c_b + K_{tr}}{d_b} = \frac{2.44 + 0.88}{1.128} = 2.94$$

FIGURE 6.8
Bar details at beam-column joint for bar development examples.

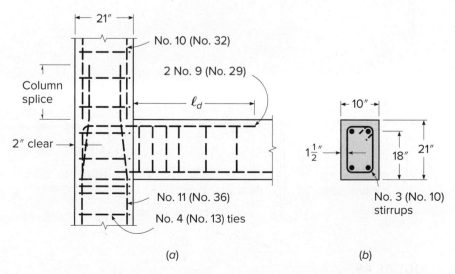

Column splice

2" clear

No. 10 (No. 32)

2 No. 9 (No. 29)

ℓ_d

No. 11 (No. 36)

No. 4 (No. 13) ties

10"

$1\frac{1}{2}$"

18"   21"

No. 3 (No. 10) stirrups

(a)   (b)

[†] If the top cover had controlled, the potential splitting plane would be vertical and one times the stirrup bar area would be used in calculating A_{tr} with $n = 1$.

This exceeds the limit value of 2.5, so $(c_b + K_{tr})/d_b$ is set to 2.5. Then from Eq. (6.5) with $\psi_s = 1.0$,

$$\ell_d = \left(\frac{3}{40} \frac{f_y}{\lambda \sqrt{f_c'}} \frac{\psi_t \psi_e \psi_s \psi_g}{\left(\dfrac{c_b + K_{tr}}{d_b} \right)} \right) d_b = \left(\frac{3}{40} \frac{60,000}{1.0\sqrt{4000}} \frac{1.3 \times 1.0 \times 1.0 \times 1.0}{(2.5)} \right) 1.128$$

$$= 37.0 \times 1.128 = 41.7 \text{ in.}$$

and the required development length is $41.7 \times 1.80/2.00 = 41.7 \times 0.90 = 37.5$ in. rather than 63 in. as before. Clearly, the use of the more accurate Eq. (6.5) permits a considerable reduction in development length. Even though its use requires much more time and effort, it is justified if the design is to be repeated many times in a structure.

6.4   ANCHORAGE OF TENSION BARS BY HOOKS

a.   Standard Dimensions

In the event that the desired tensile stress in a bar cannot be developed by bond along the length of the bar alone, it is necessary to provide special anchorage at the ends of the bar, usually by means of a 90° or a 180° hook or a headed bar (the latter is discussed in Section 6.5). The dimensions and bend radii for hooks have been standardized in ACI Code 25.3.1 and 25.3.2 as follows (see Figs. 6.9 and 6.10 and Table 6.2):

1. A 90° bend plus an extension of at least 12 bar diameters at the free end of the bar, or
2. A 180° bend plus an extension of at least 4 bar diameters, but not less than $2\frac{1}{2}$ in. at the free end of the bar, or

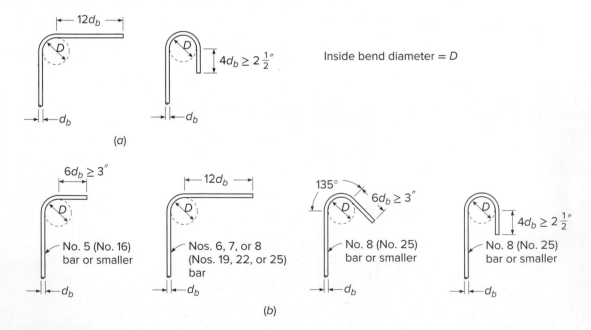

FIGURE 6.9
Standard bar hooks: (*a*) main reinforcement and (*b*) stirrups and ties. ACI Code 25.3.2 requires that standard hooks for stirrups and ties enclose longitudinal reinforcement.

FIGURE 6.10

Bar details for development of standard hooks.

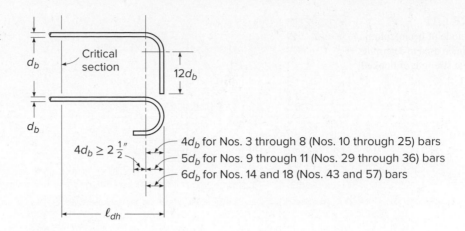

TABLE 6.2

Minimum inside bend diameters for standard hooks

Bar Size	Minimum Diameter
Nos. 3 through 8 (Nos. 10 through 25)	6 bar diameters[a]
Nos. 9, 10, and 11 (Nos. 29, 32, and 36)	8 bar diameters
Nos. 14 and 18 (Nos. 43 and 57)	10 bar diameters

[a] Nos. 3, 4, and 5 (Nos. 10, 13, and 16): 4 bar diameters for stirrups and ties.

3. For stirrup and tie anchorage only:
   (a) For Nos. 3, 4, 5 (Nos. 10, 13, and 16) bars, a 90° bend plus an extension of at least 6 bar diameters, but not less than 3 in., at the free end of the bar, or
   (b) For Nos. 6, 7, and 8 (Nos. 19, 22, and 25) bars, a 90° bend plus an extension of at least 12 bar diameters at the free end of the bar, or
   (c) For No. 8 (No. 25) bars and smaller, a 135° bend plus an extension of at least 6 bar diameters, but not less than 3 in., at the free end of the bar, or
   (d) For No. 8 (No. 25) bars and smaller, a 180° bend plus an extension of at least 4 bar diameters, but not less than 2.5 in., at the free end of the bar.

The minimum diameter of bend, measured on the inside of the bar, for standard hooks other than for stirrups or ties in sizes Nos. 3 through 5 (Nos. 10 through 16), should be not less than the values shown in Table 6.2. For stirrup and tie hooks, for bar sizes No. 5 (No. 16) and smaller, the inside diameter of bend is reduced to 4 bar diameters, according to ACI Code 25.3.2.

When welded wire reinforcement (smooth or deformed wires) is used for stirrups or ties, the inside diameter of bend should not be less than 4 wire diameters for deformed wire larger than D6 and 2 wire diameters for all other wires. Bends with an inside diameter of less than 8 wire diameters should not be less than 4 wire diameters from the nearest welded intersection.

b. **Behavior of Hooked Bars**

Hooked bars resist pullout by the combined actions of bond along the straight length of the bar leading to the hook and anchorage provided by the hook. Tests indicate that the failure of hooked bars in tension is accompanied by breakout of the concrete in

FIGURE 6.11
Failure mode of beam-column
test specimen used to determine
anchorage strength of hooked
bars (Ref. 6.25).

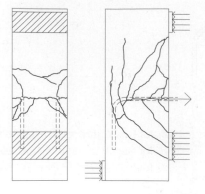

FIGURE 6.12
Beam-column test specimens
containing hooked bars at
failure: (*a*) breakout failure
and (*b*) side splitting failure
(Ref. 6.25).

(*a*)   (*b*)

the direction of the tensile force and, to a lesser degree, splitting of the concrete parallel to the plane of the hook. Splitting failure becomes more prevalent as the cover decreases and the bar size increases (Ref. 6.25). These modes of failure are illustrated in Figs. 6.11 and 6.12.

Tests demonstrate that the anchorage strength of hooked bars:

1. Increases with the compressive strength of the concrete, but to a power of f_c' close to ¼ rather than to ½, as traditionally represented by $\sqrt{f_c'}$.
2. Increases with the center-to-center spacing s of hooked bars up to about $6d_b$ and then remains constant for higher values of s.
3. Increases with confinement provided by the surrounding concrete. Hooks located within a column core with at least 2.5 in. of side cover or in other members with at least $6d_b$ of side cover on the outer hook have about 25 percent greater anchorage strength than hooked bars with lower levels of confinement.
4. Increases with confinement provided in the form of closed stirrups, ties, or other reinforcement enclosing the hook. Based on the ACI Code, the confining reinforcement must extend at least $0.75\ell_{dh}$ in the direction of the bar in tension, as shown in

FIGURE 6.13

Confining reinforcement that contributes to the anchorage strength of hooked bars: (*a*) confining reinforcement parallel to bar in tension and (*b*) confining reinforcement perpendicular to bar in tension.

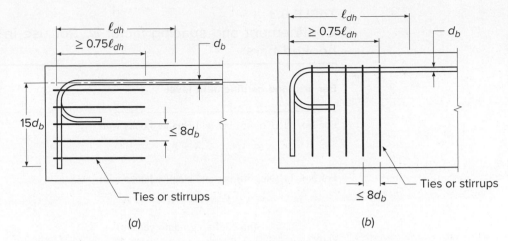

(*a*)   (*b*)

Fig. 6.13, where ℓ_{dh} is the development length of the hooked bar. The confining reinforcement may be parallel to ℓ_{dh}, enclosing the hooks within $15d_b$ of the centerline of the straight portion of the hooked bars (Fig. 6.13*a*), or perpendicular to ℓ_{dh} enclosing the hooked bars along ℓ_{dh} (Fig. 6.13*b*). The contribution of confining reinforcement to anchorage strength is based on the ratio of the total area of the confining reinforcement A_{th} (all legs of stirrups or ties) to the total area of the enclosed hooked bars A_{hs}.

The ACI Code includes other criteria, which are described in Section 6.4c. The contribution of confining reinforcement is greatest for closely spaced hooks, becoming less effective as the spacing between the hooked bars increases. Thus, the effects of increased spacing and increased confinement are not strictly additive.

The provisions in ACI Code 25.4.3 for hooked bars in tension are based on research summarized in Refs. 6.25 to 6.29. Based on that research, the development length of hooked bars ℓ_{dh} was shown to be a function of the bar diameter to the 1.5 power, as shown in Eq. (6.7):

$$\ell_{dh} = \left(\frac{f_y \psi_{cs} \psi_o}{500 f_c'^{0.25}}\right) d_b^{1.5} \tag{6.7}$$

where ψ_{cs} = factor that increases with the level of the confinement provided by stirrups or ties enclosing the hooked bars and the spacing of the hooked bars, as shown in Table 6.3.

ψ_o = 1.0 for hooked bars terminating inside a column core with side cover normal to the plane of hook $\geq$ 2.5 in., or with a side cover normal to the plane of hook $\geq 6d_b$;

= 1.25 otherwise.

Based on the limited test results for higher-strength concretes, an upper limit of 16,000 psi is placed on f_c' in Eq. (6.7). Due to a lack of experimental data, the effect of confining reinforcement is not considered for hooked bars larger than No. 11 (No. 36), as shown in Table 6.3 (Refs. 6.25 to 6.29).

c.   **Development Length and Modification Factors for Hooked Bars**

The development length ℓ_{dh} is defined as shown in Fig. 6.10 and is measured from the critical section to the farthest point on the bar, parallel to the straight part of the bar. The ACI Code requirements for the development length of hooked bars are based on Eq. (6.7) but with a number of modifications, as shown in Eq. (6.8). Rather than

TABLE 6.3
Confinement and spacing factor ψ_{cs} for use in Eq. (6.7) for hooked bars[†]

Bar size and confinement level	ψ_{cs}	
	$s/d_b{}^{\ddagger} = 2$	$s/d_b{}^{\ddagger} \geq 6$
For No. 11 (No. 36) bar and smaller hooks with $\dfrac{A_{th}}{A_{hs}} \geq 0.4$	1.0	5/6
For No. 11 (No. 36) bar and smaller hooks with $\dfrac{A_{th}}{A_{hs}} = 0$ (no confining reinforcement)	5/3	1.0
Hooked bars larger than No. 11 (No. 36) with any value of A_{th}/A_{hs}	5/3	1.0

[†]Interpolation permitted.
[‡]d_b = diameter of hooked bar.

represent the effect of concrete compressive strength as $f_c'^{0.25}$, as it is in Eq. (6.7), the ACI Code approximates $f_c'^{0.25}$ (for concrete compressive strengths below 6000 psi) as $\sqrt{f_c'}/\psi_c$, where ψ_c is defined following Eq. (6.8). Above 6000 psi, $\psi_c = 1.0$. The ACI Code also simplifies ψ_{cs} by replacing it with ψ_r, also defined following Eq. (6.8). Like Eq. (6.5) for straight bar development, Eq. (6.8) includes a term for epoxy-coated reinforcement ψ_e. In accordance with ACI Code 25.4.3, the development length of hooked bars in tension is

$$\ell_{dh} = \left(\frac{f_y \psi_e \psi_r \psi_o \psi_c}{55\lambda\sqrt{f_c'}} \right) d_b^{1.5} \tag{6.8}$$

In Eq. (6.8), the terms are defined as follows:

ψ_e = epoxy coating factor
    For epoxy-coated or zinc and epoxy dual-coated reinforcement: 1.2
    For uncoated and zinc-coated (galvanized) reinforcement: 1.0

ψ_r = confining reinforcement factor
    For No. 11 (No. 36) and smaller bars with $A_{th} \geq 0.4A_{hs}$ or
      minimum spacing between hooked bars $s \geq 6d_b$ 1.0
    Other: 1.6

ψ_o = location factor
    For No. 11 (No. 36) and smaller diameter hooked bars
      (1) terminating inside a column core with side cover normal to
      the plane of hook ≥ 2.5 in., or (2) with side cover normal to the
      plane of the hook $\geq 6d_b$ 1.0
    Other: 1.25

ψ_c = concrete strength factor
    For $f_c' < 6000$ psi $f_c'/15{,}000 + 0.6$
    Other   1.0

λ = lightweight concrete factor
    For lightweight concrete: 0.75
    Other: 1.0

FIGURE 6.14

Transverse reinforcement requirements at discontinuous ends of members with small cover distances.

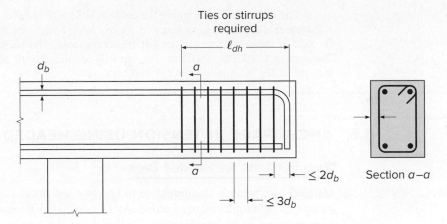

In any case, $\sqrt{f_c'}$ may not exceed 100 psi in Eq. (6.8), and ℓ_{dh} may not be less than $8d_b$ and not less than 6 in. Finally, unlike straight reinforcement, ℓ_{dh} may not be reduced by the ratio (A_s required/A_s provided).

Transverse confinement steel is essential if the full bar strength must be developed with minimum concrete confinement, such as when hooks may be required at the ends of a simply supported beam or where a beam in a continuous structure frames into an end column and does not extend past the column or when bars must be anchored in a short cantilever, as shown in Fig. 6.14 (Ref. 6.11). According to ACI Code 25.4.3.4, for bars hooked at the discontinuous ends of members with both side cover and top or bottom cover less than $2\frac{1}{2}$ in., hooks *must* be enclosed with closed stirrups or ties along the full development length, as shown in Fig. 6.14. The spacing of the confinement steel must not exceed 3 times the diameter of the hooked bar d_b, and the first stirrup or tie must enclose the bent portion of the hook within a distance equal to $2d_b$ of the outside of the bend.

EXAMPLE 6.2   **Development of hooked bars in tension.**   Referring to the beam-column joint shown in Fig. 6.8, the No. 9 (No. 29) negative bars are to be extended into the column and terminated in a standard 90° hook, keeping 2 in. clear to the outside face of the column. The column width in the direction of beam width is 16 in. Find the minimum length of embedment of the hook past the column face, and specify the hook details.

SOLUTION.   The development length for hooked bars, measured from the critical section along the bar to the far side of the vertical hook, is given by Eq. (6.8). In this case, the bars are uncoated, giving an epoxy coating factor of $\psi_e = 1.0$. The No. 9 (No. 29) bars are anchored within the column core and the side cover exceeds 2.5 in., so the location factor $\psi_o = 1.0$. From Example 6.1, the spacing between the bars is 5.11 in., or less than $6d_b$, but the confining reinforcement consists of two No. 4 (No. 13) closed ties with a total area $A_{th} = 4 \times 0.4$ in^2 = 0.8 in^2, giving $A_{th}/A_{hs} = 0.4$; thus, $\psi_r = 1.0$. For $f_c' = 4000$ psi, $\psi_c = f_c'/15{,}000 + 0.6 = 4000/15{,}000 + 0.6 = 0.867$. This gives

$$\ell_{dh} = \left(\frac{f_y \psi_e \psi_r \psi_o \psi_c}{55\lambda\sqrt{f_c'}}\right)d_b^{1.5} = \left(\frac{60{,}000 \times 1.0 \times 1.0 \times 1.0 \times 0.867}{55 \times 1.0\sqrt{4000}}\right)1.128^{1.5} = 17.9 \text{ in.}$$

The available length within the column is $21 - 2 = 19$ in. and can thus accommodate ℓ_{dh}. Note that although ℓ_{dh} is less that the available length, the hook should be placed to the far side of the column to use the full 19 in. available. The hook will be bent to a minimum diameter of $8 \times 1.128 = 9.02$ in. The bar will continue for 12 bar diameters, or 13.5 in. past the end of the bend in the vertical direction.

6.5   ANCHORAGE IN TENSION USING HEADED BARS

a.   Requirements for Headed Bars

Headed bars provide an alternative to hooks when the desired tensile stress in the bar cannot be developed by bond alone. ACI Code 20.2.1.6 requires that headed deformed bars conform to the requirements for HA heads in Annex A1 of ASTM A970. Both ASTM A970 and ACI Code 25.4.4.1 require that bearing area of the head A_{brg} be equal to at least 4 times the area of the bar A_b. For headed bars with obstructions, A_{brg} is taken as the gross area of the head minus the maximum area of the obstruction. Figure 6.15*a* shows the maximum dimensions of an obstruction as permitted by ASTM A970. When an obstruction has a gap adjacent to the head, as shown in Fig. 6.15*b*, A_{brg} is taken as the gross area of the head minus the area of the obstruction adjacent to the bearing face provided that the gap has a width at least equal to the larger of $\frac{3}{8}$ in. (10 mm) and $\frac{3}{8}d_b$, the depth of the gap does not exceed the width of the gap, and the obstruction everywhere within the gap falls inside a straight line connecting the outer dimension of the obstruction at the initiation of the gap with the dimension of the obstruction at the bearing face of the head. Test results (Refs. 6.29 to 6.32) show that small obstructions, no larger than the dimensions shown in Fig. 6.16, do not affect that anchorage strength of headed bars and, as a result, according to ASTM A970 are not considered to detract from the net bearing area of the head.

FIGURE 6.15
Headed deformed reinforcing bar: (*a*) maximum dimensions for obstruction of the deformations and (*b*) details of gap in obstruction adjacent to a head [gap width $\geq$ larger of $\frac{3}{8}$ in. (10 mm) and $\frac{3}{8}d_b$, depth of gap $\leq$ width of gap, obstruction within gap falls inside straight line connecting outer dimension of obstruction at initiation of gap with dimension of obstruction at bearing face of head].

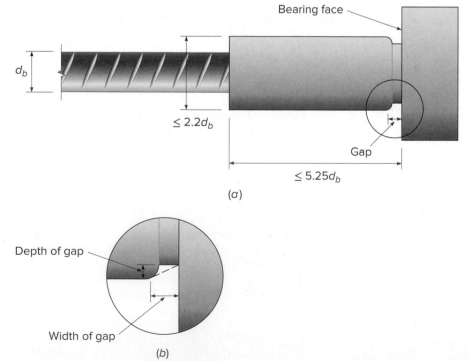

FIGURE 6.16
Headed deformed bar with an
obstruction not considered to
detract from the net bearing
area of the head.

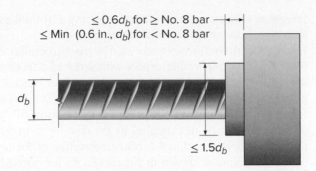

$\leq 0.6d_b$ for $\geq$ No. 8 bar

$\leq$ Min (0.6 in., d_b) for $<$ No. 8 bar

d_b

$\leq 1.5d_b$

b. Behavior of Headed Bars

Headed deformed bars resist pullout by a combination of bond along the straight length of the bar leading to the head and bearing resistance provided by the head. Many aspects of the anchorage behavior of headed bars are similar to those of hooked bars, but with some important differences. Tests indicate that failure of headed bars in tension is most often accompanied by breakout of the concrete in the direction of the tensile force, sometimes accompanied by side-face blowout. Two types of breakout failure surface were observed in the beam-column test specimens that served as the basis for the ACI Code provisions: a cone-shaped failure, as shown in Fig. 6.17a, and a similar failure where a crack extended above the joint region toward the top of the column, as shown in Fig. 6.17b.

Tests demonstrate that the anchorage strength of deformed headed bars:

1. Increases with the compressive strength of the concrete to a power of f_c' close to $\frac{1}{4}$, as observed for hooked bars.

FIGURE 6.17
Breakout failure in beam-
column test specimens
containing headed deformed
bars: (a) cone-shaped and
(b) back cover splitting
(Ref. 6.29).

(a)   (b)

2. Increases with the center-to-center spacing s of hooked bars up to about $8d_b$ and then remains constant for higher values of s.

3. Increases with confinement provided by the surrounding concrete. Headed deformed bars located within a column core with at least 2.5 in. of side cover or in other members with at least $8d_b$ of side cover on the outer headed bar have about 25 percent greater anchorage strength than headed bars with lower levels of confinement.

4. Increases with confinement provided in the form of "parallel tie reinforcement" (closed stirrups or ties parallel to the development length ℓ_{dt}) in beam-column joints and located within $8d_b$ of the centerline of the headed bar toward the middle of the joint, as shown in Fig. 6.18. As for hooked bars, the contribution of parallel tie reinforcement to anchorage strength is based on the ratio of the total area of the parallel tie reinforcement A_{tt} (all legs of stirrups or ties) to the total area of the enclosed headed bars A_{hs}. Unlike hooked bars, however, the anchorage strength of headed bars is not increased by confining reinforcement perpendicular to the development length. Because of a lack of test data, A_{tt} is not considered for anchorages other than beam-column joints.

The ACI Code includes other criteria that are described in Section 6.5c. As with hooked bars, the contribution of confining reinforcement (for headed bars in the form of parallel ties) is greatest for closely spaced headed bars, becoming less effective as the spacing between the hooked bars increases. Thus, the effects of increased spacing and increased confinement are not strictly additive.

The provisions in ACI Code 25.4.4 for headed deformed bars in tension are based on research summarized in Refs. 6.29 to 6.32. Like hooked bars, the development length of headed deformed bars ℓ_{dt} is a function of the bar diameter to the 1.5 power, as shown in Eq. (6.9):

$$\ell_{dt} = \left(\frac{f_y \psi_{cs} \psi_o}{800 f_c'^{0.25}} \right) d_b^{1.5} \tag{6.9}$$

where ψ_{cs} = factor that increases with the level of the confinement provided by parallel ties and spacing of headed deformed bars, as shown in Table 6.4.

$\psi_o = 1.0$ for headed deformed bars terminating inside a column core with side cover to the bar ≥ 2.5 in., or in any member with a side cover to the bar $\geq 8d_b$;

$= 1.25$ otherwise.

Because of a lack of experimental data, no recommendations were made for headed deformed bars larger than No. 11 (No. 36).

FIGURE 6.18

Parallel ties located within $8d_b$ of the centerline of the headed bar toward the middle of the joint contribute to the anchorage strength of headed bars in beam-column joints.

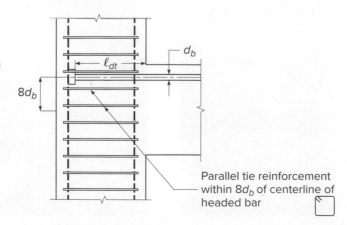

Parallel tie reinforcement within $8d_b$ of centerline of headed bar

TABLE 6.4

Confinement and spacing factor ψ_{cs} for use in Eq. (6.9) for deformed headed bars[†]

Bar size and confinement level	ψ_{cs}	
	$s/d_b{}^{‡} = 2$	$s/d_b{}^{‡} \geq 8$
$\dfrac{A_{th}}{A_{hs}} \geq 0.3$	1.2	0.8
$\dfrac{A_{th}}{A_{hs}} = 0$ (no confining reinforcement)	2.0	1.0

[†]Interpolation permitted.
[‡]d_b = diameter of headed deformed bar.

c.   Development Length and Modification Factors for Headed Bars

The development length ℓ_{dt} is defined as shown in Fig. 6.19 and is measured from the bearing face of the head to the critical section. The ACI Code requirements for development length of headed bars are based on Eq. (6.9) but with the modifications shown in Eq. (6.10). As in the case of hooked bars, rather than represent the effect of concrete compressive strength as $f_c'^{0.25}$, as shown in Eq. (6.9), the ACI Code approximates $f_c'^{0.25}$ as $\sqrt{f_c'}/\psi_c$, where ψ_c is defined following Eqs. (6.8) and (6.10). Also, as with hooked bars, the ACI Code simplifies ψ_{cs} by replacing it with another term, in this case ψ_p, defined following Eq. (6.10), and although the limits on ψ_{cs} are based on values of s/d_b between 2 and 8, the limits on ψ_p are based on values of s/d_b between 3 and 6. Equation (6.10) includes a term for epoxy-coated reinforcement ψ_e but does not include a term for lightweight concrete because no tests have been performed to evaluate the anchorage strength of headed deformed bars in lightweight concrete, and for that reason, their use is limited to normalweight concrete. In accordance with ACI Code 25.4.4, the development length of headed deformed bars in tension is

$$\ell_{dt} = \left(\frac{f_y \psi_e \psi_p \psi_o \psi_c}{75\sqrt{f_c'}} \right) d_b^{1.5} \tag{6.10}$$

In Eq. (6.10), the terms are defined as follows:

ψ_e = epoxy coating factor

For epoxy-coated or zinc and epoxy dual-coated reinforcement:	1.2
For uncoated and zinc-coated (galvanized) reinforcement:	1.0

FIGURE 6.19
Development length of headed deformed bars.

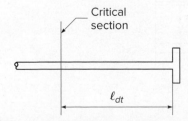

Critical section

ℓ_{dt}

ψ_p = parallel tie reinforcement factor
For No. 11 (No. 36) and smaller bars with $A_{th} \geq 0.3A_{hs}$ or minimum
    spacing between hooked bars $s \geq 6d_b$ 1.0
    Other: 1.6

ψ_o = location factor
For headed bars (1) terminating inside a column core with side
    cover to bar ≥ 2.5 in., or (2) with side cover to bar $\geq 6d_b$ 1.0
    Other: 1.25

ψ_c = concrete strength factor
For $f_c' < 6000$ psi $f_c'/15,000 + 0.6$
    Other: 1.0

The design provisions are restricted to No. 11 (No. 36) and smaller headed bars. The bars, as distinct from the heads, must have a clear cover of at least $2d_b$ and a center-to-center spacing between bars of at least $3d_b$. The value of $\sqrt{f_c'}$ may not exceed 100 psi in Eq. (6.10), and ℓ_{dt} may not be less than $8d_b$ and not less than 6 in. Like ℓ_{dh}, ℓ_{dt} may not be reduced by the ratio (A_s required/A_s provided).

When headed bars from a member, such as a beam or slab, terminate in a supporting member, such as the column shown in Fig. 6.20, ACI Commentary 25.4.4.2 recommends that the bar be extended "through the joint of the far face of the confined core of the supporting member, allowing for cover and avoidance of interference with column reinforcement, even though the resulting anchorage length may exceed ℓ_{dt}." Doing so helps adequately anchor the compressive forces that are developed at the face of the head and improves the performance of the beam column connection.

d.   Mechanical Anchorage

In cases where headed bars do not meet the requirements specified in ACI Code 25.4.4.1 or in cases where bars are terminated by mechanisms such as welded plates or other manufactured devices, ACI Code 25.4.5 allows such devices to be used to develop the reinforcement if the adequacy of the devices is established by tests. In such cases, the development of the reinforcement may consist of the combined contributions of bond along the length of the bar leading to the critical section, plus that of the mechanical anchorage, much in the way that the total resistance of headed bars is provided.

FIGURE 6.20
Headed deformed bar
extended to far side of
column with anchorage
length that exceeds ℓ_{dt}.

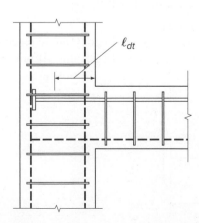

EXAMPLE 6.3   **Development of headed deformed bars in tension.**   Two No. 7 (No. 22) bars serve as top reinforcement for a bracket framing into a 16×16 in. column (Fig. 6.21). The bracket projects 15 in. from the column and is the same width as the column. The top cover to the center of the bars is 3 in., and the side cover to the center of the bars is 3.5 in. The bars are spaced laterally at 4.9 in. These dimensions are inadequate for straight development length or for standard hooks. Based on other reinforcement, cover requirements, and head thickness, total development lengths for headed bars of 13.5 in. in the column and 12.5 in. in the bracket are available. The reinforcing bars have $f_y = 60,000$ psi, and the concrete is normalweight with $f'_c = 5000$ psi. Determine if a bar with heads at both ends can be used in this application.

SOLUTION.   The minimum head size is $A_{brg} = 4A_b = 2.4$ in.^{2}. The smaller available anchorage length in the bracket governs. The bars are uncoated, giving an epoxy coating factor of $\psi_e = 1.0$. The No. 7 (No. 22) bars are not anchored in a column core and have less than $6d_b$ of side cover, so the location factor $\psi_o = 1.25$. The center-to-center bar spacing between the bars is 9 in., which is greater than $6d_b$; thus, $\psi_p = 1.0$. For $f'_c = 5000$ psi, $\psi_p = f'_c/15,000 + 0.6 = 5000/15,000 + 0.6 = 0.933$. The development length ℓ_{dt} calculated using Eq. (6.10) is

$$\ell_{dt} = \left(\frac{f_y \psi_e \psi_p \psi_o \psi_c}{75\sqrt{f'_c}} \right) d_b^{1.5} = \left(\frac{60,000 \times 1.0 \times 1.0 \times 1.25 \times 0.93}{75\sqrt{5000}} \right) 0.875^{1.5} = 10.8 \text{ in.}$$

which must be checked against the minimum values for ℓ_{dt}, which are

$$\ell_{dt} \geq 8d_b = 7 \text{ in.}$$
$$\ell_{dt} \geq 6 \text{ in.}$$

The value of ℓ_{dt} obtained using Eq. (6.10) governs and is less than the available anchorage lengths in both the column and the bracket. Thus, a bar with heads at both ends can be used. As shown in Fig. 6.21, the heads are located at the far faces of the column and the bracket, with a distance between the faces of the heads of 26 in.

FIGURE 6.21
Column and bracket for headed deformed bar development example.

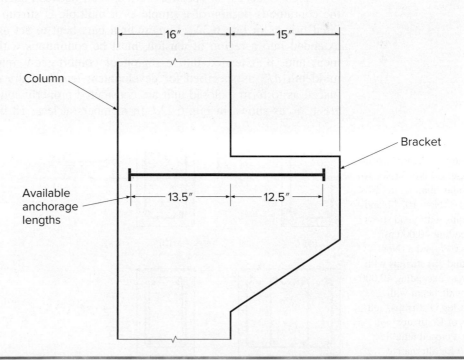

6.6   ANCHORAGE REQUIREMENTS FOR WEB REINFORCEMENT

Stirrups should be carried as close as possible to the compression and tension faces of a beam, and special attention must be given to proper anchorage. The truss model (see Section 5.8 and Fig. 5.18) for design of shear reinforcement indicates the development of diagonal compressive struts, the thrust from which is equilibrated, near the top and bottom of the beam, by the tension web members (i.e., the stirrups). Thus, at the factored load, the tensile strength of the stirrups must be developed for almost their full height. Clearly, it is impossible to do this by development length. For this reason, stirrups normally are provided with 90°, 135°, or 180° hooks at their upper end (see Fig. 6.9*b* for standard hook details) and at their lower end are bent 90° to pass around the longitudinal reinforcement. In simple spans, or in the positive bending region of continuous spans, where no top bars are required for flexure, stirrup support bars must be used. These are usually about the same diameter as the stirrups themselves, and they not only provide improved anchorage of the hooks but also facilitate fabrication of the reinforcement cage, holding the stirrups in position during placement of the concrete.

ACI Code 25.7.1 includes special provisions for anchorage of web reinforcement. The ends of single-leg, simple-U, or multiple-U stirrups are to be anchored by one of the following means:

1. For No. 5 (No. 16) bars and smaller, and for Nos. 6, 7, and 8 (Nos. 19, 22, and 25) bars with f_{yt} of 40,000 psi or less, a standard hook around longitudinal reinforcement, as shown in Fig. 6.22*a*.
2. For Nos. 6, 7, and 8 (Nos. 19, 22, and 25) stirrups with f_{yt} greater than 40,000 psi, a standard hook around a longitudinal bar, plus an embedment between midheight of the member and the outside end of the hook equal to or greater than $0.014d_bf_{yt}/\lambda\sqrt{f_c'}$ in., as shown in Fig. 6.22*b*.

ACI Code 25.7.1 specifies further that, between anchored ends, each bend in the continuous portion of a simple-U or multiple-U stirrup must enclose a longitudinal bar, as in Fig. 6.22*c*. Longitudinal bars bent to act as shear reinforcement, if extended into a region of tension, must be continuous with longitudinal reinforcement and, if extended into a region of compression, must be anchored beyond middepth $d/2$ as specified for development length. Pairs of U stirrups or ties so placed as to form a closed unit are considered properly spliced when length of laps are $1.3\ell_d$ as shown in Fig. 6.22*d*. In members at least 18 in. deep, such splices are

FIGURE 6.22
ACI requirements for stirrup anchorage: (*a*) No. 5 (No. 16) stirrups and smaller, and Nos. 6, 7, and 8 (Nos. 19, 22, and 25) stirrups with yield stress not exceeding 40,000 psi; (*b*) Nos. 6, 7, and 8 (Nos. 19, 22, and 25) stirrups with yield stress exceeding 40,000 psi; (*c*) wide beam with multiple-leg U stirrups; and (*d*) pairs of U stirrups forming a closed unit. See Fig. 6.9*b* for alternative standard hook details.

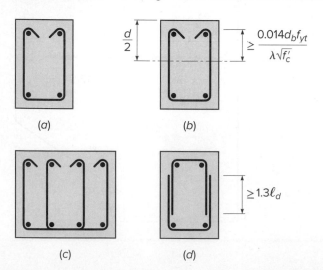

considered adequate if $A_b f_{yt} \leq 9000$ lb and the stirrup legs extend the full depth of the member. As will be discussed in Sections 6.11 and 8.5e, respectively, pairs of U stirrups may not be used in perimeter beams or for torsional reinforcement.

Other provisions are contained in ACI Code 25.7.1 relating to the use of welded wire reinforcement, which is sometimes used for web reinforcement in precast and prestressed concrete beams.

6.7   WELDED WIRE REINFORCEMENT

Tensile steel consisting of welded wire reinforcement, with either deformed or smooth wires, is commonly used in one-way and two-way slabs and certain other types of members (see Section 2.15). For *deformed* wire reinforcement, some of the development is assigned to the welded cross wires and some to the embedded length of the deformed wire. According to ACI Code 25.4.6, the development length of welded deformed wire reinforcement measured from the point of the critical section to the end of the wire is computed as the product of the development length ℓ_d from Table 6.1 or from the more accurate Eq. (6.5) and the appropriate modification factor or factors related to those equations, except that the development length may not to be less than 8 in. For welded deformed wire reinforcement with at least one cross wire within the development length and not less than 2 in. from the point of the critical section, as shown in Fig. 6.23a, a *deformed wire factor* ψ_w equal to the greater of

$$\frac{f_y - 35{,}000}{f_y} \tag{6.11a}$$

and

$$\frac{5d_b}{s} \tag{6.11b}$$

is applied, where s is the lateral spacing of the wires being developed; but this factor need not exceed 1.0. When ψ_w from Eq. (6.11a) or (6.11b) is used, the epoxy coating factor ψ_e is taken as 1.0. For welded deformed wire reinforcement with no cross wires

FIGURE 6.23
Development of (*a*) welded deformed wire reinforcement and (*b*) welded plain wire reinforcement.

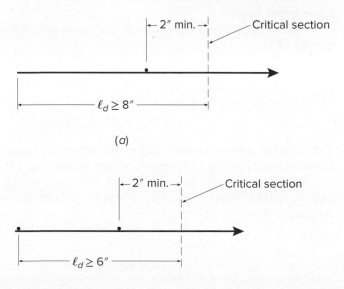

(*a*)

(*b*)

within the development length or with a single cross wire less than 2 in. from the point of the critical section, the wire fabric factor is taken to be equal to 1.0 and the development length determined as for the deformed wire.

For welded *plain* wire reinforcement, development is considered to be provided by embedment of two cross wires, with the closer wire not less than 2 in. from the critical section, as shown in Fig. 6.23b. However, the development length measured from the critical section to the outermost cross wire is not to be less than

$$\ell_d = 0.27 \left(\frac{f_y}{\lambda \sqrt{f_c'}} \right) \left(\frac{A_b}{s} \right) \tag{6.12}$$

according to ACI Code 25.4.7, where A_b is the cross-sectional area of an individual wire to be developed or spliced. The modification factor for excess reinforcement may be applied, but ℓ_d is not to be less than 6 in. for the welded plain wire reinforcement.[†]

6.8 DEVELOPMENT OF BARS IN COMPRESSION

Reinforcement may be required to develop its compressive strength by embedment under various circumstances, for example, where bars transfer their share of column loads to a supporting footing or where lap splices are made of compression bars in a column (see Section 6.13). In the case of bars in compression, a part of the total force is transferred by bond along the embedded length, and a part is transferred by end bearing of the bars on the concrete. Because the surrounding concrete is relatively free of cracks and because of the beneficial effect of end bearing, shorter basic development lengths are permissible for compression bars than for tension bars. If transverse confinement steel is present, such as spiral column reinforcement, special spiral steel around an individual bar, or column ties with a minimum spacing, the required development length is further reduced. Hooks and heads such as are shown in Figs. 6.9 and 6.15 are *not* effective in transferring compression from bars to concrete, and, if present for other reasons, should be disregarded in determining required embedment length.

According to ACI Code 25.4.9, the development length in compression is the greater of

$$\ell_{dc} = \left(\frac{f_y \psi_r}{50 \lambda \sqrt{f_c'}} \right) d_b \tag{6.13a}$$

and

$$\ell_{dc} = 0.0003 \, f_y \, \psi_r \, d_b \tag{6.13b}$$

The factor ψ_r is based on confining reinforcement and along with the factor for excess reinforcement is given in Table 6.5. ψ_r may be taken as 1.0.

In no case is ℓ_{dc} to be less than 8 in., according to the ACI Code. Basic and modified compressive development lengths are given in Table A.11 of Appendix A.

[†] The ACI Code offers no explanation as to why $\ell_{d,min} = 6$ in. for welded plain wire reinforcement, but 8 in. for welded deformed wire reinforcement, but two cross wires are required for welded plain wire reinforcement versus one cross wire for welded deformed wire reinforcement.

TABLE 6.5
Development lengths for deformed bars in compression

A. Basic development length ℓ_{dc}

$$\begin{cases} \geq \left(\dfrac{f_y \psi_r}{50 \lambda \sqrt{f_c'}} \right) d_b \\[2mm] \geq 0.0003\, f_y\, \psi_r d_b \end{cases}$$

B. Modification factors to be applied to ℓ_{dc}

ψ_r Reinforcement enclosed within a spiral, a circular continuously wound tie with $d_b \geq \frac{1}{4}$ in. and a pitch of 4 in., No. 4 (No. 13) bar ties spaced ≤ 4 in. on center, or hoops (closed tie or continuously wound tie with seismic hooks at both ends) spaced ≤ 4 in. on center	0.75
Other	1.0
Reinforcement in excess of that required by analysis	$\dfrac{A_s \text{ required}}{A_s \text{ provided}}$

6.9 BUNDLED BARS

It was pointed out in Section 4.5c that it is sometimes advantageous to "bundle" tensile reinforcement in large beams, with two, three, or four bars in contact, to provide for improved placement of concrete around and between bundles of bars. Bar bundles are typically triangular or L-shaped for three bars, and square for four and must be enclosed in transverse reinforcement. If the bars are in compression, the transverse reinforcement must be at least No. 4 (No. 13) bars. When bars are cut off in a bundled group, the cutoff points must be staggered at least 40 diameters. According to ACI Code 25.6.1, the development length of individual bars within a bundle, for both tension and compression, is that of the individual bar increased by 20 percent for a three-bar bundle and by 33 percent for a four-bar bundle, to account for the probable deficiency of bond at the inside of the bar group.

A unit of bundled bars is treated as a single bar with a diameter d_b derived from the equivalent total area and having a centroid that coincides with that of the bar group (1) to determine the appropriate spacing limitations and cover requirements based on d_b, (2) for use in Table 6.1, (3) when calculating the confinement term K_{tr} in Eq. (6.5), and (4) when selecting the epoxy coating factor ψ_e.

6.10 BAR CUTOFF AND BEND POINTS IN BEAMS

Chapter 4 deals with moments, flexural stresses, concrete dimensions, and longitudinal bar areas at the critical moment sections of beams. These critical moment sections are generally at the face of the supports (negative bending) and near the middle of the span (positive bending). Occasionally, haunched members having variable depth or width are used so that the concrete flexural capacity will agree more closely with the variation of bending moment along a span or series of spans. Usually, however, prismatic beams with constant concrete cross-sectional dimensions are used to simplify formwork and thus to reduce cost.

The steel requirement, on the other hand, is easily varied in accordance with requirements for flexure, and it is common practice either to cut off bars where they are no longer needed to resist stress or, sometimes in the case of continuous beams, to bend up the bottom steel (usually at 45°) so that it provides tensile reinforcement at the top of the beam over the supports.

a. Theoretical Points of Cutoff or Bend

The tensile force to be resisted by the reinforcement at any cross section is

$$T = A_s f_s = \frac{M}{jd}$$

where M is the value of bending moment at that section and jd is the internal lever arm of the resisting moment. The lever arm jd varies only within narrow limits and is never less than the value at the maximum-moment section. Consequently, the tensile force can be taken with good accuracy directly proportional to the bending moment. Since it is desirable to design so that the steel everywhere in the beam is as nearly fully stressed as possible, it follows that the required steel area is very nearly proportional to the bending moment.

To illustrate, the moment diagram for a uniformly loaded simple-span beam shown in Fig. 6.24a can be used as a steel requirement diagram. At the

FIGURE 6.24
Bar cutoff points from moment diagrams.

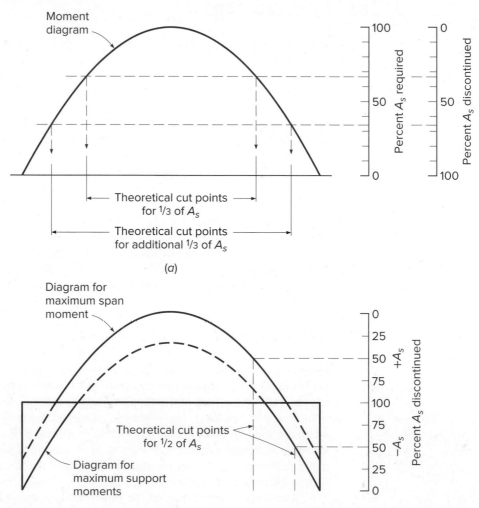

maximum-moment section, 100 percent of the tensile steel is required (0 percent can be discontinued or bent), while at the supports, 0 percent of the steel is theoretically required (100 percent can be discontinued or bent). The percentage of bars that could be discontinued elsewhere along the span is obtainable directly from the moment diagram, drawn to scale. To facilitate the determination of cutoff or bend points for simple spans, Graph A.2 of Appendix A has been prepared. It represents a half-moment diagram for a uniformly loaded simple span.

To determine cutoff or bend points for continuous beams, the moment diagrams resulting from loading for maximum span moment and maximum support moment are drawn. A moment envelope results that defines the range of values of moment at any section. Cutoff or bend points can be found from the appropriate moment curve as for simple spans. Figure 6.24b illustrates, for example, a continuous beam with moment envelope resulting from alternate loadings to produce maximum span and maximum support moments. The locations of the points at which 50 percent of the bottom and top steel may theoretically be discontinued are shown.

According to ACI Code 6.5, uniformly loaded, continuous reinforced concrete beams of fairly regular span may be designed using moment coefficients (see Table 11.1). These coefficients, analogous to the numerical constant in the expression $\frac{1}{8}wL^2$ for simple-beam bending moment, give a conservative approximation of span and support moments for continuous beams. When such coefficients are used in design, cutoff and bend points may conveniently be found from Graph A.3 of Appendix A. Moment curves corresponding to the various span and support-moment coefficients are given at the top and bottom of the chart, respectively.

Alternatively, if moments are found by frame analysis rather than from ACI moment coefficients, the location along the span where bending moment reduces to any particular value (e.g., as determined by the bar group after some bars are cut off), or to zero, is easily computed by statics.

b. Practical Considerations and ACI Code Requirements

Actually, in no case should the tensile steel be discontinued exactly at the theoretically described points. As described in Section 5.3 and shown in Fig. 5.8, when diagonal tension cracks form, an internal redistribution of forces occurs in a beam. Prior to cracking, the steel tensile force at any point is proportional to the moment at a vertical section passing through the point. However, after the crack has formed, the tensile force in the steel at the crack is governed by the moment at a section nearer midspan, which may be much larger. Furthermore, the actual moment diagram may differ from that used as a design basis, due to approximation of the real loads, approximations in the analysis, or the superimposed effect of settlement or lateral loads. In recognition of these facts, ACI Code 7.7.3.3 and 9.7.3.3, covering one-way slabs and beams, respectively, require that every bar be continued at least a distance equal to the effective depth of the beam or 12 bar diameters (whichever is larger) beyond the point at which it is theoretically no longer required to resist stress, except at supports of simple spans and at the free end of cantilevers.

In addition, it is necessary that the calculated stress in the steel at each section be developed by adequate embedded length or end anchorage, or a combination of the two. For the usual case, with no special end anchorage, this means that the full development length ℓ_d must be provided beyond critical sections at which peak stress exists in the bars. These critical sections are located at points of maximum moment and at points where adjacent terminated reinforcement is no longer needed to resist bending.[†]

Further reflecting the possible change in peak stress location, ACI Code 7.7.3.8 and 9.7.3.8 require that at least one-third of the positive-moment steel (one-fourth in continuous spans) be continued uninterrupted along the same face of the beam a distance at least 6 in. into the support. When a flexural member is a part of a primary lateral load resisting system, positive-moment reinforcement required to be extended into the support must be anchored to develop the yield strength of the bars at the face of support to account for the possibility of reversal of moment at the supports. According to ACI Code 7.7.3.8 and 9.7.3.8, at least one-third of the total reinforcement provided for negative moment at the support must be extended beyond the extreme position of the point of inflection a distance not less than one-sixteenth the clear span, or d, or $12d_b$, whichever is greatest.

Requirements for bar cutoff or bend point locations are summarized in Fig. 6.25. If negative bars L are to be cut off, they must extend a full development length ℓ_d beyond the face of the support. In addition, they must extend a distance d or $12d_b$ beyond the theoretical point of cutoff defined by the moment diagram. The remaining negative bars M (at least one-third of the total negative area) must extend at least ℓ_d beyond the theoretical point of cutoff of bars L and in addition must extend d, $12d_b$, or $\ell_n/16$ (whichever is greatest) past the point of inflection of the negative-moment diagram.

If the positive bars N are to be cut off, they must project ℓ_d past the point of theoretical maximum moment, as well as d or $12d_b$ beyond the cutoff point from the positive-moment diagram. The remaining positive bars O must extend ℓ_d past the theoretical point of cutoff of bars N and must extend at least 6 in. into the face of the support.

When bars are cut off in a tension zone, there is a tendency toward the formation of premature flexural and diagonal tension cracks in the vicinity of the cut end. This may result in a reduction of shear capacity and a loss in overall ductility of the beam. ACI Code 7.7.3.5 and 9.7.3.5 require special precautions, specifying that no flexural bar may be terminated in a tension zone unless *one* of the following conditions is satisfied:

1. The shear is not over two-thirds of the design strength ϕV_n.
2. For No. 11 (No. 36) or smaller bars, continuing bars provide twice the area required for flexure at that point, and the shear does not exceed three-quarters of the design strength ϕV_n.
3. Stirrups in excess of those normally required are provided over a distance along each terminated bar from the point of cutoff equal to $\frac{3}{4}d$. These "binder" stirrups shall provide an area $A_v \geq 60b_w s/f_{yt}$. In addition, the stirrup spacing must not exceed $d/8\beta_b$, where β_b is the ratio of the area of bars cut off to the total area of bars at the section.

As an alternative to cutting off the steel, tension bars may be anchored by bending them across the web and making them continuous with the reinforcement on the opposite face. Although this leads to some complication in detailing and placing the steel, thus adding to construction cost, some engineers prefer the arrangement because added insurance is provided against the spread of diagonal tension cracks. In some cases, particularly for relatively deep beams in which a large percentage of the total bottom

---

[†] The ACI Code is ambiguous as to whether or not the extension length d or $12d_b$ is to be added to the required development length ℓ_d. The Code Commentary presents the view that these requirements need not be superimposed, and Fig. 6.25 has been prepared on that basis. However, the argument just presented regarding possible shifts in moment curves or steel stress distribution curves leads to the conclusion that these requirements should be superimposed. In such cases, each bar should be continued a distance ℓ_d plus the greater of d or $12d_b$ beyond the peak stress location.

FIGURE 6.25
Bar cutoff requirements of
the ACI Code.

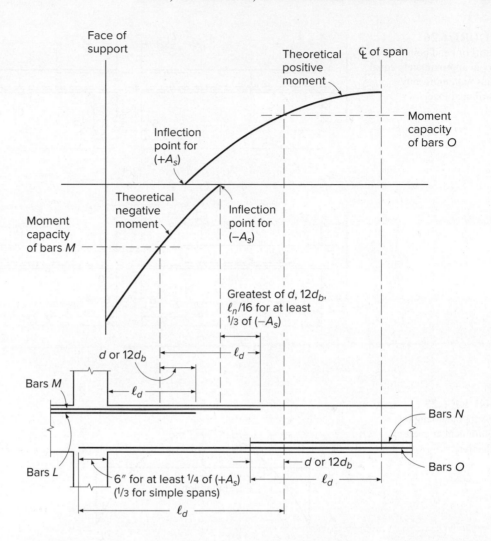

steel is to be bent, it may be impossible to locate the bend-up point for bottom bars far enough from the support for the same bars to meet the requirements for top steel. The theoretical points of bend should be checked carefully for both bottom and top steel.

Because the determination of cutoff or bend points may be rather tedious, particularly for frames that have been analyzed by elastic methods rather than by moment coefficients, many designers specify that bars be cut off or bent at more or less arbitrarily defined points that experience has proved to be safe. For nearly equal spans, uniformly loaded, in which not more than about one-half the tensile steel is to be cut off or bent, the locations shown in Fig. 6.26 are satisfactory. Note, in Fig. 6.26, that the beam at the exterior support at the left is shown to be simply supported. If the beam is monolithic with exterior columns or with a concrete wall at that end, details for a typical interior span could be used for the end span as well.

c. Special Requirements near the Point of Zero Moment

While the basic requirement for flexural tensile reinforcement is that a full development length ℓ_d be provided beyond the point where the bar is assumed fully stressed to f_y, this requirement may *not* be sufficient to ensure safety against bond distress.

FIGURE 6.26
Cutoff or bend points for
bars in approximately equal
spans with uniformly
distributed loads.

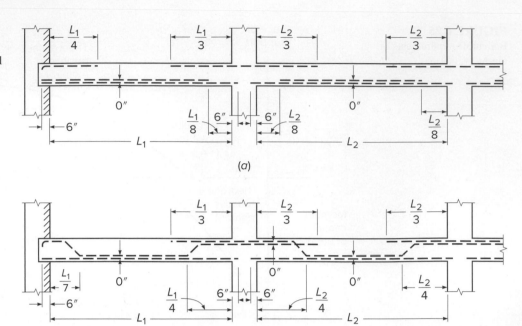

FIGURE 6.27
Development length
requirement at point of
inflection.

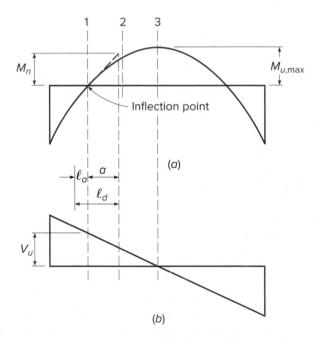

Figure 6.27 shows the moment and shear diagram representative of a uniformly
loaded continuous beam. Positive bars provided to resist the maximum moment at 3
are required to have a full development length beyond the point 3, measured in the
direction of decreasing moment. Thus, ℓ_d in the limiting case could be exactly equal
to the distance from point 3 to the point of inflection. However, if that requirement
were exactly met, then at point 2, halfway from 3 to the point of inflection, those bars

would have only one-half their development length remaining, whereas the moment would be three-quarters of that at point 3, and three-quarters of the bar force must yet be developed. This situation arises whenever the moments over the development length are greater than those corresponding to a linear reduction to zero. Therefore, the problem is a concern in the positive-moment region of continuous uniformly loaded spans, but not in the negative-moment region.

As discussed in Section 6.1, the bond force U per unit length along the tensile reinforcement in a beam is $U = dT/dx$, where dT is the change in bar tension in the length dx. Since $dT = dM/jd$, this can be written

$$U = \frac{dM}{jd\ dx} \qquad (a)$$

that is, the bond force per unit length of bar, generated by bending, is proportional to the slope of the moment diagram. In reference to Fig. 6.27a, the maximum bond force U in the positive-moment region would therefore be at the point of inflection, and U would gradually diminish along the beam toward point 3. Clearly, a conservative approach in evaluating adequacy in bond for those bars that are continued as far as the point of inflection (not necessarily the full A_s provided for M_u at point 3) would be to require that the bond resistance, which is assumed to increase linearly along the bar from its end, be governed by the maximum rate of moment increase, that is, the maximum slope dM/dx of the moment diagram, which for positive bending is seen to occur at the inflection point.

Because the slope of the moment diagram at any point is equal to the value of the shear force at that point, the slope of the moment diagram at the point of inflection is V_u. With reference to Fig. 6.27, a dashed line may therefore be drawn tangent to the moment curve at the point of inflection having the slope equal to the value of shear force V_u. Then if M_n is the nominal flexural strength provided by those bars that extend to the point of inflection, and if the moment diagram were conservatively assumed to vary linearly along the dashed line tangent to the actual moment curve, from the basic relation that $M_n/a = V_u$, a distance a is established:

$$a = \frac{M_n}{V_u} \qquad (b)$$

If the bars in question were fully stressed at a distance a to the right of the point of inflection, and if the moments diminished linearly to the point of inflection, as suggested by the dashed line, then bond failure would not occur if the development length ℓ_d did not exceed the distance a. The actual moments are less than indicated by the dashed line, so the requirement is on the safe side.

If the bars extend past the point of inflection toward the support, as is always required, then the extension can be counted as contributing toward satisfying the requirement for embedded length. Arbitrarily, according to ACI Code 7.7.3.8 and 9.7.3.8, a length past the point of inflection not greater than the larger of the beam depth d or 12 times the bar diameter d_b may be counted toward satisfying the requirement. Thus, the requirement for tensile bars at the point of inflection is that

$$\ell_d \le \frac{M_n}{V_u} + \ell_a \qquad (6.14)$$

where M_n = nominal flexural strength assuming all reinforcement at section to be stressed to f_y

V_u = factored shear force at section

ℓ_a = embedded length of bar past point of zero moment, but not to exceed the greater of d or $12d_b$

A corresponding situation occurs near the supports of simple spans carrying uniform loads, and similar requirements must be imposed. However, because of the beneficial effect of vertical compression in the concrete at the end of a simply supported span, which tends to prevent splitting and bond failure along the bars, the value M_n/V_u may be increased 30 percent for such cases, according to ACI Code 7.7.3.8 and 9.7.3.8. Thus, at the ends of a simply supported span, the requirement for tension reinforcement is

$$\ell_d \leq 1.3 \, \frac{M_n}{V_u} + \ell_a \tag{6.15}$$

The consequence of these special requirements at the point of zero moment is that, in some cases, smaller bar sizes must be used to obtain smaller ℓ_d, even though requirements for development past the point of maximum stress are met.

It may be evident from review of Sections 6.10b and 6.10c that the determination of cutoff or bend points in flexural members is complicated and can be extremely time-consuming in design. It is important to keep the matter in perspective and to recognize that the overall cost of construction will be increased very little if some bars are slightly longer than absolutely necessary, according to calculation, or as dictated by ACI Code provisions. In addition, simplicity in construction is a desired goal, and can, in itself, produce compensating cost savings. Accordingly, many engineers in practice continue *all* positive reinforcement into the face of the supports the required 6 in. and extend *all* negative reinforcement the required distance past the points of inflection, rather than using staggered cutoff points.

6.11 STRUCTURAL INTEGRITY PROVISIONS

Experience with structures that have been subjected to damage to a major supporting element, such as a column, owing to accident or abnormal loading has indicated that total collapse can be prevented through relatively minor changes in bar detailing. If some reinforcement, properly confined, is carried continuously through a support, then even if that support is damaged or destroyed, catenary action of the beams can prevent total collapse. In general, if beams have bottom and top steel meeting or exceeding the requirements summarized in Sections 6.10b and 6.10c, and if binding steel is provided in the form of properly detailed stirrups, then that catenary action can usually be ensured.

According to ACI Code 9.7.7.1, beams at the perimeter of the structure (spandrel beams) must have continuous reinforcement passing through the region bounded by the longitudinal reinforcement of the columns consisting of at least one-sixth of the tension reinforcement required for negative moment at the support, but not less than two bars, and at least one-quarter of the tension reinforcement required for positive moment at midspan, but not less than two bars. At noncontinuous supports, the reinforcement must be anchored using a standard hook or a headed deformed bar to develop f_y at the face of the support. The continuous reinforcement must be enclosed in accordance with ACI Code 25.7.1.6 by closed stirrups perpendicular to the axis of the member, hoops as used in seismic design (see Section 20.4), or a closed cage of welded wire reinforcement with transverse wires perpendicular to the axis of the member. This transverse reinforcement must be anchored by a 135°

FIGURE 6.28
Two-piece stirrup meeting
the requirements of ACI
Code 25.7.1.6 for
confinement of longitudinal
integrity reinforcement in
perimeter beams. The 90°
hook must be placed adjacent
to the slab.

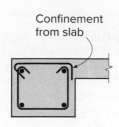

Confinement
from slab

standard hook (Fig. 6.9*b*) around a longitudinal bar, or where the concrete surrounding the anchorage is restrained against spalling by a flange or slab, by either a 90° or 135° standard hook around a longitudinal bar.

Figure 6.28 shows a two-piece stirrup that meets the requirements of ACI Code 25.7.1.6. Although the spacing of these stirrups is not specified, the requirements for minimum shear steel given in Section 5.5b provide guidance in regions where shear does not require closer spacing. The stirrups need not be extended through the joints. Overlapping pairs of U stirrups of the type shown in Fig. 6.22*d* are not permitted in perimeter beams because damage to the side cover concrete may cause both the stirrups and top longitudinal reinforcement to tear out of the concrete, thus preventing the longitudinal reinforcement from acting as a catenary.

The required continuity of longitudinal steel can be provided using top reinforcement spliced at midspan and bottom reinforcement spliced at or near the supports using Class B tension splices, or mechanical or welded splices (see Section 6.13).

In other than perimeter beams, ACI Code 9.7.7.2 requires that at least one-quarter of the positive-moment reinforcement required at midspan, but not less than two bars, must pass through the column longitudinal reinforcement and must be continuous. The requirements for anchoring this longitudinal reinforcement at noncontinuous supports and for splicing the bars to provide continuity are the same as for perimeter beams.

For one-way slabs, ACI Code 7.7.7 requires that at least one-quarter of the maximum positive-moment reinforcement be continuous and, at noncontinuous supports, the reinforcement be anchored to develop f_y at the face of the support. If splices are necessary in the structural integrity reinforcement, the reinforcement must be spliced near supports using Class B tension splices, or mechanical or welded splices.

Note that these provisions require very little additional steel in the structure. At least one-quarter of the bottom bars must be extended 6 in. into the support by other ACI Code provisions; the structural integrity provisions merely require that these bars be made continuous or spliced. Similarly, ACI Code 9.7.3.8 requires that at least one-third of the negative bars in beams be extended a certain minimum distance past the point of inflection; the structural integrity provisions for perimeter beams require only that one-half of those bars (that is, one-sixth of the negative bars at the face of the support) be further extended and spliced at midspan.

6.12   INTEGRATED BEAM DESIGN EXAMPLE

In this and in the preceding chapters, the several aspects of the design of reinforced concrete beams have been studied more or less separately: first the flexural design, then design for shear, and finally for bond and anchorage. The following example is presented to show how the various requirements for beams, which are often in some respects conflicting, are satisfied in the overall design of a representative member.

EXAMPLE 6.4 **Integrated design of T beam.** A floor system consists of single-span T beams 8 ft on centers, supported by 12 in. masonry walls spaced at 25 ft between inside faces. The general arrangement is shown in Fig. 6.29a. A 5 in. monolithic slab carries a uniformly distributed service live load of 165 psf. The T beams, in addition to the slab load and their own weight, must carry two 16,000 lb equipment loads applied over the stem of the T beam 3 ft from the span

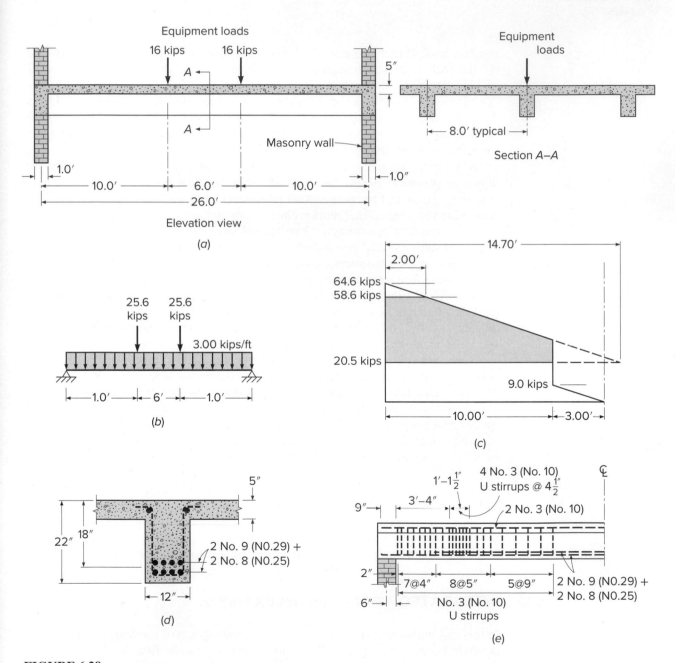

FIGURE 6.29
T beam design for Example 6.4.

centerline as shown. A complete design is to be provided for the T beams, using concrete of 4000 psi strength and bars with 60,000 psi yield stress. (*Note*: Because normalweight concrete and uncoated reinforcement with a yield strength of 60,000 psi are used, λ, ψ_e, and $\psi_g = 1.0$. λ is dropped from the calculations for shear and bond, and ψ_e and ψ_g are dropped from the calculations for bond.)

SOLUTION. According to the ACI Code, the span length is to be taken as the clear span plus the beam depth, but need not exceed the distance between the centers of supports. The latter provision controls in this case, and the effective span is 26 ft. Estimating the beam web dimensions to be 12 × 24 in., the calculated and factored dead loads are as follows:

Slab:

$$\frac{5}{12} \times 150 \times 7 = 440 \text{ lb/ft}$$

Beam:

$$\frac{12 \times 24}{144} 150 = 300 \text{ lb/ft}$$

$$w_d = 740 \text{ lb/ft}$$

$$1.2w_d = 890 \text{ lb/ft}$$

The uniformly distributed live load is

$$w_l = 165 \times 8 = 1320 \text{ lb/ft}$$

$$1.6w_l = 2110 \text{ lb/ft}$$

The factored live load is

$$w_u = 1.2w_d + 1.6w_l = 890 + 2110 = 3000 \text{ lb/ft} = 3.0 \text{ kips/ft}$$

Live load factors are applied to the two concentrated loads to obtain $P_u = 16{,}000 \times 1.6 = 25{,}600$ lb. Factored loads are summarized in Fig. 6.29*b*.

In lieu of other controlling criteria, the beam web dimensions will be selected on the basis of shear. The left and right reactions under factored load are $25.6 + 3.00 \times 13 = 64.6$ kips. With the effective beam depth estimated to be 20 in., the maximum shear that need be considered in design is $64.6 - 3.00(0.50 + 1.67) = 58.1$ kips. Although the ACI Code permits V_s as high as $8\sqrt{f_c'}b_wd$, this would require very heavy web reinforcement. A lower limit of $4\sqrt{f_c'}b_wd$ will be adopted. With $V_c = 2\sqrt{f_c'}b_wd$ this results in a maximum $V_n = 6\sqrt{f_c'}b_wd$. Then $b_w d = V_u/(6\phi\sqrt{f_c'}) = 58{,}100/(6 \times 0.75\sqrt{4000}) = 204$ in^2. Cross-sectional dimensions $b_w = 12$ in. and $d = 18$ in. are selected, providing a total beam depth of 22 in. The assumed dead load of the beam need not be revised.

According to the Code, the effective flange width b is the smallest of the three quantities

$$16h_f + b_w = 80 + 12 = 92 \text{ in.}$$

$$\frac{2\ell_n}{8} + b_w = \frac{26 \times 12}{4} + 12 = 90 \text{ in.}$$

Centerline beam spacing = 96 in.

The second controls in this case. The maximum moment is at midspan, where

$$M_u = \left(\frac{1}{8} \times 3.00 \times 26^2\right) + (25.6 \times 10) = 510 \text{ ft-kips}$$

Assuming for trial that the stress-block depth will equal the slab thickness leads to

$$A_s = \frac{M_u}{\phi f_y(d - a/2)} = \frac{510 \times 12}{0.90 \times 60 \times 15.5} = 7.31 \text{ in}^2$$

Then

$$a = \frac{A_s f_y}{0.85 f_c' b} = \frac{7.31 \times 60}{0.85 \times 4 \times 90} = 1.43 \text{ in.}$$

The stress-block depth is seen to be less than the slab depth; rectangular beam equations are valid. An improved determination of A_s is

$$A_s = \frac{510 \times 12}{0.90 \times 60 \times 17.28} = 6.56 \text{ in}^2$$

A check confirms that this is well below the maximum permitted reinforcement ratio. Four No. 9 (No. 29) plus four No. 8 (No. 25) bars will be used, providing a total area of 7.14 in². They will be arranged in two rows, as shown in Fig. 6.29d, with No. 9 (No. 29) bars at the outer end of each row. Beam width b_w is adequate for this bar arrangement.

While the ACI Code permits discontinuation of two-thirds of the longitudinal reinforcement for simple spans, in the present case it is convenient to discontinue only the upper layer of steel, consisting of one-half of the total area. The moment capacity of the member after the upper layer of bars has been discontinued is then found:

$$a = \frac{3.57 \times 60}{0.85 \times 4 \times 78} = 0.81 \text{ in.}$$

$$\phi M_n = \phi A_s f_y\left(d - \frac{a}{2}\right) = 0.90 \times 3.57 \times 60 \times 18.66 \times \frac{1}{12} = 300 \text{ ft-kips}$$

For the present case, with a moment diagram resulting from combined distributed and concentrated loads, the point at which the applied moment is equal to this amount must be calculated. (In the case of uniformly loaded beams, Graphs A.2 and A.3 in Appendix A are helpful.) If x is the distance from the support centerline to the point at which the moment is 300 ft-kips, then

$$64.6x - \frac{3.00x^2}{2} = 300$$
$$x = 5.30$$

The upper bars must be continued at least $d = 1.50$ ft or $12d_b = 1.13$ ft beyond this theoretical point of cutoff. In addition, the full development length ℓ_d must be provided past the maximum-moment section at which the stress in the bars to be cut is assumed to be f_y. Because of the heavy concentrated loads near the midspan, the point of peak stress will be assumed to be at the concentrated load rather than at midspan. For the four upper bars, assuming 1.50 in. clear cover to the outside of the No. 3 (No. 10) stirrups, the clear side cover is $1.50 + 0.38 = 1.88$ in., or $1.66d_b$. Assuming equal clear spacing between all four bars, that clear spacing is $[12.00 - 2 \times (1.50 + 0.38 + 1.13 + 1.00)]/3 = 1.33$ in., or $1.18d_b$. Noting that the ACI Code requirements for minimum stirrups are met, it is clear that all restrictions for the use of the simplified equation for development length are met. From Table 6.1 (Section 6.3), the required development length is

$$\ell_d = \frac{60,000}{20\sqrt{4000}} 1.13 = 47 \times 1.13 = 53 \text{ in.}$$

or 4.42 ft. Thus, the bars must be continued at least $3.00 + 4.42 = 7.42$ ft past the midspan point, but in addition they must continue to a point $5.30 - 1.50 = 3.80$ ft from the support centerline. The second requirement controls and the upper layer of the bars will be terminated, as shown in Fig. 6.29e, 3.30 ft from the support face. The bottom layer of bars will be extended to a point 3 in. from the end of the beam, providing 5.55 ft embedment past the critical section for cutoff of the upper bars. This exceeds the development length of the lower set of bars, confirming that cutoff and extension requirements are met.

Note that a simpler design, using very little extra steel, would result from extending all eight positive bars into the support. Whether or not the more elaborate calculations and more complicated placement are justified would depend largely on the number of repetitions of the design in the total structure.

Checking by Eq. (6.15) to ensure that the continued steel is of sufficiently small diameter determines that

$$\ell_d \leq 1.3 \, \frac{333 \times 12}{64.6} + 3 = 83 \text{ in.}$$

The actual ℓ_d of 53 in. meets this restriction.

Since the cut bars are located in the tension zone, special binding stirrups will be used to control cracking; these will be selected after the normal shear reinforcement has been determined.

The shear diagram resulting from application of factored loads is shown in Fig. 6.29c. The shear contribution of the concrete is

$$\phi V_c = 0.75 \times 2\sqrt{4000} \times 12 \times 18 = 20{,}500 \text{ lb}$$

Thus, web reinforcement must be provided for that part of the shear diagram shown shaded.

No. 3 (No. 10) stirrups are selected. The maximum spacings must not exceed $d/2 = 9$ in., 24 in., or $A_v f_{yt}/(0.75\sqrt{f_c'} b_w) = 0.22 \times 60{,}000/(0.75\sqrt{4000} \times 12) = 23$ in. $\leq A_v f_{yt}/50b_w = 0.22 \times 60{,}000/(50 \times 12) = 22$ in. The first criterion controls here. For reference, from Eq. (5.14a) the hypothetical stirrup spacing at the support is

$$s_0 = \frac{0.75 \times 0.22 \times 60 \times 18}{64.6 - 20.5} = 4.04 \text{ in.}$$

and at 2 ft intervals along the span,

$$s_2 = 4.68 \text{ in.}$$
$$s_4 = 5.55 \text{ in.}$$
$$s_6 = 6.83 \text{ in.}$$
$$s_8 = 8.87 \text{ in.}$$
$$s_{10} = 12.64 \text{ in.}$$

The spacing need not be closer than that required 2.00 ft from the support centerline ($d = 18$ in. at the *face* of the support). In addition, stirrups are not required past the point of application of concentrated load, since beyond that point the shear is less than one-half of ϕV_c. The final spacing of vertical stirrups selected is

$$
\begin{aligned}
1 \text{ space at 2 in.} &= 2 \text{ in.} \\
7 \text{ spaces at 4 in.} &= 28 \text{ in.} \\
8 \text{ spaces at 5 in.} &= 40 \text{ in.} \\
5 \text{ spaces at 9 in.} &= \underline{45 \text{ in.}}
\end{aligned}
$$

Total $= 115$ in. $= 9$ ft 7 in. from the face of the support (121 in. $= 10$ ft 1 in. from the support centerline)

Two No. 3 (No. 10) longitudinal bars will be added to meet anchorage requirements and fix the top of the stirrups.

In addition to the shear reinforcement just specified, it is necessary to provide extra web reinforcement over a distance equal to $\frac{3}{4}\,d$, or 13.5 in., from the cut ends of the discontinued steel.

The spacing of this extra web reinforcement must not exceed $d/8\beta_b = 18/(8 \times \frac{1}{2}) = 4.5$ in. In addition, the area of added steel within the distance s must not be less than $60\hat{b}_w s/f_{yt} = 60 \times 12 \times 4.5/60,000 = 0.054$ in^2. For convenience, No. 3 (No. 10) stirrups will be used for this purpose also, providing an area of 0.22 in^2 in the distance s. The placement of the four extra stirrups is shown in Fig. 6.29e.

6.13   BAR SPLICES

In general, reinforcing bars are stocked by suppliers in lengths of 60 ft for bars from No. 5 to No. 18 (No. 16 to No. 57) and in 20 or 40 ft lengths for smaller sizes. For this reason, and because it is often more convenient to work with shorter bar lengths, it is frequently necessary to splice bars in the field. Splices in reinforcement at points of maximum stress should be avoided, and when splices are used, they should be staggered, although neither condition is practical, for example, in compression splices in columns.

Splices for No. 11 (No. 36) bars and smaller are usually made simply by lapping the bars a sufficient distance to transfer stress by bond from one bar to the other. The lapped bars are usually placed in contact and lightly wired so that they stay in position as the concrete is placed. Alternatively, splicing may be accomplished by welding or by sleeves or mechanical devices. ACI Code 25.5.1.5 requires that for spliced reinforcement with $f_y \geq 80,000$ psi spaced closer than 6 in. on center, transverse reinforcement must be provided so that K_{tr} is not smaller than $0.5d_b$. As pointed out in Section 6.3a, however, ACI Code 9.7.1.4 and 10.7.1.3 require that K_{tr} be at least $0.5d_b$ in beams and columns where reinforcement with $f_y \geq 80,000$ psi is developed or spliced, independent of the bar spacing. ACI Code 25.5.1.1 prohibits use of lapped splices for bars larger than No. 11 (No. 36), except that No. 14 and No. 18 (No. 43 and No. 57) bars may be lapped in compression with No. 11 (No. 36) and smaller bars per ACI Code 16.3.5.4 and 25.5.5.3. For bars that carry only compression, it is possible to transfer load by end bearing of square cut ends, if the bars are accurately held in position by a sleeve or other device. If bars of different sizes must be spliced, the splice length must equal or exceed the greater of the splice length of the smaller bar and the development length of the larger bar.

Lap splices of bars in bundles are based on the lap splice length required for individual bars within the bundle but must be increased in length by 20 percent for three-bar bundles and by 33 percent for four-bar bundles because of the reduced effective perimeter. Individual bar splices within a bundle should not overlap, and entire bundles must not be lap-spliced.

According to ACI Code 25.5.7.1, welded splices must develop at least 125 percent of the specified yield strength of the bar. The same requirement applies to full mechanical connections. This ensures that an overloaded spliced bar would fail by ductile yielding in the region away from the splice, rather than at the splice where brittle failure is likely.

a.   Lap Splices in Tension

The required length of lap for tension splices ℓ_{st} is stated in terms of the development length ℓ_d. In the process of calculating ℓ_d, the usual modification factors are applied, except that the reduction factor for excess reinforcement should not be applied because that factor is already accounted for in the splice specification.

Two different classifications of lap splices are established, corresponding to the minimum length of lap required: a Class A splice requires a lap of $1.0\ell_d$, and a Class B splice requires a lap of $1.3\ell_d$. In either case, a minimum length of 12 in. applies. For Class B splices, the 12 in. minimum applies to $1.3\ell_d$, not to the value of ℓ_d used to calculate the lap length. Lap splices, in general, must be Class B splices, according to ACI Code 25.5.2.1, except that Class A splices are allowed when the area of reinforcement provided is at least twice that required by analysis over the entire length of the splice *and* when one-half or less of the total reinforcement is spliced within the required lap length. The effect of these requirements is to encourage designers to locate splices away from regions of maximum stress, to a location where the actual steel area is at least twice that required by analysis, and to stagger splices.

Spiral reinforcement is spliced with a lap of $48d_b$ for uncoated bars and $72d_b$ for epoxy-coated bars, in accordance with ACI Code 25.7.3.6. The lap for epoxy-coated bars is reduced to $48d_b$ if the bars are anchored with a standard stirrup or tie hook.

b. Compression Splices

Reinforcing bars in compression are spliced mainly in columns, where bars are most often terminated just above each floor or every other floor. This is done partly for construction convenience, to avoid handling and supporting very long column bars, but it is also done to permit column steel area to be reduced in steps, as loads become lighter at higher floors.

Compression bars may be spliced by lapping, by direct end bearing, or by welding or mechanical devices that provide positive connection. The minimum length of lap for compression splices is set according to ACI Code 25.5.5.1:

$$\text{For bars with } f_y \le 60{,}000 \text{ psi} \qquad \ell_{sc} = 0.0005 f_y\, d_b$$

$$\text{For bars with } 60{,}000 \text{ psi} < f_y \le 80{,}000 \text{ psi} \qquad \ell_{sc} = (0.0009 f_y - 24)d_b$$

$$\text{For bars with } f_y > 80{,}000 \text{ psi} \qquad \ell_{sc} = \text{longer of } (0.0009 f_y - 24)d_b \text{ and } \ell_{st}$$

but not less than 12 in. For f_c' less than 3000 psi, the required lap is increased by one-third. When bars of different size are lap-spliced in compression, the splice length is to be the larger of the development length of the larger bar and the splice length of the smaller bar. In exception to the usual restriction on lap splices for large-diameter bars, No. 14 and No. 18 (No. 43 and No. 57) bars *may* be lap-spliced to No. 11 (No. 36) and smaller bars.

Direct end bearing of the bars has been found by test and experience to be an effective means for transmitting compression. In such a case, the bars must be held in proper alignment by a suitable device. The bar ends must terminate in flat surfaces within 1.5° of a right angle, and the bars must be fitted within 3° of full bearing after assembly, according to ACI Code 25.5.6. End bearing splices are limited to members where ties, closed stirrups, or spirals are used.

c. Column Splices

Lap splices, butt-welded splices, mechanical connections, or end-bearing splices may be used in columns, with certain restrictions. Reinforcing bars in columns may be subjected to compression or tension, or, for different load combinations, both tension and compression. Accordingly, column splices must conform in some cases to the requirements for compression splices only or tension splices only or to requirements for both. ACI Code 10.7.5.2 requires that a minimum tension capacity be provided in each face of all columns, even where analysis indicates compression only. Ordinary compressive lap splices provide sufficient tensile resistance, but end-bearing splices may require additional bars for tension, unless the splices are staggered.

For lap splices, where the bar stress due to factored loads is compression, column lap splices must conform to the requirements presented in Section 6.13b for compression splices. Where the stress is tension and does not exceed $0.5f_y$, ACI Code 10.7.5.2 requires that lap splices must be Class B if more than one-half the bars are spliced at any section, or Class A if one-half or fewer are spliced and alternate lap splices are staggered by ℓ_d. If the stress is tension and exceeds $0.5f_y$, then lap splices must be Class B, according to ACI Code 10.7.5.2.

If lateral ties are used throughout the splice length having an effective area of at least $0.0015hs$ in both directions, where s is the spacing of ties and h is the overall thickness of the member, the required splice length may be multiplied by 0.83 but must be at least 12 in. The tie legs perpendicular to the dimension h are used to calculate the effective tie area. If spiral reinforcement confines the splice, the length required may be multiplied by 0.75 but again must be at least 12 in.

End-bearing splices, as described above, may be used for column bars stressed in compression, if the splices are staggered or additional bars are provided at splice locations. The continuing bars in each face must have a tensile strength of not less than $0.25f_y$ times the area of reinforcement in that face.

As mentioned in Section 6.13b, column splices are commonly made just above a floor. However, for frames subjected to lateral loads, a better location is within the center half of the column height, where the moments due to lateral loads are much lower than at floor level. Such placement is mandatory for columns in "special moment frames" designed for seismic loads, as will be discussed in Chapter 20.

EXAMPLE 6.5   **Compression splice of column reinforcement.**   In reference to Fig. 6.8, four No. 11 (No. 36) column bars from the floor below are to be lap-spliced with four No. 10 (No. 32) column bars from above, and the splice is to be made just above a construction joint at floor level. The column, measuring 12×21 in. in cross section, will be subject to compression only for all load combinations. Transverse reinforcement consists of No. 4 (No. 13) ties at 16 in. spacing. All vertical bars may be assumed to be fully stressed. Calculate the required splice length. Material strengths are $f_y = 60,000$ psi and $f_c' = 4000$ psi.

SOLUTION.   The length of the splice must be the larger of the development length ℓ_{dc} of the No. 11 (No. 36) bars and the splice length ℓ_{sc} of the No. 10 (No. 32) bars. For the No. 11 (No. 36) bars, the development length is equal to the larger of the values obtained with Eqs. (6.10a) and (6.10b):

$$\ell_{dc} = \left(\frac{f_y \psi_r}{50 \lambda \sqrt{f_c'}} \right) d_b = \left(\frac{60,000 \times 1.0}{50 \times 1.0 \times \sqrt{4000}} \right) 1.41 = 27 \text{ in.}$$

$$\ell_{dc} = 0.0003 f_y d_b = 0.0003 \times 60,000 \times 1.41 = 25 \text{ in.}$$

The first criterion controls. No modification factors apply. For the No. 10 (No. 32) bars, the compression splice length is

$$\ell_{sc} = 0.0005 f_y d_b = 0.0005 \times 60,000 \times 1.27 = 38 \text{ in.}$$

In the check for use of the modification factor for tied columns, the critical column dimension is 21 in., and the required effective tie area is thus $0.0015 \times 21 \times 16 \times 0.50 \text{ in}^2$. The No. 4 (No. 13) ties provide an area of only $0.20 \times 2 = 0.40$ in^2, so the reduction factor of 0.83 cannot be applied to the splice length. Thus, the compression splice length of 38 in., which exceeds the development length of 27 in. for the No. 11 (No. 36) bars, controls, and a lap splice of 38 in. is required. Note that if the spacing of the ties at the splice were reduced to 12.8 in. or less (say 12 in.), the required lap would be reduced to $38 \times 0.83 = 32$ in. This would save steel, and, although placement cost would increase slightly, would probably represent the more economical design.

REFERENCES

6.1. R. M. Mains, "Measurement of the Distribution of Tensile and Bond Stresses along Reinforcing Bars," *J. ACI*, vol. 23, no. 3, 1951, pp. 225–252.

6.2. A. H. Nilson, "Internal Measurement of Bond Slip," *J. ACI*, vol. 69, no. 7, 1972, pp. 439–441.

6.3. Y. Goto, "Cracks Formed in Concrete around Deformed Tension Bars," *J. ACI*, vol. 68, no. 4, 1971, pp. 244–251.

6.4. L. A. Lutz and P. Gergely, "Mechanics of Bond and Slip of Deformed Bars in Concrete," *J. ACI*, vol. 64, no. 11, 1967, pp. 711–721.

6.5. P. M. Ferguson and J. N. Thompson, "Development Length of High Strength Reinforcing Bars in Bond," *J. ACI*, vol. 59, no. 7, 1962, pp. 887–922.

6.6. R. G. Mathey and D. Watstein, "Investigation of Bond in Beam and Pullout Specimens with High-Strength Reinforcing Bars," *J. ACI*, vol. 32, no. 9, 1961, pp. 1071–1090.

6.7. ACI Committee 408, "Bond Stress—The State of the Art," *J. ACI*, vol. 63, no. 11, 1966, pp. 1161–1190.

6.8. ACI Committee 408, "Suggested Development, Splice, and Standard Hook Provisions for Deformed Bars in Tension," *Concr. Intl.*, vol. 1, no. 7, 1979, pp. 44–46.

6.9. J. O. Jirsa, L. A. Lutz, and P. Gergely, "Rationale for Suggested Development, Splice, and Standard Hook Provisions for Deformed Bars in Tension," *Concr. Intl.*, vol. 1, no. 7, 1979, pp. 47–61.

6.10. C. O. Orangun, J. O. Jirsa, and J. E. Breen, "A Reevaluation of the Test Data on Development Length and Splices," *J. ACI*, vol. 74, no. 3, 1977, pp. 114–122.

6.11. L. A. Lutz, S. A. Mirza, and N. K. Gosain, "Changes to and Applications of Development and Lap Splice Length Provisions for Bars in Tension," *ACI Struct. J.*, vol. 90, no. 4, 1993, pp. 393–406.

6.12. D. Darwin, M. L. Tholen, E. K. Idun, and J. Zuo, "Splice Strength of High Relative Rib Area Reinforcing Bars," *ACI Struct. J.*, vol. 93, no. 1, 1996, pp. 95–107.

6.13. D. Darwin, J. Zuo, M. L. Tholen, and E. K. Idun, "Development Length Criteria for Conventional and High Relative Rib Area Reinforcing Bars," *ACI Struct. J.*, vol. 93, no. 3, 1996, pp. 347–359.

6.14. J. Zuo and D. Darwin, "Splice Strength of Conventional and High Relative Rib Area Bars in Normal and High Strength Concrete," *ACI Struct. J.*, vol. 97, no. 4, 2000, pp. 630–641.

6.15. ACI Committee 408, *Bond and Development of Straight Reinforcement in Tension*, ACI 408R-03, American Concrete Institute, Farmington Hills, MI, 2003.

6.16. P. M. Ferguson, "Small Bar Spacing or Cover—A Bond Problem for the Designer," *J. ACI*, vol. 74, no. 9, 1977, pp. 435–439.

6.17. P. R. Jeanty, D. Mitchell, and M. S. Mirza, "Investigation of Top Bar Effects in Beams," *ACI Struct. J.*, vol. 85, no. 3, 1988, pp. 251–257.

6.18. B. B. Brettmann, D. Darwin, and R. C. Donahey, "Bond of Reinforcement to Superplasticized Concrete," *J. ACI*, vol. 83, no. 1, 1986, pp. 98–107.

6.19. R. G. Mathey and J. R. Clifton, "Bond of Coated Reinforcing Bars in Concrete," *J. Struct. Div.*, ASCE, vol. 102, no. ST1, 1976, pp. 215–228.

6.20. R. A. Treece and J. O. Jirsa, "Bond Strength of Epoxy-Coated Reinforcing Bars," *ACI Matls. J.*, vol. 86, no. 2, 1989, pp. 167–174.

6.21. B. S. Hamad, J. O. Jirsa, and N. I. dePaulo, "Anchorage Strength of Epoxy-Coated Hooked Bars," *ACI Struct. J.*, vol. 90, no. 2, 1993, pp. 210–217.

6.22. H. H. Ghaffari, O. C. Choi, D. Darwin, and S. L. McCabe, "Bond of Epoxy-Coated Reinforcement: Cover, Casting Position, Slump, and Consolidation," *ACI Struct. J.*, vol. 91, no. 1, 1994, pp. 59–68.

6.23. C. J. Hester, S. Salamizavaregh, D. Darwin, and S. L. McCabe, "Bond of Epoxy-Coated Reinforcement: Splices," *ACI Struct. J.*, vol. 90, no. 1, 1993, pp. 89–102.

6.24. D. Darwin and J. Zuo, "Discussion of Proposed Changes to ACI 318" in *ACI 318-02 Discussion and Closure*, *Concr. Intl.*, vol. 24, no. 1, 2002, pp. 91, 93, 97–101.

6.25. J. Sperry, S. Yasso, N. Searle, M. DeRubeis, D. Darwin, M. O'Reilly, A. Matamoros, L. Feldman, A. Lepage, R. Lequesne, and A. Ajaam, "Conventional and High-Strength Hooked Bars—Part 1: Anchorage Tests," *ACI Struct. J.*, vol. 114, no. 1, 2017, pp. 255–266.

6.26.   J. Sperry, D. Darwin, M. O'Reilly, R. Lequesne, S. Yasso, A. Matamoros, L. Feldman, and A. Lepage, "Conventional and High-Strength Hooked Bars—Part 2: Data Analysis," *ACI Struct. J.*, vol. 114, no. 1, 2017, pp. 267–276.

6.27.   J. Sperry, D. Darwin, M. O'Reilly, A. Lepage, R. Lequesne, A. Matamoros, L. Feldman, S. Yasso, N. Searle, M. DeRubeis, and A. Ajaam, "Conventional and High-Strength Hooked Bars: Detailing Effects," *ACI Struct. J.*, vol. 115, no. 1, 2018, pp. 247–257.

6.28.   A. Ajaam, S. Yasso, D. Darwin, M. O'Reilly, and J. Sperry, "Anchorage Strength of Closely-Spaced Hooked Bars," *ACI Struct. J.*, vol. 115, no. 4, 2018, pp. 1143–1152.

6.29.   Y. Shao, D. Darwin, M. O'Reilly, R. D. Lequesne, K. Ghimire, and Hano, M., "Anchorage of Conventional and High-Strength Headed Reinforcing Bars," SM Report No. 117, Univ. of Kansas Center for Research, Lawrence, Kansas, 2016, 234 pp.

6.30.   K. Ghimire, D. Darwin, and M. O'Reilly, "Anchorage of Headed Reinforcing Bars," SM Report No. 127, Univ. of Kansas Center for Research, Lawrence, KS, 2018, 278 pp.

6.31.   K. P. Ghimire, Y. Shao, D. Darwin, and M. O'Reilly, "Conventional and High-Strength Headed Bars—Part 1: Anchorage Tests," *ACI Struct. J.*, vol. 116, no. 4, 2019, pp. 255–264.

6.32.   K. P. Ghimire, Y. Shao, D. Darwin, and M. O'Reilly, "Conventional and High-Strength Headed Bars—Part 2: Data Analysis," *ACI Struct. J.*, vol. 116, no. 4, 2019, pp. 265–272.

PROBLEMS

6.1.   For the beam cross section shown in Fig. 4.16, what are the development lengths of the top No. 7 (No. 22) bars and bottom No. 9 (No. 29) bars for No. 4 (No. 13) stirrups with $1\frac{1}{2}$ in. clear side cover spaced at 6 in. using Eqs. (6.5) and (6.6)? Normalweight concrete, $f_c' = 4000$ psi, $f_y = 60,000$ psi. Comment.

6.2.   For the beam cross section shown in Fig. 4.16, what are the development lengths of the top No. 7 (No. 22) bars and bottom No. 9 (No. 29) bars for No. 4 (No. 13) stirrups with $1\frac{1}{2}$ in. clear side cover spaced at 4 in. using Eqs. (6.5) and (6.6)? Lightweight concrete, $f_c' = 4000$ psi, $f_y = 80,000$ psi. Comment.

6.3.   For the beam cross section shown in Fig. 4.16, what are the development lengths of the top No. 7 (No. 22) bars and bottom No. 9 (No. 29) bars for No. 4 (No. 13) stirrups with $1\frac{1}{2}$ in. clear side cover spaced at 6 in. using Eqs. (6.5) and (6.6)? Normalweight concrete, $f_c' = 8000$ psi, $f_y = 60,000$ psi. Comment.

6.4.   For the beam cross section shown in Fig. 4.16, what are the development lengths of the top No. 7 (No. 22) bars and bottom No. 9 (No. 29) bars for No. 4 (No. 13) stirrups with $1\frac{1}{2}$ in. clear side cover spaced at 6 in. using Eqs. (6.5) and (6.6)? The bars are epoxy coated. Lightweight concrete, $f_c' = 8000$ psi, $f_y = 60,000$ psi. Comment.

6.5.   For the beam cross section shown in Fig. 4.16, what are the development lengths of the top No. 7 (No. 22) bars and bottom No. 9 (No. 29) bars for No. 4 (No. 13) stirrups with $1\frac{1}{2}$ in. clear side cover spaced at 3 in. using Eqs. (6.5) and (6.6)? Normalweight concrete, $f_c' = 4000$ psi, $f_y = 80,000$ psi. Comment.

6.6.   For the beam cross section shown in Fig. 4.16, what are the development lengths of the top No. 7 (No. 22) bars and bottom No. 9 (No. 29) bars for No. 4 (No. 13) stirrups with $1\frac{1}{2}$ in. clear side cover spaced at 6 in. using Eqs. (6.5) and (6.6)? Normalweight concrete, $f_c' = 12,000$ psi, $f_y = 60,000$ psi. Comment.

6.7.   Compare the development lengths of No. 8 (No. 25) hooked (180° bend) and headed bars cast in a beam-column joint with $2\frac{1}{2}$ in. clear cover on the bars and 4 in. clear spacing between the bars. $A_{th} > 0.4A_{hs}$ and $A_{tt} > 0.3A_{hs}$. Normalweight concrete, $f_c' = 4000$ psi, $f_y = 60,000$ psi. Comment.

6.8.   Compare the development lengths of No. 11 (No. 36) hooked (90° bend) and headed bars cast in a beam-column joint with 3 in. clear cover on the bars and 6 in. clear spacing between the bars. $A_{th} = 0.2A_{hs}$ and $A_{tt} = 0.15A_{hs}$. Normalweight concrete, $f_c' = 7000$ psi, $f_y = 60,000$ psi. Comment.

6.9.   The short beam shown in Fig. P6.9 cantilevers from a supporting column at the left. The beam must carry a calculated dead load of 2.0 kips/ft including its own weight and a service live load of 2.6 kips/ft. Tensile flexural reinforcement consists of two No. 10 (No. 32) bars at a 21 in. effective depth. Transverse No. 3 (No. 10) U stirrups with 1.5 in. cover are provided at the following spacings from the face of the column: 4 in., 3 at 8 in., 5 at 10.5 in. Within the beam-column joint, No. 4 (No. 13) closed ties are provided at an 8 in. spacing, with the first tie located 2 in. below the centerline of the hooked bar.

(a) If the flexural and shear steel use $f_y = 60,000$ psi and if the beam uses lightweight concrete having $f_c' = 4000$ psi, check to see if proper development length can be provided for the No. 10 (No. 32) bars. Use the simplified development length equations, Eq. (6.6a) or (6.6b), as appropriate.

(b) Recalculate the required development length for the beam bars using the basic Eq. (6.5). Comment on your results.

(c) If the column material strengths are $f_y = 60,000$ psi and $f_c' = 5000$ psi (normalweight concrete), check to see if adequate embedment can be provided within the column for the No. 10 (No. 32) bars. If hooks are required, specify detailed dimensions.

FIGURE P6.9

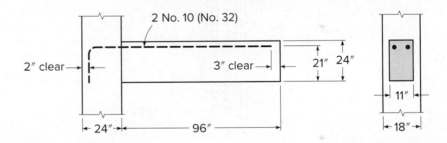

6.10.   The beam shown in Fig. P6.10 is simply supported with a clear span of 26 ft and is to carry a distributed dead load of 1.05 kips/ft including its own weight and live load of 1.62 kips/ft, unfactored, in service. The reinforcement consists of five No. 10 (No. 32) bars at a 16 in. effective depth, two of which are to be discontinued where no longer needed. Material strengths specified are $f_y = 60,000$ psi and $f_c' = 5000$ psi. No. 3 (No. 10) stirrups are used with a cover of 1.5 in. at spacing less than ACI Code maximum.

(a) Calculate the point where two bars can be discontinued.

(b) Check to be sure that adequate embedded length is provided for continued and discontinued bars.

(c) Check special requirements at the support, where $M_u = 0$.

(d) If No. 3 (No. 10) bars are used for transverse reinforcement, specify special reinforcing details in the vicinity where the No. 10 (No. 32) bar is cut off.

(e) Comment on the practical aspects of the proposed design. Would you recommend cutting off the steel as suggested? Could three bars be discontinued rather than two?

FIGURE P6.10

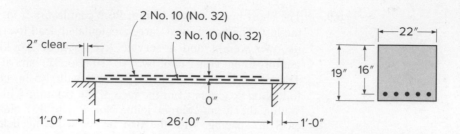

6.11. Figure P6.11 shows the column reinforcement for a 16 in. diameter concrete column, with $f_y = 80,000$ psi and $f_c' = 8000$ psi. Analysis of the building frame indicates a required $A_s = 7.30$ in^2 in the lower column and 5.80 in^2 in the upper column. Spiral reinforcement consists of a $\frac{3}{8}$ in. diameter rod with a 2 in. pitch. Column bars are to be spliced in compression just above the construction joint at the floor level, as shown in the sketch. Calculate the minimum permitted length of splice.

FIGURE P6.11

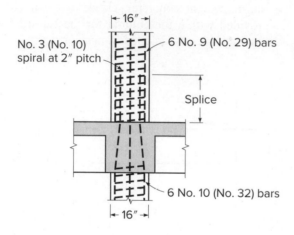

6.12. The short cantilever shown in Fig. P6.12 carries a heavy concentrated load 6 in. from its outer end. Flexural analysis indicates that three No. 8 (No. 25) bars are required, suitably anchored in the supporting wall and extending to a point no closer than 2 in. from the free end. The bars will be fully stressed to f_y at the fixed support. Investigate the need for hooks and transverse confinement steel at the right end of the member. Material strengths are $f_y = 60,000$ psi and $f_c' = 4000$ psi. If hooks and transverse steel are required, show details in a sketch.

FIGURE P6.12

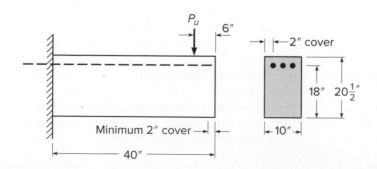

6.13. A continuous-strip wall footing is shown in cross section in Fig. P6.13. It is proposed that tensile reinforcement be provided using No. 8 (No. 25) bars at 16 in. spacing along the length of the wall, to provide a bar area of 0.59 in^2/ft. The bars have strength $f_y = 60,000$ psi, and the footing concrete has $f_c' = 4000$ psi. The critical section for bending is assumed to be at the face of the supported wall, and the effective depth to the tensile steel is 12 in. Check to ensure that sufficient development length is available for the No. 8 (No. 25) bars, and if hooks are required, sketch details of the hooks, giving dimensions.

Note: If hooks are required for the No. 8 (No. 25) bars, prepare an alternate design using bars having the same area per foot but of smaller diameter such that hooks could be eliminated; use the largest bar size possible to minimize the cost of steel placement.

FIGURE P6.13

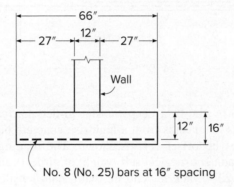

No. 8 (No. 25) bars at 16″ spacing

6.14. A closure strip is to be used between two 8 in. precast slabs (Fig. P6.14). The slabs contain No. 5 (No. 16) bars spaced at 10 in. Determine the minimum width of the closure strip for use with headed bars spliced within the strip. $A_{brg} = 4A_b$. Material strengths are $f_y = 60,000$ psi and $f_c' = 5000$ psi. The maximum size aggregate = $\frac{3}{4}$ in. Assume head thickness = 0.5 in.

FIGURE P6.14

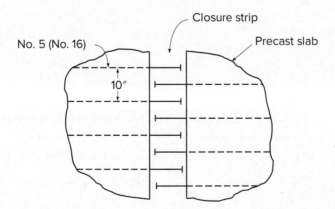

6.15. The continuous beam shown in Fig. P6.15 has been designed to carry a service dead load of 2.25 kips/ft including self-weight and service live load of 3.25 kips/ft. Flexural design has been based on ACI moment coefficients of $\frac{1}{11}$ and $\frac{1}{16}$ at the face of support and midspan, respectively, resulting in a concrete section with $b = 14$ in. and $d = 22$ in. Negative reinforcement at the support face is provided by four No. 10 (No. 32) bars, which will be cut off in pairs where no longer required by the ACI Code. Positive bars consist of four No. 8 (No. 25) bars, which will also be cut off in pairs. Specify the exact point of cutoff for all negative and positive steel. Specify also any supplementary web reinforcement that may be required. Check for satisfaction of ACI Code requirements at the point of inflection, and suggest modifications of reinforcement if appropriate. Material strengths are $f_y = 60{,}000$ psi and $f_c' = 4000$ psi.

FIGURE P6.15

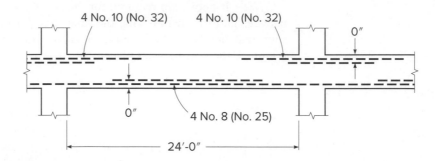

6.16. Figure P6.16 shows a deep transfer girder that carries two heavy column loads at its outer ends from a high-rise concrete building. Ground-floor columns must be offset 8 ft as shown. The loading produces an essentially constant moment (neglect self-weight of girder) calling for a concrete section with $b = 22$ in. and $d = 50$ in., with main tensile reinforcement at the top of the girder comprised of 12 No. 11 (No. 36) bars in three layers of four bars each. The maximum available bar length is 60 ft, so tensile splices must be provided. Design and detail all splices, following ACI Code provisions. Splices will be staggered, with no more than four bars spliced at any section. Also, investigate the need for special anchorage at the outer ends of main reinforcement, and specify details of special anchorage if required. Material strengths are $f_y = 60{,}000$ psi and $f_c' = 5000$ psi.

FIGURE P6.16

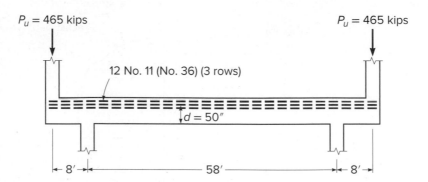

7 Serviceability

7.1 INTRODUCTION

Chapters 4, 5, and 6 have dealt mainly with the strength design of reinforced concrete beams. Methods have been developed to ensure that beams will have a proper safety margin against failure in flexure or shear, or due to inadequate bond and anchorage of the reinforcement. The member has been assumed to be at a hypothetical overload state for this purpose.

It is also important that member performance in normal service be satisfactory, when loads are those actually expected to act, that is, when load factors are 1.0. This is not guaranteed simply by providing adequate strength. Service load deflections under full load may be excessively large, or long-term deflections due to sustained loads may cause damage. Tension cracks in beams may be wide enough to be visually disturbing, and in some cases may reduce the durability of the structure. These and other questions, such as vibration or fatigue, require consideration.

Serviceability studies are carried out based on elastic theory, with stresses in both concrete and steel assumed to be proportional to strain. The concrete on the tension side of the neutral axis may be assumed uncracked, partially cracked, or fully cracked, depending on the loads and material strengths (see Section 4.2).

In early reinforced concrete designs, questions of serviceability were dealt with indirectly, by limiting the stresses in concrete and steel at service loads to the rather conservative values that had resulted in satisfactory performance. In contrast, with current design methods that permit more slender members through more accurate assessment of capacity, and with higher-strength materials further contributing to the trend toward smaller member sizes, such indirect methods no longer work. The current approach is to investigate service load cracking and deflections specifically, after proportioning members based on strength requirements.

In this chapter, methods will be developed to ensure that the cracks associated with flexure of reinforced concrete beams are narrow and well distributed, and that short- and long-term deflections at loads up to the full service load are not objectionably large.

7.2 CRACKING IN FLEXURAL MEMBERS

All reinforced concrete beams crack, generally starting at loads well below service level, and possibly even prior to loading due to restrained shrinkage. Flexural cracking due to loads is not only inevitable but actually necessary for the reinforcement to be used effectively. Prior to the formation of flexural cracks, the steel stress is no more than n times the stress in the adjacent concrete, where n is the modular ratio E_s/E_c. For materials common in current practice, n is approximately 8. Thus, when the concrete is close to its modulus of rupture of about 500 psi, the steel stress will be only $8 \times 500 = 4000$ psi, far too low to be very effective as reinforcement. At normal service loads, steel stresses 8 or 9 times that value can be expected.

In a well-designed beam, flexural cracks are fine, so-called hairline cracks, almost invisible to the casual observer, and they permit little if any corrosion of the reinforcement. As loads are gradually increased above the cracking load, both the number and the width of cracks increase, and at service load level a maximum width of crack of about 0.016 in. is typical. If loads are further increased, crack widths increase further, although the number of cracks is more or less stable.

Cracking of concrete is a random process, highly variable and influenced by many factors. Because of the complexity of the problem, present methods for predicting crack widths are based primarily on test observations. Most equations that have been developed predict the *probable maximum crack width*, which usually means that about 90 percent of the crack widths in the member are below the calculated value. However, isolated cracks exceeding twice the computed width can sometimes occur (Ref. 7.1).

a.   Variables Affecting Width of Cracks

In the discussion of the importance of a good bond between steel and concrete in Section 6.1, it was pointed out that if proper end anchorage is provided, a beam will not fail prematurely, even though the bond is destroyed along the entire span. However, crack widths will be greater than for an otherwise identical beam in which good resistance to slip is provided along the length of the span. In general, beams with smooth round bars will display a relatively small number of rather wide cracks in service, while beams with good slip resistance ensured by proper surface deformations on the bars will show a larger number of very fine, almost invisible cracks. Because of this improvement, reinforcing bars in current practice are always provided with surface deformations, the maximum spacing and minimum height of which are established by ASTM Specifications A615, A706, A996, and A1035.

A second variable of importance is the stress in the reinforcement. Studies by Gergely and Lutz and others (Refs. 7.2 to 7.4) have confirmed that crack width is proportional to f_s^n, where f_s is the steel stress and n is an exponent that varies in the range from about 1.0 to 1.4. For steel stresses in the range of practical interest, say from 20 to 40 ksi, n may be taken equal to 1.0. The steel stress is easily computed based on elastic cracked-section analysis (Section 4.2b). Alternatively, f_s may be taken equal to $\frac{2}{3}f_y$ according to ACI Code 24.3.2.

Experiments by Broms (Ref. 7.5) and others have shown that both crack spacing and crack width are related to the concrete cover distance d_c, measured from the center of the bar to the face of the concrete. In general, increasing the cover increases the spacing of cracks and also increases crack width. Furthermore, the distribution of the reinforcement in the tension zone of the beam is important. Generally, to control cracking, it is better to use a larger number of smaller-diameter bars to provide the required A_s than to use the minimum number of larger bars, and the bars should be well distributed over the tensile zone of the concrete. For deep flexural members, this includes additional reinforcement on the sides of the web to prevent excessive surface crack widths above or below the level of the main flexural reinforcement.

FIGURE 7.1
Geometric basis of crack
width calculations.

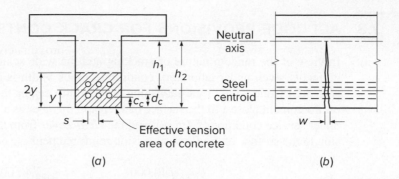

(a)   (b)

b. Equations for Crack Width

A number of expressions for maximum crack width have been developed based on
the statistical analysis of experimental data. Two expressions that have figured prom-
inently in the development of the crack control provisions in the ACI Code are those
developed by Gergely and Lutz (Ref. 7.2) and Frosch (Ref. 7.4) for the maximum
crack width at the tension face of a beam. They are, respectively,

$$w = 0.076\beta f_s \sqrt[3]{d_c A} \tag{7.1}$$

and

$$w = 2000 \frac{f_s}{E_s} \beta \sqrt{d_c^2 + \left(\frac{s}{2}\right)^2} \tag{7.2}$$

where w = maximum width of crack, thousandth inches
f_s = steel stress at load for which crack width is to be determined, ksi
E_s = modulus of elasticity of steel, ksi

The geometric parameters are shown in Fig. 7.1 and are as follows:

d_c = thickness of concrete cover measured from tension face to center of bar
closest to that face, in.
β = ratio of distances from tension face and from steel centroid to neutral axis,
equal to h_2/h_1
A = concrete area surrounding one bar, equal to total effective tension area of
concrete surrounding reinforcement and having same centroid, divided by
number of bars, in.^{2}
s = maximum bar spacing, in.

Equations (7.1) and (7.2), which apply only to beams in which deformed bars are
used, include all the factors just named as having an important influence on the width
of cracks: steel stress, concrete cover, and the distribution of the reinforcement in the
concrete tensile zone. In addition, the factor β is added to account for the increase in
crack width with distance from the neutral axis (see Fig. 7.1b).

c. Cyclic and Sustained Load Effects

Both cyclic and sustained loading account for increasing crack width. While there is a
large amount of scatter in test data, results of fatigue tests and sustained loading tests
indicate that a doubling of crack width can be expected with time (Ref. 7.1). Under
most conditions, the spacing of cracks does not change with time at constant levels of
sustained stress or cyclic stress range.

7.3   ACI CODE PROVISIONS FOR CRACK CONTROL

In view of the random nature of cracking and the wide scatter of crack width measurements, even under laboratory conditions, crack width is controlled in accordance with ACI Code 24.3.2 by establishing a maximum center-to-center spacing s for the reinforcement closest to the surface of a tension member as a function of the bar stress under service conditions f_s (in psi) and the *clear cover* from the nearest surface in tension to the surface of the flexural tension reinforcement c_c, as shown in Fig. 7.1.

$$s = 15 \left(\frac{40{,}000}{f_s} \right) - 2.5c_c \leq 12 \left(\frac{40{,}000}{f_s} \right) \tag{7.3}$$

The choice of clear cover c_c, rather than the cover to the center of the bar d_c, was made to simplify design, since this allows s to be independent of bar size. As a consequence, maximum crack widths will be somewhat greater for larger bars than for smaller bars.

As shown in Eq. (7.3), the ACI Code sets an upper limit on s of $12(40{,}000/f_s)$. The stress f_s is calculated by dividing the service load moment by the product of the area of reinforcement and the internal moment arm, as shown in Eq. (4.4). Alternatively, the ACI Code permits f_s to be taken as two-thirds of the specified yield strength f_y. For members with only a single bar, the width of the extreme tension face may not exceed s, in accordance with ACI Code 24.3.3.

Figure 7.2a compares the values of spacing s obtained using Eqs. (7.1) and (7.2) for a beam containing No. 8 (No. 25) reinforcing bars, for $f_s = 40{,}000$ psi (corresponding to $f_y = 60{,}000$ psi), $\beta = 1.2$, and a maximum crack width $w = 0.016$ in., to the values calculated using Eq. (7.3). Equations (7.1) and (7.2) give identical spacings for two values of clear cover, but significantly different spacings for other values of c_c. Equation (7.3) provides a practical representation of the values of s that are calculated using the two experimentally based expressions. The equation is plotted in Fig. 7.2b for $f_s = 26{,}667$, $40{,}000$, and $53{,}333$ psi, corresponding to $\frac{2}{3} f_y$ for Grade 40, 60, and 80 bars, respectively.

ACI Code 24.3.5 points out that the spacing s in structures subject to fatigue, designed to be watertight, or exposed to corrosion should be based on "investigations or precautions specific to those conditions." These include the use of expressions such as Eqs. (7.1) and (7.2) to determine the probable maximum crack width. Further guidance is given in Ref. 7.1. In any case, s should not exceed the value obtained using Eq. (7.3).

When concrete T beam flanges are in tension, as in the negative-moment region of continuous T beams, concentration of the reinforcement over the web may result in excessive crack width in the overhanging slab, even though cracks directly over the web are fine and well distributed. To prevent this, the tensile reinforcement should be distributed over the width of the flange, rather than concentrated. However, because of shear lag, the outer bars in such a distribution would be considerably less highly stressed than those directly over the web, producing an uneconomical design. As a reasonable compromise, ACI Code 24.3.4 requires that the tension reinforcement in such cases be distributed over the effective flange width or a width equal to one-tenth the span, whichever is smaller. If the effective flange width exceeds one-tenth of the span, some longitudinal reinforcement must be provided in the outer portions of the flange. The amount of such additional reinforcement is left to the discretion of the designer; it should at least be the equivalent of temperature reinforcement for the slab (see Section 12.3) and is often taken as twice that amount.

FIGURE 7.2
Maximum bar spacing versus clear cover: (*a*) Comparison of Eqs. (7.1), (7.2), and (7.3) for $w_c = 0.016$ in., $f_s = 40,000$ psi, $\beta = 1.2$, bar size = No. 8 (No. 25) and (*b*) Eq. (7.3) for $f_s = 26,667$, 40,000, and 53,333 psi, corresponding to $\frac{2}{3}f_y$ for Grades 40, 60, and 80 reinforcement, respectively. [*Part (a) after Ref. 7.6.*]

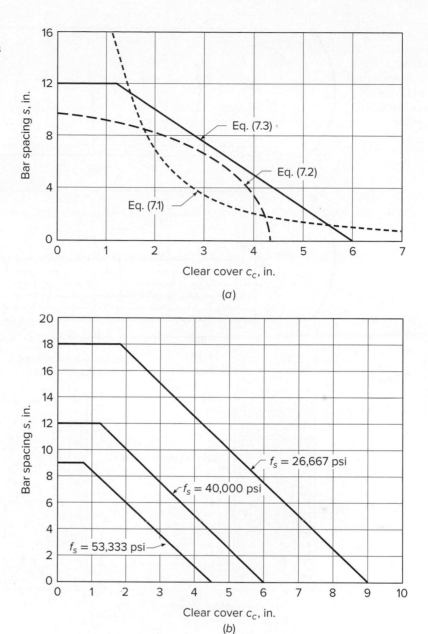

For beams with relatively deep webs, some "skin" reinforcement should be placed near the vertical faces of the web to control the width of cracks in the concrete tension zone above the level of the main reinforcement. Without such steel, well-distributed cracks at the level of the main bars tend to consolidate into a smaller number of wider cracks closer to the neutral axis (Ref. 7.7), as shown in Fig. 7.3*a*. With such steel, the cracks do not consolidate, and remain narrow, as shown in Fig. 7.3*b*. According to ACI Code 9.7.2.3, if the total depth of the beam *h* exceeds 36 in., longitudinal skin reinforcement must be uniformly distributed along both side faces of the member for a distance $h/2$ from the tension face, as shown in Fig. 7.4. The spacing *s* between longitudinal bars or wires is as specified in Eq. (7.3). The size of the bars or wires is not specified, but as indicated in ACI Commentary 9.7.2.3, No. 3 to No. 5 (No. 10 to No. 16) bars or welded wire reinforcement with a minimum area

FIGURE 7.3
Cracking in beams with
relatively deep webs:
(*a*) beam with only flexural
reinforcement and (*b*) beam
with both flexural and skin
reinforcement. (*After
Ref. 7.7.*)

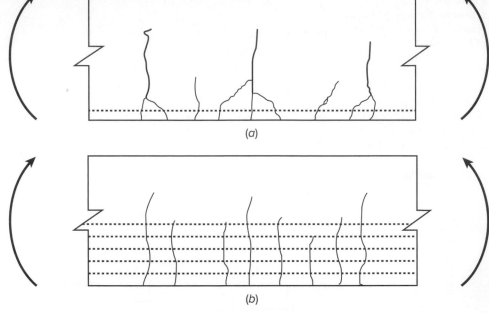

(*a*)

(*b*)

FIGURE 7.4
Skin reinforcement for
flexural members with total
depth *h* greater than 36 in.

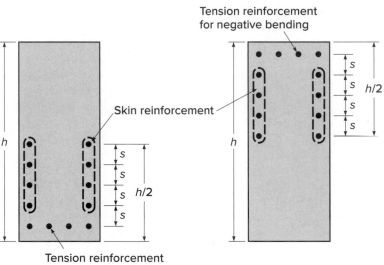

of 0.1 in² per foot of depth are typically used. The contribution of the skin reinforcement to flexural strength is usually disregarded, although it may be included in the strength calculations if a strain compatibility analysis is used to establish the stress in the skin steel at the flexural failure load.

Figure 7.2*b* provides a convenient design aid for determining the maximum center-to-center bar spacing as a function of clear cover for the usual case used in design, $f_s = \frac{2}{3} f_y$. From a practical point of view, it is even more helpful to know the minimum number of bars across the width of a beam stem that is needed to satisfy the ACI Code requirements for crack control. That number depends on side cover, as well as clear cover to the tension face, and is dependent on bar size. Table A.8 in Appendix A gives the minimum number of Grade 60 bars across a beam stem for two common cases, 2 in. clear cover on the sides and bottom, which corresponds to using No. 3 or No. 4 (No. 10 or No. 13) stirrups, and $1\frac{1}{2}$ in. clear cover on the sides and bottom, representing beams in which no stirrups are used.

EXAMPLE 7.1   **Check crack control criteria.**   Figure 7.5 shows the main flexural reinforcement at midspan for a T girder in a high-rise building that carries a service load moment of 8630 in-kips. The clear cover on the side and bottom of the beam stem is $2\frac{1}{4}$ in. $f_y = 60$ ksi. Determine if the beam meets the crack control criteria in the ACI Code.

SOLUTION.   Since the depth of the beam equals but does not exceed 36 in., skin reinforcement is not needed. To check the bar spacing criteria, the steel stress can be estimated closely by taking the internal lever arm equal to the distance $d - h_f/2$:

$$f_s = \frac{M_s}{A_s\,(d - h_f/2)} = \frac{8630}{7.9 \times 29.25} = 37.3 \text{ ksi}$$

(Alternately, the ACI Code permits using $f_s = \frac{2}{3}f_y$, giving 40.0 ksi.)

Using $f_s = 40.0$ ksi in Eq. (7.3) gives

$$s = 15\left(\frac{40{,}000}{f_s}\right) - 2.5c_c = 15\left(\frac{40{,}000}{37{,}300}\right) - 2.5 \times 2.25 = 10.5 \text{ in.}$$

By inspection, it is clear that this requirement is satisfied for the beam. If the results had been unfavorable, a redesign using a larger number of smaller-diameter bars would have been indicated.

FIGURE 7.5
T beam for crack width determination in Example 7.1.

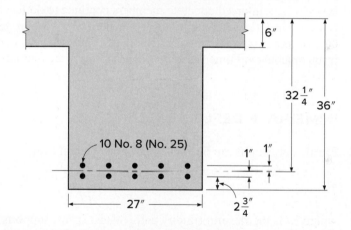

7.4   CONTROL OF DEFLECTIONS

In addition to limitations on cracking, described in the preceding sections, it is usually necessary to impose certain controls on deflections of beams to ensure serviceability. Excessive deflections can lead to cracking of supported walls and partitions, ill-fitting doors and windows, poor roof drainage, misalignment of sensitive machinery and equipment, or visually offensive sag. It is important, therefore, to maintain control of deflections, in one way or another, so that members designed mainly for strength at prescribed overloads will also perform well in normal service.

There are presently two approaches to deflection control. The first is indirect and consists in setting suitable upper limits on the span-depth ratio. This is simple, and it is satisfactory in many cases where spans, loads and load distributions, and member sizes and proportions fall in the usual ranges. Otherwise, it is essential to calculate deflections and to compare those predicted values with specific limitations that may be imposed by codes or by special requirements.

It will become clear, in the sections that follow, that calculations can, at best, provide a guide to probable actual deflections. This is so because of uncertainties regarding material properties, effects of cracking, and load history for the member under consideration. Extreme precision in the calculations, therefore, is never justified, because highly accurate results are unlikely. However, it is generally sufficient to know, for example, that the deflection under load will be about $\frac{1}{2}$ in. rather than 2 in., while it is relatively unimportant to know whether it will actually be $\frac{5}{8}$ in. rather than $\frac{1}{2}$ in.

The deflections of concern are generally those that occur during the normal service life of the member. In service, a member sustains the full dead load, plus some fraction or all of the specified service live load. Safety provisions of the ACI Code and similar design specifications ensure that, under loads up to the full service load, stresses in both steel and concrete remain within the elastic ranges. Consequently, deflections that occur at once upon application of load, the *immediate deflections*, can be calculated based on the properties of the uncracked elastic member, the cracked elastic member, or some combination of these (see Section 4.2).

It was pointed out in Sections 2.8 and 2.11, however, that in addition to concrete deformations that occur immediately when load is applied, there are other deformations that take place gradually over an extended time. These time-dependent deformations are chiefly due to concrete creep and shrinkage. As a result of these influences, reinforced concrete members continue to deflect with the passage of time. Long-term deflections continue over a period of several years, and may eventually be 2 or more times the initial elastic deflections. Clearly, methods for predicting both instantaneous and time-dependent deflections are essential.

7.5   IMMEDIATE DEFLECTIONS

Elastic deflections can be expressed in the general form

$$\Delta = \frac{f(\text{loads, spans, supports})}{EI}$$

where EI is the flexural rigidity and $f(\text{loads, spans, supports})$ is a function of the particular load, span, and support arrangement. For instance, the deflection of a uniformly loaded simple beam is $5wl^4/384EI$, so that $f = 5wl^4/384$. Similar deflection equations have been tabulated or can easily be computed for many other loadings and span arrangements, simple, fixed, or continuous, and the corresponding f functions can be determined. The particular problem in reinforced concrete structures is therefore the determination of the appropriate flexural rigidity EI for a member consisting of two materials with properties and behavior as widely different as steel and concrete.

If the maximum moment in a flexural member is so small that the tensile stress in the concrete does not exceed the modulus of rupture f_r, no flexural tension cracks will occur. The full, uncracked section is then available for resisting stress and providing rigidity. This stage of loading was analyzed in Section 4.2a based on the uncracked, transformed section (see Fig. 4.2). In agreement with this analysis, the effective moment of inertia for this low range of loads is that of the uncracked transformed section I_{ut}, and E is the modulus of concrete E_c as given by Eq. (2.3). Correspondingly, for this load range,

$$\Delta_{iu} = \frac{f}{E_c\, I_{ut}} \tag{a}$$

At higher loads, flexural tension cracks form. In addition, if shear stresses exceed v_{cr} [see Eq. (5.3)] and web reinforcement is employed to resist them, diagonal cracks can exist at service loads. In the region of flexural cracks, the position of the neutral axis varies: directly at each crack, it is located at the level calculated for the cracked transformed section (see Section 4.2b); midway between cracks, it dips to a location closer to that calculated for the uncracked transformed section. Correspondingly, flexural-tension cracking causes the effective moment of inertia to be that of the cracked transformed section (Fig. 4.4) in the immediate neighborhood of flexural-tension cracks and closer to that of the uncracked transformed section midway between cracks, with a gradual transition between these extremes.

The value of the local moment of inertia varies in those portions of the beam in which the bending moment exceeds the cracking moment of the section

$$M_{cr} = \frac{f_r I_{ut}}{y_t} \tag{7.4}$$

where y_t is the distance from the neutral axis to the tension face and f_r is the modulus of rupture. The exact variation of I depends on the shape of the moment diagram and on the crack pattern and is difficult to determine. This makes an exact deflection calculation impossible.

However, extensively documented studies (Refs. 7.8 to 7.11) have shown that deflections Δ_{ic} occurring in a beam after the maximum moment M_a has reached and exceeded the cracking moment M_{cr} can be calculated by using an effective moment of inertia I_e; that is,

$$\Delta_{ic} = \frac{f}{E_c I_e} \tag{b}$$

For many years, I_e was approximated as

$$I_e = \left(\frac{M_{cr}}{M_a}\right)^3 I_{ut} + \left[1 - \left(\frac{M_{cr}}{M_a}\right)^3\right] I_{cr} \leq I_{ut} \tag{7.5}$$

where I_{cr} is the moment of inertia of the cracked transformed section.

More recent studies (Refs 7.9 to 7.11) have demonstrated that Eq. (7.5) underestimates the deflection of beams and slabs with reinforcement ratios ρ below 0.01 and members where cracking occurs for $M_a < M_{cr}$ because the member has been subjected to tensile stresses caused by restrained shrinkage. In such cases, a more appropriate expression for reinforced concrete flexural members is

$$I_e = \frac{I_{cr}}{1 - \left(\frac{(2/3)M_{cr}}{M_a}\right)^2 \left(1 - \frac{I_{cr}}{I_{ut}}\right)} \leq I_{ut} \tag{7.6}$$

The term $(2/3)M_{cr}$ approximates the effective cracking moment under combined bending and restraint. Equations (7.5) and (7.6) produce similar results for members with $\rho \geq 0.01$ and $M_a \geq 2M_{cr}$. Equation (7.5) remains the basis in the ACI Code for calculating I_e for prestressed concrete members.

In Fig. 7.6, the effective moment of inertia, given by Eq. (7.6), is plotted as a function of the ratio M_a/M_{cr} (the reciprocal of the moment ratio used in the equation). As shown in the figure for values of maximum moment M_a less than the two-thirds of the cracking moment M_{cr} (that is, $M_a/M_{cr} < \frac{2}{3}$), $I_e = I_{ut}$. With increasing values of M_a, I_e approaches I_{cr}; for values of M_a/M_{cr} of 3 or more, I_e is almost the same as I_{cr}. Typical values of M_a/M_{cr} at full service load range from about 1.5 to 3.

FIGURE 7.6
Variation of I_e with moment ratio.

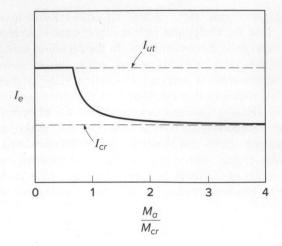

FIGURE 7.7
Deflection of a reinforced concrete beam using Eq. (7.6).

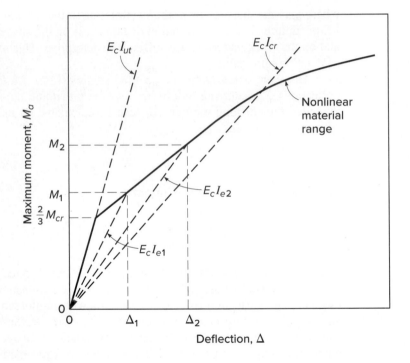

Figure 7.7 shows the growth of deflections with increasing moment for a simple-span beam and illustrates the use of Eq. (7.6). For moments no larger than $\frac{2}{3} M_{cr}$, deflections are practically proportional to moments, and the deflection at which cracking begins (assuming that the member is subject to restraint of shrinkage) is obtained from Eq. (*a*) with $M = \frac{2}{3} M_{cr}$. At larger moments, the effective moment of inertia I_e becomes progressively smaller, according to Eq. (7.6), and deflections are found by Eq. (*b*) for the load level of interest.

The moment M_2 might correspond to the full service load, for example, while the moment M_1 would represent the dead load moment for a typical case. A moment-deflection curve corresponding to the line $E_c I_{cr}$ represents an upper bound for deflections, consistent with Fig. 7.6, except that at loads somewhat beyond the service load, the nonlinear response of steel, concrete, or both causes a further non-linear increase in deflections.

Note that to calculate the increment of deflection due to live load, causing a moment increase $M_2 - M_1$, a two-step computation is required: the first for deflection Δ_2 due to live and dead load and the second for deflection Δ_1 due to dead load alone, each with the appropriate value of I_e. Then the deflection increment due to live load is found, equal to $\Delta_2 - \Delta_1$.

Most reinforced concrete spans are continuous, not simply supported. The concepts just introduced for simple spans can be applied, but the moment diagram for a given span will include both negative and positive regions, reflecting the rotational restraint provided at the ends of the spans by continuous frame action. The effective moment of inertia for a continuous span can be found by a simple averaging procedure, according to the ACI Code, that will be described in Section 7.7c.

A fundamental problem for continuous spans is that although the deflections are based on the moment diagram, the moment diagram depends, in turn, on the flexural rigidity EI for each member of the frame. The flexural rigidity depends on the extent of cracking, as has been demonstrated. Cracking, in turn, depends on the moments, which are to be found. The circular nature of the problem is evident.

One could use an iterative procedure, initially basing the frame analysis on uncracked concrete members, determining the moments, calculating effective EI terms for all members, then recalculating moments, adjusting the EI values, etc. The process could be continued for as many iterations as needed, until changes are not significant. However, such an approach would be expensive and time-consuming, even with computer use.

Usually, a very approximate approach is adopted. Member flexural stiffnesses for the frame analysis are based simply on properties of uncracked rectangular concrete cross sections. This can be defended by noting that the moments in a continuous frame depend only on the *relative* values of EI in its members, not the *absolute* values. Hence, if a consistent assumption, that is, uncracked section, is used for all members, the results should be valid. Although cracking is certainly more prevalent in beams than in columns, thus reducing the relative EI for the beams, this is compensated to a large extent, in typical cases, by the stiffening effect of the flanges in the positive bending regions of continuous T beam construction. This subject is discussed at greater length in Section 11.5.

7.6 DEFLECTIONS DUE TO LONG-TERM LOADS

Initial deflections are increased significantly if loads are sustained over a long period of time, due to the effects of shrinkage and creep. These two effects are usually combined in deflection calculations. Creep generally dominates, but for some types of members, shrinkage deflections are large and should be considered separately (see Section 7.8).

It was pointed out in Section 2.8 that creep deformations of concrete are directly proportional to the compressive stress up to and beyond the usual service load range. They increase asymptotically with time and, for the same stress, are larger for low-strength than for high-strength concretes. The ratio of additional time-dependent strain to initial elastic strain is given by the creep coefficient C_{cu} (see Table 2.2).

For a reinforced concrete beam, the long-term deformation is much more complicated than for an axially loaded cylinder, because while the concrete creeps under sustained load, the steel does not. The situation in a reinforced concrete beam is illustrated by Fig. 7.8. Under sustained load, the initial strain ε_i at the top face of the beam increases, due to creep, by the amount ε_t, while the strain ε_s in the steel

FIGURE 7.8

Effect of concrete creep on curvature: (*a*) beam cross section; (*b*) strains; and (*c*) stresses and forces. (*Adapted from Ref. 7.8.*)

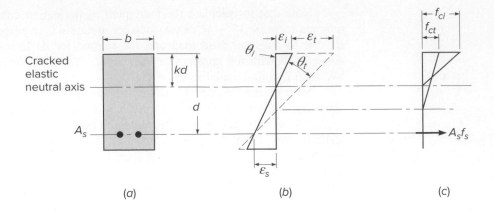

(*a*)   (*b*)   (*c*)

is essentially unchanged. Because the rotation of the strain distribution diagram is therefore about a point at the level of the steel, rather than about the cracked elastic neutral axis, the neutral axis moves down as a result of creep, and

$$\frac{\theta_t}{\theta_i} < \frac{\varepsilon_t}{\varepsilon_i} \tag{a}$$

demonstrating that the usual creep coefficients cannot be applied to initial curvatures to obtain creep curvatures (hence deflections).

The situation is further complicated. Due to the lowering of the neutral axis associated with creep (see Fig. 7.8*b*) and the resulting increase in compression area, the compressive stress required to produce a given resultant C to equilibrate $T = A_s f_s$ is less than before, in contrast to the situation in a creep test of a compressed cylinder, because the beam creep occurs at a gradually diminishing stress. On the other hand, with the new lower neutral axis, the internal lever arm between compressive and tensile resultant forces is less, calling for an increase in both resultants for a constant moment. This, in turn, will require a small increase in stress, and hence strain, in the steel; thus, ε_s is not constant as assumed originally.

Because of such complexities, it is necessary in practice to calculate additional, time-dependent deflections of beams due to creep (and shrinkage) using a simplified, empirical approach by which the initial elastic deflections are multiplied by a factor λ_Δ to obtain the additional long-time deflections. Values of λ_Δ for use in design are based on long-term deflection data for reinforced concrete beams (Refs. 7.12 to 7.15). Thus,

$$\Delta_t = \lambda_\Delta \, \Delta_i \tag{7.7}$$

where Δ_t is the *additional* long-term deflection due to the combined effect of creep and shrinkage and Δ_i is the initial elastic deflection calculated by the methods described in Section 7.5.

The coefficient λ_Δ depends on the duration of the sustained load. It also depends on whether the beam has only reinforcement A_s on the tension side or whether additional longitudinal reinforcement A'_s is provided on the compression side. In the latter case, the long-term deflections are much reduced. This is so because when no compression reinforcement is provided, the compression concrete is subject to unrestrained creep and shrinkage. On the other hand, since steel is not subject to creep, if additional bars are located close to the compression face, they will resist and thereby reduce the amount of creep and shrinkage and the corresponding deflection (Ref. 7.15). Compression steel may be included for this reason alone. Specific values of λ_Δ, used to account for the influence of creep and compression reinforcement, will be given in Section 7.7.

If a beam carries a certain sustained load W (such as the dead load plus the average traffic load on a bridge) and is subject to a short-term heavy live load P (such as the weight of an unusually heavy vehicle), the maximum total deflection under this combined loading is obtained as follows:

1. Calculate the instantaneous deflection Δ_{iw} caused by the sustained load W by methods given in Section 7.5.
2. Calculate the additional long-term deflection caused by W, that is,

$$\Delta_{tw} = \lambda_\Delta \, \Delta_{iw}$$

3. Then the total deflection caused by the sustained part of the load is

$$\Delta_w = \Delta_{iw} + \Delta_{tw}$$

4. In calculating the additional instantaneous deflection caused by the short-term load P, account must be taken of the fact that the load-deflection relation after cracking is nonlinear, as illustrated by Fig. 7.7. Hence,

$$\Delta_{ip} = \Delta_{i(w+p)} - \Delta_{iw}$$

where $\Delta_{i(w+p)}$ is the total instantaneous deflection that would be obtained if W and P were applied simultaneously, calculated by using I_e determined for the moment caused by $W + P$.

5. Then the total deflection under the sustained load plus heavy short-term load is

$$\Delta = \Delta_w + \Delta_{ip}$$

In calculations of deflections, careful attention must be paid to the load history, that is, the time sequence in which loads are applied, as well as to the magnitude of the loads. The short-term peak load on the bridge girder just described might be applied early in the life of the member, before time-dependent deflections had taken place. Similarly, for buildings, heavy loads such as stacked material are often placed during construction. These temporary loads may be equal to, or even greater than, the design live load. The state of cracking will correspond to the *maximum load* that was carried, and the sustained load deflection, on which the long-term effects are based, would correspond to that cracked condition. I_e for the maximum load reached should be used to recalculate the sustained load deflection before calculating long-term effects.

This will be illustrated referring to Fig. 7.9, showing the load-deflection plot for a building girder that is designed to carry a specified dead and live load. Assume first that the dead and live loads increase monotonically. As the full dead load W_d is applied, the load deflection curve follows the path 0-1, and the dead load deflection Δ_d is found using I_{e1} calculated from Eq. (7.6), with $M_a = M_d$. The time-dependent effect of the dead load would be $\lambda_\Delta \Delta_d$. As live load is then applied, path 1-2 would be followed. Live load deflection Δ_l would be found in two steps, as described in Section 7.5, first finding Δ_{d+l} based on I_{e2}, with M_a in Eq. (7.6) equal to M_{d+l}, and then subtracting dead load deflection Δ_d.

If, on the other hand, short-term construction loads were applied, then removed, the deflection path 1-2-3 would be followed. Then, under dead load only, the resulting deflection would be Δ_d'. Note that this deflection can be found in one step using W_d, but with I_{e2} corresponding to the maximum load reached. The long-term deflection now would be $\lambda_\Delta \Delta_d'$, significantly *larger* than before. Should the full design *live* load then be applied, the deflection would follow path 3-4, and the live load deflection would be *less* than for the first case. It, too, can be calculated by a simple one-step calculation using W_l alone, in this case, and with moment of inertia equal to I_{e2}.

FIGURE 7.9
Effect of load history on
deflection of a building
girder.

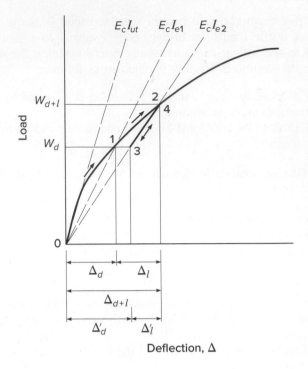

Clearly, in calculating deflections, the engineer must anticipate, as nearly as possible, both the magnitude and the time sequence of the loadings. Although long-term deflections are often calculated assuming monotonic loading, with both immediate and long-term effects of dead load occurring before application of live load, in many cases this is not realistic.

7.7   ACI CODE PROVISIONS FOR CONTROL OF DEFLECTIONS

a.   Minimum Depth-Span Ratios

As pointed out in Section 7.4, two approaches to deflection control are in current use, both acceptable under the provisions of the ACI Code, within prescribed limits. The simpler of these is to impose restrictions on the minimum member depth h, relative to the span l, to ensure that the beam will be sufficiently stiff that deflections are unlikely to cause problems in service. Deflections are greatly influenced by support conditions (for example, a simply supported uniformly loaded beam will deflect 5 times as much as an otherwise identical beam with fixed supports), so minimum depths must vary depending on conditions of restraint at the ends of the spans.

According to ACI Code 7.3.1 and 9.3.1, the minimum depths of Table 7.1 apply to one-way construction *not* supporting or attached to partitions or other construction likely to be damaged by large deflections, unless computation of deflections indicates that a lesser depth can be used without adverse effects. The values given in Table 7.1 are to be used directly for normalweight concrete with $w_c = 145$ pcf and reinforcement with $f_y = 60,000$ psi. For members using lightweight concrete with density in the range from 90 to 115 pcf, the values of Table 7.1 should be multiplied by $1.65 - 0.005w_c \geq 1.09$. For reinforcing steel with yield strengths other than 60,000 psi, the values should be multiplied by $0.4 + f_y/100,000$.

TABLE 7.1

Minimum thickness of nonprestressed beams or one-way slabs unless deflections are computed

Member	Minimum Thickness h			
	Simply Supported	One End Continuous	Both Ends Continuous	Cantilever
	Members Not Supporting or Attached to Partitions or Other Construction Likely to Be Damaged by Large Deflections			
Solid one-way slabs	$\ell/20$	$\ell/24$	$\ell/28$	$\ell/10$
Beams or ribbed one-way slabs	$\ell/16$	$\ell/18.5$	$\ell/21$	$\ell/8$

b. Calculation of Immediate Deflections

When there is need to use member depths shallower than are permitted by Table 7.1, when members support construction that is likely to be damaged by large deflections, or for prestressed members, deflections must be calculated and compared with limiting values (see Section 7.7e). The calculation of deflections, when required, proceeds along the lines described in Sections 7.5 and 7.6. For design purposes, the moment of the uncracked transformed section I_{ut} can be replaced by that of the gross concrete section I_g, neglecting reinforcement, without serious error. With this simplification, Eqs. (7.4) and (7.6) are replaced by the following:

$$M_{cr} = \frac{f_r I_g}{y_t} \tag{7.8}$$

and

$$I_e = \frac{I_{cr}}{1 - \left(\frac{(2/3)M_{cr}}{M_a}\right)^2 \left(1 - \frac{I_{cr}}{I_g}\right)} \leq I_g \tag{7.9}$$

The modulus of rupture is to be taken equal to

$$f_r = 7.5\lambda\sqrt{f_c'} \tag{7.10}$$

As explained in Section 5.5a, in accordance with ACI Code 19.2.4, λ may be taken as 1.0 for normalweight concrete and 0.75 for lightweight concrete. Values between 0.75 and 1.0 may also be used by applying linear interpolation between $\lambda = 0.75$ and 1.0 for concretes with unit weights of $w_c \leq 100$ pcf and $w_c \geq 135$ pcf, respectively. Alternatively, $\lambda = 0.75$ for all-lightweight concrete and 0.85 for sand-lightweight concrete, using linear interpolation between 0.75 and 0.85 when a portion of the lightweight fine aggregate is replaced by normalweight fine aggregate and between 0.85 and 1.0 for concretes containing normalweight fine aggregate and a blend of lightweight and normalweight coarse aggregate.

c. Continuous Spans

For continuous spans, ACI Code 24.2.3.6 calls for a simple average of values obtained from Eq. (7.9) for the critical positive- and negative-moment sections, that is,

$$I_e = 0.50I_{em} + 0.25 (I_{e1} + I_{e2}) \tag{7.11a}$$

where I_{em} is the effective moment of inertia for the midspan section and I_{e1} and I_{e2} are those for the negative-moment sections at the respective beam ends, each calculated from Eq. (7.9) using the applicable value of M_a. It is shown in Ref. 7.16 that a somewhat improved result can be had for continuous prismatic members using a weighted average for beams with both ends continuous of

$$I_e = 0.70I_{em} + 0.15\ (I_{e1} + I_{e2}) \qquad (7.11b)$$

and for beams with one end continuous and the other simply supported of

$$I_e = 0.85I_{em} + 0.15I_{e1} \qquad (7.11c)$$

where I_{e1} is the effective moment of inertia at the continuous end. As an option, ACI Code 24.2.3.7 permits I_e for continuous prismatic beams to be taken equal to the value obtained from Eq. (7.9) at midspan; for cantilevers, I_e calculated at the support section may be used.

After I_e is found, deflections may be computed with due regard for rotations of the tangent to the elastic curve at the supports. In general, in computing the maximum deflection, the loading producing the maximum positive moment may be used, and the midspan deflection may normally be used as an acceptable approximation of the maximum deflection. For members where supports may be considered fully fixed or hinged, handbook equations for deflections may be used.

d.   Long-Term Deflection Multipliers

On the basis of empirical studies (Refs. 7.8, 7.13, and 7.14), ACI Code 24.2.4.1 specifies that *additional* long-term deflections Δ_t due to the combined effects of creep and shrinkage be calculated by multiplying the immediate deflection Δ_i by the factor

$$\lambda_\Delta = \frac{\xi}{1 + 50\rho'} \qquad (7.12)$$

where $\rho' = A'_s/bd$ and ξ is a time-dependent coefficient that varies as shown in Fig. 7.10. In Eq. (7.12), the quantity $1/(1 + 50\rho')$ is a reduction factor that is essentially a section property, reflecting the beneficial effect of compression reinforcement A'_s in reducing long-term deflections, whereas ξ is a material property depending on creep and shrinkage characteristics. For simple and continuous spans, the value of ρ' used in Eq. (7.12) should be that at the midspan section, according to the ACI Code, or that at the support for cantilevers. Equation (7.12) and the values of ξ given by Fig. 7.10 apply to both normalweight and lightweight concrete beams. The additional, time-dependent deflections are thus found using values of λ_Δ from Eq. (7.12) in Eq. (7.7).

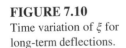

FIGURE 7.10
Time variation of ξ for long-term deflections.

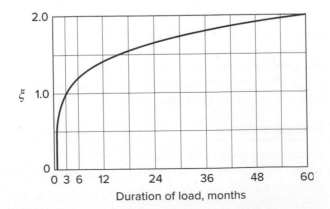

Values of ξ given in the ACI Code and Commentary are satisfactory for ordinary beams and one-way slabs, but may result in underestimation of time-dependent deflections of two-way slabs, for which Branson has suggested a 5-year value of $\xi = 3.0$ (Ref. 7.8).

Research described in Ref. 7.17 indicates that Eq. (7.12) does not properly reflect the reduced creep that is characteristic of higher-strength concretes. As indicated in Table 2.2, the creep coefficient for high-strength concrete may be as low as one-half the value for normal concrete. Clearly, the long-term deflection of high-strength concrete beams under sustained load, expressed as a ratio of immediate elastic deflection, correspondingly will be less. This suggests a lower value of the material modifier ξ in Eq. (7.12) and Fig. 7.10. On the other hand, in high-strength concrete beams, the influence of compression steel in reducing creep deflections is less pronounced, requiring an adjustment in the section modifier $1/(1 + 50\rho')$ in that equation.

Based on long-term tests involving six experimental programs, the following modified form of Eq. (7.12) is recommended (Ref. 7.17):

$$\lambda_\Delta = \frac{\mu \xi}{1 + 50\,\mu\rho'} \tag{7.13}$$

in which

$$\mu = 1.4 - f_c'\,/10{,}000$$

$$0.4 \le \mu \le 1.0 \tag{7.14}$$

The proposed equation gives results identical to Eq. (7.12) for concrete strengths of 4000 psi and below, and much improved predictions for concrete strengths between 4000 and 12,000 psi.

e. Permissible Deflections

To ensure satisfactory performance in service, ACI Code 24.2.2 imposes certain limits on deflections calculated according to the procedures just described. These limits are given in Table 7.2. Limits depend on whether or not the member supports or is attached

TABLE 7.2
Maximum permissible calculated deflections

Member	Condition		Deflection to Be Considered	Deflection Limitation
Flat roofs	Not supporting or attached to nonstructural elements likely to be damaged by large deflections		Immediate deflection due to maximum of roof live load L_r, snow load S, and rain load R	$\dfrac{\ell}{180}$
Floors			Immediate deflection due to live load L	$\dfrac{\ell}{360}$
Roof or floors	Supporting or attached to nonstructural elements	Likely to be damaged by large deflections	That part of the total deflection occurring after attachment of nonstructural elements, which is the sum of the time-dependent deflection due to all sustained loads and the immediate deflection due to any additional live load	$\dfrac{\ell}{480}$
		Not likely to be damaged by large deflections		$\dfrac{\ell}{240}$

to other nonstructural elements, and whether or not those nonstructural elements are likely to be damaged by large deflections. When long-term deflections are computed, that part of the deflection that occurs before attachment of the nonstructural elements may be deducted; information from Fig. 7.10 is useful for this purpose. The last two limits of Table 7.2 may be exceeded under certain conditions, according to the ACI Code.

EXAMPLE 7.2 **Deflection calculation.** The beam shown in Fig. 7.11 is a part of the floor system of an apartment house and is designed to carry a dead load w_d of 1.65 kips/ft and a service live load w_l of 3.3 kips/ft. Of the total live load, 20 percent is sustained in nature, while 80 percent will be applied only intermittently over the life of the structure. Under full dead and live load, the moment diagram is as shown in Fig. 7.11c. The beam will support nonstructural partitions that would be damaged if large deflections were to occur. They will be installed shortly after construction shoring is removed and dead loads take effect, but before significant creep occurs. Calculate that part of the total deflection that would adversely affect the partitions, that is, the sum of long-time deflection due to dead and partial live load plus the immediate deflection due to the nonsustained part of the live load. Material strengths are $f_c' = 4000$ psi and $f_y = 60$ ksi.

SOLUTION. For the specified materials, $E_c = 57{,}000\sqrt{4000} = 3.60 \times 10^6$ psi, and with $E_s = 29 \times 10^6$ psi, the modular ratio $n = 8$. The modulus of rupture $f_r = 7.5 \times 1.0\sqrt{4000} = 474$ psi. The effective moment of inertia will be calculated for the moment diagram shown in Fig. 7.11c corresponding to the full service load, on the basis that the extent of cracking will be governed by the full service load, even though that load is intermittent. In the positive-moment region, the centroidal axis of the uncracked T section of Fig. 7.11b is found by taking moments about the top surface, to be at 7.66 in. depth, and $I_g = 33{,}160$ in^4. By similar means, the centroidal axis of the cracked transformed T section shown in Fig. 7.11d is located 3.73 in. below the top of the slab and $I_{cr} = 10{,}860$ in^4. The cracking moment is then found by means of Eq. (7.8):

$$M_{cr} = \frac{f_r I_g}{y_t} = \frac{474 \times 33{,}160}{16.84} \times \frac{1}{12{,}000} = 78 \text{ ft-kips}$$

With $M_{cr}/M_a = 78/162 = 0.481$, the effective moment of inertia in the positive bending region is found from Eq. (7.9) to be

$$I_e = \frac{I_{cr}}{1 - \left(\dfrac{(2/3)M_{cr}}{M_a}\right)^2\left(1 - \dfrac{I_{cr}}{I_g}\right)} = \frac{10{,}860}{1 - \left(\dfrac{2}{3} \times 0.481\right)^2\left(1 - \dfrac{10{,}860}{33{,}160}\right)} = 11{,}667 \text{ in}^4$$

In the negative bending region, the gross moment of inertia will be based on the rectangular section shown in Fig. 7.11b. For this area, the centroid is 12.25 in. from the top surface and $I_g = 17{,}200$ in^4. For the cracked transformed section shown in Fig. 7.11e, the centroidal axis is found, taking moments about the bottom surface, to be 8.65 in. from that level, and $I_{cr} = 11{,}366$ in^4. Then

$$M_{cr} = \frac{474 \times 17{,}200}{12.25} \times \frac{1}{12{,}000} = 55.5 \text{ ft-kips}$$

giving $M_{cr}/M_a = 55.5/225 = 0.247$. Thus, for the negative-moment regions,

$$I_e = \frac{11{,}366}{1 - \left(\dfrac{2}{3} \times 0.247\right)^2\left(1 - \dfrac{11{,}366}{17{,}260}\right)} = 11{,}472 \text{ in}^4$$

FIGURE 7.11

Continuous T beam for
deflection calculations in
Example 7.2. The uncracked
section is shown in (b), the
cracked transformed section
in the positive-moment
region is shown in (d), and
the cracked transformed
section in the negative-
moment region is shown
in (e).

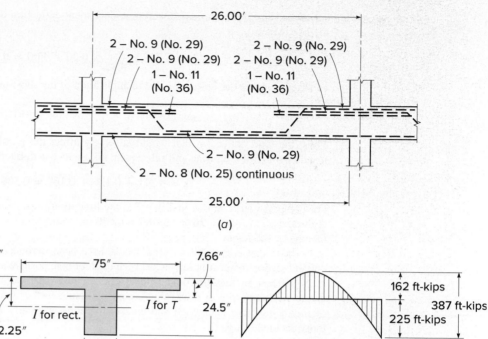

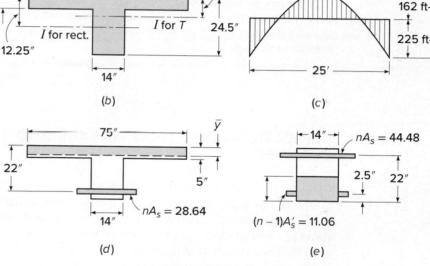

The average value of I_e to be used in calculation of deflection is

$$I_{e,av} = \frac{1}{2}\,(11{,}667 + 11{,}472) = 11{,}570 \text{ in}^4$$

It is next necessary to find the sustained-load deflection multiplier given by Eq. (7.12) and
Fig. 7.10. For the positive bending zone, with no compression reinforcement, $\lambda_{\Delta\,pos} = 2.00$.

For convenient reference, the deflection of the member under full dead plus live load
of 4.95 kips/ft, corresponding to the moment diagram of Fig. 7.11c, will be found. Making
use of the moment-area principles,

$$\Delta_{d+\ell} = \frac{1}{EI}\left[\left(\frac{2}{3}\times 387 \times 12.5 \times \frac{5}{8}\times 12.5\right) - (225 \times 12.5 \times 6.25)\right] = \frac{7620}{EI}$$

$$= \frac{7620 \times 1728}{3600 \times 11{,}570} = 0.316 \text{ in.}$$

Using this figure as a basis, the time-dependent portion of dead load deflection (the only part of
the total that would affect the partitions) is

$$\Delta_d = 0.316 \times \frac{1.65}{4.95}\times 2.00 = 0.211 \text{ in.}$$

while the sum of the immediate and time-dependent deflection due to the sustained portion of the live load is

$$\Delta_{0.20l} = 0.316 \times \frac{3.3}{4.95} \times 0.20 \times 3.00 = 0.126 \text{ in.}$$

and the instantaneous deflection due to application of the short-term portion of the live load is

$$\Delta_{0.80l} = 0.316 \times \frac{3.3}{4.95} \times 0.80 = 0.169 \text{ in.}$$

Thus, the total deflection that would adversely affect the partitions, from the time they are installed until all long-time and subsequent instantaneous deflections have occurred, is

$$\Delta = 0.211 + 0.126 + 0.169 = 0.506 \text{ in.}$$

For comparison, as shown in Table 7.2, the limitation imposed by the ACI Code in such circumstances is $\ell/480 = 26 \times 12/480 = 0.650$ in., indicating that the stiffness of the proposed member is sufficient.

Note that essentially no error would have been introduced in the above solution if the cracked-section moment of inertia had been used for both positive and negative sections rather than I_e. Significant savings in computational effort would have resulted. If M_{cr}/M_a is less than $\frac{1}{3}$, use of I_{cr} would almost always be acceptable. Note further that computation of the moment of inertia for both uncracked and cracked sections is greatly facilitated by design aids like those included in Ref. 7.18.

7.8 DEFLECTIONS DUE TO SHRINKAGE AND TEMPERATURE CHANGES

Concrete shrinkage will produce compressive stress in the longitudinal reinforcement in beams and slabs and equilibrating tensile stress in the concrete. If, as usual, the reinforcement is not symmetrically placed with respect to the concrete centroid, then shrinkage will produce curvature and corresponding deflection. The deflections will be in the same direction as those produced by the loads, if the reinforcement is mainly on the side of the member subject to flexural tension.

Shrinkage deflection is not usually calculated separately, but is combined with creep deflection, according to ACI Code procedures (see Section 7.7d). However, there are circumstances where a separate and more accurate estimation of shrinkage deflection may be necessary, particularly for thin, lightly loaded slabs. Compression steel, while it has only a small effect in reducing immediate elastic deflections, contributes significantly in reducing deflections due to shrinkage (as well as creep), and is sometimes added for this reason.

Curvatures due to shrinkage of concrete in an unsymmetrically reinforced concrete member can be found by the fictitious tensile force method (Ref. 7.8). Figure 7.12a shows the member cross section, with compression steel area A_s' and tensile steel area A_s, at depths d' and d, respectively, from the top surface. In Fig. 7.12b, the concrete and steel are imagined to be temporarily separated, so that the concrete can assume its free shrinkage strain ε_{sh}. Then a fictitious compressive force $T_{sh} = (A_s + A_s')$ $\varepsilon_{sh} E_s$ is applied to the steel, at the centroid of all the bars, a distance e below the concrete centroid, such that the steel shortening will exactly equal the free shrinkage strain of the concrete. The equilibrating tension force T_{sh} is then applied to the

FIGURE 7.12

Shrinkage curvature of a reinforced concrete beam or slab: (*a*) cross section; (*b*) free shrinkage strain; and (*c*) shrinkage curvature.

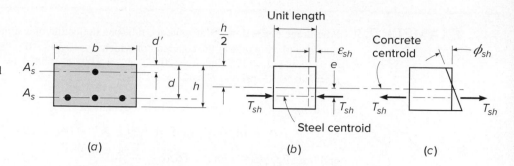

(*a*)   (*b*)   (*c*)

recombined section, as in Fig. 7.12*c*. This produces a moment $T_{sh}e$, and the corresponding shrinkage curvature is

$$\phi_{sh} = \frac{T_{sh}e}{EI}$$

The effects of concrete cracking and creep complicate the analysis, but comparisons with experimental data (Ref. 7.8) indicate that good results can be obtained using e_g and I_g for the uncracked gross concrete section and by using a reduced modulus E_{ct} equal to $\frac{1}{2}E_c$ to account for creep. Thus,

$$\phi_{sh} = \frac{2T_{sh}e_g}{E_c I_g} \tag{7.15}$$

where E_c is the usual value of concrete modulus given by Eq. (2.3).

Empirical methods are also used, in place of the fictitious tensile force method, to calculate shrinkage curvatures. These methods are based on the simple but reasonable proposition that the shrinkage curvature is a direct function of the free shrinkage and steel percentage, and an inverse function of the section depth (Ref. 7.8). Branson suggests that for steel percentage $p - p' \leq 3$ percent (where $p = 100A_s/bd$ and $p' = 100A_s'/bd$),

$$\phi_{sh} = 0.7 \frac{\varepsilon_{sh}}{h} (p - p')^{1/3} \left(\frac{p - p'}{p}\right)^{1/2} \tag{7.16a}$$

and for $p - p' > 3$ percent,

$$\phi_{sh} = \frac{\varepsilon_{sh}}{h} \tag{7.16b}$$

With shrinkage curvature calculated by either method, the corresponding member deflection can be determined by any convenient means such as the moment-area or conjugate-beam method. If steel percentages and eccentricities are constant along the span, the deflection ε_{sh} resulting from the shrinkage curvature can be determined from

$$\Delta_{sh} = K_{sh} \, \phi_{sh} \ell^2 \tag{7.17}$$

where K_{sh} is a coefficient equal to 0.500 for cantilevers, 0.125 for simple spans, 0.065 for interior spans of continuous beams, and 0.090 for end spans of continuous beams (Ref. 7.8).

EXAMPLE 7.3   **Shrinkage deflection.**   Calculate the midspan deflection of a simply supported beam of 20 ft span due to shrinkage of the concrete for which $\varepsilon_{sh} = 780 \times 10^{-6}$. With reference to Fig. 7.12a, $b = 10$ in., $d = 17.5$ in., $h = 20$ in., $A_s = 3.00$ in^2, and $A'_s = 0$. The elastic moduli are $E_c = 3.6 \times 10^6$ psi and $E_s = 29 \times 10^6$ psi.

SOLUTION.   By the fictitious tensile force method,

$$T_{sh} = (A_s + A'_s)\, \varepsilon_{sh}\, E_s = 3.00 \times 780 \times 10^{-6} \times 29 \times 10^6 = 67{,}900 \text{ lb}$$

and from Eq. (7.15) with $I_g = 6670$,

$$\phi_{sh} = \frac{2T_{sh}\, e_g}{E_c\, I_g} = \frac{2 \times 67{,}900 \times 7.5}{3.6 \times 10^6 \times 6670} = 42.4 \times 10^{-6}$$

while from Eq. (7.17) with $K_{sh} = 0.125$ for the simple span,

$$\Delta_{sh} = K_{sh}\phi_{sh}\ell^2 = 0.125 \times 42.4 \times 10^{-6} \times 240^2 = 0.305 \text{ in.}$$

Alternatively, by Branson's approximate Eq. (7.16a) with $p = 100 \times 3/175 = 1.7$ percent and $p' = 0$,

$$\phi_{sh} = 0.7\, \frac{\varepsilon_{sh}}{h}\, (p - p')^{1/3} \left(\frac{p - p'}{p}\right)^{1/2} = 0.7\, \frac{780 \times 10^{-6}}{20}\, (1.7)^{1/3}\, (1)^{1/2} = 32.5 \times 10^{-6}$$

compared with 42.4×10^{-6} obtained by the equivalent tensile force method. Considering the uncertainties such as the effects of cracking and creep, the approximate approach can usually be considered satisfactory.

Deflections will be produced as a result of differential temperatures varying from top to bottom of a member also. Such variation will result in a strain variation with member depth that may usually be assumed to be linear. For such cases, the deflection due to differential temperature can be calculated using Eq. (7.17) in which ϕ_{sh} is replaced by $\alpha\Delta T/h$, where the thermal coefficient α for concrete may be taken as 5.5×10^{-6} per °F and ΔT is the temperature differential in degrees Fahrenheit from one side to the other. The presence of the reinforcement has little influence on curvatures and deflections resulting from differential temperatures, because the thermal coefficient for the steel (6.5×10^{-6}) is very close to that for concrete.

7.9   MOMENT VS. CURVATURE FOR REINFORCED CONCRETE SECTIONS

Although it is not needed explicitly in ordinary design and is not a part of ACI Code procedures, the relation between moment applied to a given beam section and the resulting curvature, through the full range of loading to failure, is important in several ways. It is basic to the study of member ductility, understanding the development of plastic hinges, and accounting for the redistribution of elastic

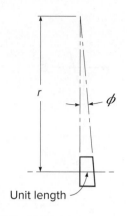

Unit length

FIGURE 7.13
Unit curvature resulting from bending of beam section.

moments that occurs in most reinforced concrete structures before collapse (see Section 11.9).

It will be recalled, with reference to Fig. 7.13, that curvature is defined as the angle change per unit length at any given location along the axis of a member subjected to bending loads:

$$\phi = \frac{1}{r} \tag{7.18}$$

where ϕ = unit curvature and r = radius of curvature. With the stress-strain relationships for steel and concrete, represented in idealized form in Fig. 7.14a and b, respectively, and the usual assumptions regarding perfect bond and plane sections, it is possible to calculate the relation between moment and curvature for a typical underreinforced concrete beam section, subject to flexural cracking, as follows.

Figure 7.15a shows the transformed cross section of a rectangular, tensile-reinforced beam in the uncracked elastic stage of loading, with steel represented by the equivalent concrete area nA_s, that is, with area $(n - 1)A_s$ added outside of the rectangular concrete section.[†] The neutral axis, a distance c_1 below the top surface of the beam, is easily found (see Section 4.2a). In the limiting case, the concrete stress at the tension face is just equal to the modulus of rupture f_r and the strain is $\varepsilon_r = f_r/E_c$. The steel is well below yield at this stage, which can be confirmed by computing, from the strain diagram, the steel strain $\varepsilon_s = \varepsilon_{cs}$, where ε_{cs} is the concrete

FIGURE 7.14
Idealized stress-strain curves: (a) steel and (b) concrete.

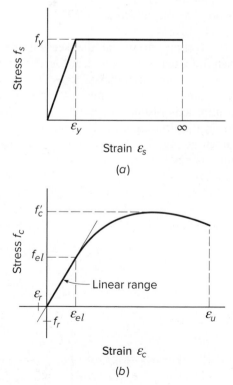

Strain ε_s

(a)

Strain ε_c

(b)

[†] Note that compression reinforcement, or multiple layers of tension reinforcement, can easily be included in the analysis with no essential complication.

FIGURE 7.15
Uncracked beam in the elastic range of loading: (*a*) transformed cross section; (*b*) strains; and (*c*) stresses and forces.

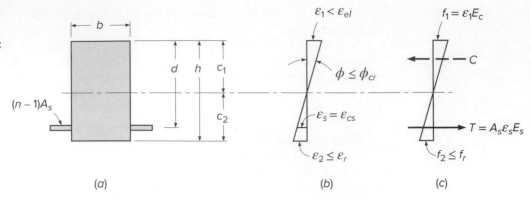

strain at the level of the steel. It is easily confirmed, also, that the maximum concrete compressive stress is well below the proportional limit. The curvature is seen, in Fig. 7.15*b*, to be

$$\phi_{cr} = \frac{\varepsilon_1}{c_1} = \frac{\varepsilon_r}{c_2} \tag{7.19}$$

and the corresponding moment is

$$M_{cr} = \frac{f_r \, I_{ut}}{c_2} \tag{7.20}$$

where I_{ut} is the moment of inertia of the uncracked transformed section. Equations (7.19) and (7.20) provide the information needed to plot point 1 of the moment-curvature graph of Fig. 7.18.

When tensile cracking occurs at the section, the stiffness is immediately reduced, and curvature increases to point 2 in Fig. 7.18 with no increase in moment. The analysis now is based on the cracked transformed section of Fig. 7.16*a*, with steel represented by the transformed area nA_s and tension concrete deleted. The cracked, elastic neutral axis distance $c_1 = kd$ is easily found by the usual methods (see Section 4.2*b*). In the limiting case, the concrete strain just reaches the proportional limit, as shown in Fig. 7.16*b*, and typically the steel is still below the yield strain. The curvature is easily computed by

$$\phi_{el} = \frac{\varepsilon_1}{c_1} = \frac{\varepsilon_{el}}{c_1} \tag{7.21}$$

and the corresponding moment is

$$M_{el} = \frac{1}{2} f_{el} \, kj \, bd^2 \tag{7.22}$$

FIGURE 7.16
Cracked beam in the elastic range of material response: (*a*) transformed cross section; (*b*) strains; and (*c*) stresses and forces.

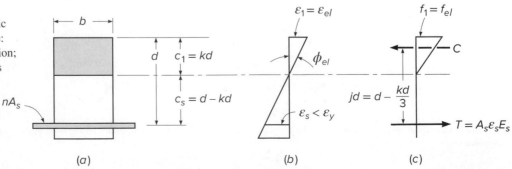

FIGURE 7.17
Cracked beam with concrete in the inelastic range of loading: (*a*) cross section; (*b*) strains; and (*c*) stresses and forces.

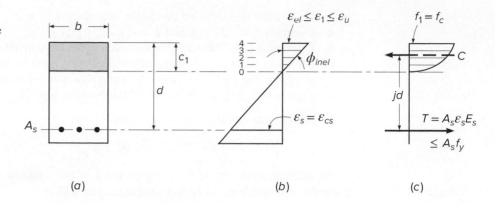

(a)   (b)   (c)

as was derived in Section 4.2b. This provides point 3 in Fig. 7.18. The curvature at point 2 can now be found from the ratio M_{cr}/M_{el}.

Next, the cracked, inelastic stage of loading is shown in Fig. 7.17. Here the concrete is well into the inelastic range, although the steel has not yet yielded. The neutral axis depth c_1 is less than the elastic kd and is changing with increasing load as the shape of the concrete stress distribution changes and the steel stress changes.

It is now convenient to adopt a numerical representation of the concrete compressive stress distribution, to find both the total concrete compressive force C and the location of its centroid, for any arbitrarily selected value of maximum concrete strain ε_1 in this range. The compressive strain diagram is divided into an arbitrary number of steps (such as, four in Fig. 7.17b), and the corresponding compressive stresses for each strain are read from the stress-strain curve of Fig. 7.14b. The stepwise representation of the actual continuous stress block is integrated numerically to find C, and its point of application is located, taking moments of the concrete forces about the top of the section. The basic equilibrium requirement $C = T$ then can be used to find the correct location of the neutral axis, for the particular compressive strain selected, following an iterative procedure.

The entire process can be summarized as follows:

1. Select any top face concrete strain ε_1 in the inelastic range, that is, between ε_{el} and ε_u.
2. Assume the neutral axis depth, a distance c_1 below the top face.

FIGURE 7.18
Moment-curvature relation for tensile-reinforced beam.

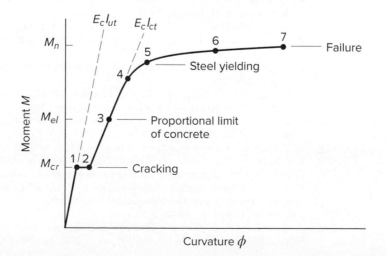

3. From the strain diagram geometry, determine $\varepsilon_s = \varepsilon_{cs}$.
4. Compute $f_s = \varepsilon_s E_s \leq f_y$ and $T = A_s f_s$.
5. Determine C by integrating numerically under the concrete stress distribution curve.
6. Check to see if $C = T$. If not, the neutral axis must be adjusted upward or downward, for the particular concrete strain that was selected in step 1, until equilibrium is satisfied. This determines the correct value of c_1.

Curvature can then be found from

$$\phi_{inel} = \frac{\varepsilon_1}{c_1} \qquad (7.23)$$

The internal lever arm jd from the centroid of the concrete stress distribution to the tensile resultant, Fig. 7.17c, is calculated, after which

$$M_{inel} = Cjd = Tjd \qquad (7.24)$$

The sequence of steps 1 through 6 is then repeated for newly selected values of concrete strain ε_1. The end result will be a series of points, such as 4, 5, 6, and 7 in Fig. 7.18. The limit of the moment-curvature plot is reached when the concrete top face strain equals ε_u, corresponding to point 7. The steel would be well past the yield strain at this loading, and at the yield stress.

It is important to be aware of the difference between a moment-unit curvature plot, such as Fig. 7.18, and a moment-rotation diagram for the hinging region of a reinforced concrete beam. The hinging region normally includes a number of discrete cracks, but between those cracks, the uncracked concrete reduces the steel strain, leading to what is termed the *tension stiffening* effect. The result is that the total rotation at the hinge is much less than would be calculated by multiplying the curvature per unit length at the cracked section by the observed or assumed length of the hinging region. Furthermore, the sharp increase in unit curvature shown in Fig. 7.18 at cracking would not be seen on the moment-rotation plot, only a small, but progressive, reduction of the slope of the diagram.

REFERENCES

7.1. ACI Committee 224, *Control of Cracking in Concrete Structures* (ACI 224R-01), American Concrete Institute, Farmington Hills, MI, 2001.

7.2. P. Gergely and L. A. Lutz, "Maximum Crack Width in Reinforced Concrete Flexural Members," in *Causes, Mechanisms, and Control of Cracking in Concrete*, SP-20, American Concrete Institute, Detroit, MI, 1968, pp. 1–17.

7.3. P. H. Kaar and A. H. Mattock, "High-Strength Bars as Concrete Reinforcement—Part 4: Control of Cracking," *Journal*, PCA Research and Development Laboratories, vol. 5, no. 1, 1963, pp. 15–38.

7.4. R. J. Frosch, "Another Look at Cracking and Crack Control in Reinforced Concrete," *ACI Struct. J.*, vol. 96, no. 3, 1999, pp. 437–442.

7.5. B. B. Broms, "Crack Width and Crack Spacing in Reinforced Concrete Members," *J. ACI*, vol. 62, no. 10, 1965, pp. 1237–1256.

7.6. ACI Committee 318, "Discussion of Proposed Revisions to Building Code Requirements for Structural Concrete (ACI 318-95) and Commentary (ACI 318R-95)," *Concr. Intl.*, vol. 21, no. 5, 1999, pp. 318-1–318-49.

7.7. G. C. Frantz and J. C. Breen, "Cracking on the Side Faces of Large Reinforced Concrete Beams," *J. ACI*, vol. 77, no. 10, 1980, pp. 307–313.

7.8. D. E. Branson, *Deformation of Concrete Structures*, McGraw-Hill, New York, 1977.

7.9. P. H. Bischoff, "Reevaluation of Deflection Prediction for Concrete Beams Reinforced with Steel and Fiber Reinforced Polymer Bars," *J. J. Struct. Engrg,* ASCE, vol. 131, no. 5, 2005, pp. 752–767.

7.10. P. H. Bischoff and A. Scanlon, "Effective Moment of Inertia for Calculating Deflections of Concrete Members Containing Steel Reinforcement and Fiber-Reinforced Polymer Reinforcement," *ACI Struct. J,* vol. 104, no. 1, 2007, pp. 68–75.

7.11. A. Scanlon and P. H. Bischoff, "Shrinkage Restraint and Loading History Effects on Deflections of Flexural Members," *ACI Struct. J,* vol. 105, no. 4, 2008, pp. 498–506.

7.12.   ACI Committee 435, *Control of Deflection in Concrete Structures* (ACI 435R-95), American Concrete Institute, Farmington Hills, MI, 1995.

7.13.   W. W. Yu and G. Winter, "Instantaneous and Long-Time Deflections of Reinforced Concrete Beams under Working Loads," *J. ACI*, vol. 57, no. 1, 1960, pp. 29–50.

7.14.   ACI Committee 209, *Prediction of Creep, Shrinkage and Temperature Effects in Concrete Structures* (ACI 209R-92), American Concrete Institute, Farmington Hills, MI, 1992.

7.15.   D. E. Branson, "Compression Steel Effect on Long-Time Deflections," *J. ACI*, vol. 68, no. 8, 1971, pp. 555–559.

7.16.   ACI Committee 435, "Proposed Revisions by Committee 435 to ACI Building Code and Commentary Provisions on Deflections," *J. ACI*, vol. 75, no. 6, June 1978, pp. 229–238.

7.17.   K. Paulson, A. H. Nilson, and K. C. Hover, "Long-Term Deflection of High-Strength Concrete Beams," *ACI Mater. J.*, vol. 88, no. 2, 1991, pp. 197–206.

7.18.   *CRSI Handbook*, 10th ed., Concrete Reinforcing Steel Institute, Schaumburg, IL, 2008.

PROBLEMS

7.1.   A rectangular beam of width $b = 15$ in., effective depth $d = 21.5$ in., and total depth $h = 24$ in. spans 18.5 ft between simple supports. It will carry a dead load of 1.08 kips/ft including self-weight, plus a service live load of 4.00 kips/ft. Reinforcement consists of four evenly spaced No. 7 (No. 22) bars in one row. The clear cover on the sides is 2 in. Material strengths are $f_y = 80,000$ psi and $f_c' = 4000$ psi.

   (*a*)   Compute the stress in the steel at full service load, and using the Gergely-Lutz equation, estimate the maximum crack width.

   (*b*)   Confirm the suitability of the proposed design based on Eq. (7.3).

7.2.   To save steel-handling costs, an alternative design is proposed for the beam in Problem 7.1, using two No. 9 (No. 29) Grade 80 bars to provide approximately the same steel strength as the originally proposed four No. 7 (No. 22) Grade 60 bars. Check to determine if the redesigned beam is satisfactory with respect to cracking according to the ACI Code. What modification could you suggest that would minimize the number of bars to reduce cost, yet satisfy requirements of crack control?

7.3.   For the beam in Problem 7.1:

   (*a*)   Calculate the increment of deflection resulting from the first application of the short-term live load.

   (*b*)   Find the creep portion of the sustained load deflection plus the immediate deflection due to live load.

   (*c*)   Compare your results with the limitations imposed by the ACI Code, as summarized in Table 7.2.

   Assume that the beam is a part of a floor system and supports cinder block partitions susceptible to cracking if deflections are excessive.

7.4.   A beam having $b = 12$ in., $d = 21.5$ in., and $h = 24$ in. is reinforced with three No. 11 (No. 36) bars. Material strengths are $f_y = 60,000$ psi and $f_c' = 4000$ psi. It is used on a 28 ft simple span to carry a total service load of 2430 lb/ft. For this member, the sustained loads include self-weight of the beam plus additional superimposed dead load of 510 lb/ft, plus 400 lb/ft representing that part of the live load that acts more or less continuously, such as furniture, equipment, and time-average occupancy load. The remaining 1220 lb/ft live load consists of short-duration loads, such as the brief peak load in the corridors of an office building at the end of a workday.

   (*a*)   Find the increment of deflection under sustained loads due to creep.

   (*b*)   Find the additional deflection increment due to the intermittent part of the live load.

In your calculations, you may assume that the peak load is applied almost immediately after the building is placed in service, then reapplied intermittently. Compare with ACI Code limits from Table 7.2. Assume that, for this long-span floor beam, construction details are provided that will avoid damage to supported elements due to deflections. If ACI Code limitations are not met, what changes would you recommend to improve the design?

7.5. A reinforced concrete beam is continuous over two equal 22 ft spans, simply supported at the two exterior supports, and fully continuous at the interior support. Concrete cross-sectional dimensions are $b = 10$ in., $h = 22$ in., and $d = 19.5$ in. for both positive and negative bending regions. Positive reinforcement in each span consists of two No. 9 (No. 29) bars, and negative reinforcement at the interior support is made up of three No. 10 (No. 32) bars. No compression steel is used. Material strengths are $f_y = 60,000$ psi and $f_c' = 5000$ psi. The beam will carry a service live load, applied early in the life of the member, of 1800 lb/ft distributed uniformly over both spans; 20 percent of this load will be sustained more or less permanently, while the rest is intermittent. The total service dead load is 1000 lb/ft including self-weight.

(a) Find the immediate deflection when shores are removed and the full dead load is applied.

(b) Find the long-term deflection under sustained load.

(c) Find the increment of deflection when the short-term part of the live load is applied.

Compare with ACI Code deflection limits; piping and brittle conduits are carried that would be damaged by large deflections. Note that midspan deflection may be used as a close approximation of maximum deflection.

7.6. Recalculate the deflections of Problem 7.5 based on the assumption that 20 percent of the live load represents the normal service condition of loading and is sustained more or less continuously, while the remaining 80 percent is a short-term peak loading that would probably not be applied until most creep deflections have occurred. Compare with your earlier results.

7.7. The tensile-reinforced rectangular beam shown in Fig. P7.7 is made using steel with $f_y = 60,000$ psi and $E_s = 29,000,000$ psi. A perfectly plastic response after yielding can be assumed. The concrete has a stress-strain curve in compression that may be approximated by the parabola $f_c = f_c' [2\varepsilon_c/\varepsilon_0 - (\varepsilon_c/\varepsilon_0)^2]$, where f_c and ε_c are the stress and strain in the concrete. The variable ε_0 is the strain at the peak stress $= 0.002$, and $f_c' = 4000$ psi. The ultimate strain in the concrete is 0.003. The concrete responds elastically in tension up to the modulus of rupture $f_r = 475$ psi. Based on this information, plot a curve relating applied moment to unit curvature at a section subjected to flexural cracking. Label points corresponding to first cracking, first yielding of steel, and peak moment.

FIGURE P7.7

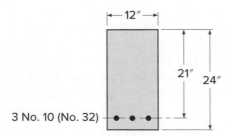

8

Analysis and Design for Torsion

8.1 INTRODUCTION

Reinforced concrete members are commonly subjected to bending moments, transverse shears associated with those bending moments, and, in the case of columns, axial forces often combined with bending and shear. In addition, torsional forces may act, tending to twist a member about its longitudinal axis. Such torsional forces seldom act alone and are almost always concurrent with bending moment and transverse shear, and sometimes with axial force as well.

For many years, torsion was regarded as a secondary effect and was not considered explicitly in design, its influence being absorbed in the overall factor of safety of rather conservatively designed structures. Current methods of analysis and design, however, have resulted in less conservatism, leading to somewhat smaller members that, in many cases, must be reinforced to increase torsional strength. In addition, there is increasing use of structural members for which torsion is a central feature of behavior; examples include curved bridge girders, eccentrically loaded box beams, and helical stairway slabs. The design procedures in the ACI Code were first proposed in Switzerland (Refs. 8.1 and 8.2) and are also included in the European and Canadian model codes (Refs. 8.3 and 8.4). Reference 8.5 provides a summary of the development of design procedures for reinforced concrete members subjected to torsion.

It is useful in considering torsion to distinguish between primary and secondary torsion in reinforced concrete structures. *Primary torsion*, sometimes called *equilibrium torsion* or *statically determinate torsion*, exists when the external load has no alternative load path but must be supported by torsion. For such cases, the torsion required to maintain static equilibrium can be uniquely determined. An example is the cantilevered slab of Fig. 8.1*a*. Loads applied to the slab surface cause twisting moments m_t to act along the length of the supporting beam. These are equilibrated by the resisting torque T provided at the columns. Without the torsional moments, the structure will collapse.

In contrast to this condition, *secondary torsion*, also called *compatibility torsion* or *statically indeterminate torsion*, arises from the requirements of continuity, that is, compatibility of deformation between adjacent parts of a structure. For this case, the torsional moments cannot be found based on static equilibrium alone. Disregard of continuity in the design will often lead to extensive cracking, but generally will not cause collapse. An internal readjustment of forces is usually possible and an alternative equilibrium of forces found. An example of secondary torsion is found in the spandrel or edge beam supporting a monolithic concrete slab, shown in Fig. 8.1*b*. If the spandrel beam is torsionally stiff and suitably reinforced, and if the columns can provide the necessary resisting torque T, then the slab moments will approximate

FIGURE 8.1
Torsional effects in
reinforced concrete:
(*a*) primary or equilibrium
torsion at a cantilevered slab;
(*b*) secondary or compatibility
torsion at an edge beam;
(*c*) slab moments if edge
beam is stiff torsionally;
and (*d*) slab moments if edge
beam is flexible torsionally.

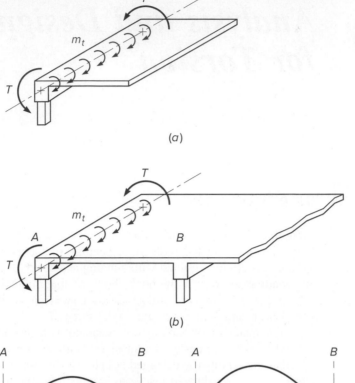

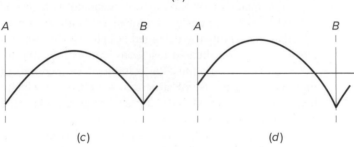

those for a rigid exterior support as shown in Fig. 8.1*c*. However, if the beam has
little torsional stiffness and inadequate torsional reinforcement, cracking will occur to
further reduce its torsional stiffness, and the slab moments will approximate those for
a hinged edge, as shown in Fig. 8.1*d*. If the slab is designed to resist the altered
moment diagram, collapse will not occur (see discussion in Section 11.10).

Although current techniques for analysis permit the realistic evaluation of torsional moments for statically indeterminate conditions as well as determinate, designers often neglect secondary torsional effects when torsional stresses are low and
alternative equilibrium states are possible. This is permitted according to the ACI
Code and many other design specifications. On the other hand, when torsional
strength is an essential feature of the design, such as for the bridge shown in Fig. 8.2,
special analysis and special torsional reinforcement are required, as described in the
remainder of this chapter.

8.2   TORSION IN PLAIN CONCRETE MEMBERS

Figure 8.3 shows a portion of a prismatic member subjected to equal and opposite
torques T at the ends. If the material is elastic, St. Venant's torsion theory indicates

FIGURE 8.2
Curved continuous beam
bridge, Las Vegas, Nevada,
designed for torsional effects.
(*Courtesy of Portland Cement
Association.*)

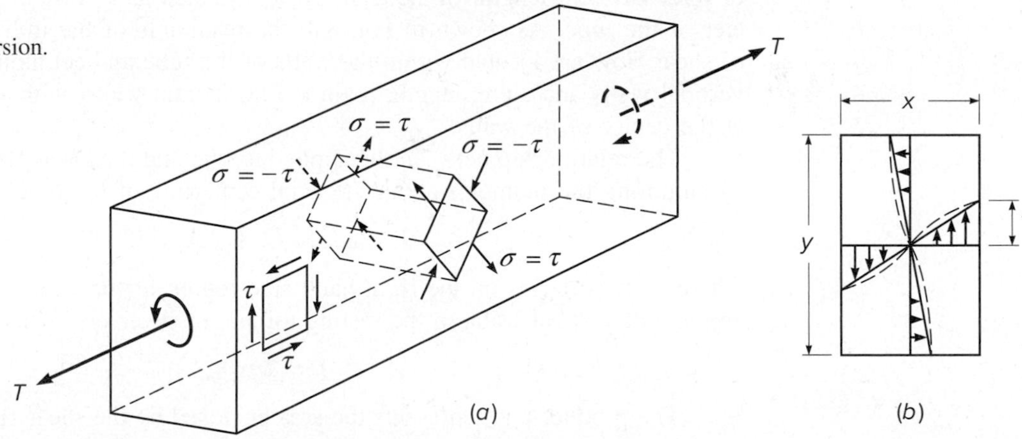

FIGURE 8.3
Stresses caused by torsion.

that torsional shear stresses are distributed over the cross section, as shown in
Fig. 8.3*b*. The largest shear stresses occur at the middle of the wide faces. If the
material deforms inelastically, as expected for concrete, the stress distribution is
closer to that shown by the dashed line.

FIGURE 8.4
Thin-walled tube under torsion.

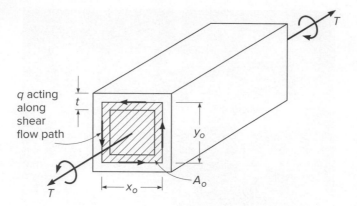

Shear stresses in pairs act on an element at or near the wide surface, as shown in Fig. 8.3a. As explained in strength of materials texts, this state of stress corresponds to equal tension and compression stresses on the faces of an element at 45° to the direction of shear. These inclined tension stresses are of the same kind as those caused by transverse shear, discussed in Section 5.2. However, in the case of torsion, since the torsional shear stresses are of opposite sign on opposing sides of the member (Fig. 8.3b), the corresponding diagonal tension stresses are at right angles to each other (Fig. 8.3a).

When the diagonal tension stresses exceed the tensile resistance of the concrete, a crack forms at some accidentally weaker location and spreads immediately across the beam. The value of torque corresponding to the formation of this diagonal crack is known as the *cracking torque* T_{cr}.

There are several ways of analyzing members subjected to torsion. The nonlinear stress distribution shown by the dotted lines in Fig. 8.3b lends itself to the use of the *thin-walled tube*, *space truss analogy*. Using this analogy, the shear stresses are treated as constant over a finite thickness t around the periphery of the member, allowing the beam to be represented by an equivalent tube, as shown in Fig. 8.4. Within the walls of the tube, torque is resisted by the shear flow q, which has units of force per unit length. In the analogy, q is treated as a constant around the perimeter of the tube. As shown in Fig. 8.4, the resultants of the individual components of shear flow are located within the walls of the tube and act along lengths y_o in the vertical walls and along lengths x_o in the horizontal walls, with y_o and x_o measured at the center of the walls.

The relationship between the applied torque and the shear flow can be obtained by summing the moments about the axial centerline of the tube, giving

$$T = 2qx_oy_o/2 + 2qy_ox_o/2 \qquad (a)$$

where the two terms on the right-hand side represent the contributions of the horizontal and vertical walls to the resting torque, respectively. Thus,

$$T = 2qx_oy_o \qquad (b)$$

The product x_oy_o represents the area enclosed by the shear flow path A_o, giving

$$T = 2qA_o \qquad (c)$$

and

$$q = \frac{T}{2A_o} \qquad (d)$$

Note that although A_o is an area, it derives from the moment calculation shown in Eq. (a) above. Thus, A_o is applicable for hollow box sections, as well as solid sections, and in such case includes the area of the central void.

For a tube wall thickness t, the unit shear stress acting within the walls of the tube is

$$\tau = \frac{q}{t} = \frac{T}{2A_o t} \tag{8.1}$$

As shown in Fig. 8.3a, the principal tensile stress $\sigma = \tau$. Thus, the concrete will crack only when $\tau = \sigma = f_t'$, the tensile strength of concrete. Considering that concrete is under biaxial tension and compression, f_t' can be conservatively represented by $4\sqrt{f_c'}$ rather than the value typically used for the modulus of rupture of concrete, which is taken as $f_r = 7.5\sqrt{f_c'}$ for normal-density concrete. Substituting $\tau = \tau_{cr} = 4\sqrt{f_c'}$ in Eq. (8.1) and solving for T give the value of the cracking torque:

$$T_{cr} = 4\sqrt{f_c'}\,(2A_o t) \tag{8.2}$$

Remembering that A_o represents the area enclosed by the shear flow path, A_o must be some fraction of the area enclosed by the outside perimeter of the full concrete cross section A_{cp}. The value of t can, in general, be approximated as a fraction of the ratio A_{cp}/p_{cp}, where p_{cp} is the perimeter of the cross section. For solid members with rectangular cross sections, t is typically one-sixth to one-fourth of the minimum width. Using a value of one-fourth for a member with a width-to-depth ratio of 0.5 yields a value of A_o approximately equal to $\frac{2}{3}A_{cp}$. For the same member, $t = \frac{3}{4}A_{cp}/p_{cp}$. Using these values for A_o and t in Eq. (8.2) gives

$$T_{cr} = 4\sqrt{f_c'}\,\frac{A_{cp}^2}{p_{cp}} \qquad \text{in.-lb} \tag{8.3}$$

It has been found that Eq. (8.3) gives a reasonable estimate of the cracking torque of solid reinforced concrete members regardless of the cross-sectional shape. For hollow sections, T_{cr} in Eq. (8.3) should be reduced by the ratio A_g/A_{cp}, where A_g is the gross cross section of the concrete, that is, not including the area of the voids (Ref. 8.6).

8.3   TORSION IN REINFORCED CONCRETE MEMBERS

To resist torsion for values of T above T_{cr}, reinforcement must consist of closely spaced stirrups and longitudinal bars. Tests have shown that longitudinal bars alone hardly increase the torsional strength, with test results showing an improvement of at most 15 percent (Ref. 8.6). This is understandable because the only way in which longitudinal steel can directly contribute to torsional strength is by dowel action, which is particularly weak and unreliable if longitudinal splitting along bars is not restrained by transverse reinforcement. Thus, the torsional strength of members reinforced only with longitudinal steel is satisfactorily, and somewhat conservatively, predicted by Eqs. (8.2) and (8.3).

When members are adequately reinforced, as in Fig. 8.5a, the concrete cracks at a torque that is equal to or only somewhat larger than in an unreinforced member, as given by Eq. (8.3). The cracks form a spiral pattern, as shown in Fig. 8.5b. Upon cracking, the torsional resistance of the concrete drops to about one-half of that of

FIGURE 8.5
Reinforced concrete beam
in torsion: (*a*) torsional
reinforcement and
(*b*) torsional cracks.

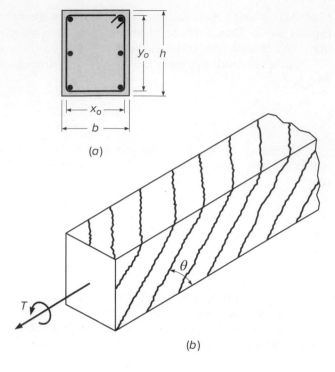

(*a*)

(*b*)

FIGURE 8.6
Torque-twist curve in
reinforced concrete member.

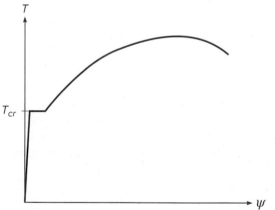

the uncracked member, the remainder being now resisted by reinforcement. This redistribution of internal resistance is reflected in the torque-twist curve (Fig. 8.6), which at the cracking torque shows continued twist at constant torque until the internal forces have been redistributed from the concrete to the steel. As the section approaches the ultimate load, the concrete outside the reinforcing cage cracks and begins to spall off, contributing progressively less to the torsional capacity of the member.

Tests show that, after cracking, the area enclosed by the shear path is defined by the dimensions x_o and y_o measured to the centerline of the outermost closed transverse reinforcement, rather than to the center of the tube walls as before. These dimensions define the gross area $A_{oh} = x_o y_o$ and the shear perimeter $p_h = 2(x_o + y_o)$ measured at the steel centerline.

Analysis of the torsional resistance of the member is aided by treating the member as a space truss consisting of spiral *concrete diagonals* that are able to take

FIGURE 8.7
Space truss analogy.

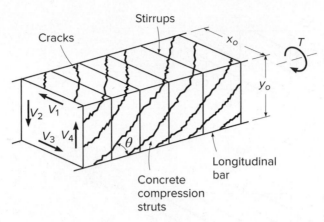

load parallel but not perpendicular to the torsional cracks, transverse *tension tie members* that are provided by closed stirrups or ties, and *tension chords* that are provided by longitudinal reinforcement. The hollow-tube, space truss analogy represents a simplification of actual behavior, since, as will be demonstrated, the calculated torsional strength is controlled by the strength of the transverse reinforcement, independent of concrete strength. Such a simplification will be used here because it aids understanding, although it greatly underestimates torsional capacity and does not reflect the higher torsional capacities obtained with higher concrete strengths (Refs. 8.7 and 8.8).

With reference to Fig. 8.7, the torsional resistance provided by a member with a rectangular cross section can be represented as the sum of the contributions of the shears in each of the four walls of the equivalent hollow tube. The contribution of the shear acting in the right-hand vertical wall of the tube to the torsional resistance, for example, is

$$T_4 = \frac{V_4 \, x_o}{2} \qquad\qquad (a)$$

Following a procedure similar to that used for analyzing the variable-angle truss shear model discussed in Section 5.8 and shown in Figs. 5.18 and 5.19, the equilibrium of a section of the vertical wall—with one edge parallel to a torsional crack with angle θ—can be evaluated using Fig. 8.8a.

Assuming that the stirrups crossing the crack are yielding, the shear in the wall under consideration is

$$V_4 = A_t f_{yt} n \qquad\qquad (b)$$

where A_t = area of one leg of a closed stirrup
f_{yt} = yield strength of transverse reinforcement
n = number of stirrups intercepted by torsional crack

Since the horizontal projection of the crack is $y_o \cot \theta$ and $n = y_o \cot \theta/s$, where θ is the slope angle of the strut and s is the spacing of the stirrups,

$$V_4 = \frac{A_t f_{yt} y_o}{s} \cot \theta \qquad\qquad (c)$$

Combining Eqs. (c) and (a) gives

$$T_4 = \frac{A_t f_{yt} y_o x_o}{2s} \cot \theta \qquad\qquad (d)$$

FIGURE 8.8
Basis for torsional design:
(*a*) vertical tension in
stirrups; (*b*) diagonal
compression in vertical wall
of beam; and (*c*) equilibrium
diagram of forces due to
shear in vertical wall.

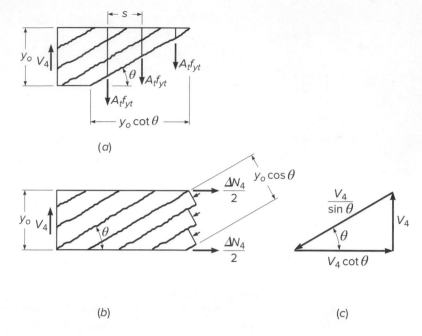

Again with reference to Fig. 8.7, the contribution of the shear in the left-hand vertical wall V_2 to the torsional resistance T_2 can be obtained by following the steps outlined for Eqs. (*a*), (*b*), and (*c*), giving a value that is identical with that for T_4 in Eq. (*d*). The contributions of the horizontal walls T_1 and T_3 can be determined in the same way, except that the internal moment arm in Eq. (*a*) becomes y_o and the horizontal projection of the crack becomes $x_o \cot \theta$, which leads to expressions for T_1 and T_3 that are also identical to that for T_4 in Eq. (*d*). Thus, summing over all four sides, the nominal capacity of the section is

$$T_n = \sum_{i=1}^{4} T_i = \frac{2A_t f_{yt} \, y_o \, x_o}{s} \cot \theta \tag{e}$$

Noting that $y_o x_o = A_{oh}$ and rearranging slightly give

$$T_n = \frac{2A_{oh} A_t f_{yt}}{s} \cot \theta \tag{8.4}$$

The diagonal compression struts that form parallel to the torsional cracks are necessary for the equilibrium of the cross section. As shown in Fig. 8.8*b* and *c*, the horizontal component of compression in the struts in the vertical wall must be equilibrated by an axial tensile force ΔN_4. Based on the assumed uniform distribution of shear flow around the perimeter of the member, the diagonal stresses in the struts must be uniformly distributed, resulting in a line of action of the resultant axial force that coincides with the midheight of the wall. Referring to Fig. 8.8*c*, the total contribution of the right-hand vertical wall to the change in axial force of the member due to the presence of torsion is

$$\Delta N_4 = V_4 \cot \theta = \frac{A_t f_{yt} y_o}{s} \cot^2 \theta$$

Again, summing over all four sides, the total increase in axial force for the member is

$$\Delta N = \sum_{i=1}^{4} \Delta N_i = \frac{A_t f_{yt}}{s} 2 (x_o + y_o) \cot^2 \theta \qquad (8.5a)$$

$$\Delta N = \frac{A_t f_{yt} p_h}{s} \cot^2 \theta \qquad (8.5b)$$

where p_h is the perimeter of the centerline of the closed stirrups.

Longitudinal reinforcement must be provided to carry the added axial force ΔN. If that steel is designed to yield, then

$$A_\ell f_y = \frac{A_t f_{yt} p_h}{s} \cot^2 \theta \qquad (8.6)$$

and

$$A_\ell = \frac{A_t}{s} p_h \frac{f_{yt}}{f_y} \cot^2 \theta \qquad (8.7)$$

where A_ℓ = total area of longitudinal reinforcement to resist torsion, in^2
f_y = yield strength of longitudinal torsional reinforcement, psi

Solving Eq. (8.4) for $A_t f_{yt}$, substituting the value in Eq. (8.7), and solving for T_n gives the limiting value of the nominal torsional capacity T_n based on A_ℓ.

$$T_n = \frac{2A_{oh} A_\ell f_y}{p_h} \tan \theta \qquad (8.8)$$

It has been found experimentally that, after cracking, the effective area enclosed by the shear flow path is somewhat less than the value of A_{oh} used in the previous development. It is recommended in Ref. 8.8 that the reduced value be taken as $A_o = 0.85A_{oh}$, where, it will be recalled, A_{oh} is the area enclosed by the centerline of the transverse reinforcement. This recommendation is incorporated in the ACI Code (see Section 8.5) and in a modified form of Eq. (8.4) with A_o substituted for A_{oh}. It has further been found experimentally that the thickness of the equivalent tube at loads near ultimate is closely approximated by $t = A_{oh}/p_h$, where p_h is the perimeter of A_{oh}.

8.4 TORSION PLUS SHEAR

Members are rarely subjected to torsion alone. The prevalent situation is that of a beam subject to the usual flexural moments and shear forces, which, in addition, must resist torsional moments. In an uncracked member, shear forces as well as torque produce shear stresses. In a cracked member, both shear and torsion increase the forces in the diagonal struts (Figs. 5.19d and 8.8b), increase the width of diagonal cracks, and increase the forces required in the transverse reinforcement (Figs. 5.19e and 8.8a).

Using the usual representation for reinforced concrete, the nominal shear stress caused by an applied shear force V is $\tau_v = V/b_w d$. The shear stress caused by torsion, given in Eq. (8.1), is $\tau_t = T/(2A_o t)$. As shown in Fig 8.9a for *hollow sections*, these stresses are directly additive on one side of the member. Thus, for a cracked concrete

FIGURE 8.9
Addition of torsional and
shear stresses: (*a*) hollow
section and (*b*) solid section.
(*Adapted from Ref. 8.8.*)

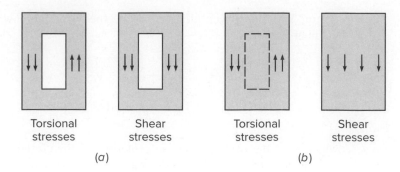

Torsional
stresses

Shear
stresses

Torsional
stresses

Shear
stresses

(*a*)   (*b*)

cross section with $A_o = 0.85A_{oh}$ and $t = A_{oh}/p_h$, the maximum shear stress can be expressed as

$$\tau = \tau_v + \tau_t = \frac{V}{b_w d} + \frac{Tp_h}{1.7A_{oh}^2} \tag{8.9}$$

For a member with a *solid section*, Fig. 8.9*b*, τ_t is predominately distributed around the perimeter, as represented by the hollow tube analogy, but the full cross section contributes to carrying τ_v. Comparisons with experimental results show that Eq. (8.9) is somewhat overconservative for solid sections and that a better representation for maximum shear stress is provided by the square root of the sum of the squares of the nominal shear stresses:

$$\tau = \sqrt{\left(\frac{V}{b_w d}\right)^2 + \left(\frac{Tp_h}{1.7A_{oh}^2}\right)^2} \tag{8.10}$$

Equations (8.9) and (8.10) serve as a measure of the shear stresses in the concrete under both service and ultimate loading.

8.5 ACI CODE PROVISIONS FOR TORSION DESIGN

The basic principles upon which ACI Code design provisions are based have been presented in the preceding sections. ACI Code 9.5.1.1 safety provisions require that

$$T_n \geq \phi T_u \tag{8.11}$$

where T_n = nominal torsional strength of member
$\quad\ \ T_u$ = required torsional strength at factored loads

The strength reduction factor $\phi = 0.75$ applies for torsion. Strength T_n as a function of the transverse reinforcement is based on Eq. (8.4) with A_o substituted for A_{oh}, thus

$$T_n = \frac{2A_o A_t f_{yt}}{s} \cot \theta \tag{8.12}$$

With the same substitution, T_n as a function of the longitudinal reinforcement, based on Eq. (8.8), is

$$T_n = \frac{2A_o A_\ell f_y}{p_h} \tan \theta \tag{8.13}$$

In accordance with ACI Code 9.4.4.3, sections located less than a distance d from the face of a support may be designed for the same torsional moment T_u as that computed at a distance d, recognizing the beneficial effects of support compression. However, if a concentrated torque is applied within this distance, the critical section must be taken at the face of the support. These provisions parallel those used in shear design. For beams supporting slabs such as are shown in Fig. 8.1, the torsional loading from the slab may be treated as being uniformly distributed along the beam.

a. T Beams and Box Sections

For T beams, a portion of the overhanging flange contributes to the cracking torsional capacity and, if reinforced with closed stirrups, to the torsional strength. According to ACI Code 9.2.4.4, the contributing width of the overhanging flange on either side of the web is equal to the smaller of (1) the projection of the beam above or below the slab, whichever is greater, and (2) 4 times the slab thickness, as shown in Fig. 8.10. As with solid sections, A_{cp} for box sections, with or without flanges, represents the area enclosed by the outside perimeter of the concrete section.

After torsional cracking, the applied torque is resisted by the portion of the section represented by A_{oh}, the area enclosed by the centerline of the outermost closed transverse torsional reinforcement. For rectangular, box, and T sections, A_{oh} is illustrated in Fig. 8.11. For sections with flanges, the Code does not require that the section used to establish A_{cp} coincide with that used to establish A_{oh}.

b. Threshold Torsion

If the value of factored torsional moment T_u is low enough, the effects of torsion may be neglected, according to ACI Code 22.7.1.1. This lower limit is ϕ times the *threshold torsion* T_{th}, which equals 25 percent of the cracking torque, given by Eq. (8.3). The presence of torsional moment at or below this limit will have a negligible effect on the flexural and shear strength of the member.

FIGURE 8.10

Portion of slab to be included with beam for torsional design.

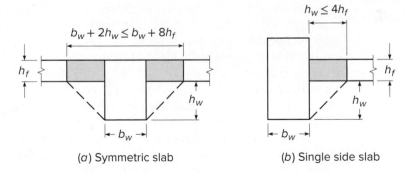

(a) Symmetric slab   (b) Single side slab

FIGURE 8.11

Definition of A_{oh}.

(*Adapted from Ref. 8.8.*)

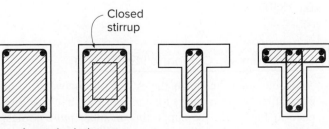

A_{oh} = shaded area

For solid cross sections without axial load, ACI Code 22.7.4 defines the threshold torsion as

$$T_{th} = \lambda \sqrt{f_c'} \left(\frac{A_{cp}^2}{p_{cp}} \right) \tag{8.14a}$$

and for members with axial load N_u and gross area A_g as

$$T_{th} = \lambda \sqrt{f_c'} \left(\frac{A_{cp}^2}{p_{cp}} \right) \sqrt{1 + \frac{N_u}{4A_g \, \lambda \sqrt{f_c'}}} \tag{8.14b}$$

For hollow cross sections without axial load, the threshold torsion is

$$T_{th} = \lambda \sqrt{f_c'} \left(\frac{A_g^2}{p_{cp}} \right) \tag{8.15a}$$

and for hollow members with axial load, the threshold torsion is

$$T_{th} = \lambda \sqrt{f_c'} \left(\frac{A_g^2}{p_{cp}} \right) \sqrt{1 + \frac{N_u}{4A_g \, \lambda \sqrt{f_c'}}} \tag{8.15b}$$

Thus, if $T_u \geq \phi T_{th}$, the member must be designed for torsion.

When calculating A_{cp}^2/p_{cp} for solid sections and A_g^2/p_{cp} for hollow sections for beams with flanges, ACI Code 9.2.4.4 requires that the flanges must be neglected when doing so produces higher values for these terms than when the flanges are included.

The value of λ is specified in ACI Code 19.2.4.2 and previously described in Section 5.5a. λ may be taken as 1.0 for normalweight concrete and 0.75 for light-weight concrete. Values between 0.75 and 1.0 may also be used by applying linear interpolation between $\lambda = 0.75$ and 1.0 for concretes with unit weights of $w_c \leq 100$ pcf and $w_c \geq 135$ pcf, respectively. Alternatively, $\lambda = 0.75$ for all-lightweight concrete and 0.85 for sand-lightweight concrete, using linear interpolation between 0.75 and 0.85 when a portion of the lightweight fine aggregate is replaced by normal-weight fine aggregate and between 0.85 and 1.0 for concretes containing normal-weight fine aggregate and a blend of lightweight and normalweight coarse aggregate.

A comparison of Eqs. (8.14) and (8.15) shows that for hollow sections (with or without axial load), A_{cp} has been replaced by the gross area of the concrete A_g to determine if torsional effects may be neglected. This has the effect of multiplying 25 percent of the cracking torque by the ratio A_g/A_{cp} twice—once to account for the reduction in cracking torque for hollow sections from the value shown in Eq. (8.3) and a second time to account for the transition from the circular interaction of combined shear and torsion stresses in Eq. (8.10) to the linear interaction represented by Eq. (8.9).

c. Equilibrium vs. Compatibility Torsion

A distinction is made in the ACI Code between equilibrium (primary) torsion and compatibility (secondary) torsion. For the first condition, described earlier with reference to Fig. 8.1a, the supporting member *must* be designed to provide the torsional resistance required by static equilibrium. For secondary torsion resulting from compatibility requirements, shown in Fig. 8.1b, it is assumed that cracking will result in a redistribution of internal forces; and according to ACI Code 22.7.3.2, the maximum torsional moment T_u may be reduced to ϕ times the cracking torque T_{cr} or $4\phi\lambda\sqrt{f_c'}(A_{cp}^2/p_{cp})$ for members not subjected to axial load and $4\phi\lambda\sqrt{f_c'}(A_{cp}^2/p_{cp}) \times \sqrt{1 + N_u/(4A_g\,\lambda\sqrt{f_c'})}$ for members subjected to axial load. In the case of hollow sections, A_{cp} is *not* replaced by A_g, and unlike the Code requirements for the threshold torsion of members with flanges, the designer is free to include or exclude flanges when calculating A_{cp}^2/p_{cp} and A_g^2/p_{cp}, but the choice should be consistent with the section properties used to calculate T_n. The design moments and shears in the supported member must be adjusted accordingly. The reduced value of T_u permitted by the ACI Code is intended to approximate the torsional cracking strength of the supporting beam, for combined torsional and flexural loading. The large rotations that occur at essentially constant torsional load result in significant redistribution of internal forces, justifying use of the reduced value for design of the torsional member and the supported elements.

d. Limitations on Shear Stress

Based largely on empirical observations, the width of diagonal cracks caused by combined shear and torsion under *service loads* can be limited by limiting the calculated shear stress under *factored shear and torsion* (Ref. 8.4) so that

$$v_{\max} \le \phi\left(\frac{V_c}{b_w d} + 8\sqrt{f_c'}\right) \tag{8.16}$$

where $v_{\max}$ in Eq. (8.16) corresponds to the upper limits on shear capacity described in Section 5.5d. Combining Eq. (8.16) with Eq. (8.9) provides limits on the cross-sectional dimensions of *hollow sections*, in accordance with ACI Code 22.7.7.1.

$$\frac{V_u}{b_w d} + \frac{T_u p_h}{1.7A_{oh}^2} \le \phi\left(\frac{V_c}{b_w d} + 8\sqrt{f_c'}\right) \tag{8.17}$$

Likewise, for *solid sections*, combining Eq. (8.16) with Eq. (8.10) gives

$$\sqrt{\left(\frac{V_u}{b_w d}\right)^2 + \left(\frac{T_u p_h}{1.7A_{oh}^2}\right)^2} \le \phi\left(\frac{V_c}{b_w d} + 8\sqrt{f_c'}\right) \tag{8.18}$$

Either member dimensions or concrete strength must be increased if the criteria in Eq. (8.17) or (8.18) are not satisfied.

ACI Code 22.7.7.1 requires that if the wall thickness varies around the perimeter of a hollow section, Eq. (8.17) be evaluated at the location where the left-hand side of the expression is a maximum. If the wall thickness is less than the assumed value of t used in the development of Eq. (8.9) A_{oh}/p_h, the actual value of t must be used in the calculation of torsional shear stress. As a result, the second term on the left-hand side of Eq. (8.17) must be taken as

$$\frac{T_u}{1.7A_{oh}t}$$

where t is the thickness of the wall of the hollow section at the location where the stresses are being checked.

e.   Reinforcement for Torsion

The nominal torsional strength based on the capacity of the transverse reinforcement is given by Eq. (8.12).

$$T_n = \frac{2A_o A_t f_{yt}}{s} \cot \theta \qquad (8.12)$$

According to ACI Code 22.7.6.1, the angle θ may assume any value between 30 and 60°, with a value of $\theta = 45°$ suggested. The area enclosed by the shear flow A_o may be determined by analysis using procedures such as suggested in Ref. 8.9, or A_o may be taken as equal to $0.85A_{oh}$. Combining Eq. (8.12) with Eq. (8.11), the required cross-sectional area of one stirrup leg for torsion is

$$A_t = \frac{T_u s}{2\phi A_o f_{yt} \cot \theta} \qquad (8.19)$$

The Code limits f_{yt} to a maximum of 60,000 psi for reasons of crack control.

The reinforcement provided for torsion must be combined with that required for shear. Based on the typical two-leg stirrup, this may be expressed as

$$\frac{A_{v+t}}{s} = \frac{A_v}{s} + 2\frac{A_t}{s} \qquad (8.20)$$

As described in Section 8.3, the transverse stirrups used for torsional reinforcement must be of a closed form to provide the required tensile capacity across the diagonal cracks of all faces of the beam. U-shaped stirrups commonly used for transverse shear reinforcement are not suitable for torsional reinforcement. On the other hand, one-piece closed stirrups make field assembly of beam reinforcement difficult, and for practical reasons, torsional stirrups are generally two-piece stirrup-ties, as shown in Fig. 8.12. A U-shaped stirrup is combined with a horizontal top bar, suitably anchored.

FIGURE 8.12
Stirrup-ties and longitudinal reinforcement for torsion:
(*a*) spandrel beam with flanges on one side;
(*b*) interior beam; (*c*) isolated rectangular beam; (*d*) wide spandrel beam; and
(*e*) T beam with torsional reinforcement in flanges.

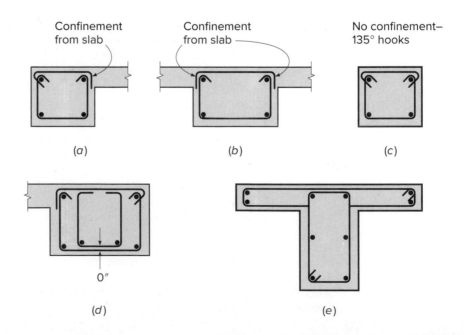

Because concrete outside the reinforcing cage tends to spall off when the member is subjected to high torque, transverse torsional reinforcement must be anchored within the concrete core (Ref. 8.10). ACI Code 25.7.1.6 and 25.7.2.5 require that stirrups or ties used for transverse longitudinal reinforcement be anchored with a 135° standard hook around a longitudinal bar, unless the concrete surrounding the anchorage is restrained against spalling by a flange or a slab, in which case 90° standard hooks may be used, as shown in Fig. 8.12a, b, and d. Overlapping U-shaped stirrups, such as shown in Fig. 6.22d, may not be used. If flanges are included in the computation of torsional strength for T- or L-shaped beams, closed torsional stirrups must be provided in the flanges, as shown in Fig. 8.12e.

The required spacing of closed stirrups, satisfying Eq. (8.20), is selected for the trial design based on standard bar sizes.

To control spiral cracking, the maximum spacing of torsional stirrups should not exceed $p_h/8$ or 12 in., whichever is smaller. In addition, for members requiring both shear and torsion reinforcement, the minimum area of closed stirrups is equal to

$$A_v + 2A_t = 0.75\sqrt{f_c'}\,\frac{b_w s}{f_{yt}} \geq 50\frac{b_w s}{f_{yt}} \tag{8.21}$$

according to ACI Code 9.6.4.2.

The area of longitudinal bar reinforcement A_ℓ required to resist T_n in Eq. (8.13) is given by Eq. (8.7)

$$A_\ell = \frac{A_t}{s}\,p_h\,\frac{f_{yt}}{f_y}\,\cot^2\theta \tag{8.7}$$

where θ must have the same value used to calculate A_t.

The term A_t/s in Eq. (8.7) should be taken as the value calculated using Eq. (8.19), not modified based on minimum transverse steel requirements. ACI Code 9.5.4.5 permits the portion of A_ℓ in the flexural compression zone to be reduced by an amount equal to $M_u/(0.9df_y)$, where M_u is the factored moment acting at the section in combination with T_u.

Based on an evaluation of the performance of reinforced concrete beam torsional test specimens, ACI Code 9.6.4.3 requires a minimum value of A_ℓ equal to the lesser of Eqs. (8.22a) and (8.22b).

$$A_{\ell,\min} = \frac{5\sqrt{f_c'}\,A_{cp}}{f_y} - \frac{A_t}{s}\,p_h\,\frac{f_{yt}}{f_y} \tag{8.22a}$$

$$A_{\ell,\min} = \frac{5\sqrt{f_c'}\,A_{cp}}{f_y} - \left(\frac{25b_w}{f_{yt}}\right)p_h\,\frac{f_{yt}}{f_y} \tag{8.22b}$$

with f_{yt} in Eq. (8.22b) in psi. As a general rule, the term $(25b_w/f_{yt})$ in Eq. (8.22b) serves as a lower bound for the term A_t/s in Eq. (8.22a).

According to ACI Code 9.7.5, the spacing of the longitudinal bars should not exceed 12 in., and they should be distributed around the perimeter of the cross section to control cracking. The bars may not be less than No. 3 (No. 10) in size or have a diameter less than 0.042 times the spacing of the transverse stirrups. At least one longitudinal bar must be placed at each corner of the stirrups. Careful attention must be paid to the anchorage of longitudinal torsional reinforcement so that it is able to develop its yield strength at the face of the supporting columns, where torsional moments are often maximum.

Reinforcement required for torsion may be combined with that required for other forces, provided that the area furnished is the sum of the individually required areas and that the most restrictive requirements of spacing and placement are met. According to ACI Code 9.7.6.3, torsional reinforcement must be provided at least a distance $b_t + d$ beyond the point theoretically required, where b_t is the width of that part of the cross section containing the closed stirrups resisting torsion. According to the provisions of the ACI Code, the point at which the torsional reinforcement is no longer required is the point at which $T_u < \phi T_{th}$, where T_{th} is the threshold torque given in Section 8.5b.

The subject of torsional design of prestressed concrete is not treated here, but as presented in the ACI Code, it differs only in certain details from the above presentation for nonprestressed reinforced concrete beams.

f. Design for Torsion

Designing a reinforced concrete flexural member for torsion involves a series of steps. The following sequence ensures that each is covered:

1. Determine if the factored torque is less than $\phi T_{th} = \phi \lambda \sqrt{f_c'}(A_{cp}^2/p_{cp})$ or $\phi \lambda \sqrt{f_c'}(A_{cp}^2/p_{cp})\sqrt{1 + N_u/(4A_g\lambda\sqrt{f_c'})}$ for members subjected to axial load. If so, torsion may be neglected. If not, proceed with the design. Note that in this step, portions of over hanging flanges, as defined in Section 8.5a, must be included in the calculation of A_{cp} and p_{cp}.

2. If the torsion is compatibility torsion, rather than equilibrium torsion, as described in Sections 8.1 and 8.5c, the maximum factored torque may be reduced to $4\phi \lambda \sqrt{f_c'}(A_{cp}^2/p_{cp})$ or $4\phi \lambda \sqrt{f_c'}(A_{cp}^2/p_{cp})\sqrt{1 + N_u/(4A_g\lambda\sqrt{f_c'})}$ for members subjected to axial load, with the moments and shears in the supported members adjusted accordingly. Equilibrium torsion cannot be adjusted.

3. Check the shear stresses in the section under combined torsion and shear, using the criteria of Section 8.5d.

4. Calculate the required transverse reinforcement for torsion using Eq. (8.19) and shear using Eq. (5.14a). Combine A_t and A_v using Eq. (8.20).

5. Check that the minimum transverse reinforcement requirements are met for both torsion and shear. These include the maximum spacing, as described in Sections 8.5e and 5.5d, and minimum area, as given in Eq. (8.21).

6. Calculate the required longitudinal torsional reinforcement A_ℓ, using the larger of the values given in Eqs. (8.7) and (8.22), and satisfy the spacing and bar size requirements given in Section 8.5e. The portion of A_ℓ in the flexural compression zone may be reduced by $M_u/(0.9df_y)$, providing that Eq. (8.22) and the spacing and bar size requirements are satisfied.

7. Continue torsional reinforcement $b_t + d$ past the point where T_u is less than $\phi \lambda \sqrt{f_c'}(A_{cp}^2/p_{cp})$ or $\phi \lambda \sqrt{f_c'}(A_{cp}^2/p_{cp})\sqrt{1 + N_u/(4A_g\lambda\sqrt{f_c'})}$ for members subjected to axial load.

EXAMPLE 8.1   **Design for torsion with shear.** The 28 ft span beam shown in Fig. 8.13a and b carries a monolithic slab cantilevering 6 ft past the beam centerline. The resulting L beam supports a live load of 900 lb/ft along the beam centerline plus 50 psf uniformly distributed over the upper slab surface. The effective depth to the flexural steel centroid is 21.5 in., and the distance from the beam surfaces to the centroid of stirrup steel is $1\frac{3}{4}$ in. Material strengths are $f_c' = 5000$ psi and $f_y = 60,000$ psi. Design the torsional and shear reinforcement for the beam.

FIGURE 8.13
Shear and torsion design
example.

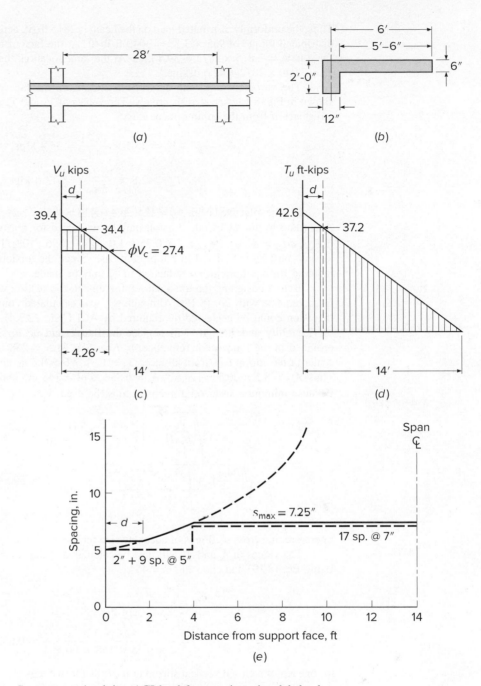

(a)

(b)

(c)

(d)

(e)

SOLUTION. Applying ACI load factors gives the slab load as

$$1.2w_d = 1.2 \times 75 \times 5.5 = 495 \text{ lb/ft}$$

$$1.6w_l = 1.6 \times 50 \times 5.5 = \underline{440 \text{ lb/ft}}$$

Total = 935 lb/ft at 3.25 ft eccentricity

while the beam carries directly

$$1.2w_d = 1.2 \times 300 = 360 \text{ lb/ft}$$

$$1.6w_l = 1.6(900 + 50) = \underline{1520 \text{ lb/ft}}$$

Total = 1880 lb/ft

Thus, the uniformly distributed load on the beam is 2815 lb/ft, acting together with a uniformly distributed torque of $935 \times 3.25 = 3040$ ft-lb/ft. At the face of the column, the design shear force is $V_u = 2.815 \times 28/2 = 39.4$ kips. At the same location, the design torsional moment is $T_u = 3.040 \times 28/2 = 42.6$ ft-kips.

The variation of V_u and T_u with distance from the face of the supporting column is shown in Fig. 8.13c and d, respectively. The values of V_u and T_u at the critical design section, a distance d from the column face, are

$$V_u = 39.4 \times \frac{12.21}{14} = 34.4 \text{ kips}$$

$$T_u = 42.6 \times \frac{12.21}{14} = 37.2 \text{ ft-kips}$$

For the effective beam, $A_{cp} = 12 \times 24 + 6 \times 18 = 396$ in^2 and $p_{cp} = 2 \times 24 + 2 \times 30 = 108$ in. According to the ACI Code, torsion may be neglected for normalweight concrete ($\lambda = 1.0$) if $T_u \le \phi T_{th} = \phi \lambda \sqrt{f_c'} \left(A_{cp}^2 / p_{cp} \right) = 0.75 \times 1.0 \sqrt{5000} \, (396^2/108)/12{,}000 = 6.4$ ft-kips. Clearly, torsion must be considered in the present case. Since the torsional resistance of the beam is required for equilibrium, no reduction in T_u may be made.

Before designing the torsional reinforcement, the section will be checked for adequacy in accordance with Eq. (8.18). Although A_{cp} was calculated considering the flange to check if torsion could be neglected (as required by ACI Code 22.7.4), subsequent calculations for serviceability and strength will neglect the flange and no torsional reinforcement will be provided in the flange. For reference, $b_w d = 12 \times 21.5 = 258$ in^2. With $1\frac{3}{4}$ in. cover to the center of the stirrup bars from all faces, $x_o = 12 - 3.5 = 8.5$ in. and $y_o = 24.0 - 3.5 = 20.5$ in. Thus, $A_{oh} = 8.5 \times 20.5 = 174$ in^2, $A_o = 0.85 \times 174 = 148$ in^2, and $p_h = 2(8.5 + 20.5) = 58$ in. Because minimum shear reinforcement must be used, $V_c = 2\sqrt{f_c'}$. Using Eq. (8.18),

$$\sqrt{\left(\frac{V_u}{b_w d} \right)^2 + \left(\frac{T_u p_h}{1.7 A_{oh}^2} \right)^2} \le \phi \left(\frac{V_c}{b_w d} + 8\sqrt{f_c'} \right)$$

$$\sqrt{\left(\frac{34.4}{258} \right)^2 + \left(\frac{37.2 \times 12 \times 58}{1.7 \times 174^2} \right)^2} \le \frac{0.75}{1000} \left(2\sqrt{5000} + 8\sqrt{5000} \right)$$

$$0.520 \text{ ksi} \le 0.530 \text{ ksi}$$

Therefore, the cross section is of adequate size for the given concrete strength.

The values of A_t and A_v will now be calculated at the column face (for reference only). Using Eq. (8.19) and choosing $\theta = 45°$,

$$A_t = \frac{T_u s}{2\phi A_o f_y \cot \theta}$$

$$= \frac{42.6 \times 12s}{2 \times 0.75 \times 148 \times 60 \times 1} = 0.0384s$$

for one leg of a closed vertical stirrup or $0.0768s$ for two legs.

The shear capacity of the concrete alone, obtained using Eq. (5.12c), is

$$\phi V_c = 0.75 \times 2\lambda \sqrt{f_c'} \, b_w d$$

$$= \frac{0.75 \times 2 \times 1.0 \sqrt{5000} \times 258}{1000} = 27.4 \text{ kips}$$

From Eq. (5.14a), the web reinforcement for transverse shear, again computed at the column face, is

$$A_v = \frac{(V_u - \phi V_c)s}{\phi f_{yt} d} = \frac{(39.4 - 27.4)s}{0.75 \times 60 \times 21.5} = 0.0124s$$

to be provided in two vertical legs.

The calculated value of A_t will decrease linearly to zero at the midspan, and the calculated value of A_v will decrease linearly to zero 4.26 ft from the face of the support, the point at which $V_u = \phi V_c$. Thus, the total area to be provided by the two vertical legs is

$$2A_t + A_v = 0.0768s\left(1 - \frac{x}{14}\right) + 0.0124s\left(1 - \frac{x}{4.26}\right)$$

for $0 \leq x \leq 4.26$ ft., where x is the distance from the face of the support, and

$$2A_t + A_v = 0.0768s\left(1 - \frac{x}{14}\right)$$

for $4.26 \leq x \leq 14$ ft.

No. 4 (No. 13) closed stirrups will provide a total area in the two legs of 0.40 in². For $2A_t + A_v = 0.40$ in², the required spacing at d and at 2 ft intervals along the span can be found using the given relationships between stirrup area and spacing:

$$s_d = 5.39 \text{ in.}$$
$$s_2 = 5.52 \text{ in.}$$
$$s_4 = 7.19 \text{ in.}$$
$$s_6 = 9.11 \text{ in.}$$
$$s_8 = 12.2 \text{ in.}$$
$$s_{10} = 18.2 \text{ in.}$$

These values of s are plotted in Fig. 8.13e. ACI provisions for maximum spacing should now be checked. For torsion reinforcement, the maximum spacing is the lesser of

$$\frac{p_h}{8} = \frac{58}{8} = 7.25 \text{ in.}$$

or 12 in., whereas for shear reinforcement, the maximum spacing is $d/2 = 10.75$ in. ≤ 24 in. The most restrictive provision is the first, and the maximum spacing of 7.25 in. is plotted in Fig. 8.13e. Stirrups between the face of the support and the distance d can be spaced at s_d. The resulting spacing requirements are shown by the solid line in the figure. These requirements are met in a practical way by No. 4 (No. 13) closed stirrups, the first placed 2 in. from the face of the column, followed by 9 at 5 in. spacing and 17 at 7 in. spacing. According to the ACI Code, stirrups may be discontinued at the point where $V_u < \phi V_c/2$ (4.9 ft from the span centerline) or $b_t + d = 2.8$ ft past the point at which $T_u < \phi \lambda \sqrt{f_c'} (A_{cp}^2/p_{cp})$. The latter point is past the centerline of the member; therefore, minimum stirrups are required throughout the span. The minimum web steel provided, 0.40 in², satisfies the ACI Code minimum $= 0.75\sqrt{f_c'} b_w s/f_{yt} = 0.75\sqrt{5000} (12) \times 7/60,000 = 0.074$ in² $\geq 50b_w s/f_{yt} = 50 \times 12 \times 7/60,000 = 0.070$ in².

The longitudinal steel required for torsion at a distance d from the column face is computed next. At that location

$$\frac{A_t}{s} = 0.0384\left(1 - \frac{1.79}{14}\right) = 0.0335$$

and from Eq. (8.7)

$$A_\ell = \frac{A_t}{s} p_h \frac{f_{yt}}{f_y} \cot^2 \theta$$

$$= 0.0335 \times 58 \times \frac{60}{60} \times 1^2 = 1.94 \text{ in}^2$$

with a total not less than the lower value from Eqs. (8.22a) and (8.22b), which are, respectively,

$$A_{\ell,min} = \frac{5\sqrt{f_c'} A_{cp}}{f_y} - \frac{A_t}{s} p_h \frac{f_{yt}}{f_y}$$

$$A_{\ell,\min} = \frac{5\sqrt{f_c'}\,A_{cp}}{f_y} - \left(\frac{25b_w}{f_{yt}}\right)p_h\frac{f_{yt}}{f_y}$$

Because $25b_w/f_{yt} = 25 \times 12/60{,}000 = 0.005$ is less than A_t/s, Eq. (8.22a) will give the lower value of $A_{\ell,\min}$.

$$A_{\ell,\min} = \frac{5\sqrt{5000} \times 396}{60 \times 1000} - 0.0335 \times 58 \times \frac{60}{60} = 0.39 \text{ in}^2$$

According to the ACI Code, the spacing must not exceed 12 in., and the bars may not be less than No. 3 (No. 10) in size or have a diameter less than $0.042s = 0.29$ in. Reinforcement will be placed at the top, middepth, and bottom of the member—each level to provide not less than $1.94/3 = 0.65$ in^2. Two No. 6 (No. 19) bars will be used at middepth, and reinforcement to be placed for flexure will be increased by 0.65 in^2 at the top and bottom of the member.

Although A_ℓ reduces in direct proportion to A_t and, hence, decreases linearly starting at d from the face of the column to the midspan, for simplicity of construction, the added bars and the increment in the flexural steel will be maintained throughout the length of the member. Although ACI Code 9.5.4.5 states that A_ℓ may be decreased in flexural compression zones by an amount equal to $M_u/(0.9df_y)$, that reduction will not be made here. Adequate embedment must be provided past the face of the column to fully develop f_y in the bars at that location.

REFERENCES

8.1.   P. Lampert and B. Thurlimann, "Ultimate Strength and Design of Reinforced Concrete Beams in Torsion and Bending," *Int. Assoc. Bridge and Struct. Eng.* Publ. 31-I, Zurich, 1971, pp. 107–131.

8.2.   B. Thurlimann, "Torsional Strength of Reinforced and Prestressed Beams—CEB Approach," in *Concrete Design: U.S. and European Practices*, SP-59, 1979, American Concrete Institute, Detroit, MI, pp. 117–143.

8.3.   *CEB-FIP Model Code 1990*, Thomas Telford, London, 1991.

8.4.   CSA Committee A23.3–14, *Design of Concrete Structures*, Canadian Standards Association, Etobicoke, Ontario, 2014, 297 pp.

8.5.   "Report on Torsion of Structural Concrete," ACI Committee 445, *ACI Manual of Concrete Practice*, Part 6, 2015.

8.6.   T. T. C. Hsu, "Torsion of Structural Concrete—Behavior of Reinforced Concrete Rectangular Members," in *Torsion of Structural Concrete*, SP-18, 1968, American Concrete Institute, Detroit, MI, pp. 261–306.

8.7.   A. H. Mattock, Disc. of "Design of Torsion" by J. G. MacGregor and M. G. Ghoneim (Ref 8.8), *ACI Struct. J.*, vol. 93, no. 1, 1996, pp. 142–143.

8.8.   J. G. MacGregor and M. G. Ghoneim, "Design of Torsion," *ACI Struct. J.*, vol. 92, no. 2, 1995, pp. 218–221.

8.9.   T. T. C. Hsu, "Shear Flow Zone in Torsion of Reinforced Concrete," *J. Struct. Eng.*, vol. 116, no. 11, 1990, pp. 3206–3226.

8.10.   D. Mitchell and M. P. Collins, "Detailing for Torsion," *J. ACI*, vol. 73, no. 9, 1976, pp. 506–511.

PROBLEMS

8.1.   A rectangular beam is 15 in. wide and 30 in. deep. For $f_c' = 4000$ psi and $f_{yt} = 60{,}000$ psi, determine the required spacing of No. 4 (No. 13) closed stirrups for a factored shear of 80 kips and factored torsional moment of 50 ft-kips. The stirrup centroid is located 1.75 in. from each concrete face. The effective depth is 26.5 in.

8.2.   A beam of rectangular cross section having $b = 22$ in. and $h = 15$ in. is to carry a total factored load of 3600 lb/ft uniformly distributed over its 26 ft span, and in addition the beam will be subjected to a uniformly distributed torsion of 1800 ft-lb/ft at factored loads. Closed stirrup-ties will be used to provide for flexural shear and torsion, placed with the stirrup steel centroid 1.75 in. from each concrete face. The corresponding flexural effective depth will be approximately 12.5 in. Design the transverse reinforcement for this beam and calculate the increment of longitudinal steel area needed to provide for torsion, using $f_c' = 4000$ psi and $f_y = 60,000$ psi.

8.3. Architectural and clearance requirements call for the use of a transfer girder, shown in Fig. P8.3, spanning 20 ft between supporting column faces. The girder must carry from above a concentrated column load of 17.5 kips at midspan, applied with eccentricity 2 ft from the girder centerline. (Load factors are already included, as is an allowance for girder self-weight.) The member is to have dimensions $b = 10$ in., $h = 20$ in., $x_o = 6.5$ in., $y_o = 16.5$ in., and $d = 17$ in. Supporting columns provide full torsional rigidity; flexural rigidity at the ends of the span can be assumed to develop 40 percent of the maximum moment that would be obtained if the girder were simply supported. Design both transverse and longitudinal steel for the beam. Material strengths are $f_c' = 5000$ psi and $f_y = 60,000$ psi.

FIGURE P8.3
Transfer girder: (*a*) top view; (*b*) front view; and (*c*) side view.

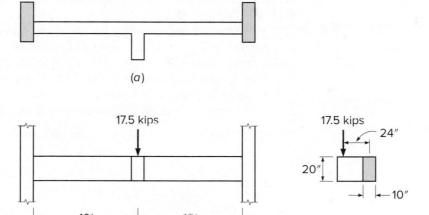

8.4. The beam shown in cross section in Fig. P8.4 is a typical interior member of a continuous building frame, with span 30 ft between support faces. At factored loads, it will carry a uniformly distributed vertical load of 3200 lb/ft, acting simultaneously with a uniformly distributed torsion of 2700 ft-lb/ft. Transverse reinforcement for shear and torsion will consist of No. 4 (No. 13) stirrup-ties, as shown, with 1.5 in. clear to all concrete faces. The effective depth to flexural steel is taken equal to 22.5 in. for both negative and positive

FIGURE P8.4

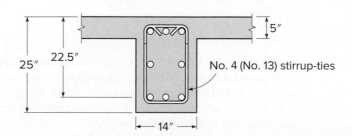

bending regions. Design the transverse reinforcement for shear and torsion, and calculate the longitudinal steel to be added to the flexural requirements to provide for torsion. Torsional reinforcement will be provided only in the web, not in the flanges. Material strengths are $f_c' = 4000$ psi and $f_y = 60,000$ psi.

8.5. The single-span T beam bridge described in Problem 4.29 is reinforced for flexure with four No. 10 (No. 32) bars in two layers, which continue uninterrupted into the supports, permitting a service live load of 1.50 kips/ft to be carried, in addition to the dead load of 0.93 kip/ft, including self-weight. Assume now that only one-half of that live load acts but that it is applied over only one-half the width of the member, entirely to the right of the section centerline. Design the transverse reinforcement for shear and torsion, and calculate the modified longitudinal steel needed for this eccentric load condition. Torsional reinforcement can be provided in the slab if needed, as well as in the web. Stirrup-ties will be No. 3 or No. 4 (No. 10 or No. 13) bars, with 1.5 in. clear to all concrete faces. Supports provide no restraint against flexural rotations but do provide full restraint against twist. Show a sketch of your final design, detailing all reinforcement. Material strengths are as given for Problem 4.29.

8.6. Design a spandrel (edge) girder for shear and torsion that is loaded with a uniform factored load of 1.1 kips/ft. In addition, beams framing into the girder apply concentrated factored vertical loads F_{u1} and F_{u2} and torsional moments T_{u1} and T_{u2}, as shown in Fig. P8.6. Girder dimensions are $h = 32$ in. and $b_w = 28$ in., and slab thickness (one side of girder only) = 6 in. An analysis of various loading combinations indicates the following results:

Case 1	$F_{u1} = F_{u2} = 80$ kips	
	$T_{u1} = T_{u2} = 160$ ft-kips	
Case 2	$F_{u1} = 83$ kips; $F_{u2} = 22$ kips	
	$T_{u1} = 160$ ft-kips; $T_{u2} = 53$ ft-kips	
Case 3	$F_{u1} = 22$ kips; $F_{u2} = 83$ kips	
	$T_{u1} = 53$ ft-kips; $T_{u2} = 160$ ft-kips	

To calculate reactions, treat the ends of the girder as fixed. Use $f_y = 60,000$ psi and $f_c' = 4000$ psi. Provide design drawings showing the transverse steel and the longitudinal steel required in addition to the flexural steel.

FIGURE P8.6

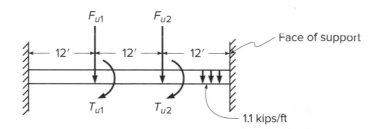

8.7. A 20 ft long rectangular beam, free-standing except for being fixed at each end against rotation, must carry a midspan live load of 35 kips. The load can be as much as 12 in. off the axis of the beam. Beam dimensions are $b = 12$ in., $d = 20$ in., and $h = 23$ in. Use $f_y = 60,000$ psi and $f_c' = 4000$ psi. Design the shear and torsion reinforcement.

9

Short Columns

9.1 INTRODUCTION: AXIAL COMPRESSION

Columns are defined as members that carry loads chiefly in compression. Usually columns carry bending moments as well, about one or both axes of the cross section, and the bending action may produce tensile forces over a part of the cross section. Even in such cases, columns are generally referred to as compression members, because the compression forces dominate their behavior. In addition to the most common type of compression member, that is, vertical elements in structures, compression members include arch ribs; rigid frame members inclined or otherwise; compression elements in trusses, shells, or portions thereof that carry axial compression; and other forms. In this chapter the term *column* will be used interchangeably with the term *compression member*, for brevity and in conformity with general usage.

Two types of reinforced concrete compression members are in use:

1. Members reinforced with longitudinal bars and transverse ties.
2. Members reinforced with longitudinal bars and continuous spirals.

The main reinforcement in columns is longitudinal, parallel to the direction of the load, and consists of bars arranged in a square, rectangular, or circular pattern, as was shown in Fig. 3.2. Figure 9.1 shows an ironworker tightening splices for the main reinforcing steel during construction of the 60-story Bank of America Corporate Center in Charlotte, North Carolina. The ratio of longitudinal steel area A_{st} to gross concrete cross section A_g is in the range from 0.01 to 0.08, according to ACI Code 10.6.1.1. The lower limit is necessary to ensure resistance to bending moments not accounted for in the analysis and to reduce the effects of creep and shrinkage of the concrete under sustained compression. Ratios higher than 0.08 not only are uneconomical but also would cause difficulty owing to congestion of the reinforcement, particularly where the steel must be spliced. Most columns are designed with ratios below 0.04. Larger-diameter bars are used to reduce placement costs and to avoid unnecessary congestion. The largest bars, No. 14 and No. 18 (No. 43 and No. 57), are produced mainly for use in columns. According to ACI Code 10.7.3.1, a minimum of four longitudinal bars is required when the bars are enclosed by spaced rectangular or circular ties, and a minimum of six bars must be used when the longitudinal bars are enclosed by a continuous spiral. A minimum of three longitudinal bars must be used when the bars are enclosed by triangular ties.

Columns may be divided into two broad categories: *short columns*, for which the strength is governed by the strength of the materials and the geometry

FIGURE 9.1
Reinforcement for primary column of 60-story Bank of America Corporate Center in Charlotte, North Carolina. (*Courtesy of Walter P. Moore and Associates.*)

of the cross section, and *slender columns*, for which the strength may be significantly reduced by lateral deflections. A number of years ago, an ACI-ASCE survey indicated that 90 percent of columns braced against sidesway and 40 percent of unbraced columns could be designed as short columns. Effective lateral bracing, which prevents relative lateral movement of the two ends of a column, is commonly provided by shear walls, elevator and stairwell shafts, diagonal bracing, or a combination of these. Although slender columns are more common now because of the wider use of high-strength materials and improved methods of dimensioning members, it is still true that most columns in ordinary practice can be considered short columns. Only short columns will be discussed in this chapter; the effects of slenderness in reducing column strength will be covered in Chapter 10.

The behavior of short, axially loaded compression members was discussed in Section 3.5 in introducing the basic aspects of reinforced concrete. It is suggested that the earlier material be reviewed at this point. In Section 3.5, it was demonstrated that, for lower loads at which both materials remain elastic, the steel carries a relatively small portion of the total load. The steel stress f_s is equal to n times the concrete stress:

$$f_s = nf_c \tag{9.1}$$

where $n = E_s/E_c$ is the modular ratio. In this range, the axial load P is given by

$$P = f_c[A_g + (n - 1)A_{st}] \tag{9.2}$$

where A_g is the gross area of the cross section, A_{st} is the total area of the reinforcing steel, and the term in brackets is the area of the transformed section (see Fig. 9.2). Equations (9.2) and (9.1) can be used to find concrete and steel stresses, respectively, for given loads, provided both materials remain elastic. Example 3.1 demonstrated the use of these equations.

In Section 3.5, it was further shown that the nominal strength of an axially loaded column can be found, recognizing the nonlinear response of both materials, by

$$P_n = 0.85f_c'A_c + f_y A_{st} \tag{9.3a}$$

where A_c = net area of concrete, or

$$P_n = 0.85f_c'(A_g - A_{st}) + f_y A_{st} \tag{9.3b}$$

that is, by summing the strength contributions of the two components of the column. At this stage, the steel carries a significantly larger fraction of the load than was the case at lower total load.

Equations (9.3a) and (9.3b) are based on the assumption that f_y in compression will be attained once the concrete reaches its limiting strain $\varepsilon_u = 0.003$. For f_y much above 80,000 psi, however, concrete will surpass a strain of 0.003 and, thus, may no longer provide an average stress of $0.85f_c'$ at a strain corresponding to f_y. For this reason, ACI Code 22.4.2 places an upper limit on f_y in compression of 80,000 psi.

The calculation of the nominal strength of an axially loaded column was demonstrated in Section 3.5.

According to ACI Code 22.4.2, the *design strength* of an axially loaded column is to be found based on Eq. (9.3b) with the introduction of certain strength reduction factors. These strength reduction factors are lower for columns than for beams, reflecting the greater effect of column failure on a structure. A beam failure would normally affect only a local region, whereas a column failure could result in the collapse of the entire structure. In addition, these factors reflect differences in the behavior of tied columns and spirally reinforced columns that will be discussed in Section 9.2. A basic ϕ factor of 0.75 is used for spirally reinforced columns and 0.65 for tied columns, vs. $\phi = 0.90$ for most beams.

A further limitation on column strength is imposed by ACI Code 22.4.2 to allow for accidental eccentricities of loading not considered in the analysis. This is done by imposing an upper limit on the axial load that is less than the calculated design strength. This upper limit is taken as 0.85 times the design strength for spirally

FIGURE 9.2

Transformed section in axial compression.

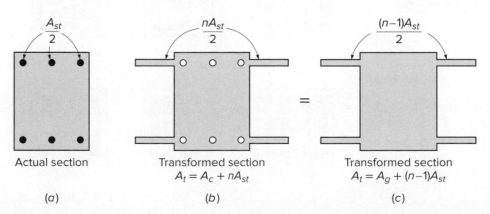

Actual section	Transformed section $A_t = A_c + nA_{st}$	Transformed section $A_t = A_g + (n-1)A_{st}$
(a)	(b)	(c)

reinforced columns and 0.80 times the calculated strength for tied columns. Thus, according to ACI Code 22.4.2, for spirally reinforced columns

$$\phi P_{n,\text{max}} = 0.85\phi\,[0.85f_c'(A_g - A_{st}) + f_y\,A_{st}] \tag{9.4a}$$

with $\phi = 0.75$. For tied columns

$$\phi P_{n,\text{max}} = 0.80\phi\,[0.85f_c'(A_g - A_{st}) + f_y\,A_{st}] \tag{9.4b}$$

with $\phi = 0.65$.

9.2   TRANSVERSE TIES AND SPIRALS

Figure 3.2 shows cross sections of the simplest types of columns, spirally reinforced or provided with transverse ties. Other cross sections frequently found in buildings and bridges are shown in Fig. 9.3. In general, in members with large axial forces and small moments, longitudinal bars are spaced more or less uniformly around the perimeter (Fig. 9.3a to d). When bending moments are large, much of the longitudinal steel is concentrated at the faces of highest compression or tension, that is, at maximum distances from the axis of bending (Fig. 9.3e to h). Specific recommended patterns for many combinations and arrangements of bars are found in Refs. 9.1 and 9.2. In heavily loaded columns with large steel percentages, the result of a large number of bars, each of them positioned and held individually by ties, is steel congestion in the forms and difficulties in placing the concrete. In such cases, bundled bars are frequently employed. Bundles consist of two to four bars tied in direct contact,

FIGURE 9.3
Tie arrangements for square and rectangular columns.

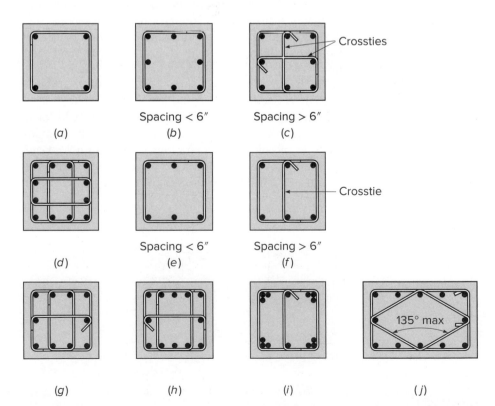

wired, or otherwise fastened together. These are usually placed in the corners, as shown in Fig. 9.3*i*. Tests have shown that adequately bundled bars act as one unit; that is, they are detailed as if a bundle constituted a single round bar of area equal to the sum of the bundled bars.

Transverse reinforcement, in the form of individual relatively widely spaced ties or a continuous closely spaced spiral, serves several functions. For one, such reinforcement is needed to hold the longitudinal bars in position in the forms while the concrete is being placed. For this purpose, longitudinal and transverse steel is wired together to form cages, which are then moved into the forms and properly positioned before placing the concrete. For another, transverse reinforcement is needed to prevent the highly stressed, slender longitudinal bars from buckling outward by bursting the thin concrete cover.

Closely spaced spirals serve these two functions. Ties, which can be arranged and spaced in various ways, must be so designed that these two requirements are met. This means that the spacing must be sufficiently small to prevent buckling between ties and that, in any tie plane, a sufficient number of ties must be provided to position and hold all bars. On the other hand, in columns with many longitudinal bars, if the column section is crisscrossed by too many ties, they interfere with the placement of concrete in the forms. To achieve adequate tying yet hold the number of ties to a minimum, ACI Code 10.7.6 and 25.7.2 give rules for tie arrangement that may be summarized as:

> Longitudinal reinforcement shall be laterally supported using ties or hoops (the latter are discussed in Chapter 20).
>
> Tie bars or wire shall be at least No. 3 (No. 10) for No. 10 (No. 32) or smaller longitudinal bars and at least No. 4 (No. 13) for No. 11 (No. 36) or larger longitudinal bars and bundled longitudinal bars.
>
> Center-to-center tie spacing shall not exceed the least of $16d_b$ of the longitudinal bar, $48d_b$ of the tie bar, and smallest dimension of member.
>
> Rectilinear ties shall be arranged so that every corner and alternate longitudinal bar shall have lateral support provided by the corner of a tie with an included angle of not more than 135°, as shown in Fig. 9.3*j*, and no bar shall be farther than 6 in. clear on each side along the tie from a laterally supported bar, as illustrated in Fig. 9.3*b*, *c*, *e*, and *f*. Intermediate lateral support can be provided by crossties, such as those shown in Fig. 9.3*c* and *e*, which must be continuous with a hook at one end with a bend not less than 135° and a standard hook at the other end, both of which must engage a longitudinal bar.
>
> Deformed wire or welded wire reinforcement of equivalent area may be used in place of ties. Circular ties may be used where longitudinal bars are located around the perimeter of a circle.

For spirally reinforced columns, ACI Code 25.7.3 gives requirements for lateral reinforcement that may be summarized as follows:

> Spirals shall consist of evenly spaced continuous bar or wire at least $\frac{3}{8}$ in. in diameter, with a clear spacing between turns of the spiral not greater than 3 in. nor less than 1 in.

In addition, a minimum ratio of spiral steel is imposed such that the structural performance of the column is significantly improved, with respect to both ultimate load and the type of failure, compared with an otherwise identical tied column.

The structural effect of a spiral is easily visualized by considering as a model a steel drum filled with sand (Fig. 9.4). When a load is placed on the sand, a lateral pressure is exerted by the sand on the drum, which causes hoop tension in the steel wall. The load on the sand can be increased until the hoop tension becomes large enough to burst the drum. The sand pile alone, if not confined in the drum, would have been able to support little load. A cylindrical concrete column, to be sure, does

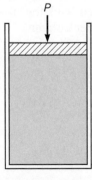

FIGURE 9.4
Model for action of a spiral.

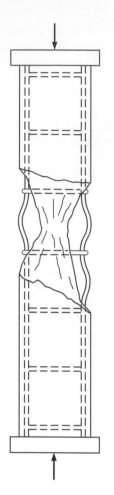

have a definite strength without any transverse confinement. As it is being loaded, it shortens longitudinally and expands laterally, depending on Poisson's ratio. A closely spaced spiral confining the column counteracts the expansion, as did the steel drum in the model. This causes hoop tension in the spiral, while the carrying capacity of the confined concrete in the core is greatly increased. Failure occurs only when the spiral steel yields, which greatly reduces its confining effect, or when it fractures.

A tied column fails at the load given by Eq. (9.3a or b). At this load, the concrete fails by crushing and shearing outward along inclined planes, and the longitudinal steel by buckling outward between ties (Fig. 9.5). In a spirally reinforced column, when the same load is reached, the longitudinal steel and the concrete within the core are prevented from moving outward by the spiral. The concrete in the outer shell, however, not being so confined, does fail; that is, the outer shell spalls off when the load P_n is reached. It is at this stage that the confining action of the spiral has a significant effect, and if sizable spiral steel is provided, the load that will ultimately fail the column by causing the spiral steel to yield or fracture can be much larger than that at which the shell spalled off. Furthermore, the axial strain limit when the column fails will be much greater than otherwise; the toughness of the column has been much increased.

In contrast to the practice in some foreign countries, it is reasoned in the United States that any excess capacity beyond the spalling load of the shell is wasted because the member, although not actually failed, would no longer be considered serviceable. For this reason, the ACI Code provides a minimum spiral reinforcement of such an amount that its contribution to the carrying capacity is just slightly larger than that of the concrete in the shell. The situation is best understood from Fig. 9.6, which compares the performance of a tied column with that of a spiral column whose spalling load is equal to the ultimate load of the tied column. The failure of the tied column is abrupt and complete. This is true, to almost the same degree, of a spiral column with a spiral so light that its strength contribution is considerably less than the strength lost in the spalled shell. With a heavy spiral the reverse is true, and with considerable prior deformation the spalled column would fail at a higher load. The "ACI spiral," its strength contribution about compensating for that lost in the spalled shell, hardly increases the ultimate load. However, by preventing instantaneous crushing of concrete and buckling of steel, it produces a more gradual and ductile failure, that is, a tougher column.

FIGURE 9.5
Failure of a tied column.

FIGURE 9.6
Behavior of spirally reinforced and tied columns.

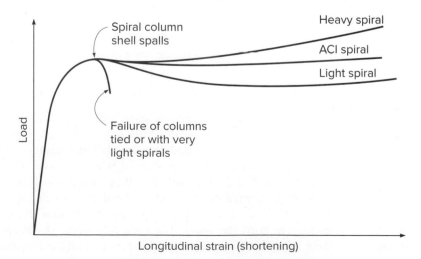

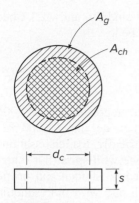

FIGURE 9.7
Confinement of core concrete
due to hoop tension.

It has been found experimentally (Refs. 9.3 to 9.5) that the increase in compressive strength of the core concrete in a column provided through the confining effect of spiral steel is closely represented by the equation

$$f_c^* - 0.85f_c' = 4.0f_2' \qquad (a)$$

where $f_c^* =$ compressive strength of spirally confined core concrete
$0.85f_c' =$ compressive strength of concrete if unconfined
$f_2' =$ lateral confinement stress in core concrete produced by spiral

The confinement stress f_2' is calculated assuming that the spiral steel reaches its yield stress f_{yt} when the column eventually fails. With reference to Fig. 9.7, a hoop tension analysis of an idealized model of a short segment of column confined by one turn of transverse steel shows that

$$f_2' = \frac{2f_{yt}\,A_{sp}}{d_c s} \qquad (b)$$

where $f_{yt} =$ yield strength of spiral steel
$A_{sp} =$ cross-sectional area of spiral wire
$d_c =$ outside diameter of spiral
$s =$ spacing or pitch of spiral wire

A *volumetric ratio* is defined as the ratio of the volume of spiral steel to the volume of core concrete:

$$\rho_s = \frac{2\pi d_c\,A_{sp}}{2}\,\frac{4}{\pi d_c^2\,s}$$

from which

$$A_{sp} = \frac{\rho_s\,d_c\,s}{4} \qquad (c)$$

Substituting the value of A_{sp} from Eq. (c) into Eq. (b) results in

$$f_2' = \frac{\rho_s f_{yt}}{2} \qquad (d)$$

To find the right amount of spiral steel, one calculates

$$\text{Strength contribution of the shell} = 0.85f_c'\,(A_g - A_{ch}) \qquad (e)$$

where A_g and A_{ch} are, respectively, the gross and core concrete areas. Then substituting the confinement stress from Eq. (d) into Eq. (a) and multiplying by the core concrete area, one finds

$$\text{Strength provided by spiral} = 2\rho_s f_{yt}\,A_{ch} \qquad (f)$$

The basis for the design of the spiral is that the strength gain provided by the spiral should be at least equal to that lost when the shell spalls, so combining Eqs. (e) and (f) yields

$$0.85f_c'\,(A_g - A_{ch}) = 2\rho_s f_{yt}\,A_{ch}$$

from which

$$\rho_s = 0.425\left(\frac{A_g}{A_{ch}} - 1\right)\frac{f_c'}{f_{yt}} \qquad (g)$$

According to the ACI Code, this result is rounded upward slightly, and ACI Code 25.7.3.3 states that the ratio of spiral reinforcement shall not be less than

$$\rho_s = 0.45 \left(\frac{A_g}{A_{ch}} - 1 \right) \frac{f_c'}{f_{yt}} \tag{9.5}$$

It is further stipulated in the ACI Code that f_{yt} should not be taken greater than 100,000 psi and that spiral reinforcement should not be spliced if f_{yt} is greater than 60,000 psi.

It follows from this development that two concentrically loaded columns designed in accordance with the ACI Code, one tied and one with spiral but otherwise identical, will fail at about the same load, the former in a sudden and brittle manner, the latter gradually with prior spalling of the shell and with more ductile behavior. This advantage of the spiral column is much less pronounced if the load is applied with significant eccentricity or when bending from other sources is present simultaneously with axial load. For this reason, while the ACI Code permits somewhat larger design loads on spiral than on tied columns when the moments are small or zero ($\phi = 0.75$ for spirally reinforced columns vs. $\phi = 0.65$ for tied), the difference is not large, and it is even further reduced for large eccentricities, for which ϕ approaches 0.90 for both.

The design of spiral reinforcement according to the ACI Code provisions is easily reduced to tabular form, as in Table A.14 of Appendix A.

9.3 COMPRESSION PLUS BENDING OF RECTANGULAR COLUMNS

Members that are axially, that is, concentrically, compressed occur rarely, if ever, in buildings and other structures. Components such as columns and arches chiefly carry loads in compression, but simultaneous bending is almost always present. Bending moments are caused by continuity, that is, by the fact that building columns are parts of monolithic frames in which the support moments of the girders are partly resisted by the abutting columns, by transverse loads such as wind forces, by loads carried eccentrically on column brackets, or in arches when the arch axis does not coincide with the pressure line. Even when design calculations show a member to be loaded purely axially, inevitable imperfections of construction will introduce eccentricities and consequent bending in the member as built. For this reason, members that must be designed for simultaneous compression and bending are very frequent in almost all types of concrete structures.

When a member is subjected to combined axial compression P and moment M, as in Fig. 9.8a, it is usually convenient to replace the axial load and moment with an equal load P applied at eccentricity $e = M/P$, as in Fig. 9.8b. The two loadings are statically equivalent. All columns may then be classified in terms of the equivalent eccentricity. Those having relatively small e are generally characterized by compression over the entire concrete section, and if overloaded, will fail by crushing of the concrete accompanied by yielding of the steel in compression on the more heavily loaded side. Columns with large eccentricity are subject to tension over at least a part of the section, and if overloaded, may fail due to tensile yielding of the steel on the side farthest from the load.

For columns, load stages below the ultimate are generally not important. Cracking of concrete, even for columns with large eccentricity, is usually not a serious problem, and lateral deflections at service load levels are seldom, if ever, a factor.

FIGURE 9.8
Equivalent eccentricity of
column load.

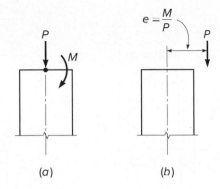

Design of columns is therefore based on the factored load, which must not exceed the design strength, as usual, that is,

$$\phi M_n \geq M_u \tag{9.6a}$$

$$\phi P_n \geq P_u \tag{9.6b}$$

9.4 STRAIN COMPATIBILITY ANALYSIS AND INTERACTION DIAGRAMS

Figure 9.9a shows a member loaded parallel to its axis by a compressive force P_n at an eccentricity e measured from the centerline. The distribution of strains at a section $a\text{-}a$ along its length, at incipient failure, is shown in Fig. 9.9b. With plane sections assumed to remain plane, concrete strains vary linearly with distance from the neutral axis, which is located a distance c from the more heavily loaded side of the member. With full compatibility of deformations, the steel strains at any location are the same as the strains in the adjacent concrete; thus, if the ultimate concrete strain is ε_u, the strain in the bars nearest the load is ε_s', while that in the tension bars at the far side is ε_s. Compression steel with area A_s' and tension steel with area A_s are located at distances d' and d, respectively, from the compression face.

The corresponding stresses and forces are shown in Fig. 9.9c. Just as for simple bending, the actual concrete compressive stress distribution is replaced by an equivalent rectangular distribution having depth $a = \beta_1 c$. A large number of tests on columns with a variety of shapes have shown that the strengths computed on this basis are in satisfactory agreement with test results (Ref. 9.6).

Equilibrium between external and internal axial forces shown in Fig. 9.9c requires that

$$P_n = 0.85 f_c' \, ab + f_s' A_s' - f_s A_s \tag{9.7}$$

Also, the moment about the centerline of the section of the internal stresses and forces must be equal and opposite to the moment of the external force P_n, so that

$$M_n = P_n e = 0.85 f_c' \, ab \left(\frac{h}{2} - \frac{a}{2} \right) + f_s' A_s' \left(\frac{h}{2} - d' \right) + f_s A_s \left(d - \frac{h}{2} \right) \tag{9.8}$$

These are the two basic equilibrium relations for rectangular eccentrically compressed members. For reasons discussed in Section 9.1, an upper limit of 80,000 psi should be placed on stress in the compression steel f_s'.

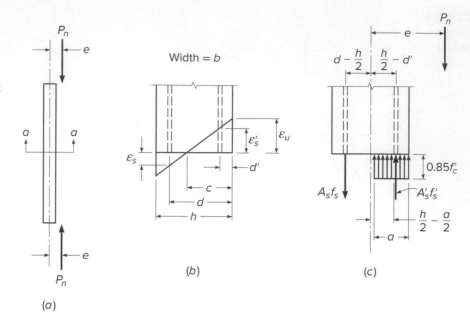

The fact that the presence of the compression reinforcement A_s' has displaced a corresponding amount of concrete of area A_s' is neglected in writing these equations. If necessary, particularly for large reinforcement ratios, one can account for this very simply. Evidently, in the above equations, a nonexistent concrete compression force of amount $A_s'\,(0.85f_c')$ has been included as acting in the displaced concrete at the level of the compression steel. This excess force can be removed in both equations by multiplying A_s' by $f_s' - 0.85f_c'$ rather than by f_s'.

For large eccentricities, failure is initiated by yielding of the tension steel A_s. Hence, for this case, $f_s = f_y$. When the concrete reaches its ultimate strain ε_u, the compression steel may or may not have yielded; this must be determined based on compatibility of strains. For small eccentricities, the concrete will reach its limit strain ε_u before the tension steel starts yielding; in fact, the bars on the side of the column farther from the load may be in compression, not tension. For small eccentricities, too, the analysis must be based on compatibility of strains between the steel and the adjacent concrete.

For a given eccentricity determined from the frame analysis (that is, $e = M_u/P_u$) it is possible to solve Eqs. (9.7) and (9.8) for the load P_n and moment M_n that would result in failure as follows. In both equations, f_s', f_s, and a can be expressed in terms of a single unknown c, the distance to the neutral axis. This is easily done based on the geometry of the strain diagram, with ε_u taken equal to 0.003 as usual, and using the stress-strain curve of the reinforcement. The result is that the two equations contain only two unknowns, P_n and c, and can be solved for those values simultaneously. However, to do so in practice would be complicated algebraically, particularly because of the need to incorporate the limit f_y on both f_s' and f_s.

A better approach, providing the basis for practical design, is to construct a *strength interaction diagram* defining the failure load and failure moment for a given column for the full range of eccentricities from zero to infinity. For any eccentricity, there is a unique pair of values of P_n and M_n that will produce the state of incipient failure. That pair of values can be plotted as a point on a graph relating P_n and M_n,

FIGURE 9.10
Interaction diagram for
nominal column strength
in combined bending and
axial load.

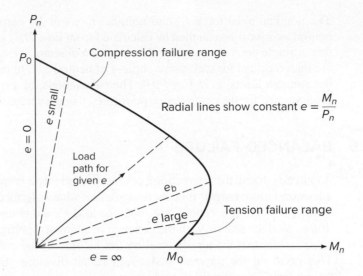

such as shown in Fig. 9.10. A series of such calculations, each corresponding to a different eccentricity, will result in a curve having a shape typically as shown in Fig. 9.10. On such a diagram, any radial line represents a particular eccentricity $e = M/P$. For that eccentricity, gradually increasing the load will define a load path as shown, and when that load path reaches the limit curve, failure will result. Note that the vertical axis corresponds to $e = 0$, and P_0 is the capacity of the column if concentrically loaded, as given by Eq. (9.3b). The horizontal axis corresponds to an infinite value of e, that is, pure bending at moment capacity M_0. Small eccentricities will produce failure governed by concrete compression, while large eccentricities give a failure triggered by yielding of the tension steel.

For a given column, selected for trial, the interaction diagram is most easily constructed by selecting successive choices of neutral axis distance c, from infinity (axial load with eccentricity 0) to a very small value found by trial to give $P_n = 0$ (pure bending). For each selected value of c, the steel strains and stresses and the concrete force are easily calculated as follows. For the tension steel,

$$\varepsilon_s = \varepsilon_u \frac{d - c}{c} \tag{9.9}$$

$$f_s = \varepsilon_u E_s \frac{d - c}{c} \leq f_y \tag{9.10}$$

while for the compression steel,

$$\varepsilon_s' = \varepsilon_u \frac{c - d'}{c} \tag{9.11}$$

$$f_s' = \varepsilon_u E_s \frac{c - d'}{c} \leq f_y \leq 80,000 \text{ psi} \tag{9.12}$$

The concrete stress block has depth

$$a = \beta_1 c \leq h \tag{9.13}$$

and consequently the concrete compressive resultant is

$$C = 0.85 f_c' ab \tag{9.14}$$

The nominal axial force P_n and nominal moment M_n corresponding to the selected neutral axis location can then be calculated from Eqs. (9.7) and (9.8), respectively, and thus a single point on the strength interaction diagram is established. The calculations are then repeated for successive choices of neutral axis to establish the curve defining the strength limits, as in Fig. 9.10. The calculations, of a repetitive nature, are easily programmed for the computer or performed using a spreadsheet.

9.5 BALANCED FAILURE

As already noted, the interaction curve is divided into a compression failure range and a tension failure range.[†] It is useful to define what is termed a *balanced failure mode* and corresponding eccentricity e_b with the load P_b and moment M_b acting in combination to produce failure, with the concrete reaching its limiting strain ε_u at precisely the same instant that the tensile steel on the far side of the column reaches yield strain. This point on the interaction diagram is the dividing point between compression failure (small eccentricities) and tension failure (large eccentricities).

The values of P_b and M_b are easily computed with reference to Fig. 9.9. For balanced failure,

$$c = c_b = d\,\frac{\varepsilon_u}{\varepsilon_u + \varepsilon_y} \tag{9.15}$$

and

$$a = a_b = \beta_1 c_b \tag{9.16}$$

Equations (9.9) through (9.14) are then used to obtain the steel stresses and the compressive resultant, after which P_b and M_b are found from Eqs. (9.7) and (9.8).

Note that, in contrast to beam design, one cannot restrict column designs such that yielding failure rather than crushing failure would always be the result of overloading. The type of failure for a column depends on the value of eccentricity e, which in turn is defined by the load analysis of the building or other structure.

It is important to observe, in Fig. 9.10, that in the region of compression failure the larger the axial load P_n, the smaller the moment M_n that the section is able to sustain before failing. However, in the region of tension failure, the reverse is true; the larger the axial load, the larger the simultaneous moment capacity. This is easily understood. In the compression failure region, failure occurs through overstraining of the concrete. The larger the concrete compressive strain caused by the axial load alone, the smaller the margin of additional strain available for the added compression caused by bending. On the other hand, in the tension failure region, yielding of the steel initiates failure. If the member is loaded in simple bending to the point at which yielding begins in the tension steel, and if an axial compression load is then added, the steel compressive stresses caused by this load will be superimposed on the previous tensile stresses. This reduces the total steel stress to a value below its yield strength. Consequently, an additional moment can now be sustained of such magnitude that the combination of the steel stress from the axial load and the increased moment again reaches the yield strength.

[†] The terms *compression failure range* and *tension failure range* are used for the purpose of general description and are distinct from *tension-controlled* and *compression-controlled* failures, as described in Chapter 4 and Section 9.9.

The typical shape of a column interaction diagram shown in Fig. 9.10 has important design implications. In the range of tension failure, a *reduction in axial load* may produce failure for a given moment. When performing a frame analysis, the designer must consider all combinations of loading that may occur, including the combination that produces minimum axial load paired with a moment because the combined loading may place the column in the tension failure range. Within this range, only the compressive load that is certain to be present should be used to calculate the capacity of the column subject to the given moment.

EXAMPLE 9.1 **Column strength interaction diagram.** A 12×20 in. column is reinforced with four No. 9 (No. 29) bars of area 1.0 in^2 each, one in each corner as shown in Fig. 9.11a. The concrete cylinder strength is $f_c' = 4000$ psi and the steel yield strength is 60 ksi. Determine (a) the load P_b, moment M_b, and corresponding eccentricity e_b for balanced failure; (b) the load and moment for a representative point in the tension failure region of the interaction curve; (c) the load and moment for a representative point in the compression failure region; and (d) the axial load strength for zero eccentricity. Then (e) sketch the strength interaction diagram for this column. Finally, (f) design the transverse reinforcement, based on ACI Code provisions.

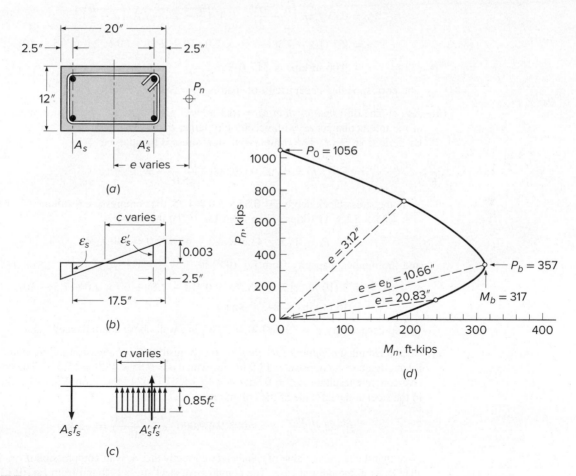

FIGURE 9.11
Column interaction diagram for Example 9.1: (a) cross section; (b) strain distribution; (c) stresses and forces; and (d) strength interaction diagram.

SOLUTION.

(*a*) The neutral axis for the balanced failure condition is easily found from Eq. (9.15) with $\varepsilon_u = 0.003$ and $\varepsilon_y = 60/29,000 = 0.0021$:

$$c_b = 17.5 \times \frac{0.003}{0.0051} = 10.3 \text{ in.}$$

giving a stress-block depth $a = 0.85 \times 10.3 = 8.76$ in. For the balanced failure condition, by definition, $f_s = f_y$. The compressive steel stress is found from Eq. (9.12):

$$f_s' = 0.003 \times 29,000 \, \frac{10.3 - 2.5}{10.3} = 65.9 \text{ ksi} \qquad \text{but} \qquad \leq 60 \text{ ksi}$$

confirming that the compression steel, too, is at the yield. The concrete compressive resultant is

$$C = 0.85 \times 4 \times 8.76 \times 12 = 357 \text{ kips}$$

The balanced load P_b is then found from Eq. (9.7) to be

$$P_b = 0.85 f_c' ab + f_s' A_s' - f_s A_s$$

$$= 357 + 60 \times 2.0 - 60 \times 2.0 = 357 \text{ kips}$$

and the balanced moment from Eq. (9.8) is

$$M_b = 0.85 f_c' ab \left(\frac{h}{2} - \frac{a}{2} \right) + f_s' A_s' \left(\frac{h}{2} - d' \right) - f_s A_s \left(d - \frac{h}{2} \right)$$

$$= 357 \, (10 - 4.38) + 60 \times 2.0 \, (10 - 2.5) + 60 \times 2.0 \, (17.5 - 10)$$

$$= 3806 \text{ in-kips} = 317 \text{ ft-kips}$$

The corresponding eccentricity of load is $e_b = 3806/357 = 10.66$ in.

(*b*) Any choice of *c smaller than* $c_b = 10.3$ in. will give a point in the tension failure region of the interaction curve, with eccentricity larger than e_b. For example, choose $c = 5.0$ in. By definition, $f_s = f_y$. The compressive steel stress is found to be

$$f_s' = 0.003 \times 29,000 \, \frac{5.0 - 2.5}{5.0} = 43.5 \text{ ksi}$$

With the stress-block depth $a = 0.85 \times 5.0 = 4.25$, the compressive resultant is $C = 0.85 \times 4 \times 4.25 \times 12 = 173$ kips. Then from Eq. (9.7), the thrust is

$$P_n = 173 + 43.5 \times 2.0 - 60 \times 2.0 = 140 \text{ kips}$$

and the moment capacity from Eq. (9.8) is

$$M_n = 173 \, (10 - 2.12) + 43.5 \times 2.0 \, (10 - 2.5) + 60 \times 2.0 \, (17.5 - 10)$$

$$= 2916 \text{ in-kips} = 243 \text{ ft-kips}$$

giving eccentricity $e = 2916/140 = 20.83$ in., well above the balanced value.

(*c*) Now selecting a *c* value *larger than* c_b to demonstrate a compression failure point on the interaction curve, choose $c = 18.0$ in., for which $a = 0.85 \times 18.0 = 15.3$ in. The compressive concrete resultant is $C = 0.85 \times 4 \times 15.3 \times 12 = 624$ kips. From Eq. (9.10) the stress in the steel at the left side of the column is

$$f_s = \varepsilon_u E_s \, \frac{d - c}{c} = 0.003 \times 29,000 \, \frac{17.5 - 18.0}{18.0} = -2 \text{ ksi}$$

Note that the negative value of f_s indicates correctly that A_s is in compression if *c* is greater than *d*, as in the present case. The compressive steel stress is found from Eq. (9.12) to be

$$f_s' = \varepsilon_u E_s \, \frac{c - d'}{c} = 0.003 \times 29,000 \, \frac{18.0 - 2.5}{18.0} = 75 \text{ ksi} \qquad \text{but} \qquad \leq 60 \text{ ksi}$$

Then the column capacity is

$$P_n = 624 + 60 \times 2.0 + 2 \times 2.0 = 748 \text{ kips}$$
$$M_n = 624 \,(10 - 7.65) + 60 \times 2.0 \,(10 - 2.5) - 2 \times 2.0 \,(17.5 - 10)$$
$$= 2336 \text{ in-kips} = 195 \text{ ft-kips}$$

giving eccentricity $e = 2336/748 = 3.12$ in.

(d) The axial strength of the column if concentrically loaded corresponds to $c = \infty$ and $e = 0$. For this case,

$$P_n = 0.85 f_c' A_g + f_y A_{st}$$
$$= 0.85 \times 4 \times 12 \times 20 + 60 \times 4.0 = 1056 \text{ kips}$$

Note that, for this as well as the preceding calculations, subtraction of the concrete displaced by the steel has been neglected. For comparison, if the deduction were made in the last calculation as done in Fig. 9.3b,

$$P_n = 0.85 f_c' (A_g - A_{st}) + f_y A_{st}$$
$$= 0.85 \times 4 \,(12 \times 20 - 4) + (60 \times 4.0) = 1042 \text{ kips}$$

The error in neglecting this deduction is only 1 percent in this case; the difference generally can be neglected, except perhaps for columns with reinforcement ratios close to the maximum of 8 percent. In the case of design aids, however, such as those presented in Refs. 9.2 and 9.7 and discussed in Section 9.10, the deduction is usually included for all reinforcement ratios.

(e) From the calculations just completed, plus similar repetitive calculations that will not be given here, the strength interaction curve of Fig. 9.11d is constructed. Note the characteristic shape, described earlier, the location of the balanced failure point as well as the "small eccentricity" and "large eccentricity" points just found, and the axial load capacity.

In the process of developing a strength interaction curve, it is possible to select the values of steel strain ε_s, as done in step a, for use in steps b and c. Selecting ε_s uniquely establishes the neutral axis depth c, as shown by Eqs. (9.9) and (9.15), and is useful in determining M_n and P_n for values of steel strain that correspond to changes in the strength reduction factor ϕ, as will be discussed in Section 9.9.

(f) The design of the column ties will be carried out following the ACI Code restrictions. For the minimum permitted tie, a No. 3 (No. 10) bar, which has a diameter of $\frac{3}{8}$ in., used with No. 9 (No. 29) longitudinal bars having a diameter of 1.128 in a column the least dimension of which is 12 in., the tie spacing is not to exceed

$$48 \times \frac{3}{8} = 18 \text{ in.}$$
$$16 \times 1.128 = 18.05 \text{ in.}$$
$$b = 12 \text{ in.}$$

The last restriction controls in this case, and No. 3 (No. 10) ties will be used at 12 in. spacing, detailed as shown in Fig. 9.11a. Note that the permitted spacing as controlled by the first and second criteria, 18 in., must be reduced because of the 12 in. column dimension.

9.6 DISTRIBUTED REINFORCEMENT

When large bending moments are present, it is most economical to concentrate all or most of the steel along the outer faces parallel to the axis of bending. Such arrangements are shown in Figs. 9.3e to h. On the other hand, with small eccentricities so that axial compression is predominant, and when a small cross section is desired, it is often advantageous to place the steel more uniformly around the perimeter, as in Fig. 9.3a to d. In this case, special attention must be paid to the intermediate bars, that is, those that are not placed along the two faces that are most highly stressed. This is so because when the ultimate load is reached, the stresses in these intermediate bars are usually below the yield point, even though the bars along one or both extreme faces may be yielding. This situation can be analyzed by a simple and obvious extension of the previous analysis based on compatibility of strains. A strength interaction diagram may be constructed just as before. A sequence of choices of neutral axis location results in a set of paired values of P_n and M_n, each corresponding to a particular eccentricity of load.

EXAMPLE 9.2   **Analysis of eccentrically loaded column with distributed reinforcement.**   The column in Fig. 9.12a is reinforced with ten No. 11 (No. 36) bars distributed around the perimeter as shown. Load P_n will be applied with eccentricity e about the strong axis. Material strengths are $f_c' = 6000$ psi and $f_y = 75$ ksi. Find the load and moment corresponding to a failure point with neutral axis $c = 18$ in. from the right face.

FIGURE 9.12
Column in Example 9.2:
(a) cross section; (b) strain distribution; and (c) stresses and forces.

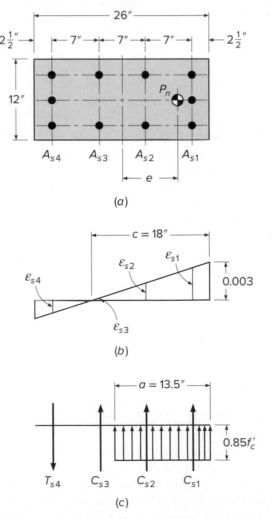

SOLUTION. When the concrete reaches its limit strain of 0.003, the strain distribution is that shown in Fig. 9.12b, the strains at the locations of the four bar groups are found from similar triangles, after which the stresses are found by multiplying strains by $E_s = 29,000$ ksi applying the limit value f_y:

$$\varepsilon_{s1} = 0.00258 \quad f_{s1} = 74.8 \text{ ksi compression}$$
$$\varepsilon_{s2} = 0.00142 \quad f_{s2} = 41.2 \text{ ksi compression}$$
$$\varepsilon_{s3} = 0.00025 \quad f_{s3} = 7.3 \text{ ksi compression}$$
$$\varepsilon_{s4} = 0.00091 \quad f_{s4} = 26.4 \text{ ksi tension}$$

For $f'_c = 6000$ psi, $\beta_1 = 0.75$ and the depth of the equivalent rectangular stress block is $a = 0.75 \times 18 = 13.5$ in. The concrete compressive resultant is $C = 0.85 \times 6 \times 13.5 \times 12 = 826$ kips, and the respective steel forces in Fig. 9.12c are

$$C_{s1} = 4.68 \times 74.8 = 350 \text{ kips}$$
$$C_{s2} = 3.12 \times 41.2 = 129 \text{ kips}$$
$$C_{s3} = 3.12 \times 7.3 \; = \; 23 \text{ kips}$$
$$T_{s4} = 4.68 \times 26.4 = 124 \text{ kips}$$

The axial load and moment that would produce failure for a neutral axis 18 in. from the right face are found by the obvious extensions of Eqs. (9.7) and (9.8):

$$P_n = 826 + 350 + 129 + 23 - 124 = 1204 \text{ kips}$$
$$M_n = 826 \, (13 - 6.75) + 350 \, (13 - 2.5) + 129 \, (13 - 9.5) - 23 \, (13 - 9.5)$$
$$+ \; 124 \, (13 - 2.5)$$
$$= 10{,}510 \text{ in-kips}$$
$$= 876 \text{ ft-kips}$$

The corresponding eccentricity is $e = 10{,}510/1204 = 8.73$ in. Other points on the interaction diagram can be computed in a similar way.

Two general conclusions can be made from this example:

1. Even with the relatively small eccentricity of about one-third of the depth of the section, only the bars of group 1 just barely reached their yield strain, and consequently their yield stress. All other bar groups of the relatively high-strength steel that was used are stressed far below their yield strength, which would also have been true for group 1 for a slightly larger eccentricity. It follows that the use of the more expensive high-strength steel is economical in symmetrically reinforced columns only for very small eccentricities, for example, in the lower stories of tall buildings.
2. The contribution of the intermediate bars of groups 2 and 3 to both P_n and M_n is quite small because of their low stresses. Again, intermediate bars, except as they are needed to hold ties in place, are economical only for columns with very small eccentricities.

9.7 UNSYMMETRICAL REINFORCEMENT

Most reinforced concrete columns are symmetrically reinforced about the axis of bending. However, for some cases, such as the columns of rigid portal frames in which the moments are uniaxial and the eccentricity is large, it is more economical to use

FIGURE 9.13

Plastic centroid of an unsymmetrically reinforced column.

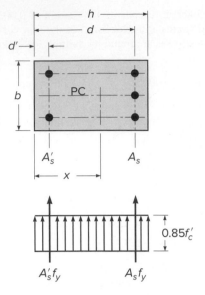

an unsymmetrical pattern of bars, with most of the bars on the tension side, as shown in Fig. 9.13. Such columns can be analyzed by the same strain compatibility approach as described above. However, for an unsymmetrically reinforced column to be loaded concentrically, the load must pass through a point known as the *plastic centroid*. The plastic centroid is defined as the point of application of the resultant force for the column cross section (including concrete and steel forces) if the column is compressed uniformly to the failure strain $\varepsilon_u = 0.003$ over its entire cross section. Eccentricity of the applied load must be measured with respect to the plastic centroid, because only then will $e = 0$ correspond to an axial load with no moment. The location of the plastic centroid for the column of Fig. 9.13 is the resultant of the three internal forces to be accounted for. Its distance from the left face is

$$x = \frac{0.85f_c' \, bh^2/2 + f_y A_s d + f_y A_s' d'}{0.85f_c' bh + f_y A_s + f_y A_s'} \tag{9.17}$$

Clearly, in a symmetrically reinforced cross section, the plastic centroid and the geometric center coincide.

9.8   CIRCULAR COLUMNS

The transverse reinforcement in circular columns may consist of ties or spirals. It was mentioned in Section 9.2 that when load eccentricities are small, spirally reinforced columns show greater toughness, that is, greater ductility, than tied columns, although this difference fades out as the eccentricity is increased. For this reason, as discussed in Section 9.2, the ACI Code provides a more favorable strength reduction factor $\phi = 0.75$ for spiral columns, compared with $\phi = 0.65$ for tied columns. Also, the maximum stipulated design load for entirely or nearly axially loaded members is larger for spirally reinforced members than for comparable tied members (see Section 9.9). It follows that spirally reinforced columns permit a somewhat more economical utilization of the materials, particularly for small calculated eccentricities, although the cost

FIGURE 9.14
Circular column with
compression plus bending.

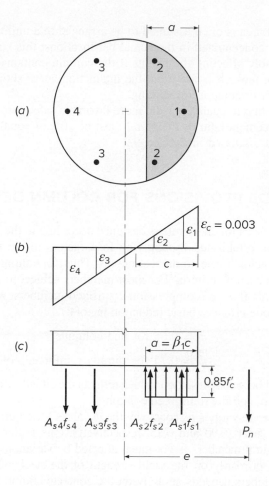

of spirals is greater than the cost of circular ties. A further advantage lies in the fact
that the circular shape is frequently desired by the architect.

Figure 9.14 shows the cross section of a circular-reinforced concrete column.
The reinforcement is as required for a spirally reinforced column, where six or
more longitudinal bars of equal size must be provided for longitudinal reinforcement. The strain distribution at the instant at which the ultimate load is reached
is shown in Fig. 9.14b. Bar groups 2 and 3 are seen to be strained to much smaller
values than groups 1 and 4. The stresses in the four bar groups are easily found.
For any of the bars with strains in excess of yield strain $\varepsilon_y = f_y/E_s$, the stress at
failure is evidently the yield stress of the bar. For bars with smaller strains, the
stress is found from $f_s = \varepsilon_s E_s$.

One then has the internal forces shown in Fig. 9.14c. Note that the situation is
analogous to that discussed in Sections 9.4 to 9.6 for rectangular columns. Calculations for P_n and M_n can be carried out exactly as in Example 9.1, except that for
circular columns the concrete compression zone subject to the equivalent rectangular
stress distribution has the shape of a segment of a circle, shown shaded in Fig. 9.14a.

Although the shape of the compression zone and the strain variation in the
different groups of bars make longhand calculations awkward, no new principles are
involved and computer solutions are easily developed.

Design or analysis of spirally reinforced columns is usually carried out by means
of design aids, such as Graphs A.13 to A.16 of Appendix A. Additional tables and
graphs are available, for example, in Ref. 9.7. In developing such design aids, the

entire steel area is often assumed to be arranged in a uniform, concentric ring, rather than being concentrated in the actual bar locations; this simplifies calculations without noticeably affecting the results if the column contains at least eight longitudinal bars. When fewer bars are used, the interaction curve should be calculated based on the weakest orientation in bending.

Note that to qualify for the more favorable safety provisions for spiral columns, the reinforcement ratio of the spiral must be at least equal to that given by Eq. (9.5) for reasons discussed in Section 9.2.

9.9 ACI CODE PROVISIONS FOR COLUMN DESIGN

For columns, as for all members designed according to the ACI Code, adequate safety margins are established by applying load factors to the service loads and strength reduction factors to the nominal strengths. Thus, for columns, $\phi P_n \geq P_u$ and $\phi M_n \geq M_u$ are the basic safety criteria. For most members subject to axial compression or compression plus flexure (compression-controlled members, as described in Chapter 4), the ACI Code provides basic reduction factors:

$$\phi = 0.65 \text{ for tied columns}$$

$$\phi = 0.75 \text{ for spirally reinforced columns}$$

The spread between these two values reflects the added safety furnished by the greater toughness of spirally reinforced columns.

There are various reasons why the ϕ values for columns are lower than those for flexure or shear (0.90 and 0.75, respectively). One is that the strength of underreinforced flexural members is not much affected by variations in concrete strength, since it depends primarily on the yield strength of the steel, while the strength of axially loaded members depends strongly on the concrete compressive strength. Because the cylinder strength of concrete under site conditions is less closely controlled than the yield strength of mill-produced steel, a larger occasional strength deficiency must be allowed for. This is particularly true for columns, in which concrete, being placed from the top down in a long, narrow form, is more subject to segregation than in horizontally cast beams. Moreover, electrical and other conduits are frequently located in building columns; this reduces their effective cross sections, often to an extent unknown to the designer, even though this is poor practice and restricted by the ACI Code. Finally, the consequences of a column failure, say in a lower story, would be more catastrophic than those of a single beam failure in the same building.

For high eccentricities, as the eccentricity increases from e_b to infinity (pure bending), the ACI Code recognizes that the member behaves progressively more like a flexural member and less like a column. As described in Chapter 4 and shown in Fig. 4.9d, this is acknowledged in ACI Code 21.2.2 by providing a linear transition in ϕ from values of 0.65 and 0.75 to 0.90 as the net tensile strain in the extreme tensile steel ε_t increases from $\varepsilon_{ty} = f_y/E_s$ (which may be taken as 0.002 for Grade 60 reinforcement) to $\varepsilon_{ty} + 0.003$.

Within the transition between tension-controlled and compression-controlled sections:

$$\phi = 0.75 + 0.15 \frac{(\varepsilon_t - \varepsilon_{ty})}{0.003} \text{ for sections with an ACI spiral} \qquad (9.18a)$$

$$\phi = 0.65 + 0.25 \frac{(\varepsilon_t - \varepsilon_{ty})}{0.003} \text{ for other sections} \qquad (9.18b)$$

FIGURE 9.15
ACI safety provisions
superimposed on column
strength interaction diagram.

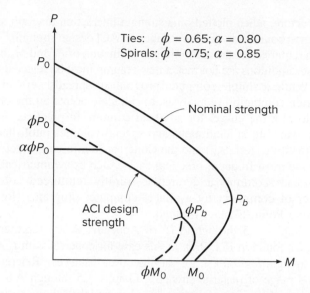

At the other extreme, for columns with very small or zero calculated eccentricities, the ACI Code recognizes that accidental construction misalignments and other unforeseen factors may produce actual eccentricities in excess of these small design values. Also, the concrete strength under high, sustained axial loads may be somewhat smaller than the short-term cylinder strength. Therefore, regardless of the magnitude of the calculated eccentricity, ACI Code 22.4.2 limits the maximum design strength to $0.80\phi P_0$ for tied columns (with $\phi = 0.65$) and to $0.85\phi P_0$ for spirally reinforced columns (with $\phi = 0.75$), where P_0 is the nominal strength of the axially loaded column with zero eccentricity [see Eq. (9.4)].

The effects of the safety provisions of the ACI Code are shown in Fig. 9.15. The solid curve labeled "nominal strength" is the same as Fig. 9.10 and represents the actual carrying capacity, as nearly as can be predicted. The smooth curve shown partially dashed, then solid, then dashed, represents the basic design strength obtained by reducing the nominal strengths P_n and M_n, for each eccentricity, by $\phi = 0.65$ for tied columns and $\phi = 0.75$ for spiral columns. The horizontal cutoff at $\alpha\phi P_0$ represents the maximum design load stipulated in the ACI Code for small eccentricities, that is, large axial loads, as just discussed. At the other end, for large eccentricities, that is, small axial loads, the ACI Code permits a linear transition of ϕ from 0.65 or 0.75, applicable for $\varepsilon_t \leq f_y/E_s$ (or 0.002 for Grade 60 reinforcement) to 0.90 at $\varepsilon_t = \varepsilon_{ty} + 0.003$. By definition, $\varepsilon_t = f_y/E_s$ at the balanced condition. The effect of the transition in ϕ is shown at the lower right end of the design strength curve.[†]

As mentioned in Section 4.5c, the spacing between longitudinal reinforcement in columns must be at least $1\frac{1}{2} d_b$, $1\frac{1}{2}$ in., and $\frac{4}{3}$ the maximum aggregate size.

9.10   DESIGN AIDS

The design of eccentrically loaded columns using the strain compatibility method of analysis described requires that a trial column be selected. The trial column is then investigated to determine if it is adequate to carry any combination of P_u and M_u that may act on it should the structure be overloaded, that is, to see if P_u and M_u from the analysis of

---

[†] While the general intent of the ACI Code safety provisions relating to eccentric columns is clear and fundamentally sound, the end result is a set of strangely shaped column design curves following no discernible physical law, as is demonstrated in Fig. 9.15. Improved column safety provisions, resulting in a smooth design curve appropriately related to the strength curve, would be simpler to use and more rational as well.

the structure, when plotted on a strength interaction diagram as in Fig. 9.15, fall within the region bounded by the curve labeled "ACI design strength." Furthermore, economical design requires that the controlling combination of P_u and M_u be close to the limit curve. If these conditions are not met, a new column must be selected for trial.

While a simple computer program or spreadsheet can be developed, based on the strain compatibility analysis, to calculate points on the design strength curve, and even to plot the curve, for any trial column, in practice, design aids are used such as are available in handbooks and special volumes published by the American Concrete Institute (Ref. 9.7) and the Concrete Reinforcing Steel Institute (Ref. 9.2). They cover the most frequent practical cases, such as symmetrically reinforced rectangular and square columns and circular spirally reinforced columns. There are also a number of commercially available computer programs (for example, spCOLUMN, Structure Point, Skokie, Illinois).

Graphs A.5 through A.16 of Appendix A are representative of column design charts (as found in Ref. 9.7), in this case for concrete with $f_c' = 4000$ psi and steel with yield strength $f_y = 60$ ksi, for varying cover distances.[†] Reference 9.7 includes charts for a broad range of material strengths. Graphs A.5 through A.8 are drawn for rectangular columns with reinforcement distributed around the column perimeter; Graphs A.9 through A.12 are for rectangular columns with reinforcement along two opposite faces. Circular columns with bars in a circular pattern are shown in Graphs A.13 through A.16.

The graphs consist of nominal strength interaction curves of the type shown in Fig. 9.15. However, instead of plotting P_n versus M_n, corresponding parameters have been used to make the charts more generally applicable, that is, load is plotted as $K_n = P_n/(f_c'A_g)$, while moment is expressed as $R_n = P_ne/(f_c'A_gh)$. Families of curves are drawn for various values of $\rho_g = A_{st}/A_g$ between 0.01 and 0.08. The graphs also include radial lines representing different eccentricity ratios e/h, as well as lines representing different ratios of stress f_s/f_y or values of strain $\varepsilon_t = 0.002$ and 0.005 in the extreme tension steel.

Charts such as these permit the direct design of eccentrically loaded columns throughout the common range of strength and geometric variables. They may be used in one of two ways as follows. For a given factored load P_u and equivalent eccentricity $e = M_u/P_u$:

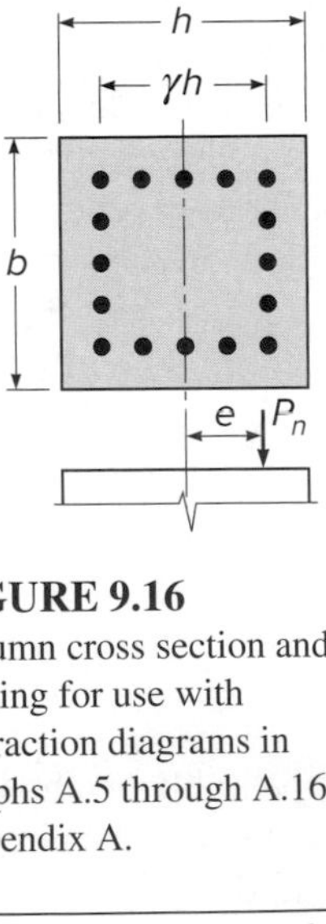

FIGURE 9.16
Column cross section and loading for use with interaction diagrams in Graphs A.5 through A.16 in Appendix A.

1. **(a)** Select trial cross-sectional dimensions b and h (refer to Fig. 9.16).
    (b) Calculate the ratio γ (see Fig. 9.16) based on required cover distances to the bar centroids, and select the corresponding column design chart.
    (c) Calculate $K_n = P_u/(\phi f_c'A_g)$ and $R_n = P_ue/(\phi f_c'A_gh)$, where $A_g = bh$.
    (d) From the graph, for the values found in (c), read the required reinforcement ratio ρ_g.
    (e) Calculate the total steel area $A_{st} = \rho_gbh$.
2. **(a)** Select the reinforcement ratio ρ_g.
    (b) Choose a trial value of h and calculate e/h and γ.
    (c) From the corresponding graph, read $K_n = P_u/(\phi f_c'A_g)$ and calculate the required A_g.
    (d) Calculate $b = A_g/h$.
    (e) Revise the trial value of h if necessary to obtain a well-proportioned section.
    (f) Calculate the total steel area $A_{st} = \rho_gbh$.

[†] Graphs A.5 through A.16 were developed for the specific bar configurations shown on the graphs. The curves exhibit changes in curvature, especially apparent near the balanced load, that result when bars within the cross section yield. The values provided in the graphs, however, are largely insensitive to the exact number of bars in the cross section and may be used for columns with similar bar configurations, but with smaller or larger numbers of bars.

Use of the column design charts will be illustrated in Examples 9.3 and 9.4.

Other design aids pertaining to ties and spirals, as well as recommendations for standard practice, will be found in Refs. 9.2 and 9.7.

EXAMPLE 9.3   **Selection of reinforcement for column of given size.**   In a three-story structure, an exterior column is to be designed for a service dead load of 222 kips, maximum live load of 297 kips, dead load moment of 136 ft-kips, and live load moment of 194 ft-kips. The minimum live load compatible with the full live load moment is 166 kips, obtained when no live load is placed on the roof but a full live load is placed on the second floor. Architectural considerations require that a rectangular column be used, with dimensions $b = 20$ in. and $h = 25$ in.

(a) Find the required column reinforcement for the condition that the full live load acts.

(b) Check to ensure that the column is adequate for the condition of no live load on the roof. Material strengths are $f_c' = 4000$ psi and $f_y = 60,000$ psi.

SOLUTION.

(a) The column will be designed initially for full load, then checked for adequacy when live load is partially removed. According to the ACI safety provisions, the column must be designed for a factored load $P_u = 1.2 \times 222 + 1.6 \times 297 = 742$ kips and a factored moment $M_u = 1.2 \times 136 + 1.6 \times 194 = 474$ ft-kips. A column 20×25 in. is specified, and reinforcement distributed around the column perimeter will be used. Bar cover is estimated to be 2.5 in. from the column face to the steel centerline for each bar. The column parameters (assuming bending about the strong axis) are

$$K_n = \frac{P_u}{\phi f_c' A_g} = \frac{742}{0.65 \times 4 \times 500} = 0.570$$

$$R_n = \frac{M_u}{\phi f_c' A_g h} = \frac{474 \times 12}{0.65 \times 4 \times 500 \times 25} = 0.175$$

With 2.5 in. cover, the parameter $\gamma = (25 - 5)/25 = 0.80$. For this column geometry and material strengths, Graph A.7 of Appendix A applies. From that figure, with the calculated values of K_n and R_n, $\rho_g = 0.024$. Thus, the required reinforcement is $A_{st} = 0.024 \times 500 = 12.00$ in^2. Twelve No. 9 (No. 29) bars will be used, one at each corner and two evenly spaced along each face of the column, providing $A_{st} = 12.00$ in^2.

(b) With the roof live load absent, the column will carry a factored load $P_u = 1.2 \times 222 + 1.6 \times 166 = 532$ kips and factored moment $M_u = 474$ ft-kips, as before. Thus, the column parameters for this condition are

$$K_n = \frac{P_u}{\phi f_c' A_g} = \frac{532}{0.65 \times 4 \times 500} = 0.409$$

$$R_n = \frac{M_u}{\phi f_c' A_g h} = \frac{474 \times 12}{0.65 \times 4 \times 500 \times 25} = 0.175$$

and $\gamma = 0.80$ as before. From Graph A.7 it is found that a reinforcement ratio of $\rho_g = 0.017$ is sufficient for this condition, less than that required in part (a), so no modification is required.

Selecting No. 3 (No. 10) ties for trial, the maximum tie spacing must not exceed $48 \times 0.375 = 18$ in., $16 \times 1.128 = 18.05$ in., or 20 in. Spacing is controlled by the diameter of the ties, and No. 3 (No. 10) ties will be used at 18 in. spacing, in the pattern shown in Fig. 9.3d.

EXAMPLE 9.4   **Selection of column size for a given reinforcement ratio.**   A column is to be designed to carry a factored load $P_u = 481$ kips and factored moment $M_u = 492$ ft-kips. Material strengths $f_y = 60,000$ psi and $f'_c = 4000$ psi are specified. Cost studies for the particular location indicate that a reinforcement ratio ρ_g of about 0.03 is optimum. Find the required dimensions b and h of the column. Bending will be about the strong axis, and an arrangement of steel with bars concentrated in two layers, adjacent to the outer faces of the column and parallel to the axis of bending, will be used.

SOLUTION.   It is convenient to select a trial column dimension h, perpendicular to the axis of bending; a value of $h = 25$ in. will be selected, and assuming a concrete cover of 2.5 in. to the bar centers, the parameter $\gamma = 0.80$. Graph A.11 of Appendix A applies. For the stated loads, the eccentricity is $e = 492 \times 12/481 = 12.3$ in., and $e/h = 12.3/25 = 0.49$. From Graph A.11 with $e/h = 0.49$ and $\rho_g = 0.03$, $K_n = P_w/\phi f'_c A_g = 0.51$. For the trial dimension $h = 25$ in., the required column width is

$$b = \frac{P_u}{\phi f'_c K_n h} = \frac{481}{0.65 \times 4 \times 0.51 \times 25} = 14.5 \text{ in.}$$

A column 15×25 in. will be used, for which the required steel area is $A_{st} = 0.03 \times 15 \times 25 = 11.25$ in.^{2}. Eight No. 11 (No. 36) bars will be used, providing $A_{st} = 12.48$ in.^{2}, arranged in two layers of four bars each, similar to the sketch shown in Graph A.11.

9.11   BIAXIAL BENDING

The methods discussed in the preceding sections permit rectangular or square columns to be designed if bending is present about only one of the principal axes. There are situations, by no means exceptional, in which axial compression is accompanied by simultaneous bending about both principal axes of the section. Such is the case, for instance, in corner columns of buildings where beams and girders frame into the columns in the directions of both walls and transfer their end moments into the columns in two perpendicular planes. Similar loading may occur at interior columns, particularly if the column layout is irregular.

The situation with respect to strength of biaxially loaded columns is shown in Fig. 9.17. Let X and Y denote the directions of the principal axes of the cross section. In Fig. 9.17a, the section is shown subject to bending about the Y axis only, with load eccentricity e_x measured in the X direction. The corresponding strength interaction curve is shown as case (a) in the three-dimensional sketch in Fig. 9.17d and is drawn in the plane defined by the axes P_n and M_{ny}. Such a curve can be established by the usual methods for uniaxial bending. Similarly, Fig. 9.17b shows bending about the X axis only, with eccentricity e_y measured in the Y direction. The corresponding interaction curve is shown as case (b) in the plane of P_n and M_{nx} in Fig. 9.17d. For case (c), which combines X and Y axis bending, the orientation of the resultant eccentricity is defined by the angle λ:

$$\lambda = \arctan \frac{e_x}{e_y} = \arctan \frac{M_{ny}}{M_{nx}}$$

Bending for this case is about an axis defined by the angle θ with respect to the X axis. The angle λ in Fig. 9.17c establishes a plane in Fig. 9.17d, passing through the vertical P_n axis and making an angle λ with the M_{nx} axis, as shown. In that plane,

FIGURE 9.17

Interaction diagram for compression plus biaxial bending: (*a*) uniaxial bending about *Y* axis; (*b*) uniaxial bending about *X* axis; (*c*) biaxial bending about diagonal axis; and (*d*) interaction surface.

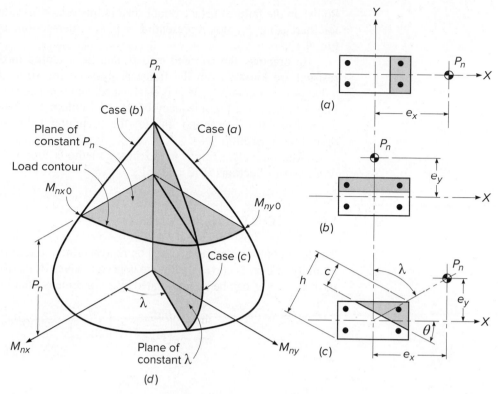

column strength is defined by the interaction curve labeled case (*c*). For other values of λ, similar curves are obtained to define a *failure surface* for axial load plus biaxial bending, as shown in Fig. 9.17*d*. The surface is exactly analogous to the *interaction curve* for axial load plus uniaxial bending. Any combination of P_u, M_{ux}, and M_{uy} falling inside the surface can be applied safely, but any point falling outside the surface would represent failure. Note that the failure surface can be described either by a set of curves defined by radial planes passing through the P_n axis, as shown by case (*c*), or by a set of curves defined by horizontal plane intersections, each for a constant P_n, defining load contours.

Constructing such an interaction surface for a given column would appear to be an obvious extension of uniaxial bending analysis. In Fig. 9.17*c*, for a selected value of θ, successive choices of neutral axis distance *c* could be taken. For each, using strain compatibility and stress-strain relations to establish bar forces and the concrete compressive resultant, then using the equilibrium equations to find P_n, M_{nx}, and M_{ny}, one can determine a single point on the interaction surface. Repetitive calculations, easily done by computer, then establish sufficient points to define the surface. The triangular or trapezoidal compression zone, as shown in Fig. 9.17*c*, is a complication, and in general the strain in each reinforcing bar will be different, but these features can be incorporated.

The main difficulty, however, is that the neutral axis will not, in general, be perpendicular to the resultant eccentricity, drawn from the column center to the load P_n. For each successive choice of neutral axis, there are unique values of P_n, M_{nx}, and M_{ny}, and only for special cases will the ratio of M_{ny}/M_{nx} be such that the eccentricity is perpendicular to the neutral axis chosen for the calculation. The result is that, for successive choices of *c* for any given θ, the value of λ in Fig. 9.17*c* and *d* will vary.

Points on the failure surface established in this way will wander up the failure surface for increasing P_n, not representing a plane intersection, as shown for case (c) in Fig. 9.17d.

In practice, the factored load P_u and the factored moments M_{ux} and M_{uy} to be resisted are known from the frame analysis of the structure. Therefore, the actual value of $\lambda = \arctan(M_{uy}/M_{ux})$ is established, and one needs only the curve of case (c), Fig. 9.17d, to test the adequacy of the trial column. An iterative computer method to establish the interaction line for the particular value of λ that applies will be described in Section 9.14.

Alternatively, simple approximate methods are widely used. These will be described in Sections 9.12 and 9.13.

9.12   LOAD CONTOUR METHOD

The load contour method is based on representing the failure surface of Fig. 9.17d by a family of curves corresponding to constant values of P_n (Ref. 9.8). The general form of these curves can be approximated by a nondimensional interaction equation

$$\left(\frac{M_{nx}}{M_{nx0}}\right)^{\alpha 1} + \left(\frac{M_{ny}}{M_{ny0}}\right)^{\alpha 2} = 1.0 \tag{9.19}$$

where

$$M_{nx} = P_n e_y$$

$$M_{nx0} = M_{nx} \qquad \text{when } M_{ny} = 0$$

$$M_{ny} = P_n e_x$$

$$M_{ny0} = M_{ny} \qquad \text{when } M_{nx} = 0$$

and α_1 and α_2 are exponents depending on column dimensions, amount and distribution of steel reinforcement, stress-strain characteristics of steel and concrete, amount of concrete cover, and size of transverse ties or spiral. When $\alpha_1 = \alpha_2 = \alpha$, the shapes of such interaction contours are as shown in Fig. 9.18 for specific α values.

Calculations reported by Bresler in Ref. 9.9 indicate that α falls in the range from 1.15 to 1.55 for square and rectangular columns. Values near the lower end of that range are the more conservative. Methods and design aids permitting a more defined estimation of α are found in Ref. 9.7.

In practice, the values of P_u, M_{ux}, and M_{uy} are known from the analysis of the structure. For a trial column section, the values of M_{nx0} and M_{ny0} corresponding to the load P_u/ϕ can easily be found by the usual methods for uniaxial bending. Then replacing M_{nx} with M_{ux}/ϕ and M_{ny} with M_{uy}/ϕ and using $\alpha_1 = \alpha_2 = \alpha$ in Eq. (9.19), or alternatively by plotting $(M_{nx}/\phi)/M_{nx0}$ and $(M_{ny}/\phi)/M_{ny0}$ in Fig. 9.18, it can be confirmed that a particular combination of factored moments falls within the load contour (safe design) or outside the contour (failure), and the design modified if necessary.

An approximate approach to the load contour method, in which the curved load contour is represented by a bilinear approximation, will be found in Ref. 9.10. It leads to a method of *trial design* in which the biaxial bending moments are represented by an equivalent uniaxial bending moment. Design charts based on this approximate approach will be found in the *ACI Design Manual* (Ref. 9.7). Trial designs arrived at in this way should be checked for adequacy by the load contour method, described above, or by the method of reciprocal loads that follows.

FIGURE 9.18

Interaction contours at constant P_n for varying α. (*Adapted from Ref. 9.8.*)

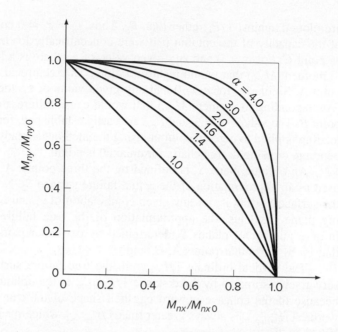

9.13 RECIPROCAL LOAD METHOD

A simple, approximate design method developed by Bresler (Ref. 9.9) has been satisfactorily verified by comparison with results of extensive tests and accurate calculations (Ref. 9.11). It is noted that the column interaction surface in Fig. 9.17d can, alternatively, be plotted as a function of the axial load P_n and eccentricities $e_x = M_{ny}/P_n$ and $e_y = M_{nx}/P_n$, as is shown in Fig. 9.19a. The surface S_1 of Fig. 9.19a can be transformed into an equivalent failure surface S_2, as shown in Fig. 9.19b, where e_x and e_y

FIGURE 9.19

Interaction surfaces for the reciprocal load method.

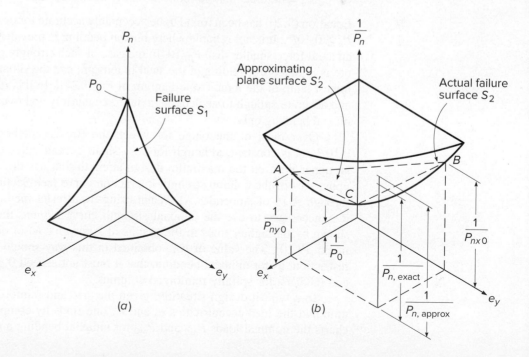

are plotted against $1/P_n$ rather than P_n. Thus, $e_x = e_y = 0$ corresponds to the inverse of the capacity of the column if it were concentrically loaded P_0, and this is plotted as point C. For $e_y = 0$ and any given value of e_x, there is a load P_{ny0} (corresponding to moment M_{ny0}) that would result in failure. The reciprocal of this load is plotted as point A. Similarly, for $e_x = 0$ and any given value of e_y, there is a certain load P_{nx0} (corresponding to moment M_{nx0}) that would cause failure, the reciprocal of which is point B. The values of P_{nx0} and P_{ny0} are easily established, for known eccentricities of loading applied to a given column, using the methods already established for uniaxial bending, or using design charts for uniaxial bending.

An oblique plane S_2' is defined by the three points: A, B, and C. This plane is used as an approximation of the actual failure surface S_2. Note that, for any point on the surface S_2 (that is, for any given combination of e_x and e_y), there is a corresponding plane S_2'. Thus, the approximation of the true failure surface S_2 involves an infinite number of planes S_2' determined by particular pairs of values of e_x and e_y, that is, by particular points A, B, and C.

The vertical ordinate $1/P_{n,\text{exact}}$ to the true failure surface will always be conservatively estimated by the distance $1/P_{n,\text{approx}}$ to the oblique plane ABC (extended), because of the concave upward eggshell shape of the true failure surface. In other words, $1/P_{n,\text{approx}}$ is always greater than $1/P_{n,\text{exact}}$, which means that $P_{n,\text{approx}}$ is always less than $P_{n,\text{exact}}$.

Bresler's reciprocal load equation derives from the geometry of the approximating plane. It can be shown that

$$\frac{1}{P_n} = \frac{1}{P_{nx0}} + \frac{1}{P_{ny0}} - \frac{1}{P_0} \tag{9.20}$$

where P_n = approximate value of nominal load in biaxial bending with eccentricities e_x and e_y
P_{ny0} = nominal load when only eccentricity e_x is present ($e_y = 0$)
P_{nx0} = nominal load when only eccentricity e_y is present ($e_x = 0$)
P_0 = nominal load for concentrically loaded column

Equation (9.20) has been found to be acceptably accurate for design purposes provided $P_n \geq 0.10P_0$. It is not reliable where biaxial bending is prevalent and accompanied by an axial force smaller than $P_0/10$. In the case of such strongly prevalent bending, failure is initiated by yielding of the steel in tension, and the situation corresponds to the lowest tenth of the interaction diagram of Fig. 9.17d. In this range, it is conservative and accurate enough to neglect the axial force entirely and to calculate the section for biaxial bending only.

Over most of the range for which the Bresler method is applicable, above $0.10P_0$, ϕ is constant, although for very small eccentricities the ACI Code imposes an upper limit on the maximum design strength that has the effect of flattening the upper part of the column strength interaction curve (see Section 9.9 and Graphs A.5 through A.16 of Appendix A). When using the Bresler method for biaxial bending, it is necessary to use the uniaxial strength curve *without* the horizontal cutoff (as shown by the lighter lines in the graphs of Appendix A) in obtaining values for use in Eq. (9.20). The value of ϕP_n obtained in this way should then be subject to the restriction, as for uniaxial bending, that it must not exceed $0.80\phi P_0$ for tied columns and $0.85\phi P_0$ for spirally reinforced columns.

In a typical design situation, given the size and reinforcement of the trial column and the load eccentricities e_y and e_x, one finds by computation or from design charts the nominal loads P_{nx0} and P_{ny0} for uniaxial bending around the X and Y axes, respectively, and the nominal load P_0 for concentric loading. Then $1/P_n$ is computed from Eq. (9.20), and from that P_n is calculated. The design requirement is that the factored load P_u not exceed ϕP_n, as modified by the horizontal cutoff mentioned above, if applicable.

EXAMPLE 9.5   **Design of column for biaxial bending.** The 12×20 in. column shown in Fig. 9.20 is reinforced with eight No. 9 (No. 29) bars arranged around the column perimeter, providing an area $A_{st} = 8.00$ in². A factored load P_u of 255 kips is to be applied with eccentricities $e_y = 3$ in. and $e_x = 6$ in., as shown. Material strengths are $f_c' = 4$ ksi and $f_y = 60$ ksi. Check the adequacy of the trial design (a) using the reciprocal load method and (b) using the load contour method.

SOLUTION.

(a) By the reciprocal load method, first considering bending about the Y axis, $\gamma = 15/20 = 0.75$ and $e/h = 6/20 = 0.30$. With the reinforcement ratio of $A_{st}/bh = 8.00/240 = 0.033$, using the average of Graphs A.6 ($\gamma = 0.70$) and A.7 ($\gamma = 0.80$),

$$\frac{P_{ny0}}{f_c' A_g} \text{ (avg)} = \frac{0.62 + 0.66}{2} = 0.64 \qquad P_{ny0} = 0.64 \times 4 \times 240 = 614 \text{ kips}$$

$$\frac{P_0}{f_c' A_g} = 1.31 \qquad\qquad P_0 = 1.31 \times 4 \times 240 = 1258 \text{ kips}$$

Then for bending about the X axis, $\gamma = \frac{7}{12} = 0.58$ (say 0.60) and $e/h = \frac{3}{12} = 0.25$. Graph A.5 of Appendix A gives

$$\frac{P_{nx0}}{f_c' A_g} = 0.65 \qquad P_{nx0} = 0.65 \times 4 \times 240 = 624 \text{ kips}$$

$$\frac{P_0}{f_c' A_g} = 1.31 \qquad P_0 = 1.31 \times 4 \times 240 = 1258 \text{ kips}$$

Substituting these values in Eq. (9.20) results in

$$\frac{1}{P_n} = \frac{1}{624} + \frac{1}{614} - \frac{1}{1258} = 0.00244$$

from which $P_n = 410$ kips. Thus, according to the Bresler method, the design load of $P_u = 0.65 \times 410 = 267$ kips can be applied safely.

FIGURE 9.20
Column cross section for
Example 9.5.

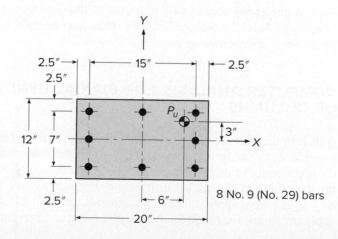

8 No. 9 (No. 29) bars

(b) By the load contour method, for Y axis bending with $P_u/(\phi f_c' A_g) = 255/(0.65 \times 4 \times 240) = 0.41$. The average from Graphs A.6 and A.7 of Appendix A is

$$\frac{M_{ny0}}{f_c' A_g h} \text{ (avg)} = \frac{0.212 + 0.235}{2} = 0.224$$

Hence, $M_{ny0} = 0.224 \times 4 \times 240 \times 20 = 4300$ in-kips. Then for X axis bending, with $P_u/(\phi f_c' A_g) = 0.41$, as before, from Graph A.5,

$$\frac{M_{nx0}}{f_c' A_g h} = 0.186$$

So $M_{nx0} = 0.186 \times 4 \times 240 \times 12 = 2140$ in-kips. The factored load moments about the Y and X axes, respectively, are

$$M_{uy} = 255 \times 6 = 1530 \text{ in-kips}$$

$$M_{ux} = 255 \times 3 = 765 \text{ in-kips}$$

Adequacy of the trial design will now be checked using Eq. (9.19) with an exponent α conservatively taken equal to 1.15. Then with $M_{nx} = M_{ux}/\phi$ and $M_{ny} = M_{uy}/\phi$, that equation indicates

$$\left(\frac{765/0.65}{2140}\right)^{1.15} + \left(\frac{1530/0.65}{4300}\right)^{1.15} = 0.502 + 0.500 = 1.002$$

This is close enough to 1.0 that the design would be considered safe by the load contour method also.

In actual practice, the values of α used in Eq. (9.19) should be checked, for the specific column, because predictions of that equation are quite sensitive to changes in α. In Ref. 9.10, it is shown that $\alpha = \log 0.5/\log \beta$, where values of β can be tabulated for specific column geometries, material strengths, and load ranges (see Ref. 9.7). For the present example, it can be confirmed from Ref. 9.7 that $\beta = 0.56$ and hence $\alpha = 1.19$, approximately as chosen.

One observes that, in Example 9.5a, an eccentricity in the Y direction equal to 50 percent of that in the X direction causes a reduction in nominal capacity of 33 percent, that is, from 614 to 410 kips. For cases in which the ratio of eccentricities is smaller, there is some justification for the frequent practice in framed structures of neglecting the bending moments in the direction of the smaller eccentricity. *In general, biaxial bending should be taken into account when the estimated eccentricity ratio approaches or exceeds 0.2.*

9.14 COMPUTER ANALYSIS FOR BIAXIAL BENDING OF COLUMNS

Although the load contour method and the reciprocal load method are widely used in practice, each has serious shortcomings. With the load contour method, selection of the appropriate value of the exponent α is made difficult by a number of factors relating to column shape and bar distribution. For many cases, the usual assumption that $\alpha_1 = \alpha_2$ is a poor approximation. Design aids are available, but they introduce further approximations, for example, the use of a bilinear representation of the load contour.

The reciprocal load method is very simple to use, but the representation of the curved failure surface by an approximating plane is not reliable in the range of large eccentricities, where failure is initiated by steel yielding.

With the general availability and wide use of computers, it is better to use simpler methods to obtain faster, and more exact, solutions to the biaxial column problem. Such a method is that developed by Ehsani (Ref. 9.12). A column strength interaction curve is established for a trial column, exactly analogous to the curve for axial load plus uniaxial bending, as described in Sections 9.3 to 9.7. However, the curve is generated for the particular value of the eccentricity angle that applies, as determined by the ratio of M_{uy}/M_{ux} from the structural frame analysis [see case (c) of Fig. 9.17d]. This is done by taking successive choices of neutral axis distance, measured in this case along one face of the column from the most heavily compressed corner, from very small (large eccentricity) to very large (small eccentricity), then calculating the axial force P_n and moments M_{nx} and M_{ny}. For each neutral axis distance, iteration is performed with successive values of the orientation angle θ, Fig. 9.17c, until $\lambda = \arctan (M_{ny}/M_{nx})$ is in agreement with the value of $\lambda = \arctan (M_{uy}/M_{ux})$ from the structural frame analysis. Thus, one point on curve (c) of Fig. 9.17d is established. The sequence of calculations is repeated: another choice of neutral axis distance is made, a value of θ is selected, the axial force and moments are calculated, λ is found, and the value of θ is iterated until λ is correct. Thus, the next point is established, and so on, until the complete strength interaction curve for that particular value of λ is complete. ACI Code safety provisions may then be imposed in the usual way, and the adequacy of the proposed design tested, for the known load and moments, against the design strength curve for the trial column.

The method is obviously impractical for manual calculation, but the iterative steps are easily and quickly performed by computer, which can also provide a graphical presentation of results. Full details will be found in Ref. 9.12.

A number of computer programs for biaxial bending are available commercially, such as spCOLUMN (Structure Point, Skokie, Illinois).

9.15   BAR SPLICING IN COLUMNS AND TIES NEAR BEAM-COLUMN JOINTS

The main vertical reinforcement in columns is usually spliced just above each floor, or sometimes at alternate floors. This permits the column steel area to be reduced progressively at the higher levels in a building, where loads are smaller, and in addition avoids handling and supporting very long column bars. Column steel may be spliced by lapping, butt welding, various types of mechanical connections, or direct end bearing, using special devices to ensure proper alignment of bars.

Special attention must be given to the problem of bar congestion at splices. Lapping the bars, for example, effectively doubles the steel area in the column cross section at the level of the splice and can result in problems placing concrete and meeting the ACI Code requirement for minimum lateral spacing of bars ($1.5d_b$ or 1.5 in.). To avoid difficulty, column steel percentages are often limited in practice to not more than about 4 percent, or the bars are extended two stories and staggered splices are used.

The most common method of splicing column steel is the simple lapped bar splice, with the bars in contact throughout the lapped length. It is standard practice to offset the lower bars, as shown in Fig. 9.21, to permit the proper positioning of the upper bars. To prevent outward buckling of the bars at the bottom bend point of such an offset, with spalling of the concrete cover, it is necessary to provide special transverse reinforcement

FIGURE 9.21
Splice details at typical interior column. Beams frame into joint from four directions.

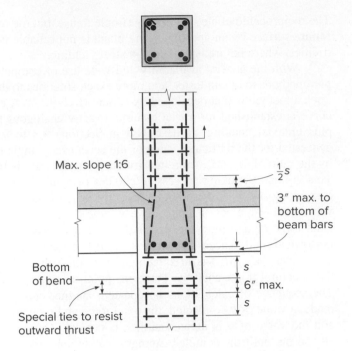

in the form of extra ties. According to ACI Code 10.7.4.1, the slope of the inclined part of an offset bar must not exceed 1 in 6. According to ACI Code 10.7.6.4, the transverse steel must be provided to resist $1\frac{1}{2}$ times the horizontal component of the computed force in the inclined part of the offset bar, and this special reinforcement must be placed not more than 6 in. from the point of bend, as shown in Fig. 9.21.

Elsewhere in the column, above and below the floor, the usual spacing requirements described in Section 9.2 apply, except that ties must be located not more than one-half the normal spacing s above the floor and must be located not farther than one-half s below the lowest horizontal reinforcement in the slab, drop panel, or shear cap, according to ACI Code 10.7.6.2. Where beams frame from four directions into a joint, as shown in Fig. 9.21, ties may be terminated within the beam-column joint. In this case, the top tie in a column must be located not more than 3 in. below the lowest reinforcement in the shallowest beam or bracket. As a result of these requirements, if beams are not present on four sides, such as for exterior columns, ties must be placed vertically at the usual spacing through the depth of the joint up to a level not more than one-half the usual spacing s below the lowest reinforcement in the slab.

Analogous requirements are found in ACI Code 10.7.6.3 and are illustrated in Ref. 9.1 for spirally reinforced columns.

As discussed in Section 6.13, in frames subjected to lateral loading, a viable alternative to splicing bars just above the floor is to splice them in the center half of the column height, where the moment due to lateral loading is much lower than at floor level. Splicing near midheight is mandatory in "special moment frames" designed for seismic loading (Chapter 20). The use of midheight splices removes the requirement for the special ties shown in Fig. 9.21 because bent bars are not used.

Column splices are mainly compression splices, although load combinations producing moderate to large eccentricity require that splices transmit tension as well. ACI Code 10.7.5 permits splicing by lapping, butt welding, mechanical connectors, or end bearing. As discussed in Section 6.13, the length of compression lap splices may be reduced in cases where ties or spiral reinforcement throughout the lap length meets specific requirements. If the column bars are in tension, Class A tension lap splices are permitted if the tensile stress does not exceed $0.5f_y$ and less than one-half of the bars are spliced at any section. Class B tension splices are required if the tensile stresses are higher than $0.5f_y$ under factored loads or where more than one-half of the reinforcement is spliced at one location. When end bearing splices are used, they must be staggered or additional reinforcement must be added so that the continuing bars on each column face possess a tensile strength not less than $0.25f_y$ times the area of the vertical reinforcement on that face, according to ACI Code 10.7.5.3.

Full requirements for both compression and tension lap splices are discussed in Section 6.13, and the design of a compression splice in a typical column is illustrated in Example 6.5.

9.16 TRANSMISSION OF COLUMN LOADS THROUGH FLOOR SYSTEMS

Quite often, the specified compressive strength of the concrete in columns will exceed that of the floor system. This is especially true for the lower stories in high-rise buildings, where high-strength concrete is used to minimize the cross-sectional area of the columns and thus maximize the usable floor space. High-strength concrete, however, is not needed for the beams and slabs that make up the floor system.

Floor systems and columns must be cast in separate placements. This is not only good construction practice to allow the concrete in the columns to settle prior to placement of the floor system but also required by ACI Code 26.5.7.2 to prevent cracking at the interface between the floor and the column that would occur if the floor and supporting members were cast at the same time. This standard practice, however, opens the possibility for placement of lower-strength concrete within the portion of a floor system that directly supports the columns above, which would, in turn, significantly reduce their capacity. The high lateral confinement provided by the floor system to the concrete in the vicinity of the column does have a mitigating effect because it places that region in triaxial compression and thus increases its usable compressive strength, as explained in Section 2.10.

To address the effects on performance of using concretes with significantly different compressive strengths in the columns and floor system, ACI Code 15.5 specifies that if f_c' of the floor system is less than $0.7f_c'$, of the column, one of three requirements must be met:

1. At the time of concrete placement in the floor system, concrete with the strength specified for the column must be placed in the floor at the column location. The concrete must extend at least 2 ft into the floor system from the face of the column for the full depth of the floor system and be integrated with the floor concrete.
2. The strength of the column through the floor system must be based on the lower compressive strength of the floor concrete. Additional reinforcement may be required.
3. For beam-column joints that are laterally supported on four sides by beams of approximately the same depth and at least three-quarters of the width of the column or slab-column joints that are supported by slabs on four sides, the strength of the column may be based on a compressive strength equal to 75 percent of the column concrete strength plus 35 percent of the floor concrete strength. The ratio of the column concrete strength to the slab concrete strength may not be taken greater than 2.5 for use in design.

9.17   SHEAR IN COLUMNS

Columns are subjected to shear due to lateral load—for example, from wind or earthquakes—as well as from bending moments transferred by beams and slabs at joints. The design of columns subjected to shear in a single direction is handled as described in Section 5.6 for members under axial load, with the column ties serving as shear reinforcement. As such, the ties must meet all requirements for shear reinforcement described in Section 5.5.

When structures are subjected to lateral load due to wind or earthquake, or columns are subjected to bending moments from orthogonal flexural members, however, columns will be subjected to biaxial shear loading. According to ACI Code 22.5.1.10, the effects of biaxial shear may be neglected if either Eq. (9.21a) or Eq. (9.21b) is satisfied:

$$\frac{V_{u,x}}{\phi V_{n,x}} \leq 0.5 \tag{9.21a}$$

$$\frac{V_{u,y}}{\phi V_{n,y}} \leq 0.5 \tag{9.21b}$$

where $V_{u,x}$ and $V_{u,y}$ are, respectively, the factored shear forces in the x and y directions, and $V_{n,x}$ and $V_{n,y}$ are, respectively, the shear strengths in the x and y directions.

If $V_{u,x}/(\phi V_{n,x}) > 0.5$ and $V_{u,y}/(\phi V_{n,y}) > 0.5$, ACI Code 22.5.1.11 requires that

$$\frac{V_{u,x}}{\phi V_{n,x}} + \frac{V_{u,y}}{\phi V_{n,y}} \leq 1.5 \tag{9.22}$$

If needed, shear reinforcement may be added to satisfy Eq. (9.22).

REFERENCES

9.1.   *ACI Detailing Manual*, SP-66, American Concrete Institute, Farmington Hills, MI, 2004.

9.2.   *CRSI Design Handbook*, 10th ed., Concrete Reinforcing Steel Institute, Schaumburg, IL, 2008.

9.3.   F. E. Richart, A. Brandtzaeg, and R. L. Brown, "A Study of the Failure of Concrete under Combined Compressive Stresses," *Univ. Ill. Eng. Exp. Stn. Bull.* 185, 1928.

9.4.   F. E. Richart, A. Brandtzaeg, and R. L. Brown, "The Failure of Plain and Spirally Reinforced Concrete in Compression," *Univ. Ill. Eng. Exp. Stn. Bull.* 190, 1929.

9.5.   S. Martinez, A. H. Nilson, and F. O. Slate, "Spirally Reinforced High Strength Concrete Columns," *J. ACI*, vol. 81, no. 5, 1984, pp. 431–442.

9.6.   A. H. Mattock, L. B. Kriz, and E. Hognestad, "Rectangular Concrete Stress Distribution in Ultimate Strength Design," *J. ACI*, vol. 32, no. 8, 1961, pp. 875–928.

9.7.   *The Reinforced Concrete Design Manual in Accordance with ACI 318-11*, SP-17(11), American Concrete Institute, Farmington Hills, MI, 2012.

9.8.   F. N. Pannell, "Failure Surfaces for Members in Compression and Biaxial Bending," *J. ACI*, vol. 60, no. 1, 1963, pp. 129–140.

9.9.   B. Bresler, "Design Criteria for Reinforced Columns under Axial Load and Biaxial Bending," *J. ACI*, vol. 32, no. 5, 1960, pp. 481–490.

9.10.   A. L. Parme, J. M. Nieves, and A. Gouwens, "Capacity of Reinforced Concrete Rectangular Members Subject to Biaxial Bending," *J. ACI*, vol. 63, no. 9, 1966, pp. 911–923.

9.11.   L. N. Ramamurthy, "Investigation of the Ultimate Strength of Square and Rectangular Columns under Biaxially Eccentric Loads," in *Symp. Reinforced Concrete Columns*, SP-13, American Concrete Institute, Detroit, MI, 1966, pp. 263–298.

9.12.   M. R. Ehsani, "CAD for Columns," *Concr. Intl.*, vol. 8, no. 9, 1986, pp. 43–47.

PROBLEMS

9.1. A 16 in. square column is reinforced with four No. 11 (No. 36) bars, one in each corner, with cover distances 3 in. to the steel center in each direction. Material strengths are $f_c' = 5000$ psi and $f_y = 60,000$ psi. Construct the interaction diagram relating axial strength P_n and flexural strength M_n. Bending will be about an axis parallel to one face. Calculate the coordinates for P_o, P_b, and at least three other representative points on the curve.

9.2. Starting with the column in Problem 9.1, perform enough additional calculations to determine the effects of increasing f_c' from 5000 to 8000 psi on column capacity at both high and low axial loads. Assuming that a compressive strength of 8000 psi is appropriate for the lower stories of a high-rise structure, would you recommend using concrete with $f_c' = 8000$ psi for the columns supporting all stories within the building? Use your analysis to support your answer.

9.3. Plot the design strength curve relating ϕP_n and ϕM_n for the column of Problem 9.1. Design and detail the tie steel required by the ACI Code. Is the column a good choice to resist a load $P_u = 540$ kips applied with an eccentricity $e = 4.44$ in.?

9.4. The short column shown in Fig. P9.4 will be subjected to an eccentric load causing uniaxial bending about the Y axis. Material strengths are $f_y = 60$ ksi and $f_c' = 4$ ksi.

FIGURE P9.4

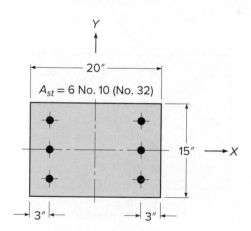

(a) Construct the nominal strength interaction curve for this column, calculating no fewer than five points, including those corresponding to pure bending, pure axial thrust, and balanced failure.

(b) Compare the calculated values with those obtained using Graph A.10 in Appendix A.

(c) Show on the same drawing the design strength curve obtained through introduction of the ACI ϕ factors.

(d) Design the transverse reinforcement for the column, giving key dimensions for ties.

9.5. The column shown in Fig. P9.5 is subjected to axial load and bending moment, causing bending about an axis parallel to that of the rows of bars. What moment M_n would cause the column to fail if the axial load P_n applied simultaneously was 1250 kips? Material strengths are $f_c' = 4000$ psi and $f_y = 60$ ksi.

FIGURE P9.5

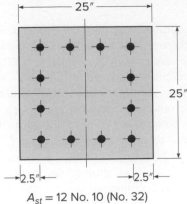

$A_{st} = 12$ No. 10 (No. 32)

9.6.   What is the strength M_n of the column of Problem 9.5 if it was loaded in pure bending (axial force $= 0$) about one principal axis?

9.7.   Construct the interaction diagram relating P_n to M_n for the building column shown in Fig. P9.7. Bending will be about the axis a-a. Calculate specific coordinates for concentric loading ($e = 0$), for P_b, and at least three other points, well chosen, on the curve. Material strengths are $f'_c = 8000$ psi and $f_y = 60,000$ psi.

FIGURE P9.7

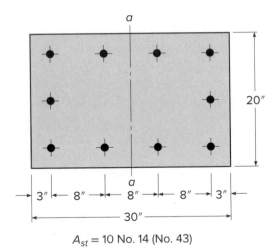

$A_{st} = 10$ No. 14 (No. 43)

9.8.   A short rectangular reinforced concrete column shown in Fig. P9.8 is to be a part of a long-span rigid frame and will be subjected to high bending moments combined with relatively low axial loads, causing bending about the strong axis. Because of the high eccentricity, steel is placed unsymmetrically as shown, with three No. 14 (No. 43) bars near the tension face and two No. 11 (No. 36) bars near the compression face. Material strengths are $f'_c = 6$ ksi and $f_y = 75$ ksi. Construct the complete strength interaction diagram, plotting P_n vs. M_n, relating eccentricities to the plastic centroid of the column (not the geometric center).

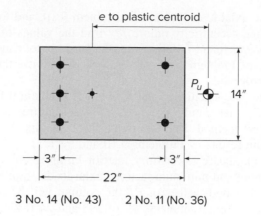

3 No. 14 (No. 43)   2 No. 11 (No. 36)

FIGURE P9.8

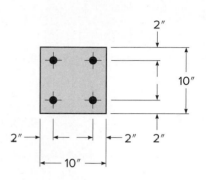

FIGURE P9.9

9.9. The square column shown in Fig. P9.9 must be designed for a factored axial load of 130 kips. Material strengths are $f_c' = 4000$ psi and $f_y = 60,000$ psi.

(a) Select the longitudinal and transverse reinforcement for an eccentricity $e_y = 2.7$ in.

(b) Select the longitudinal and transverse reinforcement for the same axial load with $e_x = e_y = 2.7$ in.

(c) Construct the strength interaction diagram and design strength curves for the column designed in part (b), given that the column will be subjected to biaxial bending with equal eccentricities about both principal axes.

9.10. The square column shown in Fig. P9.10 is a corner column subject to axial load and biaxial bending. Material strengths are $f_y = 60,000$ psi and $f_c' = 4000$ psi.

(a) Find the unique combination of P_n, M_{nx}, and M_{ny} that will produce incipient failure with the neutral axis located as in the figure. The compressive zone is shown shaded. Note that the actual neutral axis is shown, not the equivalent rectangular stress block limit; however, the rectangular stress block may be used as the basis of calculations.

(b) Find the angle between the neutral axis and the eccentricity axis, the latter defined as the line from the column center to the point of load.

FIGURE P9.10

$A_{st} = 4$ No. 14 (No. 43)

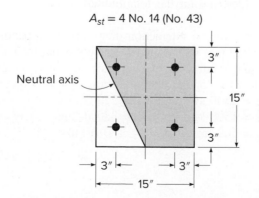

9.11. For the axial load P_n found in Problem 9.10, and for the same column, with the same eccentricity ratio e_y/e_x, find the values of M_{nx} and M_{ny} that would produce incipient failure, using the load contour method. Compare with the results of Problem 9.10. Take $\alpha = 1.30$, and use the graphs in Appendix A, as appropriate.

9.12. For the eccentricities e_x and e_y found in Problem 9.10, find the value of axial load P_n that would produce incipient failure, using the reciprocal load (Bresler) method. Use the graphs in Appendix A, as appropriate. Compare with the results of Problems 9.10 and 9.11.

9.13. A 20 in. square lower-story interior building column must be designed for maximum and minimum factored axial loads P_u of 880 and 551 kips, respectively. For both values of P_u, the column will be subjected to simultaneous factored bending moments M_u of 295 and 24 ft-kips about the Y and X axes, respectively (Fig. P9.13). Material strengths are $f_y = 60,000$ psi and $f_c' = 4000$ psi. Using equal reinforcement on all sides, design the longitudinal and transverse reinforcement for this column.

FIGURE P9.13

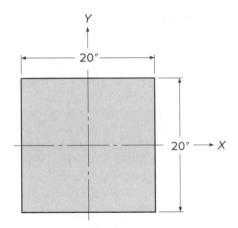

9.14. A 16 in. square lower-story corner column in the building described in Problem 9.13 will be subjected to maximum and minimum factored axial loads P_u of 209 and 130 kips, respectively. For both values of P_u, the columns must be designed for simultaneous factored bending moments M_u of 110 and 104 ft-kips about the Y and X axes, respectively. Using equal reinforcement on all sides, design the longitudinal and transverse reinforcement for this column.

9.15. Using the column interaction diagrams in Appendix A ($f_c' = 4000$ psi and $f_y = 60,000$ psi) with ρ_t approximately equal to 0.02, design square columns with equal reinforcement on all sides to carry each of the following loads and select the longitudinal and transverse reinforcement:

(a) $P_u = 2500$ kips and $M_u = 220$ ft-kips
(b) $P_u = 1500$ kips and $M_u = 330$ ft-kips
(c) $P_u = 600$ kips and $M_u = 180$ ft-kips

10 *Slender Columns*

10.1 INTRODUCTION

The material presented in Chapter 9 pertained to concentrically or eccentrically loaded *short columns*, for which the strength is governed entirely by the strength of the materials and the geometry of the cross section. Most columns in present-day practice fall in that category. However, with the increasing use of high-strength materials and improved methods of dimensioning members, it is now possible, for a given value of axial load, with or without simultaneous bending, to design a much smaller cross section than in the past. This clearly makes for more slender members. It is because of this, together with the use of more innovative structural concepts, that rational and reliable design procedures for slender columns have become increasingly important.

A column is said to be *slender* if its cross-sectional dimensions are small compared with its length. The degree of slenderness is generally expressed in terms of the *slenderness ratio* ℓ/r, where ℓ is the unsupported length of the member and r is the radius of gyration of its cross section, equal to $\sqrt{I/A}$. For square or circular members, the value of r is the same about either axis; for other shapes, r is smallest about the minor principal axis, and it is generally this value that must be used in determining the slenderness ratio of a freestanding column.

It has long been known that a member of great slenderness will collapse under a smaller compression load than a stocky member with the same cross-sectional dimensions. When a stocky member, say with $\ell/r = 10$ (such as a square column of length equal to about 3 times its cross-sectional dimension h), is loaded in axial compression, it will fail at the load given by Eq. (9.3), because at that load both concrete and steel are stressed to their maximum carrying capacity and give way, respectively, by crushing and by yielding. If a member with the same cross section has a slenderness ratio $\ell/r = 100$ (such as a square column hinged at both ends and of length equal to about 30 times its section dimension), it may fail under an axial load equal to one-half or less of that given by Eq. (9.3). In this case, collapse is caused by buckling, that is, by sudden lateral displacement of the member between its ends, with consequent overstressing of steel and concrete by the bending stresses that are superimposed on the axial compressive stresses.

Most columns in practice are subjected to bending moments as well as axial loads, as discussed in Chapter 9. These moments produce lateral deflection of a member between its ends and may also result in relative lateral displacement of joints. Associated with these lateral displacements are *secondary moments* that add to the primary moments and that may become very large for slender columns, leading to failure. A practical definition of a slender column is one for which there is a significant reduction in axial load capacity because of these secondary moments. In the development of

ACI Code column provisions, for example, any reduction greater than about 5 percent is considered significant, requiring consideration of slenderness effects.

The ACI Code and Commentary contain detailed provisions governing the design of slender columns. ACI Code 6.6.4 presents approximate methods for accounting for slenderness through the use of *moment magnification factors*. The provisions are quite similar to those used for many years for steel columns designed under the American Institute of Steel Construction (AISC) Specification. Alternatively, in ACI Code 6.7 and 6.8, a more fundamental approach is endorsed, in which the effect of lateral displacements is accounted for directly in the frame analysis. The latter approach, known as *second-order analysis*, is often incorporated as a feature in commercially available structural analysis software.

As noted, most columns in practice continue to be short columns. Simple expressions are included in the ACI Code to determine whether slenderness effects must be considered. These will be presented in Section 10.4 following the development of background information in Sections 10.2 and 10.3 relating to column buckling and slenderness effects.

10.2   CONCENTRICALLY LOADED COLUMNS

The basic information on the behavior of straight, concentrically loaded slender columns was developed by Euler more than 250 years ago. In generalized form, it states that such a member will fail by buckling at the critical load

$$P_c = \frac{\pi^2 E_t I}{(k\ell)^2} \tag{10.1}$$

where $k\ell$ is the *effective length* of the column.

It is seen that the buckling load for a given column cross section decreases rapidly as the effective length increases (Ref. 10.1).

For the simplest case of a column hinged at both ends and made of elastic material, E_t simply becomes Young's modulus and the effective length $k\ell$ is equal to the actual length ℓ of the column. At the load given by Eq. (10.1), the originally straight member buckles into a half sine wave, as shown in Fig. 10.1a. In this bent configuration, bending moments Py act at any section such as a; y is the deflection at that section. These deflections continue to increase until the bending stress caused by the increasing moment, together with the original compression stress, overstresses and fails the member.

If the stress-strain curve of a short piece of the given member has the shape shown in Fig. 10.2a, as it would be for reinforced concrete columns, E_t is equal to Young's modulus, provided that the buckling stress P_c/A is below the proportional limit f_p. If the strain is larger than f_p, buckling occurs in the inelastic range. In this case, in Eq. (10.1), E_t is the tangent modulus, that is, the slope of the tangent to the stress-strain curve. As the stress increases, E_t decreases. A plot of the buckling load vs. the slenderness ratio, the so-called column curve, therefore has the shape given in Fig. 10.2b, which shows the reduction in buckling strength with increasing slenderness. For very stocky columns, the value of the buckling load, calculated from Eq. (10.1), exceeds the direct crushing strength of the stocky column P_n, given by Eq. (9.3). This is also shown in Fig. 10.2b. Correspondingly, there is a limiting slenderness ratio $(k\ell/r)_{\text{lim}}$. For values smaller than this, failure occurs by simple crushing, regardless of $k\ell/r$; for values larger than $(k\ell/r)_{\text{lim}}$, failure occurs by buckling, the buckling load or stress decreasing for greater slenderness.

FIGURE 10.1
Buckling and effective length
of axially loaded columns.

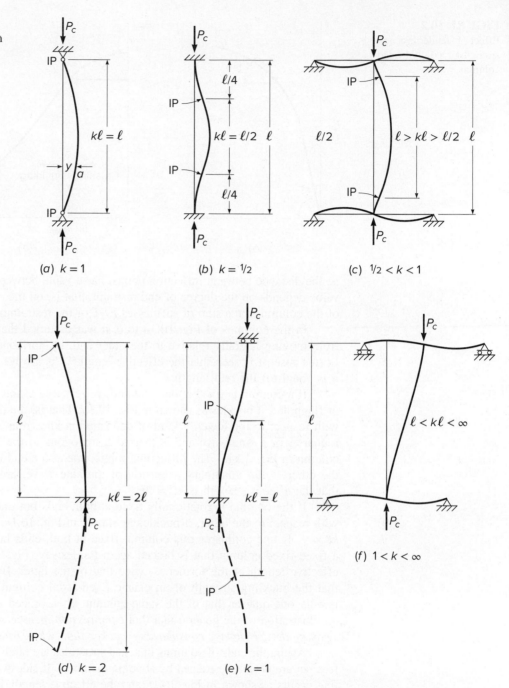

(a) $k = 1$ (b) $k = \frac{1}{2}$ (c) $\frac{1}{2} < k < 1$

(d) $k = 2$ (e) $k = 1$ (f) $1 < k < \infty$

If a member is fixed against rotation at both ends, it buckles in the shape of Fig. 10.1*b*, with inflection points (IPs) as shown. The portion between the inflection points is in precisely the same situation as the hinge-ended column of Fig. 10.1*a*, and thus, the *effective length* $k\ell$ of the fixed-fixed column, that is, the distance between inflection points, is seen to be $k\ell = \ell/2$. Equation (10.1) shows that an elastic column fixed at both ends will carry 4 times as much load as when hinged.

Columns in real structures are rarely either hinged or fixed but have ends partially restrained against rotation by abutting members. This is shown schematically in Fig. 10.1*c*, from which it is seen that for such members the effective length $k\ell$, that

FIGURE 10.2
Effect of slenderness on strength of axially loaded columns.

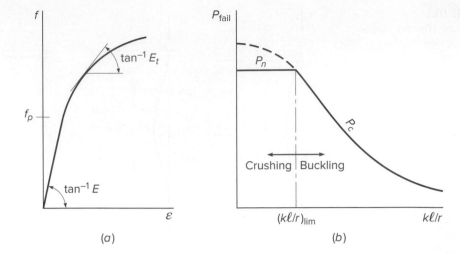

(a)   *(b)*

is, the distance between inflection points, has a value between ℓ and $\ell/2$. The precise value depends on the degree of end restraint, that is, on the ratio of the stiffness EI/ℓ of the column to the sum of stiffnesses EI/ℓ of the restraining members at both ends.

In the columns of Fig. 10.1a to c, it was assumed that one end was prevented from moving laterally relative to the other end, by horizontal bracing or otherwise. In this case, it is seen that the effective length $k\ell$ is always smaller than (or at most it is equal to) the real length ℓ.

If a column is fixed at one end and entirely free at the other (cantilever column or flagpole), it buckles as shown in Fig. 10.1d. That is, the upper end moves laterally with respect to the lower, a kind of deformation known as *sidesway*. It buckles into a quarter of a sine wave and is therefore analogous to the upper half of the hinged column in Fig. 10.1a. The inflection points, one at the end of the actual column and the other at the imaginary extension of the sine wave, are a distance 2ℓ apart, so that the effective length is $k\ell = 2\ell$.

If the column is rotationally fixed at both ends but one end can move laterally with respect to the other, it buckles as shown in Fig. 10.1e, with an effective length $k\ell = \ell$. If one compares this column, fixed at both ends but free to sidesway, with a fixed-fixed column that is braced against sidesway (Fig. 10.1b), one sees that the effective length of the former is twice that of the latter. By Eq. (10.1), this means that the buckling strength of an elastic fixed-fixed column that is free to sidesway is only one-quarter that of the same column when braced against sidesway. This is an illustration of the general fact that *compression members free to buckle in a sidesway mode are always considerably weaker than when braced against sidesway.*

Again, the ends of columns in actual structures are rarely hinged, fixed, or entirely free but are usually restrained by abutting members. If sidesway is not prevented, buckling occurs as shown in Fig. 10.1f, and the effective length, as before, depends on the degree of restraint. If the cross beams are very rigid compared with the column, the case of Fig. 10.1e is approached and $k\ell$ is only slightly larger than ℓ. On the other hand, if the restraining members are extremely flexible, a hinged condition is approached at both ends. Evidently, a column hinged at both ends and free to sidesway is unstable. It will simply topple, being unable to carry any load whatever.

In reinforced concrete structures, one is rarely concerned with single members but rather with rigid frames of various configurations. The manner in which the relationships just described affect the buckling behavior of frames is illustrated by the simple portal frame shown in Fig. 10.3, with loads applied concentrically to the columns. If sidesway is prevented, as indicated schematically by the brace in

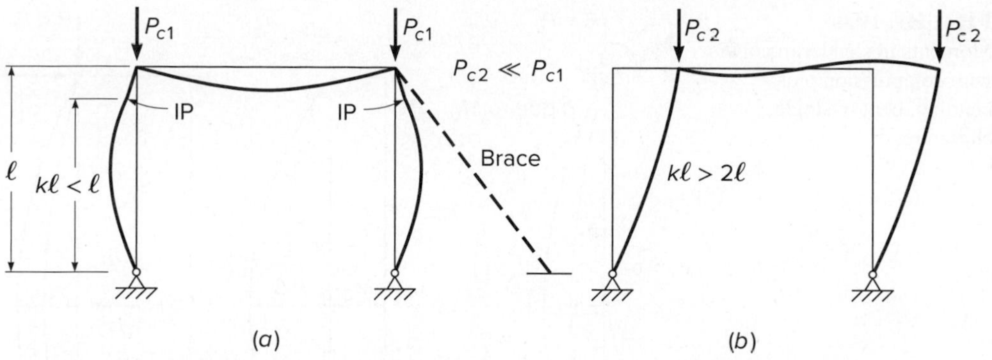

FIGURE 10.3
Rigid-frame buckling:
(*a*) laterally braced and
(*b*) unbraced.

(*a*)

(*b*)

Fig. 10.3*a*, the buckling configuration will be as shown. The buckled shape of the column corresponds to that in Fig. 10.1*c*, except that the lower end is hinged. It is seen that the effective length $k\ell$ is smaller than ℓ. On the other hand, if no sidesway bracing is provided to an otherwise identical frame, buckling occurs as shown in Fig. 10.3*b*. The column is in a situation similar to that shown in Fig. 10.1*d*, upside down, except that the upper end is not fixed but only partially restrained by the girder. It is seen that the effective length $k\ell$ exceeds 2ℓ by an amount depending on the degree of restraint. The buckling strength depends on $k\ell/r$ in the manner shown in Fig. 10.2*b*. As a consequence, even though they are dimensionally identical, the unbraced frame will buckle at a radically smaller load than the braced frame.

In summary, the following can be noted:

1. The strength of concentrically loaded columns decreases with increasing slenderness ratio $k\ell/r$.
2. In columns that are *braced against sidesway* or that are parts of frames braced against sidesway, the effective length $k\ell$, that is, the distance between inflection points, falls between $\ell/2$ and ℓ, depending on the degree of end restraint.
3. The effective lengths of columns that are *not braced against sidesway* or that are parts of frames not so braced are always larger than ℓ, the more so the smaller the end restraint. In consequence, the buckling load of a frame not braced against sidesway is always substantially smaller than that of the same frame when braced.

10.3 COMPRESSION PLUS BENDING

Most reinforced concrete compression members are also subject to simultaneous flexure, caused by transverse loads or by end moments owing to continuity. The behavior of members subject to such combined loading also depends greatly on their slenderness.

Figure 10.4*a* shows such a member, axially loaded by P and bent by equal end moments M_e. If no axial load were present, the moment M_0 in the member would be constant throughout and equal to the end moments M_e. This is shown in Fig. 10.4*b*. In this situation, that is, in simple bending without axial compression, the member deflects as shown by the dashed curve of Fig. 10.4*a*, where y_0 represents the deflection at any point caused by bending only. When P is applied, the moment at any point increases by an amount equal to P times its lever arm. The increased moments cause additional deflections, so that the deflection curve under the simultaneous action of P and M_0 is the solid curve of Fig. 10.4*a*. At any point, then, the total moment is now

$$M = M_0 + Py \tag{10.2}$$

FIGURE 10.4
Moments in slender members
with compression plus
bending, bent in single
curvature.

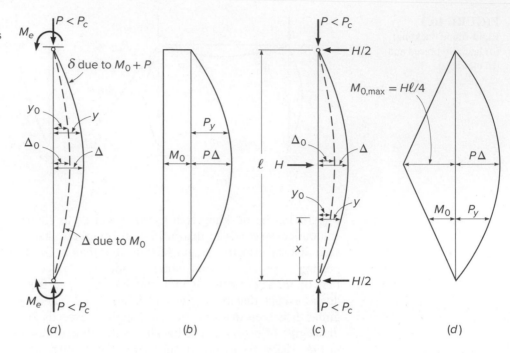

that is, the total moment consists of the moment M_0 that acts in the presence of P and the additional moment caused by P, equal to P times the deflection. This is one illustration of the so-called P-Δ effect.

A similar situation is shown in Fig. 10.4c, where bending is caused by the transverse load H. When P is absent, the moment at any point x is $M_0 = Hx/2$, with a maximum value at midspan equal to $H\ell/4$. The corresponding M_0 diagram is shown in Fig. 10.4d. When P is applied, additional moments Py are caused again, distributed as shown, and the total moment at any point in the member consists of the same two parts as in Eq. (10.2).

The deflections y of elastic columns of the type shown in Fig. 10.4 can be calculated from the deflections y_0, that is, from the deflections of the corresponding beam without axial load, using the following expression (see, for example, Ref. 10.1).

$$y = y_0 \frac{1}{1 - P/P_c} \tag{10.3}$$

If Δ is the deflection at the point of maximum moment $M_{\max}$, as shown in Fig. 10.4, $M_{\max}$ can be calculated using Eqs. (10.2) and (10.3).

$$M_{\max} = M_0 + P\Delta = M_0 + P\Delta_0 \frac{1}{1 - P/P_c} \tag{10.4}$$

It can be shown (Ref. 10.2) that Eq. (10.4) can be written as

$$M_{\max} = M_0 \frac{1 + \psi P/P_c}{1 - P/P_c} \tag{10.5}$$

where ψ is a coefficient that depends on the type of loading and varies between about ± 0.20 for most practical cases. Because P/P_c is always significantly smaller than 1,

FIGURE 10.5
Effect of slenderness on column moments.

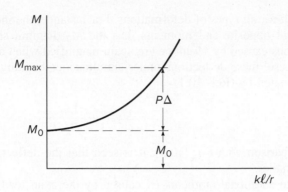

the second term in the numerator of Eq. (10.5) is small enough to be neglected. Doing so, one obtains the simplified design equation

$$M_{max} = M_0 \frac{1}{1 - P/P_c} \tag{10.6}$$

where $1/(1 - P/P_c)$ is known as the *moment magnification factor*, which reflects the amount by which the moment M_0 is magnified by the presence of a simultaneous axial force P.

Since P_c decreases with increasing slenderness ratio, it is seen from Eq. (10.6) that the moment M in the member increases with the slenderness ratio $k\ell/r$. The situation is shown schematically in Fig. 10.5. It indicates that, for a given transverse loading (that is, a given value of M_0), an axial force P causes a larger additional moment in a slender member than in a stocky member.

In the two members in Fig. 10.4, the largest moment caused by P, namely $P\Delta$, adds directly to the maximum value of M_0; for example,

$$M_0 = \frac{H\ell}{4}$$

in Fig. 10.4d. As P increases, the maximum moment at midspan increases at a rate faster than that of P in the manner given by Eqs. (10.2) and (10.6) and shown in Fig. 10.6. The member will fail when the simultaneous values of P and M become equal to P_n and M_n, the nominal strength of the cross section at the location of maximum moment.

This direct addition of the maximum moment caused by P to the maximum moment caused by the transverse load, clearly the most unfavorable situation, does

FIGURE 10.6
Effect of axial load on column moments.

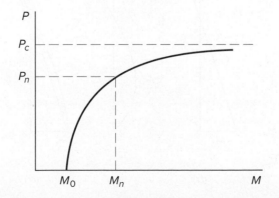

not result for all types of deformations. For instance, the member in Fig. 10.7a, with equal and opposite end moments, has the M_0 diagram shown in Fig. 10.7b. The deflections caused by M_0 alone are again magnified when an axial load P is applied. In this case, these deflections under simultaneous bending and compression can be approximated by (Ref. 10.1)

$$y = y_0 \frac{1}{1 - P/4P_c} \tag{10.7}$$

By comparison with Eq. (10.3), it is seen that the deflection magnification here is much smaller.

Y The additional moments Py caused by the axial load are distributed as shown in Fig. 10.7c. Although the M_0 moments are largest at the ends, the Py moments are seen to be largest at some distance from the ends. Depending on their relative magnitudes, the total moments $M = M_0 + Py$ are distributed as shown in either Fig. 10.7d or e. In the former case, the maximum moment continues to act at the end and to be equal to M_e; the presence of the axial force, then, does not result in any increase in the maximum moment. Alternatively, in the case of Fig. 10.7e, the maximum moment is located at some distance from the end; at that location, M_0 is significantly smaller than its maximum value M_e, and for this reason, the added moment Py increases the maximum moment to a value only moderately greater than M_e.

Y Comparing Figs. 10.4 and 10.7, one can generalize as follows. The moment M_0 will be magnified most strongly when the location where M_0 is largest coincides with that where the deflection y_0 is largest. This occurs in members bent into single curvature by symmetrical loads or equal end moments. If the two end moments of Fig. 10.4a are unequal but of the same sign, that is, producing single curvature, M_0 will still be strongly magnified, though not quite so much as for equal end moments. On the other hand, as evident from Fig. 10.7, there will be little or possibly no magnification if the end moments are of opposite sign and produce an inflection point along the member.

FIGURE 10.7
Moments in slender members with compression plus bending, bent in double curvature.

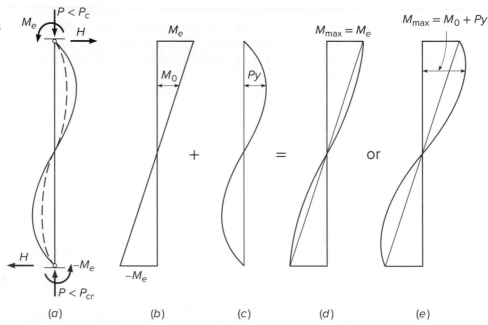

It can be shown (Ref. 10.2) that the way in which moment magnification depends on the relative magnitude of the two end moments (as in Figs. 10.4*a* and 10.7*a*) can be expressed by a modification of Eq. (10.6):

$$M_{\max} = M_0 \frac{C_m}{1 - P/P_c} \tag{10.8}$$

where

$$C_m = 0.6 - 0.4 \frac{M_1}{M_2} \geq 0.4 \tag{10.9}$$

Here M_1 is the numerically smaller and M_2 the numerically larger of the two end moments; hence, by definition, $M_0 = M_2$. The fraction M_1/M_2 is defined as negative if the end moments produce single curvature and positive if they produce double curvature. It is seen that when $M_1 = M_2$, as in Fig. 10.4*a*, $C_m = 1$, so that Eq. (10.8) becomes Eq. (10.6), as it should. Note that Eq. (10.9) *applies only to members braced against sidesway.* As will become apparent from the discussion that follows, for members not braced against sidesway, maximum moment magnification usually occurs, that is, $C_m = 1$.

Members that are braced against sidesway include columns that are parts of structures in which sidesway is prevented in one of various ways: by walls sufficiently strong and rigid in their own planes to effectively prevent horizontal displacement, by special bracing in vertical planes, in buildings by designing the utility core to resist horizontal loads and furnish bracing to the frames, or by bracing the frame against some other essentially immovable support.

If no such bracing is provided, *sidesway can occur only for the entire frame simultaneously,* not for individual columns in the frame. If this is the case, the combined effect of bending and axial load is somewhat different from that in braced columns. As an illustration, consider the simple portal frame of Fig. 10.8*a* subject to a horizontal load H, such as a wind load, and compression forces P, such as from gravity loads. The moments M_0 caused by H alone, in the absence of P, are shown in Fig. 10.8*b*; the corresponding deformation of the frame is given in dashed curves. When P is added, horizontal moments are caused that result in the magnified deformations shown in solid curves and in the moment diagram of Fig. 10.8*d*. It is seen that the maximum values of M_0, both positive and negative, and the maximum values of the additional moments M_P of the same sign occur at the same locations, namely, at the ends of the columns. They are therefore fully additive, leading to a large moment magnification. In contrast, if the frame in Fig. 10.8 is laterally braced and vertically

FIGURE 10.8
Fixed portal frame, laterally unbraced.

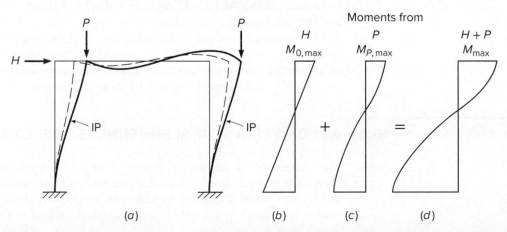

FIGURE 10.9
Fixed portal frame, laterally
braced.

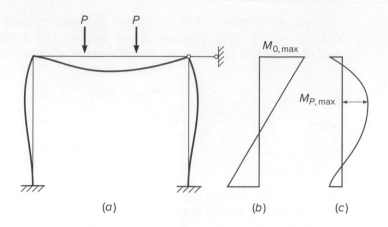

loaded, Fig. 10.9 shows that the maximum values of the two different moments occur in different locations; the moment magnification, if any, is therefore much smaller, as correctly expressed by C_m.

The moments that cause a frame to sidesway need not be caused by horizontal loads as in Fig. 10.8. Asymmetries, of frame configuration, vertical loading, or both, also result in sidesway displacements. In this case, the presence of axial column loads again results in the increased deflection and moment magnification.

In summary, it can be stated as follows:

1. In flexural members, the presence of axial compression causes additional deflections and additional moments Py. Other things being equal, the additional moments increase with increasing slenderness ratio $k\ell/r$.
2. In *members braced against sidesway* and bent in single curvature, the maxima of both types of moments, M_0 and Py, occur at the same or at nearby locations and are fully additive; this leads to large moment magnifications. If the M_0 moments result in double curvature (that is, in the occurrence of an inflection point), the opposite is true and less or no moment magnification occurs.
3. In *members in frames not braced against sidesway*, the maximum moments of both kinds, M_0 and Py, almost always occur at the same locations, the ends of the columns; they are fully additive, regardless of the presence or absence of an inflection point. Here, too, other things being equal, the additional deflections and the corresponding moments increase with increasing $k\ell/r$.

This discussion is a simplified presentation of a fairly complex subject. The provisions of the ACI Code regarding slender columns are based on the behavior and the corresponding equations that have just been presented. They take account, in an approximate manner, of the additional complexities that arise from the fact that concrete is not an elastic material, tension cracking changes the moment of inertia of a member, and under sustained load, creep increases the short-term deflections and, thereby, the moments caused by these deflections.

10.4 ACI CRITERIA FOR SLENDERNESS EFFECTS IN COLUMNS

The procedure of designing slender columns is inevitably lengthy, particularly because it involves a trial-and-error process. At the same time, studies have shown that most columns in existing buildings are sufficiently stocky that slenderness effects reduce their capacity by only a few percent. As stated in Chapter 9, an ACI-ASCE survey indicated that 90 percent of columns braced against sway, and 40 percent of unbraced columns, could be designed as short columns; that is, they could develop essentially the full cross-sectional strength with little or no reduction from slenderness (Ref. 10.3). Furthermore, lateral bracing is usually provided by shear walls, elevator shafts, stairwells, or other elements for which resistance to lateral deflection is much greater than for the columns of the building frame. It can be concluded that in most cases in reinforced concrete buildings, slenderness effects may be neglected.

To permit the designer to dispense with the complicated analysis required for slender column design for these ordinary cases, ACI Code 6.2.5 provides limits below which the effects of slenderness are insignificant and may be neglected. These limits are adjusted to result in a maximum unaccounted reduction in column capacity of no more than 5 percent. Separate limits are applied to braced and unbraced structures, alternately described in the ACI Code as *nonsway* and *sway* frames, respectively. For the purpose of determining if slenderness effects may be neglected, ACI Code 6.2.5 permits compression members to be considered as braced against sidesway if the total stiffness of the bracing elements resisting lateral movement of a story is at least 12 times the stiffness of all columns in that story. The Code provisions are as follows:

1. For compression members braced against sidesway (that is, in nonsway structures), the effects of slenderness may be neglected when $k\ell_u/r \le 34 + 12M_1/M_2$ and $k\ell_u/r \le 40$.
2. For compression members not braced against sidesway (that is, in sway structures), the effects of slenderness may be neglected when $k\ell_u/r$ is less than 22.

In these provisions, k is the effective length factor (see Section 10.2); ℓ_u is the unsupported length, taken as the clear distance between floor slabs, beams, or other members providing lateral support; M_1 is the smaller factored end moment on the compression member; M_2 is the larger factored end moment on the compression member (if transverse loading occurs between supports, M_2 is the largest moment in member); and M_1/M_2 is negative if the member is bent in single curvature and positive if bent in double curvature.

The radius of gyration r for rectangular columns may be taken as $0.30h$, where h is the overall cross-sectional dimension in the direction in which stability is being considered. For circular members, it may be taken as 0.25 times the diameter. For other shapes, r may be computed for the gross concrete section.

The effective length factor k may be conservatively taken as 1.0 for compression members that are braced against sidesway if a more accurate value is not determined by analysis. By necessity, k must be determined by analysis for compression members that are not braced against sidesway. The ACI criteria for determining k for both braced and unbraced columns are discussed in Section 10.6.

If slenderness effects must be considered, ACI Code 6.2.6 requires that the design of columns, beams restraining those columns, and other supporting members in the structure be based on a second-order analysis. The analysis may be elastic (ACI Code 6.7) or inelastic (ACI Code 6.8), or may be in accordance with the ACI moment magnifier procedure (ACI Code 6.6.4). Finite element analysis (ACI Code 6.9) is also permitted. To limit the potential for excessive moment magnification, the total factored moment including second-order effects in compression members may not exceed 1.4 times the factored moment due to first-order effects. In addition, second-order effects must be considered along the length of a member to cover cases in which the maximum moment may occur away from the ends. If a second-order analysis program is used, checking along the length of a member will require subdividing the member when it is modeled. In lieu of doing so, the ACI moment magnification method may be used. ACI Code

6.6.4, 6.7, and 6.8 require that the dimensions of all members used in the analysis be within 10 percent of the final dimensions. If not, the structure must be reanalyzed.

Elastic and inelastic second-order analyses, which are covered in ACI Code 6.7 and 6.8, are discussed in Section 10.8. The ACI moment magnification method of second-order analysis is discussed next.

10.5 ACI CRITERIA FOR NONSWAY VS. SWAY STRUCTURES

The discussion of Section 10.3 clearly shows important differences in the behavior of slender columns in nonsway (braced) structures and corresponding columns in sway (unbraced) structures. ACI Code provisions and Commentary guidelines for the approximate design of slender columns reflect this, and there are separate provisions in each relating to the important parameters in nonsway vs. sway structures, including moment magnification factors and effective length factors.

In practice, a structure is seldom either completely braced or completely unbraced. It is necessary, therefore, to determine in advance if bracing provided by shear walls, elevator and utility shafts, stairwells, or other elements is adequate to restrain the structure against significant sway effects. Both the ACI Code and Commentary provide guidance.

As suggested in ACI Commentary 6.6.4.1, a compression member can be assumed braced if it is located in a story in which the bracing elements (shear walls, etc.) have a stiffness substantial enough to limit lateral deflection to the extent that the column strength is not substantially affected. Such a determination can often be made by inspection. If not, ACI Code 6.6.4.3 provides two alternate criteria for determining if columns and stories are treated as nonsway or sway.

To be considered as a nonsway or braced column, the first criterion requires that the increase in column end moments due to second-order effects not exceed 5 percent of the first-order end moments. The designer is free to select the method for such a determination.

As an alternative, ACI Code 6.6.4.3 allows a story to be considered nonsway when the *stability index*

$$Q = \frac{\Sigma P_u \Delta_o}{V_{us}\ell_c} \tag{10.10}$$

for a story is not greater than 0.05, where ΣP_u and V_{us} are the total factored vertical load and story shear, respectively, for the story; Δ_o is the first-order relative deflection between the top and the bottom of the story due to V_{us}; and ℓ_c is the length of the compressive member measured center to center of the joints in the frame. ACI Commentary 6.6.4.3 provides the guidance that ΣP_u should be based on the lateral loading that maximizes the value of ΣP_u; the case of $V_{us} = 0$ is not included. In most cases, this calculation involves the combinations of load factors in Table 1.2 for wind, earthquake, or soil pressure (for example, $1.2D + 1.0W + 1.0L + 0.5L_r$).

As shown in Refs. 10.3 and 10.4, for Q not greater than 0.6, the stability index closely approximates the ratio P/P_c used in the calculation of the moment magnification factor, so that $1/(1 - P/P_c)$ can be replaced by $1/(1 - Q)$. Thus, for $Q = 0.05$, $M_{\max} \approx 1.05M_0$.[†]

[†] The near equivalence of Q to P/P_c for reinforced concrete columns can be demonstrated using a single sway column with ends fixed against rotation, as shown in Fig. 10.1e. For this column, $Q = P_u\Delta_o/V_{us}\ell_c$. Since V_{us}/Δ_o = the lateral stiffness of the column = $12EI/\ell_c^3$, the stability index can be expressed as $Q = P_u/(12EI/\ell_c^2)$. For an unsupported length of the column (the length used to calculate P_c) $\ell_u = 0.9\ell_c$ and $P = P_u$, $Q = P_u/(9.72EI/\ell_u^2)$ compared to $P/P_c = P_u/(\pi^2EI/\ell_u^2) = P_u/(9.87EI/\ell_u^2)$.

The section properties of the frame members used to calculate Q need to account for the effects of axial loads, cracked regions along the length of the member, and the duration of the loads. ACI Code 6.6.3 provides useful guidance that is appropriate for first-order as well as second-order analysis. According to ACI Code 6.6.3.1, section properties may be represented using the modulus of elasticity E_c given in Eq. (2.3) and the following section properties:

Moments of inertia

Beams	$0.35I_g$
Columns	$0.70I_g$
Walls—uncracked	$0.70I_g$
—cracked	$0.35I_g$
Flat plates and flat slabs	$0.25I_g$

Area $1.0A_g$

where I_g and A_g are based on the gross concrete cross section, neglecting reinforcement. As discussed in Section 11.5, I_g for T beams can be closely approximated as 2 times I_g for the web. The reduced values of I given above take into account the effect of nonlinear material behavior on the effective stiffness of the members. Reference 10.3 shows that these values for moments of inertia underestimate the true moments of inertia and conservatively overestimate second-order effects by 20 to 25 percent for reinforced concrete frames.

Based on work described in Refs. 10.5 and 10.6, ACI Code 6.6.3.1 indicates that the moments of inertia I of compression members and flexural members may also be computed using alternative expressions. For compression members,

$$I = \left(0.80 + 25\frac{A_{st}}{A_g}\right) \left(1 - \frac{M_u}{P_u h} - 0.5\frac{P_u}{P_o}\right) I_g \leq 0.875I_g \qquad (10.11)$$

where P_u and M_u are based on the load combination under consideration, or the combination of P_u and M_u resulting in the smallest value of I. The value of I calculated using Eq. (10.11) need not be taken less than $0.35I_g$.

For flexural members,

$$I = (0.10 + 25\rho) \left(1.2 - 0.2\frac{b_w}{d}\right) I_g \leq 0.5I_g \qquad (10.12)$$

The value of I calculated using Eq. (10.12) need not be taken less than $0.25I_g$. For continuous flexural members, I may be taken as the average value of I calculated at critical positive and negative moment locations along the length of the beam. The Code requires that the member dimensions and reinforcement ratios used in Eqs. (10.11) and (10.12) be within 10 percent of the final values.

To account for the effects of creep on Δ_o in Eq. (10.10) when sustained lateral loads act, the moments of inertia for compression members must be divided by $(1 + \beta_{ds})$, where β_{ds} is the ratio of the maximum factored sustained shear within a story to the maximum factored shear in that story associated with the same load combination.

10.6 ACI MOMENT MAGNIFIER METHOD FOR NONSWAY FRAMES

A slender reinforced concrete column reaches the limit of its strength when the combination of P and M at the most highly stressed section causes that section to fail. In general, P is essentially constant along the length of the member. This means that the column approaches failure when, at the most highly stressed section, the axial force P combines with a moment $M = M_{max}$, as given by Eq. (10.8), so that this combination becomes equal to P_n and M_n, which will cause the section to fail. This is easily visualized by means of Fig. 10.10.

For a column of given cross section, Fig. 10.10 presents a typical interaction diagram. For simplicity, suppose that the column is bent in single curvature with equal eccentricities at both ends. For this eccentricity, the strength of the cross section is given by point A on the interaction curve. If the column is stocky enough for the moment magnification to be negligibly small, then $P_{n,stocky}$ at point A represents the member strength of the column under the simultaneous moment $M_{n,stocky} = e_0 P_{n,stocky}$.

On the other hand, if the same column is sufficiently slender, significant moment magnification will occur with increasing P. Then the moment at the most highly stressed section is M_{max}, as given by Eq. (10.8), with $C_m = 1$ because of equal end eccentricities. The solid curve in Fig. 10.10 shows the nonlinear increase of M_{max} as P increases. The point where this curve intersects the interaction curve, that is, point B, defines the member strength $P_{n,slender}$ of the slender column, combined with the simultaneously applied end moments $M_0 = e_0 P_{n,slender}$. If end moments are unequal, the factor C_m will be less than 1, as discussed in Section 10.3.

For slender column design, the axial load and end moments in a column are first determined using conventional frame analysis (see Chapter 11), typically using the section properties given in Section 10.5. The member is then designed for that axial load and a simultaneous magnified column moment.

For a nonsway frame, the ACI Code equation for magnified moment, acting with the factored axial load P_u, is written as

$$M_c = \delta \, M_2 \qquad (10.13)$$

where the moment magnification factor is

$$\delta = \frac{C_m}{1 - P_u/0.75P_c} \geq 1 \qquad (10.14)$$

FIGURE 10.10
Effect of slenderness on carrying capacity.

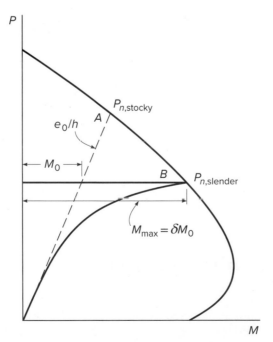

The 0.75 term in Eq. (10.14) is a *stiffness reduction factor*, designed to provide a conservative estimate of P_c. The critical load P_c, in accordance with Eq. (10.1), is given as

$$P_c = \frac{\pi^2 (EI)_{eff}}{(k\ell_u)^2} \tag{10.15}$$

where $(EI)_{eff}$ is defined later in this section and ℓ_u is defined as the unsupported length of the compression member. The value of k in Eq. (10.15) should be set equal to 1.0, unless calculated using the values of E_c and I given in Section 10.5 and procedures described later in this section.

In Eq. (10.14), the value of C_m is as previously given in Eq. (10.9):

$$C_m = 0.6 - 0.4 \frac{M_1}{M_2} \geq 0.4 \tag{10.9}$$

for columns braced against sidesway and without transverse loads between supports. Here M_2 is the larger of the two end moments, and M_1/M_2 is negative when the end moments produce single curvature and positive when they produce double curvature. The variation of C_m with M_1/M_2 is shown in Fig. 10.11. $C_m = 1.0$ for columns with transverse loads applied between the supports. In Eq. (10.14), when the calculated value of δ is smaller than 1, it indicates that the larger of the two end moments, M_2, is the largest moment in the column, a situation depicted in Fig. 10.7*d*.

In this way, the ACI Code provides for the capacity-reducing effects of slenderness in nonsway frames by means of the moment magnification factor δ. However, it is well known that for columns with no or very small applied moments, that is, axially or nearly axially loaded columns, increasing slenderness also reduces the column strength. For this situation, ACI Code 6.6.4.5 provides that the factored moment M_2 in Eq. (10.13) not be taken less than

$$M_{2,\min} = P_u(0.6 + 0.03h) \tag{10.16}$$

about each axis separately, where 0.6 and h are in inches. For members in which $M_{2,\min}$ exceeds M_2, the value of C_m in Eq. (10.9) is taken equal to 1.0 or is based on the ratio of the computed end moments M_1 and M_2.

The value of $(EI)_{eff}$ used in Eq. (10.15) to calculate P_c for an individual member must be both accurate and reasonably conservative to account for the greater variability inherent in the properties of individual columns, as compared to the

FIGURE 10.11

Values of C_m for slender columns in sway and nonsway frames.

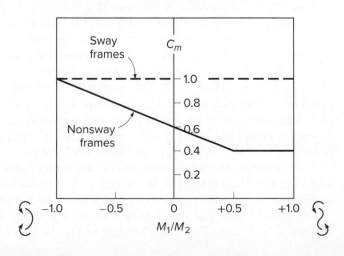

properties of the reinforced concrete frame, as a whole. The values of EI provided in Section 10.5 are adequate for general frame analysis but not for establishing P_c for individual columns.

In homogeneous elastic members, such as steel columns, EI is easily obtained from Young's modulus and the usual moment of inertia. Reinforced concrete columns, however, are nonhomogeneous, since they consist of both steel and concrete. Whereas steel is substantially elastic, concrete is not and is, in addition, subject to creep and to cracking if tension occurs on the convex side of the column. All these factors affect the effective value of EI for a reinforced concrete member. It is possible by computer methods to calculate fairly realistic effective section properties, taking account of these factors. Even these calculations are no more accurate than the assumptions on which they are based. On the basis of elaborate studies, both analytical and experimental (Ref. 10.7), the ACI Code requires that $(EI)_{eff}$ be determined using Eq. (10.11) or by either

$$(EI)_{eff} = \frac{0.2E_cI_g + E_sI_{se}}{1 + \beta_{dns}} \qquad (10.17)$$

or the simpler expression

$$(EI)_{eff} = \frac{0.4E_cI_g}{1 + \beta_{dns}} \qquad (10.18)$$

where E_c = modulus of elasticity of concrete, psi
  I_g = moment of inertia of gross section of column, in^4
  E_s = modulus of elasticity of steel = 29,000,000 psi
  I_{se} = moment of inertia of reinforcement about centroidal axis of member cross section, in^4
  β_{dns} = ratio of maximum factored sustained axial load to maximum factored axial load associated with same load combination (this definition differs from that used in Section 10.5 to calculate Δ_o)

The factor β_{dns} approximately accounts for the effects of creep. That is, the larger the sustained loads, the larger are the creep deformations and corresponding curvatures. Consequently, the larger the sustained loads relative to the temporary loads, the smaller the effective rigidity, as correctly reflected in Eqs. (10.17) and (10.18). Because, of the two materials, only concrete is subject to creep, and reinforcing steel as ordinarily used is not, the argument can be made that the creep parameter $1 + \beta_{dns}$ should be applied only to the term $0.2E_cI_g$ in Eq. (10.17). However, as explained in ACI Commentary 6.6.4.4, the creep parameter is applied to both terms because of the potential for premature yielding of steel in columns under sustained loading.

Both Eqs. (10.17) and (10.18) are conservative lower limits for large numbers of actual members (Ref. 10.3). The simpler but more conservative Eq. (10.18) is not unreasonable for lightly reinforced members, but it greatly underestimates the effect of reinforcement for more heavily reinforced members, that is, for the range of higher ρ_g values. Equation (10.17) is more reliable for the entire range of ρ_g and definitely preferable for medium and high ρ_g values (Ref. 10.8).

An accurate determination of the effective length factor k is essential in connection with Eqs. (10.13) and (10.15). In Section 10.2, it was shown that, for frames braced against sidesway (nonsway frames), k varies from $\frac{1}{2}$ to 1, whereas for laterally unbraced frames (sway frames), it varies from 1 to ∞, depending on the degree of rotational restraint at both ends. This was illustrated in Fig. 10.1. For frames, it is seen that this degree of rotational restraint depends on whether the stiffnesses of the beams framing into the column at top and bottom are large or small compared with the stiffness of the column itself. An approximate but generally satisfactory way of determining k is by means of *alignment charts* based on isolating the given column plus all members framing into it at top and bottom, as shown in Fig. 10.12. The *degree of end restraint* at each end is $\psi = \Sigma(EI/\ell_c$ of columns$) \div \Sigma(EI/\ell$ of floor members$)$. Only floor members that are in a plane at either end of the column are to be included. The value of k can be read directly from the chart of Fig. 10.13, as illustrated by the dashed lines.[†]

It is seen that k must be known before a column in a frame can be dimensioned. Yet k depends on the stiffness EI/ℓ of the members to be dimensioned, as well as on that of the abutting members. Thus, the dimensioning process necessarily involves iteration; that is, one assumes member sizes, calculates member stiffnesses and corresponding k values, and then calculates the critical buckling load and more accurate member sizes on the basis of these k values until assumed and final member sizes coincide or are satisfactorily close. The stiffness EI/ℓ should be calculated based on the values of E_c and I given in Section 10.5, and the span lengths of the members ℓ_c and ℓ should be measured center to center of the joints.

An outline of the separate steps in the analysis/design procedure for nonsway stories or frames follows along these lines:

1. Select a trial column section to carry the factored axial load P_u and moment $M_u = M_2$ from the elastic first-order frame analysis, assuming short column behavior and following the procedures of Chapter 9.
2. Determine if the frame should be considered as nonsway or sway, using the criteria of Section 10.5.
3. Find the unsupported length ℓ_u.
4. For the trial column, check for consideration of slenderness effects, using the criteria of Section 10.4 with $k = 1.0$.
5. If slenderness is tentatively found to be important, refine the calculation of k based on the alignment chart in Fig. 10.13a, with member stiffnesses EI/ℓ (Section 10.5)

FIGURE 10.12
Section of rigid frame including column to be designed.

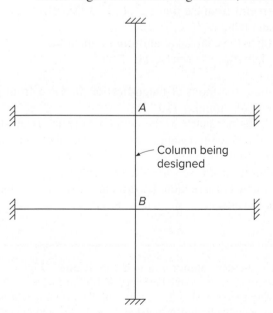

Column being designed

[†] Equations for the determination of effective length factors k, more convenient than charts for developing computer solutions, are presented in Refs. 10.9 through 10.12. The expressions in Ref. 10.12 are the most accurate.

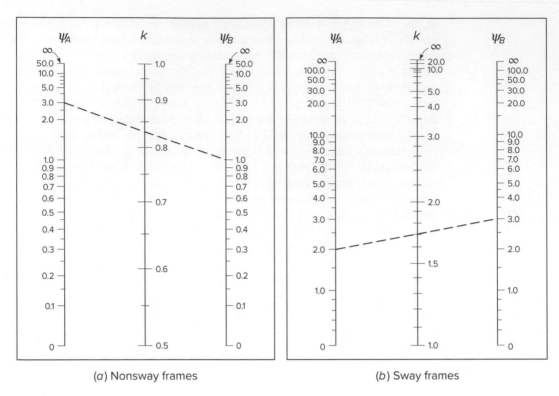

FIGURE 10.13
Alignment charts for effective length factors k.

and rotational restraint factors ψ based on trial member sizes. Recheck against the slenderness criteria.

6. If moments from the frame analysis are small, check to determine if the minimum moment from Eq. (10.16) controls.

7. Calculate the equivalent uniform moment factor C_m from Eq. (10.9).

8. Calculate β_{dns}, EI from Eq. (10.17) or (10.18), and P_c from Eq. (10.15) for the trial column.

9. Calculate the moment magnification factor δ from Eq. (10.14) and magnified moment M_c from Eq. (10.13).

10. Check the adequacy of the column to resist axial load and magnified moment, using the column design charts of Appendix A in the usual way. Revise the column section and reinforcement if necessary.

11. If column dimensions are altered, repeat the calculations for k, ψ, and P_c based on the new cross section. Determine the revised moment magnification factor and check the adequacy of the new design.

EXAMPLE 10.1   **Design of a slender column in a nonsway frame.**   Figure 10.14 shows an elevation view of a multistory concrete frame building, with 48 in. wide × 12 in. deep beams on all column lines, carrying two-way slab floors and roof. The clear height of the columns is 13 ft. Interior columns are tentatively dimensioned at 18 × 18 in., and exterior columns at 16 × 16 in. The frame is effectively braced against sway by stair and elevator shafts having concrete walls that are monolithic with the floors, located in the building corners (not shown in the figure).

FIGURE 10.14
Concrete building frame for
Example 10.1.

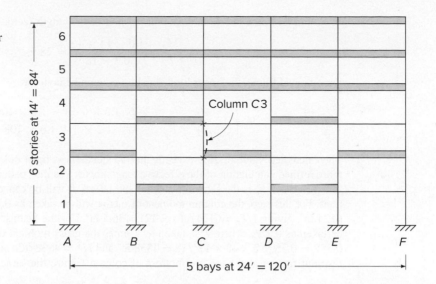

The structure will be subjected to vertical dead and live loads. Trial calculations by first-order analysis indicate that the pattern of live loading shown in Fig. 10.14, with full load distribution on roof and upper floors and a checkerboard pattern adjacent to column $C3$, produces maximum moments with single curvature in that column, at nearly maximum axial load. Dead loads act on all spans. Service load values of dead and live load axial force and moments for the typical interior column $C3$ are as follows:

Dead load   *Live load*
$P = 230$ kips $P = 173$ kips
$M_2 = 2$ ft-kips $M_2 = 108$ ft-kips
$M_1 = -2$ ft-kips $M_1 = 100$ ft-kips

The column is subjected to double curvature under dead load alone and single curvature under live load.

Design column $C3$, using the ACI moment magnifier method. Use $f_c' = 4000$ psi and $f_y = 60,000$ psi.

SOLUTION. The column will first be designed as a short column, assuming no slenderness effect. With the application of the usual load factors,

$$P_u = 1.2 \times 230 + 1.6 \times 173 = 553 \text{ kips}$$

$$M_u = 1.2 \times 2 + 1.6 \times 108 = 175 \text{ ft-kips}$$

For an 18×18 in. column, with the 1.5 in. clear to the outside steel, No. 3 (No. 10) stirrups, and (assumed) No. 10 (No. 32) longitudinal steel:

$$\gamma = (18.00 - 2 \times 1.50 - 2 \times 0.38 - 1.27)/18 = 0.72$$

Graph A.6 for $\gamma = 0.70$, with bars arranged around the column perimeter, will be used. Then

$$\frac{P_u}{\phi f_c' A_g} = \frac{553}{0.65 \times 4 \times 324} = 0.656$$

$$\frac{M_u}{\phi f_c' A_g h} = \frac{175 \times 12}{0.65 \times 4 \times 324 \times 18} = 0.138$$

and from the graph $\rho_g = 0.02$. This is low enough that an increase in steel area could be made, if necessary, to allow for slenderness, and the 18×18 in. concrete dimensions will be retained.

For an initial check on slenderness, an effective length factor $k = 1.0$ will be used. Then

$$\frac{k\ell_u}{r} = \frac{1.0 \times 13 \times 12}{0.3 \times 18} = 28.9$$

For a braced frame, the upper limit for short column behavior is

$$34 + 12 \frac{M_1}{M_2} = 34 + 12\left(-\frac{1.2 \times (-2) + 1.6 \times 100}{1.2 \times 2 + 1.6 \times 108}\right) = 23.2$$

The calculated value of 28.9 exceeds this, so slenderness must be considered in the design. A more refined calculation of the effective length factor k is thus called for.

Because E_c is the same for column and beams, it will be canceled in the stiffness calculations. For this step, the column moment of inertia will be taken as $0.7I_g = 0.7 \times 18 \times 18^3/12 = 6124$ in^4, giving $I/\ell_c = 6124/(14 \times 12) = 36.5$ in^3. For the beams, the moment of inertia will be taken as $0.35I_g$, where I_g is taken as 2 times the gross moment of inertia of the web. Thus, $0.35I_g = 0.35 \times 2 \times 48 \times 12^3/12 = 4838$ in^4 and $I/\ell = 4838/(24 \times 12) = 16.8$ in^3. Rotational restraint factors at the top and bottom of column $C3$ are the same and are

$$\psi_a = \psi_b = \frac{36.5 + 36.5}{16.8 + 16.8} = 2.17$$

From Fig. 10.13a for the braced frame, the value of k is 0.87, rather than 1.0 as used previously. Consequently,

$$\frac{k\ell_u}{r} = \frac{0.87 \times 13 \times 12}{0.3 \times 18} = 25.1$$

This is still above the limit value of 23.3, confirming that slenderness must be considered.

A check will now be made of minimum moment. According to Eq. (10.16), $M_{2,\min} = P_u(0.6 + 0.03h) = 553 \times (0.6 + 0.03 \times 18)/12 = 53$ ft-kips. It is seen that this does not control.

The coefficient C_m can now be found from Eq. (10.9) with $M_1 = 1.2 \times (-2) + 1.6 \times 100 = 158$ ft-kips and $M_2 = 1.2 \times 2 + 1.6 \times 108 = 175$ ft-kips:

$$C_m = 0.6 - 0.4 \frac{M_1}{M_2} = 0.6 - 0.4\left(-\frac{158}{175}\right) = 0.96$$

Next the factor β_{dns} will be found based on the ratio of the maximum factored axial sustained load (the factored dead load in this case) to the maximum factored axial load:

$$\beta_{dns} = \frac{1.2 \times 230}{1.2 \times 230 + 1.6 \times 173} = 0.50$$

For a relatively low reinforcement ratio, one estimated to be in the range of 0.02 to 0.03, the more approximate Eq. (10.18) for EI will be used, and

$$(EI)_{eff} = \frac{0.4E_cI_g}{1 + \beta_{dns}} = \frac{0.4 \times 3.60 \times 10^6 \times 18 \times 18^3/12}{1 + 0.50} = 8.40 \times 10^9 \text{ in}^2\text{-lb}$$

The critical buckling load is found from Eq. (10.15) to be

$$P_c = \frac{\pi^2 (EI)_{eff}}{(k\ell_u)^2} = \frac{\pi^2 \times 8.40 \times 10^9}{(0.87 \times 13 \times 12)^2} = 4.50 \times 10^6 \text{ lb}$$

The moment magnification factor can now be found from Eq. (10.14).

$$\delta = \frac{C_m}{1 - P_u/0.75P_c} = \frac{0.96}{1 - 553/(0.75 \times 4500)} = 1.15$$

Thus, the required axial strength of the column is $P_u = 553$ kips (as before), while the magnified design moment is $M_c = \delta M_2 = 1.15 \times 175 = 201$ ft-kips. As described in Section 10.4,

FIGURE 10.15
Cross section of column $C3$, Example 10.1.

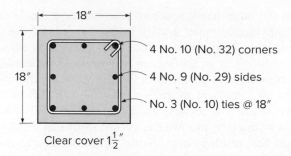

18″

18″

4 No. 10 (No. 32) corners

4 No. 9 (No. 29) sides

No. 3 (No. 10) ties @ 18″

Clear cover $1\frac{1}{2}$″

ACI Code 6.2.6 limits the magnified moment to 1.4 times the moment due to first-order effects. This limitation is clearly satisfied. With reference again to the column design chart A.6 with

$$\frac{P_u}{\phi f_c' A_g} = \frac{553}{0.65 \times 4 \times 324} = 0.656$$

$$\frac{M_u}{\phi f_c' A_g h} = \frac{201 \times 12}{0.65 \times 4 \times 324 \times 18} = 0.159$$

it is seen that the required reinforcement ratio is increased from 0.020 to 0.026 because of slenderness. The steel area now required is

$$A_{st} = 0.026 \times 324 = 8.42 \text{ in}^2$$

which can be provided using four No. 10 (No. 32) and four No. 9 (No. 29) bars ($A_{st} = 9.08$ in^2), arranged as shown in Fig. 10.15. No. 3 (No. 10) ties will be used at a spacing not to exceed the least dimension of the column (18 in.), 48 tie diameters (18 in.), or 16 bar diameters (18 in.). Single ties at 18 in. spacing, as shown in the figure, will meet requirements of the ACI Code.

Further refinements in the design could, of course, be made by recalculating the critical buckling load using Eq. (10.17). This extra step is not justified here because the column slenderness is barely above the upper limit for short column behavior and the moment magnification is not great.

10.7 ACI MOMENT MAGNIFIER METHOD FOR SWAY FRAMES

The important differences in behavior between columns braced against sidesway and columns for which sidesway is possible were discussed in Sections 10.2 and 10.3. The critical load for a column P_c depends on the effective length $k\ell_u$, and although the effective length factor k falls between 0.5 and 1.0 for braced columns, it is between 1.0 and ∞ for columns that are unbraced (see Figs. 10.1 and 10.13). Consequently, an unbraced column will buckle at a much smaller load than will a braced column that is otherwise identical.

Columns subject to sidesway do not normally stand alone but are part of a structural system including floors and roof. A floor or roof is normally very stiff in its own plane. Consequently, all columns at a given story level in a structure are subject to essentially identical sway displacements; that is, sidesway of a particular story can occur only by simultaneous lateral motion of all columns of that story. Clearly, all columns at a given level must be considered together in evaluating slenderness effects relating to sidesway.

On the other hand, it is also possible for a single column in a sway frame to buckle individually under gravity loads, the ends of the column being held against relative lateral movement by other, stiffer columns at the same floor level. This possibility, resulting in magnification of nonsway moments due to gravity loads, must also be considered in the analysis and design of slender columns in unbraced frames.

The ACI moment magnifier approach can still be used for frames subject to sidesway, but it is necessary, according to ACI Code 6.6.4.6, to separate the loads acting on a structure into two categories: loads that result in no appreciable sidesway and loads that result in appreciable sidesway. Clearly two separate frame analyses are required, one for loads of each type. In general, gravity loads acting on reasonably symmetrical frames produce little sway, and the effects of gravity load may therefore be placed in the first category. This is confirmed by tests and analyses in Ref. 10.13 that show that the sway magnification of gravity moments by the sway multiplier is unwarranted.

The maximum magnified moments caused by sway loading occur at the ends of the column, but those due to gravity loads may occur somewhere in the midheight of the column, the exact location of the latter varying depending on the end moments. Because magnified gravity moments and magnified sway moments do not occur at the same location, the argument can be made that, in most cases, no magnification should be applied to the nonsway moments when sway moments are considered; that is, it is unlikely that the actual maximum moment will exceed the sum of the nonmagnified gravity moment and the magnified sway moment. Consequently, for cases involving sidesway, Eq. (10.13) is replaced by

$$M_1 = M_{1ns} + \delta_s M_{1s} \qquad (10.19)$$

$$M_2 = M_{2ns} + \delta_s M_{2s} \qquad (10.20)$$

where M_1 = smaller factored end moment on compression member
M_2 = larger factored end moment on compression member
M_{1ns} = factored end moment on compression member at end at which M_1 acts, due to loads that cause no appreciable sidesway, calculated using a first-order elastic frame analysis
M_{2ns} = factored end moment on compression member at end at which M_2 acts, due to loads that cause no appreciable sidesway, calculated using a first-order elastic frame analysis
M_{1s} = factored end moment on compression member at end at which M_1 acts, due to loads that cause appreciable sidesway, calculated using a first-order elastic frame analysis
M_{2s} = factored end moment on compression member at end at which M_2 acts, due to loads that cause appreciable sidesway, calculated using a first-order elastic frame analysis
δ_s = moment magnification factor for frames not braced against sidesway, to reflect lateral drift resulting from lateral (and sometimes gravity) loads

ACI Code 6.6.4.6 provides two alternate methods for calculating the moment magnification factor for frames not braced against sidesway δ_s.

With the first alternative, the moment magnification factor is calculated as

$$\delta_s = \frac{1}{1 - Q} \geq 1 \qquad (10.21)$$

where Q is the stability index calculated using Eq. (10.10). The ACI Code limits application of Eq. (10.21) to values of $\delta_s = 1/(1 - Q) \leq 1.5$. An elastic second-order analysis, as described in ACI Code 6.7, or the second alternative described in ACI Code 6.6.4.6 must be used for higher values of δ_s.

For the second alternative, the moment magnification factor is calculated as

$$\delta_s = \frac{1}{1 - \Sigma P_u / 0.75 \Sigma P_c} \geq 1 \tag{10.22}$$

in which ΣP_u is the total axial load on all columns and ΣP_c is the total critical buckling load for all columns in the story under consideration. As with Eq. (10.14), the 0.75 factor in Eq. (10.22) is a stiffness reduction factor to provide a conservative estimate of the critical buckling loads P_c. The individual values of P_c are calculated using Eq. (10.15) with effective length factors k for unbraced frames (Fig. 10.13b) and values of EI from Eq. (10.17) or (10.18).

When calculating δ_s, the factor β_{ds} is defined differently than β_{dns} is for nonsway frames. As described earlier, in Section 10.5, β_{ds} is the ratio of the maximum factored sustained shear within a story to the maximum factored shear in that story. Thus, for most applications, $\beta_{ds} = 0$ for the purpose of calculating δ_s. In unusual situations, $\beta_{ds} \neq 0$ will occur, such as a building located on a sloping site that is subjected to soil pressure on a single side (Refs. 10.14 and 10.15).

The sequence of design steps for slender columns in sway frames is similar to that outlined in Section 10.6 for nonsway frames, except for the requirement that loads be separated into gravity loads, which are assumed to produce no sway, and horizontal loads producing sway. Separate frame analyses are required, and different equivalent length factors k and creep coefficients β_{dns} and β_{ds} must be applied. Note that according to ACI Code 5.3.1 (see also Table 1.2), if wind effects W are included in the design, four possible factored load combinations are to be applied:

$$U = 1.2D + 1.6L$$

$$U = 1.2D + 1.6 \, (L_r \text{ or } S \text{ or } R) + 0.5W$$

$$U = 1.2D + 1.0W + 1.0L + 0.5 \, (L_r \text{ or } S \text{ or } R)$$

$$U = 0.9D + 1.0W$$

Similar provisions are included for cases where earthquake loads are to be considered. This represents a significant complication in the sway frame analysis; however, the factored loads can be separated into gravity effects and sway effects, as required, and a separate analysis can be performed for each.

It is important to realize that, for sway frames, *the beams must be designed for the total magnified end moments of the compression members at the joint.* Even though the columns may be very rigid, if plastic hinges were to form in the restraining beams adjacent to the joints, the effective column length would be greatly increased and the critical column load much reduced.

The choice of which of the methods to use for calculating δ_s depends upon the desired level of accuracy and the available analytical tools.

Second-order analysis (discussed in greater detail in Section 10.8) provides the most accurate estimate of the magnified sway moments but requires more sophisticated techniques. The extra effort required for second-order analysis, however, usually produces a superior design. The first alternative, Eq. (10.21), will in most cases be the easiest to apply, since matrix analysis is used for virtually all frames to determine member forces under gravity and lateral loading. Such an analysis automatically generates the value of Δ_o, the first-order relative deflection within a story, allowing Q to be calculated for each story within a structure. The second alternative, Eq. (10.22), is retained with minor modifications from previous versions of the ACI Code. As will be demonstrated in the following example, calculations using Eq. (10.22) are more tedious than those needed for Eq. (10.21) but do not require knowledge of Δ_o. Application of Eq. (10.21) is limited by the Code to values of $\delta_s \leq 1.5$. For $\delta_s > 1.5$, application of Eq. (10.22) is mandatory if a second-order analysis is not used.

EXAMPLE 10.2   **Design of a slender column in a sway frame.**   Consider now that the concrete building frame of Example 10.1 acts as a sway frame, without the stairwells or elevator shafts described earlier. An initial evaluation is carried out using the member dimensions and reinforcement given in Example 10.1. The reinforcement for the interior 18 × 18 in. columns, shown in Fig. 10.15, consists of four No. 10 (No. 32) bars at the corners and four No. 9 (No. 29) bars at the center of each side. Reinforcement for the exterior 16 × 16 in. columns consists of eight No. 8 (No. 25) bars distributed in a manner similar to that shown for the longitudinal reinforcement in Fig. 10.15. The building will be subjected to gravity dead and live loads and horizontal wind loads. Elastic first-order analysis of the frame at service loads (all load factors = 1.0) using the values of E and I defined in Section 10.5 gives the following results at the third story:

	Cols. *A3 and F3*	Cols. *B3 and E3*	Cols. *C3 and D3*
P_{dead}	115 kips	230 kips	230 kips
P_{live}	90 kips	173 kips	173 kips
P_{wind}	± 48 kips	± 29 kips	± 10 kips
V_{wind}	9 kips	18 kips	18 kips
$M_{2,dead}$			2 ft-kips
$M_{2,live}$			108 ft-kips
$M_{2,wind}$			± 126 ft-kips
$M_{1,dead}$			−2 ft-kips
$M_{1,live}$			100 ft-kips
$M_{1,wind}$			± 112 ft-kips

To simplify the analysis in this example, roof loads will not be considered. The relative lateral deflection for the third story under total wind shear $V_{wind} = 90$ kips is 1.25 in.

Column *C3* is to be designed for the critical loading condition, using $f'_c = 4000$ psi and $f_y = 60,000$ psi as before.

SOLUTION.   The column size and reinforcement must satisfy requirements for each of the four load conditions noted above.

Initially, a check is made to see if a sway frame analysis is required. The factored shear $V_{us} = V_{wind} = 90$ kips. The corresponding deflection $\Delta_o = 1.25$ in. The total factored axial force on the story is obtained using the load table.

Columns *A3* and *F3*: $P_u = 1.2 \times 115 + 1.0 \times 90 = 228$ kips

Columns *B3*, *C3*, *D3*, and *E3*: $P_u = 1.2 \times 230 + 1.0 \times 173 = 449$ kips

Note that in this case the values of P_{wind} in the columns are not considered since they cancel out for the floor as a whole, that is, $\Sigma P_{wind} = 0$. Thus, $\Sigma P_u = 2 \times 228 + 4 \times 449 = 2252$ kips, and the stability index is

$$Q = \frac{\Sigma P_u \Delta_o}{V_{us}\ell_c} = \frac{2252 \times 1.25}{90 \times 14 \times 12} = 0.19$$

Since $Q > 0.05$, sway frame analysis is required for this story.

(*a*) **Gravity loads only.** All columns in sway frames must first be considered as braced columns under gravity loads acting alone, that is, for $U = 1.2D + 1.6L$. This check has already been made for column *C3* in Example 10.1.

(b) **Gravity plus wind loads.** Three additional load combinations must be considered when wind effects are included: $U = 1.2D + 1.6(L_r$ or S or $R) + 0.5W$, $U = 1.2D + 1.0W + 1.0L + 0.5(L_r$ or S or $R)$, and $U = 0.9D + 1.0W$. By inspection, the second combination will control for this case, and the others will not be considered further. From Example 10.1, $\psi_a = \psi_b = 2.17$. With reference to the alignment chart in Fig. 10.13b, the effective length factor for an unbraced frame $k = 1.64$ and

$$\frac{k\ell_u}{r} = \frac{1.64 \times 13 \times 12}{0.3 \times 18} = 47.4$$

This is much above the limit value of 22 for short column behavior in an unbraced frame. (This should be no surprise since $k\ell_u/r = 25.1$ for column $C3$ in the braced condition.) For sway frame analysis, the loads must be separated into gravity loads and sway loads, and the appropriate magnification factor must be computed and applied to the sway moments. The factored end moments resulting from the nonsway loads on column $C3$ are

$$M_{1ns} = 1.2 \times (-2) + 1.0 \times 100 = 98 \text{ ft-kips}$$

$$M_{2ns} = 1.2 \times 2 + 1.0 \times 108 = 110 \text{ ft-kips}$$

The sway effects will amplify the moments:

$$M_{1s} = -112 \text{ ft-kips}$$

$$M_{2s} = 126 \text{ ft-kips}$$

For the purposes of comparison, the magnified sway moments will be calculated based on both Q [Eq. (10.21)] and $\Sigma P_u/\Sigma P_c$ [Eq. (10.22)].

Using Eq. (10.21),

$$\delta_s = \frac{1}{1 - Q} = \frac{1}{1 - 0.19} = 1.23$$

giving

$$\delta_s M_{1s} = 1.23 \times (-112) = -138 \text{ ft-kips}$$

$$\delta_s M_{2s} = 1.23 \times 126 = 155 \text{ ft-kips}$$

To use Eq. (10.22), the critical loads must be calculated for each of the columns as follows. For columns $A3$ and $F3$,

Columns: $I = 0.7I_g = 0.7 \times 16 \times 16^3/12 = 3823 \text{ in}^4$
            and $I/\ell_c = 3823/(14 \times 12) = 22.8 \text{ in}^3$

Beams: $I = 4838 \text{ in}^4$ and $I/\ell_c = 16.8 \text{ in}^3$

Rotational restraint factors for this case, with two columns and one beam framing into the joint, are

$$\psi_a = \psi_b = \frac{22.8 + 22.8}{16.8} = 2.71$$

which, with reference to the alignment chart for unbraced frames, gives $k = 1.77$. For wind load, $\beta_{ds} = 0$. Since reinforcement has been initially selected for one column, EI will be calculated using Eq. (10.17).

$$(EI)_{eff} = 0.2E_cI_g + E_sI_{se} = 0.2 \times 3.6 \times 10^6 \times 16 \times 16^3/12 + 29 \times 10^6 \times 6 \times 0.79 \times 5.6^2$$
$$= 8.24 \times 10^9 \text{ in}^2\text{-lb}$$

Then the critical load is

$$P_c = \frac{\pi^2(EI)_{eff}}{(k\ell_u)^2} = \frac{\pi^2 \times 8.24 \times 10^9}{(1.77 \times 13 \times 12)^2} = 1.07 \times 10^6 \text{ lb}$$

For columns $B3$, $C3$, $D3$, and $E3$, from earlier calculations for column $C3$, $k = 1.64$ for the sway loading case. For these columns,

$$(EI)_{eff} = 0.2 \times 3.6 \times 10^6 \times 18 \times 18^3/12 + 29 \times 10^6 \, (4 \times 1.27 \times 6.4^2 + 2 \times 1.0 \times 6.5^2)$$

$$= 14.8 \times 10^9 \text{ in}^2\text{-lb}$$

$$P_c = \frac{\pi^2 \times 14.8 \times 10^9}{(1.64 \times 13 \times 12)^2} = 2.23 \times 10^6 \text{ lb}$$

Thus, for all the columns at this level of the structure,

$$\Sigma P_c = 2 \times 1070 + 4 \times 2230 = 11{,}060 \text{ kips}$$

The sway moment magnification factor is

$$\delta_s = \frac{1}{1 - \Sigma P_u/0.75\Sigma P_c} = \frac{1}{1 - 2252/(0.75 \times 11{,}060)} = 1.37$$

and the magnified sway moments for the top and bottom of column $C3$ are

$$\delta_s M_{1s} = 1.37 \times (-112) = -153 \text{ ft-kips}$$

$$\delta_s M_{2s} = 1.37 \times 126 = 173 \text{ ft-kips}$$

The values of $\delta_s M_s$ are higher based on $\Sigma P_u/\Sigma P_c$ than they are based on Q (173 ft-kips vs. 155 ft-kips for $\delta_s M_{2s}$), emphasizing the conservative nature of the moment magnifier approach based on Eq. (10.22). The design will proceed using the less conservative value of $\delta_s M_s$.

The total magnified moments are

$$M_1 = 98 - 138 = -40 \text{ ft-kips}$$

$$M_2 = 110 + 155 = 265 \text{ ft-kips}$$

The values do not exceed the upper limit of 1.4 times the moments due to first-order effects and will now be combined with factored axial load $P_u = 459$ kips (now including P_{wind}). In reference to Graph A.6 with column parameters

$$\frac{P_u}{\phi f_c' A_g} = \frac{459}{0.65 \times 4 \times 324} = 0.545$$

$$\frac{M_u}{\phi f_c' A_g h} = \frac{265 \times 12}{0.65 \times 4 \times 324 \times 18} = 0.211$$

it is seen that $\rho_g = 0.038$. This is considerably higher than the value of 0.026 required for column $C3$ in a braced frame. The required steel area of

$$A_{st} = 0.038 \times 324 = 12.31 \text{ in}^2$$

will be provided using eight No. 11 (No. 36) bars ($A_{st} = 12.48$ in^2), arranged as shown in Fig. 10.16. Spacing of No. 4 (No. 13) ties must not exceed the least dimension of the column, 48 tie diameters, or 16 main bar diameters. The first criterion controls, and No. 4 (No. 13) ties at 18 in. spacing will be used in the pattern shown in Fig. 10.16.

FIGURE 10.16
Cross section of column $C3$, Example 10.2.

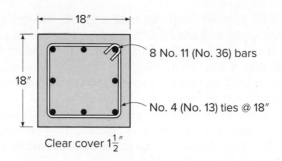

18″

18″

8 No. 11 (No. 36) bars

No. 4 (No. 13) ties @ 18″

Clear cover $1\frac{1}{2}″$

10.8 SECOND-ORDER ANALYSIS FOR SLENDERNESS EFFECTS

It may be evident from the preceding examples that although the ACI moment magnifier method works well enough for nonsway frames, its application to sway frames is complicated, with many opportunities for error, especially when Eq. (10.22) is used to calculate δ_s.

With the universal availability of computers in design offices, and because of the complexity of the moment magnifier method, it is advantageous to apply rational second-order frame analysis, or P-Δ analysis, in which the effects of lateral deflection on moments, axial forces, and, in turn, lateral deflections are computed directly. The resulting moments and deflections include the effects of slenderness, and so the problem is strictly nonlinear, whether the model used for the analysis is elastic (ACI Code 6.7) or inelastic (ACI Code 6.8).

Elastic second-order analysis in accordance with ACI Code 6.7 must consider the effects of axial loads, cracked regions within the members, and load duration, and although elastic models are simpler to implement than nonlinear models, as pointed out in Ref. 10.16, the key requirement for EI values for second-order frame analysis, whether elastic or inelastic, is that they be representative of member stiffness just prior to failure. The values of E and I in Section 10.5, which are taken from ACI Code 6.6.3.1, meet that requirement and include a stiffness reduction factor of 0.875 (Ref. 10.16). The value of the stiffness reduction factor and the moments of inertia in Section 10.5 are higher than the factor 0.75 in Eqs. (10.14) and (10.21) and the effective values of I in Eqs. (10.17) and (10.18), respectively, because of the inherently lower variability in the total stiffness of a frame compared to that of an individual member.

Inelastic second-order analysis in accordance with ACI Code 6.8 must account for the effects of material nonlinearity, load duration, shrinkage, and creep, as well as "satisfy equilibrium in the deformed configuration." ACI Code 6.8 requires that the second-order analysis procedure provide strength and deformation predictions that are in "substantial agreement" with physical tests of reinforced concrete components or structural systems that exhibit responses that are consistent with those expected in the structure under design. ACI Commentary 6.8 suggests that a prediction within 15 percent of the test results is satisfactory.

As pointed out in Section 10.5, the member dimensions used in any second-order analysis must be within 10 percent of the final dimensions. Otherwise, the frame must be reanalyzed.

A rational second-order analysis gives a better approximation of actual moments and forces than the moment magnifier method. Differences are particularly significant for irregular frames, frames subject to significant sway forces, and for lightly braced frames. There may be important economies in the resulting design.

Practical methods for performing a full second-order analysis are described in the literature (Refs. 10.3 and 10.17 through 10.20, to name a few), and general-purpose programs that perform a full nonlinear analysis including sway effects are commercially available. Linear first-order analysis programs are also available, but must include an iterative approach to produce acceptable results. This iterative approach can be summarized as follows.

Figure 10.17*a* shows a simple frame subject to lateral loads H and vertical loads P. The lateral deflection Δ is calculated by ordinary first-order analysis. As the frame is displaced laterally, the column end moments must equilibrate the lateral loads and a moment equal to $(\Sigma P)\Delta$:

$$\Sigma(M_{\text{top}} + M_{\text{bot}}) = H\ell_c + \Sigma P\Delta \tag{10.23}$$

where Δ is the lateral deflection of the top of the frame with respect to the bottom, and ΣP is the sum of the vertical forces acting. The moment $\Sigma P\Delta$ in a given story can be represented by equivalent shear forces $(\Sigma P)\Delta/\ell_c$, where ℓ_c is the story height, as shown in Fig. 10.17*b*. These shears give an overturning moment equal to that of the loads P acting at a displacement Δ.

Figure 10.17*c* shows the story shears acting in a three-story frame. The algebraic sum of the story shears from the columns above and below a given floor

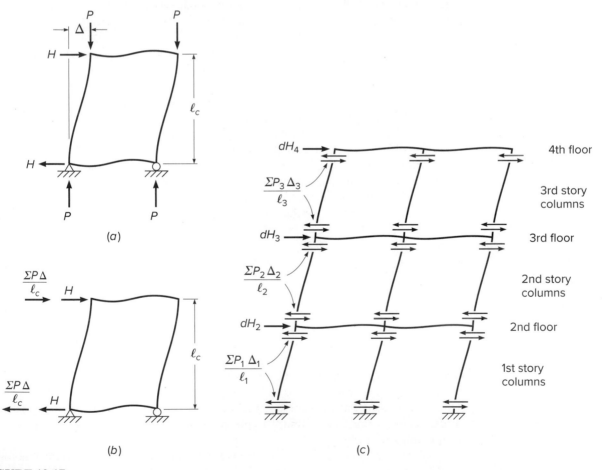

(a)

(b)

(c)

FIGURE 10.17
Basis for iterative P-Δ analysis: (*a*) vertical and lateral loads on rectangular frame; (*b*) real lateral forces H and fictitious sway forces dH; and (*c*) three-story frame subject to sway forces. (*Adapted from Ref. 10.17.*)

corresponds in effect to a sway force dH acting on that floor. For example, at the second floor the sway force is

$$dH_2 = \frac{\Sigma P_1 \Delta_1}{\ell_1} - \frac{\Sigma P_2 \Delta_2}{\ell_2} \qquad (10.24)$$

The sway forces must be added to the applied lateral force H at any story level, and the structure is then reanalyzed, giving new deflections and increased moments. If the lateral deflections increase significantly (say more than 5 percent), new dH sway forces are computed, and the structure is reanalyzed for the sum of the applied lateral forces and the new sway forces. Iteration is continued until changes are insignificant. Generally one or two cycles of iteration are adequate for structures of reasonable lateral stiffness (Ref. 10.3).

It is noted in Ref. 10.17 that a correction must be made in the analysis to account for the differences in shape between the $P\Delta$ moment diagram that has the same shape as the deflected column and the moment diagram associated with the $P\Delta/\ell$ forces, which is linear between the joints at the column ends. The area of the actual $P\Delta$ moment diagram is larger than the linear equivalent representation, and consequently lateral deflections will be larger. The difference will vary depending on the relative stiffnesses of the column and the beams framing into the joints. In Ref. 10.17, it is suggested that the increased deflection can be accounted for by taking the sway forces dH as 15 percent greater than the calculated value for each iteration. Iteration and the 15 percent increase in deflection are not required if the program performs a full nonlinear geometric analysis, since the $P\Delta$ moments are calculated in full.

The accuracy of the results of a $P\text{-}\Delta$ analysis will be strongly influenced by the values of member stiffness used, by foundation rotations, if any, and by the effects of concrete creep. In connection with creep effects, lateral loads causing significant sway are usually wind or earthquake loads of short duration, so creep effects are minimal. In general, the use of sway frames to resist *sustained* lateral loads, for example, from earth or liquid pressures, is not recommended, and it would be preferable to include shear walls or other elements to resist these loads.

REFERENCES

10.1. S. P. Timoshenko and J. M. Gere, *Theory of Elastic Stability*, 3d ed., McGraw-Hill, New York, 1969.

10.2. T. V. Galambos (ed.), *Guide to Stability Design Criteria for Metal Structures*, 5th ed., John Wiley & Sons, New York, 1998.

10.3. J. G. MacGregor and S. E. Hage, "Stability Analysis and Design of Concrete Frames," *J. Struct. Div.*, ASCE, vol. 103, no. ST10, 1977, pp. 1953–1970.

10.4. S.-M. A. Lai and J. G. MacGregor, "Geometric Nonlinearities in Unbraced Multistory Frames," *J. Struct. Eng.*, ASCE, vol. 109, no. 11, 1983, pp. 2528–2545.

10.5. M. Khuntia and S. K. Ghosh, "Flexural Stiffness of Reinforced Concrete Columns and Beams: Analytical Approach," *ACI Struct. J.*, vol. 101, no. 3, 2004, pp. 351–363.

10.6. M. Khuntia and S. K. Ghosh, "Flexural Stiffness of Reinforced Concrete Columns and Beams: Experimental Verification," *ACI Struct. J.*, vol. 101, no. 3, 2004, pp. 364–374.

10.7. J. G. MacGregor, J. E. Breen, and E. O. Pfrang, "Design of Slender Concrete Columns," *J. ACI*, vol. 67, no. 1, 1970, pp. 6–28.

10.8. J. G. MacGregor, V. H. Oelhafen, and S. E. Hage, "A Reexamination of the *EI* Value for Slender Columns," *Reinforced Concrete Columns*, American Concrete Institute, Detroit, MI, 1975, pp. 1–40.

10.9. *Code of Practice for the Structural Use of Concrete*, Part 1, "Design Materials and Workmanship" (CP110: Part 1, 1972), British Standards Institution, London, 1972.

10.10. W. B. Cranston, "Analysis and Design of Reinforced Concrete Columns," *Research Report* No. 20, Paper 41.020, Cement and Concrete Association, London, 1972.

10.11. R. W. Furlong, "Column Slenderness and Charts for Design," *J. ACI*, vol. 68, no. 1, 1971, pp. 9–18.

10.12. M. Valley and P. Dumonteil, Disc. of "*K*-Factor Equation to Alignment Charts for Column Design," by L. Duan, W.-S. King, and W.-F. Chen, *ACI Struct. J.*, vol. 91, no. 2, Mar.–Apr. 1994, pp. 229–230.

10.13. J. S. Ford, D. C. Chang, and J. E. Breen, "Design Indications from Tests of Unbraced Multipanel Concrete Frames," *Concr. Intl.*, vol. 3, no. 3, 1981, pp. 37–47.

10.14. *Building Code Requirements for Structural Concrete*, ACI 318-14, American Concrete Institute, Farmington Hills, MI, 2014.

10.15. *Commentary on Building Code Requirements for Structural Concrete*, ACI 318R-14, American Concrete Institute, Farmington Hills, MI, 2014 (published as part of Ref. 10.14).

10.16. J. G. MacGregor, "Design of Slender Concrete Columns—Revisited," *ACI Struct. J.*, vol. 90, no. 3, 1993, pp. 302–309.

10.17. J. G. MacGregor, *Reinforced Concrete*, 3d ed., Prentice-Hall, Upper Saddle River, NJ, 1997.

10.18. B. R. Wood, D. Beaulieu, and P. F. Adams, "Column Design by *P*-Delta Model," *Proc. ASCE*, vol. 102, no. ST2, 1976, pp. 487–500.

10.19. B. R. Wood, D. Beaulieu, and P. F. Adams, "Further Aspects of Design by *P*-Delta Model," *J. Struct. Div.*, ASCE, vol. 102, no. ST3, 1976, pp. 487–500.

10.20. R. W. Furlong, "Rational Analysis of Multistory Concrete Structures," *Concr. Intl.*, vol. 3, no. 6, 1981, pp. 29–35.

PROBLEMS

10.1. The 15×15 in. column shown in Fig. P10.1 must extend from footing level to the second floor of a braced frame structure with an unsupported length of 20.5 ft. Exterior exposure requires 2 in. clear cover for the outermost steel. Analysis indicates that the critical loading corresponds with the following service loads: (*a*) from dead loads, $P = 180$ kips, $M_{\text{top}} = 30$ ft-kips, $M_{\text{bot}} = 15$ ft-kips; (*b*) from live loads, $P = 110$ kips, $M_{\text{top}} = 60$ ft-kips, $M_{\text{bot}} = 30$ ft-kips, with the column bent in double curvature as shown. The effective length factor k determined using Fig. 10.13*a* is 0.90. Material strengths are $f'_c = 4000$ psi and $f_y = 60,000$ psi. Using the ACI moment magnifier method, determine whether the column is adequate to resist these loads.

FIGURE P10.1

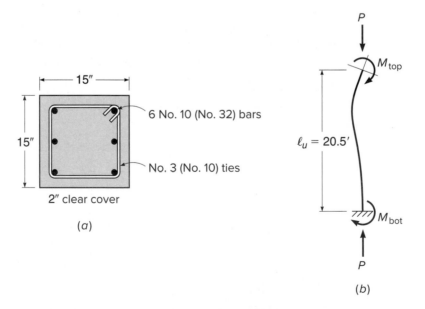

6 No. 10 (No. 32) bars

No. 3 (No. 10) ties

2″ clear cover

(*a*)

$\ell_u = 20.5'$

(*b*)

10.2. The structure shown in Fig. P10.2*a* requires tall slender columns at the left side. It is fully braced by shear walls on the right. All columns are 16×16 in., as shown in Fig. P10.2*b*, and all beams are 24×18 in. with 6 in. monolithic floor slab, as in Fig. P10.2c. Trial calculations call for column reinforcement as shown. Alternate load analysis indicates the critical condition with column *AB* bent in single curvature, and service loads and moments as follows: from dead loads, $P = 139$ kips, $M_{top} = 61$ ft-kips, $M_{bot} = 41$ ft-kips; from live load, $P = 93$ kips, $M_{top} = 41$ ft-kips, $M_{bot} = 27$ ft-kips. Material strengths are $f'_c = 4000$ psi and $f_y = 60,000$ psi. Is the proposed column, reinforced as shown, satisfactory for this load condition? Use Eq. (10.18) to calculate *EI* for the column.

FIGURE 10.2

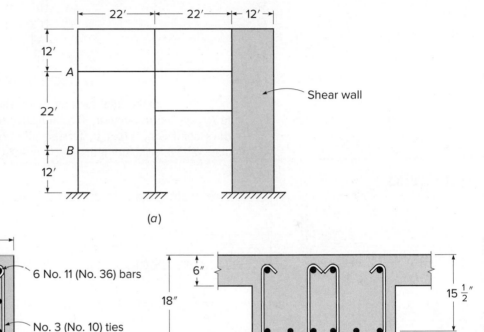

(a)

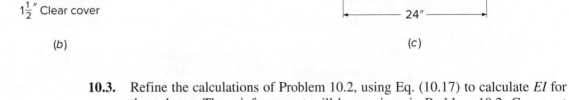

(b)

(c)

10.3. Refine the calculations of Problem 10.2, using Eq. (10.17) to calculate *EI* for the column. The reinforcement will be as given in Problem 10.2. Comment on your results.

10.4. An interior column in a braced frame has an unsupported length of 20 ft and carries the following service load forces and moments: (*a*) from dead loads, $P = 180$ kips, $M_{top} = 28$ ft-kips, $M_{bot} = -28$ ft-kips; (*b*) from live loads, $P = 220$ kips, $M_{top} = 112$ ft-kips, $M_{bot} = 112$ ft-kips, with the signs of the moments representing double curvature under dead load and single curvature under live load. Rotational restraint factors at the top and bottom may be taken equal to 1.0. Design a square tied column to resist these loads, with a reinforcement ratio of about 0.02. Use $f'_c = 4000$ psi and $f_y = 60,000$ psi.

10.5. The first three floors of a multistory building are shown in Fig. P10.5. The lateral load resisting frame consists of 20 × 20 in. exterior columns, 24 × 24 in. interior columns, and 36 in. wide × 24 in. deep girders. The center-to-center column height is 16 ft. For the second-story columns, the service gravity dead and live loads and the horizontal wind loads based on an elastic first-order analysis of the frame are

	Cols. *A2 and E2*	Cols. *B2 and D2*	Col. *C2*
P_{dead}	348 kips	757 kips	688 kips
P_{live}	137 kips	307 kips	295 kips
P_{wind}	±30 kips	±14 kips	0 kips
V_{wind}	10.5 kips	21.5 kips	21.5 kips
$M_{2,dead}$		31 ft-kips	
$M_{2,live}$		161 ft-kips	
$M_{2,wind}$		168 ft-kips	
$M_{1,dead}$		−34 ft-kips	
$M_{1,live}$		108 ft-kips	
$M_{1,wind}$		−157 ft-kips	

A matrix analysis for the total factored wind shear of 85.5 kips, using values of E and I specified in Section 10.5, indicates that the relative lateral deflection of the second story is 0.38 in. Design columns *B2* and *D2* using Eq. (10.21) to calculate δ_s. Material strengths are $f_c' = 4000$ psi and $f_y = 60,000$ psi.

FIGURE P10.5

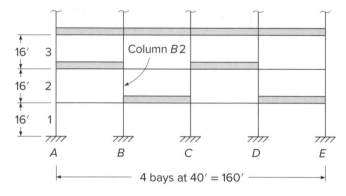

10.6. Repeat Problem 10.5, using Eq. (10.22) to calculate δ_s. Comment on your results.

10.7. Redesign column *C3* from Example 10.2 for a story height of 16 ft, a column unsupported length of 15 ft, and a relative lateral displacement of the third story of 1.80 in. Loads and other dimensions remain unchanged.

10.8. The first four floors of a multistory building are shown in Fig. P10.8. The lateral load-resisting frame consists of 22 × 22 in. exterior columns, 26 × 26 in.

FIGURE P10.8

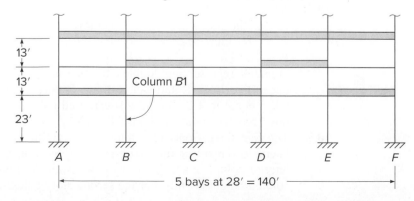

interior columns, and 33 in. wide × 18 in. deep girders. The foundation is at ground level, supported on drilled piers, and may be considered as fully fixed against rotation. The first-story columns have a clear height to the girder soffit of 21 ft 6 in., giving a floor-to-floor height of 23 ft. The upper floors have a center-to-center spacing of 13 ft. For the first-story columns, the service gravity dead and live loads plus the horizontal and vertical wind loads based on an elastic first-order analysis of the frame are

	Cols. $A1$ *and* $F1$	Cols. $B1$ *and* $E1$	Cols. $C1$ *and* $D1$
P_{dead}	495 kips	1090 kips	989 kips
P_{live}	99 kips	206 kips	188 kips
P_{wind}	±51 kips	±30 kips	±10 kips
V_{wind}	18 kips	35 kips	35 kips
$M_{2,dead}$		4 ft-kips	
$M_{2,live}$		70 ft-kips	
$M_{2,wind}$		384 ft-kips	
$M_{1,dead}$		−2 ft-kips	
$M_{1,live}$		−35 ft-kips	
$M_{1,wind}$		−384 ft-kips	

A matrix analysis for the total factored wind shear of 176 kips, using values of E and I specified in Section 10.5, indicates that the relative lateral deflection of the second story is 0.64 in. Design columns $B1$ and $E1$, using Eq. (10.21) to calculate δ_s. Material strengths are $f'_c = 4000$ psi and $f_y = 60,000$ psi.

10.9. Repeat Problem 10.8, using Eq. (10.22) to calculate δ_s. Comment on your results.

11

Analysis, Idealization, and Preliminarily Design of Reinforced Concrete Beams and Frames

11.1 CONTINUITY

The individual members that compose a steel or timber structure are fabricated or cut separately and joined together by rivets, bolts, welds, or nails. Unless the joints are specially designed for rigidity, they are too flexible to transfer moments of significant magnitude from one member to another. In contrast, in reinforced concrete structures, as much of the concrete as is practical is placed in one single operation. Reinforcing steel is not terminated at the ends of a member but is extended through the joints into adjacent members. At construction joints, special care is taken to bond the new concrete to the old by carefully cleaning the latter, by extending the reinforcement through the joint, and by other means. As a result, reinforced concrete structures usually represent monolithic, or continuous, units. A load applied at one location causes deformation and stress at all other locations. Even in precast concrete construction, which resembles steel construction in that individual members are brought to the job site and joined in the field, connections are often designed to provide for the transfer of moment as well as shear and axial load, producing at least partial continuity.

The effect of continuity is most simply illustrated by a continuous beam, as shown in Fig. 11.1*a*. With simple spans, such as provided in many types of steel construction, only the loaded member *CD* would deform, and all other members of the structure would remain straight. But with continuity from one member to the next through the support regions, as in a reinforced concrete structure, the distortion caused by a load on one single span is seen to spread to all other spans, although the magnitude of deformation decreases with increasing distance from the loaded member. All members of the six-span structure are subject to curvature, and thus also to bending moment, as a result of loading span *CD*.

Similarly, for the rigid-jointed frame of Fig. 11.1*b*, the distortion caused by a load on the single member *GH* spreads to all beams and all columns, although, as before, the effect decreases with increasing distance from the load. All members are subject to bending moment, even though they may carry no transverse load.

If horizontal forces, such as forces caused by wind or seismic action, act on a frame, it deforms as illustrated by Fig. 11.1*c*. Here, too, all members of the frame distort, even though the forces act only on the left side; the amount of distortion is seen to be the same for all corresponding members, regardless of their distance from the points of loading, in contrast to the case of vertical loading. A member such as *EH*, even without a directly applied transverse load, will experience deformations and associated bending moment.

In *statically determinate structures*, such as simple-span beams, the deflected shape and the moments and shears depend only on the type and magnitude of the

FIGURE 11.1
Deflected shape of
continuous beams and
frames.

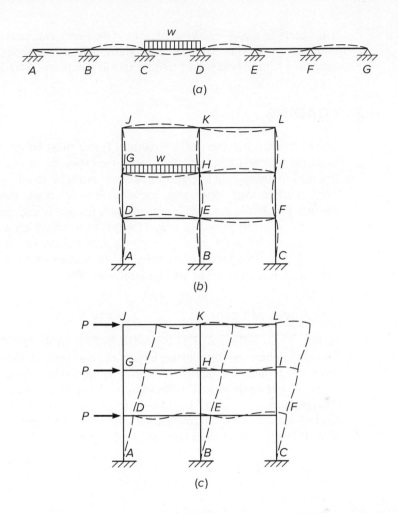

loads and the dimensions of the member. In contrast, inspection of the *statically indeterminate structures* in Fig. 11.1 shows that the deflection curve of any member depends not only on the loads but also on the joint rotations, whose magnitudes in turn depend on the distortion of adjacent, rigidly connected members. For a rigid joint such as joint *H* in the frame shown in Fig. 11.1*b* or *c*, all the rotations at the near ends of all members framing into that joint must be the same. For a correct design of continuous beams and frames, it is evidently necessary to determine moments, shears, and thrusts considering the effect of continuity at the joints.

The determination of these internal forces in continuously reinforced concrete structures is usually based on *elastic analysis* of the structure at factored loads with methods that will be described in Sections 11.2 through 11.5. Such analysis requires knowledge of the cross-sectional dimensions of the members. Member dimensions are initially estimated during preliminary design, which is described in Section 11.6 along with guidelines for establishing member proportions. For checking the results of more exact analysis, the approximate methods of Section 11.7 are useful. For many structures, a full elastic analysis is not justified, and the ACI coefficient method of analysis described in Section 11.8 provides an adequate basis for design moments and shears.

Before failure, reinforced concrete sections are usually capable of considerable inelastic rotation at nearly constant moment, as was described in Section 7.9.

This permits a *redistribution of elastic moments* and provides the basis for *plastic analysis* of beams, frames, and slabs. Plastic analysis will be developed in Section 11.9 for beams and frames and in Chapters 23 and 24 for slabs.

11.2 LOADING

The individual members of a structural frame must be designed for the worst combination of loads that can reasonably be expected to occur during its useful life. Internal moments, shears, and thrusts are brought about by the combined effect of dead and live loads, plus other loads, such as wind and earthquake, as discussed in Section 1.7. While dead loads are constant, live loads such as floor loads from human occupancy can be placed in various ways, some of which will result in larger effects than others. In addition, the various combinations of factored loads specified in Table 1.2 must be used to determine the load cases that govern member design. The subject of load placement will be addressed first.

a. Placement of Loads

In Fig. 11.2a only span *CD* is loaded by live load. The distortions of the various frame members are seen to be largest in, and immediately adjacent to, the loaded span and to decrease rapidly with increasing distance from the load. Since bending moments are proportional to curvatures, the moments in more remote members are correspondingly smaller than those in, or close to, the loaded span. However, the loading shown in Fig. 11.2a does not produce the maximum possible positive moment in *CD*. In fact, if additional live load were placed on span *AB*, this span would bend

FIGURE 11.2
Alternate live loadings for maximum effects.

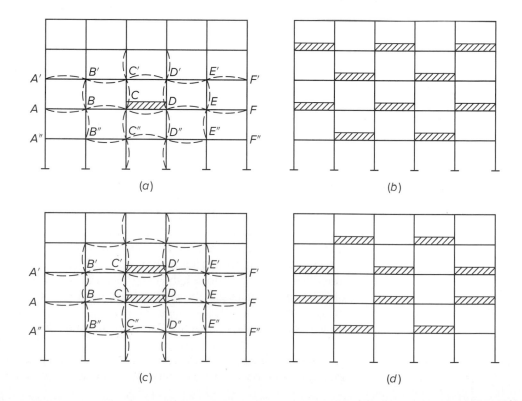

down, *BC* would bend up, and *CD* itself would bend down in the same manner, although to a lesser degree, as it is bent by its own load. Hence, the positive moment in *CD* is increased if *AB* and, by the same reasoning, *EF* are loaded simultaneously. By expanding the same reasoning to the other members of the frame, it is easy to see that the checkerboard pattern of live load shown in Fig. 11.2*b* produces the largest possible positive moments, not only in *CD* but also in all loaded spans. Hence, two such checkerboard patterns are required to obtain the maximum positive moments in all spans.

In addition to maximum span moments, it is often necessary to investigate minimum span moments. Dead load, acting as it does on all spans, usually produces only positive span moments. However, live load, placed as in Fig. 11.2*a*, and even more so in Fig. 11.2*b*, is seen to bend the unloaded spans upward, that is, to produce negative moments in the span. If these negative live load moments are larger than the generally positive dead load moments, a given girder, depending on load position, may be subject at one time to positive span moments and at another to negative span moments. It must be designed to withstand both types of moments; that is, it must be furnished with tensile steel at both top and bottom. Thus, the loading of Fig. 11.2*b*, in addition to giving maximum span moments in the loaded spans, gives minimum span moments in the unloaded spans.

Maximum negative moments at the supports of the girders are obtained, on the other hand, if loads are placed on the two spans adjacent to the particular support, as will be illustrated in Section 11.3, and in a corresponding pattern on the more remote girders. A separate loading scheme of this type is then required for each support for which maximum negative moments are to be computed.

In each column, the largest moments occur at the top or bottom. While the loading shown in Fig. 11.2*c* results in large moments at the ends of columns *CC'* and *DD'*, the reader can easily be convinced that these moments are further augmented if additional loads are placed as shown in Fig. 11.2*d*.

It is seen from this brief discussion that to calculate the maximum possible moments at all critical points of a frame, live load must be placed in a great variety of different schemes. In most practical cases, however, consideration of the relative magnitude of effects will permit limitation of analysis to a small number of significant cases.

b. Load Combinations

The ACI Code requires that structures be designed for a number of load combinations, as discussed in Section 1.7. Thus, for example, factored load combinations might include (1) dead plus live load; (2) three possible combinations that include dead, live, and wind load; and (3) two combinations that include dead load, live load, and earthquake load, with some of the combinations including snow, rain, and roof live load. While each of the combinations may be considered as an individual loading condition, experience has shown that the most efficient technique involves separate analyses for each of the basic loads without load factors, that is, a full analysis for unfactored dead load only, separate analyses for the various live load distributions described in Section 11.2*a*, and separate analyses for each of the other loads (wind, snow, etc.). Once the separate analyses are completed, it is a simple matter to combine the results using the appropriate load factor for each type of load. This procedure is most advantageous because, for example, live load may require a load factor of 1.6 for one combination, a value of 1.0 for another, and a value of 0.5 for yet another. Once the forces have been calculated for each combination, the combination of loads that governs for each member can usually be identified by inspection.

11.3   SIMPLIFICATIONS IN FRAME ANALYSIS

Considering the complexity of many practical building frames and the need to account for the possibility of alternative loadings, there is evidently a need to simplify. This can be done by means of certain approximations that allow the determination of moments with reasonable accuracy while substantially reducing the amount of computation.

Numerous trial computations have shown that, for building frames with a reasonably regular outline, not involving unusual asymmetry of loading or shape, the influence of sidesway caused by vertical loads can be neglected. In that case, moments due to vertical loads are determined with sufficient accuracy by dividing the entire frame into simpler subframes. Each of these consists of one continuous beam, plus the top and bottom columns framing into that particular beam. Placing the live loads on the beam in the most unfavorable manner permits sufficiently accurate determination of all beam moments, as well as the moments at the top ends of the bottom columns and the bottom ends of the top columns. For this partial structure, the far ends of the columns are considered fixed, except for such first-floor or basement columns where soil and foundation conditions dictate the assumption of hinged ends. Such an approach is explicitly permitted by ACI Code 6.3.1.2 and 6.4, which allow the following assumptions for floor and roof members under gravity load:

1. To calculate moments and shears in columns, beams, and slabs, the structural model may be limited to the members in the level being considered and the columns above and below that level; the far ends of columns built integrally with the structure may be considered fixed.
2. The maximum positive moment near midspan occurs with the factored live load on the span and on alternate spans, and the maximum negative moment at a support occurs with the factored live load on adjacent spans only.

When investigating the maximum negative moment at any joint, negligible error will result if the joints second removed in each direction are considered to be completely fixed. Similarly, in determining maximum or minimum span moments, the joints at the far ends of the adjacent spans may be considered fixed. Thus, individual portions of a frame of many members may be investigated separately.

Figure 11.3 demonstrates the application of the ACI Code requirements for live load on a three-span subframe. The loading in Fig. 11.3a results in maximum positive moments in the exterior spans, the minimum positive moment in the center span, and the maximum negative moments at the interior faces of the exterior columns. The loading shown in Fig. 11.3b results in the maximum positive moment in the center span and minimum positive moments in the exterior spans. The loading in Fig. 11.3c results in maximum negative moment at both faces of the interior columns. Since the structure is symmetrical, values of moment and shear obtained for the loading shown in Fig. 11.3c apply to the right side of the structure as well as the left. Due to the simplicity of this structure, joints away from the spans of interest are not treated as fixed.

Moments and shears used for design are determined by combining the moment and shear diagrams for the individual load cases to obtain the maximum values along each span length. The resulting *envelope moment and shear diagrams* are shown in Fig. 11.3d and e, respectively. The moment and shear envelopes (note the range of

FIGURE 11.3
Subframe loading as required by ACI Code 6.4: Loading for (*a*) maximum positive moments in the exterior spans, the minimum positive moment in the center span, and the maximum negative moments at the interior faces of the exterior columns; (*b*) maximum positive moment in the center span and minimum positive moments in the exterior spans; (*c*) maximum negative moment at both faces of the interior columns; (*d*) envelope moment diagram; and (*e*) envelope shear diagram. (DL and LL represent factored dead and live loads, respectively.)

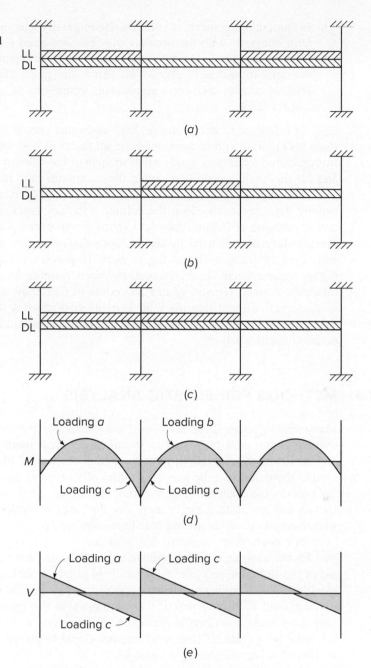

positions for points of inflection and points of zero shear) are used not only to design the critical sections but also to determine cutoff points for flexural reinforcement and requirements for shear reinforcement.

In regard to columns, the ACI Code indicates

1. The factored axial load and factored moment occurring simultaneously for each applicable factored load combination shall be considered (ACI Code 10.4.2.1).
2. For frames or continuous construction, consideration shall be given to the effect of floor and roof load patterns on the transfer of moment to exterior and interior columns and of eccentric loading due to other causes (ACI Code 6.6.2.2).

3. In computing moments in columns due to gravity loading, the far ends of columns built integrally with the structure may be considered fixed (ACI Code 6.3.1.2).
4. Floor or roof level moments shall be resisted by distributing the moment between columns immediately above and below the given floor in proportion to the relative column stiffnesses considering conditions of restraint (ACI Code 6.5.5 and 6.6.2.1).

In reference to item 1 above, load cases that should be specifically considered include (1) those with factored loads on all floors or roof and the maximum moment from factored loads on a single adjacent span of the floor or roof under consideration and (2) the loading condition giving the maximum ratio of moment to axial load.

Although it is not addressed in the ACI Code, axial loads on columns are usually determined based on the column tributary areas, which are defined based on the midspan of flexural members framing into each column. The axial load from the tributary area is used in design, with the exception of first interior columns, which are typically designed for an extra 10 percent axial load to account for the higher shear expected in the flexural members framing into the exterior face of first interior columns. The use of this procedure to determine axial loads due to gravity is conservative (note that the total vertical load exceeds the factored loads on the structure) and is adequately close to the values that would be obtained from a more detailed frame analysis.

11.4 METHODS FOR ELASTIC ANALYSIS

Many methods have been developed over the years for the elastic analysis of continuous beams and frames. The so-called classical methods (Ref. 11.1), such as application of the theorem of three moments, the method of least work (Castigliano's second theorem), and the general method of consistent deformation, will prove useful only in the analysis of continuous beams having few spans or of very simple frames and are, in fact, rarely used. For the cases generally encountered in practice, such methods prove exceedingly tedious, and alternative approaches are preferred.

For many years, moment distribution (Ref. 11.1) provided the basic analytical tool for the analysis of indeterminate concrete beams and frames, originally with the aid of the slide rule and later with handheld programmable calculators. For relatively small problems, moment distribution may still provide the most rapid results, and it is often used in current practice, for example, in the Equivalent Frame Method of design for slabs described in Section 13.6. However, with computers, manual methods have been replaced largely by computational techniques that provide rapid solutions with a high degree of accuracy.

Approximate methods of analysis, based either on careful sketches of the shape of the deformed structure under load or on moment coefficients, still provide a means for rapid estimation of internal forces and moments (Ref. 11.2). Such estimates are useful in preliminary design and in checking more exact solutions for gross errors that might result from input errors. In structures of minor importance, approximations may even provide the basis for final design.

In view of the number of excellent texts now available that treat methods of analysis, the present discussion will be confined to an evaluation of the usefulness of several of the more important of these, with particular reference to the analysis of reinforced concrete structures. Certain idealizations and approximations that facilitate the solution in practical cases will be described in more detail.

a. Moment Distribution

In 1932, Hardy Cross developed the method of moment distribution to solve problems in frame analysis that involve many unknown joint displacements and rotations. For the next three decades, moment distribution provided the standard means in engineering offices for the analysis of indeterminate frames. Even now, it serves as the basic analytical tool if computer facilities are not available.

In the moment distribution method (Ref. 11.1), the fixed-end moments for each member are modified in a series of cycles, each converging on the precise final result, to account for rotation and translation of the joints. The resulting series can be terminated whenever one reaches the degree of accuracy required. After member end moments are obtained, all member stress resultants can be obtained from the laws of statics.

It has been found by comparative analyses that, except in unusual cases, building-frame moments found by modifying fixed-end moments by only two cycles of moment distribution will be sufficiently accurate for design purposes (Ref. 11.3).

b. Computer Software

Computer software programs provide highly sophisticated tools for analysis of concrete structures. The simplest programs can be used to conduct linear elastic analyses of indeterminate structures based on either service or factored loads. The factored axial forces, moments, shear, and torsional moments from these analyses are used to design the structural members. This software requires input of member sizes and material properties. The approximate and simplified analyses presented in this chapter allow rapid determination of initial member sizes. To obtain reasonable service level information, particularly deflections, an appropriate idealization of the structure, as discussed in this chapter, is essential.

Advanced software can conduct three-dimensional structural analyses that include first- and second-order stability effects and material degradation. These programs automatically conduct multiple iterations, with each iteration incorporating changes in deflection and material behavior from the previous cycle. Software, such as found in the library at the Pacific Earthquake Engineering Research (PEER) Center, developed under the auspices of the National Science Foundation, provides powerful tools for analyzing the earthquake response of buildings. Some tall building designs are subjected to a suite of simulated earthquakes as the basis for performance-based building design (Ref. 11.4).

Building information modeling (BIM) software incorporates three-dimensional architectural visualization, structural analysis, and other digital representations into interchangeable files that can be shared by professionals working on the project (Ref. 11.5). These three-dimensional models capture important design information and can identify interferences between structural, mechanical, and electrical systems. Autodesk's *Revit* and Bentley Systems' *Open Site Designer* are just two examples of integrated BIM software.

11.5 IDEALIZATION OF THE STRUCTURE

It is seldom possible for the engineer to analyze an actual complex redundant structure. Almost without exception, certain idealizations must be made in devising an analytical model, so that the analysis will be practically possible. Thus, three-dimensional members are represented by straight lines, generally coincident with the actual centroidal axis. Supports are idealized as rollers, hinges, or rigid joints.

Loads actually distributed over a finite area are assumed to be point loads. In three-dimensional framed structures, analysis is often limited to plane frames, each of which is assumed to act independently.

In the idealization of reinforced concrete frames, certain questions require special comment. The most important of these pertain to effective span lengths, effective moments of inertia, and conditions of support.

a.   Effective Span Length

In elastic frame analysis, a structure is usually represented by a simple line diagram, based dimensionally on the centerline distances between columns and between floor beams. Actually, the depths of beams and the widths of columns (in the plane of the frame) amount to sizable fractions of the respective lengths of these members; their clear lengths are therefore considerably smaller than their centerline distances between joints.

It is evident that the usual assumption in frame analysis that the members are prismatic, with constant moment of inertia between centerlines, is not strictly correct. A beam intersecting a column may be prismatic up to the column face, but from that point to the column centerline it has a greatly increased depth, with a moment of inertia that could be considered infinite compared with that of the remainder of the span. A similar variation in width and moment of inertia is obtained for the columns. Thus, to be strictly correct, the actual variation in member depth should be considered in the analysis. Qualitatively, this would increase beam support moments somewhat and decrease span moments. In addition, it is apparent that the critical section for design for negative bending would be at the face of the support, and not at the centerline, since for all practical purposes an unlimited effective depth is obtained in the beam across the width of the support.

It will be observed that, in the case of the columns, the moment gradient is not very steep, so that the difference between centerline moment and the moment at the top or bottom face of the beam is small and in most cases can be disregarded. However, the slope of the moment diagram for the beam is usually quite steep in the region of the support, and there will be a substantial difference between the support centerline moment and face moment. If the former were used in proportioning the member, an unnecessarily large section would result. It is desirable, then, to reduce support moments found by elastic analysis to account for the finite width of the supports.

In Fig. 11.4, the change in moment between the support centerline and the support face will be equal to the area under the shear diagram between those two points. For knife edge supports, this shear area is seen to be very nearly equal to $Va\ell/2$. Actually, however, the reaction is distributed in some unknown way across the width of the support. This will have the effect of modifying the shear diagram as shown by the dashed line; it has been proposed that the reduced area be taken as equal to $Va\ell/3$. The fact that the reaction is distributed will modify the moment diagram as well as the shear diagram, causing a slight rounding of the negative moment peak, as shown in the figure, and the reduction of $Va\ell/3$ is properly applied to the moment diagram after the peak has been rounded. This will give nearly the same face moment as would be obtained by deducting the amount $Va\ell/2$ from the peak moment.

Another effect is present, however: the modification of the moment diagram due to the increased moment of inertia of the beam at the column. This effect is similar to that of a haunch, and it will mean slightly increased negative moment and slightly decreased positive moment. For ordinary values of the ratio a, this shift in the moment curve will be on the order of $Va\ell/6$. Thus, it is convenient simply to deduct the amount $Va\ell/3$ from the unrounded peak moment obtained from elastic

FIGURE 11.4
Reduction of negative and
positive moments in a frame.

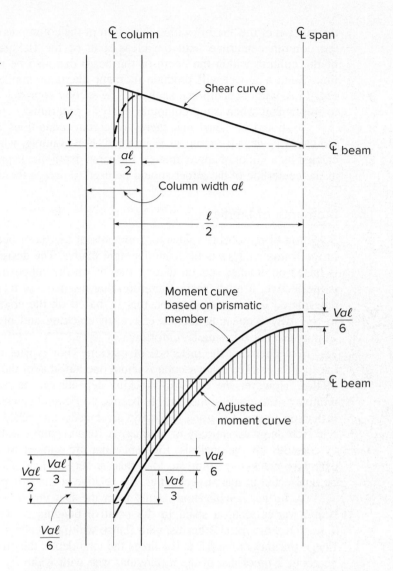

analysis. This allows for (1) the actual rounding of the shear diagram and the negative moment peak due to the distributed reaction and (2) the downward shift of the moment curve due to the haunch effect at the supports. The consistent reduction in positive moment of $Va\ell/6$ is illustrated in Fig. 11.4.

With this said, there are two other approaches that are often used by structural designers. The first is to analyze the structure based on the simple line diagram and to reduce the moment from the column centerline to the face of the support by $Va\ell/2$ without adjusting for the higher effective stiffness within the thickness width of the column. The moment diagram, although somewhat less realistic than represented by the lower curve in Fig. 11.4, still satisfies statics and requires less flexural reinforcement at the face of the support. As a consequence, there is less congestion in the beam-column joint, a location where it is often difficult to place concrete because of the high quantity of reinforcing steel from the flexural members framing into the column (usually from two different directions) and from the column itself. The somewhat higher percentage of reinforcement required at midspan usually causes little difficulty in concrete placement. The second approach involves representing the portion of the "beam" within the width of the column as a rigid link that connects the column centerline with the clear span of the flexural member. The portion of the column within the depth of the beam can also be represented using a rigid link. Such a model will produce moment diagrams similar to the lower curve in Fig. 11.4, without additional analysis. The second approach is both realistic and easy to implement when using computer analysis programs.

It should be noted that there are certain conditions of support for which no reduction in negative moment is justified. For example, when a continuous beam is carried by a girder of approximately the same depth, the negative moment in the beam at the centerline of the girder should be used to design the negative reinforcing steel.

b.   Moments of Inertia

Selection of reasonable values for moments of inertia of beams and columns for use in the frame analysis is far from a simple matter. The design of beams and columns is based on cracked section theory, that is, on the supposition that tension concrete is ineffective. It might seem, therefore, that moments of inertia to be used should be determined in the same manner, that is, based on the cracked transformed section, in this way accounting for the effects of cracking and presence of reinforcement. Things are not this simple, unfortunately.

Consider first the influence of cracking. For typical members, the moment of inertia of a cracked beam section is about one-half that of the uncracked gross concrete section. However, the extent of cracking depends on the magnitude of the moments relative to the cracking moment. In beams, no flexural cracks would be found near the inflection points. Columns, typically, are mostly uncracked, except for those having relatively large eccentricity of loading. A fundamental question, too, is the load level to consider for the analysis. Elements that are subject to cracking will have more extensive cracks near ultimate load than at service load. Compression members will be unaffected in this respect. Thus, the relative stiffness depends on load level.

A further complication results from the fact that the effective cross section of beams varies along a span. In the positive bending region, a beam usually has a T section. For typical T beams, with flange width typically 4 to 6 times web width and flange thickness from 0.2 to 0.4 times the total depth, the gross moment of inertia will be about 2 times that of the rectangular web with width b_w and depth h. However, in the negative bending region near the supports, the bottom of the section is in compression. The T flange is cracked, and the effective cross section is therefore rectangular.

The amount and arrangement of reinforcement are also influential. In beams, if bottom bars are continued through the supports, as is often done, this steel acts as compression reinforcement and stiffens the section. In columns, reinforcement ratios are generally much higher than in beams, adding to the stiffness.

Given these complications, it is clear that some simplifications are necessary. It is helpful to note that, in most cases, it is only the *ratio* of member stiffnesses that influences the final result, not the absolute value of the stiffnesses. The stiffness ratios may be but little affected by different assumptions in computing moment of inertia if there is consistency for all members.

In practice, it is generally sufficiently accurate to base stiffness calculations for frame analysis on the gross concrete cross section of the columns. In continuous T beams, cracking will reduce the moment of inertia to about one-half that of the uncracked section. Thus, the effect of the flanges and the effect of cracking may nearly cancel in the positive bending region. In the negative moment regions, there are no flanges; however, if bottom bars continue through the supports to serve as compression steel, the added stiffness tends to compensate for lack of compression flange. Thus, for beams, generally a constant moment of inertia can be used, based on the rectangular cross-sectional area $b_w h$.

ACI Code 6.3.1.1 states only that the relative stiffness be selected based on reasonable assumptions and that the assumptions be consistent for each analysis. ACI Commentary 6.3.1.1 explains that separate analyses may be used, for example, to check different conditions, such as serviceability and strength. ACI Commentary 6.3.1.1 notes that *relative* values of stiffness are important and that a common assumption is to use one-half of the gross EI for beams and the gross EI for the columns. Additional guidance is given in Section 10.5 and ACI Code 6.6.3.1, which specifies the section properties to be used for frames subject to sidesway: 35 percent of the *gross* moment of inertia for beams and 70 percent for columns. This differs from the guidance provided in ACI Commentary 6.3.1.1 but, except for a factor of 0.70, matches the guidance provided in the earlier discussion.

c. Conditions of Support

For purposes of analysis, many structures can be divided into a number of two-dimensional frames. Even for such cases, however, there are situations in which it is impossible to predict with accuracy what the conditions of restraint might be at the ends of a span; yet moments are frequently affected to a considerable degree by the choice made. In many other cases, it is necessary to recognize that structures may be three-dimensional. The rotational restraint at a joint may be influenced or even governed by the characteristics of members framing into that joint at right angles. Adjacent members or frames parallel to the one under primary consideration may likewise influence its performance.

If floor beams are cast monolithically with reinforced concrete walls (frequently the case when first-floor beams are carried on foundation walls), the moment of inertia of the wall about an axis parallel to its face may be so large that the beam end can be considered completely fixed for all practical purposes. If the wall is relatively thin or the beam particularly massive, the moment of inertia of each should be calculated, that of the wall being equal to $bt^3/12$, where t is the wall thickness and b the wall width tributary to one beam.

If the outer ends of concrete beams rest on masonry walls, as is sometimes the case, an assumption of zero rotational restraint (that is, a hinged support) is probably closest to the actual case.

For columns supported on relatively small footings, which in turn rest on compressible soil, a hinged end is generally assumed, since such soils offer but little resistance to rotation of the footing. If, on the other hand, the footings rest on solid rock, or if a cluster of piles is used with their upper portion encased by a concrete cap, the effect is to provide almost complete fixity for the supported column, and this should be assumed in the analysis. Columns supported by a continuous foundation mat should likewise be assumed fixed at their lower ends.

If members framing into a joint in a direction perpendicular to the plane of the frame under analysis have sufficient torsional stiffness, and if their far ends are fixed or nearly so, their effect on joint rigidity should be included in the computations. The torsional stiffness of a member of length ℓ is given by the expression GJ/ℓ, where G is the shear modulus of elasticity of concrete (approximately $E_c/2.2$) and J is the torsional stiffness factor of the member. For beams with rectangular cross sections or with sections made up of rectangular elements, J can be taken equal to $\Sigma(hb^3/3 - b^4/5)$, in which h and b are the cross-sectional dimensions of each

FIGURE 11.5
Slab, beam, and girder floor
system.

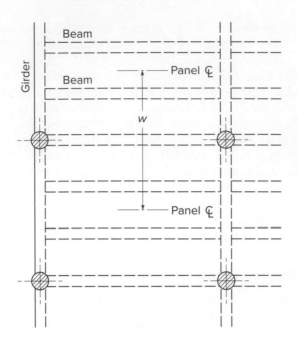

rectangular element, b being the lesser dimension in each case. In moment distribution, when the effect of torsional rigidity is included, it is important that the absolute flexural stiffness $4EI/\ell$ be used rather than relative I/ℓ values.

A common situation in beam-and-girder floors and concrete joist floors is illustrated in Fig. 11.5. The sketch shows a beam-and-girder floor system in which longitudinal beams are placed at the third points of each bay, supported by transverse girders, in addition to the longitudinal beams supported directly by the columns. If the transverse girders are quite stiff, it is apparent that the flexural stiffness of all beams in the width w should be balanced against the stiffness of one set of columns in the longitudinal bent. If, on the other hand, the girders have little torsional stiffness, there would be ample justification for making two separate longitudinal analyses, one for the beams supported directly by the columns, in which the rotational resistance of the columns would be considered, and a second for the beams framing into the girders, in which case hinged supports would be assumed. In most cases, it would be sufficiently accurate to consider the girders stiff torsionally and to add directly the stiffness of all beams tributary to a single column. This has the added advantage that all longitudinal beams will have the same cross-sectional dimensions and the same reinforcing steel, which will greatly facilitate construction. Plastic redistribution of loads upon overloading would generally ensure nearly equal restraint moments on all beams before collapse as assumed in design. Torsional moments should not be neglected in designing the girders.

11.6   PRELIMINARY DESIGN AND GUIDELINES FOR PROPORTIONING MEMBERS

In making an elastic analysis of a structural framework, it is necessary to know at the outset the cross-sectional dimensions of the members, so that moments of inertia and stiffnesses can be calculated. Yet the determination of these same cross-sectional dimensions is the precise purpose of the elastic analysis. In terms of load, the dead load on a structure is often dominated by the weight of the slab. Obviously, a preliminary estimate of member sizes must be one of the first steps in the analysis. Subsequently, with the results of the analysis at hand, members are proportioned, and the resulting dimensions compared with those previously assumed. If necessary, the assumed section properties are modified, and the analysis is repeated. Since the procedure may become quite laborious, it is obviously advantageous to make the best possible original estimate of member sizes, in the hope of avoiding repetition of the analysis.

In this connection, it is worth repeating that in ordinary frame analysis, one is concerned with relative stiffnesses only, not the absolute stiffnesses. If, in the original estimate of member sizes, the stiffnesses of all beams and columns are overestimated or underestimated by about the same amount, correction of these estimated sizes after the first analysis will have little or no effect. Consequently, no revision of the analysis would be required. If, on the other hand, a nonuniform error in estimation is made, and relative stiffnesses differ from assumed values by more than about 30 percent, a new analysis should be made.

The experienced designer can estimate member sizes with surprising accuracy. Those with little or no experience must rely on trial calculations or arbitrary rules, modified to suit particular situations. In building frames, the depth of one-way slabs (discussed at greater length in Chapter 12) is often controlled by either deflection requirements or the negative moments at the faces of the supporting beams. Minimum depth criteria are reflected in Tables 7.1 and 12.1, and negative moments at the face of the support can be estimated using coefficients described in Section 11.8. A practical minimum thickness of 4 in. is often used, except for joist construction meeting the requirements of ACI Code 9.8 (see Section 19.2d).

Beam sizes are usually governed by the negative moments and the shears at the supports, where their effective section is rectangular. Moments can be approximated by the fixed-end moments for the particular span or by using the ACI moment coefficients (see Section 11.8). In most cases, shears will not differ greatly from simple beam shears. Alternatively, many designers prefer to estimate the depth of beams at about $\frac{3}{4}$ in. per foot of span, with the width equal to about one-half the depth.

For most construction, wide, relatively shallow beams and girders are preferred to obtain minimum floor depths, and using the same depth for all flexural members allows the use of simple, low-cost forming systems. Such designs can significantly reduce forming costs, while incurring only small additional costs for concrete and reinforcing steel. It is often wise to check the reinforcement ratio ρ based on the assumed moments to help maintain overall economy. A value of $\rho \approx 0.012$ in preliminary design will give $\rho \approx 0.01$ in a final design, if a more exact analysis is used. Obviously, member dimensions are subject to modification, depending on the type and magnitude of the loads, methods of design, and material strength.

Column sizes are governed primarily by axial loads, which can be estimated quickly based on the column tributary areas, which are defined based on the midspan of flexural members framing into each column. As pointed out in Section 11.3, first interior columns are typically designed for an extra 10 percent axial load to account for the higher shear expected in the flexural members framing into the exterior face of first interior columns. The presence of moments in the columns is also cause for some increase of the area as determined by axial loads. For interior columns, in which unbalanced moments will not be large, a 10 percent increase may be sufficient, while for exterior columns, particularly for upper stories, an increase of 50 percent in area may be appropriate. In deciding on these estimated increases, the following factors should be considered. Moments are larger in exterior than in interior columns, since in the latter, dead load moments from adjacent spans will largely balance, in contrast

FIGURE 11.6

Subframe for estimating
moments in lower-story
columns of lateral
load–resisting frames.

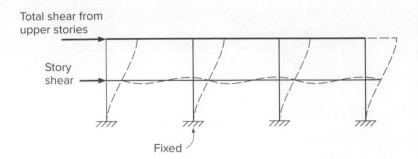

to the case in exterior columns. In addition, the influence of moments, compared
with that of axial loads, is larger in upper-floor than in lower-floor columns, because
the moments are usually of about the same magnitude, while the axial loads are
larger in the latter than in the former.

For minimum forming costs, it is highly desirable to use the same column dimensions throughout the height of a building. This can be accomplished by using higher-
strength concrete on the lower stories (for high-rise buildings, this should be the
highest-strength concrete available) and reducing concrete strength in upper stories, as
appropriate. For columns in *braced frames*, the preliminary design of the lower-story
columns may be based on zero eccentricity using $0.80\phi P_o = P_u$. A total reinforcement
ratio $\rho_g \approx 0.02$ should be used for the column with the highest axial load. With a
value of $\rho_g \approx 0.01$ for the column with the lowest axial load on higher stories, the
column size is maintained, reducing f_c' when ρ_g drops below 1 percent. Although ACI
Code 10.6.1.1 limits ρ_g to a range of 1 to 8 percent, the effective minimum value of
ρ_g is 0.005 based on ACI Code 10.3.1.2, which allows the minimum reinforcement to
be calculated based on a reduced effective area A_g not less than one-half the total area
(this provision cannot be used in regions of high seismic risk). For columns in lateral
load–resisting frames, a subframe may be used to estimate the factored bending
moments due to lateral load on the lower-story columns. The subframe illustrated in
Fig. 11.6 consists of the lower two stories in the structure, with the appropriate level
of fixity at the base. The upper flexural members in the subframe are treated as rigid.
Factored lateral loads are applied to the structure. The subframe can be analyzed using
matrix analysis or the portal frame method described in Section 11.7. Judicious consideration of factors such as those just discussed, along with simple models, as appropriate, will enable a designer to produce a reasonably accurate preliminary design,
which in most cases will permit a satisfactory analysis to be made on the first trial.

11.7   APPROXIMATE ANALYSIS

In spite of the development of refined methods for the analysis of beams and frames,
increasing attention is being paid to various approximate methods of analysis (Ref. 11.2).
There are several reasons for this. Prior to performing a complete analysis of an
indeterminate structure, it is necessary to estimate the proportions of its members to
determine their relative stiffness, upon which the analysis depends. These dimensions
can be obtained on the basis of approximate analysis. Also, even with the availability of computers, most engineers find it desirable to make a rough check of results,
using approximate means, to detect gross errors. Further, for structures of minor
importance, it is often satisfactory to design on the basis of results obtained by rough
calculation. For these reasons, many engineers at some stage in the design process

estimate the values of moments, shears, and thrusts at critical locations, using approximate sketches of the structure deflected by its loads.

Provided that points of inflection (locations in members at which the bending moment is zero and there is a reversal of curvature of the elastic curve) can be located accurately, the stress resultants for a framed structure can usually be found on the basis of static equilibrium alone. Each portion of the structure must be in equilibrium under the application of its external loads and the internal stress resultants.

For the fixed-end beam in Fig. 11.7a, for example, the points of inflection under uniformly distributed load are known to be located 0.211ℓ from the ends of the span. Since the moment at these points is zero, imaginary hinges can be placed there without modifying the member behavior. The individual segments between hinges can be analyzed by statics, as shown in Fig. 11.7b. Starting with the center segment, shears equal to 0.289$w\ell$ must act at the hinges. These, together with the transverse load, produce a midspan moment of 0.0417$w\ell^2$. Proceeding next to the outer segments, a downward load is applied at the hinge representing the shear from the center segment. This,

FIGURE 11.7

Analysis of fixed-end beam by locating inflection points.

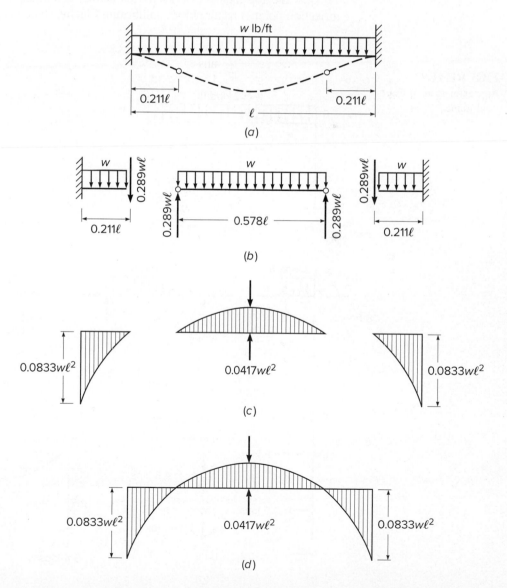

together with the applied load, produces support moments of $0.0833w\ell^2$. Note that, for this example, since the correct position of the inflection points was known at the start, the resulting moment diagram of Fig. 11.7c agrees exactly with the true moment diagram for a fixed-end beam shown in Fig. 11.7d. In more practical cases, inflection points must be estimated, and the results obtained will only approximate the true values.

The use of approximate analysis in determining stress resultants in frames is illustrated by Fig. 11.8. Figure 11.8a shows the geometry and loading of a two-member rigid frame. In Fig. 11.8b an exaggerated sketch of the probable deflected shape is given, together with the estimated location of points of inflection. On this basis, the central portion of the girder is analyzed by statics, as shown in Fig. 11.8d, to obtain girder shears at the inflection points of 7 kips, acting with an axial load P (still not determined). Similarly, the requirements of statics applied to the outer segments of the girder in Fig. 11.8c and e give vertical shears of 11 and 13 kips at B and C, respectively, and end moments of 18 and 30 ft-kips at the same locations. Proceeding then to the upper segment of the column, shown in Fig. 11.8f, with known axial load of 11 kips and top moment of 18 ft-kips acting, a horizontal shear of 4.5 kips at the inflection point is required for equilibrium. Finally, static analysis of the lower part of

FIGURE 11.8
Approximate analysis of
rigid frame.

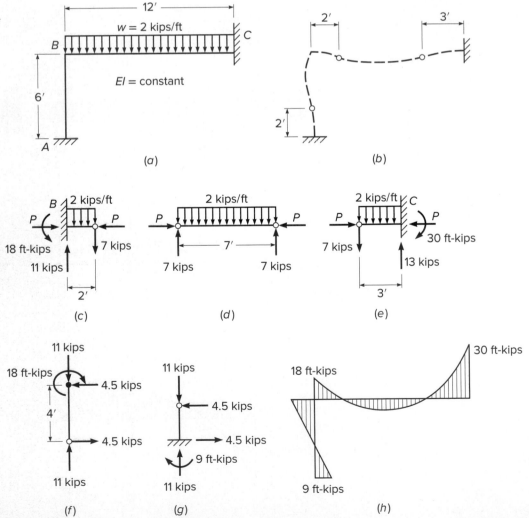

the column indicates a requirement of 9 ft-kips moment at A, as shown in Fig. 11.8g. The value of P equal to 4.5 kips is obtained by summing horizontal forces at joint B.

The moment diagram resulting from approximate analysis is shown in Fig. 11.8h. For comparison, an exact analysis of the frame indicates member end moments of 8 ft-kips at A, 16 ft-kips at B, and 28 ft-kips at C. The results of the approximate analysis would be satisfactory for design in many cases; if a more exact analysis is to be made, a valuable check is available on the magnitude of results.

A specialization of the approximate method described, known as the *portal method*, is commonly used to estimate the effects of sidesway due to lateral forces acting on multistory building frames. For such frames, it is usual to assume that horizontal loads are applied at the joints only. If this is true, moments in all members vary linearly and, except in hinged members, have opposite signs close to the midpoint of each member.

For a simple rectangular portal frame having three members, the shear forces are the same in both legs and are each equal to one-half the external horizontal load. If one of the legs is more rigid than the other, it will require a larger horizontal force to displace it horizontally the same amount as the more flexible leg. Consequently, the portion of the total shear resisted by the stiffer column is larger than that of the more flexible column.

In multistory building frames, moments and forces in the girders and columns of each individual story are distributed in substantially the same manner as just discussed for single-story frames. The portal method of computing approximate moments, shears, and axial forces from horizontal loads is, therefore, based on the following three simple propositions:

1. The total horizontal shear in all columns of a given story is equal and opposite to the sum of all horizontal loads acting above that story.
2. The horizontal shear is the same in both exterior columns; the horizontal shear in each interior column is twice that in an exterior column.
3. The inflection points of all members, columns and girders, are located midway between joints.

Although the last of these propositions is commonly applied to all columns, including those of the bottom floor, the authors prefer to deal with the latter separately, depending on conditions of foundation. If the actual conditions are such as practically to prevent rotation (foundation on rock, massive pile foundations, etc.), the inflection points of the bottom columns are above midpoint and may be assumed to be at a distance $2h/3$ from the bottom. If little resistance is offered to rotation, for example, for relatively small footings on compressible soil, the inflection point is located closer to the bottom and may be assumed to be at a distance $h/3$ from the bottom, or even lower. (With ideal hinges, the inflection point is at the hinge, that is, at the very bottom.) Since shears and corresponding moments are largest in the bottom story, a judicious evaluation of foundation conditions as they affect the location of inflection points is of considerable importance.

The first of the three cited propositions follows from the requirement that horizontal forces be in equilibrium at any level. The second takes account of the fact that in building frames interior columns are generally more rigid than exterior ones because (1) the larger axial loads require a larger cross section and (2) exterior columns are restrained from joint rotation only by one abutting girder, while interior columns are so restrained by two such members. The third proposition is very nearly true because, except for the top and bottom columns and, to a minor degree, for the exterior girders, each member in a building frame is restrained about equally at both ends. For this reason, members deflect under horizontal loads in an antisymmetrical manner, with the inflection point at midlength.

The actual computations in this method are extremely simple. Once column shears are determined from propositions 1 and 2 and inflection points located from proposition 3, all moments, shears, and forces are simply computed by statics. The process is illustrated in Fig. 11.9a.

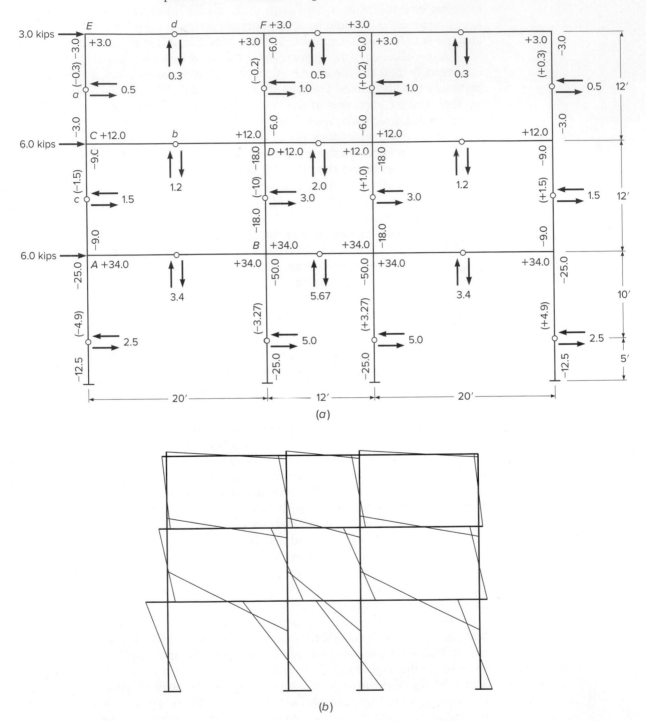

(a)

(b)

FIGURE 11.9
Portal method for determining moments from wind load in a building frame: (a) moments, shears, and thrusts and (b) variations of moments.

Consider joints C and D. The total shear in the second story is $3 + 6 = 9$ kips. According to proposition 2, the shear in each exterior column is $9/6 = 1.5$ kips, and in each interior column $2 \times 1.5 = 3.0$ kips. The shears in the other floors, obtained in the same manner, act at the hinges as shown. Consider the equilibrium of the rigid structure between hinges a, b, and c; the column moments, 3.0 and 9.0 ft-kips, respectively, are obtained directly by multiplying the shears by their lever arms, 6 ft. The girder moment at C, to produce equilibrium, is equal and opposite to the sum of the column moments. The shear in the girder is obtained by recognizing that its moment (that is, shear times one-half the girder span) must be equal to the girder moment at C. Hence, this shear is $12.0/10 = 1.2$ kips. The moment at end D is equal to that at C, since the inflection point is at midspan. At D, column moments are computed in the same manner from the known column shears and lever arms. The sum of the two girder moments, to produce equilibrium, must be equal and opposite to the sum of the two column moments, from which the girder moment to the right of D is $18.0 + 6.0 - 12.0 = 12.0$ ft-kips. Axial forces in the columns also follow from statics. Thus, for the rigid body aEd, a vertical shear of 0.3 kip is seen to act upward at d. To equilibrate it, a tensile force of -0.3 kip is required in the column CE. In the rigid body abc, an upward shear of 1.2 kips at b is added to the previous upward tension of 0.3 kip at a. To equilibrate these two forces, a tension force of -1.5 kips is required in column AC. If the equilibrium of all other partial structures between hinges is considered in a similar manner, all moments, forces, and shears are rapidly determined.

In the present case, relatively flexible foundations were assumed, and the location of the lowermost inflection points was estimated to be at $h/3$ from the bottom. The general character of the resulting moment distribution is shown in Fig. 11.9b.

11.8 ACI MOMENT COEFFICIENTS

ACI Code 6.5 includes expressions that may be used for the approximate calculation of maximum moments and shears in continuous beams and one-way slabs. The expressions for moment take the form of a coefficient multiplied by $w_u \ell_n^2$, where w_u is the total factored load per unit length on the span and ℓ_n is the clear span from face to face of supports for positive moment, or the average of the two adjacent clear spans for negative moment. Shear is taken equal to a coefficient multiplied by $w_u \ell_n / 2$. The coefficients, found in ACI Code 6.5.2 and 6.5.4, are shown in Table 11.1 and summarized in Fig. 11.10.

The ACI moment coefficients were derived by elastic analysis, considering alternative placement of live load to yield maximum negative or positive moments at the critical sections, as was described in Section 11.2. They are applicable within the following limitations:

1. Members are prismatic.
2. Loads are uniformly distributed.
3. The unfactored live load does not exceed 3 times the unfactored dead load.
4. There are two or more spans.
5. The longer of two adjacent spans does not exceed the shorter by more than 20 percent.

As discussed in Section 11.3 for more general loading conditions, the alternative loading patterns considered in applying the Code moment coefficients result in an envelope of maximum moments, as illustrated in Fig. 11.11 for one span of a continuous frame. For maximum positive moment, that span would carry dead and live loads,

TABLE 11.1

Moment and shear values using ACI coefficient[†]

Positive moment	
End spans	
If discontinuous end is integral with the support	$\frac{1}{14}\,w_u\ell_n^2$
If discontinuous end is unrestrained	$\frac{1}{11}\,w_u\ell_n^2$
Interior spans	$\frac{1}{16}\,w_u\ell_n^2$
Negative moment at interior faces of exterior supports for members built integrally with their supports	
Where the support is a spandrel beam or girder	$\frac{1}{24}\,w_u\ell_n^2$
Where the support is a column	$\frac{1}{16}\,w_u\ell_n^2$
Negative moment at exterior face of first interior support	
Two spans	$\frac{1}{9}\,w_u\ell_n^2$
More than two spans	$\frac{1}{10}\,w_u\ell_n^2$
Negative moment at other faces of interior supports	$\frac{1}{11}\,w_u\ell_n^2$
Negative moment at face of all supports for (1) slabs with spans not exceeding 10 ft and (2) beams and girders where ratio of sum of column stiffness to beam stiffness exceeds 8 at each end of the span	$\frac{1}{12}\,w_u\ell_n^2$
Shear in end members at first interior support	$1.15\,\dfrac{w_u\ell_n}{2}$
Shear at all other supports	$\dfrac{w_u\ell_n}{2}$

[†]w_u = total factored load per unit length of beam or per unit area of slab.

  ℓ_n = clear span for positive moment and shear and the average of the two adjacent clear spans for negative moment.

while adjacent spans would carry dead load only, producing the diagram of Fig. 11.11*a*. For maximum negative moment at the left support, dead and live loads would be placed on the given span and that to the left, while the adjacent span on the right would carry only dead load, with the result shown in Fig. 11.11*b*. Figure 11.11*c* shows the corresponding results for maximum moment at the right support.

    The composite moment diagram formed from the controlling portions of those just developed (Fig. 11.11*d*) provides the basis for design of the span. As observed in Section 11.3, there are a range of positions for the points of inflection resulting from alternate loadings. The extreme locations, required to determine bar cutoff points, can be found with the aid of Graph A.3 of Appendix A. In the region of the inflection point, it is evident from Fig. 11.11*d* that there may be a reversal of moments for alternative load patterns. However, within the stated limits for use of the coefficients, there should be no reversal of moments at the critical design sections near midspan or at the support faces.

    Comparison of the moments found using the ACI coefficients with those calculated by more exact analysis will usually indicate that the coefficient moments are quite conservative. Actual elastic moments may be considerably smaller. Consequently, in many reinforced concrete structures, significant economy can be achieved by making a more precise analysis. This is mandatory for beams and slabs with spans differing by more than 20 percent, sustaining loads that are not uniformly distributed, or carrying live loads greater than 3 times the dead load.

FIGURE 11.10
Summary of ACI moment coefficients: (a) beams with more than two spans; (b) beams with two spans only; (c) slabs with spans not exceeding 10 ft; and (d) beams in which the sum of column stiffnesses exceeds 8 times the sum of beam stiffnesses at each end of the span.

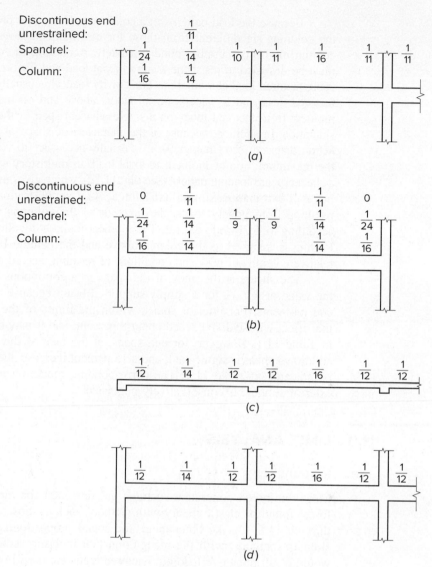

FIGURE 11.11
Maximum moment diagrams and moment envelope for a continuous beam: (a) maximum positive moment; (b) maximum negative moment at left end; (c) maximum negative moment at right end; and (d) composite moment envelope.

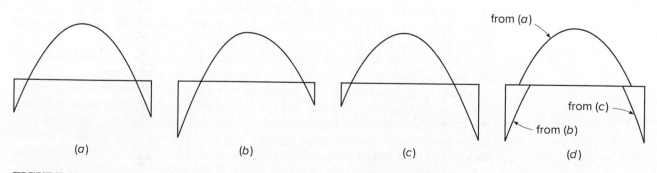

Because the load patterns in a continuous frame that produce critical moments in the columns are different from those for maximum negative moments in the beams, column moments must be found separately. According to ACI Code 10.4.2, columns must be designed to resist the factored axial load and factored moment for each applicable factored load combination. For gravity load, this usually involves axial load from factored dead and live loads on all floors above and on the roof plus the maximum moment from factored loads on a single adjacent span of the floor or roof under consideration. In addition, because of the characteristic shape of the column strength interaction diagram (see Chapter 9), it is usually necessary to consider the case that gives the maximum ratio of moment to axial load. In multistory structures, this results from a checkerboard loading pattern (see Fig. 11.2*d*), which gives maximum column moments but at a less-than-maximum axial force. As a simplification, in computing moments resulting from gravity loads, the far ends of the columns may be considered fixed, according to ACI Code 6.3.1.2. The moment found at a column-beam joint for a given loading is assigned to the column above and the column below in proportion to the relative column stiffness and conditions of restraint, according to ACI Code 6.5.5.

The shears at the ends of the spans in a continuous frame are modified from the value of $w_u \ell_n / 2$ for a simply supported beam because of the usually unbalanced end moments. For interior spans, within the limits of the ACI coefficient method, this effect will seldom exceed about 8 percent, and it may be neglected, as suggested in Table 11.1. However, for end spans, at the face of the first interior support, the additional shear is significant, and a 15 percent increase above the simple beam shear is indicated in Table 11.1. The corresponding reduction in shear at the face of the exterior support is conservatively neglected.

11.9   LIMIT ANALYSIS

a.   Introduction

Most reinforced concrete structures are designed for moments, shears, and axial forces found by elastic theory with methods such as those described in Sections 11.1 through 11.8. On the other hand, the actual proportioning of members is done by strength methods, with the recognition that inelastic section and member response would result upon overloading. Factored loads are used in the elastic analysis to find moments in a continuous beam, for example, after which the critical beam sections are designed with the knowledge that the steel would be well into the yield range and the concrete stress distribution very nonlinear before final collapse. Clearly this is an inconsistent approach to the total analysis-design process, although it can be shown to be both safe and conservative. A beam or frame so analyzed and designed will not fail at a load lower than the value calculated in this way.[†]

On the other hand, it is known that a continuous beam or frame normally will not fail when the nominal moment capacity of just one critical section is reached. A *plastic hinge* will form at that section, permitting large rotation to occur at essentially constant resisting moment and thus transferring load to other locations along the span where the limiting resistance has not yet been reached. Normally in a continuous beam or frame, excess capacity will exist at those other locations because they would have been reinforced for moments resulting from different load distributions selected to produce maximum moments at those other locations.

[†]See the discussion of upper and lower bound theorems of the theory of plasticity in Section 23.2 for an elaboration on this point.

FIGURE 11.12
Three-span continuous beam after the formation of plastic hinges at the interior supports. (*Photograph by Arthur H. Nilson*)

As loading is further increased, additional plastic hinges may form at other locations along the span and eventually result in collapse of the structure, but only after a significant *redistribution of moments* has occurred. The ratio of negative to positive moments found from elastic analysis is no longer correct, for example, and the true ratio after redistribution depends upon the flexural strengths actually provided at the hinging sections.

Recognition of redistribution of moments can be important because it permits a more realistic appraisal of the actual load-carrying capacity of a structure, thus leading to improved economy. In addition, it permits the designer to modify, within limits, the moment diagrams for which members are to be designed. Certain sections can be deliberately underreinforced if moment resistance at adjacent critical sections is increased correspondingly. Adjustment of design moments in this way enables the designer to reduce the congestion of reinforcement that often occurs in high-moment areas, such as at the beam-column joints.

The formation of plastic hinges is well established by tests such as that pictured in Fig. 11.12. The three-span continuous beam illustrates the inelastic response typical of heavily overloaded members. It was reinforced in such a way that plastic hinges would form at the interior support sections before the limit capacity of sections elsewhere was reached. The beam continued to carry increasing load well beyond the load that produced first yielding at the supports. The extreme deflections and sharp changes in slope of the member axis that are seen here were obtained only slightly before final collapse.

The *inconsistency* of the present approach to the total analysis-design process, the possibility of using the *reserve strength* of concrete structures resulting from moment redistribution, and the opportunity to *reduce steel congestion* in critical regions have motivated considerable interest in limit analysis for reinforced concrete based on the concepts just described. For beams and frames, ACI Code 6.6.5 permits limited redistribution of moments, depending upon the strain in the tensile steel ε_t. For slabs, which generally use very low reinforcement ratios and consequently have great ductility, plastic design methods are especially suitable.

b. Plastic Hinges and Collapse Mechanisms

If a short segment of a reinforced concrete beam is subjected to a bending moment, curvature of the beam axis will result, and there will be a corresponding rotation of one face of the segment with respect to the other. It is convenient to express this in terms of an angular change per unit length of the member. The relation between moment and angle change per unit length of beam, or curvature, at a reinforced concrete beam section subject to tensile cracking was developed in Section 7.9. Methods were presented there by which the theoretical moment-curvature graph might be drawn for a given beam cross section, as in Fig. 7.18.

The actual moment-curvature relationship measured in beam tests differs somewhat from that shown in Fig. 7.18, mainly because, from tests, curvatures are calculated from average strains measured over a finite gage length, usually about equal to the effective depth of the beam. In particular, the sharp increase in curvature upon concrete cracking shown in Fig. 7.18 is not often seen because the crack occurs at only one discrete location along the gage length. Elsewhere, the uncracked concrete shares in resisting flexural tension, resulting in what is known as *tension stiffening*. This tends to reduce curvature. Furthermore, the exact shape of the moment-curvature relation depends strongly upon the reinforcement ratio as well as upon the exact stress-strain curves for the concrete and steel.

Figure 11.13 shows a somewhat simplified moment-curvature diagram for an actual concrete beam section having a tensile reinforcement ratio equal to about one-half the balanced value. The diagram is linear up to the cracking moment M_{cr}, after which a nearly straight line of somewhat flatter slope is obtained. At the moment

FIGURE 11.13
Plastic hinge characteristics in a reinforced concrete member: (*a*) typical moment-curvature diagram; (*b*) strains and stresses at start of yielding; and (*c*) strains and stresses at incipient failure.

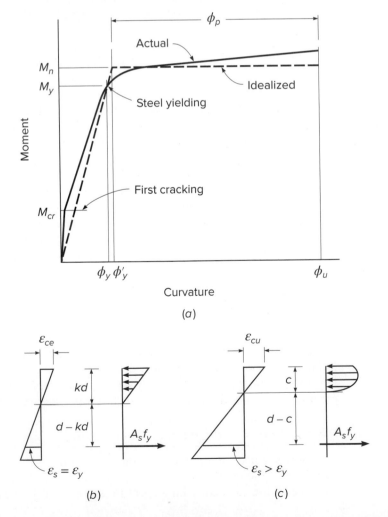

that initiates yielding M_y, the curvature starts to increase disproportionately. Further increase in applied moment causes extensive inelastic rotation until, eventually, the compressive strain limit of the concrete is reached at the ultimate rotation ϕ_u. The maximum moment is often somewhat above the calculated flexural strength M_n, due largely to strain hardening of the reinforcement.

The effect of inelastic concrete response prior to steel yielding is small for typically underreinforced sections, as is indicated in Fig. 7.18, and the yield moment can be calculated based on the elastic concrete stress distribution shown in Fig. 11.13b:

$$M_y = A_s f_y \left(d - \frac{kd}{3} \right) \tag{11.1}$$

where kd is the distance from the compression face to the cracked elastic neutral axis (see Section 4.2b). The nominal moment capacity M_n, based on Fig. 11.13c, is calculated by the usual expression

$$M_n = A_s f_y \left(d - \frac{a}{2} \right) = A_s f_y \left(d - \frac{\beta_1 c}{2} \right) \tag{11.2}$$

For purposes of limit analysis, the $M-\phi$ curve is usually idealized, as shown by the dashed line in Fig. 11.13a. The slope of the elastic portion of the curve is obtained with satisfactory accuracy using the moment of inertia of the cracked transformed section. After the nominal moment M_n is reached, continued plastic rotation is assumed to occur with no change in applied moment. The elastic curve of the beam will show an abrupt change in slope at such a section. The beam behaves as if there were a hinge at that point. However, the hinge will not be "friction-free," but will have a constant resistance to rotation.

If such a plastic hinge forms in a determinate structure, as shown in Fig. 11.14, uncontrolled deflection takes place, and the structure will collapse. The resulting system is referred to as a *mechanism*, an analogy to linkage systems in mechanics. Generalizing, one can say that a statically determinate system requires the formation of only one plastic hinge to become a mechanism.

This is not so for indeterminate structures. In this case, stability may be maintained even though hinges have formed at several cross sections. The formation of such hinges in indeterminate structures permits a redistribution of moments within the beam or frame. It will be assumed for simplicity that the indeterminate beam of Fig. 11.15a is symmetrically reinforced, so that the negative bending capacity is the same as the positive. Let the load P be increased gradually until the elastic moment at the fixed support, $\frac{3}{16}P\ell$, is just equal to the plastic moment capacity of the section M_n. This load is

$$P = P_{el} = \frac{16}{3} \frac{M_n}{\ell} = 5.33 \frac{M_n}{\ell} \tag{a}$$

FIGURE 11.14

Statically indeterminate member after the formation of plastic hinge.

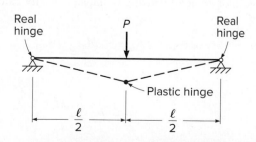

FIGURE 11.15
Indeterminate beam with
plastic hinges at support and
midspan.

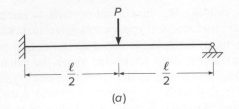

(a)

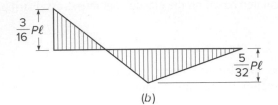

(b)

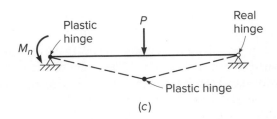

(c)

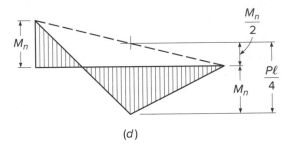

(d)

At this load, the positive moment under the load is $\frac{5}{32}P\ell$, as shown in Fig. 11.15*b*. The beam still responds elastically everywhere but at the left support. At that point the actual fixed support can be replaced for purposes of analysis with a plastic hinge offering a known resisting moment M_n. Because a redundant reaction has been replaced by a known moment, the beam is now determinate.

The load can be increased further until the moment under the load also becomes equal to M_n, at which load the second hinge forms. The structure is converted into a mechanism, as shown in Fig. 11.15*c*, and collapse occurs. The moment diagram at collapse load is shown in Fig. 11.15*d*.

The magnitude of load causing collapse is easily calculated from the geometry of Fig. 11.15*d*:

$$M_n + \frac{M_n}{2} = \frac{P\ell}{4}$$

from which

$$P = P_n = \frac{6M_n}{\ell} \tag{b}$$

By comparison of Eqs. (*b*) and (*a*), it is evident that an increase in *P* of 12.5 percent is possible, beyond the load that caused the formation of the first plastic hinge, before the beam will actually collapse. Due to the formation of plastic hinges, a redistribution of moments has occurred such that, at failure, the ratio between the positive moment and negative moment is equal to that assumed in reinforcing the structure.

c.   Rotation Requirement

It may be evident that there is a direct relation between the amount of redistribution desired and the amount of inelastic rotation at the critical sections of a beam required to produce the desired redistribution. In general, the greater the modification of the elastic moment ratio, the greater the required rotation capacity to accomplish that change. To illustrate, if the beam of Fig. 11.15*a* had been reinforced according to the elastic moment diagram of Fig. 11.15*b*, no inelastic-rotation capacity at all would be required. The beam would, at least in theory, yield simultaneously at the left support and at midspan. On the other hand, if the reinforcement at the left support had been deliberately reduced (and the midspan reinforcement correspondingly increased), inelastic rotation at the support would be required before the strength at midspan could be realized.

The amount of rotation required at plastic hinges for any assumed moment diagram can be found by considering the requirements of compatibility. The member must be bent, under the combined effects of elastic moment and plastic hinges, so that the correct boundary conditions are satisfied at the supports. Usually, zero support deflection is to be maintained. Moment-area and conjugate-beam principles are useful in quantitative determination of rotation requirements (Ref. 11.6). In deflection calculations, it is convenient to assume that plastic hinging occurs at a point, rather than being distributed over a finite *hinging length*, as is actually the case. Consequently, in loading the conjugate beam with unit rotations, plastic hinges are represented as concentrated loads.

Calculation of rotation requirements will be illustrated by the two-span continuous beam shown in Fig. 11.16*a*. The elastic moment diagram resulting from a single concentrated load is shown in Fig. 11.16*b*. The moment at support *B* is $0.096P\ell$, while that under the load is $0.182P\ell$. If the deflection of the beam at support *C* were calculated using the unit rotations equal to M/EI, based on this elastic moment diagram, a zero result would be obtained.

Figure 11.16*c* shows an alternative, statically admissible moment diagram that was obtained by arbitrarily increasing the support moment from $0.096P\ell$ to $0.150P\ell$. If the beam deflection at *C* were calculated using this moment diagram as a basis, a nonzero value would be obtained. This indicates the necessity for inelastic rotation at one or more points to maintain geometric compatibility at the right support.

If the beam were reinforced according to Fig. 11.16*c*, increasing loads would produce the first plastic hinge at *D*, where the beam has been deliberately made under strength. Continued loading would eventually result in formation of the second plastic hinge at *B*, creating a mechanism and leading to collapse of the structure.

Limit analysis requires calculation of rotation at all plastic hinges up to, but not including, the last hinge that triggers actual collapse. Figure 11.16*d* shows the M/EI load to be imposed on the conjugate beam of Fig. 11.16*e*. Also shown is the concentrated angle change θ_d, which is to be evaluated. Starting with the left span, taking moments of the M/EI loads about the internal hinge of the conjugate

FIGURE 11.16
Moment redistribution in a
two-span beam: (*a*) loaded
beam; (*b*) elastic moments;
(*c*) modified moments;
(*d*) *M/EI* loads; (*e*) conjugate
beam; and (*f*) deflection
curve.

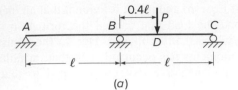

(*a*)

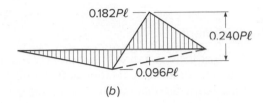

(*b*)

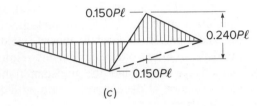

(*c*)

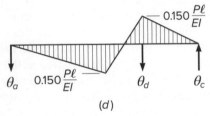

(*d*)

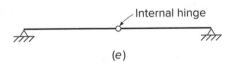

(*e*)

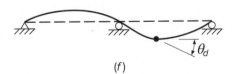

(*f*)

beam at *B*, one obtains the left reaction of the conjugate beam (equal to the slope
of the real beam):

$$\theta_a = 0.025 \frac{P\ell^2}{EI}$$

With that reaction known, moments are taken about the support *C* of the conjugate
beam and set equal to zero to obtain

$$\theta_d = 0.060 \frac{P\ell^2}{EI}$$

This represents the necessary discontinuity in the slope of the elastic curve shown in Fig. 11.16f to restore the beam to zero deflection at the right support. The beam must be capable of developing at least that amount of plastic rotation if the modified moment diagram assumed in Fig. 11.16c is to be valid.

d. Rotation Capacity

The capacity of concrete structures to absorb inelastic rotations at plastic-hinge locations is not unlimited. The designer adopting full limit analysis in concrete must calculate not only the amount of rotation required at critical sections to achieve the assumed degree of moment redistribution but also the rotation capacity of the members at those sections to ensure that it is adequate.

Curvature at initiation of yielding is easily calculated from the elastic strain distribution shown in Fig. 11.13b.

$$\phi_y = \frac{\varepsilon_y}{d(1-k)} \tag{11.3}$$

in which the ratio k establishing the depth of the elastic neutral axis is found from Eq. (4.8). The curvature corresponding to the nominal moment can be obtained from the geometry of Fig. 11.13c:

$$\phi_u = \frac{\varepsilon_{cu}}{c} \tag{11.4}$$

Although it is customary in flexural strength analysis to adopt $\varepsilon_{cu} = 0.003$, for purposes of limit analysis a more refined value is needed. Extensive experimental studies (Refs. 11.7 and 11.8) indicate that the ultimate strain capacity of concrete is strongly influenced by the beam width b, the moment gradient, and the presence of additional reinforcement in the form of compression steel and confining steel (that is, web reinforcement). The last parameter is conveniently introduced by means of a reinforcement ratio ρ'', defined as the ratio of the volume of one stirrup plus the compressive steel volume within the concrete volume along the length of the member corresponding to one stirrup. On the basis of empirical studies, the ultimate flexural strain at a plastic hinge is

$$\varepsilon_{cu} = 0.003 + 0.02\frac{b}{z} + \left(\frac{\rho''f_y}{14.5}\right)^2 \tag{11.5}$$

where z is the distance between points of maximum and zero moment. Based on Eqs. (11.3) to (11.5), the inelastic curvature for the idealized relation shown in Fig. 11.13a is

$$\phi_p = \phi_u - \phi_y \frac{M_n}{M_y} \tag{11.6}$$

This plastic rotation is not confined to one cross section but is distributed over a finite length referred to as the *hinging length*. The experimental studies upon which Eq. (11.5) is based measured strains and rotations in a length equal to the effective depth d of the test members. Consequently, ε_{cu} is an *average* value of ultimate strain over a finite length, and ϕ_p, given by Eq. (11.6), is an *average* value of curvature. The total inelastic rotation θ_p can be found by multiplying the average curvature by the hinging length:

$$\theta_p = \left(\phi_u - \phi_y \frac{M_n}{M_y}\right)\ell_p \tag{11.7}$$

On the basis of current evidence, it appears that the hinging length ℓ_p in support regions, on either side of the support, can be approximated by the expression

$$\ell_p = 0.5d + 0.05z \tag{11.8}$$

in which z is the distance from the point of maximum moment to the nearest point of zero moment.

e. Moment Redistribution under the ACI Code

Full use of the plastic capacity of reinforced concrete beams and frames requires an extensive analysis of all possible mechanisms and an investigation of rotation requirements and capacities at all proposed hinge locations. The increase in design time may not be justified by the gains obtained. On the other hand, a restricted amount of redistribution of elastic moments can safely be made without complete analysis, yet may be sufficient to obtain most of the advantages of limit analysis.

A limited amount of redistribution is permitted by ACI Code 6.6.5, depending upon a rough measure of available ductility, without explicit calculation of rotation requirements and capacities. The net tensile strain in the extreme tension steel at nominal strength ε_t, given in Eq. (4.25), is used as an indicator of rotation capacity. Accordingly, ACI Code 6.6.5 provides as follows:

> Except where approximate values for moments are used in accordance with ACI Code 6.5 [moment coefficients described in Section 11.8], where moments have been calculated in accordance with ACI Code 6.8 [inelastic analysis], or where moments in two-way slabs are determined using pattern loading specified in ACI Code 6.4.3.3 [using 75 percent of the factored live load, as described in Section 13.6d], reduction of moments at sections of maximum negative or maximum positive moment calculated by elastic theory shall be permitted for any assumed loading arrangement if (a) and (b) are satisfied: (a) Flexural members are continuous; (b) $\varepsilon_t \geq 0.0075$ at the section at which moment is reduced. Redistribution shall not exceed the lesser of $1000\varepsilon_t$ percent and 20 percent. The reduced moment shall be used to calculate redistributed moments at all other sections within the spans such that static equilibrium is maintained after redistribution of moments for each loading arrangement. Shears and support reactions shall be calculated in accordance with static equilibrium considering the redistributed moments for each loading arrangement.

Redistribution for values of $\varepsilon_t < 0.0075$ is conservatively prohibited. The ACI Code provisions are shown graphically in Fig. 11.17. The value of ρ corresponding to a given value of ε_t, and thus a given percentage change in moment, can be calculated using Eq. (4.26a) from Section 4.3d.

To demonstrate the advantage of moment redistribution when alternative loadings are involved, consider the concrete beam of Fig. 11.18. A three-span continuous beam is shown, with dead load of 1 kip/ft and live load of 2 kips/ft. To obtain maximum moments at all critical design sections, it is necessary to consider three alternative loadings. Case a, with live and dead load over exterior spans and dead load only over the interior span, will produce the maximum positive moment in the exterior spans. Case b, with dead load on exterior spans and dead and live load on the interior span, will produce the maximum positive moment in the interior span. The maximum negative moment over the interior support is obtained by placing dead and live load on the two adjacent spans and dead load only on the far exterior span, as shown in case c.

It will be assumed for simplicity that a 10 percent adjustment of maximum negative and positive moments is permitted throughout, provided that other span moments

FIGURE 11.17
Allowable moment
redistribution under the
ACI Code.

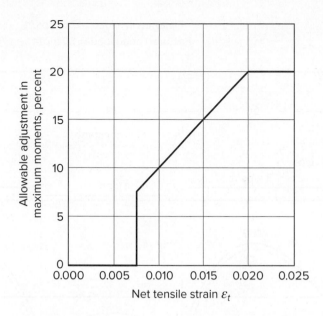

are modified accordingly. An overall reduction in design moments through the entire three-span beam may be possible. Case *a*, for example, produces an elastic maximum span moment in the exterior spans of 109 ft-kips. Corresponding to this is an elastic negative moment of 80 ft-kips at the interior support. Adjusting the maximum positive moment downward by 10 percent, one obtains a positive moment of 98 ft-kips, which results in an upward adjustment of the support moment to 102 ft-kips.

Now consider case *b*. By a similar redistribution of moments, the middle-span moment is reduced from 70 to 63 ft-kips, accompanied by an increase in the support moment from 80 to 87 ft-kips.

The moment obtained at the first interior support for loading case *c*, 134 ft-kips, can be decreased by 10 percent to 121 ft-kips. To limit the increase in the controlling span moment of the interior span, the right interior support moment is not decreased. The positive moments in the left exterior span and in the interior span corresponding to the modified moment at the left interior support are 90 and 57 ft-kips, respectively.

It will be observed that the reduction obtained for the span moments in cases *a* and *b* was achieved at the expense of increasing the moment at the first interior support. However, the increased support moment in each case was less than the moment for which that support would have to be designed based on the loading *c*, which produced the maximum support moment. Similarly, the reduction in support moment in case *c* was taken at the expense of an increase in span moments in the two adjacent spans. However, in each case, the increased span moments were less than the maximum span moments obtained for other loading conditions. The final design moments at all critical sections are underlined in Fig. 11.18. It can be seen, then, that the net result is a reduction in design moments over the entire beam. This modification of moments does not mean a reduction in safety factor below that implied in code safety provisions; rather, it means a reduction of the *excess* strength that would otherwise be present in the structure because of the actual redistribution of moments that would occur before failure. It reflects the fact that the maximum design moments are obtained from alternative load patterns, which could not exist concurrently. The end result is a more realistic appraisal of the actual collapse load of the indeterminate structure.

FIGURE 11.18
Redistribution of moments in a three-span continuous beam. The final design moments are underlined.

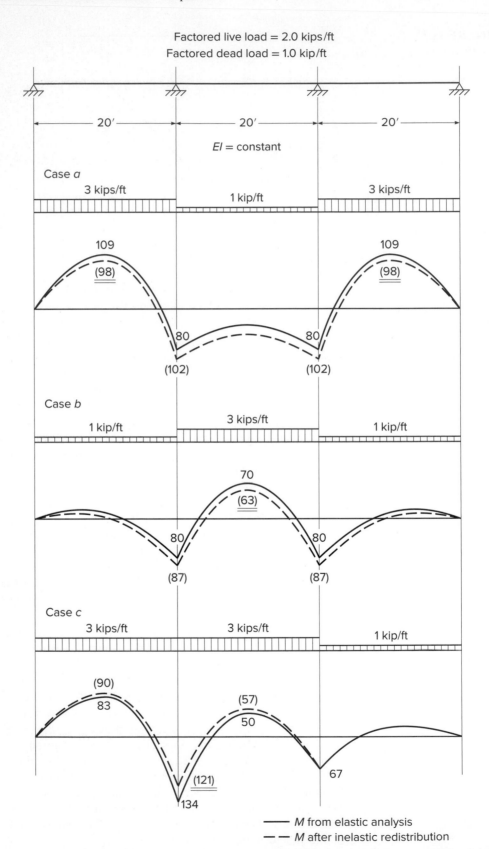

11.10 CONCLUSION

The problems associated with analysis of reinforced concrete structures are many. The engineer not only must accept the uncertainties of load placement, magnitude, and duration typical of any structural analysis but also must cope with other complications that are unique to reinforced concrete. These are mainly associated with estimation of moment of inertia of the reinforced concrete sections and with the influence of concrete creep. They may be summarized briefly as follows: (1) effective moments of inertia change depending on the sign of the bending moment; (2) moments of inertia depend not only on the effective concrete section but also on the steel, a part of which may be discontinuous; (3) moments of inertia depend on cracking, which is both location-dependent and load-dependent; and (4) the concrete is subject to creep under sustained loads, reducing its effective modulus. In addition, joint restraints and conditions of support for complex structures are seldom completely in accordance with the idealization. The student may well despair of accurate calculation of the internal forces for which the members of a reinforced concrete frame must be designed.

It may be reassuring to know that reinforced concrete has a remarkable capacity to adapt to the assumptions of the designer. This has been pointed out by a number of outstanding engineers. *Luigi Nervi*, the renowned Italian architect-engineer, has stated it eloquently as follows:

> Mainly because of plastic flow, a concrete structure tries with admirable docility to adapt itself to our calculations—which do not always represent the most logical and spontaneous answer to the request of the forces at play—and even tries to correct our deficiencies and errors. Sections and regions too highly stressed yield and channel some of their loads to other sections or regions, which accept this additional task with commendable spirit of collaboration, within the limits of their own strength.[†]

Hardy Cross, best known for his development of the moment distribution method of analysis (see Section 11.4), noted the beneficial effects of concrete creep, by which a structure can adapt to support settlements, which, on the basis of elastic analysis, cause forces and movements sufficient to fail the structure. *Halvard Birkeland*, one of the pioneers in the development of prestressed concrete in the United States, referred to the "wisdom of the structure," noting that ". . . the structure, in many instances, will accept our rash assumptions and our imperfect mathematical models . . . the structure will exhaust all means of standing before it decides to fall."[‡]

Thus, it may be of some comfort to know that *a reinforced concrete structure will tend to act as the engineer has assumed it will act*. Reasonable assumptions in the analysis may safely be made. But a corollary to this important principle is the acceptance of its limits: *the general pattern of forces and moments must be recognized, and at least one reasonable load path provided*. Too great a deviation from the actual distribution of internal forces can result in serviceability problems associated with cracking and deflection and can even result in premature failure. It is for this reason that methods of limit analysis for reinforced concrete include restrictions on the amount of redistribution of elastic moments (see Section 11.9). But it is reassuring to know that if good judgment is used in assigning internal forces to critical sections, the *wisdom of the structure* will prevail.

[†]P. L. Nervi, *Structures*, F. W. Dodge Corp., New York, 1956.
[‡]H. L. Birkeland, "The Wisdom of the Structure," *J. ACI*, April 1978, pp. 105–111.

REFERENCES

11.1. J. C. McCormac, *Structural Analysis Using Classical and Matrix Methods*, 4th ed., Wiley, Hoboken, New Jersey, 2007.

11.2. K. M. Leet, C.-M. Uang, J. Lanning, and A. Gilbert, *Fundamentals of Structural Analysis*, 5th ed., McGraw-Hill, New York, 2018.

11.3. *Continuity in Concrete Building Frames*, 4th ed., Portland Cement Association, Skokie, IL, 1959.

11.4. J. A. Fry, J. D. Hooper, and R. Klemencic, "Core Wall Case Study Design for Pacific Earthquake Engineering Research/California Seismic Safety Commission," *Structural Design of Tall and Special Buildings*, vol. 19, no. 1–2, 2010, pp. 61–75.

11.5. K. Kensek and D. Noble, *Building Information Modeling: BIM in Current and Future Practice*, John Wiley & Sons, Hoboken, New Jersey, 2014.

11.6. G. C. Ernst, "A Brief for Limit Design," *Trans. ASCE*, vol. 121, 1956, pp. 605–632.

11.7. A. H. Mattock, "Rotation Capacity of Hinging Regions in Reinforced Concrete Frames," *Proc. Int. Symp. Flexural Mech. Reinforced Concrete*, ACI Publication SP-12, 1964.

11.8. J. S. Ford, D. C. Chang, and J. E. Breen, "Design indications from Tests of Unbraced Multi-panel Concrete Frames," *Concr. Intl.*, vol. 3, no. 3, 1981, pp. 37–47.

PROBLEMS

11.1. Complete the preliminary design of the four-story heavy storage facility shown in Fig. P11.1. The floor live load is 250 psf, the roof live load is 12 psf, and the dead load on all floors and the roof consists of the structure self-weight plus 10 psf for utilities. The building is enclosed in a self-supporting curtain wall that also carries the lateral load on the structure. Beams are spaced at 12 ft; girders are spaced at 27 ft. The minimum clear space between floors is 11 ft, and the floor depth should not exceed 30 in. The column cross sections should be maintained from floor to floor. Use $f_y = 60{,}000$ psi and $f_c' = 4000$ psi for the floors. Concrete with f_c' up to 8000 psi is available for the columns. The preliminary design should include the initial dimensions of the structural slab, beams, girders, and columns for a typical floor.

FIGURE P11.1

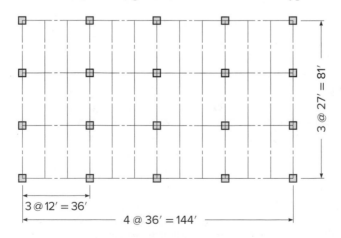

11.2. A concrete beam with $b = 12$ in., $h = 26.5$ in., and $d = 24$ in., having a span of 24 ft, can be considered fully fixed at the left support and supported vertically but with no rotational restraint (for example, roller) at the right end. It is reinforced for positive bending with a combination of bars giving $A_s = 2.45$ in^2, and for negative bending at the left support with $A_s = 2.88$ in^2. Positive bars are carried 6 in. into the face of the left support, according to the ACI Code requirements, but lack the embedded length to be considered effective as compression steel. No. 3 (No. 10) closed hoop stirrups are provided at 9 in. spacing over the full span. The factored load consists of a single concentrated force of 63.3 kips at midspan. Self-weight of the beam may be neglected in the calculations. Calculate the rotation requirement at the first plastic hinge to form (*a*) if the beam is reinforced according to the description above; (*b*) if, to reduce bar congestion at the left support, that steel area is reduced by 12.5 percent, with an appropriate increase in the positive steel area; and (*c*) if the steel area at the left support is reduced by 25 percent, compared with the original description, with an appropriate increase in the positive steel area. Also calculate the rotation capacity of the critical section, for comparison with the requirements of (*a*), (*b*), and (*c*). Comment on your results and compare with the approach to moment redistribution presented in the ACI Code. Material strengths are $f_y = 60$ ksi and $f_c' = 4$ ksi.

11.3. A 12-span continuous reinforced concrete T beam is to carry a calculated dead load of 900 lb/ft including self-weight, plus a service live load of 1400 lb/ft on uniform spans measuring 26.5 ft between centers of supporting columns (25 ft clear spans). The slab thickness is 6 in., and the effective flange width is 75 in. Web proportions are $b_w = 0.6d$, and the maximum reinforcement ratio will be set at 0.011. All columns will be 18 in. square. Material strengths are $f_c' = 4000$ psi and $f_y = 60,000$ psi.

(*a*) Find the factored moments for the exterior and first interior span based on the ACI Code moment coefficients of Table 11.1.

(*b*) Find the factored moments in the exterior and first interior span by elastic frame analysis, assuming the floor-to-floor height to be 10 ft. Note that alternative live loadings should be considered (see Section 11.2a) and that moments can be reduced to account for the support width (see Section 11.5a). Compare your results with those obtained using the ACI moment coefficients.

(*c*) Adjust the factored negative and positive moments, taking advantage of the redistribution provisions of the ACI Code. Assume that a 10 percent minimum redistribution is possible.

(*d*) Design the exterior and first interior spans for flexure and shear, finding concrete dimensions and bar requirements, basing your design on the assumptions and modified moments in part (*c*).

11.4. A continuous reinforced concrete frame consists of a two-span rectangular beam *ABC*, with center-to-center spans *AB* and *BC* of 24 ft. Columns measuring 14 in. square are provided at *A*, *B*, and *C*. The columns may be considered fully fixed at the floors above and below for purposes of analysis. The beam will carry a service live load of 1200 lb/ft and a calculated dead load of 1000 lb/ft, including self-weight. Floor-to-floor height is 12 ft. Material strengths are $f_y = 60,000$ psi and $f_c' = 4000$ psi.

(*a*) Perform an elastic analysis of the two-span frame, considering alternate live loadings to maximize the bending moment at all critical sections. Design the beams, using a maximum reinforcement ratio of 0.012 and $d = 2b$. Find the required concrete section and required steel areas at positive and negative bending sections. Select the reinforcement. Cutoff points can be determined according to Fig. 6.25. Note that negative design moments are at the face of supports, not support centerlines.

(*b*) Take maximum advantage of the redistribution provisions of ACI Code 8.4 (see Section 11.9e) to reduce design moments at all critical sections and redesign the steel for the beams. Keep the concrete section unchanged. Select reinforcement and determine cutoff points.

(c) Comment on your two designs with regard to the amount of steel required and the possible congestion of steel at the critical bending sections. You may assume that the shear reinforcement is unchanged in the redesigned beam.

11.5. Complete the preliminary design of a three-story library building shown in Fig. P11.5. The first floor (slab on grade) will be used as the reading room (live load = 60 lb/ft^2). The second and third floors will be stack areas (live load = 150 lb/ft^2). The dead load on each floor consists of the structure weight plus 20 lb/ft^2 for floor finish, lighting, and partitions. The roof live load and the snow load are both 20 lb/ft^2. Assume a load of 70 lb/ft^2 on the roof for the mechanical penthouse. The building is enclosed in a self-supporting curtain wall that also carries the lateral load on the structure. The column cross sections should be maintained from floor to floor. Beams are spaced at 8 ft; girders are spaced at 20 ft. The minimum clear height between floors is 12 ft, and the floor depth may not exceed 24 in. Use $f_y = 60,000$ psi and $f'_c = 4000$ psi. The preliminary design should include the structural slab, beams, girders, and columns for a typical floor.

FIGURE P11.5

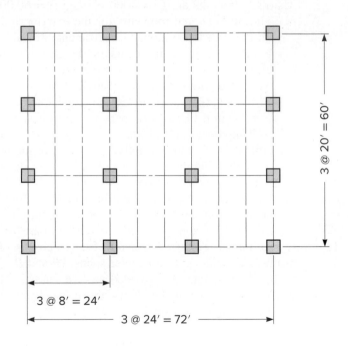

12

Analysis and Design of One-Way Slabs

12.1 TYPES OF SLABS

In reinforced concrete construction, slabs are used to provide flat, useful surfaces. A reinforced concrete slab is a broad, flat plate, usually horizontal, with top and bottom surfaces parallel or nearly so. It may be supported by reinforced concrete beams (and is usually cast monolithically with such beams), by masonry or reinforced concrete walls, by structural steel members, directly by columns, or continuously by the ground.

Slabs may be supported on two opposite sides only, as shown in Fig. 12.1*a*, in which case the structural action of the slab is essentially *one-way*, the loads being carried by the slab in the direction perpendicular to the supporting beams. There may be beams on all four sides, as shown in Fig. 12.1*b*, so that *two-way* slab action is obtained. Intermediate beams, as shown in Fig. 12.1*c*, may be provided. If the ratio of length to width of one slab panel is larger than about 2, most of the load is carried in the short direction to the supporting beams and one-way action is obtained in effect, even though supports are provided on all sides.

Concrete slabs in some cases may be carried directly by columns, as shown in Figs. 1.3 and 12.1*d*, without the use of beams or girders. Such slabs are described as *flat plates* and are commonly used where spans are not large and loads not particularly heavy. *Flat slab* construction, shown in Figs. 1.4 and 12.1*e*, is also beamless but incorporates a thickened slab region in the vicinity of the column and often employs flared column tops. They are referred to as *drop panels* and *column capitals*, respectively. Both devices increase the shear capacity around columns. Drop panels increase bending capacity, as well. *Shear caps*, a smaller version of drop panels, can be used to increase shear but not bending capacity. Closely related to the flat plate slab is the two-way joist, also known as a *grid* or *waffle slab*, shown in Fig. 12.1*f*. To reduce the dead load of solid-slab construction, voids are formed in a rectilinear pattern through use of metal or fiberglass form inserts. A two-way ribbed construction results. Usually, form inserts are omitted near the columns, so a solid slab is available to resist the higher moments and shears in these areas.

In addition to the column-supported types of construction shown in Fig. 12.1, many slabs are supported continuously on the ground, as for highways, airport runways, and warehouse floors. In such cases, a well-compacted layer of crushed stone or gravel is usually provided to ensure uniform support and to allow for proper subgrade drainage.[†]

[†] Design guidance for slabs-on-ground, including the effects of deformation of both the slab and the subgrade, can be found in *Design of Slabs-on-Ground* reported by ACI Committee 360 (Ref. 12.1) and PTI DC 10.5-1 (Ref 12.2).

FIGURE 12.1
Types of structural slabs.

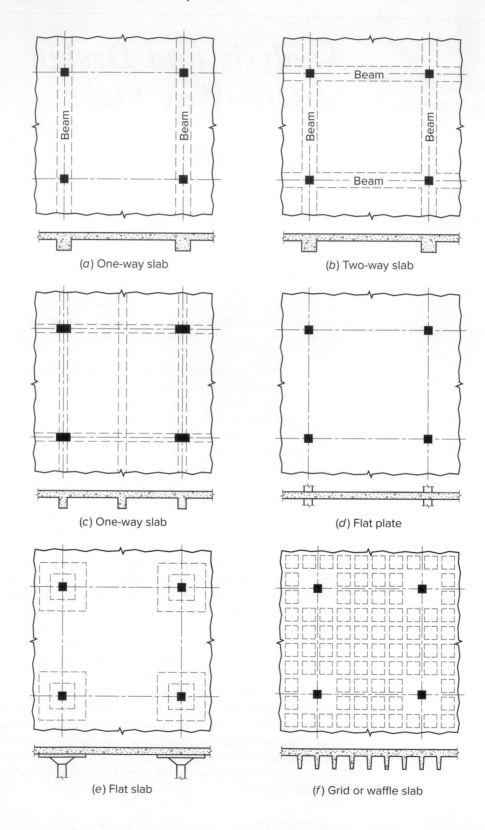

(*a*) One-way slab

(*b*) Two-way slab

(*c*) One-way slab

(*d*) Flat plate

(*e*) Flat slab

(*f*) Grid or waffle slab

Reinforcing steel for slabs is primarily parallel to the slab surfaces. Straight bar reinforcement is generally used, although in continuous slabs bottom bars are sometimes bent up to serve as negative reinforcement over the supports. Welded wire reinforcement is commonly employed for slabs on the ground. Bar mats are available for the heavier reinforcement sometimes needed in highway slabs and airport runways. Slabs may also be prestressed using bonded or unbonded post-tensioning tendons.

Reinforced concrete slabs of the types shown in Fig. 12.1 are usually designed for loads assumed to be uniformly distributed over one entire slab panel, bounded by supporting beams or column centerlines. Minor concentrated loads can be accommodated through two-way action of the reinforcement (two-way flexural steel for two-way slab systems or one-way flexural steel plus lateral distribution steel for one-way systems). Heavy concentrated loads generally require supporting beams.

One-way edge-supported slabs, such as shown in Fig. 12.1*a*, are discussed in Sections 12.2 and 12.3. Two-way beamless systems, such as shown in Fig. 12.1*d*, *e*, and *f*, as well as two-way edge-supported slabs (Fig. 12.1*b* and *c*), are covered in Chapter 13. Special methods based on limit analysis at the overload state, applicable to all types of slabs, are presented in Chapters 23 and 24.

12.2   DESIGN OF ONE-WAY SLABS

The structural action of a one-way slab may be visualized in terms of the deformed shape of the loaded surface. Figure 12.2 shows a rectangular slab, simply supported along its two opposite long edges and free of any support along the two opposite short edges. If a uniformly distributed load is applied to the surface, the deflected shape is shown by the solid lines. Curvatures, and consequently bending moments, are the same in all strips *s* spanning in the short direction between supported edges, whereas there is no curvature, hence no bending moment, in the long strips *l* parallel to the supported edges. The surface is approximately cylindrical.

For purposes of analysis and design, a unit strip of such a slab cut out at right angles to the supporting beams, as shown in Fig. 12.3, may be considered as a rectangular beam of unit width, with a depth *h* equal to the thickness of the slab and a span ℓ_a equal to the distance between supported edges. This strip can then be analyzed by the methods that were used for rectangular beams, the bending moment being computed for the strip of unit width. The load per unit area on the slab becomes the load

FIGURE 12.2
Deflected shape of uniformly loaded one-way slab.

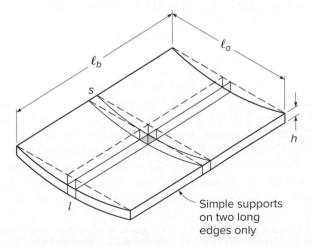

Simple supports on two long edges only

FIGURE 12.3
Unit strip basis for flexural
design.

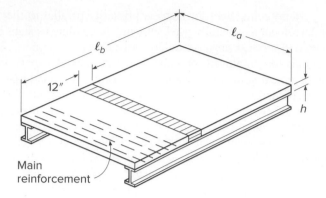

per unit length on the slab strip. Since all of the load on the slab must be transmitted to the two supporting beams, it follows that all of the reinforcement should be placed at right angles to these beams, with the exception of any bars that may be placed in the orthogonal direction to control shrinkage and temperature cracking. A one-way slab, thus, consists of a set of rectangular beams side by side.

This simplified analysis, which assumes Poisson's ratio to be zero, is slightly conservative. Actually, flexural compression in the concrete in the direction of ℓ_a results in lateral expansion in the direction of ℓ_b unless the compressed concrete is restrained. In a one-way slab, this lateral expansion is resisted by adjacent slab strips, which tend to expand also. The result is a slight strengthening and stiffening in the span direction, but this effect is small and can be disregarded.

The reinforcement ratio for a slab can be determined by dividing the area of one bar by the area of concrete between two successive bars, the latter area being the product of the depth to the center of the bars and the distance between them, center to center. The reinforcement ratio can also be determined by dividing the average area of steel per foot of width by the effective area of concrete in a 1 ft strip. The average area of steel per foot of width is equal to the area of one bar times the average number of bars in a 1 ft strip (12 divided by the spacing in inches), and the effective area of concrete in a 1 ft (or 12 in.) strip is equal to 12 times the effective depth d.

To illustrate the latter method of obtaining the reinforcement ratio ρ, assume a 5 in. thick slab with an effective depth of 4 in., with No. 4 (No. 13) bars spaced $4\frac{1}{2}$ in. center to center. The average number of bars in a 12 in. strip of slab is $12/4.5 = 2\frac{2}{3}$ bars, and the average steel area in a 12 in. strip is $2\frac{2}{3} \times 0.20 = 0.533$ in². Hence $\rho = 0.533/(12 \times 4) = 0.0111$. By the other method,

$$\rho = \frac{0.20}{4.5 \times 4} = 0.0111$$

The spacing of bars that is necessary to furnish a given area of steel per foot of width is obtained by dividing the number of bars required to furnish this area into 12 in. For example, to furnish an average area of 0.46 in²/ft, with No. 4 (No. 13) bars, requires $0.46 \div 0.20 = 2.3$ bars per foot; the bars must be spaced not more than $12/2.3 = 5.2$ in. center to center. The determination of slab steel areas for various combinations of bars and spacings is facilitated by Table A.3 of Appendix A.

Factored moments and shears in one-way slabs can be found by computer or elastic analysis or through the use of the same coefficients as used for beams (see Chapter 11). In general, center-to-center distances should be used in continuous slab analysis, but a reduction is allowed in negative moments to account for support width as discussed in Chapter 11. For slabs with clear spans not more than 10 ft that are built integrally with their supports, ACI Code 6.6.2.3 permits analysis as a continuous slab on knife edge supports with spans equal to the clear spans and the width of the beams otherwise neglected. If moment and shear coefficients are used, computations should be based on clear spans.

One-way slabs are normally designed with tensile reinforcement ratios well below the maximum practical value of ρ_{max}. Typical reinforcement ratios range from about 0.004 to 0.008. This is partially for reasons of economy, because the saving in reinforcement associated with increasing the effective depth more than compensates for the cost of the additional concrete, and partially because very thin slabs with high reinforcement ratios would likely result in large deflections. Thus, flexural design may start with selecting a relatively low reinforcement ratio, say about $0.3\rho_{max}$, setting $M_u = \phi M_n$ in Eq. (4.34), and solving for the required effective depth d, given that $b = 12$ in. for the unit strip. Alternatively, Table A.5 or Graph A.1 of Appendix A may be used. Table A.9 is also useful. The required steel area per 12 in. strip $A_s = \rho b d$ is then easily found.

ACI Code 7.3.1 specifies the minimum thickness in Table 12.1[†] for nonprestressed slabs of normalweight concrete ($w_c = 145$ pcf) using Grade 60 reinforcement and typical building loads, provided that the slab is not supporting or attached to construction that is likely to be damaged by large deflections. Lesser thicknesses may be used if calculation of deflections indicates no adverse effects. For concretes having unit weight w_c in the range from 90 to 115 pcf, the tabulated values should be multiplied by $1.65 - 0.005w_c$, but not less than 1.09. For reinforcement having a yield stress f_y other than 60,000 psi, the tabulated values should be multiplied by $0.4 + f_y/100,000$. Slab deflections may be calculated, if required, by the same methods as for beams (see Section 7.7).

Shear seldom controls the design of one-way slabs, particularly if low tensile reinforcement ratios are used. The shear capacity of the concrete ϕV_c is usually above the required shear strength V_u at factored loads.

The total slab thickness h is usually rounded to the next higher $\frac{1}{2}$ in. Best economy is often achieved when the slab thickness is selected to match standardized form or nominal lumber dimensions. The concrete protection below the reinforcement should follow the requirements of ACI Code 20.5.1, calling for $\frac{3}{4}$ in. below the bottom of the steel (see Fig. 4.13b). In a typical slab, 1 in. below the center of the steel may be assumed. The lateral spacing of the bars, except those used only to control shrinkage and temperature cracks (see Section 12.3), should not exceed 3 times the thickness h or 18 in., whichever is less, according to ACI Code 7.7.2.3. Generally, bar size should be selected so that the actual spacing is not less than about 1.5 times the slab thickness, to avoid excessive cost for bar fabrication and handling. Also, to reduce cost, straight bars are usually used for slab reinforcement, cut off where permitted as described for beams in Section 6.10.

TABLE 12.1

Minimum thickness h of nonprestressed one-way slabs

Simply supported	$\ell/20$
One end continuous	$\ell/24$
Both ends continuous	$\ell/28$
Cantilever	$\ell/10$

[†] This table first appeared in the 1963 ACI Code and may not fully reflect current loads, materials, and design practice. The authors recommend a check of all member deflections.

12.3 CONSIDERATIONS FOR ONE-WAY SLAB DESIGN

a. Temperature and Shrinkage Reinforcement

Concrete shrinks as it cures and dries, as was pointed out in Section 2.11. It is advisable to minimize such shrinkage by using concretes with the smallest possible amounts of water and cement compatible with other requirements, such as strength and workability, and by thorough moist-curing of sufficient duration. However, no matter what precautions are taken, a certain amount of shrinkage is unavoidable. If a slab of moderate dimensions rests freely on its supports, it can contract to accommodate the shortening of its length produced by shrinkage. Usually, however, slabs and other members are joined rigidly to other parts of the structure and cannot contract freely. This results in tension stresses known as *shrinkage stresses*. A decrease in temperature relative to that at which the slab was cast, particularly in outdoor structures such as parking garages and bridges, may have an effect similar to shrinkage. That is, the slab tends to contract and, if restrained from doing so, becomes subject to tensile stresses.

Since concrete is weak in tension, these temperature and shrinkage stresses are likely to result in cracking. Cracks of this nature are not detrimental, provided their size is limited to what are known as *hairline cracks*. This can be achieved by placing reinforcement in the slab to counteract contraction and distribute the cracks uniformly. As the concrete tends to shrink, such reinforcement resists the contraction and consequently becomes subject to compression. The total shrinkage in a slab so reinforced is less than that in one without reinforcement; in addition, whatever cracks do occur will be of smaller width and more evenly distributed by virtue of the reinforcement.

In one-way slabs, the reinforcement provided for resisting the bending moments has the desired effect of reducing shrinkage and distributing cracks. However, as contraction takes place equally in all directions, it is necessary to provide special reinforcement for shrinkage and temperature contraction in the direction perpendicular to the main reinforcement. This added steel is known as *shrinkage and temperature reinforcement*, or *distribution steel*.

Reinforcement for shrinkage and temperature stresses normal to the principal reinforcement should be provided in a structural slab in which the principal reinforcement extends in one direction only. ACI Code 24.4.3.2 specifies a minimum ratio of reinforcement area to *gross concrete area,* that is, based on the total depth of the slab, of 0.0018, and in no case may such reinforcing bars be placed farther apart than 5 times the slab thickness or more than 18 in.

The steel required by the ACI Code for shrinkage and temperature crack control also represents the minimum permissible reinforcement in the span direction of one-way slabs; the usual minimums for flexural steel do not apply.

b. Concrete Shear Capacity

The concrete shear capacity of a one-way slab is specified in ACI Code 22.5.5.1 for slabs with and without shear reinforcement. Most one-way slabs are governed by flexural requirements, and consequently, no shear reinforcement is required. The concrete shear capacity for slabs without shear reinforcement is given in Eq. (5.12a), which is repeated here:

$$V_c = 8\lambda_s\lambda(\rho_w)^{1/3}\sqrt{f_c'}\,b_w d \tag{12.1}$$

For slabs with minimum flexural reinforcement, this results in a maximum concrete shear strength contribution of about $V_c = 1.0\lambda\sqrt{f_c'}\,b_w d$. The concrete shear capacity is additionally reduced by the size effect described in Chapter 5 and given in Eq. (5.12b). The size effect only applies to slabs with an effective depth greater than 10 in. These two effects result in a lower concrete shear contribution than used prior to 2019. Short, thick, or highly loaded one-way slabs should be checked for the shear capacity before proceeding with the flexural design. These effects are especially noted in *podium slabs* that transfer loads from upper portions of a structure over short spans to supporting members and footings when shear controls the overall thickness.

In cases where the concrete shear capacity is insufficient, the options for increasing the shear capacity include increasing the slab thickness, increasing the longitudinal reinforcement, adding shear reinforcement, or using higher strength concrete. Each option has cost implications and must be evaluated based on local conditions.

c. Lateral Distribution of Shear Reinforcement

Wide one-way slabs are sometimes used as girders or podium slabs to transfer loads from upper portions of a structure to lower portions to minimize disruption to the floor height. These wide slabs are deeper and more heavily loaded than ordinary floor slabs. An example is an apartment building where upper-story wood framing is transferred to the concrete structure below. In these instances, ACI Code 9.7.6.2 requires that any shear reinforcement be distributed along the length at a spacing of not more than $d/2$ and across the width at a spacing of not more than d if the required shear in the stirrups is less than or equal to $4\sqrt{f_c'}\,b_w d$ and $d/4$ along the length and $d/2$ across the width if the shear demand exceeds this amount. This transverse distribution provides a more uniform transfer of shear stress across the slab.

d. Structural Integrity

ACI Code Section 7.7.7 requires that one quarter of the positive reinforcement extend to the support and be fully developed at the face of the support. Satisfying this provision may require hooked bars at simple supports. If the slab is continuous and splices are needed, splices of structural integrity reinforcement should be placed near the supports. This placement would be away from any maximum positive moment region.

e. Vibrations

Vibrations are rarely of concern for cast-in-place one-way slabs, provided that the minimum thicknesses in Table 12.1 is maintained. If thinner slabs are used, ACI Code Commentary R24.1 recommends investigation of vibration performance. No specific direction is provided, but several references are cited for guidance, including Refs. 12.3 and 12.4.

EXAMPLE 12.1   **One-way slab design.**   A reinforced concrete slab is built integrally with its supports and consists of two equal spans, each with a clear span of 15 ft. The service live load is 100 psf, and 4000 psi concrete is specified for use with steel with a yield stress equal to 60,000 psi. Design the slab, following the provisions of the ACI Code.

SOLUTION. The thickness of the slab is first estimated, based on the minimum thickness from Table 12.1, $\ell/28 = 15 \times 12/28 = 6.43$ in. A trial thickness of 6.50 in. is used, for which the weight is $150 \times 6.50/12 = 81$ psf. The specified live load and computed dead load are multiplied by the ACI load factors:

$$
\begin{aligned}
\text{Dead load} &= 81 \times 1.2 = \,97 \text{ psf}\\
\text{Live load} &= 100 \times 1.6 = \underline{160 \text{ psf}}\\
\text{Total} &= 257 \text{ psf}
\end{aligned}
$$

For this case, factored moments at critical sections may be found using the ACI moment coefficients (see Table 11.1):

At interior support: $-M = \frac{1}{9} \times 0.257 \times 15^2 = 6.43$ ft-kips

At midspan: $+M = \frac{1}{14} \times 0.257 \times 15^2 = 4.13$ ft-kips

At exterior support: $-M = \frac{1}{24} \times 0.257 \times 15^2 = 2.41$ ft-kips

The maximum practical reinforcement ratio is, according to Eq. (4.26d),

$$
\rho_{max} = 0.85\beta_1 \frac{f_c'}{f_y} \frac{\varepsilon_u}{\varepsilon_u + \varepsilon_{min}} = (0.85^2)\frac{4}{60}\frac{0.003}{0.003 + 0.005} = 0.0181
$$

If this value of ρ were actually used, the minimum required effective depth, controlled by negative moment at the interior support, would be found from Eq. (4.34) to be

$$
\begin{aligned}
d^2 &= \frac{M_u}{\phi\rho f_y\, b(1 - 0.59\rho f_y/f_c')}\\[2mm]
&= \frac{6.43 \times 12}{0.90 \times 0.0181 \times 60 \times 12[1 - 0.59 \times 0.0181 \times (60/4)]} = 7.83 \text{ in}^2\\[2mm]
d &= 2.80 \text{ in.}^\dagger
\end{aligned}
$$

This is less than the effective depth of $6.50 - 1.00 = 5.50$ in. resulting from application of Code restrictions, and $h = 6.5$ in. is adopted. At the interior support, if the stress-block depth $a = 1.00$ in., the area of steel required per foot of width in the top of the slab is [Eq. (4.33)]

$$
A_s = \frac{M_u}{\phi f_y(d - a/2)} = \frac{6.43 \times 12}{0.90 \times 60 \times (5.50 - 1.00/2)} = 0.29 \text{ in}^2
$$

Checking the assumed depth a by Eq. (4.28), one gets

$$
a = \frac{A_s f_y}{0.85f_c' b} = \frac{0.29 \times 60}{0.85 \times 4 \times 12} = 0.43 \text{ in.}
$$

A second trial is made with $a = 0.43$ in. Then

$$
A_s = \frac{6.43 \times 12}{0.90 \times 60 \times (5.50 - 0.43/2)} = 0.27 \text{ in}^2
$$

for which $a = 0.43 \times 0.27/0.29 = 0.40$ in. No further revision is necessary. At other critical-moment sections, it will be satisfactory to use the same lever arm to determine steel areas, and

At midspan: $\qquad A_s = \dfrac{4.13 \times 12}{0.90 \times 60 \times (5.50 - 0.40/2)} = 0.17 \text{ in}^2$

At exterior support: $\quad A_s = \dfrac{2.41 \times 12}{0.90 \times 60 \times (5.50 - 0.40/2)} = 0.10 \text{ in}^2$

The minimum reinforcement is that required for control of shrinkage and temperature cracking. This is

$$
A_s = 0.0018 \times 12 \times 6.50 = 0.14 \text{ in}^2
$$

per 12 in. strip. This requires a small increase in the amount of steel used at the exterior support.

---

† The depth is more easily found using Graph A.1 of Appendix A. For $\rho = \rho_{max}$, $M_u/\phi bd^2 = 913$, from which $d = 2.80$ in. Table A.5a may also be used.

Using Table 11.1 for the factored shear, the factored shear force at a distance d from the face of the interior support is

$$V_u = 1.15 \times \frac{257 \times 15}{2} - 257 \times \frac{5.50}{12} = 2100 \text{ lb}$$

The slab is less than 10 in. thick, so $\lambda_s = 1.0$. The reinforcement ratio is $\rho = 0.14/12 \times 5.5 = 0.0021$. By Eq. (12.1), the nominal shear strength of the concrete slab is

$$V_c = 8\lambda_s\lambda(\rho_w)^{1/3}\sqrt{f_c'}\, b_w d = 8 \times 1.0 \times 1.0 \times (0.0021)^{1/3}\sqrt{4000} \times 12 \times 5.5 = 4290 \text{ lb}$$

Thus, the design strength of the concrete slab, $\phi V_c = 0.75 \times 4290 = 3220$ lb, is above the required factored shear of $V_u = 2100$ lb.

The required tensile steel areas may be provided in a variety of ways, but whatever the selection, due consideration must be given to the actual placing of the steel during construction. The arrangement should be such that the steel can be placed rapidly with the minimum of labor costs even though some excess steel is necessary to achieve this end.

Two possible arrangements are shown in Fig. 12.4. In Fig. 12.4a, bent bars are used, while in Fig. 12.4b all bars are straight, consistent with current practice.

For the arrangement shown in Fig. 12.4a, No. 4 (No. 13) bars at 10 in. furnish 0.24 in^2 of steel at midspan, slightly more than required. If two-thirds of these bars are bent upward for negative reinforcement over the interior support, the average spacing of such bent bars at the interior support will be $(10 + 20)/2 = 15$ in. Since an identical pattern of bars is bent upward from the other side of the support, the effective spacing of the No. 4 (No. 13) bars over the interior support is $7\frac{1}{2}$ in. This pattern satisfies the required steel area of 0.27 in^2 per foot width of slab over the support. The bars bent at the interior support will also be bent

FIGURE 12.4

One-way slab design example.

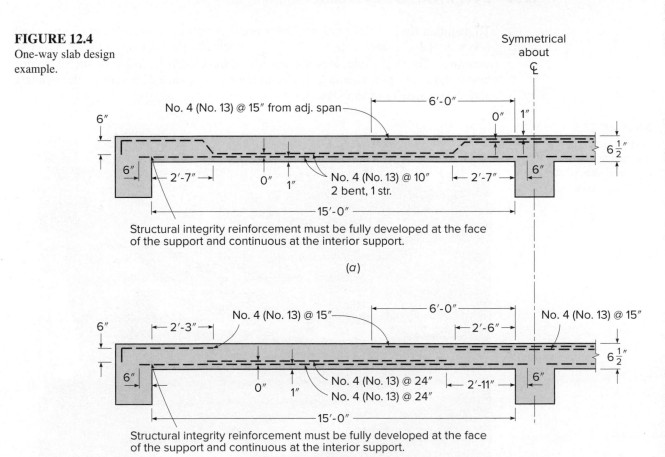

(a)

(b)

upward for negative reinforcement at the exterior support, providing reinforcement equivalent to No. 4 (No. 13) bars at 15 in., or 0.16 in^2 of steel.

Note that it is not necessary to achieve uniform spacing of reinforcement in slabs, and that the steel provided can be calculated safely on the basis of average spacing, as in the example. Care should be taken to satisfy requirements for both minimum and maximum spacing of principal reinforcement, however.

The locations of bend and cutoff points shown in Fig. 12.4a were obtained using Graph A.3 of Appendix A, as explained in Section 6.10, and Table A.10 (see also Fig. 6.20).

The arrangement shown in Fig. 12.4b uses only straight bars. Although it is satisfactory according to the ACI Code (since the shear stress does not exceed two-thirds of that permitted), cutting off the shorter positive and negative bars as shown leads to an undesirable condition at the ends of those bars, where there are concentrations of stress in the concrete. The design would be improved if the negative bars were cut off 3 ft from the face of the interior support rather than 2 ft 6 in. as shown, and if the positive steel were cut off at 2 ft 2 in. rather than at 2 ft 11 in. This would result in an overlap of approximately $2d$ of the cut positive and negative bars. Figure 6.26a suggests a somewhat simpler arrangement that would also prove satisfactory.

The required area of steel to be placed normal to the main reinforcement for purposes of temperature and shrinkage crack control is 0.14 in^2. This is provided by No. 4 (No. 13) bars at 16 in. spacing, placed directly on top of the main reinforcement in the positive-moment region and below the main steel in the negative-moment zone.

12.4   INTERNAL DUCTWORK

To maintain floor-to-floor heights, ductwork is sometimes embedded in one-way and two-way slabs, as shown in Fig. 12.5. The ducts are used for air circulation and as wireways. The ACI Code does not directly address the installation of such systems; however, the designer should be aware of the effect of such ductwork on the moment and shear capacity of the slab.

FIGURE 12.5
Internal slab ductwork installed in a two-way post-tensioned slab. (*Courtesy of ECCO Manufacturing.*)

REFERENCES

12.1. ACI Committee 360, *Guide to Design of Slabs-on-Ground* (ACI 360R-10), American Concrete Institute, Farmington Hills, MI, 2010.

12.2. PTI, *DC10.5-12: Standard Requirements for Design and Analysis of Shallow Post-Tensioned Concrete Foundations on Expansive Soils*, Post-Tensioning Institute, Farmington Hills, MI, 2012, 48 pp.

12.3. D. A. Fanella and M. Mota, *Design Guide for Vibrations of Reinforced Concrete Floor Systems*, 10-DG-Vibration, Concrete Reinforcing Steel Institute, Schaumburg, IL, 2014, 60 pp.

12.4. M. R. Wilford and P. Young, *A Design Guide for Footfall Induced Vibration of Structures*, The Concrete Centre, Surrey, UK, 2006, 83 pp.

PROBLEMS

12.1. A footbridge is to be built, consisting of a one-way solid slab spanning 16 ft between masonry abutments, as shown in Fig. P12.1. A service live load of 100 psf must be carried. In addition, a 2000 lb concentrated service load, assumed to be uniformly distributed across the bridge width, may act at any location on the span. A 2 in. asphalt wearing surface is used, weighing 20 psf, including the curb weight. Precast curbs are attached so as to be nonstructural. Prepare a design for the slab, using material strengths $f_y = 60,000$ psi and $f_c' = 4000$ psi, and summarize your results in the form of a sketch showing all concrete dimensions and reinforcement.

FIGURE P12.1

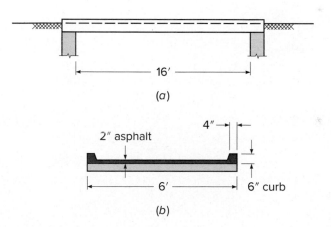

12.2. A reinforced concrete building floor system consists of a continuous one-way slab built monolithically with its supporting beams, as shown in cross section in Fig. P12.2. Service live load is 125 psf. Dead loads include a 10 psf allowance for nonstructural lightweight concrete floor fill and surface, and a 10 psf allowance for suspended loads, plus the self-weight of the floor. Using ACI coefficients from Chapter 11, calculate the design moments and shears and design the slab, using a maximum tensile reinforcement ratio of 0.006. Use all straight bar reinforcement. One-half of the positive-moment bars will be discontinued where no longer required; the other half will be continued into the supporting beams as specified by the ACI Code. All negative steel will be discontinued at the same distance from the support face in each case.

Summarize your design with a sketch showing concrete dimensions, and size, spacing, and cutoff points for all reinforcement. Material strengths are $f_y = 60,000$ psi and $f_c' = 3000$ psi.

FIGURE P12.2

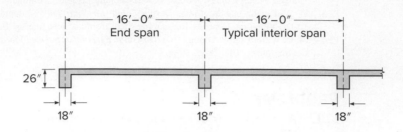

13

Analysis and Design of Two-Way Slabs

13.1 TWO-WAY EDGE-SUPPORTED SLABS

The slabs discussed in Sections 12.2 and 12.3 deform under load into an approximately cylindrical surface. The main structural action is one-way in such cases, in the direction normal to supports on two opposite edges of a rectangular panel. In many cases, however, rectangular slabs are of such proportions and are supported in such a way that two-way action results. When loaded, such slabs bend into a dished surface rather than a cylindrical one. This means that at any point the slab is curved in both principal directions, and since bending moments are proportional to curvatures, moments also exist in both directions. To resist these moments, the slab must be reinforced in both directions, by at least two layers of bars perpendicular, respectively, to two pairs of edges. The slab must be designed to take a proportionate share of the load in each direction.

Types of reinforced concrete construction that are characterized by two-way action include slabs supported by walls or beams on all sides (Fig. 12.1*b*), flat plates (Fig. 12.1*d*), flat slabs (Fig. 12.1*e*), and waffle slabs (Fig. 12.1*f*).

The simplest type of two-way slab action is that represented by Fig. 12.1*b*, where the slab, or slab panel, is supported along its four edges by relatively deep, stiff, monolithic concrete beams or by walls or steel girders. If the concrete edge beams are shallow or are omitted altogether, as they are for flat plates and flat slabs, deformation of the floor system along the column lines significantly alters the distribution of moments in the slab panel itself (Ref. 13.1). Two-way systems of this type are considered separately, beginning in Section 13.2. The present discussion pertains to two-way slabs in which edge supports are stiff enough to be considered unyielding.

Such a slab is shown in Fig. 13.1*a*. To visualize its flexural performance, it is convenient to think of it as consisting of two sets of parallel strips, in each of the two directions, intersecting each other. Evidently, part of the load is carried by one set and transmitted to one pair of edge supports, and the remainder by the other.

Figure 13.1*a* shows the two center strips of a rectangular plate with short span ℓ_a and long span ℓ_b. If the uniform load is q per square foot of slab, each of the two strips acts approximately as a simple beam, uniformly loaded by its share of q. Because these imaginary strips actually are part of the same monolithic slab, their deflections at the intersection point must be the same. Equating the center deflections of the short and long strips gives

$$\frac{5q_a\ell_a^4}{384EI} = \frac{5q_b\ell_b^4}{384EI} \qquad (a)$$

397

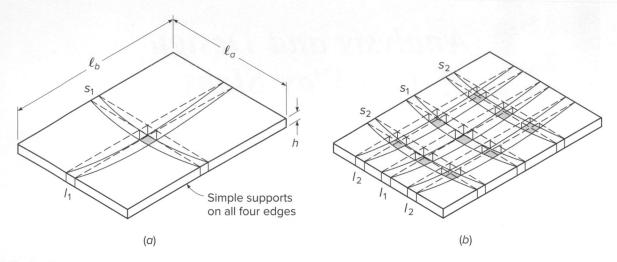

FIGURE 13.1
Two-way slab on simple edge supports: (*a*) bending of center strips of slab and (*b*) grid model of slab.

where q_a is the share of the load q carried in the short direction and q_b is the share of the load q carried in the long direction. Consequently,

$$\frac{q_a}{q_b} = \frac{\ell_b^4}{\ell_a^4} \tag{b}$$

One sees that the larger share of the load is carried in the short direction, the ratio of the two portions of the total load being inversely proportional to the fourth power of the ratio of the spans.

This result is approximate because the actual behavior of a slab is more complex than that of the two intersecting strips. An understanding of the behavior of the slab itself can be gained from Fig. 13.1*b*, which shows a slab model consisting of two sets of three strips each. It is seen that the two central strips s_1 and l_1 bend in a manner similar to that shown in Fig. 13.1*a*. The outer strips s_2 and l_2, however, are not only bent but also twisted. Consider, for instance, one of the intersections of s_2 with l_2. It is seen that at the intersection the exterior edge of strip l_2 is at a higher elevation than the interior edge, while at the nearby end of strip l_2 both edges are at the same elevation; the strip is twisted. This twisting results in torsional stresses and torsional moments that are seen to be most pronounced near the corners. Consequently, the total load on the slab is carried not only by the bending moments in two directions but also by the twisting moments. For this reason, bending moments in elastic slabs are smaller than would be computed for sets of unconnected strips loaded by q_a and q_b. For instance, for a simply supported square slab, $q_a = q_b = q/2$. If only bending were present, the maximum moment in each strip would be

$$\frac{(q/2)\ell^2}{8} = 0.0625q\ell^2 \tag{c}$$

The exact theory of bending of elastic plates shows that actually the maximum moment in such a square slab is only $0.048q\ell^2$, so that in this case the twisting moments relieve the bending moments by about 25 percent.

The largest moment occurs where the curvature is sharpest. Figure 13.1*b* shows this to be the case at midspan of the short strip s_1. Suppose the load is increased until this location is overstressed, so that the steel at the middle of strip s_1 is yielding. If the strip were an isolated beam, it would now fail. Considering the slab as a whole, however, one sees that no immediate failure will occur. The neighboring strips (those parallel as well as those perpendicular to s_1), being actually monolithic with it, will take over any additional load that strip s_1 can no longer carry until they, in turn, start yielding. This inelastic redistribution will continue until in a rather large area in the central portion of the slab all the steel in both directions is yielding. Only then will the entire slab fail. From this reasoning, which is confirmed by tests, it follows that slabs need not be designed for the absolute maximum moment in each of the two directions (such as $0.048q\ell^2$ in the example given in the previous paragraph), but only for a smaller average moment in each of the two directions in the central portion of the slab. For instance, one of the several analytical methods in general use permits a square slab to be designed for a moment of $0.036q\ell^2$. By comparison with the actual elastic maximum moment $0.048q\ell^2$, it is seen that, owing to inelastic redistribution, a moment reduction of 25 percent is provided.

The largest moment in the slab occurs at midspan of the short strip s_1 of Fig. 13.1b. It is evident that the curvature, and hence the moment, in the short strip s_2 is less than at the corresponding location of strip s_1. Consequently, a variation of short-span moment occurs in the long direction of the span. This variation is shown qualitatively in Fig. 13.2. The short-span moment diagram in Fig. 13.2a is valid only along the center strip at 1-1. Elsewhere, the maximum-moment value is less, as shown. Other moment ordinates are reduced proportionately. Similarly, the long-span moment diagram in Fig. 13.2 applies only at the longitudinal centerline of the slab; elsewhere, ordinates are reduced according to the variation shown. These variations in maximum moment across the width and length of a rectangular slab are accounted for in an approximate way in most practical design methods by designing for a reduced moment in the outer quarters of the slab span in each direction.

It should be noted that only slabs with side ratios less than about 2 need be treated as two-way slabs. From Eq. (b) above, it is seen that for a slab of this proportion, the share of the load carried in the long direction is only on the order of one-sixteenth of that in the short direction. Such a slab acts almost as if it were spanning in the short direction only. Consequently, rectangular slab panels with an aspect ratio of 2 or more may be reinforced for one-way action, with the main steel perpendicular to the long edges.

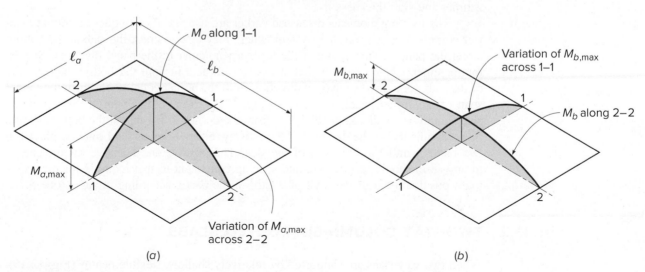

FIGURE 13.2
Moments and moment variations in a uniformly loaded slab with simple supports on four sides.

FIGURE 13.3
Special reinforcement at exterior corners of a beam-supported two-way slab.

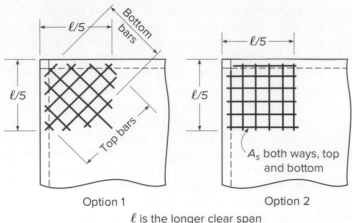

Option 1   Option 2

ℓ is the longer clear span

Consistent with the assumptions of the analysis of two-way edge-supported slabs, the main flexural reinforcement is placed in an orthogonal pattern, with reinforcing bars parallel and perpendicular to the supported edges. As the positive steel is placed in two layers, the effective depth d for the upper layer is smaller than that for the lower layer by one bar diameter. Because the moments in the long direction are the smaller ones, it is economical to place the steel in that direction on top of the bars in the short direction. The stacking problem does not exist for negative reinforcement perpendicular to the supporting edge beams except at the corners, where moments are small.

Either straight bars, cut off where they are no longer required, or bent bars may be used for two-way slabs, but economy of bar fabrication and placement generally favors all straight bars. The precise locations of inflection points (or lines of inflection) are not easily determined, because they depend upon the side ratio, the ratio of live to dead load, and continuity conditions at the edges. The standard cutoff and bend points for beams, summarized in Fig. 6.26, may be used for edge-supported slabs as well.

According to ACI Code 8.6.1, the minimum reinforcement near the tension face in each direction for two-way slabs is 0.0018 times the gross section area A_g and is the value required for shrinkage and temperature crack control. For two-way systems, the spacing of flexural reinforcement at critical sections must not exceed 2 times the slab thickness h.

The twisting moments discussed earlier are usually of consequence only at exterior corners of a two-way slab system, where they tend to crack the slab at the bottom along the panel diagonal, and at the top perpendicular to the panel diagonal. Special reinforcement should be provided at exterior corners in both the bottom and the top of the slab, for a distance in each direction from the corner equal to one-fifth the longer span of the corner panel, as shown in Fig. 13.3. The reinforcement at the top of the slab should be parallel to the diagonal from the corner, while that at the bottom should be perpendicular to the diagonal. Alternatively, either layer of steel may be placed in two bands parallel to the sides of the slab. The positive and negative reinforcement, in any case, should be of a size and spacing equivalent to that required for the maximum positive moment (per foot of width) in the panel, according to ACI Code 8.7.3.

13.2   TWO-WAY COLUMN-SUPPORTED SLABS

When two-way slabs are supported by relatively shallow, flexible beams (Fig. 12.1b), or if column-line beams are omitted altogether, as for flat plates (Fig. 12.1d), flat slabs (Fig. 12.1e), or two-way joist systems (Fig. 12.1f), then a number of new considerations are introduced. Figure 13.4*a* shows a portion of a floor system in which a rectangular slab panel is supported by relatively shallow beams on four sides. The beams, in turn, are carried by columns at the intersection of their centerlines. If a surface load q is applied, that load is shared between imaginary slab strips ℓ_a in the short direction and ℓ_b in the long direction, as described in Section 13.1. The portion of the load that is carried by the long strips ℓ_b is delivered to the beams B1 spanning in the short direction of the panel. The portion carried by the beams B1 plus that carried directly in the short direction by the slab strips ℓ_a sums up to 100 percent of the load applied to the panel. Similarly, the short-direction slab strips ℓ_a deliver a part of the load to long-direction beams B2. That load, plus the load carried directly in the long direction by the slab, includes 100 percent of the applied load. It is clearly a requirement of statics that, for column-supported construction, *100 percent of the applied load must be carried in each direction*, jointly by the slab and its supporting beams (Ref. 13.2).

A similar situation is obtained in the flat plate floor shown in Fig. 13.4*b*. In this case, beams are omitted. However, broad strips of the slab centered on the column lines in each direction serve the same function as the beams of Fig. 13.4*a*; for

FIGURE 13.4
Column-supported two-way slabs: (*a*) two-way slab with beams and (*b*) two-way slab without beams.

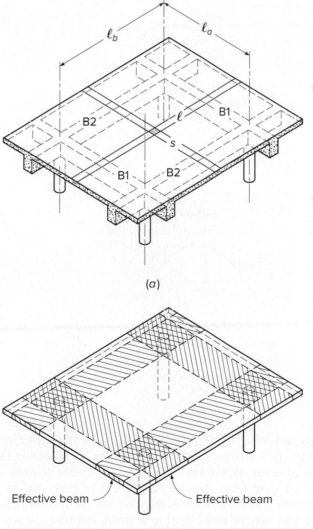

(*a*)

(*b*)

FIGURE 13.5
Moment variation in column-supported two-way slabs: (*a*) critical-moment sections; (*b*) moment variation along a span; and (*c*) moment variation across the width of critical sections.

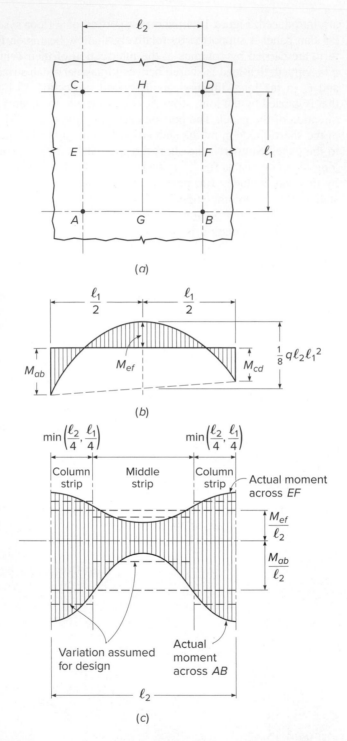

this case, also, the full load must be carried in each direction. The presence of drop panels or column capitals (Fig. 12.1*e*) or shear caps in the double-hatched zone near the columns does not modify this requirement of statics.

Figure 13.5*a* shows a flat plate floor supported by columns at *A*, *B*, *C*, and *D*. Figure 13.5*b* shows the moment diagram for the direction of span ℓ_1. In this direction, the slab may be considered as a broad, flat beam of width ℓ_2. Accordingly, the load per foot of span is $q\ell_2$. In any span of a continuous beam, the sum of the midspan positive moment and the average of the negative moments at adjacent supports is equal to the midspan positive moment of a corresponding simply supported beam. In terms of the slab, this requirement of statics may be written as

$$\tfrac{1}{2}(M_{ab} + M_{cd}) + M_{ef} = \tfrac{1}{8}\, q\ell_2\ell_1^2 \qquad\qquad (a)$$

A similar requirement exists in the perpendicular direction, leading to the relation

$$\tfrac{1}{2}(M_{ac} + M_{bd}) + M_{gh} = \tfrac{1}{8}\, q\ell_1\ell_2^2 \qquad\qquad (b)$$

These results disclose nothing about the relative magnitudes of the support moments and span moments. The proportion of the total static moment that exists at each critical section can be found from an elastic analysis that considers the relative span lengths in adjacent panels, the loading pattern, and the relative stiffness of the supporting beams, if any, and that of the columns. Alternatively, empirical methods that have been found to be reliable under restricted conditions may be adopted.

The moments across the width of critical sections such as AB or EF are not constant but vary as shown qualitatively in Fig. 13.5c. The exact variation depends on the presence or absence of beams on the column lines, the existence of drop panels, column capitals, and shear caps, as well as on the intensity of the load. For design purposes, it is convenient to divide each panel as shown in Fig. 13.5c into column strips, having a width of one-fourth the panel width, on each side of the column centerlines, and middle strips in the one-half panel width between two column strips. Moments may be considered constant within the bounds of a middle strip or column strip, as shown, unless beams are present on the column lines. In the latter case, while the beam must have the same curvature as the adjacent slab strip, the beam moment will be larger in proportion to its greater stiffness, producing a discontinuity in the moment-variation curve at the lateral face of the beam. Since the total moment must be the same as before, according to statics, the slab moments must be correspondingly less.

Chapter 8 of the ACI Code deals in a unified way with all such two-way systems. Its provisions apply to slabs supported by beams and to flat plates and flat slabs, as well as to two-way joist slabs. As described in Section 12.1, a distinction is made between flat plate (see Figs. 1.3 and 12.1d) and flat slab construction (Figs. 1.4 and 12.1e), where the former incorporates a slab of constant thickness and the latter incorporates column capitals or a thickened slab, using either drop panels or shear caps. Column capitals, shown along with drop panels in Figs. 1.4 and 12.1e, increase the shear capacity around columns. Drop panels are used to increase shear capacity, reduce the minimum required thickness of the remaining slab, and reduce the quantity of reinforcement needed for the negative moment near the column. According to ACI Code 8.2.4, a drop panel must project below the soffit of the slab at least one-quarter of the adjacent panel thickness and extend outward at least one-sixth of the center to center of supports. Shear caps are used to increase the critical section for shear at a slab-column joint. According to ACI Code 8.2.5, a shear cap must project below the slab soffit and extend horizontally from the face of the column a distance at least equal to the thickness of the projection below the slab soffit.

While ACI Code 6.2 permits analysis using most computer and finite element programs, and ACI Code 8.2.1 permits design "by any procedure satisfying the conditions of equilibrium and geometrical compatibility," specific reference is made in ACI Code 8.2.1 to two alternative approaches: a semiempirical *direct design method* and an approximate elastic analysis known as the *equivalent frame method*.

FIGURE 13.6

Portion of slab to be included
with beam.

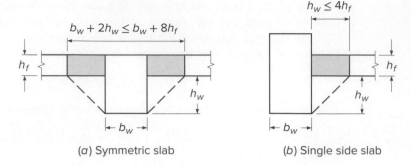

(a) Symmetric slab   (b) Single side slab

The details of both alternatives have been eliminated from the 2019 ACI Code but
may be found in earlier editions of the ACI Code.

A typical panel is divided, for purposes of design, into *column strips* and
middle strips. A column strip is defined in the ACI Code as a strip of slab having
a width on each side of the column centerline equal to one-fourth the smaller of the
panel dimensions ℓ_1 and ℓ_2. Such a strip includes column-line beams, if present. A
middle strip is a design strip bounded by two column strips. In all cases, ℓ_1 is defined
as the span in the direction of the moment analysis and ℓ_2 as the span in the lateral
direction measured center to center of the support. In the case of monolithic construction, beams are defined to include that part of the slab on each side of the beam
extending a distance equal to the projection of the beam above or below the slab h_w
(whichever is greater) but not greater than 4 times the slab thickness (see Fig. 13.6).
In practice when commercial analysis programs are used, the effective locations of
column and middle strips may deviate somewhat from those defined by the ACI
Code, providing the engineer with some flexibility to revise the definition of column
and middle strips to match the structural response.

a.   Lateral Distribution of Moments

For design purposes, it is convenient to consider the moments constant within the
bounds of a middle strip or column strip unless there is a beam present on the column
line. In the latter case, because of its greater stiffness, a beam within a column strip
tends to take a larger share of the column-strip moment than the adjacent slab. The
distribution of total negative or positive moment between slab middle strips, slab
column strips, and beams depends upon the ratio ℓ_2/ℓ_1, the relative stiffness of the
beam and the slab, and the degree of torsional restraint provided by the edge beam.

A convenient parameter defining the relative stiffness of the beam and slab
spanning in either direction is

$$\alpha_f = \frac{E_{cb} I_b}{E_{cs} I_s} \tag{13.1}$$

in which E_{cb} and E_{cs} are the moduli of elasticity of the beam and slab concrete
(usually the same) and I_b and I_s are the moments of inertia of the effective beam
and the slab. Subscripted parameters α_{f1} and α_{f2} are used to identify α computed for
the directions of ℓ_1 and ℓ_2, respectively. By definition, both I_b and, thus, α_f equal
zero when no beam is present.

The flexural stiffnesses of the beam and slab may be based on the gross concrete
section, neglecting reinforcement and possible cracking, and variations due to column
capitals and drop panels may be neglected. For the beam, if present, I_b is based on the

effective cross section defined as in Fig. 13.6. For the slab, I_s is taken equal to $bh^3/12$, where b in this case is the width between panel centerlines on each side of the beam.

The relative restraint provided by the torsional resistance of the effective transverse edge beam is reflected by the parameter β_t, defined as

$$\beta_t = \frac{E_{cb}C}{2E_{cs}I_s} \tag{13.2}$$

where I_s, as before, is calculated for the slab spanning in direction ℓ_1 and having width bounded by panel centerlines in the ℓ_2 direction. The constant C pertains to the torsional rigidity of the effective transverse beam, which is defined as the largest of the following:

1. A portion of the slab having a width equal to that of the column, bracket, or capital in the direction in which moments are taken.
2. The portion of the slab specified in 1 plus that part of any transverse beam above and below the slab.
3. The transverse beam defined as in Fig. 13.6.

The constant C is calculated by dividing the section into its component rectangles, each having smaller dimension x and larger dimension y, and summing the contributions of all the parts by means of the equation

$$C = \sum \left(1 - 0.63\frac{x}{y}\right) \frac{x^3 y}{3} \tag{13.3}$$

The subdivision can be done in such a way as to maximize C.

With these parameters defined, the negative and positive moments are distributed between column strips and middle strips, assigning to the column strips the portions of positive and negative moments shown in Table 13.1. Linear interpolations are to be made between the values shown.

Implementation of these provisions is facilitated by the interpolation charts of Graph A.4 of Appendix A. The portion of the total interior negative and positive moments assigned to the column strip can be read directly from the charts for known

TABLE 13.1

Column-strip moment, portion of total moment at critical section

		ℓ_2/ℓ_1		
		0.5	**1.0**	**2.0**
Interior negative moment				
$\alpha_{f1}\ell_2/\ell_1 = 0$		0.75	0.75	0.75
$\alpha_{f1}\ell_2/\ell_1 \geq 1.0$		0.90	0.75	0.45
Exterior negative moment				
$\alpha_{f1}\ell_2/\ell_1 = 0$	$\beta_t = 0$	1.00	1.00	1.00
	$\beta_t \geq 2.5$	0.75	0.75	0.75
$\alpha_{f1}\ell_2/\ell_1 \geq 1.0$	$\beta_t = 0$	1.00	1.00	1.00
	$\beta_t \geq 2.5$	0.90	0.75	0.45
Positive moment				
$\alpha_{f1}\ell_2/\ell_1 = 0$		0.60	0.60	0.60
$\alpha_{f1}\ell_2/\ell_1 \geq 1.0$		0.90	0.75	0.45

values of ℓ_2/ℓ_1 and $\alpha_{f1}\ell_2/\ell_1$. For exterior negative moment, the parameter β_t requires an additional interpolation, facilitated by the auxiliary diagram on the right side of the charts in Graph A.4. To illustrate its use for $\ell_2/\ell_1 = 1.55$ and $\alpha_{f1}\ell_2/\ell_1 = 0.6$, the dotted line indicates values of 1.00 for $\beta_t = 0$ and 0.65 for $\beta_t = 2.5$. Projecting to the right as indicated by the arrow to find the appropriate vertical scale of 2.5 divisions for an intermediate value of β_t, say 1.0, then upward and finally to the left, one reads the corresponding value of 0.86 on the main chart.

The column-line beam spanning in the direction ℓ_1 is to be proportioned to resist 85 percent of the column-strip moment if $\alpha_{f1}\ell_2/\ell_1$ is equal to or greater than 1.0. For values between 1 and 0, the proportion to be resisted by the beam may be obtained by linear interpolation. Concentrated or linear loads applied directly to such a beam should be accounted for separately.

The portion of the moment not resisted by the column strip is proportionately assigned to the adjacent half-middle strips. Each middle strip is designed to resist the sum of the moments assigned to its two half-middle strips. A middle strip adjacent and parallel to a wall is designed for twice the moment assigned to the half-middle strip corresponding to the first row of interior supports.

b.   Shear in Slab Systems with Beams

Special attention must be given to providing the proper resistance to shear, as well as to moment, when designing slabs with beams. Beams with $\alpha_{f1}\ell_2/\ell_1 \geq 1.0$ must be proportioned to resist the shear caused by loads on a tributary area defined as shown in Fig. 13.7. For values of $\alpha_{f1}\ell_2/\ell_1$ between 1 and 0, the proportion of load carried by beam shear is found by linear interpolation. The remaining fraction of the load on the shaded area is assumed to be transmitted directly by the slab to the columns at the four corners of the panel, and the shear stress in the slab computed accordingly (see Section 13.7).

FIGURE 13.7
Tributary areas for shear calculation.

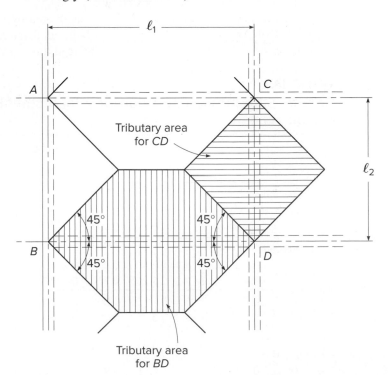

13.3 FLEXURAL REINFORCEMENT FOR COLUMN-SUPPORTED SLABS

Consistent with the assumptions made in analysis, flexural reinforcement in two-way slab systems is placed in an orthogonal grid, with bars parallel to the sides of the panels. Bar diameters and spacings are described in Section 12.2. Straight bars are generally used throughout. To provide for local concentrated loads, as well as to ensure that tensile cracks are narrow and well distributed, a maximum bar spacing at critical sections of 2 times the total slab thickness or 18 in. is specified by ACI Code 8.7.2.2 for two-way slabs. At least the minimum steel required for temperature and shrinkage crack control (see Section 12.3) must be provided. For protection of the steel against damage from fire or corrosion, at least $\frac{3}{4}$ in. concrete cover must be maintained.

Because of the stacking that results when bars are placed in perpendicular layers, the inner steel has an effective depth 1 bar diameter less than the outer steel. For flat plates and flat slabs, the issue of stacking relates to middle-strip positive steel and column-strip negative bars. In two-way slabs with beams on the column lines, stacking occurs for the middle-strip positive steel, and in the column strips is important mainly for the column-line beams, because slab moments are usually very small in the region where column strips intersect.

In the discussion of reinforcement stacking for two-way slabs supported by walls or stiff edge beams, in Section 13.1 it was pointed out that, because curvatures and moments in the short direction are greater than in the long direction of a rectangular panel, short-direction bars are normally placed closer to the top or bottom surface of the slab, with the larger effective depth d, and long-direction bars are placed inside these, with the smaller d. For two-way beamless flat plates, or slabs with relatively flexible edge beams, things are not so simple.

Consider a rectangular interior panel of a flat plate floor. If the slab column strips provided unyielding supports for the middle strips spanning in the perpendicular direction, the short-direction middle-strip curvatures and moments would be the larger. In fact, the column strips deflect downward under load, and this softening of the effective support greatly reduces curvatures and moments in the supported middle strip.

For the entire panel, including both middle strips and column strips in each direction, the moments in the long direction will be larger than those in the short direction, as is easily confirmed by calculating the static moment $M_o = q\ell_2\ell_n^2/8$ in each direction for a rectangular panel. Noting that the apportioning of M_o first to negative- and positive-moment sections, and then laterally to column and middle strips, is done by applying exactly the same ratios in each direction to the corresponding section, it is clear that the middle-strip positive moments (for example) are larger in the long direction than the short direction, exactly the opposite of the situation for the slab with stiff edge beams. In the column strips, positive and negative moments are larger in the long than in the short direction. On this basis, the designer is led to place the long-direction negative and positive bars, in both middle and column strips, closer to the top or bottom surface of the slab, respectively, providing a larger effective depth.

If column-line beams are added, and if their stiffness is progressively increased for comparative purposes, it will be found that the short-direction slab moments gradually become dominant, although the long-direction beams carry larger moments than the short-direction beams. This will be clear from a careful study of Table 13.1.

The situation is further complicated by the influence of the ratio of short to long side dimensions of a panel, and by the influence of varying conditions of edge restraint (such as, corner vs. typical exterior vs. interior panel). While no firm rules can be given, the best guide in specifying steel placement order in areas where stacking occurs is the relative magnitudes of design moments obtained from analysis for a particular case, with maximum d provided for the bars resisting the largest moment. For square slab panels, many designers calculate the required steel area based on the average effective depth, thus obtaining the same bar size and spacing in each direction. This is slightly conservative for the outer layer and slightly unconservative for the inner steel. Redistribution of loads and moments before failure would provide for the resulting differences in capacities in the two directions.

Reinforcement cutoff points could be calculated from moment envelopes if available; however, when either the equivalent frame or the direct design method is used, moment envelopes and lines of inflection are not found explicitly. In such a case, standard bar cutoff points from Fig. 13.8 are used, as recommended in the ACI

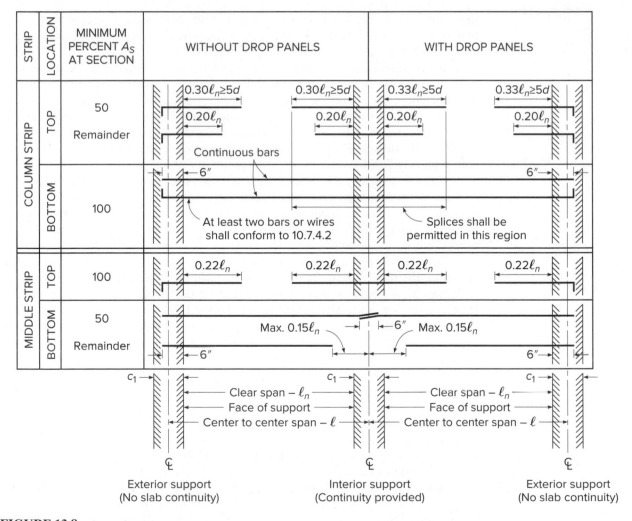

FIGURE 13.8
Minimum length of slab reinforcement in a slab without beams.

Code. The requirement that at least 50 percent of the negative reinforcement extend at least $5d$ from the face of the exterior support in slabs without drop panels is to ensure that a shear crack in a deep slab does not pass beyond the end of the terminated reinforcement.

The structural integrity provisions of ACI Code 8.7.4 require that all bottom bars within the column strip in each direction be continuous or spliced with Class B splices (see Section 6.13a) or mechanical or welded splices located as shown in Fig. 13.8. At least two of the column strip bars in each direction must pass within the column core and must be anchored at exterior supports. The continuous column strip bottom steel is intended to provide some residual ability to carry load to adjacent supports by catenary action if a single support should be damaged or destroyed. The two continuous bars through the column can be considered to be "integrity steel" and are provided to give the slab some residual capacity following a single punching shear failure.

The need for special reinforcement at the exterior corners of two-way beam-supported slabs was described in Section 13.1, and typical corner reinforcement is shown in Fig. 13.3. According to ACI Code 8.7.3, such reinforcement is required for slabs with beams between supporting columns if the value of α_f given by Eq. (13.1) is greater than 1.0.

13.4 DEPTH LIMITATIONS OF THE ACI CODE

To ensure that slab deflections in service will not be troublesome, the recommended approach is to calculate deflections for the total load or load component of interest and to compare the calculated deflections with limiting values. Methods have been developed that are both simple and acceptably accurate for predicting deflections of two-way slabs. A method for calculating the deflection of two-way column-supported slabs is presented in Section 13.10.

Alternatively, in accordance with ACI Code 8.3.1, deflection control can be achieved indirectly by adhering to more or less arbitrary limitations on minimum slab thickness, limitations developed from review of test data and study of the observed deflections of actual structures.[†] As a result of efforts to improve the accuracy and generality of the limiting equations, they have become increasingly complex.

ACI Code 8.3.1 establishes minimum thicknesses for two-way construction. Simplified criteria are included pertaining to slabs without interior beams (flat plates and flat slabs with or without edge beams), while more complicated limit equations are applied to slabs with beams spanning between the supports on all sides. In both cases, minimum thicknesses less than the specified value may be used if calculated deflections are within Code-specified limits, as quoted in Table 7.2.

a. Slabs without Interior Beams

The minimum thickness of two-way slabs without interior beams, according to ACI Code 8.3.1, must not be less than provided by Table 13.2. Edge beams, often provided even for two-way slabs otherwise without beams to improve moment and shear transfer at the exterior supports, permit a reduction in minimum thickness of about

---

[†] These limitations have been in the Code for many years and may not fully reflect current loads, material properties, and design practice. The authors recommend a check of member deflections in all cases.

TABLE 13.2

Minimum thickness of slabs without interior beams

Yield Stress f_y, psi	Without Drop Panels			With Drop Panels		
	Exterior Panels		Interior Panels	Exterior Panels		Interior Panels
	Without Edge Beams	With Edge Beams[a]		Without Edge Beams	With Edge Beams[a]	
40,000	$\ell_n/33$	$\ell_n/36$	$\ell_n/36$	$\ell_n/36$	$\ell_n/40$	$\ell_n/40$
60,000	$\ell_n/30$	$\ell_n/33$	$\ell_n/33$	$\ell_n/33$	$\ell_n/36$	$\ell_n/36$
80,000	$\ell_n/27$	$\ell_n/30$	$\ell_n/30$	$\ell_n/30$	$\ell_n/33$	$\ell_n/33$

[a] Slabs with beams along exterior edges. The value of α_f for the edge beam shall not be less than 0.8.

10 percent in exterior panels. In all cases, the minimum thickness of slabs without interior beams must not be less than the following:

For slabs without drop panels   5 in.
For slabs with drop panels   4 in.

b. Slabs with Beams on All Sides

The parameter used to define the relative stiffness of the beam and slab spanning in either direction is α_f, calculated from Eq. (13.1) of Section 13.2a. Then α_{fm} is defined as the average value of α_f for all beams on the edges of a given panel. According to ACI Code 8.3.1.2, for α_{fm} equal to or less than 0.2, the minimum thicknesses of Table 13.2 shall apply.

For α_{fm} greater than 0.2 but not greater than 2.0, the slab thickness must not be less than

$$h = \frac{\ell_n (0.8 + f_y/200,000)}{36 + 5\beta(\alpha_{fm} - 0.2)} \tag{13.4a}$$

and not less than 5.0 in.

For α_{fm} greater than 2.0, the thickness must not be less than

$$h = \frac{\ell_n (0.8 + f_y/200,000)}{36 + 9\beta} \tag{13.4b}$$

and not less than 3.5 in.,

where ℓ_n = clear span in long direction, in.
α_{fm} = average value of α_f for all beams on edges of a panel [see Eq. (13.1)]
β = ratio of clear span in long direction to clear span in short direction

At discontinuous edges, an edge beam must be provided with a stiffness ratio α_f not less than 0.8; otherwise, the minimum thickness provided by Eq. (13.4a) or (13.4b) must be increased by at least 10 percent in the panel with the discontinuous edge.

In all cases, slab thickness less than the stated minimum may be used if it can be shown by calculation that deflections do not exceed the limit values of Table 7.2.

FIGURE 13.9
Parameter F governing
minimum thickness
of two-way slabs;
minimum thickness
$h = \ell_n (0.8 + f_y/200,000)/F$.

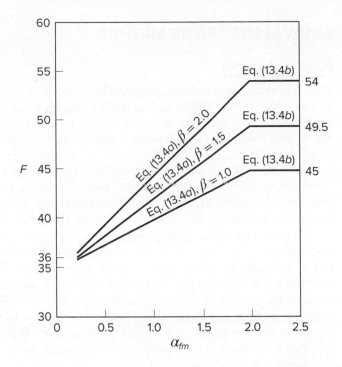

Equations (13.4a) and (13.4b) can be restated in the general form

$$h = \frac{\ell_n (0.8 + f_y/200,000)}{F} \qquad (13.4c)$$

where F is the value of the denominator in each case. Figure 13.9 shows the value of F as a function of α_{fm}, for comparative purposes, for three panel aspect ratios β:

1. Square panel, with $\beta = 1.0$
2. Rectangular panel, with $\beta = 1.5$
3. Rectangular panel, with $\beta = 2.0$, the upper limit of applicability of Eqs. (13.4a) and (13.4b)

Note that, for α_{fm} less than 0.2, column-line beams have little effect, and minimum thickness is given by Table 13.2. For stiff, relatively deep edge beams, with α_{fm} of 2 or greater, Eq. (13.4b) governs. Equation (13.4a) provides a transition for slabs with shallow column-line beams having α_{fm} in the range from 0.2 to 2.0.

13.5 DIRECT DESIGN METHOD

The direct design method is a semiempirical approach to finding moments and shears in two-way slab structures. The method is highly restrictive, requiring at least three continuous, nearly equal spans in each direction; minor variations in column placement; and only gravity loadings. The method is not allowed for the design of footings or for prestressed concrete slabs. Because the direct design method is seldom used in practice, it is not included in this volume; however, the details of this method are found in the 15th and earlier editions of this book and in the ACI Code prior to 2019.

13.6   EQUIVALENT FRAME METHOD

a.   Basis of Analysis

The equivalent frame method, proposed by Peabody in 1948 and later updated based on laboratory testing (Refs. 13.3 to 13.11), was incorporated in subsequent editions of the ACI Code as *design by elastic analysis*. The method was greatly expanded and refined based on research in the 1960s (Refs. 13.12 and 13.13) as the *equivalent frame method*.

The equivalent frame method was derived with the assumption that the analysis would be done using the moment distribution method (see Chapter 11). If analysis is done by computer using a standard frame analysis program, special modeling devices are necessary. This point is discussed further in Section 13.6e.

By the equivalent frame method, the structure is divided, for analysis, into continuous frames centered on the column lines and extending both longitudinally and transversely, as shown by the shaded strips in Fig. 13.10. Each frame is composed of a row of columns and a broad continuous beam. The beam, or slab beam, includes the portion of the slab bounded by panel centerlines on either side of the columns, together with column-line beams or drop panels, if used. For vertical loading, each floor with its columns may be analyzed separately, with the columns assumed to be fixed at the floors above and below. In calculating bending moment at a support, it is convenient and sufficiently accurate to assume that the continuous frame is completely fixed at the support with two panels removed from the given support, provided the frame continues past that point.

FIGURE 13.10
Building idealization for equivalent frame analysis.

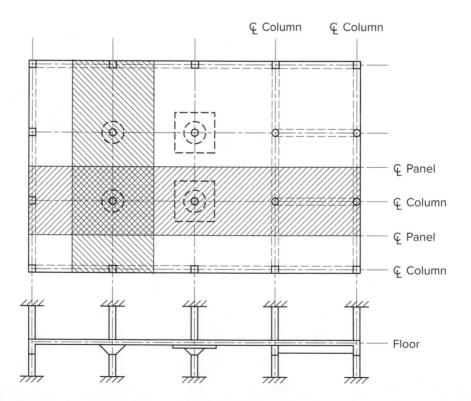

b. Moment of Inertia of Slab Beam

Moments of inertia used for analysis may be based on the concrete cross section, neglecting reinforcement, but variations in cross section along the member axis should be accounted for.

For the slab-beam strips, the first change from the midspan moment of inertia normally occurs at the edge of drop panels, if they are used. The next occurs at the edge of the column or column capital. While the stiffness of the slab-beam strip could be considered infinite within the bounds of the column or capital, at locations close to the panel centerlines (at each edge of the slab-beam strip), the stiffness is much less. From the center of the column to the face of the column or capital, the moment of inertia of the slab beam is taken equal to the value at the face of the column or capital, divided by the quantity $(1 - c_2/\ell_2)^2$, where c_2 and ℓ_2 are the size of the column or capital and the panel width, respectively, both measured transverse to the direction in which moments are being determined.

Accounting for these changes in moments of inertia results in a member, for analysis, in which the moment of inertia varies in a stepwise manner. The stiffness factors, carryover factors, and uniform-load fixed-end moment factors needed for moment distribution analysis (see Chapter 11) are given in Table A.13*a* of Appendix A for a slab without drop panels and in Table A.13*b* for a slab with drop panels with a depth equal to 1.25 times the slab depth and a total length equal to one-third the span length.

c. The Equivalent Column

In the equivalent frame method of analysis, the columns are considered to be attached to the continuous slab beam by torsional members that are transverse to the direction of the span for which moments are being found; the torsional member extends to the panel centerlines bounding each side of the slab beam under study. Torsional deformation of these transverse supporting members reduces the effective flexural stiffness provided by the actual column at the support. This effect is accounted for in the analysis by use of what is termed an *equivalent column* having stiffness less than that of the actual column.

The action of a column and the transverse torsional member is easily explained with reference to Fig. 13.11, which shows, for illustration, the column and transverse beam at the exterior support of a continuous slab-beam strip. From Fig. 13.11, it is

FIGURE 13.11
Torsion at a transverse supporting member illustrating the basis of the equivalent column.

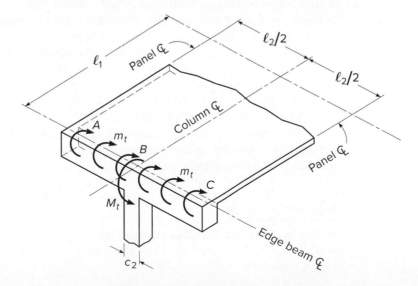

clear that the rotational restraint provided at the end of the slab spanning in the direction ℓ_1 is influenced not only by the flexural stiffness of the column but also by the torsional stiffness of the edge beam AC. With distributed torque m_t applied by the slab and resisting torque M_t provided by the column, the edge-beam sections at A and C will rotate to a greater degree than the section at B, owing to torsional deformation of the edge beam. To allow for this effect, the actual column and beam are replaced by an equivalent column, so defined that the total flexibility (inverse of stiffness) of the equivalent column is the sum of the flexibilities of the actual column and beam. Thus,

$$\frac{1}{K_{ec}} = \frac{1}{\Sigma K_c} + \frac{1}{K_t} \tag{13.5}$$

where K_{ec} = flexural stiffness of equivalent column
K_c = flexural stiffness of actual column
K_t = torsional stiffness of edge beam

all expressed in terms of moment per unit rotation. In computing K_c, the moment of inertia of the actual column is assumed to be infinite from the top of the slab to the bottom of the slab beam, and I_g is based on the gross concrete section elsewhere along the length. Stiffness factors for such a case are given in Table A.13c.

The effective cross section of the transverse torsional member, which may or may not include a beam web projecting below the slab, as shown in Fig. 13.11, is the same as defined earlier in Section 13.3a. The torsional constant C is calculated by Eq. (13.3) based on the effective cross section so determined. The torsional stiffness K_t can then be calculated by the expression

$$K_t = \sum \frac{9E_{cs}C}{\ell_2 (1 - c_2/\ell_2)^3} \tag{13.6}$$

where E_{cs} = modulus of elasticity of slab concrete
c_2 = size of rectangular column, capital, or bracket in direction ℓ_2
C = cross-sectional constant [see Eq. (13.3)]

The summation applies to the typical case in which there are slab beams (with or without edge beams) on both sides of the column. The length ℓ_2 is measured center to center of the supports and thus may have different values in each of the summation terms in Eq. (13.6), if the transverse spans are unequal.

If a panel contains a beam parallel to the direction in which moments are being determined, the value of K_t obtained from Eq. (13.6) leads to values of K_{ec} that are too low. Accordingly, in such cases, the value of K_t found by Eq. (13.6) should be multiplied by the ratio of the moment of inertia of the slab with such a beam to the moment of inertia of the slab without it.

The concept of the equivalent column, illustrated with respect to an exterior column, is employed at *all* supporting columns for each continuous slab beam, according to the equivalent frame method.

d.   Moment Analysis

With the effective stiffness of the slab-beam strip and the supports found as described, the analysis of the equivalent frame can proceed by moment distribution (see Chapter 11).

In keeping with the requirements of statics (see Section 13.2), equivalent beam strips in each direction must each carry 100 percent of the load. If the unfactored live load does not exceed three-quarters of the unfactored dead load, maximum moment may be assumed to occur at all critical sections when the full factored live load (plus factored dead load) is on the entire slab, according to ACI Code 6.4.3. Otherwise, pattern loadings must be used to maximize positive and negative moments. Maximum positive moment is calculated with three-quarters factored live load on the panel and on alternate panels, while maximum negative moment at a support is calculated with three-quarters factored live load on the adjacent panels only. Use of three-quarters live load rather than the full value recognizes that maximum positive and negative moments cannot occur simultaneously (since they are found from different loadings) and that redistribution of moments to less highly stressed sections will take place before failure of the structure occurs. It also recognizes, as explained in Section 13.1, that two-way systems have an inherently greater load capacity than represented by the simple approach used in the equivalent frame method in which the slab is treated as carrying load in only one direction. Factored moments must not be taken less than those corresponding to full factored live load on all panels, however.

Negative moments obtained from that analysis apply at the centerlines of supports. Since the support is not a knife edge but a rather broad band of slab spanning in the transverse direction, some reduction in the negative design moment is proper (see also Section 11.5a). At interior supports, the critical section for negative bending, in both column and middle strips, may be taken at the face of the supporting column or capital, but in no case at a distance greater than $0.175\ell_1$ from the center of the column. To avoid excessive reduction of negative moment at the exterior supports (where the distance to the point of inflection is small) for the case where columns are provided with capitals, the critical section for negative bending in the direction perpendicular to an edge should be taken at a distance from the face of support not greater than one-half the projection of the capital beyond the face of the support.

With positive and negative design moments obtained as just described, it still remains to distribute these moments across the widths of the critical sections. For design purposes, the total strip width is divided into column strip and adjacent half-middle strips, defined previously in Section 13.2, and moments are assumed constant within the bounds of each. The distribution of moments to column and middle strips is done using the percentages given in Table 13.1 and by the interpolation charts of Graph A.4 of Appendix A.

The distribution of moments and shears to column-line beams, if present, is in accordance with the procedures in Section 13.2a. If beams are used on the column lines, the relative stiffness of the beams in the two perpendicular directions, given by the ratio $\alpha_{f1}\ell_2^2/\alpha_{f2}\ell_1^2$, must be between 0.2 and 5.0.

EXAMPLE 13.1   **Design of flat plate floor by equivalent frame method.**   An office building is planned using a flat plate floor system with the column layout as shown in Fig. 13.12. No beams, drop panels, or column capitals are permitted. Specified live load is 100 psf, and the dead load includes the weight of the slab plus an allowance of 20 psf for finish floor plus suspended loads. The columns are 18 in. square, and the floor-to-floor height of the structure is 12 ft. Design the interior panel C, using material strengths $f_y = 60,000$ psi and $f_c' = 4000$ psi. Straight-bar reinforcement is selected.

FIGURE 13.12
Two-way flat plate floor.

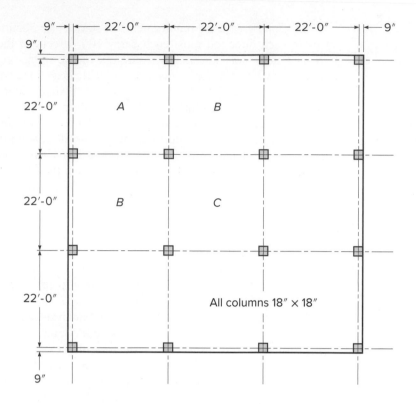

SOLUTION. Minimum thickness h for a flat plate, according to the ACI Code, may be found from Table 13.2.[†] For the present example, the minimum h for the exterior panel is

$$h = \frac{\ell_n}{30} = \frac{20.5 \times 12}{30} = 8.20 \text{ in.}$$

This is rounded up for practical reasons, with calculations based on a trial thickness of 8.5 in. for all panels. Thus, the dead load of the slab is $150 \times 8.5/12 = 106$ psf, to which the superimposed dead load of 20 psf must be added. The factored design loads are

$$1.2q_d = 1.2(106 + 20) = 151 \text{ psf}$$
$$1.6q_l = 1.6 \times 100 = 160 \text{ psf}$$

The structure is identical in each direction, permitting the design for one direction to be used for both (an average effective depth to the tensile steel will be used in the calculations). Moments will be found by the method of moment distribution.

For flat plate structures, it is usually acceptable to calculate stiffnesses as if all members were prismatic, neglecting the increase in stiffness within the joint region, as it generally has negligible effect on design moments and shears. Then, for the slab spans,

$$K_s = \frac{4E_c I_c}{\ell}$$
$$= \frac{4E_c\,(264 \times 8.5^3)}{12 \times 264} = 205E_c$$

and the column stiffnesses are

$$K_c = \frac{4E_c\,(18 \times 18^3)}{12 \times 144} = 243E_c$$

[†] In many flat plate floors, the minimum slab thickness is controlled by requirements for shear transfer at the supporting columns, and h is determined either to avoid supplementary shear reinforcement or to limit the excess shear to a reasonable margin above that which can be carried by the concrete. Design for shear in flat plates and flat slabs is treated in Section 13.7.

TABLE 13.3
Moments in flat plate floor, ft-kips

Panel	B		C		B	
Joint	1	2	2	3	3	4
(a) 311 psf all panels						
Fixed-end moments	+276	−276	+276	−276	+276	−276
Final moments	+125	−323	+295	−295	+323	−125
Span moment in C			119			
(b) 151 psf panels B and 271 psf panel C						
Fixed-end moments	+134	−134	+240	−240	+134	−134
Final moments	+50	−200	+224	−224	+200	−50
Span moment in C			137			
(c) 271 psf panels B (left) and C and 151 psf panel B (right)						
Fixed-end moments	+240	−240	+240	−240	+134	−134
Final moments	+107	−290	+274	−207	+191	−52
Span moment in C			120			

Calculation of the equivalent column stiffness requires consideration of the torsional deformation of the transverse strip of slab that functions as the supporting beam. Applying the criteria of the ACI Code establishes that the effective torsional member has width 18 in. and depth 8.5 in. For this section, the torsional constant C from Eq. (13.3) is

$$C = \Sigma\left(1 - 0.63\,\frac{x}{y}\right)\frac{x^3 y}{3} = \left(1 - 0.63 \times \frac{8.5}{18}\right)8.5^3 \times \frac{18}{3} = 2590 \text{ in}^4$$

and the torsional stiffness, from Eq. (13.6), is

$$K_t = \Sigma\frac{9E_{cs}C}{\ell_2(1 - c_2/\ell_2)^3} = \frac{9E_c \times 2590}{264(1 - 1.5/22)^3} = 109E_c$$

From Eq. (13.5), accounting for two columns and two torsional members at each joint,

$$\frac{1}{K_{ec}} = \frac{1}{\Sigma K_c} + \frac{1}{K_t} = \frac{1}{2 \times 243E_c} + \frac{1}{2 \times 109E_c}$$

from which $K_{ec} = 151E_c$. Distribution factors at each joint are then calculated in the usual way.

For the present example, the ratio of service live load to dead load is $100/126 = 0.79$, and because this exceeds 0.75, according to ACI Code 6.4.3 maximum positive and negative moments must be found based on pattern loadings, with full factored dead load in place and three-quarters factored live load positioned to cause the maximum effect. In addition, the design moments must not be less than those produced by full factored live and dead loads on all panels. Thus, three load cases must be considered: (a) full factored dead and live load, 311 psf, on all panels; (b) factored dead load of 151 psf on all spans plus three-quarters factored live load, 120 psf, on panel C; and (c) full factored dead load on all spans and three-quarters live load on first and second spans. Fixed-end moments and final moments obtained from moment distribution are summarized in Table 13.3. The results indicate that load case a controls the slab design in the support region, while load case b controls at the midspan of panel C. Moment diagrams for the two controlling cases are shown in Fig. 13.13a. As described in Section 13d, the critical section at interior supports may be taken at the face of supports, but not greater than $0.175\ell_1$ from the column centerline. The former criterion controls here, and the negative design moment is calculated by subtracting the area under the shear diagram between the centerline and face of support, for load case a, from the negative moment at the support centerline. The shear diagram for load case a is given in Fig. 13.13b, with the adjusted design moments shown in Fig. 13.13a.

FIGURE 13.13
Design moments and shears
for flat plate floor interior
panel C: (a) moments and
(b) shears.

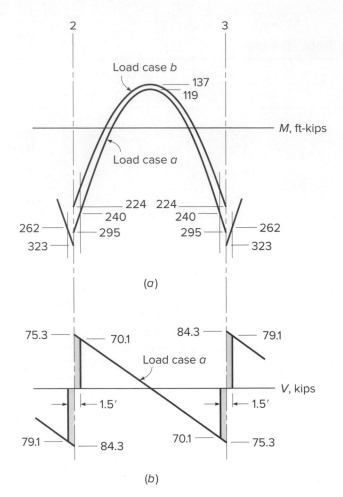

TABLE 13.4
Design of flat plate reinforcement

(1)	(2)	(3)	(4)	(5)	(6)	(7)	(8)	(9)
	Location	M_u, ft-kips	b, in.	d, in.	$M_u \times 12/b$, ft-kips/ft	ρ	A_s, in²	Number and Size of Bars
Column strip	Negative	196	132	7	17.82	0.0075	6.93	16 No. 6 (No. 19)
	Positive	82	132	7	7.45	0.0029	2.68	9 No. 5 (No. 16)
Two half-middle strips	Negative	66	132	7	6.00	0.0023	2.13	8 No. 5 (No. 16)[a]
	Positive	55	132	7	5.00	0.0020	1.85	8 No. 5 (No. 16)[a]

[a] Number of bars controlled by maximum spacing requirement.

Because the effective depth for all panels is the same, and because the negative steel for panel C continues through the support region to become the negative steel for panels B, the larger negative moment found for the panels B controls. Accordingly, the design negative moment is 262 ft-kips and the design positive moment is 137 ft-kips.

Moments are distributed laterally across the slab width according to Table 13.1, which indicates that 75 percent of the negative moment is assigned to the column strip and 60 percent of the positive moment is assigned to the column strip. The design of the slab reinforcement is summarized in Table 13.4.

Other important aspects of the design of flat plates include design for punching shear at the columns, which may require supplementary shear reinforcement, and transfer of unbalanced moments to the columns, which may require additional flexural reinforcement in the negative bending region of the column strips or adjustment of spacing of negative steel. These considerations are of special importance at exterior columns and corner columns, such as shown in Fig. 13.12. Shear and moment transfer at the columns is discussed in Sections 13.7 and 13.8, respectively.

e. Equivalent Frame Analysis by Computer

It is clear that the equivalent frame method, as just described, is oriented toward analysis using the method of moment distribution. Most design offices make use of computers, and frame analysis is done using general-purpose programs. Plane frame analysis programs can be used for slab analysis based on the concepts of the equivalent frame method, but the frame must be specially modeled. Variable moments of inertia along the axis of slab-beams and columns require nodal points (continuous joints) between sections where I is to be considered constant (that is, in the slab at the junction of slab and drop panel, drop panel and capital, and in the columns at the bottom of the capitals). In addition, it is necessary to compute K_{ec} for each column and then to compute the equivalent value of the moment of inertia for the column.

Alternately, a three-dimensional frame or finite element analysis may be used, in which the torsional properties of the transverse supporting beams may be included directly. A third option is to make use of specially written computer programs, one such being spSlab by StructurePoint (Skokie, Illinois).

13.7 SHEAR DESIGN IN FLAT PLATES AND FLAT SLABS

When two-way slabs are supported directly by columns, as in flat slabs and flat plates, or when slabs carry concentrated loads, as in footings, shear near the columns is of critical importance. Tests of flat plate structures indicate that, in most practical cases, the capacity is governed by shear (Ref. 13.14).

a. Slabs without Special Shear Reinforcement

Two kinds of shear may be critical in the design of flat slabs, flat plates, or footings. The first is the familiar beam-type shear leading to diagonal tension failure. Applicable particularly to long narrow slabs or footings, this analysis considers the slab to act as a wide beam, spanning between supports provided by the perpendicular column strips. A potential diagonal crack extends in a plane across the entire width ℓ_2 of the slab. The critical section is taken a distance d from the face of the column or capital. As for beams, the design shear strength ϕV_c must be at least equal to the required strength V_u at factored loads. The nominal shear strength V_c should be calculated using Eq. (5.12a), with b_w equal to the panel width ℓ_2 in this case.

Alternatively, failure may occur by *punching shear*, with the potential diagonal crack following the surface of a truncated cone or pyramid around the column, capital, or drop panel, as shown in Fig. 13.14a. The failure surface extends from the bottom of the slab, at the support, diagonally upward to the top surface. The angle of inclination with the horizontal θ (see Fig. 13.14b) depends upon the nature and amount of reinforcement in the slab. It may range between 20 and 45°. The critical section for shear is taken perpendicular to the plane of the slab and a distance $d/2$

FIGURE 13.14
Failure surface defined by
punching shear.

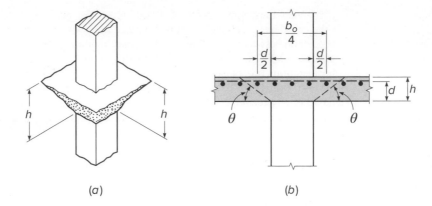

(a)   (b)

from the periphery of the support, as shown. The shear force V_u to be resisted can be calculated as the total factored load on the area bounded by panel centerlines around the column less the load applied within the area defined by the critical shear perimeter, unless significant moments must be transferred from the slab to the column (see Section 13.8).

At such a section, in addition to the shear stresses and horizontal compressive stresses due to negative bending moment, vertical or somewhat inclined compressive stress is present, owing to the reaction of the column. The simultaneous presence of vertical and horizontal compression increases the shear strength of the concrete. For slabs supported by columns having a ratio of long to short sides not greater than 2, tests indicate that due to two-way action, the shear stress on the critical section at the nominal shear strength may be taken equal to

$$v_c = 4\lambda_s \lambda \sqrt{f_c'} \qquad (13.7a)$$

according to ACI Code 22.6.5, where λ is the lightweight concrete factor (see Section 5.5a) and λ_s is the size factor given in Eq. (5.12b). The nominal shear strength of the concrete $V_c = v_c b_o d$, where b_o = the perimeter of the critical section.

However, for slabs supported by very rectangular columns, the shear strength predicted by Eq. (13.7a) has been found to be unconservative. According to tests reported in Ref. 13.15, the value of v_c approaches $2\lambda \sqrt{f_c'}$ as β, the ratio of long to short sides of the column, becomes very large. Reflecting these test data, ACI Code 22.6.5 states further that the punching shear stress v_c shall not be taken greater than

$$v_c = \left(2 + \frac{4}{\beta}\right)\lambda_s \lambda \sqrt{f_c'} \qquad (13.7b)$$

The variation of the shear strength coefficient, as governed by Eqs. (13.7a) and (13.7b), is shown in Fig. 13.15 as a function of β.

Further tests, reported in Ref. 13.16, have shown that the shear stress v_c decreases as the ratio of critical perimeter to slab depth b_o/d increases. Accordingly, ACI Code 22.6.5 states that v_c in punching shear must not be taken greater than

$$v_c = \left(\frac{\alpha_s d}{b_o} + 2\right)\lambda_s \lambda \sqrt{f_c'} \qquad (13.7c)$$

where α_s is 40 for interior columns, 30 for edge columns, and 20 for corner columns, that is, columns having critical sections with 4, 3, or 2 sides, respectively.

Thus, according to the ACI Code, the stress corresponding to the nominal punching shear strength provided by concrete for slabs and footings is to be taken

FIGURE 13.15

Shear strength coefficient for flat plates as a function of ratio β of long side to short side of support.

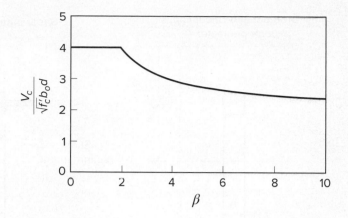

as the smallest of the values of v_c given by Eqs. (13.7a), (13.7b), and (13.7c). The design strength is taken as ϕv_c as usual, where $\phi = 0.75$ for shear. The basic requirement is then $v_u \leq \phi v_c$.

For columns with nonrectangular cross sections, the ACI Code indicates that the perimeter b_o must be of minimum length, but need not approach closer than $d/2$ to the perimeter of the reaction area. The manner of defining the critical perimeter b_o and the ratio β for such irregular support configurations is illustrated in Fig. 13.16.

The ACI Code allows the concrete shear strength given in Eqs. (13.7a), (13.7b), and (13.7c) for shear stress at a distance $d/2$ from the column face. As the critical section moves away from the column face, the confinement that allows this higher shear stress is reduced. Equations (13.7b) and (13.7c) imply that the lower limit of the concrete contribution to shear for the conditions corresponding to these equations is

$$v_c = 2\lambda_s \lambda \sqrt{f_c'} \tag{13.8}$$

Equation (13.8) exceeds the shear stress corresponding to the shear strength of beams given in Eq. (5.12a) because the proximity to the column contributing to two-way action still provides some confinement. Equation (13.8) is used in the examples and homework in this text. The nominal concrete shear strength of a supporting beam

FIGURE 13.16

Punching shear for columns of irregular shape.

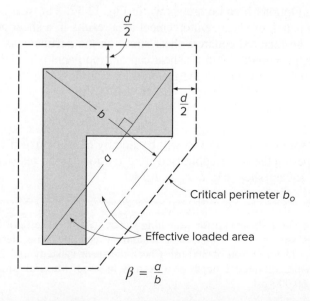

FIGURE 13.17
Critical outer shear
perimeters. (Flexural
reinforcement not shown.)

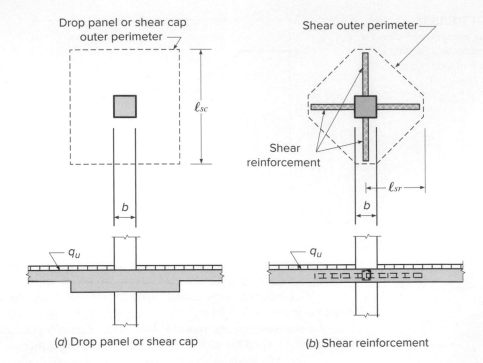

(*a*) Drop panel or shear cap

(*b*) Shear reinforcement

(Fig. 13.7) is not defined in the ACI Code. The engineer may use Eq. (13.8) or shear strength may be based on one-way behavior, as given in Eq. (5.12*a*).

This leads to several options for the shear design of two-way slabs:

a. If v_u is less than that calculated by Eqs. (13.7*a*), (13.7*b*), and (13.7*c*), no further action is needed and the shear is carried by the concrete.

If v_u is greater than that calculated by Eq. (13.7*a*), (13.7*b*), or (13.7*c*),

b. Add a shear cap or drop panel,
c. Use shear reinforcement, or
d. Increase the concrete strength.

Options **b** and **c** are shown in Fig. 13.17. The size of the projected shear cap, drop panel, or shear reinforcement area results in a shear at the outer perimeter such that the factored concrete shear stress on the perimeter is less than or equal to ϕv_c. An approximate size for a shear cap or drop panel is found by solving for the perimeter needed to support the factored load, giving

$$b_o = \frac{V_u}{\phi v_c d} \tag{13.9}$$

The solution in Eq. (13.9) will be slightly conservative if the load on the slab inside the perimeter is not subtracted from the load to be resisted. The length of one side of a square shear cap is $\ell_{sc} = b_o/4$.

EXAMPLE 13.2   **Design of a shear cap for punching shear.**   A flat plate floor has a thickness of $h = 7\frac{1}{2}$ in. and is supported by 18 in. square columns spaced 20 ft on centers each way. The floor carries a total factored load of 300 psf. Check the shear capacity and design a shear cap if needed. An average effective depth of 6 in. may be used. Material strengths are $f_c' = 4000$ psi and $f_y = 60,000$ psi.

Solution: The first critical section for punching shear is $d/2 = 3$ in. from the column face, providing a shear perimeter of $b_o = 24 \times 4 = 96$ in. The factored shear is

$$V_u = 0.300 \, (20^2 - 2^2) = 118.8 \text{ kips}$$

For normalweight concrete and a slab depth less than 10 in., $\lambda = 1.0$ and $\lambda_s = 1.0$. If no shear reinforcement is used, the design strength of the slab, controlled by Eq. (13.7a), is

$$\phi V_c = \phi \lambda_s \lambda 4 \sqrt{f_c'} b_o d = 0.75 \times 1 \times 1 \times 4\sqrt{4000} \times 96 \times 6/1000 = 109.3 \text{ kips}$$

confirming that a shear cap is needed. The necessary depth of the shear cap with no reinforcement is determined by

$$d_{\text{cap}} = \frac{V_u}{\phi \lambda_s \lambda 4 \sqrt{f_c'} b_o} = \frac{118{,}800}{0.75 \times 1 \times 1 \times 4\sqrt{4000} \times 96} = 6.5 \text{ in.}$$

which is rounded up to $d_{\text{cap}} = 7.0$ in., giving $h = 8.5$ in. It is noted that b_o was not corrected for the increased shear cap depth in recognition that the depth would be rounded to a nominal dimension.

To check the design, the depth of the slab at the edge of the shear cap is used and, away from the column face, the allowable concrete shear stress is taken as

$$\phi v_c = \phi \lambda_s \lambda 2 \sqrt{f_c'} = 0.75 \times 1 \times 1 \times 2\sqrt{4000} = 94.9 \text{ psi}$$

The required perimeter from Eq. (13.9) is

$$b_o = \frac{V_u}{\phi v_c d} = \frac{118{,}800}{0.75 \times 94.9 \times 6} = 278 \text{ in.}$$

This is rounded up to 70 in. per side for a square cap and clearly meets the requirement that the cap extends out more than the depth of the slab. Checking the design capacity at a perimeter of 4×53 in. is

$$\phi V_n = \phi v_c b_o d = 0.75 \times 94.9 \times 4 \times 70 \times 6/1000 = 119.6 \text{ kips}$$

which exceeds the factored load, indicating that the design is acceptable.

b. Shear Reinforcement

Shear reinforcement consists of either *integral beams* or *headed shear stud reinforcement,* the latter sometimes referred to as *stud rails,* as shown in Fig. 13.18. Integral beams, illustrated in Fig. 13.18a, consist of vertical stirrups used in conjunction with supplementary horizontal bars radiating outward in two perpendicular directions from the column and are contained entirely within the slab thickness. These beams act in the same general way as shear caps. Adequate anchorage of the stirrups is difficult in slabs thinner than about 10 in. ACI Code 22.6.7 requires the slab effective depth d to be at least 6 in. but not less than 16 times the diameter of the shear reinforcement. In all cases, closed hoop stirrups should be used, with a larger diameter horizontal bar at each bend point, and the stirrups must be terminated with a standard hook (Ref. 13.17).

Headed shear stud reinforcement, shown in Fig. 13.18b, is governed by ACI Code 8.7.7 and 22.6.8. This reinforcement consists of large-headed studs welded to steel strips. The strips are supported on chairs during construction to maintain the required concrete cover to the bottom of the slab below the strip, and the usual cover is maintained over the top of the head. According to tests, the positive anchorage provided by the stud head and the steel strip makes these devices more effective than integral beam reinforcement (Refs. 13.18 and 13.19). In addition, headed stud

FIGURE 13.18
Integral beam and headed
shear stud shear reinforcement.

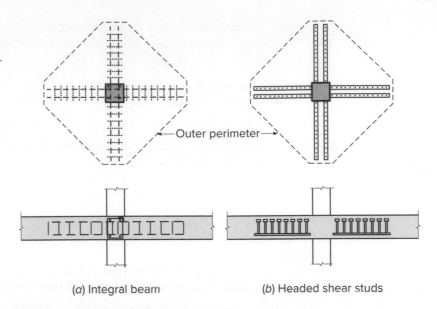

(*a*) Integral beam   (*b*) Headed shear studs

reinforcement can be placed more easily, with less interference with other reinforcement, than an integral beam.

Differences in test performance between integral beams and headed shear studs result in different values of concrete shear stress v_c. Table 13.5 summarizes the nominal two-way shear strength provided by concrete, v_c, for these applications. From Table 13.5, it is seen that when reinforcement is added, v_c is reduced below the value $4\lambda_s\lambda\sqrt{f_c'}$, resulting in the reinforcement having to provide additional capacity to account for drop in shear strength provided by concrete. As with shear caps and drop panels, v_c on the outer perimeter decreases to the value of Eq. (13.8) and further decreases to that of a one-way slab perpendicular to a supporting beam.

ACI Code 22.6.4 requires that the area enclosed by the outer perimeter of a reinforced slab b_o be minimized. For a square column, this usually results in an

TABLE 13.5

Stress corresponding to nominal two-way shear strength provided by concrete v_c for two-way members with shear reinforcement

Type of Shear Reinforcement	Critical Sections	v_c	
Stirrups	All	$2\lambda_s\lambda\sqrt{f_c'}$	
Headed shear stud reinforcement	At $d/2$ from the face of column or concentrated load	Least of (*a*), (*b*), and (*c*)	$3\lambda_s\lambda\sqrt{f_c'}$ (*a*)
			$\left(2+\dfrac{4}{\beta}\right)\lambda_s\lambda\sqrt{f_c'}$ (*b*)
			$\left(2+\dfrac{\alpha_s d}{b_o}\right)\lambda_s\lambda 2\sqrt{f_c'}$ (*c*)
	Critical section at $d/2$ beyond the outer-most reinforcement	$2\lambda_s\lambda\sqrt{f_c'}$	

Notes: (1) β is the ratio of the long to short side of the column, concentrated load, or reaction area.

(2) λ_s is the size effect factor given in Eq. (5.12*b*).

(3) α_s is 40 for interior columns, 30 for edge columns, and 20 for corner columns.

octagonal shape, as indicated in Fig. 13.18b, rather than a rectangle similar to a shear cap. Determination of the length of the extended arm of the shear reinforcement from the center of the column may be estimated by using the square perimeter in Fig. 13.17a, rotating it 45 degrees and, ignoring the small triangles at the corners, dividing the side of the square ℓ_{sc} by $\sqrt{2}$. Thus, the approximate length of the shear arm ℓ_{sr} from the center of the column is approximately

$$\ell_{sr} = \frac{\ell_{sc}}{\sqrt{2}} \qquad (13.10)$$

c. Design of Integral Beams with Vertical Stirrups

Shear reinforcement using vertical stirrups in *integral beams* is shown in Fig. 13.18a. The first critical section for shear design in the slab is taken at $d/2$ from the column face, as usual, and the stirrups, if needed, are extended outward from the column in four directions for the typical interior case (three or two directions for exterior or corner columns, respectively) until the concrete alone can carry the shear, with $v_c = 2\lambda_s\lambda\sqrt{f_c'}$ at the second critical section. Within the region adjacent to the column, where shear resistance is provided by a combination of concrete and steel, the nominal shear strength v_n must not exceed $6\sqrt{f_c'}$ according to ACI Code 26.6.6. In this region, the concrete contribution is reduced to $v_c = 2\lambda_s\lambda\sqrt{f_c'}$. The second critical section crosses each integral beam at a distance $d/2$ measured outward from the last stirrup and is located so that its perimeter b_o is a minimum (that is, for the typical case, defined by 45° lines between the integral beams). The required spacing of the vertical stirrups s is found using Eq. (5.14a), but according to ACI Core 8.7.6 must not exceed $d/2$, with the first line of stirrups not more than $d/2$ from the column face. The spacing of the stirrup legs (measured parallel to the face of the column) in the first line of shear reinforcement must not exceed $2d$.

The problem of anchorage of the shear reinforcement in shallow flat plates is critical, and closed hoop stirrups, terminating in standard hooks, always should be provided with interior corner bars to improve pullout resistance.

EXAMPLE 13.3   **Design of an integral beam with vertical stirrups.**   The flat plate slab with 7.5 in. total thickness and 6 in. effective depth shown in Fig. 13.19 is carried by 12 in. square columns 25 ft on centers in each direction. A factored load of $q_u = 190$ psf must be transmitted from the slab to a typical interior column. Concrete and steel strengths used are, respectively, $f_c' = 4000$ psi and $f_y = 60,000$ psi. Determine if shear reinforcement is required for the slab; and if so, design integral beams with stirrups to carry the excess shear.

SOLUTION.   The factored load to be transferred is $V_u = q_u(\ell^2 - b^2) = 120$ kips. The design shear strength of the concrete alone at the critical section $d/2$ from the face of the column, by the controlling Eq. (13.7a), is

$$\phi V_c = 0.75 \times 4 \times 1 \times 1\sqrt{4000} \times 72 \times 6/1000 = 82.0 \text{ kips}$$

This is less than $V_u = 120$ kips, indicating that shear reinforcement is required. The effective depth $d = 6$ in. just satisfies the minimum allowed to use stirrup reinforcement, as described in Section 13.7b. In this case, the maximum design strength allowed by the ACI Code is

$$\phi V_n = 0.75 \times 6\sqrt{4000} \times 72 \times 6/1000 = 122.9 \text{ kips}$$

FIGURE 13.19
Vertical stirrup shear
reinforcement for slab in
Example 13.3.

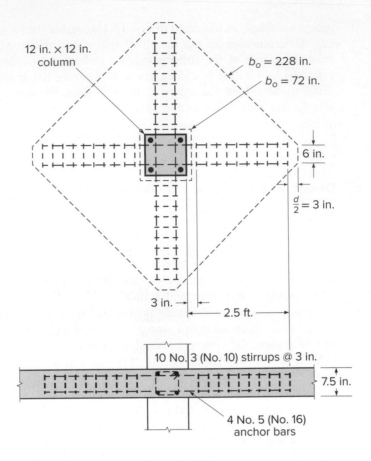

satisfactorily above the actual V_u. When shear is resisted by combined action of concrete and bar reinforcement, the concrete contribution is reduced to

$$\phi V_c = 0.75 \times 1 \times 1 \times 2\sqrt{4000} \times 72 \times 6/1000 = 41.0 \text{ kips}$$

No. 3 (No. 10) vertical closed hoop stirrups are used since d must be ≥ 16 times the stirrup diameter $\left(d/16 = \frac{3}{8} \text{ in.}\right)$ and arranged along four integral beams as shown in Fig. 13.19. Thus, the A_v provided is $4 \times 2 \times 0.11 = 0.88 \text{ in}^2$ at the first critical section, a distance $d/2$ from the column face, and the required spacing can be found from Eq. (5.14a):

$$s = \frac{\phi A_v f_y\, d}{V_u - \phi V_c} = \frac{0.75 \times 0.88 \times 60 \times 6}{120 - 41.0} = 3.01 \text{ in.}$$

However, the maximum spacing of $d/2 = 3$ in. controls here, and No. 3 (No. 10) stirrups at a constant spacing of 3 in. are used. In other cases, stirrup spacing might be increased with distance from the column, as excess shear is less, although this would complicate placement of the reinforcement and generally save little steel.

At the outer perimeter, the stress corresponding nominal shear strength provided by concrete is

$$\phi v_c = \phi 2\lambda_s \lambda \sqrt{f_c'} = 0.75 \times 2 \times 1 \times 1 \times \sqrt{4000} = 94.8 \text{ psi}$$

The required perimeter is

$$b_o = \frac{V_u}{\phi v_c d} = \frac{120}{0.0948 \times 6} = 211 \text{ in.}$$

or an extended arm of $\ell_{sr} = 211/4\sqrt{2} = 37.3$ in. $= 3.11$ ft.

Select 3.25 ft, giving a length over which the shear reinforcement must be placed of 3.25 ft less $b/2$ and $d/2 = 2.5$ ft. The perimeter is $b_o = 4 \times (6 \text{ in.} + (\ell_{sr} - 3 \text{ in.})\sqrt{2}) = 228$ in., resulting in a nominal shear capacity of

$$V_n = \phi v_c b_o d = 0.0948 \times 228 \times 6 = 130 \text{ kips}$$

which exceeds the factored load and is therefore acceptable. Placing stirrups at $d/2 = 3$ in. on center, with the first stirrup placed 1.5 in. from the column, results in 10 stirrups on each side of the column, as shown in Fig. 13.19. A more precise solution obtained by removing the load on the portion of the slab inside the perimeter results in an arm of 3.01 ft, hardly enough different to warrant the additional calculations.

Four longitudinal No. 5 (No. 16) bars are provided inside the corners of each closed hoop stirrup, as shown, to provide for proper anchorage of the shear reinforcement.

d. Design of Headed Shear Stud Reinforcement

Slab shear reinforcement consisting of integral beams with stirrups, as described in Section 13.7c, is widely used. However, the cage that is formed by the stirrups and longitudinal anchor bars may be difficult to install. Also, the slab-column joint region is somewhat congested, with top and bottom slab steel running in two perpendicular directions, with vertical bars in the column, and with the stirrups. Congestion can become critical when the slab has openings, which are frequently required, at or near the column faces.

Shear stud reinforcing strips, as shown in Figs. 13.18b and 13.20a and b, are widely used in Germany, Switzerland, and Canada (Refs. 13.17 and 13.18). Their use in the United States is based on extensive testing (Refs. 13.19 and 13.20), and is incorporated in ACI Code 22.6.6 and 22.6.8. The studs have a minimum yield strength f_{yt} of 51,000 psi and are available in diameters of 0.375, 0.500, 0.625, and 0.750 in., in accordance with ASTM specification A1044.

These devices are composed of vertical bars with anchor heads at their top, welded to a steel strip at the bottom. Multiple strips are arranged in two perpendicular directions for square and rectangular columns or usually in radial directions for circular columns. They are secured in position in the forms before the top and bottom flexural steel is in place. The steel strip rests on bar chairs to maintain the needed concrete cover below the steel and is held in position by nails through holes in the strip.

Headed shear studs are placed perpendicular to the plane of the slab. The overall height of the shear stud assembly may not be less than the thickness of the member less the sum of (1) the concrete cover over the top reinforcement, (2) the concrete cover on the base rail, and (3) one-half the diameter of the tension flexural reinforcement. Two critical shear sections exist. The first is located a distance $d/2$ from the face of the column and the second is located a distance $d/2$ from the outermost peripheral line of studs, as shown in Fig. 13.21 for a typical interior column. As with the integral beams with vertical stirrups described in Section 13.7c, the studs are extended outward from the column until the concrete alone can carry the shear; but in the case of slabs reinforced with headed shear studs, the shear stress due to the factored shear force and any unbalanced moment (see Section 13.8) may not exceed $\phi 2\lambda_s \lambda \sqrt{f_c'}$ on the second critical section.

The nominal shear capacity of the headed shear stud assembly V_n is the sum of the concrete contribution V_c and the shear stud contribution V_s. In the region adjacent to the column, the concrete contribution V_c is reduced to $3\lambda_s \lambda \sqrt{f_c'} b_o d$, as

FIGURE 13.20a
Shear stud reinforcement for concrete slabs: shear stud assembly. (*Courtesy of Amin Ghali and Walter H. Dilger.*)

FIGURE 13.20b
Shear stud reinforcement for concrete slabs: shear reinforcement installed in forms for prestressed concrete slab. (*Courtesy of Amin Ghali and Walter H. Dilger.*)

shown in Table 13.5, and the total nominal capacity V_n may not exceed $8\sqrt{f_c'}b_o d$. The shear stud contribution is $A_v f_{yt} d/s$, where A_v is the area of the studs on a peripheral line and s is the spacing between the peripheral lines, as shown in Fig. 13.20. The value of the shear stud contribution, expressed as a stress on the critical section as $A_v f_{yt}/b_o s$, must be at least $2\sqrt{f_c'}$ in accordance with ACI Code 22.6.6. The ACI Code requires the use of a minimum perimeter for the shear critical section outside

FIGURE 13.21
Arrangement of headed shear
studs and critical sections for
a typical interior column.

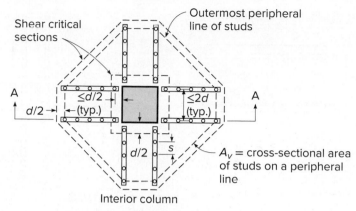

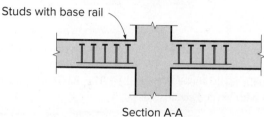

Section A-A

the last stirrups or studs, as shown in Fig. 13.20. A perimeter formed by enlarging the rectangular shear critical section around the column to a distance $d/2$ past the last stirrups or studs does not meet the Code intent.

The spacing of the studs between the column face and the first peripheral line of studs should not exceed $d/2$, and the spacing of the concentric peripheral lines of studs s should be based on the combined effects of shear and any unbalanced moment on the critical section adjacent to the column face and should not exceed $0.75d$ when the shear stress due to factored loads is less than or equal to $6\phi\sqrt{f_c'}$ or $0.5d$ when the shear stress exceeds $6\phi\sqrt{f_c'}$. Lastly, the spacing between the shear stud rails should not exceed $2d$.

EXAMPLE 13.4   **Design of headed stud reinforcement.**   Repeat Example 13.3, using headed stud reinforcement. No. 5 (No. 16) bars are used as negative flexural reinforcement. The yield strength is $f_{yt} = 51,000$ psi for studs.

SOLUTION.   The minimum height of the shear stud assembly equals the thickness of the slab less the cover over the rail, the cover over the reinforcement, and one-half the reinforcement bar diameter. Thus, from Fig 13.22a the minimum height is

$$7.5 - 0.75 - 0.75 - 0.5 \times 0.625 = 5.68 \text{ in.}$$

A stud height of 6 in. is selected, which is consistent with the effective depth and exceeds the minimum requirement. As in Example 13.3, the design shear strength of the concrete alone at the critical section from the face of the column based on an effective depth of 6 in., by the controlling Eq. (13.7a), is

$$\phi V_c = 0.75 \times 4 \times 1 \times 1\sqrt{4000} \times 72 \times 6/1000 = 82.0 \text{ kips}$$

which is less than $V_u = 120$ kips, indicating that shear reinforcement is required. The maximum design strength allowed by the ACI Code when headed stud reinforcement is used is

$$\phi V_n = 0.75 \times 8\sqrt{4000} \times 72 \times 6/1000 = 163.9 \text{ kips}$$

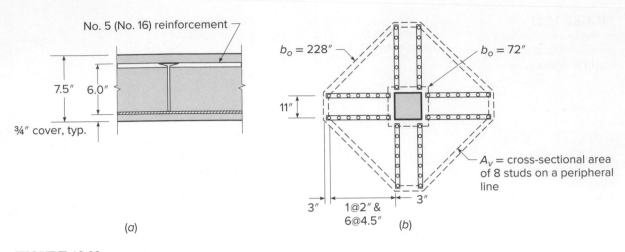

FIGURE 13.22
Headed shear stud arrangement and detail of structural depth requirements for Example 13.4.

which is satisfactorily above the actual V_u. The maximum concrete strength allowed by ACI in conjunction with headed shear studs is

$$\phi V_c = 0.75 \times 3 \times 1 \times 1\sqrt{4000} \times 72 \times 6/1000 = 61.5 \text{ kips}$$

The maximum spacing between the stud rails must be less than $2d$, so two lines of studs are needed for a 12 in. square column; a center-to-center spacing of 11 in. is selected. The shear stress in the slab at the first critical section is approximately $4.4\sqrt{f_c'}$, which is below $\phi 6\sqrt{f_c'} = 4.5\sqrt{f_c'}$, giving a maximum stud spacing of $0.75d$. A spacing of 4.5 in., equal to the maximum, is selected. The area of the studs is found from Eq. (5.14a):

$$A_v = \frac{(V_u - \phi V_c)s}{\phi f_{yt}d} = \frac{(120 - 61.5) \times 4.5}{0.75 \times 51 \times 6} = 1.15 \text{ in}^2$$

A peripheral line of studs contains eight studs, requiring a cross-sectional area of 0.14 in^2 per stud, so 0.500 in. diameter studs with a cross-sectional area of 0.20 in^2 per stud are selected.

The required perimeter of the second critical section, at which the concrete alone can carry the shear, is based on a maximum shear stress of $\phi 2\lambda_s \lambda \sqrt{f_c'}$. Thus,

$$\phi V_c = 0.75 \times 2 \times 1 \times 1\sqrt{4000} \times b_o \times 6 = 120,000 \text{ lb}$$

from which the minimum perimeter $b_o = 211$ in., as in Example 13.3. The first stud is placed at 2 in. ($\leq d/2$) from the column face. Six studs at a spacing of 4.5 in. provide a minimum perimeter of 228 in. (Fig. 13.22b).

13.8 TRANSFER OF MOMENTS AT COLUMNS

The analysis for punching shear in flat plates and flat slabs presented in Section 13.7 assumed that the shear force V_u was resisted by shearing stresses uniformly distributed around the perimeter b_o of the critical section, a distance $d/2$ from the face of the supporting column. The stress corresponding to the nominal shear strength v_c was given by Eqs. (13.7a), (13.7b), and (13.7c).

If significant moments are to be transferred from the slab to the columns, as would result from unbalanced gravity loads on either side of a column or from horizontal loading due to wind or seismic effects, the shear stress on the critical section is no longer uniformly distributed.

FIGURE 13.23
Transfer of moment from slab to column: (*a*) forces resulting from vertical load and unbalanced moment; (*b*) critical section for an interior column; (*c*) shear stress distribution for an interior column; (*d*) critical section for an edge column; and (*e*) shear stress distribution for an edge column.

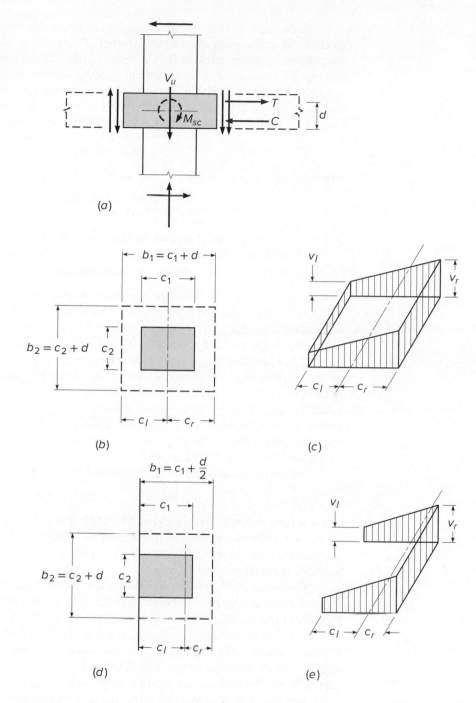

(*a*)

(*b*)

(*c*)

(*d*)

(*e*)

The situation can be modeled as shown in Fig. 13.23*a*. Here, V_u represents the total vertical reaction to be transferred to the column and M_{sc} represents the unbalanced moment to be transferred, both at factored loads. The vertical force V_u causes shear stress distributed more or less uniformly around the perimeter of the critical section as assumed earlier, represented by the inner pair of vertical arrows, acting downward. The unbalanced moment M_{sc} causes additional loading on the joint, represented by the outer pair of vertical arrows, which add to the shear stresses otherwise present on the right side, in the sketch, and subtract on the left side.

Tests indicate that for square columns about 60 percent of the unbalanced moment is transferred by flexure (forces T and C in Fig. 13.23a) and about 40 percent by shear stresses on the faces of the critical section (Ref. 13.21). For rectangular columns, it is reasonable to suppose that the portion transferred by flexure increases as the width of the critical section that resists the moment increases, that is, as $c_2 + d$ becomes larger relative to $c_1 + d$ in Fig. 13.23b. According to ACI Code 8.4.2.2, the slab moment M_{sc} considered to be transferred by flexure is

$$M_{ub} = \gamma_f M_{sc} \tag{13.11a}$$

where

$$\gamma_f = \frac{1}{1 + \frac{2}{3}\sqrt{b_1/b_2}} \tag{13.11b}$$

and b_1 = width of critical section for shear measured in direction of span for which moments are determined

$\quad\ \ b_2$ = width of critical section for shear measured in direction perpendicular to b_1

The value of γ_f may be modified if certain conditions are met in accordance with ACI Code Table 8.4.2.2.4: For unbalanced moments about an axis parallel to the edge of exterior supports, γ_f may be increased to 1.0, provided that the factored shear V_u at the edge support does not exceed $0.75\phi V_c$ or at a corner support does not exceed $0.5\phi V_c$. For unbalanced moments at interior supports and about an axis perpendicular to the edge at exterior supports, γ_f may be increased up to 1.25 times the value in Eq. (13.11b), provided that $V_u \leq 0.4\phi V_c$. In all of these cases, the net tensile strain ε_t calculated for the section within $1.5h$ on either side of the column or column capital must be at least $\varepsilon_{ty} + 0.003$ for corner columns and for edge columns for spans perpendicular to the edge, and $\varepsilon_{ty} + 0.008$ for other cases, where $\varepsilon_{ty} = f_y/E_s$, as defined following Eq. (4.26c).

The moment assumed to be transferred by shear, by ACI Code 8.4.4, is

$$M_{uv} = (1 - \gamma_f)M_{sc} = \gamma_v M_{sc} \tag{13.11c}$$

For a square column, Eqs. (13.11a), (13.11b), and (13.11c) indicate that 60 percent of the unbalanced moment is transferred by flexure and 40 percent by shear, in accordance with the available data. If b_2 is very large relative to b_1, nearly all of the moment is transferred by flexure.

The moment M_{ub} can be accommodated by concentrating a suitable fraction of the slab column-strip reinforcement near the column. According to ACI Code 8.4.2.2, this steel must be placed within the effective slab width b_{slab}, defined by the distance on either side of a column or capital equal to the lesser of (1) $1.5h$ of the slab and (2) the distance to the edge for *slabs without a drop panel or shear cap* or the lesser of (1) $1.5h$ of the drop or cap and (2) the distance to the edge of the drop or cap plus $1.5h$ of the slab for *slabs with a drop panel or shear cap.*

The moment M_{uv}, together with the vertical reaction delivered to the column, causes shear stresses assumed to vary linearly with distance from the centroid of the critical section, as indicated for an interior column by Fig. 13.23c. The stresses can be calculated from

$$v_l = \frac{V_u}{A_c} - \frac{M_{uv}c_l}{J_c} \tag{13.12a}$$

$$v_r = \frac{V_u}{A_c} + \frac{M_{uv}c_r}{J_c} \tag{13.12b}$$

where A_c = area of critical section = $2d[(c_1 + d) + (c_2 + d)]$

c_l, c_r = distances from centroid of critical section to left and right faces of section, respectively

J_c = property of critical section analogous to polar moment of inertia

For an interior column, the quantity J_c is

$$J_c = \frac{2d(c_1 + d)^3}{12} + \frac{2(c_1 + d)d^3}{12} + 2d(c_2 + d)\left(\frac{c_1 + d}{2}\right)^2 \qquad (13.13)$$

Note the implication, in the use of the parameter J_c in the form of a polar moment of inertia, that shear stresses indicated on the near and far faces of the critical section in Fig. 13.23c have horizontal as well as vertical components.

According to ACI Code 8.5.1.1, the maximum shear stress calculated by Eq. (13.12) must not exceed ϕv_n. For slabs without shear reinforcement, $\phi v_n = \phi V_c/b_o d = \phi v_c$, where v_c is the smallest value given by Eqs. (13.7a), (13.7b), or (13.7c). For slabs with shear reinforcement, $\phi v_n = \phi(V_c + V_s)/b_o d$, where V_c and V_s are as established in Section 13.7c or d. Even though the critical sections for direct shear transfer and shear due to moment transfer differ, they coincide or are in close proximity at the column corners where failures initiate, and it is conservative to take the maximum shear as the sum of the two components.

Equations similar to those above can be derived for the edge columns shown in Fig. 13.23d and e or for a corner column. Although the centroidal distances c_l and c_r are equal for the interior column, this is not true for the edge column of Fig. 13.23d or for a corner column.

The application of moment to a column from a slab or beam introduces shear to the column also, as is clear from Fig. 13.23a. This shear must be considered in the design of lateral column reinforcement.

As pointed out in Section 13.7, most flat plate structures, if they are overloaded, fail in the region close to the column, where large shear and bending forces must be transferred. There has been much research aimed at developing improved design details for this region. The design engineer should consult Refs. 13.21 through 13.23 for additional specific information.

13.9 TRANSFER COLUMN LOADS THROUGH SLABS

As discussed in Section 9.16, it is common in building construction to cast the column up to the soffit of the beam or slab of the next floor. The floor is then cast, and the next column is formed and cast. Figure 1.16 shows the floor formwork for this sequence. Using this construction sequence means that the concrete strength in the slab can differ from the concrete strength in the column. As described in Chapter 2, high-strength concrete is sometimes used in building columns to reduce their size and increase the usable floor area. If high-strength columns are used in conjunction with normal-strength concrete in a two-way slab, the slab may not be able to transfer the load to the column below.

The slab concrete surrounding the column provides confinement to allow loads to be safely transferred through concrete that is slightly lower in strength than the column concrete. Should there be a large difference in concrete strength, additional measures are required to assure the integrity of the load transfer. The requirements in these circumstances are covered in Section 9.16. When lower-strength concrete is the basis for design, vertical dowels and transverse reinforcement are required to achieve the design strength. Joints with slabs and beams framing into the column are addressed in Chapter 18.

13.10   OPENINGS IN SLABS

Almost invariably, slab systems must include openings. These may be of substantial size, as required by stairways and elevator shafts, or they may be of smaller dimensions, such as those needed to accommodate heating, plumbing, and ventilating risers; floor and roof drains; and access hatches.

Relatively small openings usually are not detrimental in *beam-supported slabs*. As a general rule, the equivalent of the interrupted reinforcement should be added at the sides of the opening. Additional diagonal bars should be included at the corners to control the cracking that will almost inevitably occur there. The importance of small openings in *slabs supported directly by columns* (flat slabs and flat plates) depends upon the location of the opening with respect to the columns. From a structural point of view, they are best located away from the columns, preferably in the area common to the slab middle strips. Unfortunately, architectural and functional considerations usually cause them to be located close to the columns. In this case, the reduction in effective shear perimeter is the major concern, because such floors are usually shear-critical.

According to ACI Code 22.6.4.3, if the opening is closer than $4h$ to a column, concentrated load, or reaction area, then that part of b_o included within the radial lines projecting from the opening to the centroid of the column, concentrated load, or reaction area should be considered ineffective. Research described in Ref. 13.24 has shown that openings located further away than $4h$ have no effect on the punching shear strength of a slab. The effect of openings is shown in Fig. 13.24, along with the effect of free edges on the perimeter of the critical section.

FIGURE 13.24
Effect of openings and free edges on the determination of the perimeter of the critical section for shear b_o.

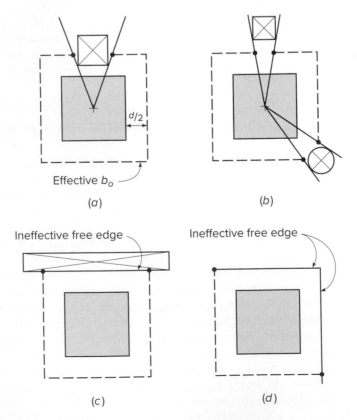

With regard to flexural requirements, the total amount of steel required by calculation must be provided regardless of openings. Any steel interrupted by holes should be matched with an equivalent amount of supplementary reinforcement on either side, properly lapped to transfer stress by bond. Concrete compression area to provide the required strength must be maintained; usually, this would be restrictive only near the columns. According to ACI Code 8.5.4, openings of any size may be located in the area common to intersecting middle strips. In the area common to intersecting column strips, not more than one-eighth of the width of the column strip in either span can be interrupted by openings. In the area common to one middle strip and one column strip, not more than one-quarter of the reinforcement in either strip may be interrupted by the opening.

ACI Code 8.5.4 permits openings of *any* size if it can be shown by analysis that the strength of the slab is at least equal to that required and that all serviceability conditions, that is, cracking and deflection limits, are met. The *strip method* of analysis and design for openings in slabs, by which specially reinforced integral beams, or *strong bands*, of depth equal to the slab depth are used to frame the openings, are described in detail in Chapter 24. Very large openings should preferably be framed by beams or slab bands of increased depth to restore, as nearly as possible, the continuity of the slab. The beams must be designed to carry a portion of the floor load, in addition to loads applied directly by partition walls, elevator support beams, or stair slabs. Section 12.4 discusses ductwork constructed within the slab thickness.

13.11 DEFLECTION CALCULATIONS

The deflection of a uniformly loaded flat plate, flat slab, or two-way slab supported by beams on column lines can be calculated by an equivalent frame method that corresponds with the method for moment analysis described in Section 13.6 (Ref. 13.25) or are the output of the computer analysis program. The definition of column and middle strips, the longitudinal and transverse moment distribution coefficients, and many other details are the same as for the moment analysis. Following the calculation of deflections by this means, they can be compared directly with limiting values like those of Table 7.2, which are applicable to slabs as well as to beams, according to the ACI Code.

A slab region bounded by column centerlines is shown in Fig. 13.25. While no column-line beams, drop panels, or column capitals are shown, the presence of any of these introduces no fundamental complication.

The deflection calculation considers the deformation of such a typical region in one direction at a time, after which the contributions from each direction are added to obtain the total deflection at any point of interest.

In reference to Fig. 13.25a, the slab is considered to act as a broad, shallow beam of width equal to the panel dimension ℓ_y and having the span ℓ_x. Initially, the slab is considered to rest on unyielding support lines at $x = 0$ and $x = \ell_x$. Because of variation of moment as well as flexural stiffness across the width of the slab, not all unit strips in the X direction will deform identically. Typically, the slab curvature in the middle-strip region will be less than that in the region of the column strips because the middle-strip moments are less. The result is as indicated in Fig. 13.25a.

Next the slab is analyzed for bending in the Y direction (Fig. 13.25b). Once again the effect of transverse variation of bending moment and flexural rigidity is seen.

FIGURE 13.25

Basis of equivalent frame
method for deflection
analysis: (*a*) *X* direction
bending; (*b*) *Y* direction
bending; and (*c*) combined
bending.

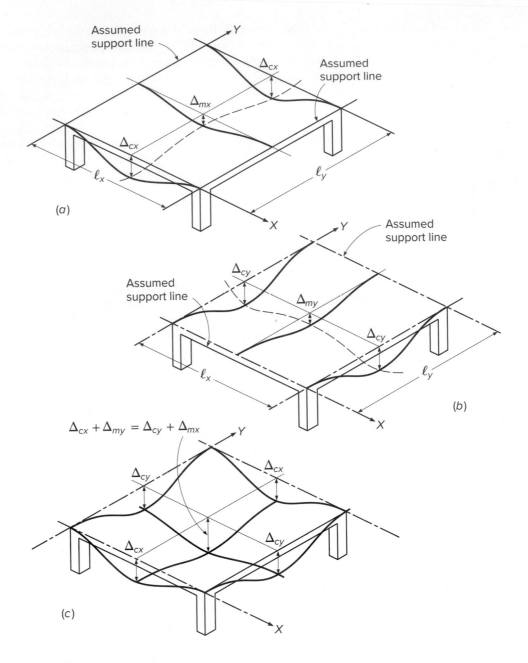

The actual deformed shape of the panel is represented in Fig. 13.25*c*. The midpanel deflection is the sum of the midspan deflection of the column strip in one direction and that of the middle strip in the other direction; that is,

$$\Delta_{\max} = \Delta_{cx} + \Delta_{my} \qquad (13.14a)$$

or

$$\Delta_{\max} = \Delta_{cy} + \Delta_{mx} \qquad (13.14b)$$

In calculations of the deformation of the slab panel in either direction, it is convenient first to assume that it deforms into a cylindrical surface, as it would if the bending moment at all sections were uniformly distributed across the panel width

FIGURE 13.26

Effective cross sections for deflection calculations: (*a*) full-width frame; (*b*) column strip; and (*c*) middle strips.

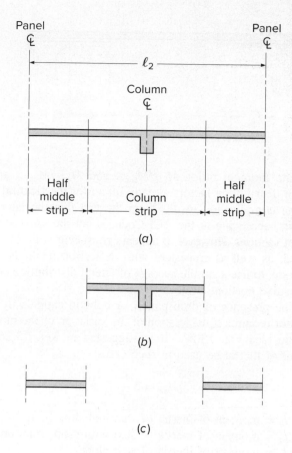

and if lateral bending of the panel were suppressed. The supports are considered to be fully fixed against both rotation and vertical displacement at this stage. Thus, a *reference deflection* is computed:

$$\Delta_{f,\text{ref}} = \frac{w\ell^4}{384E_c I_{\text{frame}}} \tag{13.15}$$

where w is the load per foot along the span of length ℓ and I_{frame} is the moment of inertia of the full-width panel (Fig. 13.26*a*) including the contribution of the column-line beam or drop panels and column capitals if present.

The effect of the actual moment variation across the width of the panel and the variation of stiffness due to beams, variable slab depth, etc., are accounted for by multiplying the reference deflection by the ratio of M/EI for the respective strips to that of the full-width frame:

$$\Delta_{f,\text{col}} = \Delta_{f,\text{ref}} \frac{M_{\text{col}}}{M_{\text{frame}}} \frac{E_c I_{\text{frame}}}{E_c I_{\text{col}}} \tag{13.16a}$$

$$\Delta_{f,\text{mid}} = \Delta_{f,\text{ref}} \frac{M_{\text{mid}}}{M_{\text{frame}}} \frac{E_c I_{\text{frame}}}{E_c I_{\text{mid}}} \tag{13.16b}$$

The subscripts relate the deflection Δ, the bending moment M, or the moment of inertia I to the full-width frame, column strip, or middle strip, as shown in Fig. 13.26*a*, *b*, and *c*, respectively.

FIGURE 13.27
Flat slab span with variable
moment of inertia.

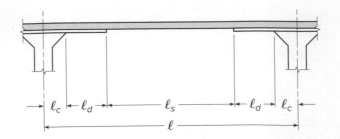

The moment ratios M_{col}/M_{frame} and M_{mid}/M_{frame} are identical to the lateral moment distribution factors already found for the flexural analysis (see Table 13.1). A minor complication results from the fact that the lateral distribution of bending moments, according to the ACI Code, is not the same at the negative and positive-moment sections. However, it appears consistent with the degree of accuracy usually required, as well as consistent with deflection methods endorsed elsewhere in the ACI Code, to use a simple average of lateral distribution coefficients for the negative and positive portions of each strip.

The presence of drop panels or column capitals in the column strip of a flat slab floor requires consideration of the variation of the moment of inertia in the span direction (see Fig. 13.27). It is suggested in Ref. 13.26 that a weighted average moment of inertia be used in such cases:

$$I_{av} = 2\,\frac{\ell_c}{\ell}\,I_c + 2\,\frac{\ell_d}{\ell}\,I_d + \frac{\ell_s}{\ell}\,I_s \tag{13.17}$$

where I_c = moment of inertia of slab including both drop panel and capital
   I_d = moment of inertia of slab with drop panel only
   I_s = moment of inertia of slab alone

Span distances are defined in Fig. 13.27.

Next it is necessary to correct for the rotations of the equivalent frame at the supports, which until now were considered fully fixed. If the ends of the columns are considered fixed at the floor above and floor below, as usual for frame analysis, the rotation of the column at the floor divided by the stiffness of the equivalent column is

$$\theta = \frac{M_{net}}{K_{ec}} \tag{13.18}$$

where θ = angle change, radians
   M_{net} = difference in floor moments to left and right of column
   K_{ec} = stiffness of equivalent column (see Section 13.6c)

In some cases, the connection between the floor slab and column transmits negligible moment, as for lift slabs; thus $K_{ec} = 0$. The flexural analysis indicates that the net moment is zero. The support rotation can be found in such cases by applying the moment-area theorems, taking moments of the M/EI area about the far end of the span and dividing by the span length.

Once the rotation at each end is known, the associated midspan deflection of the equivalent frame can be calculated. It is easily confirmed that the midspan deflection of a member experiencing an end rotation of θ rad, the far end being fixed, is

$$\Delta_\theta = \frac{\theta\ell}{8} \tag{13.19}$$

Thus, the total deflection at midspan of the column strip or middle strip is the sum of the three parts

$$\Delta_{\text{col}} = \Delta_{f,\text{col}} + \Delta_{\theta l} + \Delta_{\theta r} \qquad (13.20a)$$

$$\Delta_{\text{mid}} = \Delta_{f,\text{mid}} + \Delta_{\theta l} + \Delta_{\theta r} \qquad (13.20b)$$

where the subscripts l and r refer to the left and right ends of the span, respectively.

The calculations described are repeated for the equivalent frame in the second direction of the structure, and the total deflection at midpanel is obtained by summing the column-strip deflection in one direction and the middle-strip deflection in the other, as indicated by Eqs. (13.14a) and (13.14b).

The midpanel deflection should be the same whether calculated by Eq. (13.14a) or Eq. (13.14b). Actually, a difference is usually obtained because of the approximate nature of the calculations. For very rectangular panels, the main contribution to midpanel deflection is that of the long-direction column strip. Consequently, the midpanel deflection is best found by summing the deflections of the long-direction column strip and the short-direction middle strip. However, for exterior panels, the important contribution is from the column strips perpendicular to the discontinuous edge, even though the long side of the panel may be parallel to that edge.

In slabs, as in beams, the effect of concrete cracking is to reduce the flexural stiffness. According to ACI Code 8.3.2, the effective moment of inertia given by Eq. (7.9) is applicable to slabs as well as beams, although other values may be used if results are in reasonable agreement with tests. In most cases, two-way slabs are essentially uncracked at service loads, and it is satisfactory to base deflection calculations on the uncracked moment of inertia I_g (see Ref. 13.25 for comparison with tests). ACI Code 8.3.1.1, however, requires that for f_y greater than 80 ksi, a reduced modulus of rupture $f_r = \sqrt{f_c'}$ be used to calculate deflections. In Ref. 13.27, Branson suggests the following refinements: (1) for slabs without beams, use I_g for all dead load deflections; for dead plus live load deflections, use I_g for middle strips and I_e for column strips; (2) for slabs with beams, use I_g for all dead load deflections; for dead plus live load deflections, use I_g for column strips and I_e for middle strips. For continuous spans, I_e can be based on the midspan positive moment without serious error.

The deflections calculated using the procedure described are short-term deflections. Long-term slab deflections can be calculated by multiplying the short-term deflections by the factor λ_Δ of Eq. (7.12), as for beams. Because compression steel is seldom used in slabs, a multiplier of 2.0 results. Test evidence and experience with actual structures indicate that this may seriously underestimate long-term slab deflections, and multipliers for long-term deflection from 2.5 to 4.0 have been recommended (Refs. 13.27 to 13.29). A multiplier of 3.0 gives acceptable results in most cases.

It should be recognized that the prediction of slab deflections, both initial elastic and long-term, is complicated by the many uncertainties associated with actual building construction. Loading history, particularly during construction, has a profound effect on final deflections (Ref. 13.30). Construction loads can equal or exceed the service live load. Such loads may include the weight of stacked building material and usually include the weight of slabs above the one cast earlier, applied through shoring and reshoring to the lower slab. Because construction loads are applied to immature concrete in the slabs, the immediate elastic deflections are large, and, upon removal of the construction loads, elastic recovery is less than the initial elastic deflection because E_c increases with age. Cracking resulting from construction loading does not disappear with removal of the temporary load and may result in live load deflections greater than expected. Creep during construction loading may be greater than expected because of the early age of the concrete when loaded. Shrinkage deflections of thin slabs are often of the same order of magnitude as the elastic deflections, and some cases must be calculated separately.

It is important to recognize that both initial and time-dependent slab deflections are subject to a high degree of variability. Calculated deflections are an estimate, at best, and considerable deviation from calculated values is to be expected in actual structures.

EXAMPLE 13.5   **Calculation of deflections.** Find the deflections at the center of the typical exterior panel of the two-way floor designed in Example 13.1 due to dead and live loads. The live load may be considered a short-term load and is distributed uniformly over all panels. The floor supports nonstructural elements that are likely to be damaged by large deflections. Take $E_c = 3.6 \times 10^6$ psi.

SOLUTION. The elastic deflection due to the self-weight of 88 psf is found, after which the additional long-term dead load deflection can be found by applying the factor $\lambda = 3.0$, and the short-term live load deflection due to 144 psf by direct proportion.

The effective concrete cross sections, upon which moment of inertia calculations are based, are shown in Fig. 13.28 for the full-width frame, the column strip, and the middle strips, for the short-span and long-span directions. Note that the width of the column strip in both directions is based on the shorter panel span, according to the ACI Code. The values of moment of inertia are as follows:

	Short Direction	Long Direction
I_{frame}	27,900 in^4	25,800 in^4
I_{col}	21,000 in^4	21,000 in^4
I_{mid}	5,150 in^4	3,430 in^4

FIGURE 13.28
Cross-sectional dimensions for deflection example: (*a*) short-span direction frame, column strip, and middle strip and (*b*) long-span direction frame, column strip, and middle strip.

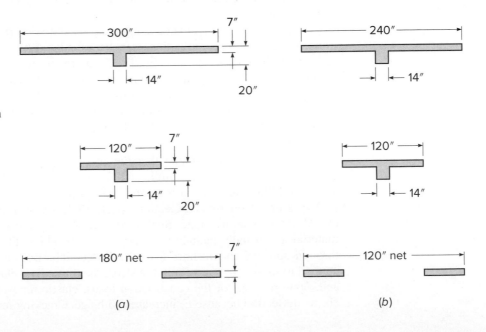

First calculating the deflections of the floor in the *short-span direction* of the panel, from Eq. (13.15) the reference deflection is

$$\Delta_{f,\text{ref}} = \frac{w\ell^4}{384 E_c I_{\text{frame}}} = \frac{88 \times 25(20 \times 12)^4}{12 \times 384 \times 3.6 \times 10^6 \times 27,900} = 0.016 \text{ in.}$$

(Note that the centerline span distance is used here, although clear span was used in the moment analysis to approximate the moment reduction due to support width, according to ACI Code procedures.) From the moment analysis in the short-span direction, it was concluded that 68 percent of the moment at both negative and positive sections was taken by the column strip and 32 percent by the middle strips. Accordingly, from Eqs. (13.16a) and (13.16b),

$$\Delta_{f,\text{col}} = \Delta_{f,\text{ref}} \frac{M_{\text{col}}}{M_{\text{frame}}} \frac{E_c I_{\text{frame}}}{E_c I_{\text{col}}} = 0.016 \times 0.68 \times \frac{27,900}{21,000} = 0.014 \text{ in.}$$

$$\Delta_{f,\text{mid}} = \Delta_{f,\text{ref}} \frac{M_{\text{mid}}}{M_{\text{frame}}} \frac{E_c I_{\text{frame}}}{E_c I_{\text{mid}}} = 0.016 \times 0.32 \times \frac{27,900}{5150} = 0.028 \text{ in.}$$

For the panel under investigation, which is fully continuous over both supports in the short direction, it may be assumed that support rotations are negligible; consequently, $\Delta_{\theta t}$ and $\Delta_{\theta r} = 0$, and from Eqs. (13.20a) and (13.20b),

$$\Delta_{\text{col}} = 0.014 \text{ in.}$$

$$\Delta_{\text{mid}} = 0.028 \text{ in.}$$

Now calculating the deformations in the *long direction* of the panel gives the reference deflection

$$\Delta_{f,\text{ref}} = \frac{88 \times 20(25 \times 12)^4}{12 \times 384 \times 3.6 \times 10^6 \times 25,800} = 0.033 \text{ in.}$$

From the moment analysis, it was found that the column strip would take 93 percent of the exterior negative moment, 81 percent of the positive moment, and 81 percent of the interior negative moment. Thus, the average lateral distribution factor for the column strip is

$$\left(\frac{0.93 + 0.81}{2} + 0.81 \right) \frac{1}{2} = 0.84$$

or 84 percent, while the middle strips are assigned 16 percent. Then from Eqs. (13.16a) and (13.16b),

$$\Delta_{f,\text{col}} = 0.033 \times 0.84 \times \frac{25,800}{21,000} = 0.034 \text{ in.}$$

$$\Delta_{f,\text{mid}} = 0.033 \times 0.16 \times \frac{25,800}{3430} = 0.040 \text{ in.}$$

While rotation at the interior column may be considered negligible, rotation at the exterior column cannot. For the dead load of the slab, the full static moment is

$$M_o = \tfrac{1}{8} \times 0.088 \times 20 \times 25^2 = 137.5 \text{ ft-kips}$$

It was found that 16 percent of the static moment, or 22.0 ft-kips, should be assigned to the exterior support section. The resulting rotation is found from Eq. (13.18). It is easily confirmed that the stiffness of the equivalent column (see Section 13.6c) is $169 \times 3.6 \times 10^6$ in-lb/rad; hence,

$$\theta = \frac{22,000 \times 12}{169 \times 3.6 \times 10^6} = 0.00043 \text{ rad}$$

From Eq. (13.19), the corresponding midpanel deflection component is

$$\Delta_\theta = \frac{0.00043 \times 25 \times 12}{8} = 0.016 \text{ in.}$$

Thus, from Eqs. (13.20*a*) and (13.20*b*), the deflections of the column and middle strips in the long direction are

$$\Delta_{col} = 0.034 + 0.016 = 0.050 \text{ in.}$$

$$\Delta_{mid} = 0.040 + 0.016 = 0.056 \text{ in.}$$

and from Eq. (13.14*a*) the short-term midpanel deflection due to self-weight is

$$\Delta_{max} = 0.050 + 0.028 = 0.078 \text{ in.}$$

The long-term deflection due to dead load is $3.0 \times 0.078 = 0.234$ in., and the short-term live load deflection is $0.078 \times 144/88 = 0.128$.

The ACI limiting value for the present case is found to be 1/480 times the span, or $20 \times 12/480 = 0.500$ in., based on the sum of the long-time deflection due to sustained load and the immediate deflection due to live load. The sum of these deflection components in the present case is

$$\Delta_{max} = 0.234 + 0.128 = 0.362 \text{ in.}$$

well below the permissible value.

13.12   ANALYSIS FOR HORIZONTAL LOADS

The methods described in the preceding sections of this chapter may be used for the analysis of two-way slab systems for gravity loads, according to ACI Code 8.2.1. These procedures, however, are not meant to apply to the analysis of buildings subject to lateral loads, such as loads caused by wind or earthquake. For lateral load analysis, the designer may select any method that is shown to satisfy equilibrium and geometric compatibility and give results that are in reasonable agreement with available test data. The results of the lateral load analysis may then be combined with those from the vertical load analysis, according to ACI Code 8.4.1.9.

Plane frame analysis, with the building assumed to consist of parallel frames each bounded laterally by the panel centerlines on either side of the column lines, has often been used in analyzing unbraced buildings for horizontal loads, as well as vertical. For vertical load analysis by the equivalent frame method, a single floor is usually studied as a substructure with attached columns assumed fully fixed at the floors above and below, but for horizontal frame analysis the equivalent frame includes all floors and columns, extending from the bottom to the top of the structure.

The main difficulty in equivalent frame analysis for horizontal loads lies in modeling the stiffness of the region at the beam-column (or slab-beam-column) connections. Transfer of forces in this region involves bending, torsion, shear, and axial load and is further complicated by the effects of concrete cracking in reducing stiffness and reinforcement in increasing it. Frame moments are greatly influenced by horizontal displacements at the floors, and a conservatively low value of stiffness should be used to ensure that a reasonable estimate of drift is included in the analysis.

While a completely satisfactory basis for modeling the beam-column joint stiffness has not been developed, at least two methods have been used in practice (Ref. 13.31). The first is based on an equivalent beam width $\alpha \ell_2$, less than the actual width, to reduce the stiffness of the slab for purposes of analysis. Figure 13.29*a* shows a plate fixed at the far edge and supported by a column of width c_2 at the near side.

FIGURE 13.29
Equivalent beam width for
horizontal load analysis.

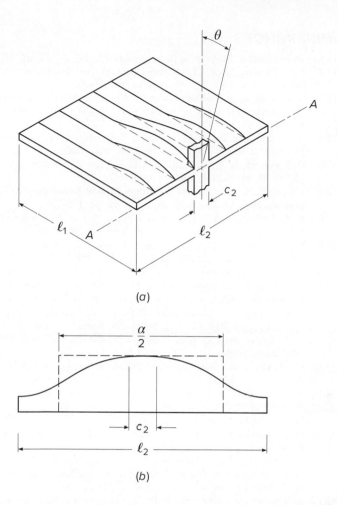

(a)

(b)

If a rotation θ is imposed at the column, the plate rotation along the axis A will vary
as shown by Fig. 13.29a, from θ at the column to smaller values away from the column. An equivalent width factor α is obtained from the requirement that the stiffness
of a prismatic beam of width $\alpha\ell_2$ must equal the stiffness of the plate of width ℓ_2.
This equality is obtained if the areas under the two rotation diagrams of Fig. 13.29b
are equal. Thus, the frame analysis is based on a reduced slab (or slab-beam) stiffness
found using $\alpha\ell_2$ rather than ℓ_2. Comparative studies indicate that, for flat plate floors,
a value for α between 0.25 and 0.50 may be used (Ref. 13.31).

Alternatively, the beam-column stiffness can be modeled based on a transverse
torsional member corresponding to that used in deriving the stiffness of the equivalent column for the vertical load analysis of two-way slabs by the equivalent frame
method (see Section 13.6c). Rotational stiffness of the joint is a function of the
flexural stiffness of the columns framing into the joint from above and below and
the torsional stiffness of the transverse strip of slab or slab beam at the column. The
equivalent column stiffness is found from Eq. (13.5) and the torsional stiffness from
Eq. (13.6), as before.

Finally, for frames in which two-way systems act as primary members resisting
lateral loads, ACI Code 8.7.4.1 requires that the lengths of reinforcement be determined by analysis because the lengths shown in Fig. 13.8 may not be adequate. The
values in Fig. 13.8, however, are retained as minimum values.

REFERENCES

13.1. W. L. Gamble, "Moments in Beam-Supported Slabs," *J. ACI*, vol. 69, no. 3, 1972, pp. 149–157.

13.2. J. R. Nichols, "Statical Limitations upon the Steel Requirement in Reinforced Concrete Flat Slab Floors," *Trans. ASCE*, vol. 77, 1914, pp. 1670–1736.

13.3. D. Peabody, Jr., "Continuous Frame Analysis of Flat Slabs," *J. Boston Society Civ. Eng.*, January 1948.

13.4. D. S. Hatcher, M. A. Sozen, and C. P. Siess, "Test of a Reinforced Concrete Flat Plate," *J. Struct. Div.*, ASCE, vol. 91, no. ST5, 1965, pp. 205–231.

13.5. S. A. Guralnick and R. W. LaFraugh, "Laboratory Study of a Forty-Five Foot Square Flat Plate Structure," *J. ACI*, vol. 60, no. 9, 1963, pp. 1107–1185.

13.6. D. S. Hatcher, M. A. Sozen, and C. P. Siess, "Test of a Reinforced Concrete Flat Slab," *J. Struct. Div.*, ASCE, vol. 95, no. ST6, 1969, pp. 1051–1072.

13.7. J. O. Jirsa, M. A. Sozen, and C. P. Siess, "Test of a Flat Slab Reinforced with Welded Wire Fabric," *J. Struct. Div.*, ASCE, vol. 92, no. ST3, 1966, pp. 199–224.

13.8. W. L. Gamble, M. A. Sozen, and C. P. Siess, "Tests of a Two-Way Reinforced Concrete Floor Slab," *J. Struct. Div.*, ASCE, vol. 95, no. ST6, 1969, pp. 1073–1096.

13.9. M. D. Vanderbilt, M. A. Sozen, and C. P. Siess, "Test of a Modified Reinforced Concrete Two-Way Slab," *J. Struct. Div.*, ASCE, vol. 95, no. ST6, 1969, pp. 1097–1116.

13.10. W. L. Gamble, "Moments in Beam-Supported Slabs," *J. ACI*, vol. 69, no. 3, 1972, pp. 149–157.

13.11. J. O. Jirsa, M. A. Sozen, and C. P. Siess, "Pattern Loadings on Reinforced Concrete Floor Slabs," *J. Struct. Div.*, ASCE, vol. 95, no. ST6, 1969, pp. 1117–1137.

13.12. W. G. Corley, M. A. Sozen, and C. P. Siess, "The Equivalent Frame Method for Reinforced Concrete Slabs," Univ. Ill. Dept. Civ. Eng. Struct. Res. Series 218, June 1961.

13.13. W. G. Corley and J. O. Jirsa, "Equivalent Frame Analysis for Slab Design," *J. ACI*, vol. 67, no. 11, 1970, pp. 875–884.

13.14. ASCE-ACI Task Committee 426, "The Shear Strength of Reinforced Concrete Members—Slabs," *J. Struct. Div.*, ASCE, vol. 100, no. ST8, 1974, pp. 1543–1591.

13.15. N. M. Hawkins, H. B. Fallsen, and R. C. Hinojosa, "Influence of Column Rectangularity on the Behavior of Flat Plate Structures," in *Cracking, Deflection, and Ultimate Load of Concrete Slab Systems*, ACI Publication SP-30, 1971, p. 127.

13.16. M. D. Vanderbilt, "Shear Strength of Continuous Plates," *J. Struct. Div.*, ASCE, vol. 98, no. ST5, 1972, pp. 961–973.

13.17. A. Ghali, "An Efficient Solution to Punching of Slabs," *Concr. Intl.*, vol. 11, no. 6, 1989, pp. 50–54.

13.18. A. A. Elgabry and A. Ghali, "Design of Stud-Shear Reinforcement for Slabs," *ACI Struct. J.*, vol. 87, no. 3, 1990, pp. 350–361.

13.19. A. S. Mokhtar, A. Ghali, and W. Dilger, "Stud Shear Reinforcement for Flat Concrete Plates," *J. ACI*, vol. 82, no. 5, 1985, pp. 676–683.

13.20. A. A. Elgabry and A. Ghali, "Tests on Concrete Slab-Column Connections with Stud Shear Reinforcement Subjected to Shear-Moment Transfer," *ACI Struct. J.*, vol. 84, no. 5, 1987, pp. 433–442.

13.21. N. M. Hawkins, A. Bao, and J. Yamazaki, "Moment Transfer from Concrete Slabs to Columns," *ACI Struct. J.*, vol. 86, no. 6, 1989, pp. 705–716.

13.22. "Recommendations for Design of Slab-Column Connections in Monolithic Reinforced Concrete Structures," reported by ACI-ASCE Committee 352, *ACI Struct. J.*, vol. 85, no. 6, 1988, pp. 675–696.

13.23. J. P. Moehle, M. E. Kreger, and R. Leon, "Background to Recommendations for Design of Reinforced Concrete Slab-Column Connections," *ACI Struct. J.*, vol. 85, no. 6, 1988, pp. 636–644.

13.24. A. S Genikomsou and M. A. Polak, "Effect of Openings on Punching Shear Strength of Reinforced Concrete Slabs—Finite Element Investigation," *ACI Structural J.*, vol. 114, no. 5, 2017, pp. 1249–1261.

13.25. A. H. Nilson and D. B. Walters, "Deflection of Two-Way Floor Systems by the Equivalent Frame Method," *J. ACI*, vol. 72, no. 5, May 1975, pp. 210–218.

13.26. E. S. Hoffman, D. P. Gustafson, and A. J. Gouwens, *Structural Design Guide to the ACI Building Code*, 4th ed., Kluwer Academic Publishers, Boston, 1998.

13.27. D. E. Branson, *Deformation of Concrete Structures*, McGraw-Hill, New York, 1977.

13.28. P. J. Taylor and J. L. Heiman, "Long-Term Deflection of Reinforced Concrete Flat Slabs and Plates," *J. ACI*, vol. 74, no. 11, 1977, pp. 556–561.

13.29. C. J. Graham and A. Scanlon, "Long-Time Multipliers for Estimating Two-Way Slab Deflections," *J. ACI*, vol. 83, no. 6, 1986, pp. 899–908.

13.30. N. J. Gardner and A. Scanlon, "Long-Term Deflections of Two-Way Slabs," *Concr. Intl.*, vol. 12, no. 1, 1990, pp. 63–67.

13.31. M. D. Vanderbilt and W. G. Corley, "Frame Analysis of Concrete Buildings," *Concr. Intl.*, vol. 5, no. 12, 1983, pp. 33–43.

PROBLEMS

13.1. For the one-way slab floor in Problem 12.2, calculate the immediate and long-term deflection due to dead loads. Assume that all dead loads are applied when the construction shoring is removed. Also determine the deflection due to application of the full-service live load. Assuming that sensitive equipment will be installed 6 months after the shoring is removed, calculate the relevant deflection components and compare the total with maximum values recommended in the ACI Code.

13.2. A monolithic reinforced concrete floor consists of rectangular bays measuring 21 × 26 ft, as shown in Fig. P13.2. The floor is designed to carry a service live load of 125 psf uniformly distributed over its surface in addition to its own weight, using a concrete strength of 5000 psi and reinforcement having $f_y = 60{,}000$ psi. Design the typical exterior panel of the floor of Example 12.2 as a part of a flat plate structure, with no beams between interior columns but with beams provided along the outside edge to stiffen the slab. No drop panels or column capitals are permitted, but shear reinforcement may be incorporated if necessary. Column size is 20 × 20 in., and the floor-to-floor height is 12 ft. Use the equivalent frame method. Summarize your design by means of a sketch showing plan and typical cross sections.

FIGURE P13.2

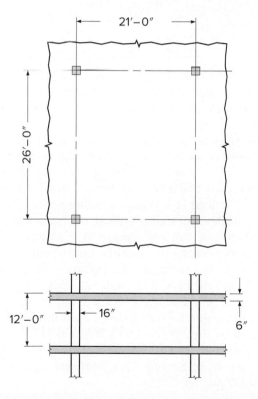

13.3. For the four-story structure shown in Figure P13.3, (*a*) select the slab thickness, (*b*) design the highlighted floor slab panel using the equivalent frame method, (*c*) prepare sketches of the steel layout, and (*d*) comment

FIGURE P13.3

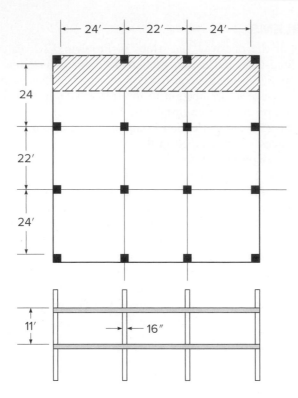

on your selection of the original thickness and what effect using shear studs might have on the design. Material strengths are $f_y = 60,000$ psi and $f'_c = 4000$ psi. Building loads are the superimposed dead load of 30 psf and live load of 50 psf.

13.4. A multistory commercial building is to be designed as a flat plate system with floors of uniform thickness having no beams or drop panels. Columns are laid out on a uniform 20 ft spacing in each direction and have a 16 in. square section and a vertical dimension 10 ft from floor to floor. Specified service live load is 100 psf including partition allowance. Using the direct design method, design a typical interior panel, determining the required floor thickness, size and spacing of reinforcing bars, and bar details including cutoff points. To simplify construction, the reinforcement in each direction is the same; use an average effective depth in the calculations. Use all straight bars. For moderate spans such as this, it has been determined that supplementary shear reinforcement would not be economical, although column capitals may be used if needed. Thus, slab thickness may be based on Eqs. (13.7a), (13.7b), and (13.7c); or column capital dimensions can be selected using those equations if slab thickness is based on the equations in Section 13.5. Material strengths are $f_y = 60,000$ psi and $f'_c = 4000$ psi.

13.5. Prepare alternative designs for shear reinforcement at the supports of the slab described in Example 13.4 (a) using integral beams with vertical stirrups similar to Fig. 13.17a, and (b) using headed shear stud reinforcement similar to Fig. 13.17b.

13.6. Figure P13.6 shows a flat plate floor designed to carry a factored load of 325 psf. The total slab thickness $h = 7\frac{1}{2}$ in. and the average effective depth $d = 6$ in. Material strengths are $f_y = 60,000$ psi and $f_c' = 4000$ psi. The design for punching shear at a typical interior column B2 provided the basis for Example 13.3. To provide a full perimeter b_o at the exterior column B1, the slab is cantilevered past the columns as shown. A total shear force $V_u = 105$ kips must be transmitted to the column, along with a bending moment $M_u = 120$ ft-kips about an axis parallel to the edge of the slab. Check for punching shear at column B1 and, if ACI Code restrictions are not met, suggest appropriate modifications in the proposed design. Edge beams are not permitted.

FIGURE P13.6

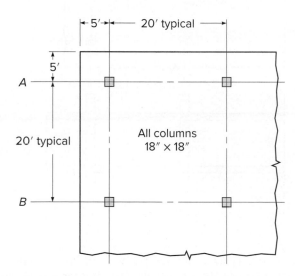

13.7. For the flat plate floor in Example 13.1, find the following deflection components at the center of panel C: (*a*) immediate deflection due to total dead load; (*b*) additional dead load deflection after a long period of time, due to total dead load; (*c*) immediate deflection due to three-quarters full live load. The moment of inertia of the cross concrete sections I_g may be used for all calculations. It may be assumed that maximum deflection is obtained for the same loading pattern that would produce maximum positive moment in the panel. Check predicted deflection against ACI limitations, assuming that nonstructural attached elements would be damaged by excessive deflections.

13.8. A parking garage is to be designed using a two-way flat slab on the column lines, as shown in Fig. P13.8. A live load of 100 psf is specified. Find the required slab thickness, using a reinforcement ratio of approximately 0.005, and design the reinforcement for a typical corner panel A, edge panel B, and interior panel C. Check shear capacity. Detail the reinforcement, showing size, spacing, and length. All straight bars are used. Material strengths are $f_y = 60,000$ psi and $f_c' = 5000$ psi. Specify the design method selected and comment on your results.

FIGURE P13.8

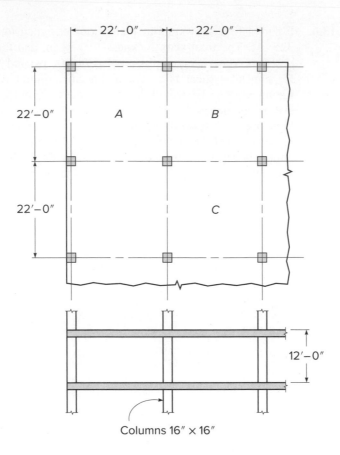

Columns 16″ × 16″

13.9. For the typical interior panel *C* of the parking garage in Problem 13.8, (*a*) compute the immediate and long-term deflections due to dead load and (*b*) compute the deflection due to the full-service live load. Compare with ACI Code maximum permissible values, given that there are no elements attached that would be damaged by large deflections.

14

Walls

14.1 INTRODUCTION

Walls serve a number of functions, ranging from separation of space in a building to restraining earth adjacent to a building or building site. As a general rule, the exterior walls of a reinforced concrete building are supported at each floor by the skeleton framework, their only function being to enclose the building. Such walls are called *panel walls*. They may be made of concrete (often precast), concrete block, brick, tile blocks, or insulated metal panels. The latter may be faced with aluminum, stainless steel, or a porcelain-enamel finish over steel, backed by insulating material and an inner surface sheathing. The thickness of each of these types of panel walls will vary according to the material, type of construction, climatological conditions, and the building requirements governing the particular locality in which the construction takes place.

Wind pressure is usually the principal load that is considered in determining the structural thickness of a wall panel, although in some cases exterior walls are used as diaphragms to transmit forces caused by horizontal loads down to the building foundations.

Curtain walls are similar to panel walls, except they are not supported at each story by the frame of the building, but are self-supporting. They are, however, often anchored to the building frame at each floor to provide lateral support.

In general, *bearing walls*, such as those shown in Fig. 14.1, may be defined as walls that carry any vertical load in addition to their own weight. Such walls may be constructed of stone masonry, brick, concrete block, or reinforced concrete. Occasional projections or pilasters add to the strength of the wall and are often used at points of load concentration. In small commercial buildings, bearing walls may be used with economy and expediency. In larger commercial and manufacturing buildings, when the element of time is an important factor, the delay necessary for the erection of the bearing wall and the attendant increased cost of construction often dictate the use of some other arrangement. In addition to vertical loads, bearing walls must often be designed to carry out-of-plane loads and bending due to eccentricity of the load, as are columns.

Shear walls resist in-plane forces and are used to provide lateral stability in systems subjected to wind and earthquake loads. Shear walls may be bearing or nonbearing.

Bearing and shear walls are considered *structural walls* by ASCE/SEI 7 (Ref. 14.1), and in either case both must be designed for a combination of all applied loads.

Figure 14.2 illustrates the in-plane and out-of-plane forces on walls and identifies general wall dimensions. *Retaining walls* resist out-of-plane soil pressure, are designed as flexural members, and are addressed in Chapter 16.

FIGURE 14.1
Bearing walls under
construction. (*Photograph by
Charles W. Dolan.*)

FIGURE 14.2
Wall forces and dimensions.

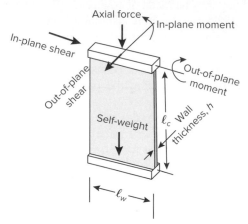

Walls may be cast-in-place, precast in a plant off-site, or precast on-site and tilted up into their final position. Walls are anchored to intersecting elements, such as adjacent walls, floors, columns, pilasters, or buttresses—and to footings unless the analysis indicates such restraint is not necessary. Walls are assumed to be laterally supported at each floor. The degree of lateral support is dependent on the overall structural framing system. Shear walls, when placed orthogonally in a building, provide the lateral restraint both to the other walls and to the columns.

14.2 GENERAL DESIGN CONSIDERATIONS

The design of structural walls has evolved through years of practice. The ACI Code allows a variety of design approaches, including computer analysis and empirical methods. For out-of-plane loading, walls may be designed as *wide columns* using the procedures discussed in Chapters 9 and 10, or they may be designed using methods presented in ACI Code 11.5.3 and 11.8 and discussed in Sections 14.3 and 14.4. Design for in-plane loading is addressed in ACI Code 11.5.4 and Section 14.5.

The effective length of a wall ℓ_w is determined by the loading. A uniformly loaded wall has an effective length equal to the overall length of the wall. Walls supporting concentrated loads have effective lengths equal to the lesser of the center-to-center spacing of the loads or the bearing length plus four times the wall thickness. The effective length is not considered to extend beyond the boundary of the wall. For cast-in-place walls with axial factored load $P_u > 0.2f_c'A_g$, where A_g is the gross area of the wall, ACI Code 14.2.4 requires that the portion of the wall within the thickness of the floor system have a specified compressive strength not less than 80 percent of the specified compressive strength of the wall.

a. Wall Thickness

In all cases involving vertical loads, wall design must consider slenderness effects. The ACI Code does not limit the minimum thickness h of bearing, exterior basement, or foundation walls, except for walls designed by a method described in ACI Code 11.5.3 as the simplified design method (see Section 14.3). For the simplified method, ACI Code 11.3.1 limits the minimum thickness of interior walls to the smallest of $h = 4$ in., $\ell_c/25$, or $\ell_w/25$ for bearing walls and 7.5 in. for exterior or foundation walls, where ℓ_c is the unsupported height of the wall and ℓ_w is the unsupported length of the wall, both measured center-to-center of the supports. The minimum thickness of nonbearing walls is limited to the smallest of $h = 4$ in., $\ell_c/30$, and $\ell_w/30$. In all cases, the minimum thickness h may be reduced if analysis indicates that a wall with lower thickness has adequate strength and stability.

b. Modeling Considerations

Modeling walls for analysis for out-of-plane loading typically assumes that the unbraced wall height ℓ_c is the center-to-center distance between supports, the wall is simply supported between floors, and the maximum out-of-plane moment occurs at midheight. Stability considerations assume that the wall is cracked when lateral loads are applied. The maximum moment on walls subjected to out-of-plane loading is increased by the effects of vertical loads. Iterative calculations may be required to determine the design load, thus:

$$M_u = M_{ua} + P_u\Delta_u \tag{14.1}$$

where M_{ua} is the maximum factored load not including the P-Δ effects. The out-of-plane deflection is calculated using cracked section properties. The strength reduction factor $\phi = 0.65$ for walls, as it is for other compression members subjected to axial loads and moment with tied reinforcement. In regions of walls reinforced by spiral reinforcement, $\phi = 0.75$.

c. Reinforcement

ACI Code 11.6.1 requires minimum reinforcement in the longitudinal (vertical) and transverse (horizontal) directions in walls, independent of loading, unless "adequate strength and stability can be demonstrated by structural analysis." The value of the required reinforcement depends upon whether the in-plane factored shear is less than or greater than

$$0.5\phi\alpha_c\lambda\sqrt{f_c'}A_{cv} \tag{14.2}$$

where α_c is the coefficient defining contribution of concrete strength to nominal wall shear strength; $\alpha_c = 3$ for $h_w/\ell_w \le 1.5$, $= 2$ for $h_w/\ell_w \ge 2.0$, and varies linearly between 3 and 2 for $1.5 < h_w/\ell_w < 2.0$; h_w and ℓ_w are, respectively, the height from base to top and length of entire wall, or the clear height and length of wall segment or wall pier considered, in.; and A_{cv} is the gross area of wall bounded by web thickness h and length of wall in the direction of shear force ℓ_w, minus area of any openings, in^2.

For walls with in-plane shear, $V_u \leq 0.5\phi\alpha_c\lambda\sqrt{f_c'}A_{cv}$, the reinforcement ratio in the longitudinal direction ρ_ℓ must be at least 0.0012 for No. 5 (No. 16) and smaller bars with $f_y \geq 60$ ksi, increasing to 0.0015 for other deformed bars. The reinforcement ratio in the transverse direction ρ_t must be at least 0.0020 for No. 5 (No. 16) and smaller bars with $f_y \geq 60$ ksi, increasing to 0.0025 for other deformed bars. For welded wire reinforcement not larger than W31 or D31, the minimum vales of ρ_ℓ and ρ_t are, respectively, 0.0012 and 0.0020.

For walls with in-plane shear $V_u > 0.5\phi\alpha_c\lambda\sqrt{f_c'}A_{cv}$, ACI Code 11.6.2 requires $\rho_t \geq 0.0025$, independent of reinforcement type. Test results indicate that for low shear walls, vertical as well as horizontal reinforcement is needed, and the ACI Code requires vertical bars or wires with area A_ℓ and spacing s_1 (discussed in Section 14.5) such that the ratio of the vertical steel to gross concrete area for a horizontal section ρ_ℓ is not less than given in Eq. (14.3)

$$\rho_\ell \geq 0.0025 + 0.5(2.5 - h_w/\ell_w)(\rho_t - 0.0025) \tag{14.3}$$

and not less than 0.0025, but need not be greater than the minimum *horizontal* reinforcement ρ_t required for in-plane shear (see Section 14.5). ACI Code 11.6.2 is somewhat problematic because, as will be apparent in Section 14.5, if the factored in-plane shear V_u is greater than $0.5\phi\alpha_c\lambda\sqrt{f_c'}A_{cv}$ but less than $\phi\alpha_c\lambda\sqrt{f_c'}A_{cv}$, no transverse reinforcement is required; as a result, ρ_ℓ would equal 0. The minimum value of ρ_t, however, would remain 0.0025. It is unlikely that $\rho_\ell = 0$ is the outcome desired by ACI Committee 318. As a result, these provisions will be interpreted as requiring a minimum value of 0.0025 for both ρ_t and ρ_ℓ when $V_u > 0.5\phi\alpha_c\lambda\sqrt{f_c'}A_{cv}$.

Although not stated in the Code, ACI Commentary 11.6.2 indicates that when applying Eq. (14.3), h_w/ℓ_w should be limited to values between 0.5 and 2.5, using 0.5 for $h_w/\ell_w < 0.5$, in which case $\rho_\ell = \rho_t$, and 2.5 for $h_w/\ell_w > 2.5$, in which case $\rho_\ell = 0.0025$. When $\rho_t = 0.0025$, the longitudinal reinforcement requirement is $\rho_\ell = 0.0025$.

Reinforcement spacing must not exceed $3h$ or 18 in., in accordance with ACI Code 11.7. If shear reinforcement is required to carry the factored load, the maximum spacing may not exceed $\ell_w/3$ for longitudinal reinforcement and $\ell_w/5$ for transverse reinforcement. For walls with thicknesses greater than 10 in., except basement and cantilever retaining walls, the reinforcement must be distributed in two layers parallel to the wall faces.

ACI Code 11.7.5 requires additional reinforcement to limit cracks that can occur near the corners of openings in walls, such as used for windows and doors. That reinforcement must consist of at least two No. 5 (No. 16) bars in walls with two layers of reinforcement in both directions or one No. 5 (No. 16) bar in walls with a single layer of reinforcement in both directions. The additional reinforcement must be anchored to develop f_y in tension at the corners of the openings and may be placed on the diagonal near each corner or oriented in both the horizontal and the vertical directions near each corner.

14.3   SIMPLIFIED DESIGN METHOD FOR AXIAL LOAD AND OUT-OF-PLANE MOMENT

ACI Code 11.5.3 permits the use of a simplified design method for solid walls with uniform thickness if the resultant of all factored vertical loads lies within the middle third of the wall thickness. The method has been shown to be conservative when compared to detailed calculations for walls with the same dimensions (Ref. 14.2). For walls meeting the restrictions, the nominal axial load capacity may be calculated as

$$P_n = 0.55 f_c' A_g \left[1 - \left(\frac{k\ell_c}{32h} \right)^2 \right] \tag{14.4}$$

where A_g is the gross area of the wall, k is the effective length factor to account for stability due to end fixity, ℓ_c is the unsupported height of the wall measured center to center of the joints as it is for columns, and h is the wall thickness. The value of k equals 0.8 for walls braced against lateral translation and restrained against rotation at one or both ends, 1.0 for walls restrained against translation and unrestrained against rotation at both ends, and 2.0 for walls not braced against translation. There is no requirement to design for bending using the simplified design method.

EXAMPLE 14.1   **Bearing wall design.**   A 12 ft long interior bearing wall carries a factored load of 15,000 plf, as shown in Fig. 14.3. The load is centered on the wall thickness. For a center-to-center floor height $h_w = \ell_c = 12$ ft-6 in., $f_c' = 4000$ psi, and $f_y = 60,000$ psi, design the wall using the simplified design method.

SOLUTION.   When designing a wall using the simplified method, the wall thickness h cannot be less than the largest of 4 in., $\ell_c/25$, or $\ell_w/25$. For $\ell_c = 12.5$ ft and $\ell_w = 12$ ft, h must be at least 6 in. A wall thickness of $h = 6$ in. gives a wall dead load of $w_d = h/12 \times 150$ pcf $\times h_w = 6/12 \times 150 \times 12.5 = 938$ plf. The factored load on the wall is

$$P_u = (1.2 \times w_d + w_u)\ell_w = (1.2 \times 938 + 15{,}000) \times 12/1000 = 195 \text{ kips}$$

The load is applied within the middle third of the wall thickness, so the simplified design method may be used. Using Eq. (14.3) with $k = 1.0$, the wall capacity is

$$\phi P_n = \phi 0.55 f_c' A_g \left[1 - \left(\frac{k\ell_c}{32h} \right)^2 \right]$$

$$= 0.65 \times 0.55 \times 4000 \times 6 \times 12 \times 12 \left[1 - \left(\frac{1.0 \times 12.5 \times 12}{32 \times 6} \right)^2 \right] = 270 \text{ kips}$$

This exceeds the factored axial load.

There is no factored shear load on the wall; thus, minimum reinforcement as described in Section 14.2b is required. For $h = 6$ in. and assuming that No. 5 or smaller bars will be used as reinforcement, the minimum longitudinal reinforcement requirement in in^2/ft is $0.0012 \times h \times 12 = 0.086$ in^2/ft, and the minimum transverse reinforcement is $0.0020 \times h \times 12 = 0.144$ in^2/ft. Since the wall is less than 10 in. thick, the reinforcement may be placed in a single layer near the center of the wall. The maximum spacing of both the longitudinal and transverse reinforcement is 18 in. or $3h = 18$ in. Reinforcement consisting of No. 4 (No. 13) bars placed at 18 in. longitudinally gives 0.13 in^2/ft and No. 4 (No. 13) bars placed at 16 in. horizontally gives 0.15 in^2/ft, thereby meeting the reinforcement requirements.

FIGURE 14.3
Factored loads on wall.

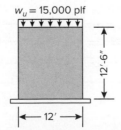

$w_u = 15{,}000$ plf

12'-6"

12'

14.4 ALTERNATIVE METHOD FOR OUT-OF-PLANE SLENDER WALL ANALYSIS

ACI Code 11.8 provides an alternative, empirical procedure for the design of slender walls subjected to axial load, out-of-plane bending, and shear. The procedure applies to all cases where the limitations of the simplified design method are not satisfied. This method addresses walls where out-of-plane deformations are calculated and the effects of secondary moments due to axial loads and out-of-plane deformations based on cracked sections are included. Results from computer analysis, including second-order effects and adjustment for cracked sections, may be used to design the wall for combined axial load and moments. The nominal strength of these walls for in-plane shear is addressed in ACI Code 11.5.4 and Section 14.5.

14.5 SHEAR WALLS

Horizontal forces acting on buildings, such as those due to wind or seismic action, can be resisted by different means. Rigid-frame resistance of the structure, augmented by the contribution of ordinary masonry walls and partitions, can provide for wind loads in many cases. However, when heavy horizontal loading is likely, such as would result from an earthquake or high winds, reinforced concrete shear walls are used. These may be added solely to resist horizontal forces, or concrete walls enclosing stairways or elevator shafts may also serve as shear walls.

Figure 14.4 shows a building with wind or seismic forces represented by arrows acting on the edge of each floor or roof. The horizontal surfaces act as deep beams to transmit loads to vertical resisting elements *A* and *B*. These shear walls, in turn, act as cantilever beams fixed at their base to carry loads down to the foundation. They are subjected to (1) variable shear, which reaches a maximum at the base; (2) bending moment, which tends to cause vertical tension near the loaded edge and

FIGURE 14.4
Building with shear walls subject to horizontal loads: (*a*) typical floor; (*b*) front elevation; and (*c*) end elevation.

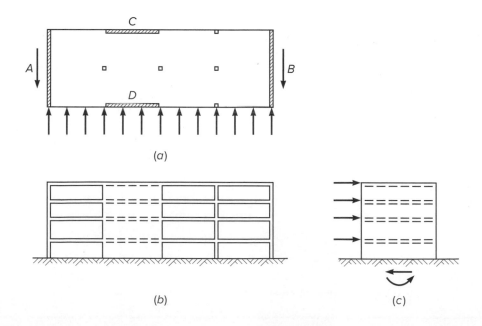

FIGURE 14.5
Geometry and reinforcement
of a typical shear wall:
(*a*) cross section and
(*b*) elevation.

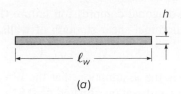

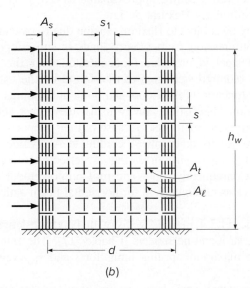

compression at the far edge; and (3) vertical compression due to ordinary gravity loading from the structure. For the building shown, additional shear walls *C* and *D* are provided to resist loads acting in the long direction of the structure.

Shear is apt to be critical for walls with a relatively low ratio of height to length. High shear walls are controlled mainly by flexural requirements.

Figure 14.5 shows a typical shear wall with total height h_w, length ℓ_w, and thickness h. It is assumed to be fixed at its base and loaded horizontally along its left edge. Vertical flexural reinforcement of area A_s is provided at the left edge, with its centroid a distance d from the extreme compression face. To allow for reversal of load, identical reinforcement is provided along the right edge. Horizontal shear reinforcement with area A_t at spacing s is provided, as well as vertical shear reinforcement with area A_ℓ at spacing s_1.

When designing shear walls, ACI Code 11.5.1 requires that

$$\phi V_n \geq V_u \tag{14.5}$$

V_n, given in ACI Code 11.5.4, is

$$V_n = (\alpha_c \lambda \sqrt{f_c'} + \rho_t f_{yt}) A_{cv} \tag{14.6}$$

where α_c and A_{cv} are defined following Eq. (14.2), λ is the lightweight concrete modification factor (see Section 5.5a), ρ_t is the reinforcement ratio in the transverse direction, and f_{yt} is the yield strength of the transverse reinforcement.

For walls subject to vertical tension force N_u,

$$\alpha_c = 2\left(1 + \frac{N_u}{500 A_g}\right) \geq 0.0 \tag{14.7}$$

where N_u is negative for tension.

To guard against diagonal compression failure (Refs. 14.3 and 14.4), an upper limit is placed on the nominal shear strength of walls.

$$V_n \le 8\sqrt{f_c'}A_{cv} \tag{14.8}$$

Inherent in Eq. (14.8) is the assumption that the effective depth of a shear wall in the plane of the wall d can be approximated as $0.8\ell_w$. This assumption is also often a useful for preliminary flexural design.

Walls may be subject to flexural tension due to overturning moment, even when the vertical compression from gravity loads is superimposed. In many but not all cases, vertical steel is provided, concentrated near the wall edges, as shown in Fig. 14.5. The required steel area can be found by the usual methods for beams.

The dual function of the floors and roofs in buildings with shear walls should be noted. In addition to resisting gravity loads, they must act as deep beams spanning between shear-resisting elements. Because of their proportions, both shearing and flexural stresses are usually quite low. According to ACI Code 5.3, the load factor for live load drops to 1.0 when wind or earthquake effects are combined with the effects of gravity loads. Consequently, floor and roof reinforcement designed for gravity loads can usually serve as reinforcement for horizontal beam action also, with no increase in bar areas.

ACI Code 11.5.4 permits walls with height-to-length ratios below 2.0 to be designed using strut-and-tie models (Chapter 17). The minimum shear reinforcement criteria and the maximum spacing limits for s and s_1 given in Section 14.2c must be satisfied.

There are special considerations and requirements for reinforced concrete walls in structures designed to resist forces associated with seismic motion. These are based on design for energy dissipation in the nonlinear range of response. This subject will be treated separately, in Chapter 20.

EXAMPLE 14.2   **Shear wall design.**   Redesign the wall from Example 14.1 for the factored vertical and lateral loads shown in Fig. 14.6. The horizontal load V_u can be applied to the left or to the right. The center-to-center floor and total wall height $h_w = \ell_c = 12$ ft-6 in., $f_c' = 4000$ psi, and $f_y = 60,000$ psi.

SOLUTION: Assume a wall thickness $h = 12$ in. giving a wall dead load of $w_d = h/12 \times 150$ pcf $\times h_w = 12/12 \times 150 \times 12.5 = 1875$ plf. The critical loads occur at the bottom corner of the wall. Thus, the loads on the wall are

$$P_u = (1.2 \times w_d + w_u)\,\ell_w = (1.2 \times 1875 + 15{,}000) \times 12/1000 = 207 \text{ kips}$$

$$M_u = 200 \text{ kips} \times 12.5 \text{ ft} = 2500 \text{ kip-ft}$$

$$V_u = 200 \text{ kips}$$

Based on the very similar axial load and the use of a wall thickness twice that used in Example 14.1, the out-of-plane effects are not of concern.

For the in-plane effects, the axial load 207 kips is less than $0.10\,f_c'\,A_g = 0.10 \times 4000 \times 12 \times 12 \times 12/1000 = 691$ kips, allowing the wall to be treated as a cantilever beam for the purposes of calculating flexural reinforcement, in accordance with ACI Code 9.5.2. The effective depth d will be taken as $0.8\ell_w = 9.6$ ft $= 115$ in.

Using the design aids in Appendix A,

$$R = \frac{M_u}{\phi b d^2} = \frac{2500 \times 12{,}000}{0.90 \times 12 \times 115^2} = 210$$

giving $\rho = 0.0036$ from Table A.5b and a required area of reinforcement $A_s = 0.0036bd = 0.0036 \times 12 \times 115 = 4.97$ in^2. This can be satisfied using five No. 9 (No. 29) bars placed at each end of the wall.

The longitudinal bars require ties with a minimum spacing equal to the lesser of 48 tie diameters, 12 longitudinal bar diameters, or the wall thickness. In this instance, 12 in. is the maximum tie spacing using No. 3 (No. 10) ties.

The factored shear on the wall $V_u = 200$ kips. $h_w/\ell_w = 12.5/12 = 1.04$. Because $h_w/\ell_w \leq 1.5$, $\alpha_c = 3$ and $0.5\phi\alpha_c\lambda\sqrt{f_c'}A_{cv} = 0.5 \times 0.75 \times 3 \times 1.0\sqrt{4000} \times 12 \times 115/1000 = 98$ kips. Thus, V_u is greater than $0.5\phi\alpha_c\lambda\sqrt{f_c'}A_{cv}$ and also slightly greater than $\phi\alpha_c\lambda\sqrt{f_c'}A_{cv} = 196$ kips, meaning that, as shown in Eq. (14.6), where $V_n = (\alpha_c\lambda\sqrt{f_c'} + \rho_t f_{yt})A_{cv}$, there is a very low requirement for transverse reinforcement. The minimum quantity of transverse reinforcement for $V_u \geq 0.5\phi\alpha_c\lambda\sqrt{f_c'}A_{cv}$, $\rho_t = 0.0025$, will easily provide the needed shear strength. As pointed out in Section 14.2c, when using Eq. (14.3) with $\rho_t = 0.0025$, $\rho_\ell = 0.0025$. Thus, the minimum required wall reinforcement in both the horizontal and the vertical directions is $0.0025 \times h \times 12$ in./ft $= 0.0025 \times 12 \times 12$ in./ft $= 0.36$ in^2/ft. The wall is greater than 10 in. thick; consequently, reinforcement must be placed in two layers. The maximum reinforcement spacing is 18 in. or $3h$, with 18 in. being the controlling value. Using No. 5 (No. 16) bars placed at 18 in. in both directions near each face gives 0.41 in^2/ft, exceeding the minimum requirements. Final detailing of the wall is shown in Fig. 14.7.

Lastly, the footing must be designed to develop the shear wall reinforcement and prevent the wall from sliding or overturning.

FIGURE 14.6
Wall dimensions and loading for Example 14.2.

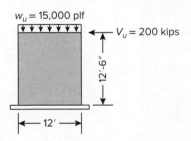

FIGURE 14.7
Wall reinforcement details for Example 14.2.

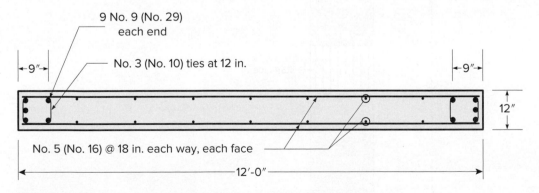

REFERENCES

14.1. *Minimum Design Loads for Buildings and Other Structures*, ASCE/SEI 7-16, American Society of Civil Engineers, Reston, VA, 2016.

14.2. K. M. Kripanarayanan, "Interesting Aspects of the Empirical Wall Design Equation," *J. ACI, Proceedings*, vol. 74, no. 5, 1977, pp. 204–207.

14.3. A. E. Cardenas, J. M. Hanson, W. G. Corley, and E. Hognestad, "Design Provisions for Shear Walls," *J. ACI*, vol. 70, no. 3, 1973, pp. 221–230.

14.4. A. E. Cardenas and D. D. Magura, "Strength of High-Rise Shear Walls: Rectangular Cross Section," pp. 119–150, in *Response of Multistory Structures to Lateral Forces*, SP-36, American Concrete Institute, Farmington Hills, MI, 1973.

PROBLEMS

14.1. A bearing wall carries a factored load of 24,000 plf centered on the wall. Design the wall using $\ell_w = 8$ ft, $\ell_c = 15$ ft, and material strengths $f'_c = 5000$ psi and $f_y = 80,000$ psi.

14.2. A shear wall has the factored lateral loads shown in Fig. P14.2. Design the first story of the wall for these factored lateral loads and a cumulative axial factored of 284 kips, excluding the wall weight. $h_w = 15$ ft and $\ell_w = 8$ ft. The material strengths are $f'_c = 5000$ psi and $f_y = 80,000$ psi.

FIGURE P14.2

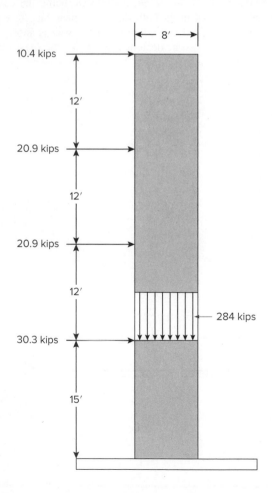

15 *Footings and Foundations*

15.1 TYPES AND FUNCTIONS

The substructure, or foundation, is the part of a structure that is usually placed below the surface of the ground and that transmits the load to the underlying soil or rock. All soils compress noticeably when loaded and cause the supported structure to settle. The two essential requirements in the design of foundations are that the total settlement of the structure be limited to a tolerably small amount and that differential settlement of the various parts of the structure be eliminated as nearly as possible. With respect to possible structural damage, the elimination of differential settlement, that is, different amounts of settlement within the same structure, is more important than limitations on uniform overall settlement. Foundations are proportioned for bearing effects, stability, overturning, and sliding at the soil-foundation interface.

To limit settlements as indicated, it is necessary (1) to transmit the load of the structure to a soil stratum of sufficient strength and (2) to spread the load over a sufficiently large area of that stratum to minimize bearing pressure. If adequate soil is not found immediately below the structure, it becomes necessary to use deep foundations such as *piles, drilled piers,* or *caissons* to transmit the load to deeper, firmer layers. If satisfactory soil directly underlies the structure, it is merely necessary to spread the load, by footings or other means. Such substructures are known as *spread* foundations. When deep foundations are used, columns loads are transferred to the piles, drilled piers, or caissons through *pile caps*. Loads on deep foundations may be axial or axial plus shear and moment. The details of loading are found in texts on geotechnical engineering, Refs. 15.1 to 15.4.

15.2 SPREAD FOOTINGS

Spread footings can be classified as wall and column footings. The horizontal outlines of the most common types are given in Fig. 15.1. A wall footing is simply a strip of reinforced concrete, wider than the wall, that distributes its pressure. Single-column footings are usually square, sometimes rectangular, and represent the simplest and most economical foundation. Their use under exterior columns meets with difficulties if property rights prevent the use of footings projecting beyond the exterior walls. In this case, combined footings or strap footings are used that enable one to design a footing that does not project beyond the wall column. Combined footings under two or more columns are also used under closely spaced, heavily loaded interior columns where single footings, if they were provided, would completely or nearly merge.

459

FIGURE 15.1
Types of spread footing.

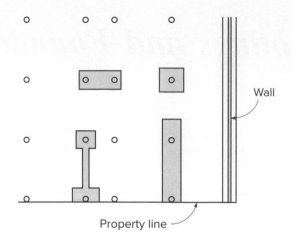

Such individual or combined column footings are the most frequently used types of spread foundations on soils of reasonable bearing capacity. If the soil is weak and/or column loads are great, the required footing areas become so large as to be uneconomical. In this case, unless a deep foundation is called for by soil conditions, a mat or raft foundation must be used. This consists of a solid reinforced concrete slab that extends under the entire building and, consequently, distributes the load of the structure over the maximum available area. Such a foundation, in view of its own rigidity, also minimizes differential settlement. It consists, in its simplest form, of a concrete slab reinforced in both directions. A form that provides more rigidity consists of an inverted girder floor. Girders are located in the column lines in each direction, and the slab is provided with two-way reinforcement, spanning between girders. Inverted flat slabs, with capitals at the bottoms of the columns, are also used for mat foundations.

15.3 DESIGN FACTORS

In ordinary construction, the load on a wall or column is transmitted vertically to the footing, which in turn is supported by the upward pressure of the soil on which it rests. If the load is symmetrical with respect to the bearing area, the bearing pressure is assumed to be uniformly distributed (Fig. 15.2a). It is known that this is only approximately true. Under footings resting on coarse-grained soils, the pressure is larger at the center of the footing and decreases toward the perimeter (Fig. 15.2b). This is so because the individual grains in such soils are somewhat mobile, so that the soil located close to the perimeter can shift very slightly outward in the direction of lower soil stresses. In contrast, in clay soils pressures are higher near the edge than at the

FIGURE 15.2
Bearing pressure distribution:
(a) as assumed; (b) actual, for
granular soils; and (c) actual,
for cohesive soils.

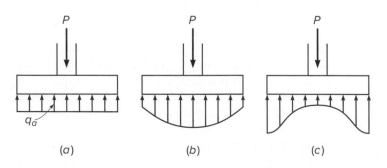

center of the footing, since in such soils the load produces a shear resistance around the perimeter that adds to the upward pressure (Fig. 15.2c). It is customary to disregard these nonuniformities (1) because their numerical amount is uncertain and highly variable, depending on types of soil, and (2) because their influence on the magnitudes of bending moments and shearing forces in the footing is relatively small.

On compressible soils, footings should be loaded concentrically to avoid tilting, which will result if bearing pressures are significantly larger under one side of the footing than under the opposite side. This means that single footings should be placed concentrically under the columns and wall footings concentrically under the walls and that, for combined footings, the centroid of the footing area should coincide with the resultant of the column loads. Eccentrically loaded footings can be used on highly compacted soils and on rock. It follows that one should count on rotational restraint of the column by a single footing only when such favorable soil conditions are present and when the footing is designed for both the column load and the restraining moment. Even then, less than full fixity should be assumed, except for footings on rock.

The accurate determination of stresses in foundation elements of all kinds is difficult, partly because of the uncertainties in determining the actual distribution of upward pressures but also because the structural elements themselves represent relatively massive blocks or thick slabs subject to heavy concentrated loads from the structure above. Design procedures for single-column footings are based largely on the results of experimental investigations by Talbot (Ref. 15.5) and Richart (Ref. 15.6). These tests and the recommendations resulting from them have been reevaluated in the light of more recent research, particularly that focusing on shear and diagonal tension (Refs. 15.7 to 15.9). Combined footings and mat foundations also can be designed by simplified methods, although increasing use is made of more sophisticated tools, such as finite element analysis and the strut-and-tie method (Ref. 15.10).

15.4 LOADS, BEARING PRESSURES, AND FOOTING SIZE

Footing sizes are determined for *unfactored* service loads and *allowable* soil pressures, in contrast to the strength design of reinforced concrete members, which uses factored loads and factored nominal strengths. This is because, for footing design, safety is provided by overall safety factors, in contrast to the separate load and strength reduction factors used to dimension concrete members. Soil capacities are usually provided as allowable stresses, while the structure is designed on a strength basis. Pressures resulting from strength design calculations may be incompatible with the soils data, requiring the structural engineer to work closely with the soils engineer to ensure a competent design.

Allowable bearing pressures are established from principles of soil mechanics, on the basis of load tests and other experimental determinations (see, for example, Refs. 15.1 to 15.4). Allowable bearing pressures q_a under service loads are usually based on a safety factor of 2.5 to 3.0 against exceeding the bearing capacity of the particular soil and to keep settlements within tolerable limits. Many local building codes contain allowable bearing pressures for the types of soils and soil conditions found in the particular locality.

For concentrically loaded footings, the required area is determined from

$$A_{\text{req}} = \frac{D + L}{q_a} \qquad (15.1)$$

In addition, most building codes, including the *International Building Code* (IBC) (Ref. 15.11), which is used throughout the United States, permit a 33 percent increase in the allowable pressure when the effects of wind W or earthquake E are included, if specific loading combinations are used for foundation design. For example, using the loading requirements in ASCE/SEI 7 (Ref 15.12),

$$A_{\text{req}} = \frac{D + 0.75L + 0.45W + 0.75\ (L_r \text{ or } S \text{ or } R)}{1.33q_a}$$

or

$$A_{\text{req}} = \frac{D + 0.75L + 0.525E + 0.75S}{1.33q_a}$$

(15.2)

where L_r = roof live load, S = snow load, and R = rain load. The factors applied to W and E in Eq. (15.2) recognize that wind and earthquake loads are strength level forces, as prescribed by Refs. 15.11 and 15.12.

The required footing area A_{req} is the larger of those determined by Eqs. (15.1) and (15.2). The loads in the numerators of Eqs. (15.1) and (15.2) must be calculated at the level of the base of the footing, that is, at the contact plane between soil and footing. This means that the weight of the footing and surcharge (that is, fill and possible liquid pressure on top of the footing) must be included. Wind loads and other lateral loads cause a tendency to overturn. In checking for overturning of a foundation, only those live loads that contribute to overturning should be included, and dead loads that stabilize against overturning should be multiplied by 0.9. A safety factor of at least 1.5 should be maintained against overturning, unless otherwise specified by the local building code (Ref. 15.8).

A footing is eccentrically loaded if the supported column is not concentric with the footing area or if the column transmits at its juncture with the footing not only a vertical load but also a bending moment. In either case, the load effects at the footing base can be represented by the vertical load P and a bending moment M. The resulting bearing pressures are again assumed to be linearly distributed. As long as the resulting eccentricity $e = M/P$ does not exceed the kern distance k of the footing area ($k = \ell/6$ for a square or rectangular footing), the usual flexure formula

$$q_{\substack{\max \\ \min}} = \frac{P}{A} \pm \frac{Mc}{I}$$

(15.3)

permits the determination of the bearing pressures at the two extreme edges, as shown in Fig. 15.3*a*. The footing area is found by trial and error from the condition $q_{\max} \leq q_a$. If the eccentricity falls outside the kern, Eq. (15.3) gives a negative value (tension) for q along one edge of the footing. Because no tension can be transmitted at the contact area between soil and footing, Eq. (15.3) is no longer valid and bearing pressures are distributed as shown in Fig. 15.3*b*. For rectangular footings of size $\ell \times b$, the maximum pressure can be found from

$$q_{\max} = \frac{2P}{3bm}$$

(15.4)

which, again, must be no larger than the allowable pressure q_a. For nonrectangular footing areas of various configurations, kern distances and other aids for calculating bearing pressures can be found in Refs. 15.1 and 15.8 and elsewhere.

Once the required footing area has been determined, the footing must then be designed to develop the necessary strength to resist all moments, shears, and other internal actions caused by the applied loads. For this purpose, the load factors of

FIGURE 15.3

Assumed bearing pressures
under eccentrically loaded
footing.

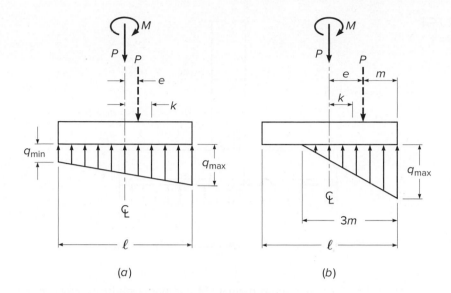

(a)   (b)

ACI Code 5.3 apply to footings as to all other structural components. Correspondingly, for strength design, the footing is dimensioned for the effects of the load combinations in Table 1.2. The most common is

$$U = 1.2D + 1.6L$$

or if wind effects are to be included,

$$U = 1.2D + 1.0W + 1.0L + 0.5 \ (L_r, \text{ or } S \text{ or } R)$$

In seismic zones, earthquake forces E must be considered according to Table 1.2. The requirement that

$$U = 0.9D + 1.0W$$

hardly ever governs the strength design of a footing, but affects overturning and stability. Lateral earth pressure H and fluid pressure F must be included if present.

These factored loads must be counteracted and equilibrated by corresponding bearing pressures in the soil. Consequently, once the footing area is determined, the bearing pressures are recalculated for the factored loads for purposes of strength calculations. These are fictitious pressures that are needed only to determine the factored loads for use in design. To distinguish them from the actual pressures q under service loads, the soil pressures that equilibrate the factored loads U is designated q_u.

15.5 WALL FOOTINGS

The simple principles of beam action apply to wall footings with only minor modifications. Figure 15.4 shows a wall footing with the forces acting on it. If bending moments were calculated from these forces, the maximum moment would be found to occur at the middle of the width. Actually, the very large rigidity of the wall modifies this situation, and the tests cited in Section 15.3 show that, for footings under concrete walls, it is satisfactory to compute the moment at the face of the wall (section 1-1). Tension cracks in these tests formed at the locations shown in Fig. 15.4, that is, under the face of the wall rather than in the middle. For footings supporting masonry walls,

FIGURE 15.4
Wall footing.

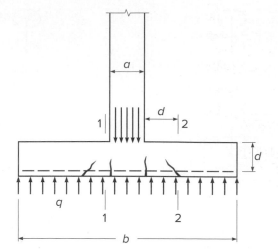

the maximum moment is computed midway between the middle and the face of the wall, because masonry is generally less rigid than concrete. The maximum bending moment in footings under concrete walls is therefore given by

$$M_u = \frac{1}{8} q_u (b - a)^2 \tag{15.5}$$

For determining shear stresses, the vertical shear force is computed on section 2-2, located, as in beams, at a distance d from the face of the wall. Thus,

$$V_u = q_u \left(\frac{b - a}{2} - d \right) \tag{15.6}$$

The calculation of development length is based on the section of maximum moment, that is, section 1-1.

Because repairs to foundations can be extremely difficult and expensive, it is desirable that the elements of the foundation remain essentially elastic during strong ground motions. Wall footings in Seismic Design Categories D, E, or F require additional attention to assure this behavior. ACI Code provisions 18.13 and 21.2.4 address these requirements.

EXAMPLE 15.1   **Design of wall footing.**   A 16 in. concrete wall supports a dead load $D = 14$ kips/ft and a live load $L = 10$ kips/ft. The allowable bearing pressure is $q_a = 4.5$ kips/ft^2 at the level of the bottom of the footing, which is 4 ft below grade. Design a footing for this wall using 4000 psi concrete and Grade 60 steel.

SOLUTION.   With a 12 in. thick footing, the footing weight per square foot is 150 psf, and the weight of the 3 ft fill on top of the footing is $3 \times 100 = 300$ psf. Consequently, the portion of the allowable bearing pressure that is available or effective for carrying the wall load is

$$q_e = 4500 - (150 + 300) = 4050 \text{ psf}$$

The required width of the footing is therefore $b = 24,000/4050 = 5.93$ ft. A 6 ft wide footing is assumed.

The bearing pressure for strength design of the footing, caused by the factored loads, is

$$q_u = \frac{1.2 \times 14 + 1.6 \times 10}{6} \times 10^3 = 5470 \text{ psf}$$

From this, the factored moment at the face of the wall, section 1-1, for strength design is

$$M_u = \frac{1}{8} \times 5470 \; (6 - 1.33)^2 \times 12 = 178{,}900 \text{ in-lb/ft}$$

and assuming $d = 10$ in., the shear at section 2-2 is

$$V_u = 5470 \left[\frac{1}{2} \; (6 - 1.33) - \frac{10}{12} \right] = 8214 \text{ lb/ft}$$

Shear often governs the depth of footings, particularly when the use of shear reinforcements in footings is uneconomical. The design shear strength per foot of width [see Eq. (5.12a), noting that the size effect is not required for one-way footings], assuming minimum flexural reinforcement ratio of $0.0018 \, A_g$ or about a reinforcement ratio of 0.0023, is

$$V_c = \left[8\lambda(\rho)^{1/3}\sqrt{f_c'} + \frac{N_u}{6A_g} \right] bd = 8 \times 1.0 \times (0.0023)^{1/3} \sqrt{4000} \times 12d = 772d$$

from which

$$d = 8214/772 = 10.6 \text{ in.}$$

ACI Code 20.5.1.3 calls for a 3 in. clear cover on bars; therefore a 14 in. thick footing is selected, giving $d = 10.5$ in. This is sufficiently close to the assumed value, and the calculations need not be revised. To determine the required steel area, $R = M_u/\phi bd^2 = 178{,}900/(0.90 \times 12 \times 10.5^2) = 150$, is used to enter Table A.5a of Appendix A. For this value, the table gives the reinforcement ratio $\rho = 0.0025$. The required steel area is then $A_s = 0.0025 \times 10.5 \times 12 = 0.32 \text{ in}^2/\text{ft}$. The minimum reinforcement for a one-way slab is $0.0018 \, A_g$, giving $A_s = 0.0018 \times 12 \times 14 = 0.30 \text{ in}^2/\text{ft}$, and does not control in this case. No. 5 (No. 16) bars, 10 in. on centers, furnish $A_s = 0.37 \text{ in}^2/\text{ft}$. The required development length according to Table A.10 of Appendix A is 24 in. This length is to be furnished from section 1-1 outward. The length of each bar, if end cover is 3 in., is $72 - 6 = 66$ in., and the available development length from section 1-1 to the nearby end is $1/2 \, (66 - 16) = 25$ in., which is more than the required development length. The reinforcement ratio of 0.0025 is close to that assumed for shear in calculation of the depth, so no further revision is needed.

Longitudinal shrinkage and temperature reinforcement, according to ACI Code 24.4, must be at least $0.0018 \times 12 \times 14 = 0.32 \text{ in}^2/\text{ft}$. No. 5 (No. 16) bars on 10 in. centers furnish $0.37 \text{ in.}^2/\text{ft}$ and are selected for placement in the longitudinal direction.

15.6 COLUMN FOOTINGS

In plan, single-column footings are usually square. Rectangular footings are used if space restrictions dictate this choice or if the supported columns have a strongly elongated rectangular cross section. In the simplest form, they consist of a single slab (Fig. 15.5a). Another type is that shown in Fig. 15.5b, where a pedestal or cap is interposed between the column and the footing slab; the pedestal provides for a more favorable transfer of load and in many cases is required to provide the

FIGURE 15.5

Types of single-column footings.

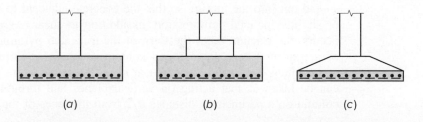

(a)   (b)   (c)

necessary development length for dowels. This form is also known as a *stepped* footing. All parts of a stepped footing must be cast at one time to provide monolithic action. Sometimes sloped footings like those shown in Fig. 15.5c are used. They require less concrete than stepped footings, but the additional labor necessary to produce the sloping surfaces (formwork, etc.) usually makes stepped footings more economical. In general, single-slab footings (Fig. 15.5a) are most economical for thicknesses up to 3 ft.

Single-column footings can be represented as cantilevers projecting out from the column in both directions and loaded upward by the soil pressure. Corresponding tension stresses are caused in both of these directions at the bottom surface. Such footings are, therefore, reinforced by two layers of steel, perpendicular to each other and parallel to the edges.

The required bearing area is obtained by dividing the total load, including the weight of the footing, by the selected bearing pressure. Weights of footings, at this stage, must be estimated and usually amount to 4 to 8 percent of the column load, the former value applying to the stronger types of soils.

In calculating bending moments and shears, only the upward pressure q_u that is caused by the factored column loads is considered. The weight of the footing proper does not cause moments or shears, just as no moments or shears are present in a book lying flat on a table. Circular or regular polygon-shaped columns may be considered as square columns of equivalent area for locating critical shear and moment sections.

a. Shear

Once the required footing area A_{req} has been established from the allowable bearing pressure q_a and the most unfavorable combination of service loads, including weight of footing and overlying fill (and such surcharge as may be present), the thickness h of the footing must be determined. In single footings, the effective depth d is mostly governed by shear. Since such footings are subject to two-way action, that is, bending in both major directions, their performance in shear is much like that of flat slabs in the vicinity of columns (see Section 13.7). However, in contrast to two-way floor and roof slabs, it is generally not economical in footings to use shear reinforcement. For this reason, only the design of footings in which all shear is carried by the concrete is discussed here. For the rare cases where the thickness is restricted so that shear reinforcement must be used, the information in Section 13.7 about slabs applies also to footings.

Two different types of shear strength are distinguished in footings: two-way, or punching shear, and one-way, or beam, shear. ACI Code 13.2.6 states that the size effect factors may be neglected for one-way shallow foundations, two-way combined footings, and mat foundations.

A column supported by the slab shown in Fig. 15.6 tends to punch through that slab because of the shear stresses that act in the footing around the perimeter of the column. At the same time, the concentrated compressive stresses from the column spread out into the footing so that the concrete adjacent to the column is in vertical or slightly inclined compression, in addition to shear. As a consequence, if failure occurs, the fracture takes the form of the truncated pyramid shown in Fig. 15.6 (or of a truncated cone for a round column), with sides sloping outward at an angle approaching 45°. The average shear stress in the concrete that fails in this manner can be taken as that acting on vertical planes laid through the footing around the column on a perimeter a distance $d/2$ from the faces of the column (vertical section

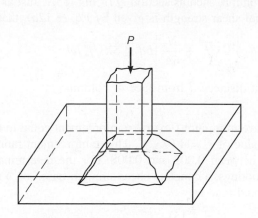

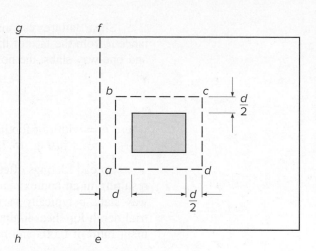

FIGURE 15.6
Punching-shear failure in single footing.

FIGURE 15.7
Critical sections for shear.

through *abcd* in Fig. 15.7). The concrete subject to this shear stress v_{u1} is also in vertical compression from the stresses spreading out from the column, and in horizontal compression in both major directions because of the biaxial bending moments in the footing. This triaxiality of stress increases the shear strength of the concrete. Tests of footings and of flat slabs have shown, correspondingly, that for punching-type failures the shear stress computed on the critical perimeter area is larger than in one-way action (such as in beams).

As discussed in Section 13.7, expanding the ACI Code equations (13.7*a*, *b*, and *c*) give the nominal punching-shear strength on this perimeter:

$$V_c = 4\lambda\sqrt{f_c'}\,b_o d \tag{15.7a}$$

except for columns of elongated cross section, for which

$$V_c = \left(2 + \frac{4}{\beta}\right)\lambda\sqrt{f_c'}\,b_o d \tag{15.7b}$$

For cases in which the ratio of critical perimeter to slab depth b_o/d is very large,

$$V_c = \left(\frac{\alpha_s d}{b_o} + 2\right)\lambda\sqrt{f_c'}\,b_o d \tag{15.7c}$$

where b_o is the perimeter *abcd* in Fig. 15.7; $\beta = a/b$ is the ratio of the long to short sides of the column cross section; and α_s is 40 for interior loading, 30 for edge loading, and 20 for corner loading of a footing. The punching-shear strength of the footing is to be taken as the smallest of the values given by Eqs. (15.7*a*), (15.7*b*), and (15.7*c*); and the design strength is ϕV_c, as usual, where $\phi = 0.75$ for shear.

The application of Eqs. (15.7) to punching shear in footings under columns with other than a rectangular cross section is shown in Fig. 13.19. For such situations, ACI Code 22.6.5 indicates that the perimeter b_o must be of minimum length but need not approach closer than $d/2$ to the perimeter of the actual loaded area. The manner of defining a and b for such irregular loaded areas is also shown in Fig. 13.19. If a moment is transferred from the column to the footing, the criteria discussed in Section 13.8 for the transfer of moment by bending and shear at slab-column connections must be satisfied.

Shear failures can also occur, as in beams or one-way slabs, at a section a distance d from the face of the column, such as section ef of Fig. 15.7. Just as in beams and one-way slabs, the nominal shear strength is given by Eq. (5.12a), that is,

$$V_c = \left[8\lambda(\rho)^{1/3}\sqrt{f_c'} + \frac{N_u}{6A_g} \right] bd \leq 5\lambda\sqrt{f_c'}\,bd \tag{15.8a}$$

where b = width of footing at distance d from face of column
$= ef$ in Fig. 15.7

Spread footings often have minimum flexural reinforcement equal to $0.0018A_g$, resulting in an approximate value of $v_c = 1.0\lambda\sqrt{f_c'}$. The reinforcement ratio for one-way slabs ρ typically varies between 0.002 and 0.008. For the determination of a trial depth for shear limited footings, a concrete shear strength based on a reinforcement ratio of 0.003 and no axial load is

$$v_c = 1.15\lambda\sqrt{f_c'} \tag{15.8b}$$

This concrete shear strength is less than that permitted in ACI Codes prior to 2019 and, consequently, it is advantageous to check one-way shear prior to designing for punching shear.

The required depth of footing d is then calculated from the usual equation

$$\phi V_c \geq V_u \tag{15.9}$$

applied separately in connection with Eqs. (15.7) and (15.8). For Eq. (15.7), $V_u = V_{u1}$ is the total upward pressure caused by q_u on the area outside the perimeter $abcd$ in Fig. 15.7. For Eq. (15.8), $V_u = V_{u2}$ is the total upward pressure on the area $efgh$ outside the section ef in Fig. 15.7. The required depth is then the larger of those calculated from either Eq. (15.7) or Eq. (15.8). For shear, $\phi = 0.75$.

Although the lightweight concrete factor λ appears in Eqs. (15.7) and (15.8), normalweight concrete ($\lambda = 1$) is almost universally used in foundations.

b.   Bearing: Transfer of Forces at Base of Column

When a column rests on a footing or pedestal, it transfers its load to only a part of the total area of the supporting member. The adjacent footing concrete provides lateral support to the directly loaded part of the concrete. This causes triaxial compressive stresses that increase the strength of the concrete that is loaded directly under the column. Based on tests, ACI Code 22.8.3 provides that when the supporting area is wider than the loaded area on all sides, the design bearing strength is

$$\phi B_n = \phi(0.85f_c' A_1)\sqrt{\frac{A_2}{A_1}} \leq \phi 2 \times (0.85f_c' A_1) \tag{15.10}$$

For bearing on concrete, $\phi = 0.65$, f_c' is the specified compressive strength of the footing concrete, which frequently is less than that of the column, and A_1 is the loaded area. A_2 is the area of the lower base of the largest frustum of a pyramid, cone, or tapered wedge contained wholly within the support and having for its upper base the loaded area and having side slopes of 1 vertical to 2 horizontal. The meaning of this definition of A_2 is clarified by Fig. 15.8. For the somewhat unusual case shown, where the top of the support is stepped, a step that is deeper or closer to the loaded area than that shown may result in reduction in the value of A_2. A footing for which the top surface is sloped away from the loaded area more steeply than 1 to 2 results in a value of A_2 equal to A_1. In most usual cases, for which the top of the footing is flat and the sides

FIGURE 15.8
Definition of areas A_1 and A_2.

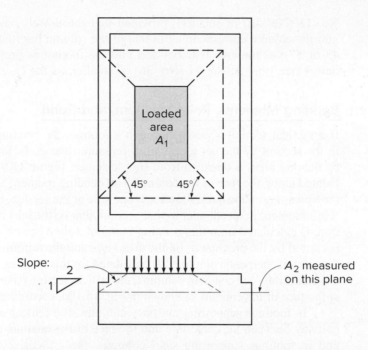

are vertical, A_2 is simply the maximum area of the portion of the supporting surface that is geometrically similar to, and concentric with, the loaded area.

All axial forces and bending moments that act at the bottom section of a column must be transferred to the footing at the bearing surface by compression in the concrete and by reinforcement. With respect to the reinforcement, this may be done either by extending the column bars into the footing or by providing dowels that are embedded in the footing and project above it. In the latter case, the column bars merely rest on the footing and in most cases are tied to dowels. This results in a simpler construction procedure than extending the column bars into the footing. To ensure the integrity of the junction between column and footing, ACI Code 16.3.4 requires that the minimum area of reinforcement that crosses the bearing surface (dowels or column bars) be 0.005 times the gross area of the supported column. The length of the dowels or bars of diameter d_b must be sufficient on both sides of the bearing surface to provide the required development length for compression bars (see Section 6.8), that is, $\ell_{dc} \geq (0.02 f_y \psi_r / \lambda \sqrt{f_c'}) d_b$ and $\geq (0.0003 f_y \psi_r) d_b$, where the confinement factor ψ_r is taken as 1.0. In addition, if dowels are used, the lapped length must be at least that required for a lap splice in compression (see Section 6.8); that is, the length of lap must not be less than the usual development length in compression and must not be less than $0.0005 f_y d_b$ or 12 in. For bars with f_y greater than 60,000 psi and less than or equal to 80,000 psi, the development length must not be less than $(0.0009 f_y - 24)$ or 12 in. Where bars of different sizes are lap-spliced, the splice length should be the larger of the development length of the larger bar or the splice length of the smaller bar, according to the ACI Code.

The two largest bar sizes, Nos. 14 (No. 43) and 18 (No. 57), are frequently used in columns with large axial forces. Under normal circumstances, the ACI Code specifically prohibits the lap splicing of these bars because tests have shown that welded splices or other positive connections are necessary to develop these heavy bars fully. A specific exception, however, is made for dowels for Nos. 14 (No. 43) and 18 (No. 57) column bars. Relying on long-standing successful use, ACI Code 25.5.5 permits these heavy bars to be spliced to dowels of lesser diameter [that is,

No. 11 (No. 36) or smaller], provided that the dowels have a development length into the column corresponding to that of the column bar [that is, Nos. 14 or 18 (Nos. 43 or 57), as the case may be] and into the footing as prescribed for the particular dowel size [that is, No. 11 (No. 36) or smaller, as the case may be].

c. Bending Moments, Reinforcement, and Bond

If a vertical section is passed through a footing, the bending moment that is caused in the section by the net upward soil pressure (that is, factored column load divided by bearing area) is obtained from simple statics. Figure 15.9 shows such a section *cd* located along the face of the column. The bending moment about *cd* is that caused by the upward pressure q_u on the area to one side of the section, that is, the area *abcd*. The reinforcement perpendicular to that section, that is, the bars running in the long direction, is calculated from this bending moment. Likewise, the moment about section *ef* is caused by the pressure q_u on the area *befg*, and the reinforcement in the short direction, that is, perpendicular to *ef*, is calculated for this bending moment. In footings that support reinforced concrete columns, the critical section *cd* or *ef* for bending is located at the face of the column, as shown in Fig. 15.10*a*, according to ACI Code 13.2.7.

In footings supporting masonry columns, the critical section *cd* or *ef* is located halfway between the centerline and the face of the column, as shown in Fig. 15.10*b*; and in footings supporting steel columns, the critical section *cd* or *ef* is located halfway between the face of the steel column and the edge of the steel base plate, as shown in Fig. 15.10*c*.

In footings with *pedestals*, the width resisting compression in sections *cd* and *ef* is that of the pedestal; the corresponding depth is the sum of the thickness of

FIGURE 15.9
Critical sections for bending and bond.

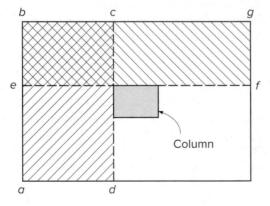

FIGURE 15.10
Critical sections *cd* or *ef* for concrete, masonry, and steel columns.

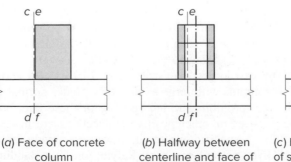

(*a*) Face of concrete column

(*b*) Halfway between centerline and face of masonry wall or column

(*c*) Halfway between face of steel column and edge of steel base plate

pedestal and footing. Further sections parallel to *cd* and *ef* are passed at the edge of the pedestal, and the moments are determined in the same manner, to check the strength at locations in which the depth is that of the footing only.

For footings with relatively small pedestals, the latter are often discounted in moment and shear computation, and bending is checked at the face of the column, with width and depth equal to that of the footing proper.

In *square footings*, the reinforcement is uniformly distributed over the width of the footing in each of the two layers; that is, the spacing of the bars is constant. The moments for which the two layers are designed are the same. However, the effective depth *d* for the upper layer is less by 1 bar diameter than that of the lower layer. Consequently, the required A_s is larger for the upper layer. Instead of using different spacings or different bar diameters in each of the two layers, it is customary to determine A_s based on average depth and to use the same arrangement of reinforcement for both layers.

In *rectangular footings*, the reinforcement *in the long direction* is again uniformly distributed over the pertinent (shorter) width. In locating the bars in the short direction, one has to consider that the support provided to the footing by the column is concentrated near the middle. Consequently, the curvature of the footing is sharpest, that is, the moment per foot largest, immediately under the column, and it decreases in the long direction with increasing distance from the column. For this reason, a larger steel area per longitudinal foot is needed in the central portion than near the far ends of the footing. ACI Code 13.3.3.3, therefore, provides the following:

> For reinforcement in the short direction, a portion of the total reinforcement $\gamma_s A_s$ shall be distributed uniformly over a band width equal to the length of short side of footing, centered on centerline of column or pedestal. The remainder of the reinforcement required in the short direction $(1 - \gamma_s)A_s$ shall be distributed uniformly outside the center band width of the footing

$$\gamma_s = \frac{\text{reinforcement in band width}}{\text{total reinforcement in short direction}} = \frac{2}{\beta + 1} \tag{15.11}$$

where β is the ratio of the long side to the short side of the footing.

According to the ACI Code 8.6.1, the usual minimum flexural reinforcement ratios of Section 4.3e need not be applied to either slabs or footings. Instead, the minimum steel requirements for shrinkage and temperature crack control for structural slabs are to be imposed, as given in Table 12.2. The maximum spacing of bars in the direction of the span is reduced to the lesser of 3 times the footing thickness *h* and 18 in., rather than 5*h* as is normal for shrinkage and temperature steel. These requirements for minimum steel and maximum spacing are to be applied to mat foundations as well as individual footings.

The critical sections for development length of footing bars are the same as those for bending. Development length may also have to be checked at all vertical planes in which changes of section or of reinforcement occur, as at the edges of pedestals or where part of the reinforcement may be terminated.

EXAMPLE 15.2   **Design of a square footing.**   A column 18 in. square, with $f'_c = 4$ ksi, reinforced with eight No. 8 (No. 25) bars of $f_y = 60$ ksi, supports an unfactored dead load of 225 kips and a live load of 175 kips. The soil (fill) has a unit weight of 100 pcf. The allowable soil pressure q_a is 5 kips/ft². Design a square footing with base 5 ft below grade, using $f'_c = 4$ ksi and $f_y = 60$ ksi.

SOLUTION.   Since the space between the bottom of the footing and the surface is occupied partly by concrete and partly by soil (fill), an average unit weight of 125 pcf is assumed. The pressure of this material at the 5 ft depth is $5 \times 125 = 625$ psf, leaving a bearing pressure of $q_e = 5000 - 625 = 4375$ psf available to carry the column service load. Hence, the required footing area $A_{req} = (225 + 175)/4.375 = 91.5$ ft^2. A base 9 ft 6 in. square is selected, furnishing a footing area of 90.3 ft^2, which differs from the required area by about 1 percent.

For strength design, the upward pressure caused by the factored column loads is $q_u = (1.2 \times 225 + 1.6 \times 175)/9.5^2 = 6.10$ kips/ft^2.

The footing depth is determined by one-way shear or by punching shear. The lower one-way concrete shear contribution usually is the governing condition. The design checks one-way shear strength first then checks the punching shear. A trial depth of 19 in. is selected. The factored one-way shear along line *e-f* is

$$V_{u1} = 6.10 \times 2.42 \times 9.5 = 140 \text{ kips}$$

And the nominal shear strength, assuming minimum flexural reinforcement using Eq. (15.8*b*) is

$$V_c = \phi \lambda 1.15 \sqrt{f_c'}\, bd = 0.75 \times 1.0 \times 1.15 \times \sqrt{4000} \times 9.5 \times 12 \times d/1000 = 6.22d$$

resulting in a required depth of $d = 140/6.22 = 22.5$ in. For concrete cast against the ground, 3 in. cover is required and allowing 1 in. for the diameter of the reinforcement to the center of the mat and rounding up gives a footing thickness of 27 in. and an effective depth $d = 27 - 4 = 23$ in.

The footing is then checked for two-way or punching shear on the critical perimeter *abcd* in Fig. 15.11. The length of the critical perimeter for a depth of 26.5 in. is

$$b_o = 4(18 + 23) = 164 \text{ in.}$$

The shear force acting on this perimeter, being equal to the total upward pressure minus that acting within the perimeter *abcd*, is

$$V_{u2} = q_u(l^2 - b^2) = 6.10 \left[9.5^2 - \left(\frac{23 + 18}{12} \right)^2 \right] = 479 \text{ kips}$$

The corresponding nominal shear strength from Eq. (15.7) is

$$V_c = 4\lambda \sqrt{f_c'}\, b_o d = 4 \times 1 \times \sqrt{4000} \times 178 \times \frac{26.5}{1000} = 1193 \text{ kips}$$

and

$$\phi V_c = 0.75 \times 1193 = 895 \text{ kips}$$

which exceeds the applied load, so punching shear capacity is adequate.

FIGURE 15.11
Critical sections for
Example 15.2.

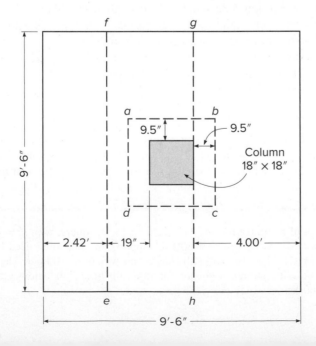

The bending moment on section gh of Fig. 15.11 is

$$M_u = 6.10 \times 9.5 \left(\frac{4.0^2}{2}\right) 12 = 5560 \text{ in-kips}$$

Because the depth required for shear is greatly in excess of that required for bending, the reinforcement ratio will be low and the corresponding depth of the rectangular stress block small. If $a = 2$ in., the required steel area is

$$A_s = \frac{M}{\phi f_y \left(d - \dfrac{a}{2}\right)} = \frac{5560}{0.90 \times 60(23 - 2/2)} = 4.68 \text{ in}^2$$

Checking the minimum reinforcement $0.0018A_g$ according to ACI Code 7.6.1.1 results in

$$A_s = 0.0018 \times 27 \times 114 = 5.54 \text{ in}^2$$

The controlling value of 5.54 in^2 is larger than the 4.68 in^2 calculated for bending. Ten No. 7 (No. 22) bars furnishing 6.00 in^2 are placed in each direction. The required development length beyond section gh is found from Table A.10 to be 41 in., which is more than adequately met by the actual length of bars beyond section gh, namely, $48 - 3 = 45$ in.

Checking for transfer of forces at the base of the column shows that the footing concrete, which has the same f_c' as the column concrete and for which the strength is enhanced according to Eq. (15.10), is clearly capable of carrying that part of the column load transmitted by the column concrete. The force in the column carried by the steel is transmitted to the footing using dowels to match the column bars. These must extend into the footing the full development length in compression, which is found from Table A.11 of Appendix A to be 19 in. for No. 8 (No. 25) bars. This is accommodated in a footing with $d = 19$ in. Above the top surface of the footing, the No. 8 (No. 25) dowels must extend into the column that same development length, but not less than the requirement for a lapped splice in compression (see Section 6.13b). The minimum lap splice length for the No. 8 (No. 25) bars is $0.0005 \times 1.0 \times 60,000 = 30$ in., which is seen to control here. Thus the bars are carried 30 in. into the column, requiring a total dowel length of 49 in. This is rounded upward for practical reasons to 4.25 ft, as shown in Fig. 15.12. It is easily confirmed that the minimum dowel steel requirement of $0.005 \times 18 \times 18 = 1.62$ in^2 does not control here.

FIGURE 15.12
Footing in Example 15.2.

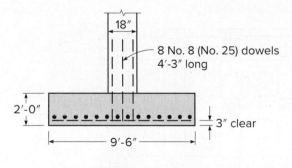

8 No. 8 (No. 25) dowels
4'-3" long

2'-0"

3" clear

9'-6"

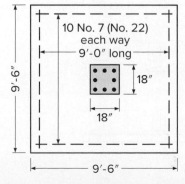

10 No. 7 (No. 22)
each way
9'-0" long

18"

18"

9'-6"

9'-6"

15.7 COMBINED FOOTINGS

Spread footings that support more than one column or wall are known as *combined footings*. They can be divided into two categories: those that support two columns and those that support more than two (generally large numbers of) columns.

Examples of the first type, that is, two-column footings, are shown in Fig. 15.1. In buildings where the allowable soil pressure is large enough for single footings to be adequate for most columns, two-column footings are seen to become necessary in two situations: (1) if columns are so close to the property line that single-column footings cannot be made without projecting beyond that line and (2) if some adjacent columns are so close to each other that their footings would merge. Both situations are shown in Fig. 15.1.

When the bearing capacity of the subsoil is low so that large bearing areas become necessary, individual footings are replaced by *continuous strip footings* that support more than two columns and usually all columns in a row. Sometimes such strips are arranged in both directions, in which case a *grid foundation* is obtained, as shown in Fig. 15.13. Strip footings can be made to develop a much larger bearing area much more economically than can be done by single footings because the individual strips represent continuous beams whose moments are much smaller than the cantilever moments in large single footings that project far out from the column in all four directions.

In many cases, the strips are made to merge, resulting in a mat foundation, as shown in Fig. 15.14. That is, the foundation consists of a solid reinforced concrete slab under the entire building. In structural action, such a mat is very similar to a flat slab or a flat plate, upside down, that is, loaded upward by the bearing pressure and downward by the concentrated column reactions. The mat foundation evidently develops the maximum available bearing area under the building. If the soil's capacity is so low that even this large bearing area is insufficient, some form of deep foundation, such as piles or caissons, must be used.

FIGURE 15.13
Grid foundation.

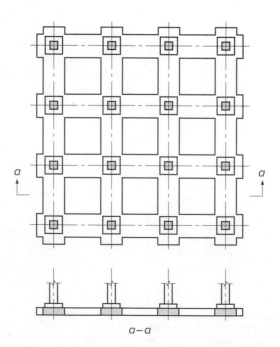

FIGURE 15.14
Mat foundation.

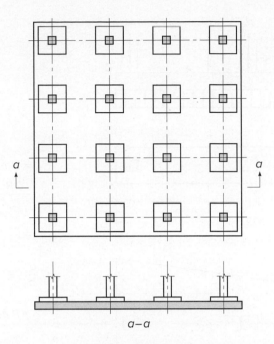

These are discussed in Section 15.10 and texts on foundation design and fall outside the scope of the present volume.

Mat foundations may be designed with the column pedestals, as shown in Figs. 15.13 and 15.14, or without them, depending on whether they are necessary for shear strength and the development length of dowels.

Apart from developing large bearing areas, another advantage of strip and mat foundations is that their continuity and rigidity help in reducing differential settlements of individual columns relative to each other, which may otherwise be caused by local variations in the quality of subsoil, or other causes. For this purpose, continuous foundations are frequently used in situations where the superstructure or the type of occupancy provides unusual sensitivity to differential settlement.

Much useful and important design information pertaining to combined footings and mats is found in Refs. 15.10 and 15.13.

15.8   TWO-COLUMN FOOTINGS

It is desirable to design combined footings so that the centroid of the footing area coincides with the resultant of the two column loads. This produces uniform bearing pressure over the entire area and forestalls a tendency for the footings to tilt. In plan, such footings are rectangular, trapezoidal, or T-shaped, the details of the shape being arranged to produce coincidence of centroid and resultant. The simple relationships shown in Fig. 15.15 facilitate the determination of the shape of the bearing area (from Ref. 15.8). In general, the distances m and n are given, the former being the distance from the center of the exterior column to the property line and the latter the distance from that column to the resultant of both column loads.

Another expedient that is used if a single footing cannot be centered under an exterior column is to place the exterior column footing eccentrically and to connect it with the nearest interior column footing by a beam or strap. This strap, being counterweighted by the interior column load, resists the tilting tendency of the eccentric

FIGURE 15.15
Two-column footing.
(*Adapted from Ref. 15.8.*)

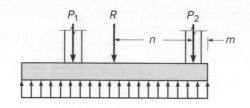

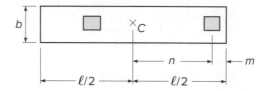

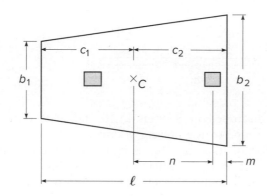

$$\ell = 2(m+n)$$

$$b = \frac{R}{q_e\ell}$$

$$\frac{b_2}{b_1} = \frac{3(n+m)-\ell}{2\ell - 3(n+m)}$$

$$(b_1 + b_2) = \frac{2R}{q_e\ell}$$

$$c_1 = \frac{\ell(b_1 + 2b_2)}{3(b_1 + b_2)}$$

$$c_2 = \frac{\ell(2b_1 + b_2)}{3(b_1 + b_2)}$$

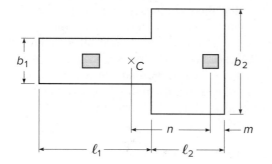

$$b_1 = \frac{R}{q_e} = \left[\frac{2(n+m)-\ell_2}{\ell_1(\ell_1 + \ell_2)}\right]$$

$$b_2 = \frac{R}{\ell_2 q_e} - \frac{\ell_1 b_1}{\ell_2}$$

$$\ell_1 b_1 + \ell_2 b_2 = \frac{R}{q_e}$$

exterior footing and equalizes the pressure under it. Such foundations are known as *strap, cantilever,* or *connected footings.*

The two examples that follow demonstrate some of the peculiarities of the design of two-column footings.

EXAMPLE 15.3   **Design of a combined footing supporting one exterior and one interior column.**   An exterior 24 × 18 in. column with service loads $D = 170$ kips and $L = 130$ kips, and an interior 24 × 24 in. column with $D = 250$ kips, $L = 200$ kips are to be supported on a combined rectangular footing whose outer end cannot protrude beyond the outer face of the exterior column (see Fig. 15.1). The distance center to center of columns is 18 ft 0 in., and the allowable bearing pressure of the soil is 6000 psf. The bottom of the footing is 7 ft below grade, and a surcharge of 100 psf is specified on the surface. Design the footing for $f_c' = 4000$ psi and $f_y = 60,000$ psi.

SOLUTION. The space between the bottom of the footing and the surface is occupied partly by concrete (footing, concrete floor) and partly by backfill. An average unit weight of 125 pcf can be assumed. Hence, the effective portion of the allowable bearing pressure that is available for carrying the column loads is $q_e = q_a -$ (weight of fill and concrete + surcharge) = $6000 - (7 \times 125 + 100) = 5025$ psf. Then the required area A_{req} = sum of column loads/$q_e = 750/5.03 = 149.3$ ft^2. The resultant of the column loads is located from the center of the exterior column a distance $450 \times 18/750 = 10.8$ ft. Hence, the length of the footing must be $2(10.8 + 0.75) = 23.1$ ft. A length of 23 ft 3 in. is selected. The required width is then $149.3/23.25 = 6.42$ ft. A width of 6 ft 6 in. is selected (see Fig. 15.16).

Longitudinally, the footing represents a one-way slab, loaded from below, spanning between columns and cantilevering beyond the interior column. Since this slab is considerably wider than the columns, the column loads are distributed crosswise by transverse slabs, one under each column. In the present relatively narrow and long footing, it is found that the required minimum depth for the transverse beams is smaller than is required for the footing in the longitudinal direction. These "slabs," therefore, are not really distinct members but merely represent transverse strips in the main body of the footing, reinforced so that they are capable of resisting the transverse bending moments and the corresponding shears. It then becomes necessary to decide how large the effective width of this transverse beam can be assumed to be. The strip directly under the column does not deflect independently and is strengthened by the adjacent parts of the footing. The effective width of the transverse beams is therefore evidently larger than that of the column. In the absence of definite rules for this case, or of research results on which to base such rules, the authors recommend conservatively that the load be assumed to spread outward from the column into the footing at a slope of 2 vertical to 1 horizontal. This means that the effective width of the transverse beam is assumed to be equal to the width of the column plus $d/2$ on either side of the column, d being the effective depth of the footing.

FIGURE 15.16
Combined footing in Example 15.3.

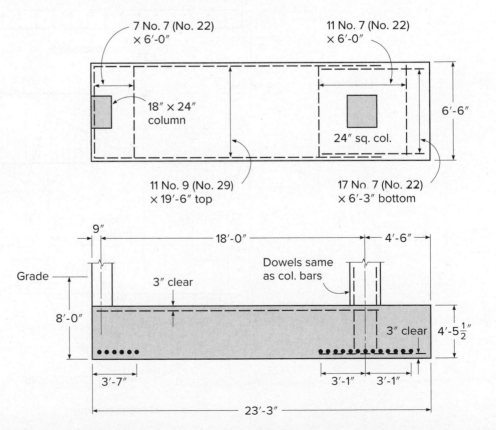

Strength design in longitudinal direction

The net upward pressure caused by the factored column loads is

$$q_u = \frac{1.2(170 + 250) + 1.6(130 + 200)}{23.25 \times 6.5} = 6.83 \text{ kips/ft}^2$$

Then the net upward pressure per linear foot in the longitudinal direction is $6.83 \times 6.5 = 44.4$ kips/ft. The maximum negative moment between the columns occurs at the section of zero shear. Let x be the distance from the outer edge of the exterior column to this section. Then (see Fig. 15.17)

$$V_u = 44,400x - 412,000 = 0$$

results in $x = 9.28$ ft. The moment at this section is

$$M_u = \left[44,400 \frac{9.28^2}{2} - 412,000(9.28 - 0.75)\right] 12 = -19,230,000 \text{ in-lb}$$

The moment at the right edge of the interior column is

$$M_u = 44,400 \left(\frac{3.5^2}{2}\right) 12 = 3,260,000 \text{ in-lb}$$

and the details of the moment diagram are as shown in Fig. 15.17. Try $d = 50.0$ in.

FIGURE 15.17
Moment and shear diagrams for footing in Example 15.3.

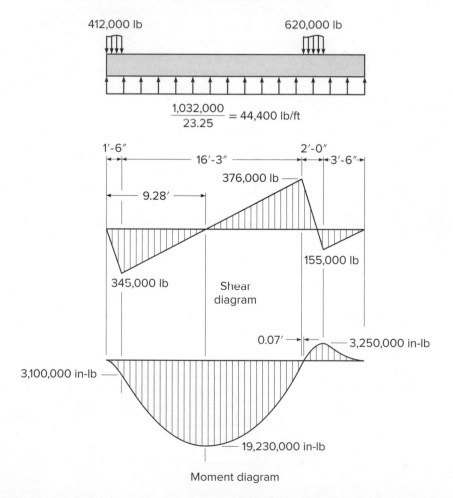

From the shear diagram in Fig. 15.17, it is seen that the critical section for flexural shear occurs at a distance d to the left of the left face of the interior column. At that point, the factored shear is

$$V_u = 376,000 - \frac{50}{12} \, 44,400 = 191,000 \text{ lb}$$

and the design shear strength assuming minimum reinforcement is

$$\phi V_c = \phi\left(1.0\sqrt{f_c'}\right)bd = 0.75(1.0 \times \sqrt{4000}) \, 78 \times d = 3700d$$

giving the required depth $d = 191,000/3700 = 56.1$ in., which is close to the trial value. Use $d = 50$ in. and confirm the concrete shear strength after longitudinal reinforcement is selected.

Additionally, as in single footings, punching shear should be checked on a perimeter section a distance $d/2$ around the column. In this case, the 50 in. depth places the critical perimeter outside the footing; therefore, only the one-way shear need be checked.

With $d = 50$ in., and with 3.5 in. cover from the center of the bars to the top surface of the footing, the total thickness is 53.5 in.

To determine the required steel area, $M_u/\phi bd^2 = 19,230,000/(0.9 \times 78 \times 50^2) = 109$ is used to enter Graph A.1b of Appendix A. For this value, the curve 60/3 gives the reinforcement ratio $\rho = 0.0028$. The required steel area is $A_s = 0.0028 \times 50 \times 78 = 10.9$ in^2. Eleven No. 9 (No. 29) bars furnish 11.00 in^2. The required development length is found to be 6.7 ft. From Fig. 15.17, the distance from the point of maximum moment to the nearer left end of the bars is seen to be $9.30 - \frac{3}{12} = 9.05$ ft, much larger than the required minimum development length. The selected reinforcement is therefore adequate for both bending and bond.

For the portion of the longitudinal beam that cantilevers beyond the interior column, the required steel area exceeds the minimum required steel area controls.

$$A_{s,\min} = 0.0018A_g = 0.0018 \times 78 \times 53.5 = 7.51 \text{ in}^2$$

Seventeen No. 7 (No. 22) bars with $A_s = 10.2$ in^2 are selected; their development length is calculated and for bottom bars is found satisfactory. The final reinforcement ratio is 0.0028 resulting in a concrete shear strength of

$$\phi V_c = \phi\left[8\lambda(\rho)^{1/3}\sqrt{f_c'}\right]bd = 0.75 \times 8 \times 1.0 \times 0.0028^{1/3} \times \sqrt{4000} \times 78 \times 50 = 208,600 \text{ lb}$$

which exceeds the factored shear of 191,000 lb. Thus, the 50 in. depth is adequate. Because the footing depth is established by the shear capacity and the final shear capacity exceeds the demand, it is possible to iterate the solution to reduce the thickness slightly. Alternatively, additional reinforcement could be added to increase the shear capacity and decrease the thickness.

Design of transverse beam under interior column

The width of the transverse beam under the interior column can now be established as previously suggested and is $24 + 2(d/2) = 24 + 2 \times 50/2 = 74$ in. The net upward load per linear foot of the transverse beam is $620,000/6.5 = 95,400$ lb/ft. The moment at the edge of the interior column is

$$M_u = 95,400 \left(\frac{2.25^2}{2}\right) 12 = 2,900,000 \text{ in-lb/ft}$$

Since the transverse bars are placed on top of the longitudinal bars (see Fig. 15.16), the actual value of d furnished is $50 - 1.0 = 49$ in. The required steel area is 7.53 in^2 and exceeds the minimum required area of controls; that is,

$$A_{s,\min} = 0.0018 \times 74 \times 50 = 6.66 \text{ in}^2$$

Eleven No. 7 (No. 22) bars are selected and placed within the 74 in. effective width of the transverse beam.

Punching shear at the perimeter a distance $d/2$ from the column has been checked before. The critical section for regular flexural shear, at a distance d from the face of the column, lies beyond the edge of the footing, and therefore no further check on shear is needed.

The design of the transverse beam under the exterior column is the same as the design of that under the interior column, except that the effective width is 43 in. The details of the calculations are not shown. It can be easily checked that seven No. 7 (No. 22) bars, placed within the 43 in. effective width, satisfy all requirements. Design details are shown in Fig. 15.16.

EXAMPLE 15.4   **Design of a strap footing.** In a strap or connected footing, the exterior footing is placed eccentrically under its column so that it does not project beyond the property line. Such an eccentric position would result in a strongly uneven distribution of bearing pressure, which could lead to tilting of the footing. To counteract this eccentricity, the footing is connected by a beam or strap to the nearest interior footing.

Both footings are so proportioned that under service load the pressure under each of them is uniform and the same under both footings. To achieve this, it is necessary, as in other combined footings, that the centroid of the combined area for the two footings coincide with the resultant of the column loads. The resulting forces are shown schematically in Fig. 15.18. They consist of the loads P_e and P_i of the exterior and interior columns, respectively, and of the net upward pressure q, which is uniform and equal under both footings. The resultants R_e and R_i of these upward pressures are also shown. Since the interior footing is concentric with the interior column, R_i and P_i are collinear. This is not the case for the exterior forces R_e and P_e where the resulting couple just balances the effect of the eccentricity of the column relative to the center of the footing. The strap proper is generally constructed so that it does not bear on the soil. This can be achieved by providing formwork not only for the sides but also for the bottom face and by withdrawing it before backfilling.

To illustrate this design, the columns in Example 15.3 will now be supported on a strap footing. Its general shape, plus dimensions as determined only subsequently by calculations, is seen in Fig. 15.19. With an allowable bearing pressure of $q_a = 6.0$ kips/ft^2 and a depth of 6 ft to the bottom of the footing as before, the bearing pressure available for carrying the external loads applied to the footing is $q_e = 5.15$ kips/ft^2. These external loads, for the strap footing, consist of the column loads and of the weight plus fill and surcharge of that part of the strap that is located between the footings. (The portion of the strap located directly on top of the footing displaces a corresponding amount of fill and therefore is already accounted for in the determination of the available bearing pressure q.) If the bottom of the strap is 6 in. above the bottom of the footings to prevent bearing on soil, the total depth to grade is 5.5 ft. If the strap width is estimated to be 2.5 ft, its estimated weight plus fill and surcharge is $2.5 \times 5.5 \times 0.125 + 0.100 \times 2.5 = 2$ kips/ft. If the gap between footings is estimated to be 8 ft, the total weight of the strap is 16 kips. Hence, for purposes of determining the required footing area, 8 kips is be added to the dead load of each column. The required total area of both footings is then $(750 + 16)/5.15 = 149$ ft^2. The distance of the resultant of the

FIGURE 15.18

Forces and reactions on the strap footing in Example 15.4.

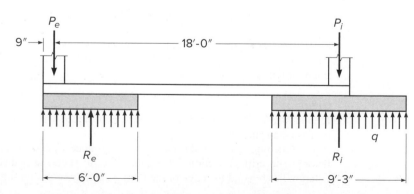

FIGURE 15.19

Strap footing in Example 15.4.

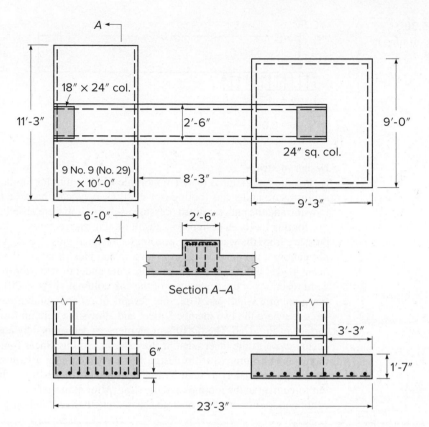

Section A–A

two column loads plus the strap load from the axis of the exterior column, with sufficient accuracy, is $458 \times 18/766 = 10.75$ ft, or 11.50 ft from the outer edge, almost identical to that calculated for Example 15.3. Trial calculations show that a rectangular footing 6 ft 0 in. $\times$ 11 ft 3 in. under the exterior column and a square footing 9 $\times$ 9 ft under the interior column have a combined area of 149 ft^2 and a distance from the outer edge to the centroid of the combined areas of $(6 \times 11.25 \times 3 + 9 \times 9 \times 18.75) \div 149 = 11.55$ ft, which is almost equal to the previously calculated distance to the resultant of the external forces.

For *strength calculations*, the bearing pressure caused by the factored external loads, including that of the strap with its fill and surcharge, is

$$q_u = \frac{1.2(170 + 250 + 16) + 1.6(130 + 200)}{149} = 7.06 \text{ kips/ft}^2$$

Design of footings

The exterior footing performs exactly like a wall footing with a length of 6 ft. Even though the column is located at its edge, the balancing action of the strap results in uniform bearing pressure, the downward load being transmitted to the footing uniformly by the strap. Hence, the design is carried out exactly as it is for a wall footing (see Section 15.5).

The interior footing, even though it merges in part with the strap, can safely be designed as an independent, square single-column footing (see Section 15.6). The main difference is that, because of the presence of the strap, punching shear cannot occur along the truncated pyramid surface shown in Fig. 15.6. For this reason, two-way or punching shear, according to Eq. (15.7), should be checked along a perimeter section located at a distance $d/2$ outward from the longitudinal edges of the strap and from the free face of the column, d being the effective depth of the footing. Flexural or one-way shear, as usual, is checked at a section a distance d from the face of the column.

FIGURE 15.20
Forces acting on strap in
Example 15.4.

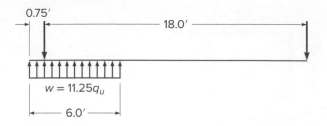

Design of strap

Even though the strap is in fact monolithic with the interior footing, the effect on the strap of the soil pressure under this footing can safely be neglected because the footing has been designed to withstand the entire upward pressure as if the strap were absent. In contrast, because the exterior footing has been designed as a wall footing that receives its load from the strap, the upward pressure from the wall footing becomes a load that must be resisted by the strap. With this simplification of the actually somewhat more complex situation, the strap represents a single-span beam loaded upward by the bearing pressure under the exterior footing and supported by downward reactions at the centerlines of the two columns (Fig. 15.20). A width of 30 in. is selected. For a column width of 24 in., this permits beam and column bars to be placed without interference where the two members meet and allows the column forms to be supported on the top surface of the strap. The maximum moment, as determined by equating the shear force to zero, occurs close to the inner edge of the exterior footing. Shear forces are large in the vicinity of the exterior column. Stirrup design is completed using a strut-and-tie model. The footing is drawn approximately to scale in Fig. 15.19, which also shows the general arrangement of the reinforcement in the footings and the strap. Minimum shear reinforcement in the strap would be continued to improve the concrete shear strength. See Problem 15.6.

15.9 STRIP, GRID, AND MAT FOUNDATIONS

As mentioned in Section 15.7, continuous foundations are often used to support heavily loaded columns, especially when a structure is located on relatively weak or uneven soil. The foundation may consist of a *continuous strip footing* supporting all columns in a given row, or of two sets of such strip footings intersecting at right angles so that they form one continuous *grid foundation* (Fig. 15.13). For even larger loads or weaker soils, the strips are made to merge, resulting in a *mat foundation* (Fig. 15.14).

For the design of such continuous foundations, it is essential that reasonably realistic assumptions be made regarding the distribution of bearing pressures that act as upward loads on the foundation. For compressible soils, it can be assumed, as a first approximation, that the deformation or settlement of the soil at a given location and the bearing pressure at that location are proportional to each other. If columns are spaced at moderate distances and if the strip, grid, or mat foundation is quite rigid, the settlements in all portions of the foundation are substantially the same. This means that the bearing pressure, also known as *subgrade reaction*, is the same, provided that the centroid of the foundation coincides with the resultant of the loads. If they do not coincide, then for such rigid foundations the subgrade reaction can be assumed to vary linearly. Bearing pressures can be calculated based on statics, as discussed for single footings (see Fig. 15.3). In this case, all loads, the downward column loads as well as the upward-bearing

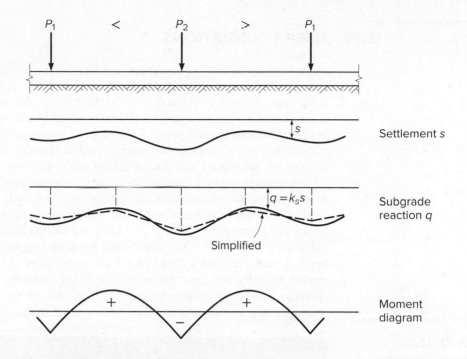

FIGURE 15.21
Strip footing. (*Adapted from Ref. 15.8.*)

pressures, are known. Hence, moments and shear forces in the foundation can be found by statics alone. Once these are determined, the design of strip and grid foundations is similar to that of inverted continuous beams and that of mat foundations to that of inverted flat slabs or plates.

On the other hand, if the foundation is relatively flexible and the column spacing large, settlements will no longer be uniform or linear. For one thing, the more heavily loaded columns cause larger settlements, and thereby larger subgrade reactions, than the lighter ones. Also, since the continuous strip or slab midway between columns deflects upward relative to the nearby columns, the soil settlement, and thereby the subgrade reaction, is smaller midway between columns than directly at the columns. This is shown schematically for a strip footing in Fig. 15.21; the subgrade reaction can no longer be assumed to be uniform. Mat foundations likewise require different approaches, depending on whether they can be assumed to be rigid when calculating the soil reaction.

Criteria have been established as a measure of the relative stiffness of the structure versus the stiffness of the soil (Refs. 15.10 and 15.13). If the relative stiffness is low, the foundation should be designed as a flexible member with a nonlinear upward reaction from the soil. For strip footings, a reasonably accurate but fairly complex analysis can be done using the theory of beams on elastic foundations (Ref. 15.14). Kramrisch (Ref. 15.8) has suggested simplified procedures, based on the assumption that contact pressures vary linearly between load points, as shown in Fig. 15.21.

For nonrigid mat foundations, great advances in analysis have been made using finite element methods, which can account specifically for the stiffnesses of both the structure and the soil. There are a large number of commercially available programs (such as spMats, StructurePoint, Skokie, Illinois) based on the finite element method, permitting quick modeling and analysis of combined footings, strip footings, and mat foundations.

15.10   DEEP FOUNDATIONS

If the bearing capacity of the upper soil layers is insufficient for a spread or shallow foundation, but firmer strata are available at greater depth, *piles* or *drilled piers*, sometimes referred to as *caissons* or *drilled shafts*, are used to transfer the loads to these deeper strata. Piles are long reinforced or precast-prestressed concrete members, driven into the soil using a pile driving hammer, Fig. 15.22. Drilled piers are constructed using an auger to create a void in the soil. Casings and reinforcement, if required, are placed and the remaining void filled with concrete.

Loadings on deep foundations are determined through principles of soil or rock mechanics in accordance with the general building code, or other requirements, in conjunction with the loads and load combinations in ASCE/SEI 7 (Ref. 15.12), which are summarized in Table 1.2. Deep foundation elements are considered in two categories: those that carry axial load alone and those that carry axial load plus bending and shear. ACI Code 13.4.2 addresses piles or drilled shafts that are supported laterally for their entire length. When bending moments are less than a moment caused by an eccentricity of 5 percent of the diameter or width of the member, they are considered to carry only axial load.

FIGURE 15.22
Prestressed pile driving operation. A prestressed pile is being loaded into the driver on the left and driven piles are seen on the right.
(*Charles Dolan*)

a.   Service Level Design and Detailing

Cast-in-place concrete piles or drilled piers may be either uncased or confined in a metal casing. To be considered a confined metal cased pile, ACI Code 13.4.2 requires the pile meet the following six conditions:

- The design cannot use the casing to resist any portion of the axial load.
- The casing has a sealed tip.
- The casing thickness must be at least standard gage No. 14 (0.068 in.) steel.
- The casing must be seamless or seams must provide the strength of the basic material and be configured to provide confinement.
- The ratio of the yield strength of the steel casing to f_c' must be at least 6, and the yield strength must not be less than 30,000 psi.
- The nominal diameter must not exceed 16 in.

For these conditions, the maximum allowable axial compressive strength for the members is given in Table 15.1. All other piles or drilled piers, including portions that are in air, water, or soil not providing lateral support, are designed as columns using the strength design principles in Section 15.10b. Higher values than those given in Table 15.1 are allowed based on load tests or if accepted by the building official.

ACI Code 13.4.5 requires precast concrete pile reinforcement to be placed in a symmetrical pattern and contain a minimum of four bars and a minimum area of reinforcement of at least $0.008A_g$. Prestressed concrete piles require a minimum compressive stress based on the length of the pile. Piles are installed using impact hammers or vibratory hammers. Stresses in the piles are sometimes monitored using a pile driving analyzer to limit the tensile stresses in the concrete. The minimum compressive stresses are intended to keep the pile from cracking due to the rebound tensile stresses during driving and are given in Table 15.2 (Refs. 15.15 and 15.16). The PCI Design Handbook recommends an effective prestress of 700 psi for all precast prestressed piles regardless of length (Ref. 15.17). The effective prestress is calculated based on an assumed loss of 30,000 psi. Chapter 22 contains more detail on the design of prestressed concrete. Minimum transverse reinforcement is required for all precast concrete piles. Table 15.3 contains both the minimum reinforcement size and spacing.

TABLE 15.1

Maximum allowable compressive strength of deep foundation members

Deep Foundation Member Type	Maximum Compressive Strength[a]
Uncased cast-in-place concrete drilled shaft	$P_a = 0.3f_c'A_g + 0.4f_yA_s$
Confined metal cased concrete pile meeting the ACI Code confinement criteria	$P_a = 0.4f_c'A_g$
Cast-in-place concrete pile in a pipe, tube, other permanent casing, or rock that does not satisfy ACI Code confinement criteria above[b]	$P_a = 0.33f_c'A_g + 0.4f_yA_s$
Precast non-prestressed piles	$P_a = 0.33f_c'A_g + 0.4f_yA_s$
Precast prestressed concrete piles	$P_a = (0.33f_c' - 0.27f_{pe})A_g$

[a]A_g is the gross cross-sectional area. If a temporary or permanent casing is used, the inside face of the casing may be used to determine the gross area.

[b]The steel casing, pipe, or tube is not counted as part of A_s. If the steel casing is to be considered composite with the concrete, design guidance is given in AISC 360.

TABLE 15.2
Minimum compressive stress in precast prestressed piles per ACI 318-19

Pile Length (ft)	Minimum Compressive Stress (psi)
Pile length $\leq$ 30	400
30 < Pile length $\leq$ 50	550
Pile length > 50	700

TABLE 15.3
Minimum transverse reinforcement

MINIMUM TRANSVERSE REINFORCEMENT SIZE		MAXIMUM TRANSVERSE REINFORCEMENT SPACING	
Least Horizontal Pile Dimension (in.)	Minimum Wire Size of Transverse Reinforcement[a]	Reinforcement Location in Pile	Maximum Center-to-Center Spacing (in.)
$h \leq 16$	W4, D4	First five ties or spirals at each end of pile	1
$16 < h < 20$	W4.5, D5	24 in. from end of pile	4
$h \geq 20$	W5.5, D5	Remainder of pile	6

[a]If bars are used, minimum No. 3 (No. 10) applies to all values of h.

Cover for drilled piers is 3 in. in accordance with the criteria for concrete cast against the ground in the ACI Code. Industry practice historically uses a cover of 2.5 in. for drilled piers when spacers are used to center the reinforcement cage. The ACI Code implies that reinforcement runs the full length of a drilled pier; however, engineers often terminate the reinforcement cage where the bending moment drops to less than that corresponding to a 5 percent eccentricity.

b. Strength Design

ACI Code 13.4.3 allows strength design for all deep foundation members and requires strength design for free-standing elements and elements in water. Strength design of deep foundations follows the principles of column design presented in Chapters 9 and 10. For axial loads without moments, compression strength reduction factors from Table 15.4 are used. The 0.55 compression strength reduction factor represents an upper bound for well-understood soil conditions with quality workmanship. A lower value may be appropriate for less-well-understood soil conditions and construction quality control. Members with moments exceeding the equivalent of an eccentricity of 5 percent are designed using the strength reduction factors in Table 1.3.

c. Pile Caps

Piles and drilled shafts are generally arranged in groups or clusters, one group under each column. The group is capped by a spread footing or cap that distributes the

TABLE 15.4
Compressive strength reduction factors for deep foundations

Deep Foundation Type	Compressive Strength Reduction Factor ϕ
Uncased cast-in-place concrete drilled pier	0.55
Cast-in-place concrete pile in rock, a pipe, tube, or other permanent casing less than 0.25 in. thick, not meeting the confinement requirements discussed in Section 15.10a	0.60
Cast-in-place concrete filled steel pipe at least 0.25 in. thick	0.70
Metal cased concrete pile meeting the confinement requirements in Section 15.10a	0.65
Precast non-prestressed concrete pile	0.65
Prestressed precast concrete pile	0.65

column load to all piles in the group. These pile caps are in most ways very similar to footings on soil, except for two features. For one, reactions on caps act as concentrated loads at the individual piles, rather than as distributed pressures. For another, if the total of all pile reactions in a cluster is divided by the area of the footing to obtain an equivalent uniform pressure (for purposes of comparison only), it is found that this equivalent pressure is considerably higher in pile caps than for spread footings. This means that moments, and particularly shears, are also correspondingly larger, which requires greater footing depths than used for a spread footing of similar horizontal dimensions. To spread the load evenly to all piles, it is in any event advisable to provide ample rigidity, that is, depth, for pile caps.

Allowable bearing capacities of piles R_a are obtained from soil exploration, pile-driving energy, and test loadings, and their determination is not within the scope of the present book (see Refs. 15.1 to 15.4). As in spread footings, the effective portion of R_a available to resist the unfactored column loads is the allowable pile reaction less the weight of footing, backfill, and surcharge per pile. That is,

$$R_e = R_a - W_f \tag{15.12}$$

where W_f is the total weight of footing, fill, and surcharge divided by the number of piles.

Once the available or effective pile reaction R_e is determined, the number of piles in a concentrically loaded cluster is the integer next larger than

$$n = \frac{D + L}{R_e}$$

As far as the effects of wind, earthquake moments at the foot of the columns, and safety against overturning are concerned, design considerations are the same as described in Section 15.4 for spread footings. These effects generally produce an eccentrically loaded pile cluster in which different piles carry different loads. The number and location of piles in such a cluster are determined by successive approximations based on the requirement that the load on the most heavily loaded pile should not exceed the allowable pile reaction R_a. With a linear distribution of pile loads due to bending, the maximum pile reaction is

$$R_{\max} = \frac{P}{n} + \frac{M}{I_{pg}/c} \tag{15.13}$$

where P is the maximum load (including weight of cap, backfill, etc.) and M the moment to be resisted by the pile group, both referred to the bottom of the cap; I_{pg} is the moment of inertia of the entire pile group about the centroidal axis about which bending occurs; and c is the distance from that axis to the extreme pile. $I_{pg} = \Sigma_1^n(1 \times y_i^2)$; that is, it is the moment of inertia of n piles, each counting as one unit and located a distance y_i from the described centroidal axis.

Piles are generally arranged in tight patterns, which minimizes the cost of the caps, but they cannot be placed closer than conditions of driving and of undisturbed carrying capacity permit. A spacing of about 3 times the butt (top) diameter of the pile but no less than 2 ft 6 in. is customary. Commonly, piles with allowable reactions of 30 to 70 tons are spaced at 3 ft 0 in. (Ref. 15.8).

The *design* of footings on piles is similar to that of single-column footings. One approach is to design the cap for the pile reactions calculated for the factored column loads. For a concentrically loaded cluster, this would give $R_u = (1.2D + 1.6L)/n$. However, since the number of piles was taken as the next-larger integral according to Eq. (15.13), determining R_u in this manner can lead to a design where the strength of the cap is less than the capacity of the pile group. It is therefore recommended that the pile reaction for strength design be taken as

$$R_u = R_e \times \text{average load factor} \tag{15.14}$$

where the average load factor $= (1.2D + 1.6L)/(D + L)$. In this manner, the cap is designed to be capable of developing the full allowable capacity of the pile group. Details of a typical pile cap are shown in Fig. 15.23.

FIGURE 15.23

Typical single-column footing on piles (pile cap).

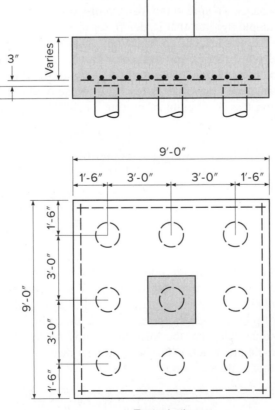

Typical pile cap

As in single-column spread footings, the depth of the pile cap is usually governed by shear. Shear design must follow the procedures for flat slabs and footings, as described in Section 15.6a. For closer spacings between piles and columns, the Code specifies either the use of the procedures described in Section 15.6a or the use of a three-dimensional strut-and-tie model (ACI Code Chapter 23) based on the principles described in Chapter 17. In the latter case, the interior struts must be designed without transverse reinforcement (Table 17.1) because of the difficulty of providing such reinforcement in a pile cap. The use of strut-and-tie methods to design pile caps is discussed in Ref. 15.18, but the work predates the 2019 Code requirements. See Problems 17.5 and 17.6.

When the procedures for flat slabs and footings are used, both punching or two-way shear and flexural or one-way shear need to be considered. The critical sections are the same as given in Section 15.6a. The difference is that shear in caps is caused by concentrated pile reactions rather than by distributed bearing pressures. This poses the question of how to calculate shear if the critical section intersects the circumference of one or more piles. For this case ACI Code 13.4.6 accounts for the fact that a pile reaction is not really a point load, but is distributed over the pile-bearing area. Correspondingly, for piles with diameters d_{pile}, the ACI Code stipulates as follows:

Calculation of shear on any section through a footing on piles shall be in accordance with the following:

(*a*) The entire reaction from any pile whose center is located $d_{pile}/2$ or more outside this section shall be considered as producing shear on that section.

(*b*) The reaction from any pile whose center is located $d_{pile}/2$ or more inside the section shall be considered as producing no shear on that section.

(*c*) For intermediate positions of the pile center, the portion of the pile reaction to be considered as producing shear on the section shall be based on straight-line interpolation between the full value at $d_{pile}/2$ outside the section and zero at $d_{pile}/2$ inside the section.

In addition to checking two-way and one-way shear, as just discussed, punching shear must also be investigated for the individual pile. Particularly in caps on a small number of heavily loaded piles, it is this possibility of a pile punching upward through the cap that may govern the required depth. The critical perimeter for this action, again, is located at a distance $d/2$ outside the upper edge of the pile. However, for relatively deep caps and closely spaced piles, critical perimeters around adjacent piles may overlap. In this case, fracture, if any, would undoubtedly occur along an outward-slanting surface around both adjacent piles. For such situations the critical perimeter is so located that its length is a minimum, as shown for two adjacent piles in Fig. 15.24.

FIGURE 15.24

Critical section for punching shear with closely spaced piles.

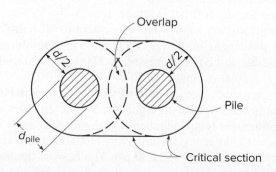

REFERENCES

15.1. R. B. Peck, W. E. Hanson, and T. H. Thornburn, *Foundation Engineering*, 2nd ed., John Wiley & Sons, New York, 1974.

15.2. K. Terzaghi, R. B. Peck, and G. Mesri, *Soil Mechanics in Engineering Practice*, 3rd ed., John Wiley & Sons, New York, 1996.

15.3. J. E. Bowles, *Foundation Analysis and Design*, 5th ed., McGraw-Hill, New York, 1996.

15.4. H.-Y. Fang, *Foundation Engineering Handbook*, 2nd ed., Springer Science and Business Media, 2013.

15.5. A. N. Talbot, "Reinforced Concrete Wall Footings and Column Footings," *Univ. Ill. Eng. Exp. Stn. Bull.* 67, 1913.

15.6. F. E. Richart, "Reinforced Concrete Wall and Column Footings," *J. ACI*, vol. 45, 1948, pp. 97 and 237.

15.7. E. Hognestad, "Shearing Strength of Reinforced Column Footings," *J. ACI*, vol. 50, 1953, p. 189.

15.8. F. Kramrisch, "Footings," chap. 5 in M. Fintel (ed.), *Handbook of Concrete Engineering*, 2nd ed., Van Nostrand Reinhold, New York, 1985.

15.9. ASCE–ACI Committee 426, "The Shear Strength of Reinforced Concrete Members—Slabs," *J. Struct. Div.*, ASCE, vol. 100, no. ST8, 1974, pp. 1543–1591.

15.10. ACI Committee 336, *Suggested Analysis and Design Procedures for Combined Footings and Mats* (Reapproved 2002), ACI 336.2R-88, American Concrete Institute, Farmington Hills, MI, 2002.

15.11. *International Building Code*, International Code Council, Washington, DC, 2020.

15.12. *Minimum Design Loads for Buildings and Other Structures*, ASCE /SEI 7-16, American Society of Civil Engineers, Reston, VA, 2016.

15.13. *Design and Performance of Mat Foundations—State of the Art Review*, SP-152, American Concrete Institute, Detroit, MI, 1995.

15.14. M. Hetenyi, *Beams on Elastic Foundations*, University of Michigan Press, Ann Arbor, 1946.

15.15. A. R. Anderson and S. E. Moustafa, "Ultimate Strength of Prestressed Concrete Piles and Columns," *ACI Journal & Proceedings*, vol 67, no. 8, 1970.

15.16. A. R. Anderson and S. E. Moustafa, "Dynamic Driving Stresses in Prestressed Concrete Piles," *Civil Engineering*, ASCE, New York, 1971.

15.17. *PCI Design Handbook*, 8th ed,. Precast and Prestressed Concrete Institute, Chicago, 2017.

15.18. P. Adebar, D. Kuchma, and M. P. Collins, "Strut-and-Tie Models for the Design of Pile Caps—An Experimental Study," *ACI Struct. J.*, vol. 87, no. 1, 1990, pp. 81–92.

PROBLEMS

15.1. A continuous strip footing is to be located concentrically under a 12 in. wall that delivers service loads $D = 25{,}000$ lb/ft and $L = 15{,}000$ lb/ft to the top of the footing. The bottom of the footing is 4 ft below the final ground surface. The soil has a density of 120 pcf and allowable bearing capacity of 8000 psf. Material strengths are $f_c' = 3000$ psi and $f_y = 60{,}000$ psi. Find (*a*) the required width of the footing, (*b*) the required effective and total depths, based on shear, and (*c*) the required flexural steel area.

15.2. An interior column for a tall concrete structure carries total service loads $D = 500$ kips and $L = 514$ kips. The column is 22×22 in. in cross section and is reinforced with 12 No. 11 (No. 36) bars centered 3 in. from the column faces (equal number of bars each face). For the column, $f_c' = 4000$ psi and $f_y = 60{,}000$ psi. The column is supported on a square footing, with the bottom of the footing 6 ft below grade. Design the footing, determining all concrete dimensions and amount and placement of all reinforcement, including length and placement of dowel steel. No shear reinforcement is permitted. The allowable soil bearing pressure is 8000 psf. Material strengths for the footing are $f_c' = 3000$ psi and $f_y = 60{,}000$ psi.

15.3. Design a single-column footing (including dowels) to support an 11 in. square column reinforced with eight No. 9 (No. 29) bars centered 2.5 in. from the column faces (equal number of bars on each face). The unfactored axial dead load = 135 kips, and the unfactored axial live load = 125 kips. For the column, $f_c' = 4000$ psi and $f_y = 60{,}000$ psi. The base of the footing is 3 ft below grade.

The allowable soil bearing pressure is 3000 lb/ft². Material strengths for the footing are $f'_c = 3000$ psi and $f_y = 60,000$ psi.

15.4. Two interior columns for a high-rise concrete structure are spaced 15 ft apart, and each carries service loads $D = 500$ kips and $L = 514$ kips. The columns are to be 22 in. square in cross section, and each is reinforced with 12 No. 11 (No. 36) bars centered 3 in. from the column faces, with an equal number of bars at each face. For the columns, $f'_c = 4000$ psi and $f_y = 60,000$ psi. The columns are supported on a rectangular combined footing with a long-side dimension twice that of the short side. The allowable soil bearing pressure is 8000 psf. The bottom of the footing is 6 ft below grade. Design the footing for these columns, using $f'_c = 4000$ psi and $f_y = 60,000$ psi. Specify all reinforcement, including length and placement of footing bars and dowel steel.

15.5. A pile cap is to be designed to distribute a concentric force from a single column to a nine-pile group, with geometry as shown in Fig. 15.23. The cap carries calculated dead and service live loads of 280 and 570 kips, respectively, from a 19 in. square concrete column reinforced with six No. 14 (No. 43) bars. The permissible load per pile at service load is 100 kips, and the pile diameter is 16 in. Find the required effective and total depths of the pile cap and the required reinforcement. Check all relevant aspects of the design, including the development length for the reinforcement and transfer of forces at the base of the column. Material strengths for the column are $f'_c = 4000$ psi and $f_y = 60,000$ psi and for the pile cap are $f'_c = 4000$ psi and $f_y = 60,000$ psi.

15.6. Complete the design of the strap footing in Example 15.4 and determine all dimensions and reinforcement.

16 *Retaining Walls*

16.1 FUNCTION AND TYPES OF RETAINING WALLS

Retaining walls are used to hold back masses of earth or other loose material where conditions make it impossible to let those masses assume their natural slopes. Such conditions occur when the width of an excavation, cut, or embankment is restricted by conditions of ownership, use of the structure, or economy. For example, in railway or highway construction the width of the right of way is fixed, and the cut or embankment must be contained within that width. Similarly, the basement walls of buildings must be located within the property and must retain the soil surrounding the basement.

Freestanding retaining walls, as distinct from those that form parts of structures, such as basement walls, are of various types, the most common of which are shown in Fig. 16.1. The gravity wall (Fig. 16.1a) retains the earth entirely by its own weight and generally contains no reinforcement. The reinforced concrete cantilever wall (Fig. 16.1b) consists of the vertical arm that retains the earth and is held in position by a footing or base slab. In this case, the weight of the fill on top of the heel, in addition to the weight of the wall, contributes to the stability of the structure. Since the arm represents a vertical cantilever, its required thickness increases rapidly with increasing height. To reduce the bending moments in vertical walls of great height, counterforts are used spaced at distances from each other equal to or slightly larger than one-half of the height (Fig. 16.1c). Property rights or other restrictions sometimes make it necessary to place the wall at the forward edge of the base slab, that is, to omit the toe. Whenever it is possible, toe extensions of one-third to one-fourth of the width of the base provide a more economical solution.

Which of the three types of walls is appropriate in a given case depends on a variety of conditions, such as local availability and price of construction materials and property rights. In general, gravity walls are economical only for relatively low walls, possibly up to about 10 ft. Cantilever walls are economical for heights from 10 to 20 ft, while counterforts are used for greater heights.

16.2 EARTH PRESSURE

In terms of physical behavior, soils and other granular masses occupy a position intermediate between liquids and solids. If sand is poured from a dump truck, it flows, but, unlike a frictionless liquid, it does not assume a horizontal surface. It maintains itself in a stable heap with sides reaching an *angle of repose*, the tangent of which is roughly equal to the coefficient of intergranular friction. If a pit is dug in clay soil, its

FIGURE 16.1
Types of retaining walls and
back drains: (*a*) gravity wall;
(*b*) cantilever wall; and
(*c*) counterfort wall.

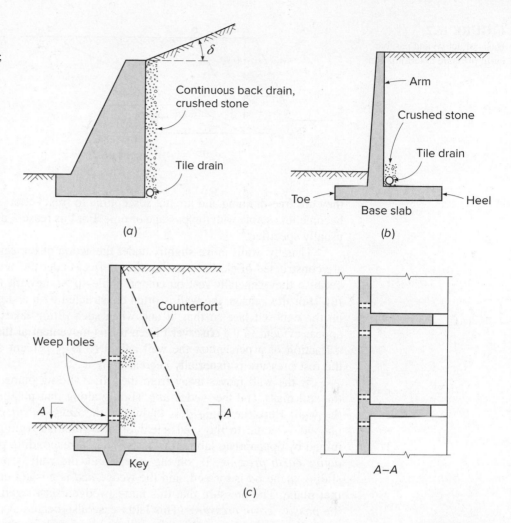

(*a*)

(*b*)

(*c*)

sides can usually be made vertical over considerable depths without support; that is, the clay behaves as a solid and retains the shape it is given. If, however, the pit is flooded, the sides will give way, and in many cases the saturated clay will be converted nearly into a true liquid. The clay is capable of maintaining its shape by means of its internal cohesion, but flooding reduces that cohesion greatly, often to zero.

If a wall is built in contact with a solid, such as a rock face, no pressure is exerted on it. If, on the other hand, a wall retains a liquid, as in a reservoir, it is subject at any level to the hydrostatic pressure $w_w h$, where w_w is the unit weight of the liquid and h is the distance from the surface. If a vertical wall retains soil, the earth pressure similarly increases proportionally to the depth, but its magnitude is

$$p_h = K_0 w h \tag{16.1}$$

where w is the unit weight of the soil and K_0 is a constant known as the *coefficient of earth pressure at rest*. The value of K_0 depends not only on the nature of the backfill but also on the method of depositing and compacting it. It has been determined experimentally that, for uncompacted noncohesive soils such as sands and gravels, K_0 ranges between 0.4 and 0.5, while it may be as high as 0.8 for the same soils in a highly compacted state (Refs. 16.1 through 16.3). For cohesive soils, K_0 may be on the order of 0.7 to 1.0. Clean sands and gravels are considered superior to all other soils because

FIGURE 16.2
Basis of active and passive
earth pressure determination.

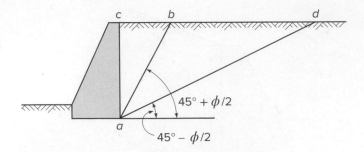

they are free-draining and are not susceptible to frost action and because they do not become less stable with the passage of time. For this reason, noncohesive backfills are usually specified.

Usually, walls move slightly under the action of the earth pressure. Since walls are constructed of elastic material, they deflect under the action of the pressure, and because they generally rest on compressible soils, they tilt and shift away from the fill. (For this reason, the wall is often constructed with a slight batter toward the fill on the exposed face so that, if and when such tilting takes place, the tilt does not appear evident to the observer.) Even if this movement at the top of the wall is only a fraction of a percent of the wall height ($\frac{1}{2}$ to $\frac{1}{10}$ percent according to Ref. 16.2), the rest pressure is materially decreased by it.

If the wall moves away from the fill, a sliding plane ab (Fig. 16.2) forms in the soil mass, and the wedge abc, sliding along that plane, exerts pressure against the wall. Here the angle ϕ is known as the *angle of internal friction*; that is, its tangent is equal to the coefficient of intergranular friction, which can be determined by appropriate laboratory tests. The corresponding pressure is known as the *active earth pressure*. If, on the other hand, the wall is pushed against the fill, a sliding plane ad is formed, and the wedge acd is pushed upward by the wall along that plane. The pressure that this larger wedge exerts against the wall is known as the *passive earth pressure*. (This latter case also occurs at the left face of the gravity wall in Fig. 16.1a when this wall yields slightly to the left under the pressure of the fill.)

The magnitude of these pressures has been analyzed by Rankine, Coulomb, and others. If the soil surface makes an angle δ with the horizontal (Fig. 16.1a), then, according to Rankine, the *coefficient for active earth pressure* is

$$K_a = \cos \delta \, \frac{\cos \delta - \sqrt{\cos^2 \delta - \cos^2 \phi}}{\cos \delta + \sqrt{\cos^2 \delta - \cos^2 \phi}} \tag{16.2}$$

and the *coefficient for passive pressure* is

$$K_p = \cos \delta \, \frac{\cos \delta + \sqrt{\cos^2 \delta - \cos^2 \phi}}{\cos \delta - \sqrt{\cos^2 \delta - \cos^2 \phi}} \tag{16.3}$$

K_a and K_p replace K_0 in Eq. (16.1) to determine soil pressure p_h under active and passive conditions, respectively.

For the frequent case of a horizontal surface, that is, $\delta = 0$ (Fig. 16.2), for active pressure,

$$K_{ah} = \frac{1 - \sin \phi}{1 + \sin \phi} \tag{16.4}$$

and for passive pressure,

$$K_{ph} = \frac{1 + \sin \phi}{1 - \sin \phi} \tag{16.5}$$

Rankine's theory is valid only for noncohesive soils such as sand and gravel but, with corresponding adjustments, can also be used successfully for cohesive clay soils.

From Eqs. (16.1) through (16.5), it is seen that the earth pressure at a given depth h depends on the inclination of the surface δ, the unit weight w, and the angle of friction ϕ. The first two of these are easily determined, while little agreement has yet been reached as to the proper values of ϕ. For the ideal case of a dry, noncohesive fill, ϕ could be determined by laboratory tests and then used in the formulas. This is impossible for clays, only part of whose resistance is furnished by intergranular friction, while the rest is due to internal cohesion. For this reason, their actual ϕ values are often increased by an arbitrary amount to account implicitly for the added cohesion. However, this is often unsafe since, as was shown by the example of the flooded pit, cohesion may vanish almost completely due to saturation and inundation.

In addition, fills behind retaining walls are rarely uniform, and, what is more important, they are rarely dry. Proper drainage of the fill is vitally important to reduce pressures (see Section 16.6), but even in a well-drained fill, the pressure temporarily increases during heavy storms or sudden thaws. This is so because even though the drainage may successfully remove the water as fast as it appears, its movement through the fill toward the drains causes additional pressure (seepage pressure). In addition, frost action and other influences may temporarily increase its value over that of the theoretical active pressure. Many walls that were designed without regard to these factors have failed, been displaced, or cracked.

It is good practice, therefore, to select conservative values for ϕ, considerably smaller than the actual test values, in all cases except where extraordinary and usually expensive precautions are taken to keep the fill dry under all conditions. An example of recommended earth pressure values, which are quite conservative, though based on extensive research and practical experience, can be found in Ref. 16.2. Less conservative values are often used in practical designs, but these should be employed (1) with caution in view of the fact that occasional trouble has been encountered with walls so designed and (2) preferably with the advice of a geotechnical engineer.

Table 16.1 gives representative values for w and ϕ often used in engineering practice. (Note that the ϕ values do not account for probable additional pressures due to porewater, seepage, frost, etc.) The table also contains values for the coefficient of

TABLE 16.1
Unit weights w, effective angles of internal friction ϕ, and coefficients of friction with concrete f

Soil	Unit Weight w, pcf	ϕ, deg	f
1. Sand or gravel without fine particles, highly permeable	110–120	33–40	0.5–0.6
2. Sand or gravel with silt mixture, low permeability	120–130	25–35	0.4–0.5
3. Silty sand, sand and gravel with high clay content	110–120	23–30	0.3–0.4
4. Medium or stiff clay	100–120	25–35[a]	0.2–0.4
5. Soft clay, silt	90–110	20–25[a]	0.2–0.3

[a] For saturated conditions, ϕ for clays and silts may be close to zero.

friction f between concrete and various soils. The values of ϕ for soils 3 through 5 may be quite unconservative; under saturated conditions, clays and silts may become entirely liquid (that is, $\phi = 0$). Soils of type 1 or 2 should be used as backfill for retaining walls wherever possible.

16.3   EARTH PRESSURE FOR COMMON CONDITIONS OF LOADING

In calculating earth pressures on walls, three common conditions of loading are most often met: (1) horizontal surface of fill at the top of the wall, (2) inclined surface of fill sloping up and back from the top of the wall, and (3) horizontal surface of fill carrying a uniformly distributed additional load (surcharge), such as from goods in a storage yard or traffic on a road.

The increase in pressure caused by uniform surcharge s (case 3) is calculated by converting its load into an equivalent, imaginary height of earth h' above the top of the wall such that

$$h' = \frac{s}{w} \tag{16.6}$$

and measuring the depth to a given point on the wall from this imaginary surface. This amounts to replacing h with $h + h'$ in Eq. (16.1).

The distributions of pressure for cases 1 to 3 are shown in Fig. 16.3. The total earth thrust P per linear foot of wall is equal to the area under the pressure distribution figure, and its line of action passes through the centroid of the pressure. Figure 16.3 gives information, calculated in this manner, on magnitude, point of action, and direction of P for these three cases.

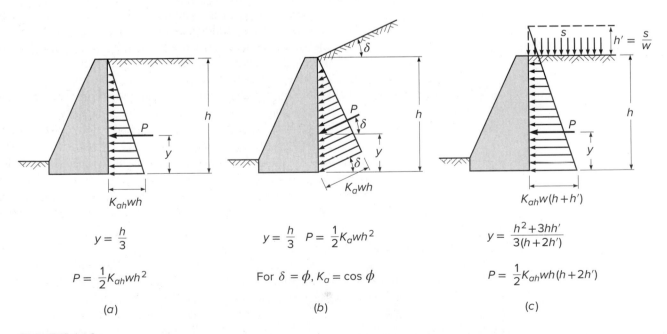

$$y = \frac{h}{3}$$

$$P = \frac{1}{2}K_{ah}wh^2$$

(a)

$$y = \frac{h}{3} \quad P = \frac{1}{2}K_awh^2$$

For $\delta = \phi$, $K_a = \cos \phi$

(b)

$$y = \frac{h^2 + 3hh'}{3(h + 2h')}$$

$$P = \frac{1}{2}K_{ah}wh(h + 2h')$$

(c)

FIGURE 16.3
Earth pressures for (a) horizontal surface; (b) sloping surface; and (c) horizontal surface with surcharge s.

Occasionally retaining walls must be built for conditions in which the groundwater level is above the base of the wall, either permanently or seasonally. In that case, the pressure of the soil *above* groundwater is determined as usual. The part of the wall *below* groundwater is subject to the sum of the water pressure and the earth pressure. The former is equal to the full hydrostatic pressure $p_w = w_w h_w$, where w_w and h_w are, respectively, the unit weight of water and the distance from the groundwater level to the point on the wall. The additional pressure of the soil below the groundwater level is calculated from Eq. (16.1), where, however, for the portion of the soil below water, w is replaced with $w - w_w$, while h, as usual, is measured from the soil surface. That is, for submerged soil, buoyancy reduces the effective weight in the indicated manner. Pressures of this magnitude, which are considerably larger than those of drained soil, also occur temporarily after heavy rainstorms or thaws in walls without provision for drainage, or if drains have become clogged.

The seeming simplicity of the determination of earth pressure, as indicated here, should not lull the designer into a false sense of security and certainty. No theory is more accurate than the assumptions on which it is based. Actual soil pressures are affected by irregularities of soil properties, porewater and drainage conditions, and climatic and other factors that cannot be expressed in formulas. This situation, on the one hand, indicates that involved refinements of theoretical earth pressure determinations, as sometimes attempted, are of little practical value. On the other hand, the design of a retaining wall is seldom a routine procedure, since the local conditions that affect pressures and safety vary from one locality to another.

16.4   EXTERNAL STABILITY

A wall may fail in two different ways: (1) its individual parts may not be strong enough to resist the acting forces, such as when a vertical cantilever wall is cracked by the earth pressure acting on it, and (2) the wall as a whole may be bodily displaced by the earth pressure, without breaking up internally. To design against the first possibility requires the determination of the necessary dimensions, thicknesses, and reinforcement to resist the moments and shears; this procedure, then, is in no way different from that of determining required dimensions and reinforcement of other types of concrete structures. The usual load factors and strength reduction factors of the ACI Code may be applied (see Section 16.5).

To safeguard the wall against solid body displacements, that is, to ensure its external stability, requires special consideration. Consistent with current practice in geotechnical engineering, the stability investigation is based on actual earth pressures (as nearly as they may be determined) and on calculated or estimated service dead and live loads, all without load factors. Calculated bearing pressures are compared with allowable values, and overall factors of safety evaluated by comparing resisting forces to maximum loads acting under service conditions.

A wall, such as that in Fig. 16.4, together with the soil mass *ijkl* that rests on the base slab, may be bodily displaced by the earth thrust P that acts on the plane *ak* by *sliding* along the plane *ab*. Such sliding is resisted by the friction between the soil and the footing along the same plane. To forestall motion, the forces that resist sliding must exceed those that tend to produce sliding; a factor of safety of 1.5 is generally assumed satisfactory in this connection.

In Fig. 16.4, the force that tends to produce sliding is the horizontal component P_h of the total earth thrust P. The resisting friction force is fR_v, where f is the coefficient of friction between concrete and soil (see Table 16.1) and R_v is the vertical

FIGURE 16.4
External stability of a
cantilever wall.

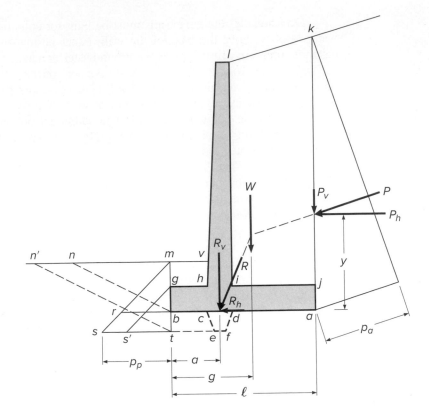

component of the total resultant R; that is, $R_v = W + P_v$ (W = weight of wall plus soil resting on the footing, P_v = vertical component of P). Hence, to provide sufficient safety,

$$f(W + P_v) \geq 1.5P_h \qquad (16.7)$$

Actually, for the wall to slide to the left, it must push with it the earth *nmb*, which gives rise to the passive earth pressure indicated by the triangle *rmb*. This passive pressure represents a further resisting force that could be added to the left side of Eq. (16.7). However, this should be done only if the proper functioning of this added resistance is ensured. For that purpose, the fill *ghvm* must be placed before the backfill *ijkl* is put in place and must be secure against later removal by scour or other means throughout the lifetime of the wall. If these conditions are not met, it is better not to count on the additional resistance of the passive pressure.

   If the required sliding resistance cannot be developed by these means, a key wall *cdfe* can be used to increase horizontal resistance. In this case, sliding, if it occurs, takes place along the planes *ad* and *tf*. While along *ad* and *ef*, the friction coefficient f applies, sliding along *te* occurs within the soil mass. The coefficient of friction that applies in this portion is consequently $\tan \phi$, where the value of ϕ may be taken from the next to last column in Table 16.1. In this situation sliding of the front soil occurs upward along *tn'* so that if the front fill is secure, the corresponding resistance from passive soil pressure is represented by the pressure triangle *stm*. If doubt exists as to the reliability of the fill above the toe, the free surface should more conservatively be assumed at the top level of the footing, in which case the passive pressure is represented by the triangle *s'tg*.

Next, it is necessary to ensure that the pressure under the footing not exceed the *permissible bearing pressure* for the particular soil. Let a (Fig. 16.4) be the distance from the front edge b to the intersection of the resultant with the base plane, and let R_v be the vertical component of R. (This intersection need not be located beneath the vertical arm, as shown, even though an economical wall generally results if it is so located.) Then the base plane ab, 1 ft wide longitudinally, is subject to a normal force R_v and to a moment about the centroid $(\ell/2 - a)R_v$. When these values are substituted in the usual formula for bending plus axial force

$$q_{\substack{max \\ min}} = \frac{N}{A} \pm \frac{Mc}{I} \tag{16.8}$$

it is found that if the resultant is located within the middle third $(a > \ell/3)$, compression acts throughout the section, and the maximum and minimum pressures can be calculated from the equations in Fig. 16.5a. If the resultant is located just at the edge of the middle third $(a = \ell/3)$, the pressure distribution is as shown in Fig. 16.5b, and Eq. (16.8) results in the formula given there.

FIGURE 16.5

Bearing pressures for different locations of resultant.

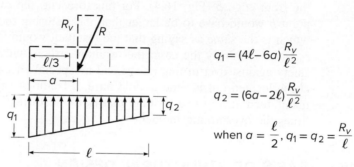

$$q_1 = (4\ell - 6a)\frac{R_v}{\ell^2}$$

$$q_2 = (6a - 2\ell)\frac{R_v}{\ell^2}$$

when $a = \dfrac{\ell}{2}$, $q_1 = q_2 = \dfrac{R_v}{\ell}$

(*a*) Resultant in middle third

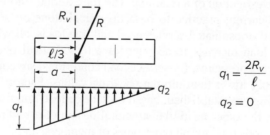

$$q_1 = \frac{2R_v}{\ell}$$

$$q_2 = 0$$

(*b*) Resultant at edge of middle third

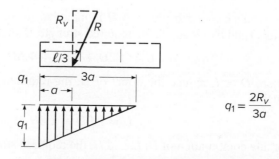

$$q_1 = \frac{2R_v}{3a}$$

(*c*) Resultant outside middle third

If the resultant were located outside the middle third ($a < \ell/3$), Eq. (16.8) would indicate tension at and near point a. Obviously, tension cannot be developed between soil and a concrete footing that merely rests on it. Hence, in this case the pressure distribution of Fig. 16.5c develops, which implies a slight lifting off the soil of the rear part of the footing. Equilibrium requires that R_v pass through the centroid of the pressure distribution triangle, from which the formula for q_1 for this case can easily be derived.

It is good practice, in general, to have the resultant located within the middle third. This not only reduces the magnitude of the maximum bearing pressure but also prevents too large a nonuniformity of pressure. If the wall is founded on a highly compressible soil, such as certain clays, a pressure distribution as in Fig. 16.5b results in a much larger settlement of the toe than of the heel, with a corresponding tilting of the wall. In a foundation on such a soil, the resultant, therefore, should strike at or very near the center of the footing. If the foundation is on very incompressible soil, such as well-compacted gravel or rock, the resultant can be allowed to fall outside the middle third (Fig. 16.5c).

A third mode of failure is the possibility of the wall *overturning* bodily around the front edge b (Fig. 16.4). For this to occur, the overturning moment yP_h about point b would have to be larger than the restoring moment $Wg + P_y\ell$ in Fig. 16.4, which is the same as saying that the resultant would have to strike outside the edge b. If, as is mostly the case, the resultant strikes within the middle third, adequate safety against overturning exists, and no special check need be made. If the resultant is located outside the middle third, a factor of safety of at least 1.5 should be maintained against overturning; that is, the restoring moment should be at least 1.5 times the overturning moment.

16.5 BASIS OF STRUCTURAL DESIGN

In the investigation of a retaining wall for external stability, described in Section 16.4, it is the current practice to base the calculations on actual earth pressures, and on calculated or estimated service dead and live loads, all with load factors of 1.0 (that is, without load increase to account for a hypothetical overload condition). Calculated soil bearing pressures, for service load conditions, are compared with allowable values set suitably lower than ultimate bearing values. Factors of safety against overturning and sliding are established, based on service load conditions.

On the other hand, the structural design of a retaining wall should be consistent with methods used for all other types of members, and thus should be based on factored loads in recognition of the possibility of an increase above service loading. ACI Code load factors relating to structural design of retaining walls are summarized as follows:

1. If resistance to earth pressure H is included in the design, together with dead loads D and live loads L, the required strength U shall be at least equal to

$$U = 1.2D + 1.6L + 1.6H$$

2. Where D or L reduce the effect of H, the required strength U shall be at least equal to

$$U = 0.9D + 1.6H$$

3. For any combination of D, L, and H, the required strength shall not be less than

$$U = 1.2D + 1.6L$$

While the ACI Code approach to load factor design is logical and relatively easy to apply to members in buildings, its application to structures that are to resist earth pressures is not so easy. Many alternative combinations of factored dead and live loads and lateral pressures are possible. Dead loads such as the weight of the concrete should be multiplied by 0.9 where they reduce design moments, such as for the toe slab of a cantilevered retaining wall, but should be multiplied by 1.2 where they increase moments, such as for the heel slab. The vertical load of the earth over the heel should be multiplied by 1.6. Obviously, no two factored load states could be obtained concurrently. For each combination of factored loads, different reactive soil pressures are produced under the structure, requiring a new determination of those pressures for each alternative combination. Furthermore, there is no reason to believe that soil pressure would continue to be linearly distributed at the overload stage, or would increase in direct proportion to the load increase; knowledge of soil pressure distributions at incipient failure is incomplete. Necessarily, a somewhat simplified view of load factor design must be adopted in designing retaining walls.

Following the ACI Code, lateral earth pressures are multiplied by a load factor of 1.6. In general, the reactive pressure of the soil under the structure at the factored load stage is taken equal to 1.6 times the soil pressure found for service load conditions in the stability analysis.[†] For cantilever retaining walls, the calculated dead load of the toe slab, which causes moments acting in the opposite sense to those produced by the upward soil reaction, is multiplied by a factor of 0.9. For the heel slab, the required moment capacity is based on the dead load of the heel slab itself and is multiplied by 1.2, while the downward load of the earth is multiplied by 1.6. Surcharge, if present, is treated as live load with a load factor of 1.6. The upward pressure of the soil under the heel slab is taken equal to zero, recognizing that for the severe overload stage a nonlinear pressure distribution is probably obtained, with most of the reaction concentrated near the toe. Similar assumptions appear to be reasonable in designing counterfort walls.

In accordance with ACI Code Chapter 13, cantilever retaining walls are designed following the flexural design provisions covered in Chapter 12, with minimum horizontal (transverse) reinforcement provided in accordance with ACI Code 7.6.1, which stipulates minimum reinforcement equal to $0.0018A_g$ for deformed bars and welded wire reinforcement, where A_g is the gross area of the wall.

16.6 DRAINAGE AND OTHER DETAILS

Such failures or damage to retaining walls as have occasionally occurred were due, in most cases, to one of two causes: overloading of the soil under the wall with consequent forward tipping or insufficient drainage of the backfill. In the latter case, hydrostatic pressure from porewater accumulated during or after rainstorms greatly increases the thrust on the wall; in addition, in subfreezing weather, ice pressure of considerable magnitude can develop in such poorly drained soils. The two causes are often interconnected, since large thrusts correspondingly increase the bearing pressure under the footing.

[†] These reactions are caused by the assumed factored load condition and have no direct relationship to ultimate soil bearing values or pressure distributions.

Allowable bearing pressures should be selected with great care. It is necessary, for this purpose, to investigate not only the type of soil immediately underlying the footing but also the deeper layers. Unless reliable information is available at the site, subsurface borings should be made to a depth at least equal to the height of the wall. The foundation must be laid below *frost depth*, which amounts to 4 to 5 ft and more in the northern states, to ensure against heaving by the freezing of soils containing moisture.

Drainage can be provided in various ways. *Weep holes* consisting of 6 or 8 in. pipe embedded in the wall, as shown in Fig. 16.1c, are usually spaced horizontally at 5 to 10 ft. In addition to the bottom row, additional rows should be provided in walls of substantial height. To facilitate drainage and prevent clogging, 1 ft³ or more of crushed stone is placed at the rear end of each weeper. Care must be taken that the outflow from the weep holes is carried off safely so as not to seep into and soften the soil underneath the wall. To prevent this, instead of weepers, *longitudinal drains* embedded in crushed stone or gravel can be provided along the rear face of the wall (Fig. 16.1b) at one or more levels; the drains discharge at the ends of the wall or at a few intermediate points. The most efficient drainage is provided by a *continuous backdrain* consisting of a layer of gravel or crushed stone covering the entire rear face of the wall (Fig. 16.1a), with discharge at the ends. Such drainage is expensive, however, unless appropriate material is cheaply available at the site. Wherever possible, the surface of the fill should be covered with a layer of low permeability and, in the case of a horizontal surface, should be laid with a slight slope away from the wall toward a gutter or other drainage.

In long walls, provision must be made against damage caused by *expansion* or *contraction* from temperature changes and shrinkage. The AASHTO *LRFD Bridge Design Specifications* (Ref. 16.4) require that for gravity walls, as well as reinforced concrete walls, expansion joints be placed at intervals of 90 ft or less, and contraction joints at not more than 30 ft (Ref. 16.4). The same specifications provide that, in reinforced concrete walls, temperature reinforcement equal to 0.0018bh in both the vertical and horizontal directions be distributed uniformly on the exposed (including end) surfaces. This AASHTO requirement is expressed as an area of reinforcement per foot on each face equal to

$$A_s \geq \frac{1.30bh}{2(b + h)f_y} \tag{16.9a}$$

$$0.11 \leq A_s \leq 0.60 \text{ in}^2/\text{ft} \tag{16.9b}$$

where b = least width of the component, h = least thickness of the component, and f_y = yield strength of the bars, expressed in ksi, ≤ 75 ksi. Similar provisions for railroad structures are found in Ref. 16.5.

16.7   EXAMPLE: DESIGN OF A GRAVITY RETAINING WALL

A gravity wall is to retain a bank 11 ft 6 in. high whose horizontal surface is subject to a live load surcharge of 400 psf. The soil is a sand and gravel mixture with a rather moderate amount of fine, silty particles. It can, therefore, be assumed to be in class 2 of Table 16.1 with the following characteristics: unit weight $w = 120$ pcf, $\phi = 30°$ (with adequate drainage to be provided), and base friction coefficient $f = 0.5$. With $\sin 30° = 0.5$, from Eqs. (16.4) and (16.5), the soil pressure coefficients are $K_{ah} = 0.333$ and $K_{ph} = 3.0$. The allowable bearing pressure is assumed to be 8000 psf. This coarse-grained soil has little compressibility, so that the resultant can be allowed to strike near the outer-third point (see Section 16.4). The weight of the concrete is $w_c = 150$ pcf.

FIGURE 16.6
Gravity retaining wall.

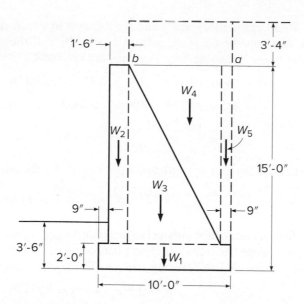

The optimum design of any retaining wall is a matter of successive approximation. Reasonable dimensions are assumed based on experience, and the various conditions of stability are checked for these dimensions. On the basis of a first trial, dimensions are readjusted, and one or two additional trials usually result in a favorable design. In the following, only the final design is analyzed in detail. The final dimensions are shown in Fig. 16.6.

The equivalent height of surcharge is $h' = 400/120 = 3.33$ ft. From Fig. 16.3c the total earth thrust is

$$P = \tfrac{1}{2} \times 0.333 \times 120 \times 15 \times 21.67 = 6500 \text{ lb}$$

and its distance from the base is $y = (225 + 150)/(3 \times 21.67) = 5.77$ ft. Hence, the overturning moment $M_o = 6500 \times 5.77 = 37{,}500$ ft-lb. To calculate the weight W and its restoring moment M_r about the edge of the toe, individual weights are taken, as shown in Fig. 16.6. With x representing the distance of the line of action of each subweight from the front edge, the following calculation results:

Component Weights	W, lb	x, ft	$M_r = xW$, ft-lb
W_1: $10 \times 2 \times 150$	3,000	5.0	15,000
W_2: $1.5 \times 13 \times 150$	2,930	1.5	4,400
W_3: $7/2 \times 13 \times 150$	6,830	4.58	31,300
W_4: $7/2 \times 13 \times 120$	5,460	6.92	37,800
W_5: $0.75 \times 13 \times 120$	1,170	9.63	11,270
Total	19,390		99,770

The distance of the resultant from the front edge is

$$a = \frac{99{,}770 - 37{,}500}{19{,}390} = 3.21 \text{ ft}$$

which is just outside the middle third. The safety factor against overturning, $99{,}770/37{,}500 = 2.66$, is ample. From Fig. 16.5c the maximum soil pressure is $q = (2 \times 19{,}390)/(3 \times 3.21) = 4030$ psf.

These calculations are made for the case in which the surcharge extends only to the rear edge of the wall, point a of Fig. 16.6. If the surcharge extends forward to point b, the following modifications are obtained:

$$W = 19{,}390 + 400 \times 7.75 = 22{,}490 \text{ lb}$$

$$M_r = 99{,}770 + 400 \times 7.75 \times 6.13 = 118{,}770 \text{ ft-lb}$$

$$a = \frac{118{,}770 - 37{,}500}{22{,}490} = 3.61 \text{ ft}$$

This is inside the middle third, and from Fig. 16.5a, the maximum bearing pressure is

$$q_1 = \frac{(40.0 - 21.7)22{,}490}{100} = 4120 \text{ psf}$$

The situation most conducive to sliding occurs when the surcharge extends only to point a, since additional surcharge between a and b would increase the total weight and the corresponding resisting friction. The friction force is

$$F = 0.5 \times 19{,}390 = 9695 \text{ lb}$$

Additionally, sliding is resisted by the passive earth pressure on the front of the wall. Although the base plane is 3.5 ft below grade, the top layer of soil cannot be relied upon to furnish passive pressure, since it is frequently loosened by roots and the like, or it could be scoured out by cloudbursts. For this reason, the top 1.5 ft is discounted in calculating the passive pressure, which then becomes

$$P_p = \tfrac{1}{2}wh^2 K_{ph} = \tfrac{1}{2} \times 120 \times 2^2 \times 3.0 = 720 \text{ lb}$$

The safety factor against sliding, $(9695 + 720)/6500 = 1.6$, is but slightly larger than the required value 1.5, indicating a favorable design. Ignoring the passive pressure gives a safety factor of 1.49, which is very close to the acceptable value.

16.8 EXAMPLE: DESIGN OF A CANTILEVER RETAINING WALL

A cantilever wall is to be designed for the situation of the gravity wall in Section 16.7. Concrete with $f_c' = 4500$ psi and steel with $f_y = 60{,}000$ psi is used.

a. Preliminary Design

To facilitate calculation of weights for checking the stability of the wall, it is advantageous first to ascertain the thickness of the arm and the footing.[†] For this purpose the thickness of the footing is roughly estimated, and then the required thickness of the arm is determined at its bottom section. With the bottom of the footing at 3.5 ft below grade and an estimated footing thickness of 1.5 ft, the free height of the arm is 13.5 ft. Hence, with respect to the bottom of the arm (see Fig. 16.3c),

$$P = \tfrac{1}{2} \times 0.333 \times 120 \times 13.5 \times 20.16 = 5440 \text{ lb}$$

$$y = \frac{183 + 135}{3 \times 20.16} = 5.25 \text{ ft}$$

$$M_u = 1.6 \times 5440 \times 5.25 = 45{,}700 \text{ ft-lb}$$

[†] Valuable guidance is provided for the designer in tabulated designs such as those found in Ref. 16.6 and by the sample calculations in Ref. 16.7.

For the given grades of concrete and steel, the maximum practical reinforcement ratio $\rho_{max} = 0.0197$. For economy and ease of bar placement, a ratio of about 40 percent of the maximum, or 0.008, is selected. Then from Graph A.1b of Appendix A,

$$\frac{M_u}{\phi b d^2} = 430$$

For a unit length of the wall ($b = 12$ in.), with $\phi = 0.90$, the required effective depth is

$$d = \sqrt{\frac{45,700 \times 12}{0.90 \times 12 \times 430}} = 10.9 \text{ in.}$$

A protective cover of 2 in. is required for concrete exposed to earth. Thus, estimating the bar diameter to be 1 in., the minimum required thickness of the arm at the base is 13.4 in. This is increased to 16 in., giving $d = 13.5$ in., because the cost of the extra concrete in such structures is usually more than balanced by the simultaneous saving in steel and ease of concrete placement. The arm is then checked for shear at the base of the wall at the intersection with the supporting slab per ACI Code 13.3.6.3, or 13.5 ft below the top of the wall:

$$P = \tfrac{1}{2} \times 0.333 \times 120 \times 13.5 \times 19.16 = 5170 \text{ lb}$$

$$V_u = 1.6 \times 5170 = 8270 \text{ lb}$$

The factored shear force is approximately $\phi 1.0\sqrt{f_c'}b_w d$. This is low enough to proceed with the design. Because the shear strength is dependent on the flexural reinforcement ratio and the size effect factor, the final check of shear strength is conducted after the flexural design.

The thickness of the base is usually the same as or slightly larger than that at the bottom of the arm. Hence, the estimated 1.5 ft need not be revised. Since the moment in the arm decreases with increasing distance from the base and is zero at the top, the arm thickness at the top is made 8 in. It is now necessary to assume lengths of heel and toe slabs and to check the stability for these assumed dimensions. Intermediate trials are omitted here, and the final dimensions are shown in Fig. 16.7a. Trial calculations have shown that safety against sliding can be achieved only by an excessively long heel or by a key. The latter, requiring the smaller concrete volume, has been adopted.

b. Stability Investigation

Weights and moments about the front edge are as follows:

Component Weights	W, lb	x, ft	M_r, ft-lb
W_1: $0.67 \times 13.5 \times 150$	1,360	4.08	5,550
W_2: $0.67 \times 0.5 \times 13.5 \times 150$	680	4.67	3,180
W_3: $9.75 \times 1.5 \times 150$	2,190	4.88	10,700
W_4: $1.33 \times 1.25 \times 150$	250	4.42	1,100
W_5: $3.75 \times 2 \times 120$	900	1.88	1,690
W_6: $0.67 \times 0.5 \times 13.5 \times 120$	540	4.86	2,620
W_7: $4.67 \times 13.5 \times 120$	7,570	7.42	56,200
Total	13,490		81,040

FIGURE 16.7
Cantilever retaining wall:
(*a*) cross section; (*b*) bearing
pressure with surcharge
to *a*; (*c*) bearing pressure
with surcharge to *b*;
(*d*) reinforcement; and
(*e*) moment variation
with height.

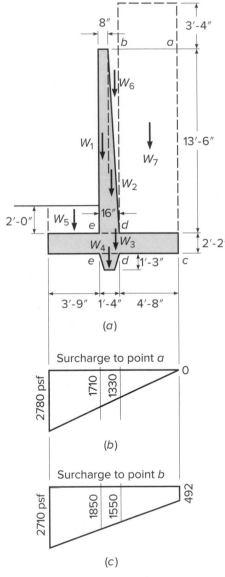

(*a*)

Surcharge to point *a*

(*b*)

Surcharge to point *b*

(*c*)

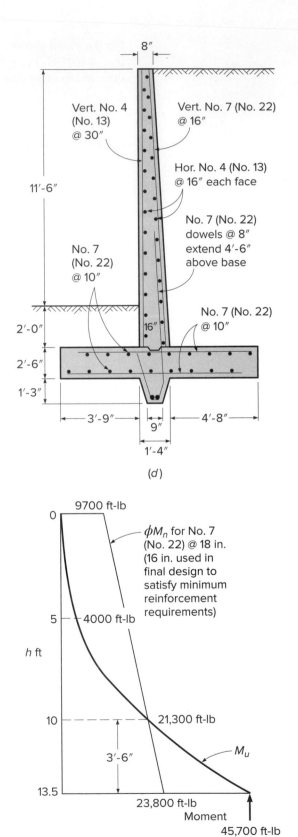

(*d*)

(*e*)

The total soil pressure on the plane *ac* is the same as for the gravity wall designed in Section 16.7; that is, $P = 6500$ lb, and the overturning moment is

$$M_o = 37,500 \text{ ft-lb}$$

The distance of the resultant from the front edge is

$$a = \frac{81,040 - 37,500}{13,490} = 3.23 \text{ ft}$$

which locates the resultant barely outside of the middle third. The corresponding maximum soil pressure at the toe, from Fig. 16.5*c*, is

$$q_1 = \frac{2 \times 13,470}{3 \times 3.23} = 2780 \text{ psf}$$

The factor of safety against overturning, $81,040/37,500 = 2.16$, is ample.

To check the safety against sliding, remember (Section 16.4) that if sliding occurs, it proceeds between concrete and soil along the heel and key (that is, length *ae* in Fig. 16.4), but takes place within the soil in front of the key (that is, along length *te* in Fig. 16.4). Consequently, the coefficient of friction that applies for the former length is $f = 0.5$, while for the latter it is equal to the internal soil friction, that is, $\tan 30° = 0.577$.

The bearing pressure distribution is shown in Fig. 16.7*b*. Since the resultant is at a distance $a = 3.23$ ft from the front, that is, nearly at the middle third, it is assumed that the bearing pressure becomes zero exactly at the edge of the heel, as shown in Fig. 16.7*b*.

The resisting force is then calculated as the sum of the friction forces of the rear and front portion, plus the passive soil pressure in front of the wall. For the latter, as in Section 16.7, the top 1.5 ft layer of soil is discounted as unreliable. Hence,

Friction, toe:	$(2780 + 1710) \times 0.5 \times 3.75 \times 0.577 = 4860$ lb
Friction, heel and key:	$1710 \times 0.5 \times 6 \times 0.5 = 2570$ lb
Passive earth pressure:	$0.5 \times 120 \times 3.25^2 \times 3.0 = \underline{1900 \text{ lb}}$
Total resistance to sliding:	$= 9330$ lb

The factor of safety against sliding, $9330/6500 = 1.44$, is only 4 percent below the recommended value of 1.5 and can be regarded as adequate.

The calculations hold for the case in which the surcharge extends from the right to point *a* above the edge of the heel. The other case of load distribution, in which the surcharge is placed over the entire surface of the fill up to point *b*, evidently does not change the earth pressure on the plane *ac*. It does, however, add to the sum of the vertical forces and increases both the restoring moment M_r and the friction along the base. Consequently, the danger of sliding or overturning is greater when the surcharge extends only to *a*, for which situation these two cases have been checked and found adequate. In view of the added vertical load, however, the bearing pressure is largest when the surface is loaded to *b*. For this case,

$$W = 13,490 + 400 \times 5.33 = 15,600 \text{ lb}$$

$$M_r = 81,040 + 400 \times 5.33 \times 7.09 = 96,200 \text{ ft-lb}$$

$$a = \frac{96,200 - 37,500}{15,600} = 3.76 \text{ ft}$$

which places the resultant inside the middle third. Hence, from Fig. 16.5a,

$$q_1 = (39.0 - 22.5)\,\frac{15,600}{9.75^2} = 2710 \text{ psf}$$

$$q_2 = (22.5 - 19.5)\,\frac{15,600}{9.75^2} = 492 \text{ psf}$$

which are far below the allowable pressure of 8000 psf. The corresponding bearing pressure distribution is shown in Fig. 16.7c.

The external stability of the wall has now been ascertained, and it remains to determine the required reinforcement and to check internal resistances.

c. Arm and Key

The moment at the bottom section of the arm has previously been determined as $M_u = 45,700$ ft-lb, and a wall thickness of 16 in. at the bottom and 8 in. at the top has been selected. With a concrete cover of 2 in. clear, $d = 16.0 - 2.0 - 0.5 = 13.5$ in. Then

$$\frac{M_u}{\phi b d^2} = \frac{45,700 \times 12}{0.90 \times 12 \times 13.5^2} = 279$$

Interpolating from Graph A.1b of Appendix A, with $f_y = 60,000$ psi and $f_c' = 4500$ psi, the required reinforcement ratio ρ is 0.0049 and $A_s = 0.0049 \times 12 \times 13.5 = 0.79$ in^2/ft. The required area of steel is provided by No. 7 (No. 22) bars at 9 in. on centers.

The bending moment in the arm decreases rapidly with increasing distance from the bottom. For this reason, only part of the main reinforcement is needed at higher elevations, and alternate bars are discontinued where no longer needed. To determine the cutoff point, the moment diagram for the arm has been drawn by calculating bending moments at two intermediate levels, 10 and 5 ft from the top. These two moments, determined in the same manner as that at the base of the arm, were found to be 21,300 and 4000 ft-lb, respectively. The resisting moment provided by alternate bars, that is, by No. 7 (No. 22) bars at 18 in. center to center, at the bottom of the arm is

$$\phi M_n = \frac{0.90 \times 0.40 \times 60,000}{12}\,(13.50 - 0.26) = 23,800 \text{ ft-lb}$$

At the top, $d = 8.0 - 2.5 = 5.5$ in., and the resisting moment of the same bars is only $\phi M_n = 23,800(5.5/13.5) = 9700$ ft-lb. Hence, the straight line drawn in Fig. 16.7e indicates the resisting moment provided at any elevation by one-half the number of main bars. The intersection of this line with the moment diagram at a distance of 3 ft 6 in. from the bottom represents the point above which alternate bars are no longer needed. ACI Code 7.7.3.3 specifies that any bar shall be extended beyond the point at which it is no longer needed to carry flexural stress for a distance equal to d or 12 bar diameters, whichever is greater. In the arm, at a distance of 3 ft 6 in. from the bottom, $d = 11.4$ in., while 12 bar diameters for No. 7 (No. 22) bars are equal to 10.5 in. Hence, one-half the bars can be discontinued 12 in. above the point where no longer needed, or a distance of 4 ft 6 in. above the base. This exceeds the required development length of 39 in. above the base.

To facilitate construction, the footing is placed first, and a construction joint is provided at the base of the arm, as shown in Fig. 16.7d. The main bars of the arm, therefore, end at the top of the base slab, and dowels are placed in the latter to be spliced with them; the integrity of the arm depends entirely on the strength of the splices used for these tension bars. Splicing all tension bars in one section by simple contact splices can easily lead to splitting of the concrete owing to the stress concentrations at the ends of the spliced bars. One way to avoid this difficulty is to weld all splices; this entails considerable extra cost.

In this particular wall, another way of placing the reinforcing offers a more economical solution. Because alternate bars in the arm can be discontinued at a distance of 4 ft 6 in. above the base, the dowels are carried up 4 ft 6 in. from the top of the base. These need not be spliced at all, because above that level only alternate No. 7 (No. 22) bars, 18 in. on centers, are needed. These latter bars are placed full length over the entire height of the arm and are spliced at the bottom with alternate shorter dowels. By this means, only 50 percent of the bars needed at the bottom of the arm are spliced; this is not objectionable.

For splices of deformed bars in tension, at sections where the ratio of steel provided to steel required is less than 2 and where no more than 50 percent of the steel is spliced, the ACI Code requires a Class B splice with a length equal to 1.3 times the development length of the bar (see Section 6.13a). The development length of the No. 7 (No. 22) bars for the given material strengths is 39 in., and so the required splice length is $1.3 \times 39 = 50.7$ in., which is less than the 4 ft 6 in. available.

According to the ACI Code, main flexural reinforcement is not to be terminated in a tension zone unless one of three conditions is satisfied: (1) shear at the cutoff point does not exceed two-thirds that permitted, (2) certain excess shear reinforcement is provided, or (3) the continuing reinforcement provides double the area required for flexure at the cutoff point and the factored shear does not exceed three-fourths of the design shear (see Section 6.10b). It is easily confirmed that the shear 4 ft 6 in. above the base is well below two-thirds the value that can be carried by the concrete; thus, main bars can be terminated as planned.

Prior to completing the design of the arm, the minimum tensile reinforcement ratio specified by the ACI Code must be checked. The actual ratio provided by the No. 7 (No. 22) bars at 18 in. spacing, with $d = 10.8$ in. just above the cutoff point, is 0.0031, about 10 percent below the minimum value of $3\sqrt{4500}/60,000 = 0.0034$. To handle this, the spacing of the No. 7 (No. 22) bars is reduced to 8 in., giving a spacing of 16 in. above the cutoff. This increases the amount of steel, but by less than would be needed if the bars were extended to a height where the decreasing value of d allowed the minimum reinforcement ratio to be satisfied. A final ACI Code requirement is that the maximum spacing of the primary flexural reinforcement exceed neither 3 times the wall thickness nor 18 in.; these restrictions are satisfied as well.

The final check of the shear capacity at the base of the wall is now made. No. 7 (No. 22) bars at 8 in. spacing have a reinforcing ratio of

$$\rho = \frac{A_s}{bd} = \frac{0.60 \times \dfrac{12}{8}}{12 \times 13.5} = 0.0056$$

The corresponding concrete shear strength is

$$v_c = 8\rho^{1/3}\sqrt{f_c'} = 8 \times (0.0056)^{1/3}\sqrt{4500} = 95.0 \text{ psi}$$

The depth of the arm exceeds 10 in. and requires an adjustment for size effects.

$$\lambda_s = \sqrt{\frac{2}{1 + \dfrac{d}{10}}} = \sqrt{\frac{2}{1 + 13.5/10}} = 0.85$$

The shear strength at the base of the wall is

$$\phi V_c = \phi \lambda \lambda_s v_c bd = 0.75 \times 1.0 \times 0.85 \times 95.0 \times 12 \times 13.5 = 9810 \text{ lb}$$

which is sufficient, exceeding the factored load of $V_u = 8270$ lb. In this case, the reduction of the bar spacing led to a higher reinforcement ratio and adequate shear strength. Had the originally calculated spacing been used, the member thickness would have to be increased.

Since the dowels had to be extended at least partly into the key to produce the necessary length of embedment, they were bent as shown to provide both reinforcement for the key and anchorage for the arm reinforcement. The exact force that the key must resist is difficult to determine, since probably the major part of the force acting on the portion of the soil in front of the key is transmitted to it through friction along the base of the footing. The relatively strong reinforcement of the key by means of the extended dowels is considered sufficient to prevent separation from the footing.

The sloping sides of the key were provided to facilitate excavation without loosening the adjacent soil. This is necessary to ensure proper functioning of the key. The hook extends toward the toe as recommended in ACI Commentary R13.6.3.3.

In addition to the main steel in the stem, reinforcement must be provided in the horizontal direction to control shrinkage and temperature cracking, in accordance with ACI Code 11.6.1. Calculations are based on the average wall thickness of 12 in. The required steel area is 0.0018 times the gross concrete area. No. 4 (No. 13) bars 16 in. on centers, each face, are selected as shown in Fig. 16.7d. Although not required by the Code for cantilever retaining walls, vertical steel equal to 0.0012 times the gross concrete area is also provided (to limit horizontal surface cracking), with at least one-half of this value provided on the exposed face, as specified for other walls under ACI Code 7.6.1. No. 4 (No. 13) bars 30 in. on centers satisfies this requirement.

d.   Toe Slab

The toe slab acts as a cantilever projecting outward from the face of the stem. It must resist the upward pressures shown in Fig. 16.7b or c and the downward load of the toe slab itself, each multiplied by appropriate load factors. The downward load of the earth fill over the toe is neglected because it is subject to possible erosion or removal. A load factor of 1.6 is applied to the service load bearing pressures. Comparison of the pressures of Fig. 16.7b and c indicates that for the toe slab, the more severe loading case results from surcharge to b. Because the self-weight of the toe slab tends to reduce design moments and shears, it is multiplied by a load factor of 0.9. Thus, the factored load moment at the outer face of the stem is

$$M_u = 1.6 \left(\frac{2710}{2} \times 3.75^2 \times \frac{2}{3} + \frac{1850}{2} \times 3.75^2 \times \frac{1}{3} \right) - 0.9 \left(225 \times 3.75^2 \times \frac{1}{2} \right)$$

$$= 25{,}800 \text{ ft-lb}$$

Because the concrete shear strength is low, the trial depth is increased to 26 in. For concrete cast against and permanently exposed to earth, a minimum protective cover for steel of 3 in. is required; if the bar diameter is about 1 in., the effective depth will be $26.0 - 3.0 - 0.5 = 22.5$ in. Thus, for a 12 in. strip of heel slab,

$$\frac{M_u}{\phi b d^2} = \frac{25{,}800 \times 12}{0.90 \times 12 \times 22.5^2} = 57$$

Graph A.1b of Appendix A shows that, for this value, the required reinforcement ratio is below the minimum of 0.0018A_g. A somewhat thinner base slab appears possible. However, moments in the heel slab are yet to be investigated, as well as shears in both the toe and heel, and the trial effective depth of 22.5 in. and overall thickness of 26 in. are retained tentatively. The required flexural steel

$$A_s = 0.0018 \times 12 \times 26 = 0.56 \text{ in}^2/\text{ft}$$

is provided by No. 7 (No. 22) bars 12 in. on centers. The required length of embedment for these bars past the exterior face of the stem is the full development length of 39 in. Thus, they are continued 39 in. past the face of the wall, as shown in Fig. 16.7d.

Shear is checked at a distance $d = 1.5$ ft from the face of the stem (2.25 ft from the end of the toe), according to the usual ACI Code procedures. The service load bearing pressure at that location (with reference to Fig. 16.7c) is 2130 psf, and the factored load shear is

$$V_u = 1.6(2710 \times \tfrac{1}{2} \times 2.54 + 2130 \times \tfrac{1}{2} \times 2.54) - 0.9(225 \times 2.54)$$

$$= 9320 \text{ lb}$$

The reinforcement ratio is $0.60/22.5 \times 12 = 0.0022$, resulting in a design shear strength of the concrete of

$$\phi V_c = \phi \lambda 8 \rho^{1/3} \sqrt{f_c'}\, bd = 0.75 \times 1.0 \times 8 \times 0.0022^{1/3} \sqrt{4500} \times 12 \times 22.5 = 14{,}200 \text{ lb}$$

which exceeds the factored load and is, therefore, satisfactory.

e. **Heel Slab**

Together, the heel and toe slabs act as a one-way shallow footing, and consequently, the size effect factor may be neglected in the calculations, as stated in ACI 13.2.6.2. The heel slab, too, acts as a cantilever, projecting in this case from the back face of the stem and loaded by surcharge, earth fill, and its own weight. The upward reaction of the soil is neglected here, for reasons given earlier. Applying appropriate load factors, the moment to be resisted is

$$M_u = 1.2 \times 225 \times 4.67^2 \times \tfrac{1}{2} + 1.6(400 \times 4.67^2 \times \tfrac{1}{2} + 1620 \times 4.67^2 \times \tfrac{1}{2})$$

$$= 38{,}200 \text{ ft-lb}$$

Thus,

$$\frac{M_u}{\phi bd^2} = \frac{38{,}200 \times 12}{0.90 \times 12 \times 22.5^2} = 84$$

Interpolating from Graph A.1b, the required reinforcement ratio is less than the minimum value based on 0.0018A_g. The required flexural steel is, thus, $A_s = 0.0018 \times 12 \times 26 = 0.56$ in^2/ft. The required steel area is provided using No. 7 (No. 22) bars 12 in. on centers, as shown in Fig. 16.7d. These bars are classified as top bars, as they have more than 12 in. of concrete below; thus, the required length of embedment to the left of the inside face of the stem is $39 \times 1.3 = 51$ in.

According to normal ACI Code procedures, the first critical section for shear would be a distance d from the face of support. However, the justification for this provision of the ACI Code is the presence, in the usual case, of vertical compressive stress near a support which tends to decrease the likelihood of shear failure in that region. However, the cantilevered heel slab is essentially hung from the bottom of the stem by the flexural tensile steel in the stem, and the concrete compression normally found near a support is absent here. Consequently, the critical section for shear in the heel slab is taken at the back face of the stem. At that location,

$$V_u = 1.2(225 \times 4.67) + 1.6(2020 \times 4.67)$$

$$= 16,350 \text{ lb}$$

The design shear strength provided by the concrete is the same as for the toe slab:

$$\phi V_c = 14,200 \text{ lb}$$

which is insufficient. Increasing the footing depth to 30 in. requires No. 7 (No. 22) bars at 10 in. An effective depth of 26.5 in. gives a reinforcement ratio of $0.60/10 \times 28.5 = 0.00226$ and a shear capacity of

$$\phi V_c = \phi \lambda 8 \rho^{1/3} \sqrt{f_c'} \, bd = 0.75 \times 1.0 \times 8 \times 0.00226^{1/3} \sqrt{4500} \times 12 \times 26.5$$

$$= 16,800 \text{ lb}$$

Thus, additional slab thickness is required. The final thickness is 30 in. Because the depth and weights have increased, it is not necessary to recalculate the stability aspects of the wall. The spacing of the toe reinforcement is reduced to 10 in. to satisfy the minimum reinforcement requirements.

The base slab is well below grade and will not be subjected to the extremes of temperature that are imposed on the stem concrete. Consequently, crack control steel in the direction perpendicular to the main reinforcement is not a major consideration. No. 4 (No. 13) bars 12 in. on centers are provided, at one face only, placed as shown in Fig. 16.7d. These bars serve chiefly as spacers for the main flexural reinforcement.

16.9 COUNTERFORT RETAINING WALLS

The external stability of a counterfort retaining wall is determined in the same manner as in the examples of Sections 16.7 and 16.8. The toe slab represents a cantilever built in along the front face of the wall, loaded upward by the bearing pressure, exactly as in the cantilever wall described in Section 16.8. Reinforcement is provided by bars a in Fig. 16.8.

A panel of the vertical wall between two counterforts is a slab acted upon by horizontal earth pressure and supported along three sides, that is, at the two counterforts and the base slab, while the fourth side, the top edge, is not supported. The earth pressure increases with distance from the free surface. The determination of moments and shears in such a slab supported on three sides and nonuniformly loaded is rather involved. It is customary in the design of such walls to disregard the support of the vertical wall by the base slab and to design it as if it were a continuous slab spanning horizontally between counterforts. This procedure is conservative because the moments obtained by this approximation are larger than those corresponding to the actual conditions of support, particularly in the lower part of the wall. Hence, for very large installations, significant savings may be achieved by a more accurate analysis. The best computational tool for this is the Hillerborg *strip method*, a plasticity-based theory for design of slabs described in detail in Chapter 24. Alternatively, results of elastic analysis are tabulated for a range of variables in Ref. 16.8.

FIGURE 16.8
Details of counterfort
retaining wall.

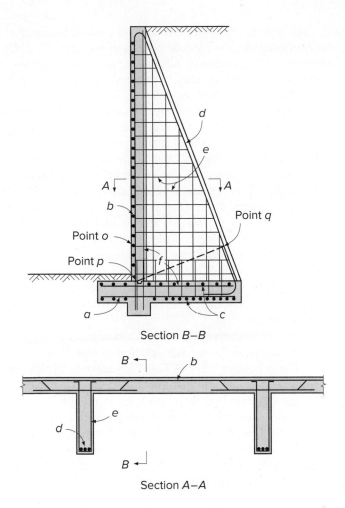

Section *B–B*

Section *A–A*

Slab moments are determined for strips 1 ft wide spanning horizontally, usually for the strip at the bottom of the wall and for three or four equally spaced additional strips at higher elevations. The earth pressure on the different strips decreases with increasing elevation and is determined using Eq. (16.1). Moment values for the bottom strips may be reduced to account for the fact that additional support is provided by the base slab. Horizontal bars *b* (Fig. 16.8) are provided, as required, with increased spacing or decreased diameter corresponding to the smaller moments. Alternate bars are bent to provide for the negative moments in the wall at the counterforts, or additional straight bars are used as negative reinforcement, as shown in Section *A–A* of Fig. 16.8.

The heel slab is supported, as is the wall slab, that is, by the counterforts and at the wall. It is loaded downward by the weight of the fill resting on it, its own weight, and such surcharge as there may be. This load is partially counteracted by the upward bearing pressure on the underside of the heel. As in the vertical wall, a simplified analysis consists in neglecting the influence of the support along the third side and in determining moments and shears for strips parallel to the wall, each strip representing a continuous beam supported at the counterforts. For a horizontal soil surface, the downward load is constant for the entire heel, whereas the upward load from the bearing pressure is usually smallest at the rear edge and increases frontward.

For this reason, the span moments are positive (compression on top) and the support moments negative in the rear portion of the heel. Near the wall, the bearing pressure often exceeds the vertical weights, resulting in a net upward load. The signs of the moments are correspondingly reversed, and steel must be placed accordingly. Bars c are provided for these moments.

The counterforts are wedge-shaped cantilevers built in at the bottom in the base slab. They support the wall slab and, therefore, are loaded by the total soil pressure over a length equal to the distance center to center between counterforts. They act as a T beam of which the wall slab is the flange and the counterfort the stem. The maximum bending moment is that of the total earth pressure, taken about the bottom of the wall slab. This moment is held in equilibrium by the force in the bars d, and hence, the effective depth for bending is the perpendicular distance pq from the center of bars d to the center of the bottom section of the wall slab. Since the moment decreases rapidly in the upper parts of the counterfort, part of the bars d can be discontinued.

In regard to shear, the authors suggest the horizontal section oq as a conservative location for checking adequacy. Modification of the customary shear calculation is required for wedge-shaped members (see Section 5.7). Usually concrete alone is sufficient to carry the shear, although bars e act as stirrups and can be used for resisting excess shear.

The main purpose of bars e is to counteract the pull of the wall slab, and they are thus designed for the full reaction of this slab.

The remaining bars of Fig. 16.8 serve as shrinkage reinforcement, except that bars f have an important additional function. It will be recalled that the wall and heel slabs are supported on three sides. Even though they were designed as if supported only by the counterforts, they develop moments where they join. The resulting tension in and near the reentrant corner should be provided for by bars f.

The question of reinforcing bar details, always important, is particularly so for corners subject to substantial bending moments, such as are present for both cantilever and counterfort retaining walls. Valuable suggestions are found in Ref. 16.9.

16.10   PRECAST RETAINING WALLS

Largely because of the high cost of forming cast-in-place retaining walls, there has been increasing use of various forms of precast concrete walls. Sections can be mass produced under controlled factory conditions using standardized forms, with excellent quality control. On-site construction time is greatly reduced, and generally only a small crew using light equipment is required. Weather becomes much less of a factor in completion of the work than for cast-in-place walls.

One type of precast wall is shown in Fig. 16.9. Reinforced, precast T-WALL® Retaining Wall System units are used, each standard unit is 2.5 to 5 ft high and 5 to 7.5 ft wide, with stems varying according to requirements from 6 to 30 ft. Individual units are stacked, using shear keys in the space created where teeth of a top and bottom unit come together. Calculations for stability against sliding and overturning and for bearing pressures are in accordance with ACI, AASHTO, and AREMA design guidelines with stability provided by the combined weight of the concrete wall and compacted select backfill. Such walls can be constructed with a vertical or battered face, with heights up to 50 ft.

Walls of the type shown have been used for highways, parking lots, commercial and industrial sites, bank stabilization, wing walls, and similar purposes.

FIGURE 16.9
Precast T-WALL® Retaining
Wall System, North Milliken
Avenue Grade Separation,
Ontario, CA. (*Courtesy of The
Neel Company, Springfield, VA.*)

REFERENCES

16.1. H-Y Fang, *Foundation Engineering Handbook*, Springer Science & Business Media, 2013.

16.2. K. Terzaghi, R. B. Peck, and G. Mesri, *Soil Mechanics in Engineering Practice*, 3rd ed., John Wiley & Sons, Inc., New York, 1996.

16.3. W. C. Huntington, *Earth Pressures and Retaining Walls*, John Wiley & Sons, Inc., New York, 1957.

16.4. *AASHTO LRFD Bridge Design Specifications*, 8th ed., American Association of State Highway and Transportation Officials (AASHTO), Washington, DC, 2017.

16.5. *Manual for Railway Engineering*, American Railway Engineering and Maintenance of Way Association (AREMA), Landover, MD, 2018.

16.6. *CRSI Design Guide Suite, Buildings*, Concrete Reinforcing Steel Institute, Shaumburg, IL.

16.7. M. Fintel, *Handbook of Concrete Engineering*, 2nd ed., Van Nostrand Reinhold Co., New York, 1985.

16.8. *Rectangular Concrete Tanks*, Publication No. IS003, Portland Cement Association, Skokie, IL, 1998.

16.9. I. H. E. Nilsson and A. Losberg, "Reinforced Concrete Corners and Joints Subjected to Bending Moment," *J. Struct. Div.*, ASCE, vol. 102, no. ST6, 1976, pp. 1229–1254.

PROBLEMS

16.1. A cantilever retaining wall is to be designed with geometry as indicated in Fig. P16.1. Backfill material is well-drained gravel having unit weight $w = 120$ pcf, internal friction angle $\phi = 25°$, and friction factor against the concrete base $= 0.60$. Backfill placed in front of the toe will have the same properties and will be well compacted. The final grade behind the wall is level with the top of the wall, with no surcharge.

(a) Check the stability of the wall.

(b) Design the reinforcement, specifying size and placement. Materials strengths are $f'_c = 5000$ psi and $f_y = 60,000$ psi. Allowable soil bearing pressure is 4000 psf.

FIGURE P16.1

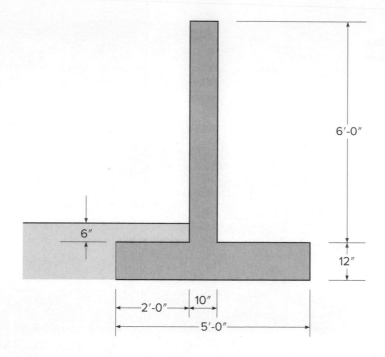

16.2. A cantilever retaining wall is to be designed with the geometry shown in Fig. P16.2 with no counterfort. Backfill material is well-drained gravel having unit weight $w = 120$ pcf, internal friction angle $\phi = 33°$, and friction factor against the concrete base $= 0.55$. Backfill placed in front of the toe will have the same properties and will be well compacted. The final grade behind the wall is level with the top of the wall, with no surcharge. At the lower level, the grade is 3 ft above the top of the base slab. To improve sliding resistance, a key is used, tentatively projecting to a depth 4 ft below the top of the base slab. (This dimension may be modified if necessary.)

(a) Based on a stability investigation, select wall geometry suitable for the specified conditions. For a first trial, place the outer face of the wall $\frac{1}{3}$ the width of the base slab back from the toe.

(b) Prepare the complete structural design, specifying size, placement, and cutoff points for all reinforcement. Materials strengths are $f_c' = 4000$ psi and $f_y = 60,000$ psi. Allowable soil bearing pressure is 5000 psf.

FIGURE P16.2

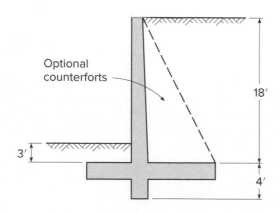

17

Strut-and-Tie Method

17.1 INTRODUCTION

Reinforced concrete beam theory is based on equilibrium, compatibility, and the constitutive behavior of the materials, steel and concrete. Of particular importance is the assumption that strain varies linearly through the depth of a member and that, as a result, plane sections remain plane. This assumption is validated by St. Venant's principle, which stipulates that strains induced by discontinuities in load or in member cross section vary in an approximately linear fashion at distances greater than or equal to the greatest cross-sectional dimension h from the point of load application. St. Venant's principle underlies the development of beam theory as presented in Chapters 1, 3, and 4.

St. Venant's principle, however, does not apply at points closer than the distance h to discontinuities in applied load or geometry. This leads to the identification of *discontinuity regions* within reinforced concrete members near concentrated loads, openings, or changes in cross section. For example, because of their geometry, the full volume of deep beams and column brackets qualify as discontinuity regions. Thus, reinforced concrete structures may be divided into regions where beam theory is valid, often referred to as *B-regions*, and regions where discontinuities affect member behavior, known as *D-regions*. A number of D-regions are illustrated in Fig. 17.1.

At low stresses, when the concrete is elastic and uncracked, the stresses within D-regions may be calculated using finite element analysis or elasticity theory. When concrete cracks, the strain field is disrupted, causing a redistribution of the internal forces. Once this happens, it is possible to represent the internal forces within discontinuity regions using a statically determinate truss, referred to as a *strut-and-tie model*. This allows a complex design problem to be greatly simplified, producing a safe solution that satisfies statics. As shown in Fig. 17.2, strut-and-tie models consist of concrete compression *struts*, steel tension *ties*, and joints that are referred to as *nodal zones*. (For consistency of presentation, struts are represented by dashed lines and ties are represented by solid lines.)

17.2 DEVELOPMENT OF THE STRUT-AND-TIE METHOD

The strut-and-tie method evolved in the early 1980s in Europe (Refs. 17.1 to 17.4). Its use is permitted by ACI Code 6.2.4.4 and defined in Chapter 23 of the Code (Ref. 17.5). As defined, two- or three-dimensional strut-and-tie models divide members into D-regions and B-regions. A D-region is that portion of a member that is within a distance equal to the member height h from a force or geometric discontinuity, as shown in Fig. 17.1.

FIGURE 17.1
Geometric and load
discontinuities for D-regions.

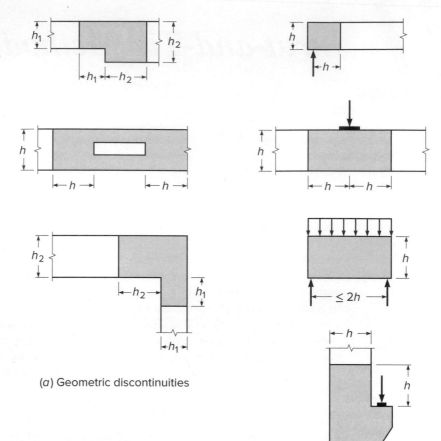

(*a*) Geometric discontinuities

(*b*) Loading discontinuities

FIGURE 17.2
Strut-and-tie model.

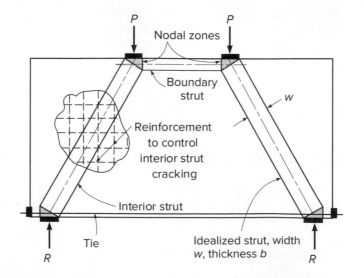

B-regions are, in general, any portions of a member outside of D-regions. The assumption is that within B-regions strain varies linearly through the member cross section and plane sections remain plane.

Strut-and-tie models are applied within D-regions. Models consist of struts and ties connected at nodal zones that are capable of transferring loads to the supports or adjacent B-regions. The cross-sectional dimensions of the struts and ties are designated as thickness and width. Thickness b is perpendicular to the plane of the truss model, and width w is measured in the plane of the model, as shown in Fig. 17.2.

a. Struts

A strut is an internal compression member. It may consist of a single element, parallel elements, or a fan-shaped compression field. Struts are classified as *boundary* struts, located at the edge of the member, or as *interior* struts, as shown in Fig. 17.2. For design purposes, a strut is typically idealized as a prismatic member between two nodes. While not preferred, a strut can also be idealized as a uniformly tapered compression member if the design criteria require different widths at the ends of the strut. The dimensions of the cross section of the strut are established by the contact area between the strut and the nodal zone. Interior struts are subjected to tensile stresses perpendicular to the axis of the strut, which may result in longitudinal splitting; boundary struts are not. The capacity of a strut is a function of the effective concrete compressive strength, which is affected by transverse stresses. Because of longitudinal splitting, interior struts are weaker than boundary struts. Transverse reinforcement, as shown in Fig. 17.2, is incorporated to control longitudinal splitting. Without lateral confinement provided by transverse or other reinforcement, the capacity of an interior strut is further reduced.

b. Ties

A tie is a tension member within a strut-and-tie model. Ties consist of reinforcement (prestressed or nonprestressed) plus a portion of the concrete that is concentric with and surrounds the axis of the tie. The surrounding concrete defines the tie area and the region available to anchor the tie. For design purposes, it is assumed that the concrete within the tie does not carry any tensile force. Even though the tensile capacity of the concrete is not used in design, it assists in reducing tie deformation at service load.

c. Nodal Zones

Nodes are locations within strut-and-tie models where the axes of struts, ties, and reactions or concentrated loads intersect. A nodal zone is the volume of concrete around a node where force transfer occurs. A nodal zone may be treated as a single region or may be subdivided into two smaller zones to equilibrate forces. For example, the nodal zone shown in Fig. 17.3a is subdivided, as shown in Fig. 17.3b, where two reactions R_1 and R_2 equilibrate the vertical components of strut forces C_1 and C_2.

For equilibrium, at least three forces must act on a node. Nodes are classified by the sign of these forces (Fig. 17.4). Thus, a *C-C-C* node resists three compressive forces, and a *C-C-T* node resists two compressive forces and one tensile force. Both tensile and compressive forces place nodes in compression because tensile forces are treated as if they pass through the node and apply a compressive force on the far side, or anchorage face. Thus, within the plane of a strut-and-tie model truss, nodal

FIGURE 17.3
Subdivision of nodal zones.

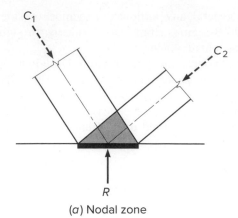

(a) Nodal zone

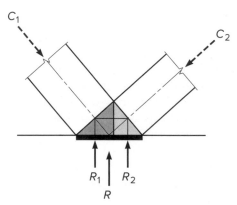

(b) Subdivided nodal zone

FIGURE 17.4
Classification of nodes.

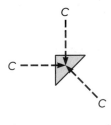

(a) C-C-C node

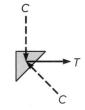

(b) C-C-T node

(c) C-T-T node

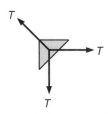

(d) T-T-T node

FIGURE 17.5
Nodal zones and extended
nodal zones.

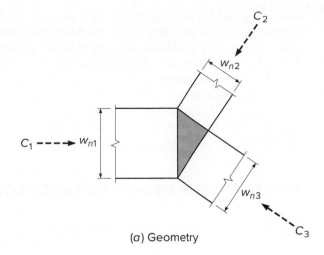

(a) Geometry

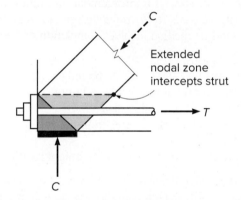

(b) Tension force anchored by plate

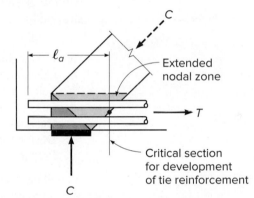

(c) Tension force anchored by bond

zones are considered to be in compression, as shown in Fig. 17.5a. If the nodal zone
dimensions w_{n1}, w_{n2}, and w_{n3} are proportional to the applied compressive forces,
the state of stress becomes one of *hydrostatic* compression. The dimension of one
side of a nodal zone is often determined based on the contact area of the load, such
as a bearing plate, column base, or beam support. If a hydrostatic state of stress is
desired, the dimensions of the remaining sides of the node are selected to maintain
a constant level of stress p within the node. By selecting nodal zone dimensions that
are proportional to the applied loads, the stresses on the faces of the nodal zone are
equal.[†] If, instead, the dimensions are determined based on preselected strut dimensions, such as minimum width, the state of stress may no longer be hydrostatic. The
decision to use a hydrostatic or a nonhydrostatic state of stress is made by the
designer, with the former being more typical because the latter results in a more
complex design.

    The length of a nodal zone is often not adequate to allow for anchorage of tie
reinforcement. For this reason, an *extended nodal zone*, defined by the intersection
of the nodal zone and the associated strut (shown in light shading in Fig. 17.5b

---

[†] The state of stress within a nodal zone is not truly hydrostatic since out-of-plane stresses are not considered; however, solutions remain valid.

and *c*), is used. An extended nodal zone may be regarded as that portion of the overlap region between struts and ties that is not already counted as part of a primary node. It increases the length within which the tensile force from the tie can be transferred to the concrete and, thus, defines the available anchorage length for ties. Ties may be developed outside of the nodal and extended nodal zones if needed, as shown to the left of the node in Fig. 17.5*c*.

In some *C-T-T* nodes, where the tension tie is continuous around the corner, the accumulated tensile forces can result in crushing of the concrete under the bend radius. These *curved bar nodes* require additional detailing attention.

17.3   STRUT-AND-TIE DESIGN METHODOLOGY

Strut-and-tie models are used in several ways during the design process. At the *conceptual* design level, sketching a strut-and-tie model provides insight into structural behavior and detailing requirements. Examples of conceptual design can be seen in the development of connection details in Chapter 18. Strut-and-tie models may be used to *validate* design details, such as for special reinforcement configurations. Finally, the strut-and-tie method may form the basis for *detailed* design of a member.

Application of the strut-and-tie method involves completion of the following steps.

a. Define and isolate the D-regions.
b. Calculate the force resultants on each D-region boundary.
c. Select a truss model to transfer the forces across a D-region.
d. Select dimensions for strut-and-tie nodal zones.
e. Design the struts and nodes to have sufficient strength.
f. Design the ties and the tie anchorage.
g. Prepare design details and check minimum reinforcement requirements.

As will be described shortly, the design process requires interaction between these steps.

According to ACI Code 23.3.1, design using a strut-and-tie model requires that

$$\text{for struts:}\quad \phi F_{ns} \geq F_{us} \tag{17.1a}$$

$$\text{for ties:}\quad \phi F_{nt} \geq F_{ut} \tag{17.1b}$$

$$\text{for nodes:}\quad \phi F_{un} \geq F_{un} \tag{17.1c}$$

where F_{ns}, F_{nt}, and F_{nn} = nominal capacity of strut, tie, and nodal zone, respectively
  F_{us}, F_{ut}, and F_{un} = factored force acting in strut, tie, and nodal zone, respectively
  ϕ = strength reduction factor

In addition to strength criteria, service level performance must be considered in design because strut-and-tie models, which are based on strength, do not necessarily satisfy serviceability requirements. To this end, the spacing of reinforcement within ties should be checked using Eq. (7.3). ACI Code 9.9.2 limits the nominal shear strength of deep beams to $10\sqrt{f_c'}\, b_w d$. This limit applies to strut-and-tie models and should be checked prior to beginning a detailed design, as described in Section 17.4*d*.

a.   D-Region

A D-region extends on both sides of a discontinuity by a distance equal to the member height *h*. At geometric discontinuities, a D-region may have different dimensions on either side of the discontinuity, as shown in Fig. 17.1.

FIGURE 17.6
Resolution of forces in a
D-region.

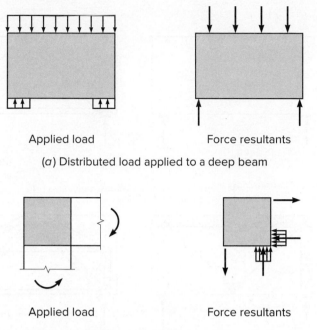

Applied load   Force resultants

(*a*) Distributed load applied to a deep beam

Applied load   Force resultants

(*b*) Moment resisting corner

b. Force Resultants on D-Region Boundaries

Once the D-region is defined, the next step involves determining the magnitude, location, and direction of the resultant forces acting on the D-region boundaries. These
forces serve as input for the strut-and-tie method and assist in establishing the geometry of the truss model. When one face of a D-region is loaded with a uniform or linearly varying stress field, or when a face is loaded by bending of a concrete section, it
may be necessary to subdivide the boundary into segments corresponding to struts or
ties and then to calculate the resultant force on each segment, as shown in Fig. 17.6.
For example, in Fig. 17.6*a*, the distributed load along the top of the deep beam is represented by four concentrated loads, and the stresses at the beam-column interface are
represented by concentrated reactions. In Fig. 17.6*b*, the moments at the faces of the
beam-column joint are represented by couples consisting of tensile and compressive
forces acting at the interfaces between the members and the joint.

c. The Truss Model

The truss representing the strut-and-tie model must fit within the envelope defined by
the D-region. The selection of struts and ties is made at the discretion of the designer,
and, therefore, multiple acceptable solutions are possible. The axes of the truss
members are chosen to coincide with the centroids of the tension and compression
fields, and the geometry so established is used to calculate the forces in the members.
The layout of a truss model is constrained by the geometric requirement that struts
must intersect only at nodal zones. Ties may cross struts. An effective model represents a minimum energy distribution through the D-region (Refs. 17.1 and 17.4);
that is, within the model, forces should follow the stiffest load path. Because struts
are typically much stiffer than ties, a model with a minimum number of tension ties
is generally preferred. Alternative truss models for a deep beam are compared in

FIGURE 17.7
Alternatives for a deep beam
truss model.

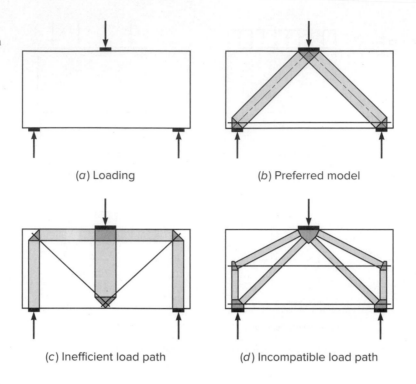

(a) Loading

(b) Preferred model

(c) Inefficient load path

(d) Incompatible load path

Fig. 17.7. Figure 17.7a shows a deep beam subjected to a concentrated load at mid-span. Figure 17.7b shows the preferred strut-and-tie model for this beam and loading condition. In this case, struts carry the load directly to nodal regions at the supports, which are, in turn, connected by a single tension tie. The model in Fig. 17.7c shows an ineffective load path, with a single strut carrying the load to a node at the bottom of the beam that is supported by two diagonal tension ties that are, in turn, supported by vertical struts over the supports. In this instance, the number of transfer points and tension ties is greater, as is the flexibility of the truss, indicating a solution that is much less effective than that shown in Fig. 17.7b. Lastly, Fig. 17.7d illustrates a model with multiple struts and ties. This particular layout not only is unduly complex but also includes an upper tension tie that will be effective only after extensive yielding and possible failure of the lower tension tie.

Theoretically, there is a unique minimum energy solution for a strut-and-tie model. Practically, any model that satisfies equilibrium and pays attention to structural stiffness proves satisfactory. Using the rationale just discussed allows the engineer to select a logical model that effectively mobilizes ties and minimizes the potential for excessive cracking. Finite element analyses and solutions based on the theory of elasticity for the full structure can provide an indication of where maximum stresses occur. A truss model that provides struts in regions of high compression and ties in regions of high tension based on these analyses will, in general, provide an efficient load path.

d.   Selecting Dimensions for Struts and Nodal Zones

The struts, ties, and nodal zones within the truss that represents a strut-and-tie model have finite widths that must be considered when selecting the dimensions of the truss. The width of each truss member depends on the magnitude of the forces and the

FIGURE 17.8
Resolution of forces in nodal
zones.

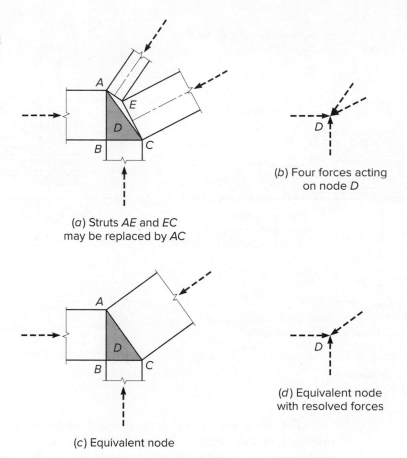

(a) Struts *AE* and *EC*
may be replaced by *AC*

(b) Four forces acting
on node *D*

(c) Equivalent node

(d) Equivalent node
with resolved forces

dimensions of the adjoining elements. An external element, such as a bearing plate or column, can serve to define a nodal zone. If the bearing area is too small, a high hydrostatic pressure results, and the corresponding width of the node or struts will not be sufficient to carry the applied load. The solution in this case is to increase the size of the bearing surface and, thus, reduce the contact pressures. Some engineers intentionally select struts and nodes that are large enough to keep the compressive stresses low; in this case, only the tension ties require detailed design. To minimize cracking and to reduce complications that may result from incompatibility in the deformations due to struts shortening and ties elongating in nearly the same plane, the angle between struts and ties at a node should be at least 25°. Engineers, however, often prefer to use an angle of at least 40° because forces in the struts and ties are often unacceptably high at lower angles.

The design of nodal zones is based on the assumption that the principal stresses within the intersecting struts and ties are parallel to the axes of these truss members. The widths of the struts and ties are, in general, proportional to the magnitude of the force in the elements. If two or more struts converge on the same face, such as shown in Fig. 17.8*a* and *b*, it is generally necessary to resolve the forces into a single force and to orient the face of the nodal zone so that it is perpendicular to the combined force, as shown in Fig. 17.8*c* and *d*. Some geometric arrangements preclude establishing a purely hydrostatic node. In these cases, the width of the strut is determined by the geometry of the bearing plate or tension tie, as shown in Fig. 17.9*a*.

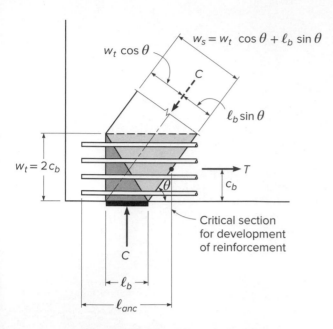

(a) C-C-T node with eccentric strut

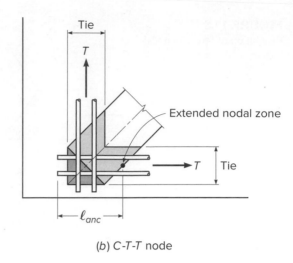

(b) C-T-T node

FIGURE 17.9
Extended nodal zone definition.

The thickness of the strut, tie, and nodal zone is typically equal to the thickness of the member. If the thickness of the bearing area is less than the thickness of the member, it may be necessary to add reinforcement perpendicular to the principal plane of the member to add confinement and prevent splitting parallel to the plane of the truss. In this instance, a strut-and-tie model perpendicular to the plane of the truss may be used to determine the requirements for transverse reinforcement in a manner that is similar to that used to reinforce interior struts.

e. Capacity of Struts

Strut capacity is based on both the strength of the strut itself and the strength of the nodal zone. If a strut does not have sufficient capacity, the design must be revised by providing compression reinforcement or by increasing the size of the nodal zone. This may, in turn, affect the size of the bearing plate or column.

f. Design of Ties and Anchorage

To control cracking in a D-region, ties are designed so that the stress in the reinforcement is below yield at service loads. The geometry of the tie must be selected so that the reinforcement fits within the tie dimensions and is fully anchored.

Anchorage for ties is provided within the nodal and extended nodal zones plus regions on the far side of the node that may be available based on the geometry of the member. Figure 17.9a illustrates an extended nodal zone and the length available for anchorage of ties ℓ_{anc}. In this case, the tie is extended to the left of the nodal zone to allow for full development of the reinforcement. The shape of the extended nodal zone is a function of the strut angle θ and the width of the tie w_t. Figure 17.9*a* illustrates the geometry and dimensions of a *C-C-T* node with a tension tie that contains multiple layers of reinforcement. Figure 17.9*b* shows a *C-T-T* nodal zone. If insufficient length is available to anchor the reinforcement within the nodal and extended nodal zones, the reinforcement must extend beyond the node or a hook or headed bar must be used to fully develop the reinforcement.

g. **Design Details and Minimum Reinforcement Requirements**

A complete design includes verification that (1) tie reinforcement can be placed in the section, (2) nodal zones are confined by compressive forces or tension ties, (3) tie anchorage is fully developed, and (4) minimum reinforcement requirements are satisfied. Reinforcement within ties must meet the ACI Code requirements for bar spacing (see Section 4.5c) and fit within the overall width and thickness of the tie. Tie details should be reviewed to ensure that ties are adequately developed on the far side of nodes by tension development length, hooks, headed bars, or other mechanical anchorage. Shear reinforcement requirements are satisfied by ensuring that the factored shear is less than the ACI Code maximum, as described in Chapter 5, longitudinal cracking of interior struts is controlled, or the minimum reinforcement requirements described in Section 17.4d are met.

17.4 ACI PROVISIONS FOR THE STRUT-AND-TIE METHOD

ACI Code Chapter 23 provides guidance for sizing struts, nodes, and ties. It addresses the performance of highly stressed compression zones that may be adjacent to or crossed by cracks in a member, the effect of stresses in nodal zones, and the requirements for anchorage of ties. The effective compressive strength of concrete $0.85f_c'$ is modified by factors that account for the effects of transverse cracks, confining reinforcement in struts, and the anchorage of ties in nodal zones, as shown in Tables 17.1 and 17.2.

The balance of this section describes the steps needed to calculate the capacity of struts, verify nodal zones, and design ties and tie anchorage. A strength reduction factor $\phi = 0.75$ is used for struts, ties, nodal zones, and bearing areas.

TABLE 17.1
Strut and node confinement modification factor β_c

Location		β_c
End of a strut connected to a node that includes a bearing surface, or	Lesser of	$\sqrt{\dfrac{A_2}{A_1}}$ where A_1 is defined by the bearing surface
Node that includes a bearing surface		2.0
Other cases		1.0

TABLE 17.2
Strut coefficient β_s

Strut Location	Strut Type	Criteria	β_s
Tension member or tension zone of members	Any	All cases	0.40
All other cases	Boundary strut	All cases	1.0
	Interior struts	Satisfying minimum distributed reinforcement criteria in Table 17.3	0.75
		Located in regions satisfying shear limitation of Eq. (17.5)	0.75
		Beam-column joints	0.75
		All other cases	0.4

a. ## Strength of Struts

The strength of a strut is limited based on the strength of the concrete in the strut and the strength of the nodal zones at the ends of the strut. The nominal compressive strength of a strut F_{ns} is given as

$$F_{ns} = f_{ce}A_{cs} \tag{17.2}$$

where f_{ce} is the effective compressive strength of the concrete in a strut and A_{cs} is the cross-sectional area at one end of the strut, which is equal to the product of the strut thickness b and the strut width w. The effective strength of concrete in a strut is

$$f_{ce} = 0.85\beta_c\beta_s f_c' \tag{17.3}$$

where β_c is a confinement factor from Table 17.1 and β_s is a factor adjusting the strength of the strut for location and transverse reinforcement (Table 17.2). The values of β_s range from 1.0 for a boundary strut to 0.4 for struts in tension members, the tension flanges of members, and members without transverse reinforcement (Table 17.1).

Compression reinforcement may be added to increase the strength of a strut, so that

$$F_{ns} = f_{ce}A_{cs} + A_s' f_s' \tag{17.4}$$

where f_s' is based on the strain in the concrete at peak stress. For Grades 40 and 60 reinforcement, $f_s' = f_y$. In accordance with ACI Code 23.6, compression reinforcement must be properly anchored, oriented parallel to the axis of the strut, located within the strut, and enclosed by ties or spirals. Transverse reinforcement spacing must be the least of the smallest dimension of the strut, 48 times the transverse reinforcement diameter, and 16 times the longitudinal bar diameter. Transverse reinforcement size is the same as required for columns in Section 9.2.

The strut-and-tie method is derived from a lower bound theorem of plasticity. (See Section 23.2 for discussion of upper and lower bound theorems.) A member designed using this method requires sufficient reinforcement to allow redistribution of internal forces in a cracked state (Ref. 17.3). Table 17.2 allows the use of a β_s value of 0.75 for struts with minimum distributed transverse reinforcement and for struts in regions of low sectional shear. This value reflects the beneficial effects of distributed reinforcement and the lower likelihood of diagonal splitting under lower shear load.

TABLE 17.3
Minimum distributed shear reinforcement

Lateral Restraint of Strut	Reinforcement Configuration	Minimum Distributed Reinforcement Ratio
Not restrained	Orthogonal grid	0.0025 in each direction
	Reinforcement in one direction crossing the strut at an angle α_1	$\dfrac{0.0025}{\sin^2 \alpha_1}$
Restrained	Reinforcement not required and $\beta_s = 0.4$	

FIGURE 17.10
Details of reinforcement crossing a strut.

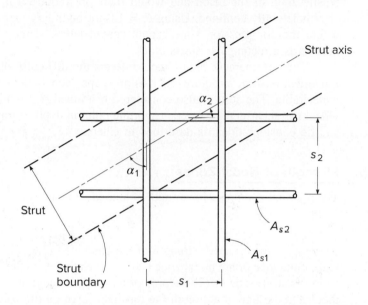

The shear strength limitation in Table 17.2 based on Eq. (17.5) is to control diagonal tension splitting of the strut when minimum reinforcement is not provided. That limitation is satisfied if the member dimensions are selected such that the shear in the member

$$V_u \leq \phi \, 5 \tan\theta \, \lambda \lambda_s \, \sqrt{f_c'} \, b_w d \qquad (17.5)$$

where λ is the factor for lightweight concrete and λ_s is the size effect modification factor. The value for λ_s may be taken as 1.0 if distributed reinforcement is provided. Otherwise, λ_s is calculated from Eq. (17.6).

$$\lambda_s = \sqrt{\dfrac{2}{\left(1 + \dfrac{d}{10}\right)}} \qquad (17.6)$$

Shear limitations in the strut-and-tie chapter of the ACI Code are sometimes overridden by other member-specific Code chapters. For example, the ACI Code has limitations on shear in walls and in diaphragms that must be adhered to.

Minimum reinforcement requirements are given in Table 17.3, and the geometry of the minimum reinforcement placement is indicated in Fig. 17.10. The distributed reinforcement in Table 17.3 must have lateral spacing not exceeding 12 in., and the angle α_1 must be at least 40°. The reinforcement must be developed beyond the boundaries of the strut.

An evaluation of a database of strut-and-tie method tests indicates that if struts are proportioned for a β_s value of 0.4, diagonal tension failures were precluded (Ref. 17.6). The same database examined the effect of depth of the members and determined that there was no significant reduction in capacity for members up to 38 in. deep when transverse reinforcement is present.

Struts are considered laterally confined if the discontinuity region is continuous perpendicular to the plane of the strut-and-tie model or the concrete restraining the strut extends beyond each face of the strut by a distance equal to at least half the width of the strut. Laterally confined struts have a β_s of 0.4 and no reinforcement requirement. For example, the strut-and-tie model seen in Fig. 18.16 occurs in the tensile zone of the beam and would therefore require a β_s of 0.4 even though the strut is laterally confined. Using $\beta_s = 1.0$ for boundary elements reflects a condition where the compression zone cannot spread laterally, thereby allowing the strut to behave as a rectangular stress block.

The confinement requirement reflects the difficulty of placing transverse reinforcement in some structures such as pile caps. Nonetheless, transverse reinforcement is desirable. The strut is also considered restrained if it is in a corner joint or a joint where the shear depth exceeds twice the column depth. Joint design, including strut-and-tie considerations, is described in Chapter 18.

b. Strength of Nodal Zones

The nominal compressive strength of a nodal zone is

$$F_{nn} = f_{ce}A_{nz} \tag{17.7}$$

where f_{ce} is the effective strength of the concrete in the nodal zone and A_{nz} is (1) the area of the face of the nodal zone taken perpendicular to the line of action of the force from the strut or tie or (2) the area of a section through the nodal zone taken perpendicular to the line of action of the resultant force on the section. The latter condition occurs when multiple struts intersect a node, as shown in Fig. 17.8.

The effective concrete strength in a nodal zone is

$$f_{ce} = 0.85\beta_c\beta_n f_c' \tag{17.8}$$

where f_c' is the compressive strength of the concrete in the nodal zone, β_c is the confinement modification factor given in Table 17.1, and β_n is a factor that reflects the degree of disruption in nodal zones due to the incompatibility of tensile strains in ties with compressive strains in struts. $\beta_n = 1.0$ for C-C-C nodes, 0.80 for C-C-T nodes, and 0.60 for C-T-T or T-T-T nodes. The values of β_n are summarized in Table 17.4. ACI Code 23.9.3 permits the strength of a node to be increased above the value given in Eq. (17.8) if the node contains confining reinforcement and the effect of that reinforcement is demonstrated by tests and analysis.

TABLE 17.4
β_n values for node strength

Nodal Zone Condition	Classification	β_n
Bounded by struts or bearing area, or both	C-C-C	1.0
Anchoring one tie	C-C-T	0.80
Anchoring two or more ties	C-T-T or T-T-T	0.60

FIGURE 17.11
Curved bar node details.
(*Adapted from Ref. 17.11.*)

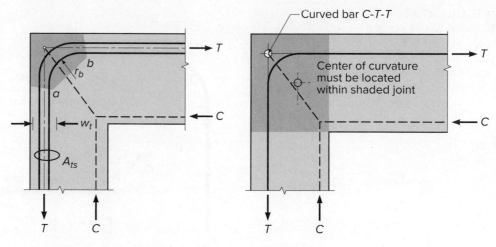

(*a*) Curved bar node with two layers of
reinforcement (node zone shaded)

(*b*) Permissible zone for center of curved
bar node at a frame

Unless compression reinforcement is used in the struts, the lower value of f_{ce}
from Eqs. (17.3) and (17.8) governs and should be used to design both the node and
the adjoining struts.

c. Curved Bar Nodes

Curved bar nodes occur where the tie reinforcement is continuous through a 90 or
180° bend. Unlike a standard hook, the tensile force is applied to both ends of the bar,
and the resulting compression stresses under the bend can split the concrete where
there is insufficient side cover. Curved bar node performance is affected by the cover
normal to the bars, the bar radius, and the total force on the bars (Ref. 17.6).

Curved bar nodes are addressed in ACI Code 23.10. If the specified cover
normal to the plane of the bend is $2d_b$ or greater, then the bend radius for bars r_b
for bends less than 180° is

$$r_b \geq \frac{2A_{ts}f_y}{b_s f_c'} \tag{17.9}$$

and for 180° bends is

$$r_b \geq \frac{1.5A_{ts}f_y}{w_t f_c'} \tag{17.10}$$

where A_{ts} is the total area of tie reinforcement, b_s is the dimension of the strut transverse
to the plane, and w_t is the effective tie width, as shown in Fig. 17.11*a*, where the ties with
180° bends are in a plane perpendicular to the plane of the truss. In no case is the radius
to be less that the radius for standard hooks. If the cover normal to the bend is less than
$2d_b$, the bend radius is multiplied by the ratio $2d_b/c_c$, where c_c is the specified clear cover
to the side face. Where curved bar nodes are formed by more than one layer of reinforcement, A_{ts} is the total area of tie reinforcement and r_b is the radius of the innermost layer
(Fig 17.11*a*). The center of the radius of the curved-bar node in a frame lies within the
area of the node (Fig. 17.11*b*). Lastly, the length of the curved portion of the bent bar
ℓ_{cb} in Fig. 17.12 should be enough to develop any difference in tie tension between the
two straight ends extending from the bent region. Curved bar radii in curved bar nodes

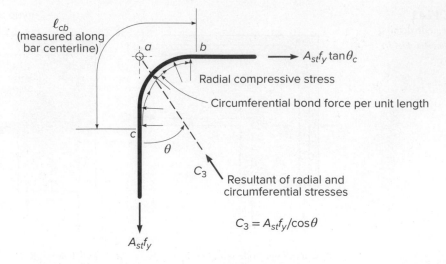

with a single layer of Grade 60 reinforcement spaced at least 8 bar diameters apart
for bar sizes No. 4 to 8 (No. 13 to 25) and 6 bar diameters for bars sizes No. 9 to 11
(No. 29 to 36) or 6 bar diameters for 180° bends are all less than the standard bend
radius for the bars. Bar spacing less than 8 or 6 bar diameters or multiple layers of
reinforcement require adjustment of the bend radius.

d. Strength of Ties

The nominal strength of ties F_{nt} is the sum of the strengths of the reinforcing steel and
prestressing steel within the tie.

$$F_{nt} = A_{ts}f_y + A_{tp}\Delta f_p \qquad (17.11)$$

where A_{ts} = area of reinforcing steel

  f_y = yield strength of reinforcing steel

  A_{tp} = area of prestressing steel, if any

  f_{pe} = effective stress in prestressing steel

  Δf_p = increase in prestressing steel stress due to factored load

A_{tp} is zero for nonprestressed members. The value of Δf_p may be found by analysis;
or, in lieu of formal analysis, ACI Code 23.7.3 allows a value 60,000 psi to be used for
bonded tendons and 10,000 psi to be used for unbonded tendons. The calculated value
of Δf_p cannot be greater than the yield stress of the prestressing reinforcement f_{py} less
the effective prestress f_{pe}.

The effective width of a tie w_t (see Fig. 17.9a) depends on the distribution of
the tie reinforcement. If the reinforcement in a tie is placed in a single layer, the
effective width of a tie may be taken as the diameter of the largest bars in the tie
plus twice the cover to the surface of the bars. Alternatively, the width of a tie may
be taken as the width of the anchor plates. The practical upper limit for tie width
$w_{t,\max}$ is equal to the width corresponding to the width of a nodal zone, given as

$$w_{t,\max} = \frac{F_{nt}}{b_s f_{ce}} \qquad (17.12)$$

where f_{ce} is the effective nodal zone compressive stress given in Eq. (17.8) and b_s is
the thickness of the strut.

Ties must be anchored before they leave the extended nodal zone, a point defined by the centroid of the bars in the tie and the extension of the outlines of either the strut or the bearing area, as shown in Fig. 17.9. If the combined lengths of the nodal zone and extended nodal zone are inadequate to provide for development of the reinforcement, additional anchorage may be obtained by extending the reinforcement beyond the nodal zone, using 90° hooks, heads, or mechanical anchors. If the tie is anchored with a 90° hook, the hooks should be confined by reinforcement extending into the beam from supporting members to avoid splitting of the concrete within the anchorage region.

e. ACI Shear Requirements for Deep Beams

Beams with clear spans less than or equal to 4 times the total member depth or with concentrated loads placed within twice the member depth of a support are classified as deep beams, according to ACI Code 9.9.[†] Examples of deep beams are shown in Fig. 17.13. ACI Code 9.9.1 allows such members to be designed either by using a nonlinear analysis or by applying the strut-and-tie method of ACI Code Chapter 23. While solutions based on nonlinear strain distributions are available (Ref. 17.7), the strut-and-tie approach allows a rational design solution.

ACI Code 9.9.2 specifies that the nominal shear in a deep beam may not exceed $10\sqrt{f'_c}\,b_w d$, where b_w is the width of the web and d is the effective depth. ACI Code 9.9.3 provides minimum steel requirements for horizontal and vertical reinforcement within a deep beam. The minimum reinforcement perpendicular to a span is

$$A_v \geq 0.0025 b_w s_1 \tag{17.13}$$

FIGURE 17.13
Deep beam D-regions.

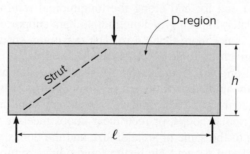

(a) Deep beam with $\ell \leq 4h$

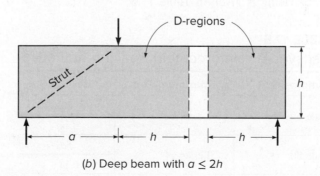

(b) Deep beam with $a \leq 2h$

where s_1 is the spacing of the reinforcement. The minimum reinforcement parallel to a span is

$$A_{vh} \geq 0.0025 b_w s_2 \tag{17.14}$$

where s_2 is the spacing of the reinforcement perpendicular to the longitudinal reinforcement. Spacings s and s_2 may not exceed $d/5$ or 12 in. ACI Code 9.9.3 allows Eq. (17.5) to be used in lieu of Eqs. (17.13) and (17.14). For strut-and-tie models, b_w equals the thickness of the element b.

f. Strut-and-Tie Method in Earthquake Applications

The strut-and-tie method applied to earthquake loading has additional detailing requirements. Strut-and-tie models in Seismic Design Categories D, E, and F must be designed in accordance with ACI Code Chapter 18. In lieu of meeting ACI Code Chapter 18 requirements, ACI Code 23.11 allows seismic effects to be multiplied by an overstrength factor not less than 2.5. Complementing this requirement, the ACI Code requires the effective compressive strength of the strut be multiplied by 0.8. The intent of the overstrength factor and reduced strut strength is to ensure that the strut remains essentially elastic during the earthquake.

ACI Code 23.11.3 contains two options for detailing struts in earthquake situations. Option one requires a minimum of four longitudinal bars, one in each corner of the strut, as required for columns of special moment frames. The longitudinal bars are required to have transverse reinforcement consisting of single or overlapping spirals, circular hoops, or rectilinear hoops engaging the peripheral longitudinal bars. Crossties of the same or smaller size and the hoops are permitted providing they are alternated end for end along the longitudinal reinforcement and are arranged such that the lateral spacing of longitudinal bars supported by the corner of a crosstie or hoop leg is not more than 14 in. The maximum spacing for the transverse reinforcement is given in Table 17.5.

Option two requires transverse reinforcement in orthogonal directions and through the thickness of the entire member cross section containing the struts instead of individual struts. The transverse reinforcement consists of single or overlapping spirals, circular hoops, or rectilinear hoops meeting the requirements of Table 17.6. Crossties of the same or smaller size and the hoops are permitted providing they are alternated end for end along the longitudinal reinforcement and are arranged such that no longitudinal bar has more than 8 in. lateral spacing. The maximum reinforcement spacing is given in Table 17.5.

TABLE 17.5
Transverse reinforcement spacing limitation

Reinforcement	Maximum Transverse Bar Spacing	
Grade 60	Lesser of	$6d_b$
		6 in.
Grade 80	Lesser of	$5d_b$
		6 in.
Grade 100	Lesser of	$4d_b$
		6 in.

TABLE 17.6
Transverse reinforcement for struts^{a,b}

Transverse Reinforcement	Applicable Expression	
A_{sh}/sb_c for rectilinear hoops	Greater of	$0.3\left(\dfrac{A_{cs}}{A_{ch}} - 1\right)\dfrac{f_c'}{f_{yt}}$
		$0.09\,\dfrac{f_c'}{f_{yt}}$

^{a} A_{ch} is measured to the outside edges of the transverse reinforcement for the strut.
^{b} It is permitted to configure hoops using two pieces of reinforcement.

Node design in models for earthquake applications is identical to design for other strut-and-tie nodes, except that the nominal concrete strength is multiplied by a factor of 0.8 to account for the deterioration of concrete strength due to cyclic loadings (Refs. 17.8 and 17.9). Earthquake tie design is the same as ordinary strut-and-tie design, except that the development length for ties in earthquake designs must be 1.25 times the standard development length to account for the strain hardening of the steel beyond the yield strength.

17.5 APPLICATIONS

While there are a number of possible applications for a strut-and-tie model, ACI Code 9.9 and 13.2 specifically allow deep beam and foundation design to be completed with this method. The following examples examine the details of deep beams and dapped beam end design by the strut-and-tie method. Additional examples of strut-and-tie modeling may be found in Chapter 18.

a. Deep Beams

Deep beams represent one of the principal applications of strut-and-tie models, since the alternative under ACI Code 9.9 is a nonlinear analysis. Two examples of deep beam design are presented, one that includes the application of concentrated loads at the upper surface of a transfer girder and one that addresses design for distributed and concentrated loads.

EXAMPLE 17.1 **Deep beam.** A transfer girder is to carry two 12 in. square columns, each with factored loads of 300 kips located at the third points of its 27 ft span, as shown in Fig. 17.14a. The beam has a thickness of 14 in. and a total height of 9 ft. Design the beam for the given loads, ignoring the self-weight, using $f_c' = 5000$ psi and $f_y = 60,000$ psi.

SOLUTION. The span-to-depth ratio for the beam is 3, thereby qualifying it as a deep beam. A strut-and-tie solution is selected.

Definition of D-region
All of the supports and loads are within h of each other or the supports, so the entire structure may be characterized as a D-region. The thickness of the struts and ties is equal to the thickness of the beam $b = 14$ in. Assuming an effective depth $d = 0.8h = 0.8 \times 9 = 7.2$ ft

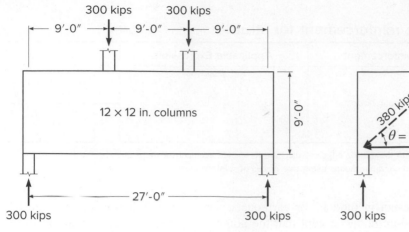

(a) Beam dimensions and loading

(b) Beam internal forces and trial truss model

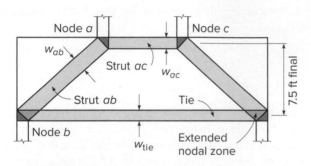

(c) Schematic showing widths of struts, ties, and nodes in final truss model

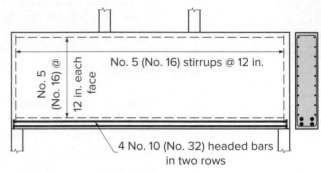

(d) Reinforcement details

FIGURE 17.14
Deep beam design for Example 17.1.

in the middle third of the beam, the maximum design shear capacity of the beam is $\phi V_n = \phi 10\sqrt{f'_c}\, b_w d = 0.75 \times 10\sqrt{5000} \times 12 \times 7.2 \times 12/1000 = 513$ kips. This is greater than $V_u = 300$ kips. Thus, the design may continue.

Force resultants on D-region boundaries
The 300 kip column loads on the upper face of the beam are equilibrated by 300 kip reactions at the supports, as shown in Fig. 17.14b. Based on an assumed center-to-center distance between the horizontal strut and the tie of $0.8h$, the trial diagonal struts form at an angle $\theta = 38.7°$ and carry a load of 380 kips. A horizontal 240 kip compression strut runs between the two column loads, and a 240 kip tension tie runs between the bottom nodes.

The truss model
Based on the beam geometry and loading, a single truss is sufficient to carry the column loads, as shown in Fig. 17.14c. The truss has a trapezoidal shape. This is an acceptable solution since the nodes are not true pins and instability within the plane of the truss is not a concern in a strut-and-tie model. The truss geometry is established by the assumed intersection of the struts and ties and used to determine θ.

Selecting dimensions for strut and nodal zones
Select a node dimension of 14 in. square under the column. The width of strut ab is then 14 in./$\sin\theta$ = 22.4 in. and strut ac is 22.4 in. × $\cos\theta$ = 17.5 in. The beam design will use

normalweight concrete and transverse reinforcement allowing λ and λ_s to be 1.0. Checking the shear capacity using Eq. (17.5.),

$$\phi V_c = \phi 5 \tan \theta \lambda \lambda_s \sqrt{f_c'} bd = 0.75 \times 5 \times \tan(38.7) \times 1 \times 1 \times \sqrt{5000} \times 14 \times 12 \times 7.2$$
$$= 257 \text{ kips}$$

which is greater than the sectional shear allowing a value of $\beta_s = 0.75$ for the interior strut, as shown in Table 17.2

Check capacity of struts and nodes
Checking the interior strut using Eq. (17.3), the capacity for $\beta_s = 0.75$ is

$$f_{ce} = \phi \beta_c \beta_s 0.85 f_c' = 0.75 \times 1.0 \times 0.75 \times 0.85 \times 5000 = 2390 \text{ psi}$$

the width of strut ac is 22.41 in. so the strut capacity is

$$F_{ac} = f_{ce} b w_{ab} = 2390 \times 14 \times 22.41 = 750 \text{ kips}$$

which exceeds the factored load of 380 kips. The width of the tie and the boundary strut $w_{ac} = w_{ab} \cos \theta = 17.5$ in. Checking the boundary strut, the strut capacity for $\beta_s = 1.0$ is

$$f_{ce} = \phi \beta_c \beta_n 0.85 f_c' = 0.75 \times 1.0 \times 1.0 \times 0.85 \times 5000 = 3190 \text{ psi}$$

and

$$F_{ac} = f_{ce} b w_{ab} = 3190 \times 14 \times 17.5 = 781 \text{ kips}$$

which exceeds the factored load of 200 kips. Thus, both struts are satisfactory.

Design ties and anchorage
The tie design consists of three steps: selection of the area of steel, design of the anchorage, and validation that the tie fits within the available tie width. The steel area is computed as $A_{ts} = F_{tu}/\phi f_y = 200/(0.75 \times 60) = 4.44 \text{ in}^2$. This is satisfied by using four No. 10 (No. 32) bars, having a total area of 5.08 in^2. The bars are placed in two layers of two bars. Headed bars are selected, and it is desired to have a center-to-center spacing of at least $6d_b = 7.6$ in. so that $\psi_p = 1.0$. For a No. 10 (No. 32) headed bar with a bearing area of $4A_b$ (gross area of $5A_b$), the diameter of the head is $2.24d_b = 2.24 \times 1.27$ in. $= 2.84$ in. With 1.5 in. cover to the heads, the center-to-center spacing of the bars is 14 in. $- 3 - 2.84$ in. $= 8.16$ in., which is greater than $6d_b$. Vertically, a center-to-center spacing of $6d_b = 7.6$ in. fits easily within the width of the tie w_{ac}. The overall structural depth of the final truss is 9 ft $- 17.5$ in./12 $= 7.5$ ft, more than the 7.2 feet assumed, which is conservative, and no revision is necessary. Using $\psi_e = 1.0$, $\psi_p = 1.0$, $\psi_o = 1.25$, and $\psi_c = 5{,}000/15{,}000 + 0.6 = 0.93$, the development length for a headed No. 10 (No. 32) bar is given by Eq. (6.10)

$$\ell_{dt} = \frac{f_y \psi_e \psi_p \psi_o \psi_c}{75 \sqrt{f_c'}} d_b^{1.5} = \frac{60{,}000 \times 1.0 \times 1.0 \times 1.25 \times 0.93}{75 \sqrt{5000}} 1.27^{1.5} = 18.8 \text{ in.}$$

The length of the node and extended nodal zone measured to the midheight of the tie is $14 + 17.5 \times (\cot 38.7°)/2 = 24.9$ in., which easily accommodates ℓ_{dt}. The bar is placed with 1½ in. cover from the beam end.

Design details and minimum reinforcement requirements
ACI Code 9.9.3.1 requires that transverse reinforcement in deep beams satisfy Eqs. (17.13) and (17.14). In addition, satisfying Eq. (17.5) ensures that sufficient reinforcement is present to control longitudinal splitting in the interior struts. Using $s_1 = s_2 = 12$ in. in Eqs. (17.13) and (17.14) will give the area of reinforcement per ft of member, which can be used, in turn, to select the bar size and spacing. Using this approach,

$$A_v = A_h = 0.0025 b_w s_1 = 0.0025 b_w s_2 = 0.0025 \times 14 \times 12 = 0.42 \text{ in}^2/\text{ft}$$

This is satisfied by No. 5 (No. 16) bars at 12 in., the maximum spacing allowed by the ACI Code, placed on each face, giving a total area of reinforcement equal to 0.62 in^2/ft. Final details are given in Fig. 17.14d.

EXAMPLE 17.2   **Deep beam with distributed loads**.   In addition to concentrated loads, the transfer girder shown in Fig. 17.15a carries a distributed factored load of 3.96 kips/ft applied along its top edge. Design for the given loads, plus the self-weight, using $f_c' = 5000$ psi and $f_y = 60,000$ psi.

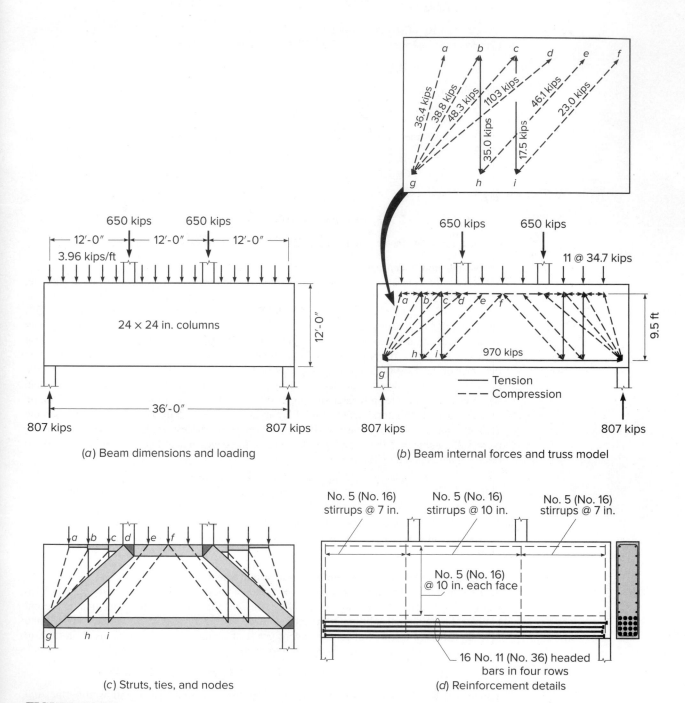

FIGURE 17.15
Deep beam with distributed loads for Example 17.2.

SOLUTION.   The factored self-weight of the beam is $1.2(12 \text{ ft} \times 2 \text{ ft} \times 0.15 \text{ kips/ft}^3) = 4.32 \text{ kips/ft}$. Thus, the total factored distributed load is $4.32 + 3.96 = 8.28 \text{ kips/ft}$, resulting in a total factored load of $8.28 \text{ kips/ft} \times 38 \text{ ft} = 314 \text{ kips}$, approximately 20 percent of the column loads. The solution follows Example 17.1 and accounts for the distributed loads. For this example, the self-weight of the beam is combined with the superimposed dead load. A more conservative solution could place the self-weight at the bottom of the beam and correspondingly increase the vertical tension tie requirements to transfer the load to the top flange. The top placement is used in this case because the self-weight is a small percentage of the total load and the concentrated forces are moved slightly toward the center of the beam for a conservative placement.

Definition of D-region
The entire beam is a D-region, as shown in Fig. 17.15a. The maximum factored shear in the beam is $V_u = 650 + 314/2 = 807 \text{ kips} < \phi 10 \sqrt{f_c'}\, b_w d = 1650 \text{ kips}$, the maximum design shear using $d = 9.6$ ft. Thus, the design can continue. A further check of the shear capacity using Eq. (17.3) with λ and λ_s equal to 1.0 for the section shows

$$V_u = \phi 5 \tan \theta \lambda \lambda_s \sqrt{f_c'}\, b_w d = 0.75 \times 5 \times \tan 38.7 \times 1.0 \times 1.0 \times \sqrt{5000} \times 24 \times 12 \times 9.5$$
$$= 581 \text{ kips}$$

which is less than the reaction, therefore $\beta_s = 0.4$ from Table 17.2.

Force resultants on D-region boundaries
The lower column reactions are equal to 807 kips. Based on the column size, stress at the beam-column interface is $p = 807/(24 \times 24) = 1.40 \text{ ksi}$.

The stress on the column node cannot exceed the effective concrete strength. For a C-C-T node,

$$p \le \phi f_{ce} = \phi 0.85 \beta_c \beta_n f_c' = 0.75 \times 1.0 \times 0.85 \times 0.80 \times 5000/1000 = 2.55 \text{ ksi}$$

Therefore, the bottom column of 24×24 in., having $p = 1.40 \text{ ksi} < 2.55 \text{ ksi}$, is adequate. The center-to-center distance between the horizontal strut and the horizontal tie at midspan is taken as 9.5 ft to compute the slope of the strut dg as $\theta = 38.7°$. The vertical dimension for struts ag, bg, and cg is assumed to be 10.5 ft because they are anchored closer to the top edge of the beam.

The total distributed load of 314 kips is represented by nine 35 kip concentrated loads placed at 3 ft centers, as shown in Fig. 17.15b. Distributed loads can be grouped at the discretion of the designer. It would be equally satisfactory to group them into 12 loads, placed one per foot, or combine some load with the column loads. The loads are combined with the column loads in this example to illustrate design for distributed loads. Using the geometric layout of the loads, strut-and-tie forces are computed and summarized in Fig. 17.15b and Table 17.7.

TABLE 17.7
Diagonal strut properties and capacities for Example 17.2

Strut	Vertical Load, kips	Slope, deg	Strut Axial Load, kips	w, in.	β_s	β_n	Node Capacity, kips	Horizontal Force, kips
dg	684.5	38.4	1102.8	22.4	0.4	0.8	1369.0	864.6
ag	35	74.1	36.4	1.1	0.4	0.8	70.0	10.0
bg	35	64.5	38.8	1.1	0.4	0.8	70.0	16.7
cg	35	46.4	48.3	1.1	0.4	0.8	70.0	33.3
bh	35	49.4	46.1	1.1	0.4	0.6	52.5	30.0
fi	17.5	49.4	23.0	0.6	0.4	0.6	26.3	15.0
Total tie force								969.6

The truss model

In addition to the struts and ties needed to carry the column loads, struts and ties to carry the distributed loads are now included in the truss. The distributed loads between the columns are carried by struts to the bottom chord; tension ties then transfer the vertical component of the load to the top chord, while the horizontal component is transferred to the bottom tie. The geometry of the struts is selected to allow tension ties to be placed vertically. The loads at nodes a, b, and c between the column and the support create a fan of compression struts to node g, as shown in Fig. 17.15b and c.

Selecting dimensions for strut and nodal zones

The forces in the "fan" struts are based on the geometry of the struts. The widths of struts ag, bg, and cg are calculated based on the strut capacity. The maximum stress in a strut with $\beta_c = 1.0$ when transverse reinforcement is used and $\beta_s = 0.4$ from above is

$$f_{ce} = \phi\beta_c\beta_s 0.85 f_c' = 0.75 \times 1.0 \times 0.4 \times 0.85 \times 5000 = 1275 \text{ psi}$$

To maintain uniform stresses in the struts, the strut width is the tensile force divided by $f_{ce} \times b$. The resulting geometries and loads are summarized in Table 17.7.

Capacity of struts and nodes

By inspection, the stresses in the fan struts ag, bg, and cg are satisfied by using the maximum allowable stress. Table 17.7 tabulates the node capacities, and in all cases, the node capacity exceeds the strut load. The total width of the struts, when using the maximum allowable stress for the struts, is less than the assumed node face. The design could be modified to match the node dimension; however, that refinement is not necessary because the struts have adequate capacity.

Design ties and anchorage

Tie design includes the horizontal thrust from the distributed loads, giving a total force of 970 kips. The required area of steel for the tie is $A_{ts} = F_{ut}/\phi f_y = 970/(0.75 \times 60) = 21.6 \text{ in}^2$ or sixteen No. 11 (No. 36) bars, which provide an area of 24.96 in². Headed bars are selected. The steel is placed in four layers of four bars. The minimum center-to-center spacing for headed bars is 3 bar diameters. For a No. 11 (No. 36) headed bar with a bearing area of $4A_b$ (gross area of $5A_b$), the diameter of the head is $2.24d_b = 2.24 \times 1.41 \text{ in.} = 3.16 \text{ in.}$ With 1.5 in. cover to the heads, the center-to-center spacing of the bars is $(24 - 3.0 - 3.16)/3 = 6.28 \text{ in.}$, which is greater than $3d_b$. At this spacing, however, $\psi_p = 1.6$. For a $3d_b$ center-to-center spacing of the bars and 1.5 in. to the heads, the depth of bottom boundary tie is $3 + 3.16 + 9 \times 1.41 = 18.9 \text{ in.}$ The length of the node and extended nodal zone measured to the midheight of the tie is $24 + 18.9 \times (\cot 38.7°)/2 = 47.6 \text{ in.}$ Using $\psi_e = 1.0$, $\psi_p = 1.6$, $\psi_o = 1.25$, and $\psi_c = 5,000/15,000 + 0.6 = 0.93$, the development length of No. 11 (No. 36) headed bars is

$$\ell_{dt} = \frac{f_y \psi_e \psi_p \psi_o \psi_c}{75 \sqrt{f_c'}} d_b^{1.5} = \frac{60,000 \times 1.25 \times 1.6 \times 1.25 \times 0.93}{75 \sqrt{5000}} 1.41^{1.5} = 44.0 \text{ in.}$$

which fits within the extended nodal zone.

The vertical tie bh carries 46.1 kips. The required area of steel for this tie distributed over a 3 ft width is $A_{ts} = F_{ut}/(3 \times \phi f_y) = 46.1/(3 \times 0.75 \times 60) = 0.34 \text{ in}^2/\text{ft}$, which is added to the transverse reinforcement.

The transverse reinforcement is 0.0025 times the gross area = $0.0025 \times 24 \times 12 \times 12 \text{ in./ft} = 0.72 \text{ in}^2/\text{ft}$. The transverse reinforcement is satisfied by No. 5 (No. 16) bars at 10 in. in each direction and No. 5 (No. 16) bars at 7 in. vertically between the center and end columns. The final details are given in Fig. 17.15d.

Design details and minimum reinforcement requirements

The minimum reinforcement requirements from Example 17.1 remain unchanged. The final details are shown in Fig. 17.15d.

FIGURE 17.16
Dapped beam ends and
strut-and-tie alternatives

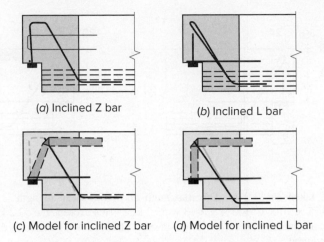

(a) Inclined Z bar   (b) Inclined L bar

(c) Model for inclined Z bar   (d) Model for inclined L bar

Examples 17.1 and 17.2 illustrate that the limiting size of the struts and nodes is determined by the element with the lowest strength, that is, the lowest value of f_{ce}. Examination of the β_s and β_n values in Tables 17.2 and 17.4 shows the variation in strength. Recognizing that using the lowest value of β (and f_{ce}) establishes the minimum usable strut and node dimensions allows the engineer to minimize the number of iterations needed to construct the truss model.

b. Dapped Beam Ends

Precast and prestressed concrete beams often have *dapped* or notched ends, as shown in Fig. 17.16, to reduce the floor-to-floor height of buildings. The recess allows structural overlap between the main beams and the floor beams. While the dapped end is advantageous in controlling building floor-to-floor height, it creates two structural problems. First, the shear at the end of the beam must be carried by a much smaller section, and second, the mechanism of load transfer through the notched zone is difficult to represent using conventional design techniques. As a result, dapped-end beams lend themselves to strut-and-tie design methods (Refs. 17.10 and 17.11).

EXAMPLE 17.3   **Design of a dapped beam end.** The 24 in. deep precast concrete T beam shown in Fig. 17.17 has an 8 in. thick web that carries factored end reactions of 53.2 kips in the vertical direction and a tensile force of 11 kips in the horizontal direction. The beam end is notched 10 in. vertically and 8 in. along the beam axis. The load is transferred to the support through a 4×8 in. bearing plate. Design the end reinforcement, using $f'_c = 5000$ psi and $f_y = 60,000$ psi.

SOLUTION. The combination of the concentrated load and the geometric discontinuity suggests the use of a strut-and-tie solution. The 14 in. nib depth is close to the 60 percent desired for successful implementation.

Definition of D-region
The D-region for this beam is approximately one structural depth in from the end of the notch. The bearing plate has longitudinal reinforcement welded to it to allow for horizontal load transfer and a vertical bar for shear-friction design of the end. The effective depth at the notch

FIGURE 17-17
Detail of dapped beam end

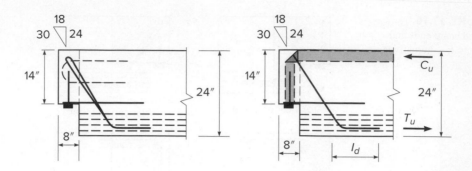

is taken as 13.0 in. The maximum allowable shear strength is $V_u \leq \phi V_n = \phi 10 \sqrt{f_c'} wd = 0.75 \times 10 \times \sqrt{5000} \times 8 \times 13 = 55.2$ kips. This exceeds the 53.2 kip applied load, so the section is adequate to proceed with the design. The load is high enough that transverse reinforcement is required.

Force resultants on D-region boundaries and the truss model
The truss model follows the inclined L configuration, as shown in Fig. 17.17. The assumed depth of the truss is 18 in. The tensile and compressive forces at the D-region interface are $T_u = C_u = V_u \times 24/18 = 70.9$ kips.

Selecting dimensions for strut and nodal zones
The nodal zone stress is established at the bearing plate. The stress at the node and the strut is $V_u/b_w d = 53.2/4 \times 8 = 1.66$ ksi.

Strength of struts
The strut strength for a boundary strut with $\beta_c = 1.00$ and $\beta_s = 1.00$ is $\phi \times 1.0 \times 1.0 \times 0.85 f_c' \sqrt{5000} = 2.39$ ksi. This exceeds the applied load of 1.66 ksi, so the strut is adequate. The horizontal strut matching the D-region boundary is assumed to be 4 in. deep and 8 in. wide. The stress at the boundary is $T_u/w \times b = 70.9/(4 \times 8) = 2.12$ ksi, which is less than the 2.39 ksi allowed.

Design ties and anchorage
The tension tie force is 30/24 times the vertical reaction or $T_u = 53.2 \times 30/24 = 66.5$ kips. The area of reinforcement required is

$$A_s = \frac{T_u}{\phi f_y} = \frac{66.5}{0.75 \times 60} = 1.48 \text{ in}^2$$

Two No. 6 (No. 19) hairpin bars provide $A_s = 4 \times 0.44 = 1.76$ in.^{2}.

The No. 6 (No. 19) bars extend into the web and have a full tension lap splice with the prestressing reinforcement. Checking for a curved bar node for the No. 6 (No. 19) bar from Eq. (17.8)

$$r_b = \frac{1.5 A_{ts} f_y}{w_t f_c'} = \frac{1.5 \times 2 \times 0.44 \times 60,000}{4 \times 5000} = 3.96 \text{ in.}$$

which is greater than the standard bend diameter, so a special bend of 8 in. diameter is required. The hairpin is to be placed with the bend vertical to fit within the web.

Design details and minimum reinforcement requirements
The horizontal force of 11 kips is taken by a bar welded to the bearing plate. The area of steel is $A_s = 11/(0.75 \times 60) = 0.24$ in.^{2}, indicating one No. 5 bar is sufficient. To prevent a shear friction failure at the bearing plate, the area of reinforcement for a friction factor of $\mu = 1.4$ is

$$A_{sv} = \frac{V_u}{\phi f_y \mu} = \frac{53.2}{0.75 \times 60 \times 1.4} = 0.84 \text{ in}^2$$

Use two No. 6 (No. 19) bars or two ¾ in. diameter headed studs.

The design is completed by detailing the inclined L bar to have a development length along the length of the strand to assure that the tensile force is transmitted into the longitudinal reinforcement.

These examples illustrate the methodology of strut-and-tie design and the importance of understanding the detailing requirements needed to transfer forces at nodes and to the corresponding reinforcement. Failure to appreciate the need to provide anchorage for the tie or to supply thrust resistance for the struts can lead to failure. In the examples, the contact area was used to establish the hydrostatic nodal pressure. As discussed, an equally acceptable solution would have been to select the maximum stress for one of the struts because β_s is typically lower than β_n is for nodes. The diagonal hairpin loop ensures anchorage at the top node, and the lower portion provides transfer to the prestress.

REFERENCES

17.1.   J. Schlaich, K. Schäfer, and M. Jennewein, "Toward a Consistent Design of Structural Concrete," *J. PCI*, vol. 32, no. 3, May–June 1987, pp. 74–150.

17.2.   P. Marti, "Truss Models in Detailing," *Concr. Intl.*, vol. 7, no. 12, 1985, pp. 66–73.

17.3.   P. Marti, "Basic Tools for Reinforced Concrete Design," *J. ACI*, vol. 82, no. 1, 1985, pp. 46–56.

17.4.   J. Schlaich and K. Schäfer, "Design and Detailing of Structural Concrete Using Strut-and-Tie Models," *Struct. Eng.*, vol. 69, no. 6, March 1991, 13 pp.

17.5.   *Building Code Requirements for Structural Concrete and Commentary*, Appendix A, ACI 318-08, American Concrete Institute, Farmington Hills, MI, 2008.

17.6.   G. J. Klein, "Curved-Bar Nodes," *Concr. Intl.*, vol. 30, no. 9, 2008, pp. 43–47.

17.7.   L. Chow, H. Conway, and G. Winter, "Stresses in Deep Beams," *Trans. ASCE*, vol. 118, 1953, p. 686.

17.8.   M. Mansour and T. Hsu, "Behavior of Reinforced Concrete Elements under Cyclic Shear. 2: Theoretical Model," *J. Struct. Eng.*, vol. 131, no. 11, 2005, pp. 1–12.

17.9.   N. H. T. To, S. Sritharan, and J. Ingham, "Strut-and-Tie Nonlinear Cyclic Analysis of Concrete Frames," *J. Struct. Eng.*, vol. 135, no. 10, 2009, pp. 1259–1268.

17.10.   W.-Y. Lu, I.-J. Lin, and H.-W. Yu , "Behavior of Reinforced Concrete Dapped-End Beams," *Mag. Concrete Res.*, vol. 64, no. 9, 2012, pp. 793–805.

17.11.   G. J. Klein, B. M. Andrews, and K. P. Holloway, *Development of Rational Design Methodologies for Dapped Ends of Prestressed Concrete Thin-Stemmed Members, Draft final report*, PCI R&D Committee, July 23, 2015, 218 pp.

PROBLEMS

17.1.   Redesign Example 17.1 with a single 300 kip column load 9 ft in from the left end of the beam. In your solution, include (*a*) a sketch of the load path and truss layout, (*b*) the sizes and geometry of the struts, ties, and nodal zones, and (*c*) a complete sketch of the final design.

17.2.   Redesign the column bracket shown in Example 18.5 using the strut-and-tie method. Your strut-and-tie model may be based on Fig. 18.23. Material properties remain the same as in Example 18.5.

17.3.   A transfer girder has an overall depth of 11 ft and spans 22 ft between column supports. In addition to its own weight, it will pick up a uniformly distributed factored load of 4.0 kips/ft from the floor above and will carry a 14 × 14 in. column delivering a concentrated factored load of 270 kips from floors above at midspan. The girder width must be equal to or less than 16 in. Design the beam for the given loads. Find the girder width and the area and geometry of tie steel, and specify the placement details. Material strengths are $f_c' = 5000$ psi and $f_y = 60{,}000$ psi.

17.4. A column transfers a factored load of 700 kips to the 9 ft square spread footing shown in Fig. P17.4, resulting in a factored uniform soil pressure of 8640 psf. Design the footing reinforcement using strut-and-tie methods. Material strengths are $f_c' = 4000$ psi and $f_y = 60,000$ psi. Because footings typically contain no shear reinforcement, your design should be based on unreinforced interior struts.

FIGURE P17.4

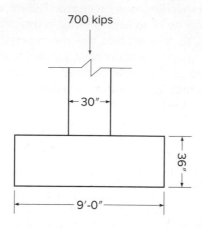

700 kips

30″

36″

9′-0″

17.5. Redesign the footing in Problem 17.4 using traditional flexure and shear methods, as described in Chapter 15. Compare your solution to the solution for Problem 17.4, and comment on your results.

17.6. Design the pile cap in Problem 15.5 for pile load transfer to column using a strut-and-tie method.

18 *Design of Reinforcement at Joints*

18.1 INTRODUCTION

Most reinforced concrete failures occur not because of any inadequacies in analysis of the structure or in design of the members but because of inadequate attention to the detailing of reinforcement. Frequently, the problem is at the connections of main structural elements (Ref. 18.1).

There is a tendency in modern structural practice for the engineer to rely upon a detailer, employed by the reinforcing bar fabricator, to provide the joint design. Certainly, in many cases, standard details such as those found in the ACI Detailing Manual (Ref. 18.2) can be followed, but only the design engineer, with the complete results of analysis of the structure at hand, can make this judgment. In many other cases, special requirements for force transfer require that joint details be fully specified on the engineering drawings, including bend configurations and cutoff points for main bars and provision of supplementary reinforcement.

The basic requirement at joints is that all of the forces existing at the ends of the members must be transmitted through the joint to the supporting members. Complex stress states exist at the junction of beams and columns, for example, that must be recognized in designing the reinforcement. Sharp discontinuities occur in the direction of internal forces, and it is essential to place reinforcing bars, properly anchored, to resist the resulting tension. Some frequently used connection details, when tested, have been found to provide as little as 30 percent of the strength required (Refs. 18.1 and 18.3).

Over the years, research has been directed toward establishing an improved basis for joint design (Refs. 18.4 and 18.5). Full-scale tests of beam-column joints have led to improved design methods such as those described in *Recommendations for Design of Beam-Column Joints in Monolithic Reinforced Concrete Structures*, reported by ACI-ASCE Committee 352 (Ref. 18.6). The recommendations are largely incorporated in Chapter 15 of the ACI Code, which covers beam-column and beam-slab joints. Such recommendations provide a basis for the safe design of beam-column joints both for ordinary construction and for buildings subject to seismic forces. Other tests have given valuable insight into the behavior of beam-girder joints, wall junctions, and other joint configurations, thus providing a sound basis for design.

The practicality of the joint design should not be overlooked. Beam reinforcement entering a beam-column joint must clear the vertical column bars, and timely consideration of this fact in selecting member widths and bar size and spacing can avoid costly delays in the field. Similarly, beam reinforcement and girder reinforcement, intersecting at right angles at a typical beam-girder-column joint, cannot be in

FIGURE 18.1
Reinforcement congestion at
beam-girder-column joint.
(*Photograph by Arthur
H. Nilson.*)

the same horizontal plane as they enter the joint. Figure 18.1 illustrates the congestion of reinforcing bars at such an intersection. Concrete placement in such a region is difficult at best, but is assisted with the use of plasticizer admixtures.

Most of this chapter treats the design of joint regions for typical continuous-frame monolithic structures that are designed according to the strength requirements of the ACI Code for gravity loads or normal wind loads. Joints connecting members that must sustain strength under reversals of deformation into the inelastic range, as in earthquakes, represent a separate category and are covered in Chapter 20. Brackets and corbels, although they are most often a part of precast buildings rather than monolithic construction, have features in common with monolithic joints, and these will be covered here.

18.2   BEAM-COLUMN JOINTS

A *beam-column joint* is defined as the portion of a column within the depth of the beams that frame into it. Formerly, the design of monolithic joints was limited to providing adequate anchorage for the reinforcement. However, the increasing use of high-strength concrete, resulting in smaller member cross sections, and the use of larger-diameter and higher-strength reinforcing bars now require that greater attention be given to joint design and detailing. Although guidance is provided by Chapter 15 of the ACI Code, the ACI-ASCE Committee 352 report *Recommendations for Design of Beam-Column Joints in Monolithic Reinforced Concrete Structures* (Ref. 18.6) provides a basis for the design of joints in both ordinary structures and structures required to resist heavy cyclic loading into the inelastic range.

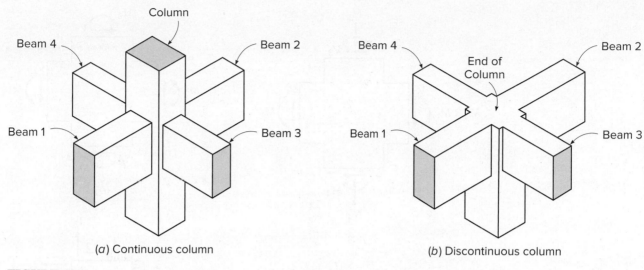

FIGURE 18.2
Typical monolithic interior beam-column joint.

a. Classification of Joints

Reference 18.6 classifies structural joints into two categories. A *Type 1* joint connects members in an ordinary structure designed on the basis of strength, according to the main body of the ACI Code, to resist gravity and normal wind load. A *Type 2* joint connects members designed to have sustained strength under deformation reversals into the inelastic range, such as members in a structure designed for earthquake motions, high winds, or blast effects. Type 1 joints are considered in ACI Code Chapter 15 and this chapter. Joints for earthquake design are addressed in ACI Code Chapter 18 and in Chapter 20.

Figure 18.2*a* shows typical *interior joints* in a monolithic reinforced concrete frame, with beams 1 and 2 framing into opposite faces of the column and beams 3 and 4 framing into the column faces in the perpendicular direction. An *exterior joint* would include beams 1, 2, and 3, or in some cases only beams 1 and 2. A *corner joint* would include only beams 1 and 3, or occasionally only a single beam 1. A joint may have beams framing into it from two perpendicular directions as shown, but for purposes of analysis and design each direction can be considered separately. Joint confinement is further affected if the column is discontinuous, as illustrated in Fig. 18.2*b*.

b. Joint Loads and Resulting Forces

The joint region must be designed to resist forces that the beams and column transfer to the joint, including axial loads, bending, torsion, and shear. Figure 18.3*a* shows joint loads acting on the free body of a typical joint of a frame subject to gravity loads, with moments M_1 and M_2 acting on opposite faces, in the opposing sense. In general, these moments will be unequal, with their difference equilibrated by the sum of the column moments M_3 and M_4. Figure 18.3*b* shows the resulting forces to be transmitted through the joint. Similarly, Fig. 18.4*a* shows the loads on a joint in a structure subjected to sidesway loading. The corresponding joint forces are shown in Fig. 18.4*b*. Only for heavy lateral loading, such as from seismic forces, would the moments acting on opposite faces of the joint act in the same sense, as shown here, producing very high horizontal shears within the joint.

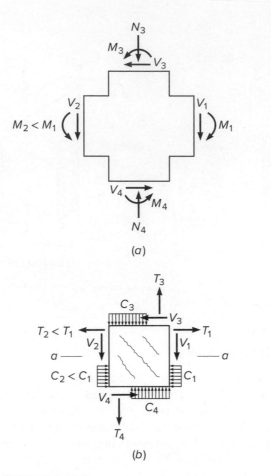

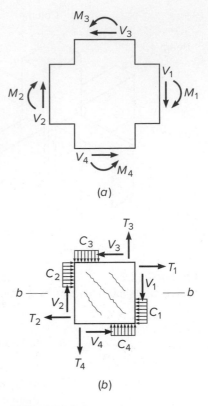

FIGURE 18.3

Joint loads and forces resulting from gravity loads: (*a*) forces and moments on the free body of a joint region and (*b*) resulting internal forces.

FIGURE 18.4

Joint loads and forces resulting from lateral loads: (*a*) forces and moments on the free body of a joint and (*b*) resulting internal forces.

According to the recommendations by Committee 352, the forces to be considered in designing joint regions are not those determined from the conventional frame analysis; rather, they are calculated based on the *nominal strengths of the members*. The column joint shear force acts on a horizontal plane midheight in the joint. The shear is based on the maximum moment transferred to the joint based on a factored load analysis. Alternatively, the shear in the direction under consideration is based on the beam nominal moment strength M_n.

Where a typical underreinforced beam meets the column face, the tension force from the negative moment reinforcement at the top of the beam is to be taken as $T = A_s f_y$, and the compression force at the face is from equilibrium $C = T$, not the nominal compressive capacity of the concrete. The design moment applied at the joint face is that corresponding to these maximum forces, $M_u = M_n = A_s f_y (d - a/2)$, rather than that from the overall analysis of the frame. Note that the inclusion of the usual strength reduction factor ϕ would be unconservative in the present case because it would reduce the forces for which the joint is to be designed; it is therefore not included in this calculation.

FIGURE 18.5
Free-body diagram of an
interior column and joint
region.

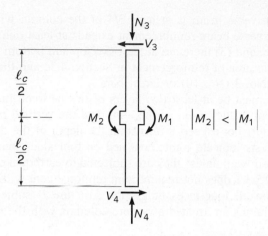

With the moment applied to each joint face found in this way, the corresponding column forces for joint design are those forces required to keep the connection in equilibrium. To illustrate, the column shears V_3 and V_4 of Figs. 18.3a and 18.4a are calculated based on the free body of the column between inflection points, as shown in Fig. 18.5. The inflection points generally can be assumed at column midheight, as shown.

c.   **Confinement and Transverse Joint Reinforcement**

The successful performance of a beam-column joint depends strongly on the lateral confinement of the joint. Confinement has two benefits: (1) The core concrete is strengthened and its strain capacity improved, and (2) the vertical column bars are prevented from buckling outward. Confinement can be provided either by the beams that frame into the joint or by special column ties provided within the joint region.

Confinement by beams is illustrated in Fig. 18.6. ACI Code 15.2.8 considers a beam-column joint confined for the direction of the joint shear under consideration if two transverse beams are provided that satisfy three requirements: (1) The width

FIGURE 18.6
Confinement of joint
concrete by beams:
(a) confinement in X and Y
directions and (b) confinement
in X direction only.

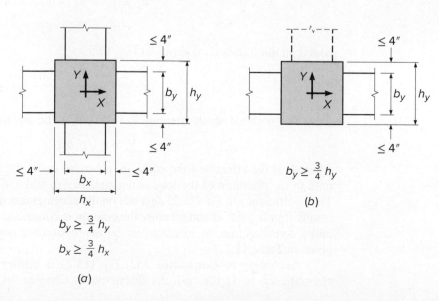

of each transverse beam is at least 3/4 of the column width into which it frames, (2) the transverse beam reinforcement extends at least one beam depth h beyond the column face, and (3) there are at least two top and bottom reinforcing bars that meet the minimum area of reinforcement in Section 4.3e and the transverse reinforcement consists of No. 3 (No. 10) or larger bars.

There must be at least two layers of ties between the top and bottom flexural reinforcement in the beams at the joint, and the vertical center-to-center spacing of these ties must not exceed 8 in. within the depth of the deepest beam framing into the joint. Tests indicate joints confined on four sides, but without shear reinforcement, perform well unless they are subjected to earthquake effects (Ref. 18.7). Thus, ACI Code 15.3.1 does not require joint reinforcement within interior joints confined by beams on all four faces unless the structure is subject to earthquake loading. Circular columns are treated as square columns with the same area.

d.   Shear Strength of a Joint

A joint subject to the forces shown in Fig. 18.3b or 18.4b will develop a pattern of diagonal cracking owing to the diagonal tensile stresses that result from the normal forces and shears, as indicated by those figures. The approach used by Committee 352 is to limit the shear force on a horizontal plane through the joint to a value established by tests. The design basis is

$$\phi V_n \geq V_u \tag{18.1}$$

where V_u is the applied shear force, V_n is the nominal shear strength of the joint, and ϕ is taken equal to 0.75.

The shear force V_u is to be calculated on a horizontal plane at midheight of the joint, such as plane a-a of Fig. 18.3b or plane b-b of Fig. 18.4b, by summing horizontal forces acting on the joint above that plane. For example, in Fig. 18.3b the joint shear on plane a-a is

$$V_u = T_1 - T_2 - V_3$$

and in Fig. 18.4b, the joint shear on plane b-b is

$$V_u = T_1 + C_2 - V_3$$
$$= T_1 + T_2 - V_3$$

Based on the free-body diagram in Fig. 18.5,

$$V_3 = V_4 = \frac{M_1 - M_2}{\ell_c}$$

The nominal shear strength V_n is given by the equation

$$V_n = \gamma \sqrt{f_c'}\, b_j h_c \tag{18.2}$$

where b_j is the effective joint width in inches, h_c is the thickness in inches of the column in the direction of the load being considered, and $\sqrt{f_c'}$ is expressed in psi units. The coefficient γ in Eq. (18.2) depends on the confinement of the joint provided by the beams framing into it and whether the column is continuous or terminates at the level under consideration, as indicated in the expression for nominal joint shear strength given in Table 18.1.

According to Committee 352, Eq. (18.2) is conservative for concretes with strengths up to 15,000 psi. As discussed in Chapter 20, ACI Code 18.8 follows

TABLE 18.1
Nominal joint shear strength V_n

Column	Beam in Direction of V_u	Confinement by Transverse Beams 15.2.8	V_n (lb)[a]
Continuous or contains an extension for reinforcement	Continuous or contains an extension for reinforcement	Confined	$24\lambda\sqrt{f_c'}A_j$
		Not confined	$20\lambda\sqrt{f_c'}A_j$
	Other	Confined	$20\lambda\sqrt{f_c'}A_j$
		Not confined	$15\lambda\sqrt{f_c'}A_j$
Other	Continuous or contains an extension for reinforcement	Confined	$20\lambda\sqrt{f_c'}A_j$
		Not confined	$15\lambda\sqrt{f_c'}A_j$
	Other	Confined	$15\lambda\sqrt{f_c'}A_j$
		Not confined	$12\lambda\sqrt{f_c'}A_j$

[a] λ is 0.75 for concrete containing lightweight aggregate and 1.0 for concrete with normalweight aggregate.

similar procedures for the design of joints in moment resistant frames, the only difference being that lower values for the coefficient γ are recommended.

For joints with beams framing in from two perpendicular directions, as for a typical interior joint, the horizontal shear should be checked independently in each direction. Although such a joint is designed to resist shear in two directions, only one classification is made for the joint in this case (that is, only one value of γ is selected based on the joint classification, and that value is used to compute V_n when checking the design shear capacity in each direction).

According to Committee 352 recommendations, the effective joint width b_j to be used in Eq. (18.2) depends on the transverse width of the beams that frame into the column as well as the transverse width of the column. With regard to the beam width b_b, if there is a single beam framing into the column in the load direction, then b_b is the width of that beam. If there are two beams in the direction of shear, one framing into each column face, then b_b is the average of the two beam widths. In reference to Fig. 18.7a, when the beam width is less than the column width, the effective joint width is equal to the smaller of the average of the beam width and column width,

$$b_j = \frac{b_b + b_c}{2} \tag{18.3}$$

and

$$b_j = b_b + \sum\frac{mh_c}{2} \tag{18.4}$$

where m is a slope that depends on the eccentricity e of the beam centerline with respect to the column centroid (Fig. 18.7b). If e is greater than $b_c/8$, $m = 0.3$; otherwise, $m = 0.5$. As shown in Fig. 18.7a, the slope m defines a width at the centroid of the column. According to Committee 352, $mh_c/2$ should not be taken as greater than the extension of the column edge beyond the edge of the beam.

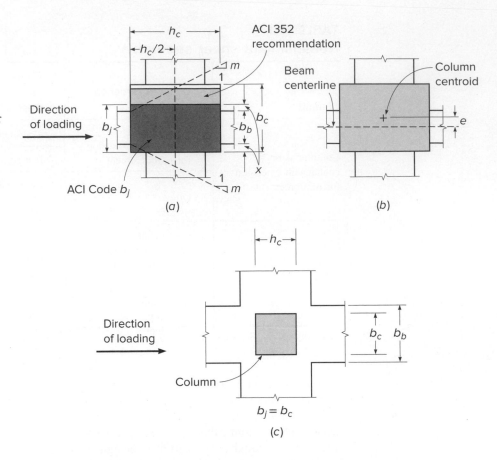

ACI Code 15.4.2 simplifies the ACI 352 requirements by specifying that b_j be the lesser of the column width or $b_j + 2x$, as shown in Fig. 18.7a. If the beam width b_b *exceeds* the column width, the effective joint width b_j is equal to the column width b_c, as shown in Fig. 18.7c.

e.   Anchorage and Development of Beam Reinforcement

For interior joints, normally the flexural reinforcement in a beam entering one face of the joint is continued through the joint to become the flexural reinforcement for the beam entering the opposite face. Therefore, for loadings associated with Type 1 joints, pullout is unlikely, and no special recommendations are made. However, for exterior or corner joints, where one or more of the beams do not continue beyond the joint, a problem of bar anchorage exists. The critical section for development of the yield strength of the beam reinforcement is at the face of the column. Column dimensions seldom permit development of the reinforcement entering the joint by straight embedment alone, and hooks are usually needed for the negative beam reinforcement. Headed bars or 90° hooks are used, with the hook extending toward and beyond the middepth of the joint. If the bottom bars entering the joint need to develop their strength $A_s f_y$ at the face of the joint, as they do if the beam is a part of a primary lateral-force-resisting system, they should also be anchored with 90° hooks, in this case turned upward to extend toward the middepth of the joint, or headed bars. Requirements for development of hooked bars given in Chapter 6 are applicable in both cases, including modification factors for concrete cover and for enclosure with ties or stirrups.

EXAMPLE 18.1   **Design of exterior Type 1 joint.**   Design the exterior joint shown in Fig. 18.8 as a part of a continuous, monolithic, reinforced concrete frame designed to resist gravity loads only. Member section dimensions $b \times h$ and reinforcements are as shown. The frame story height is 12 ft. Material strengths are $f_c' = 4000$ psi and $f_y = 60,000$ psi.

SOLUTION.   First the joint geometry must be carefully laid out, to be sure that beam bars and column bars do not interfere with one another and that placement and vibration of the concrete are practical. In this case, bar layout is simplified by making the column 4 in. wider than the beams. Column reinforcement is placed with the usual 1.5 in. of concrete outside of the No. 4 (No. 13) ties. Beam top and bottom bars are placed just inside the outer column bars. The slight offset of the center top beam bars to avoid the center column bars is of no concern. Top bars of the spandrel beams are placed just under the top normal beam bars, except for the outer spandrel bar, which is above the hook shown in Fig. 18.8*b*. Bottom bars enter the joint at different levels without interference.

Confinement is provided in the direction of the spandrel beams by the beams themselves because the spandrel width of 18 in. exceeds $\frac{3}{4}$ of the column width in the direction of the normal beam. However, because the joint is not confined on all four sides, column ties that satisfy Eq. (5.13) must be provided within the joint. The ties must be distributed over a column height not less than the deepest beam framing into the column with a spacing that does not exceed one-half of the depth of the shallowest beam. Solving Eq. (5.13) for the spacing s gives Eq. (5.15*a*). Using No. 4 (No. 13) bars as ties

$$s_{max} = \frac{A_{vt}f_{yt}}{0.75\sqrt{f_c'}b} = \frac{0.40 \times 60,000}{0.75\sqrt{4000} \times 18} = 28.1 \text{ in.}$$

$$\leq \frac{A_{vt}f_{yt}}{50b} = \frac{0.40 \times 60,000}{50 \times 18} = 26.7 \text{ in.}$$

FIGURE 18.8

Exterior beam-column joint for Example 18.1: (*a*) plan view; (*b*) cross section through spandrel beam; and (*c*) cross section through normal beam. Note that beam stirrups and column ties outside of the joint are not shown.

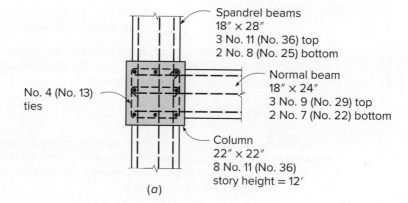

Spandrel beams
18″ × 28″
3 No. 11 (No. 36) top
2 No. 8 (No. 25) bottom

Normal beam
18″ × 24″
3 No. 9 (No. 29) top
2 No. 7 (No. 22) bottom

No. 4 (No. 13) ties

Column
22″ × 22″
8 No. 11 (No. 36)
story height = 12′

(*a*)

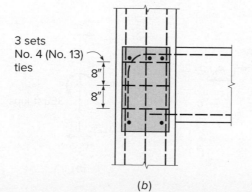

3 sets
No. 4 (No. 13) ties

8″

8″

(*b*)

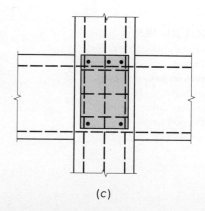

(*c*)

Clearly, the 8 in. maximum spacing criterion of ACI Code 15.3.1.4 controls. Three sets of No. 4 (No. 13) ties spaced at 8 in. are provided, as shown in Fig. 18.8. The clear distance between column bars is 5.89 in. here, less than 6 in., so strictly speaking the single-leg crosstie is not required. However, it will improve the joint confinement, guard against outward buckling of the central No. 11 (No. 36) column bar, and add little to the cost of construction, so crossties will be specified, as shown in Fig. 18.8a. Note that a 90° hook at one end, rather than the 135° bend shown, meets ACI Code tie anchorage requirements and facilitates reinforcement fabrication.

No anchorage problems exist for the spandrel beam top reinforcement, which is continuous through the joint. However, the normal beam top reinforcement must be provided with hooks to develop its yield strength at the face of the column. Being inside the column bars, the beam top bars have side cover of $1.5 + 0.5 + 1.4 = 3.4$ in. The center-to-center spacing of the No. 9 (No. 29) hooked bars $= (22 - 2 \times 3.4 - 1.128)/2 = 7.04$ in., which exceeds $6d_b$, giving $\psi_r = 1.0$. The side cover to the hooked bars exceeds 2.5 in., so the modification factor $\psi_o = 1.0$, while $\psi_e = 1.0$. For 4000 psi concrete, $\psi_c = f_c'/15{,}000 + 0.6 = 4000/15{,}000 + 0.6 = 0.87$ Referring to Eq. (6.8), the required hook development length is

$$\ell_{dh} = \left(\frac{f_y \psi_e \psi_r \psi_o \psi_c}{55\lambda\sqrt{f_c'}}\right)d_b^{1.5} = \left(\frac{60{,}000 \times 1.0 \times 1.0 \times 1.0 \times 0.87}{55 \times 1.0 \times \sqrt{4000}}\right)1.128^{1.5} = 18.0 \text{ in.}$$

If the hooked bars are carried down just inside the column ties, the actual embedded length is $22.0 - 1.5 - 0.5 = 20.0$ in., exceeding 18.0 in., so development is ensured. None of the beams are a part of the primary, lateral-force-resisting system of the frame, so the bottom bars simply can be carried 6 in. into the face of the joint and stopped.

Next the shear strength of the joint must be checked. In the direction of the spandrel beams, moments applied to the joint will be about the same and acting in the opposite sense, so very little joint shear is expected in that direction. However, the normal beam will subject the joint to horizontal shears. In reference to Fig. 18.9a, which shows a free-body sketch of the top half of the joint, the maximum force from the beam top reinforcement is

$$A_s f_y = 3.00 \times 60 = 180 \text{ kips}$$

The joint moment is calculated based on this tensile force. The normal beam effective depth is $d = 24.0 - 1.5 - 0.5 - 1.128/2 = 21.4$ in. and with stress block depth $a = A_s f_y/0.85f_c'b_w = 180/(0.85 \times 4 \times 18) = 2.94$ in., the design moment is

$$M_u = M_n = A_s f_y\left(d - \frac{a}{2}\right) = \frac{180}{12}\left(21.4 - \frac{2.94}{2}\right) = 299 \text{ ft-kips}$$

Column shears corresponding to this joint moment are found based on the free body of the column between assumed midheight inflection points, as shown in Fig. 18.9b: $V_{\text{col}} = 299/12 = 24.9$ kips. Then summing horizontal forces on the joint above the middepth plane a-a, the joint shear in the direction of the normal beam is

$$V_u = 180 - 24.9 = 155 \text{ kips}$$

FIGURE 18.9

Basis of column shear for Example 18.1: (a) horizontal forces on joint free-body sketch and (b) free-body sketch of column between inflection points.

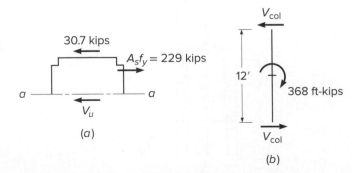

(a)

(b)

The joint width $b_j = b + 2x = 18 + 4 = 22$ in. $= h$. The joint does not meet the bar extension requirements but does meet the confinement criteria. From Table 18.1, the nominal shear strength is $15\lambda\sqrt{f_c'}\,A_j$.

Using $b_j = 22$ in., the nominal and design shear strengths of the joint are, respectively,

$$V_n = \gamma\sqrt{f_c'}\,b_j h = 15 \times 1.0\,\sqrt{4000} \times 22 \times \frac{22}{1000} = 459 \text{ kips}$$

$$\phi V_n = 0.75 \times 455 = 341 \text{ kips}$$

The applied shear $V_u = 155$ kips does not exceed the design strength, so shear is satisfactory.

EXAMPLE 18.2   **Design of interior Type 1 joint.**   Figure 18.10 shows a proposed interior joint of a reinforced concrete building, with beam and column dimensions and reinforcement as indicated. The building frame is to carry gravity loads and normal wind loads. Design and detail the joint reinforcement.

SOLUTION.   Because the joint is to be a part of the primary, lateral-force-resisting system, beam bottom bars as well as top bars are carried straight through the joint for anchorage. In such cases, it is usually convenient to lap splice the bottom reinforcement near the point of inflection of the beams.

In Fig. 18.10a and b, top and bottom beam bars entering the joint in one direction must pass, respectively, under and over the corresponding bars in the perpendicular direction. It will be assumed that this has been recognized by adjusting the effective depths in designing

FIGURE 18.10

Interior beam-column joint for Example 18.2: (a) plan view and (b) section through beam.

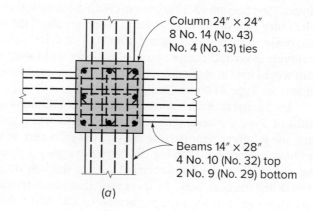

Column 24″ × 24″
8 No. 14 (No. 43)
No. 4 (No. 13) ties

Beams 14″ × 28″
4 No. 10 (No. 32) top
2 No. 9 (No. 29) bottom

(a)

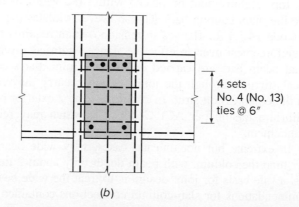

4 sets
No. 4 (No. 13)
ties @ 6″

(b)

the beams. Because the column is 10 in. wider than the beams, the outer beam bars can be passed inside the corner column bars without interference. Four bars are used for the beam top reinforcement in order to avoid interference with the center column bar.

Even the combination of normal wind loading with gravity loads should not produce large unbalanced moment on opposite faces of this interior column, and it can be safely assumed that joint shear will not be critical. However, confinement of the joint region by the beams is considered inadequate because the beam width of 14 in. is less than $\frac{3}{4}$ the column width of 24 in. Consequently, transverse column ties must be added within the joint for confinement. For the 24 in. square column, the spacing between the vertical bars exceeds 6 in., so it is necessary, according to the ACI Code, to provide ties to support the intermediate bars as well as the corner bars. Three ties are used per set, as shown in Fig. 18.10a. Since the joint is a part of the lateral-force-resisting system, the vertical spacing of these tie sets, 6 in., meets the maximum spacing requirement of 8 in. Four sets within the joint, as indicated in Fig. 18.10b, are adequate to satisfy this requirement.

f. Wide-Beam Joints

In multistory buildings, to reduce the construction depth of each floor and to reduce the overall building height, wide shallow beams are sometimes used. Joint design in cases where the beams are wider than the column introduces some important concepts not addressed in the Committee 352 report. It is important to equilibrate all of the forces applied to the joint. The tension from the top bars in the usual case, with beam width no greater than the column, are equilibrated by the horizontal component of a diagonal compression strut within the joint. The diagonal compression at the ends of the strut, in turn, is equilibrated by the beam compression and the thrust from the column. (See Section 18.3 for a more complete description of the strut-and-tie model.) If the outer bars of the normal beam pass *outside* of the column, as they often do in wide-beam designs, the diagonal strut will also be outside of the column, with no equilibrating vertical compression at its upper and lower ends. The outer parts of the beam would tend to shear off, resulting in premature failure. The problem is of special concern for Type 2 connections.

To minimize the problem, Committee 352 suggests that satisfactory performance of Type 2 connections with wide beams will result if, to provide satisfactory bond, the reinforcement passing outside the joint core is selected so that the ratio of the column depth h_c to the bar diameter d_b is greater than or equal to 24 and that at least one-third of the reinforcement passes through or is anchored in the column between the vertical bars. In the event that these restrictions cannot be met, two possibilities exist to improve performance. The first solution requires that all of the beam top reinforcement be placed within the width of the column, and preferably inside the outer column bars. This, however, violates the crack control provisions of ACI Code 24.3.4 for flanges in tension, which requires the distribution of some of the reinforcement over an effective flange width, as discussed in Section 7.3. If the normal beam bars are carried outside the joint, the second solution is to provide vertical stirrups through the joint region to carry the vertical component of thrust from the compression strut. In addition, Type 2 exterior beams must be designed for equilibrium torsion per ACI Code 22.7, which may require additional transverse reinforcement.

In extreme but not unusual cases, very wide beams are used, several times wider than the column, with beam depth only about 2 times the slab depth. In such cases, a safe basis for joint design is to treat the wide beam as a slab and follow the recommendations for slab-column connections contained in Chapter 13.

EXAMPLE 18.3   **Design of exterior Type 1 joint with wide beams.**   Figure 18.11 shows a typical exterior joint in the floor of a wide-beam structure, designed to resist gravity loads. Here the beams in each direction are 8 in. wider than the corresponding column dimension. Check the proposed joint geometry and shear strength, and design the transverse joint reinforcement. Material strengths are $f_c' = 4000$ psi and $f_y = 60,000$ psi. Story height is 12 ft.

SOLUTION.   For the present case, all normal beam top reinforcement is passed inside the core of the joint, terminating in 90° hooks at the outside of the column. Top reinforcement in the spandrel beams is continuous through the joint and is also carried inside the joint core. Bottom beam bars, in each case, can be spread across the width of the beam, and they are carried only 6 in. into the joint for the normal beam because the joint is not a part of the primary, lateral-force-resisting system. The bottom spandrel beam bars are continued to provide structural integrity (ACI Code 9.7). Beam stirrups outside of the joint, not shown in Fig. 18.11, would be carried outside of the outer bottom bars and bent up. They would require small-diameter horizontal bars inside the hooks for proper anchorage at the upper ends of their vertical legs.

Transverse confinement reinforcement must be provided in the direction of the normal beam, between the top and bottom bars of the normal beam, with spacing not to exceed one-half of the 20-in. beam depth., or 10 in. or the code limit of 8 in. As described in conjunction with calculation of the hook development length, five sets of No. 4 (No. 13) column ties will be used, as shown in Fig. 18.11. In addition to the hoop around the outside bars, a single-leg crosstie is required for the middle column bars because the clear distance between column bars exceeds 6 in. The ties satisfy Eq. (5.13a).

FIGURE 18.11
Exterior beam-column joint for Example 18.3: (*a*) plan view; (*b*) section through spandrel beam; and (*c*) section through normal beam.

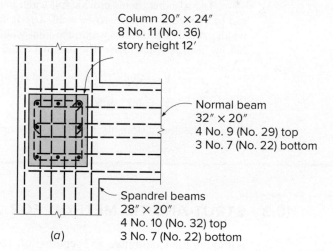
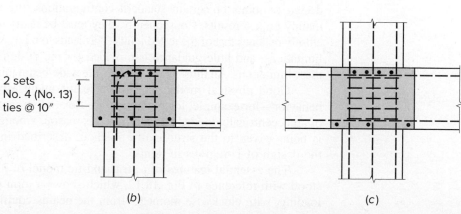

Checking the required hook development length of the No. 9 (No. 29) top bars of the normal beam, shows that the spacing between the hooked bars is less than $6d_b$. The only way to anchor the hooked bars within the available length is to increase the confining reinforcement to $0.4 \times A_{hs} = 0.4 \times 4$ in$^2 = 1.6$ in^2 that must be located within $15d_b$ or 16.9 in. of the centerline of the straight portion of the hooked bars. Doing so gives $\psi_r = 1.0$. The required confining reinforcement is provided by four ties spaced at 4 in. Five ties in all will be used through the depth of the joint.

$$\ell_{dh} = \left(\frac{f_y\psi_e\psi_r\psi_o\psi_c}{55\lambda\sqrt{f_c'}}\right)d_b^{1.5} = \left(\frac{60{,}000 \times 1.0 \times 1.0 \times 1.0 \times 0.87}{55 \times 1.0 \times \sqrt{4000}}\right)1.128^{1.5} = 18.0 \text{ in.}$$

If the hooks are carried down in the plane of the outer column bars, the available embedment is $20.0 - 1.5 - 0.5 = 18.0$ in., equaling the minimum required embedment.

Moments from the spandrels on either side of the joint will be about equal, so no joint shear problem exists in that direction. In the direction of the normal beam, shear must be checked. The tensile force applied by the top bars is $A_sf_y = 4.00 \times 60 = 240$ kips. The depth of the beam compressive stress block is $a = A_s f_y/0.85f_c' b_w = 240/(0.85 \times 4 \times 32) = 2.51$ in., and the corresponding moment is

$$M_u = M_n = A_sf_y\left(d - \frac{a}{2}\right) = \frac{240}{12}\left(17.4 - \frac{2.51}{2}\right) = 322 \text{ ft-kips}$$

Column shears are based on a free body corresponding to that of Fig. 18.9b and are equal to $V_{\text{col}} = 322/12 = 26.9$ kips. Thus, the joint shear at middepth is $V_u = 240 - 26.9 = 213$ kips.

The spandrel beams provide full-width joint confinement in their direction, and the joint can be classed as exterior, so $V_n = 20\lambda\sqrt{f_c'} A_j$. In the perpendicular direction, when the beam width exceeds the column width, the joint width b_j is to be taken equal to the column width (24 in. in the present case). The nominal and design shear strengths are, respectively,

$$V_n = 20\lambda\sqrt{f_c'} \ b_jh = 20 \times 1.0\sqrt{4000} \times 24 \times 20/1000 = 607 \text{ kips}$$

$$\phi V_n = 0.75 \times 607 = 455 \text{ kips}$$

Because the design strength is well above the applied shear of 213 kips, the shear requirement is met.

18.3   STRUT-AND-TIE METHOD FOR JOINT BEHAVIOR

Although the Committee 352 report (Ref. 18.6) is an important contribution to the safe design of joints of certain standard configurations, the recommendations are based mainly on test results. Consequently, they must be restricted to joints whose geometry closely matches that of the tested joints. This leads to many seemingly arbitrary geometric limitations, and little guidance is provided for the design of joints that may not meet these limitations. An illustration of this is the wide-beam joint discussed in Section 18.2f.

Good physical models are available for many aspects of reinforced concrete behavior—for example, for predicting the flexural strength of a beam or the strength of an eccentrically loaded column. For this reason, among others, increasing attention is being given to the strut-and-tie method, described in Chapter 17, as a basis for the design of D-regions in joints.

The essential features of a strut-and-tie model of joint behavior may be understood with reference to Fig. 18.12, which shows a joint of a frame subject to lateral loading, with clockwise moments from the beams equilibrated by counterclockwise

FIGURE 18.12
Strut-and-tie model for behavior of a beam-column joint.

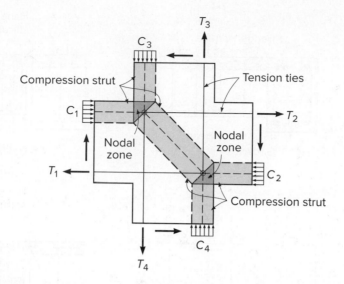

moments from the columns. The line of action of the horizontal forces C_1 and T_2 intersects that of the vertical forces C_3 and T_4 at a *nodal zone*, where the resultant force is equilibrated by a diagonal *compression strut* within the joint. At the lower end of the strut, the diagonal compression equilibrates the resultant of the horizontal forces T_1 and C_2 and the vertical forces T_3 and C_4. The tension bars must be well anchored by extension into and through the joint, or in the case of discontinuous bars (such as the top beam reinforcement in an exterior joint) by headed bars or hooks. The concrete within the nodal zone is subjected to a biaxial or, in many cases, a triaxial state of stress.

With this simple model, the flow of forces in a joint is easily visualized, satisfaction of the requirements of equilibrium is confirmed, and the need for proper anchorage of bars is emphasized. In a complete strut-and-tie model analysis, through proper attention to deformations within the joint, serviceability is ensured through control of cracking.

According to the strut-and-tie method, the main function of the column ties required within the joint region by conventional design procedures, in addition to preventing outward buckling of the vertical column bars, is to confine the concrete in the compression strut, thereby improving both its strength and ductility, and to control the cracking that may occur owing to diagonal tension perpendicular to the axis of the compression strut. The main load is carried by the uniaxially loaded struts and ties.

The strut-and-tie method not only provides valuable insights into the behavior of ordinary beam-column joints but also represents an important tool for the design of joints that fall outside of the limited range of those considered in Ref. 18.6. In the sections of this chapter that follow, a number of types of joints are considered that occur commonly in reinforced concrete structures, for which the strut-and-tie models provide essential aid in developing proper bar details.

18.4 BEAM-TO-GIRDER JOINTS

Commonly in concrete construction, secondary floor beams are supported by primary girders, as shown in Fig. 18.13a and b. It is often assumed that the reaction from the floor beam is more or less uniformly distributed through the depth of the interface

FIGURE 18.13

FIGURE 18.13

Main girder supporting secondary beam: (*a*) cross section through girder showing hanger stirrups; (*b*) cross section through beam; (*c*) truss model showing transfer of beam load to girder at load near ultimate; and (*d*) truss model showing transfer of load into the girder.

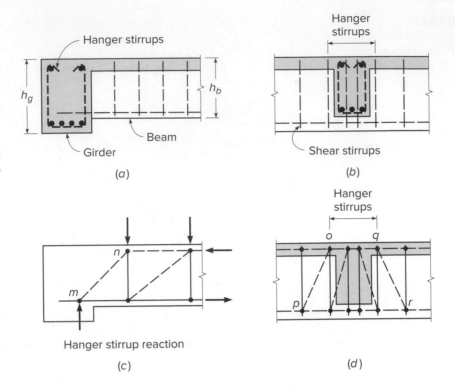

between beam and girder. This incorrect assumption is perhaps encouraged by the ACI Code "$V_c + V_s$" approach to shear design, which makes use of a nominal average shear stress in the concrete, $v_c = V_c/b_w d$, suggesting a uniform distribution of shear stress through the beam web.

The actual behavior of a diagonally cracked beam, as indicated by tests, is quite different, and the flow of forces can be represented, in somewhat simplified form, by the truss model of the beam shown in Fig. 18.13*c* (Ref. 18.7). The main reaction is delivered from beam to girder by a diagonal compression strut *mn*, which applies its thrust near the bottom of the carrying girder. Failure to provide for this thrust may result in splitting off the concrete at the bottom of the girder followed by collapse of the beam. A graphic example of lack of support for diagonal compression at the junction of a beam and its supporting girder is shown in Fig. 18.14. Tests indicate that no action is required if the shear stress to be transferred is less than $3\sqrt{f_c'}$ or if the beam is in the top half of the girder depth (Ref. 18.8). ACI Commentary R9.7.6.2 recognizes this behavior and leaves the design to the engineer.

Proper detailing of reinforcement in the region of such a joint requires the use of well-anchored *hanger* stirrups in the girder, as shown in Fig. 18.13*a* and *b*, to provide for the downward thrust of the compression strut at the end of the beam (Refs. 18.9 and 18.10). These stirrups serve as tension ties to transmit the reaction of the beam to the compression zone of the girder, where it can be equilibrated by diagonal compression struts in the girder. The hanger stirrups, which are required *in addition* to the normal girder stirrups required for shear, can be designed based on equilibrating part or all of the reaction from the beam, with the hanger stirrups assumed to be stressed to their yield stress f_y at the factored load stage.

The strut-and-tie model allows visualization of the transfer of the beam load along the girder as seen in Fig. 18.13*d*. The compression struts *op* and *qr* complete

FIGURE 18.14
Failure due to lack of support for diagonal compression in beam-girder joint. (*Courtesy of M. P. Collins, University of Toronto.*)

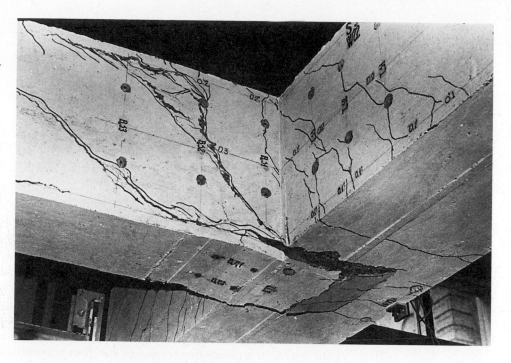

the shear transfer into the girder. The orientation of these compression struts depends on the location of the beam relative to the girder end.

If the beam and girder are the same depth, the hangers should take the full reaction. However, if the beam depth is much less than that of the girder, hangers may prove unnecessary. It is suggested in Ref. 18.8 that hanger stirrups be placed to resist a downward force of V_s^*, where

$$V_s^* = \frac{h_b}{h_g} V \tag{18.5}$$

Here h_b is the depth of the beam, h_g is the depth of the carrying girder, as indicated by Fig. 18.13, and V is the end reaction received from the beam.

Hangers will also be unnecessary if the factored beam shear is less than ϕV_c (as is usually the case for one-way joists, for example), because in such a case diagonal cracks would not form in the supported member. The predictions of the truss model would thus not be valid, and the reaction would be more nearly uniform through the depth.

The hanger stirrups should pass around the main flexural reinforcement of the girder, as shown in Fig. 18.13. If the beam and girder have the same depth, the main flexural bars in the girder should pass below those entering the connection from the beam to provide the best possible reaction platform for the diagonal compression strut.

18.5 LEDGE GIRDERS

Frequently in precast concrete construction, an L or inverted T girder is used to provide a seat, or ledge, to support precast beams framing into the carrying girder from the perpendicular direction. Typical ledge girder cross sections are shown in Fig. 18.15. The end reaction of the beams introduces a heavy concentrated load near the bottom of such girders, requiring special reinforcement in the projecting ledge and in the girder web.

FIGURE 18.15
Ledge girders carrying precast T beams: (*a*) L girder providing exterior support for T beam and (*b*) inverted T girder carrying two T beam reactions.

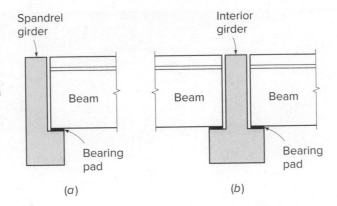

The design of such reinforcement is facilitated through use of a strut-and-tie model, as illustrated in Fig. 18.16. The downward reaction of the supported beam creates a compression fan in the ledge that distributes the reaction along a length greater than that of the bearing plate, as shown in Fig. 18.16*b*. The horizontal components of the fan are equilibrated by a compression strut along the lower flange of the girder.

In the cross section view of Fig. 18.16*a*, the downward thrust under the bearing plate is equilibrated by a diagonal compression strut, with the outward thrust at the top of the strut causing tension in the upper horizontal leg of closed hoop stirrups in the lower part of the girder. In many cases, a short structural steel angle is used just under the bearing plate, and the main tie at the top of the ledge is welded to the angle to ensure positive anchorage. At the bottom of the diagonal strut, the horizontal component of thrust is equilibrated by the opposing thrust from the other side, and the vertical component causes tension in stirrups that extend to the top of the girder. These stirrups are used in addition to those required for girder shear. Proper anchorage at the nodes is ensured by passing longitudinal bars inside the bends of both sets of stirrups.

FIGURE 18.16
Strut-and-tie model for behavior of inverted T ledge girder: (*a*) girder cross section and (*b*) side elevation.

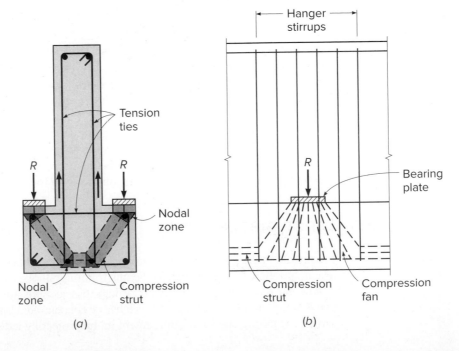

EXAMPLE 18.4   **Inverted T beam connection design.**   The inverted T beam shown in Fig. 18.17a supports 40 ft long, 12 ft wide double T beams. The width of the double T stem is 4.75 in., which is supported by an elastomeric pad (not shown) on a 6 in. long steel bearing plate. The dead load of the double T is 71 psf, including self-weight, and the beam carries a live load of 40 psf. In accordance with ACI Code 16.2.2.3(b) for connections on bearing pads, a factored horizontal force must be taken equal to 20 percent of the sustained unfactored vertical reaction multiplied by a load factor of 1.6. Design the reinforcement in the inverted T at the double T bearing point. Material properties are $f_c' = 6000$ psi and $f_y = 60,000$ psi.

SOLUTION.   The factored loads on the beam stem for a 6-ft tributary width are

$$q_u = 1.2\, R_d + 1.6\, R_l = 1.2 \times 71 + 1.6 \times 40 = 149 \text{ psf}$$

$$R_u = q_u \times b \times L/2 = 0.149 \text{ psf} \times 6 \text{ ft} \times 40 \text{ ft}/2 = 17.9 \text{ kips}$$

and treating the dead load of the double T beams as the sustained load,

$$N_{uc} = (1.6 \times 0.2 \times 71 \times 6 \text{ ft} \times 40 \text{ ft}/2)/1000 = 2.7 \text{ kips}$$

The bearing area under the double T leg is 6 in. by 4.75 in. = 28.5 in^2, giving a nodal bearing stress of

$$f_n = \frac{17.9}{28.5} = 0.63 \text{ ksi}$$

which is well below the nominal capacity of the nodes and interior or boundary struts even with $\beta_s = 0.4$, as required for a strut in a tension zone. The low stress is used to demonstrate an alternative solution methodology. By using the low stress, the node and strut capacities are adequate by inspection; however, the size of strut cd must be confirmed. Solving for the geometry and forces in Fig. 18.17b, $T_{ab} = 12.3$ kips, $T_{df} = 17.9$ kips, and strut cd carries $F_{cd} = 12.3$ kips. The thickness of the strut is assumed as 4.75 in., the same as the bearing pad. Therefore, the width of strut cd is

$$w_{cd} = \frac{12.3}{0.63 \times 4.75} = 4.11 \text{ in.}$$

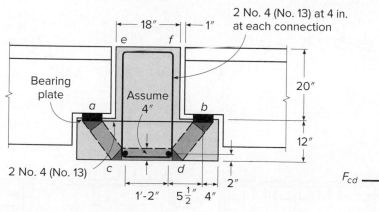

(a) Strut-and-tie model   (b) Truss detail

FIGURE 18.17
Strut-and-tie model for Example 18.4.

This is slightly more than the 4 in. assumed. A minor modification to the bearing stress would make this acceptable; therefore, the design continues with the selected geometry. The required area of reinforcement for tie *ab* is

$$A_{ts} = \frac{T_{ab} + N_{uc}}{\phi f_y} = \frac{15.0}{0.75 \times 60} = 0.33 \text{ in}^2$$

which is satisfied by using two No. 4 (No. 13) bars welded to each bearing plate. For tie *df*,

$$A_{ts} = \frac{T_{df}}{\phi f_y} = \frac{17.9}{0.75 \times 60} = 0.40 \text{ in}^2$$

which is also met using two No. 4 (No. 13) closed stirrups at 4 in. on center at each load point.

18.6 CORNERS AND T JOINTS

In many common types of reinforced concrete structures, moments and other forces must be transmitted around corners. Some examples, shown in Fig. 18.18, include gable frames, retaining walls, liquid storage tanks, and large box culverts. Reinforcement detailing at the corners is rarely obvious. A comprehensive experimental study of such joints by Nilsson and Losberg (Ref. 18.3) showed that many commonly used joint details will transmit only a small fraction of their assumed strength. Ideally, the joint should resist a moment at least as large as the calculated failure moment of the members framing into it (that is, the *joint efficiency* should be at least 100 percent). Tests have shown that, for common reinforcing details, joint efficiency may be as low as 30 percent.

Corner joints may be subjected to opening moments, causing flexural tension on the inside of the joint, or closing moments, causing tension on the outside. Generally, the first case is the more difficult to detail properly.

Consider, for example, a corner joint subjected to opening moments, such as an exterior corner of the liquid storage tank shown in the plan view in Fig. 18.18*d*.

FIGURE 18.18
Structures with corners subject to opening or closing moments: (*a*) gable frame; (*b*) earth-retaining wall; (*c*) liquid storage tank; (*d*) plan view of multicell liquid storage tank; and (*e*) large box culvert.

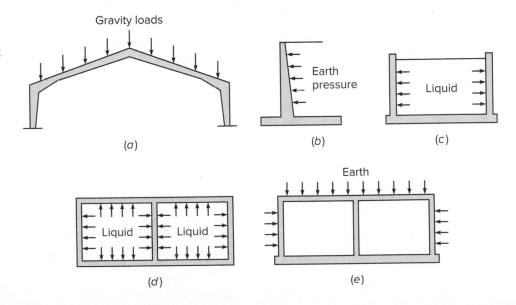

FIGURE 18.19
Corner joint subject to
opening moments:
(*a*) cracking in an improperly
designed joint and (*b*) strut-
and-tie model of joint
behavior.

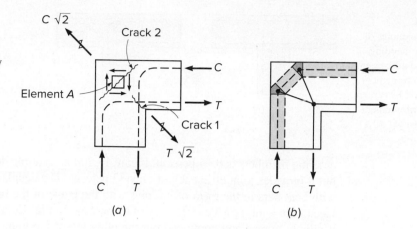

Figure 18.19*a* shows the system of forces acting on such a corner. The reinforcing bar pattern shown is *not* recommended. Formation of crack 1, radiating inward from the corner, is perhaps obvious. Crack 2, which may lead to splitting off the outside corner, may not be so obvious. However, the resultant of the two compressive forces C, having a magnitude $C\sqrt{2}$, is equilibrated by the resultant tension $T\sqrt{2}$. These two forces, one applied near the outer corner and one near the inner corner, require high tensile stress between the two, leading to formation of crack 2 as shown. The same conclusion is reached considering a small concrete element A in the corner. It is subjected to the shearing forces shown as a result of the forces C and T from the entering members. The resultant of these shearing stresses is 45° principal tension across the corner, confirming formation of crack 2.

One may, at first, be tempted to add an L-shaped bar around the outside of the corner in an attempt to confine the outer concrete. Such a bar would serve little purpose, however, because the bar would be in compression and may actually assist in pushing the corner off. The strut-and-tie model of Fig. 18.19*b* provides valuable insight into the needed reinforcement, indicating that, in addition to well-anchored tensile bars to transmit the forces T into the joint, some form of radial reinforcement is required to permit the compressive forces C to "turn the corner."

Test results for a large number of joints with alternative bar details are reported in Ref. 18.3. Comparative efficiencies for some specific details, relating the maximum moment transmitted by the corner joint to the flexural capacity of the entering members, are summarized in Fig. 18.20. In all cases, the reinforcement ratio of the

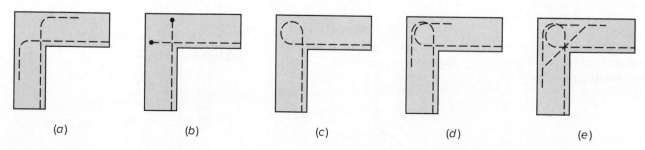

FIGURE 18.20
Efficiencies of corner joints subject to opening moments for various reinforcing details: (*a*) 32 percent; (*b*) 68 percent; (*c*) 77 percent; (*d*) 87 percent; and (*e*) 115 percent. (*Adapted from Ref. 18.3.*)

FIGURE 18.21
Comparative efficiencies of
T joints subject to bending
moment: (*a*) 24 to 40 percent
depending on reinforcement
ratio and (*b*) 82 to 110 percent
depending on reinforcement
ratio. (*Adapted from Ref. 18.3.*)

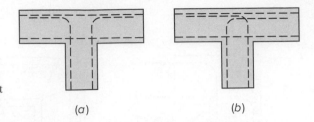

(*a*)   (*b*)

entering members is 0.75 percent. Figure 18.20*a* is a simple detail, probably often used, but it provides joint efficiency of only 32 percent. The details in Fig. 18.20*b*, reinforced with bent bars in the form of hairpins with the plane of the hooks parallel to the inside face of the joint, provide efficiency of 68 percent. In Fig. 18.20*c*, the main reinforcement is simply looped and continued out the other leg of the joint, resulting in an efficiency of 77 percent. The somewhat similar detail shown in Fig. 18.20*d*, in which the bars entering the joint are terminated with separate loops, gives an efficiency of 87 percent. The best performance results from the detail shown in Fig. 18.20*e*—the same as in Fig. 18.20*d* except for the addition of a diagonal bar. This improves joint efficiency to 115 percent, so that the joint is actually stronger than the design strength of the members framing into it. It was determined experimentally that the area of the diagonal bar should be about one-half that of the main reinforcement.

The joints between the vertical wall and horizontal base slab of retaining walls (see Fig. 18.18*b*) are also subjected to opening moments. Tests of such joints confirm the benefit of placing a diagonal bar similar to Fig. 18.20*e*. Retaining wall bar details are discussed further in Chapter 16.

T joints also may be subjected to bending moments, such as if only one cell of the multiple-cell liquid storage tank of Fig. 18.18*d* were filled. Tests of such joints, reported in Ref. 18.3, again indicate the importance of proper detailing. The reinforcing bar arrangement shown in Fig. 18.21*a*, which is sometimes seen, permits a joint efficiency of only 24 to 40 percent, but the simple rearrangement shown in Fig. 18.21*b* improves the efficiency to between 82 and 110 percent. In both cases, efficiency depends upon the main reinforcement ratio in the entering members, with highest efficiency corresponding to the lowest tensile reinforcement ratio.

A strut-and-tie model for the T joint confirms the research results presented above. Figure 18.22*a* shows that a clockwise moment applied to the stem of the T is resisted by shear forces at the inflection points of the T top. The strut-and-tie model in Fig. 18.22*b* clearly shows that the stem reinforcement must hook to the left for the joint to be effective, as shown in Fig. 18.21*b*.

FIGURE 18.22
T joint behavior subjected to
moment: (*a*) bending moment
and resulting shear forces and
(*b*) strut-and-tie model.

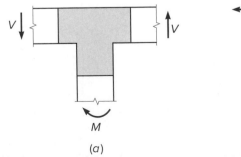
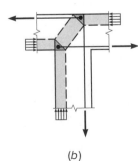

(*a*)   (*b*)

Joints subjected to closing moments, with main reinforcement passing around the corner close to the outside face, cause few detailing problems because the main tension reinforcement from the entering members can be carried around the outside of the corner. There is, however, a risk of splitting the concrete in the plane of the bend, or concrete crushing inside the bend. Therefore, the bend radius, as determined by a curved bar node assessment from Section 17.4c, is needed to properly detail the joint.

18.7 BRACKETS AND CORBELS

Brackets such as shown in Fig. 18.23a are widely used in precast construction for supporting precast beams at the columns. When they project from a wall, rather than from a column, they are properly called corbels, although the two terms are often used interchangeably. Brackets are designed mainly to provide for the vertical reaction V_u at the end of the supported beam, but unless special precautions are taken to avoid horizontal forces caused by restrained shrinkage, creep (in the case of prestressed beams), or temperature change, they must also resist a horizontal force N_{uc}.

Steel bearing plates or angles are generally used on the top surface of the brackets, as shown, to provide a uniform contact surface and to distribute the reaction. A corresponding steel bearing plate or angle may be provided at the lower corner of the supported member, or a bearing pad installed. If the two plates are welded together, horizontal forces clearly must be allowed for in the design. Even with Teflon or elastomeric bearing pads, frictional forces will develop due to volumetric change.

The structural performance of a bracket can be visualized using the strut-and-tie model shown in Fig. 18.23b. The downward thrust of the load V_u is equilibrated by the vertical component of the reaction from the diagonal compression strut that carries the load down into the column. The outward thrust at the top of the strut is balanced by the tension in the horizontal tie bars across the top of the bracket; these also take the tension, if any, imparted by the horizontal force N_{uc}. At the left end of the horizontal tie, the tension is equilibrated by the horizontal component of thrust from the second compression strut shown. The vertical component of this

FIGURE 18.23

Typical reinforced concrete bracket: (a) loads and reinforcement and (b) strut-and-tie model for internal forces.

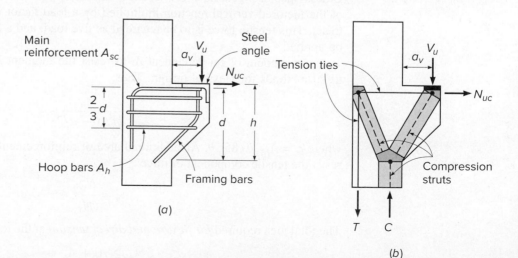

thrust requires the tensile force shown acting downward at the left side of the supporting column.

The reinforcement required, according to the strut-and-tie model, is shown in Fig. 18.23a. The main bars A_{sc} must be carefully anchored because they need to develop their full yield strength f_y directly under the load V_u, and for this reason they are usually welded to the underside of the bearing angle and a 90° hook is provided for anchorage at the left side. Closed hoop bars with area A_h confine the concrete in the two compression struts and resist a tendency for splitting in a direction parallel to the thrust. The framing bars shown are usually of about the same diameter as the stirrups and serve mainly to improve the stirrup anchorage at the outer face of the bracket.

The bracket may also be considered as a very short cantilevered beam, with flexural tension at the column face resisted by the top bars A_{sc}. Either concept will result in about the same area of main reinforcement.

A second possible mode of failure is by direct shear along a plane more or less flush with the vertical face of the main part of the column. Shear-friction reinforcement crossing such a crack (see Section 5.9) would include the area A_{sc} previously placed in the top tie and the area A_h from the hoops below it. Other failure modes include flexural tension failure, with yielding of the top bars followed by crushing of the concrete at the bottom of the bracket; crushing of the concrete under the bearing angle (particularly if end rotation of the supported beam causes the force V_u to be applied too close to the outer corner of the bracket); and direct tension failure, if the horizontal force N_{uc} is larger than anticipated.

The provisions of ACI Code 16.5 for the design of brackets and corbels have been developed mainly based on tests (Refs. 18.10, 18.11, and 18.12) and relate to the flexural model of bracket behavior. They apply to brackets and corbels with a shear span ratio a_v/d of 1.0 or less (see Fig. 18.23a). Brackets and corbels with a_v/d less than 2 may be designed using strut-and-tie models, as described in Chapter 17. The distance d is measured at the column face, and the depth at the outside edge of the bearing area must not be less than $0.5d$. The usual design basis is employed, that is, $\phi M_n \geq M_u$ and $\phi V_n \geq V_u$, and for brackets and corbels (for which shear dominates the design), ϕ is to be taken equal to 0.75 for all strength calculations, including flexure and direct tension as well as shear.

The section at the face of the supporting column must simultaneously resist the shear V_u, the moment $M_u = V_u a_v + N_{uc}(h - d)$, and the horizontal tension N_{uc}. Unless special precautions are taken, a horizontal tension not less than 20 percent of the factored vertical reaction multiplied by a load factor of 1.6 must be assumed to act. This tensile force is to be regarded as live load, and a load factor of 1.6 should be applied.

An amount of reinforcement A_f to resist the moment M_u can be found by the usual methods for flexural design. Thus,

$$A_f = \frac{M_u}{\phi f_y(d - a/2)} \tag{18.6}$$

where $a = A_f f_y / 0.85 f'_c b$. An additional area of reinforcement A_n must be provided to resist the tensile component of force:

$$A_n = \frac{N_{uc}}{\phi f_y} \tag{18.7}$$

The total area required *for flexure and direct tension* at the top of the bracket is thus

$$A_{sc} \geq A_f + A_n \tag{18.8}$$

Design for shear is based on the shear-friction method of Section 5.9, and the total shear-friction reinforcement A_{vf} is found by

$$A_{vf} = \frac{V_u}{\phi \mu f_y} \tag{18.9}$$

where the friction factor μ for monolithic construction is 1.4λ, where $\lambda = 1.0$ for normalweight concrete, 0.75 for all lightweight concrete, and is otherwise calculated based on volumetric proportions of lightweight and normalweight aggregate not to exceed 0.85 in accordance with ACI Code 22.9.4.2. The value of $V_n = V_u/\phi$ must not exceed the smallest of $0.2f_c'\, b_w d$, $(480 + 0.08f_c')b_w d$, and $1600b_w d$ at the support face for normalweight concrete or the smaller of $(0.2 - 0.07a_v/d)f_c'\, b_w d$ and $(800 - 280a_v/d)b_w d$ for lightweight concrete. Then, according to ACI Code 16.5, the total area required *for shear plus direct tension* at the top of the bracket is

$$A_{sc} \geq \frac{2}{3} A_{vf} + A_n \tag{18.10}$$

with the remaining part of A_{vf} placed in form of closed hoops having area A_h in the lower part of the bracket, as shown in Fig. 18.23a.

Thus, the total reinforcement area A_{sc} required at the top of the bracket is equal to the larger of the values given by $A_f + A_n$ or by Eq. (18.10). An additional restriction, that A_{sc} not be less than $0.04(f_c'/f_y)bd$, is intended to avoid the possibility of sudden failure upon formation of a flexural tensile crack at the top of the bracket.

According to the ACI Code, closed hoop stirrups having area A_h (see Fig. 18.23a) not less than $0.5(A_{sc} - A_n)$ must be provided and be uniformly distributed within two-thirds of the effective depth adjacent to and parallel to A_{sc}.

EXAMPLE 18.5   **Design of column bracket.**   A column bracket on a 16 in. square column having the general features shown in Fig. 18.24 is to be designed to carry the end reaction from a long-span precast girder. Vertical reactions from service dead and live loads are 17 and 33 kips, respectively, applied at $a_v = 5.5$ in. from the column face. An elastomeric bearing pad is provided for the girder, which will rest directly on a $5 \times 3 \times \frac{3}{8}$ in. steel angle anchored at the outer corner of the bracket. Bracket reinforcement includes main reinforcement A_{sc} welded to the underside of the steel angle, closed hoop stirrups having total area A_h distributed appropriately through the bracket depth, and framing bars in a vertical plane near the outer face. Select appropriate concrete dimensions, and design and detail all reinforcement. Material strengths are $f_c' = 5000$ psi and $f_y = 60,000$ psi. The column width and depth is 16 in.

SOLUTION.   The vertical factored load to be carried is

$$V_u = 1.2\, R_d + 1.6 \times R_l = 1.2 \times 17 + 1.6 \times 33 = 73.2 \text{ kips}$$

In the absence of a roller or low-friction support pad, a horizontal tensile force of

$$N_{uc} = 1.6 \times 0.2 \times R_d = 1.6 \times 0.20 \times 17 = 5.4 \text{ kips}$$

is required based on the unfactored sustained load. According to the shear friction provisions of the ACI Code, the nominal shear strength V_n must not exceed $0.2f_c'\, bd$, $(480 + 0.08f_c')bd$, or $1600bd$. With $f_c' = 5000$ psi, the second limit controls. Then, with $V_u = \phi V_n$ and with the column width $b = 12$ in.,

$$73.2 = 0.75 \times 0.880 \times 12d$$

FIGURE 18.24
Column bracket design
example.

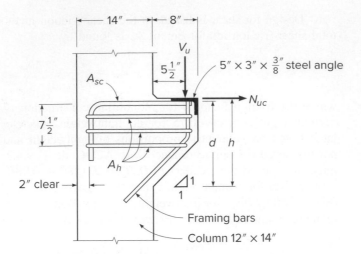

from which $d = 9.24$ in. A total depth of 16 in. is selected, and estimating 1 in. from the center of the main reinforcement to the top surface of the bracket, d is approximately equal to 15 in., the exact value depending on the bar diameter chosen for A_{sc}. If a 45° slope is used, as indicated in Fig. 18.24, the bracket depth at the outside of the bearing area is 8 in. This is not less than $0.5d = 7.5$ in., as required. For the bracket geometry selected, $a_v/d = 5.5/15 = 0.36$. This does not exceed the 1.0 limit imposed by the ACI Code.

The total shear friction reinforcement is found from Eq. (18.9) and for monolithic concrete with $\mu = 1.4$:

$$A_{vf} = \frac{V_u}{\phi \mu f_y} = \frac{73.2}{0.75 \times 1.4 \times 60} = 1.16 \text{ in}^2$$

The bending moment to be resisted is

$$M_u = V_u a_v + N_{uc}(h - d)$$
$$= 73.2 \times 5.5 + 5.4 \times 1 = 408 \text{ in.-kips}$$

The depth of the flexural compression stress block will be estimated to be 2 in., so, from Eq. (18.6),

$$A_f = \frac{M_u}{\phi f_y(d - a/2)} = \frac{408}{0.75 \times 60(15 - 2.0/2)} = 0.65 \text{ in}^2$$

Checking the stress block depth gives

$$a = \frac{A_f f_y}{0.85 f_c' \, b} = \frac{0.65 \times 60}{0.85 \times 5 \times 12} = 0.76 \text{ in.}$$

so the revised reinforcement area is

$$A_f = \frac{408}{0.75 \times 60(15 - 0.76/2)} = 0.62 \text{ in}^2$$

The tensile force of 5.4 kips requires an additional reinforcement area, from Eq. (18.7), of

$$A_n = \frac{N_{uc}}{\phi f_y} = \frac{5.4}{0.75 \times 60} = 0.12 \text{ in}^2$$

Thus, from Eqs. (18.8) and (18.10), respectively, the total reinforcement area at the top of the bracket must not be less than

$$A_{sc} \geq A_f + A_n = 0.62 + 0.12 = 0.74 \text{ in}^2$$

or less than

$$A_{sc} \geq \tfrac{2}{3} A_{vf} + A_n = \tfrac{2}{3} \times 1.16 + 0.12 = 0.89 \text{ in}^2$$

The second requirement controls here. The minimum reinforcement requirement of

$$A_{sc,\min} = 0.04 \frac{f_c'}{f_y} \, bd = 0.04 \times \frac{5}{60} \times 12 \times 15 = 0.60 \text{ in}^2$$

is seen not to control. A total of three No. 6 (No. 19) bars, providing $A_{sc} = 1.32 \text{ in}^2$, are selected.

Closed hoop reinforcement having a total area A_h not less than $0.5(A_{sc} - A_n)$ must be provided. Thus,

$$A_h \geq 0.5A_f = 0.5 \times 0.62 = 0.31 \text{ in}^2$$

and

$$A_h \geq 0.5 \times \tfrac{2}{3} A_{vf} = \tfrac{1}{3} \times 1.16 = 0.39 \text{ in}^2$$

The second requirement controls. Three No. 3 (No. 10) closed hoops are provided, to cover the entire bracket giving total area $A_h = 0.66 \text{ in}^2$. These must be placed within $\tfrac{2}{3}$ of the effective depth of the main reinforcement. A spacing of 2.5 in. will be satisfactory, as indicated in Fig. 18.24. A pair of No. 3 (No. 10) framing bars are added at the inside corner of the hoops to improve anchorage, as shown.

The final step is to check the required hook development length of the No. 6 (No. 19) top bars of the normal beam. The spacing between the bars is less than $6d_b$. Thus, ψ_e and $\psi_r = 1.0$, and $\psi_o = 1.25$. The concrete compressive strength is less than 6000 psi, so $\psi_c = 5000/15{,}000 + 0.6 = 0.93$. The development length of the three No. 6 (No. 19) hooked bars is

$$\ell_{dh} = \left(\frac{f_y \psi_e \psi_r \psi_o \psi_c}{55 \sqrt{f_c'}} \right) d_b^{1.5} = \left(\frac{60{,}000 \times 1.0 \times 1.0 \times 1.25 \times 0.93}{55 \sqrt{5000}} \right) 0.75^{1.5} = 11.7 \text{ in.}$$

which fits in the 16 in. overall dimension of the column. For the hoop bars, a standard 135° hook, as shown in Fig. 6.9, is specified.

REFERENCES

18.1. "Reinforced Concrete Design Includes Approval of Details," CRSI Engineering Practice Committee, *Concr. Int.*, vol. 10, no. 1, 1988, pp. 21–22.

18.2. *ACI Detailing Manual*, ACI Special Publication SP66, American Concrete Institute, Farmington Hills, MI, 2004.

18.3. I. H. E. Nilsson and A. Losberg, "Reinforced Concrete Corners and Joints Subjected to Bending Moment," *J. Struct. Div.*, ASCE, vol. 102, no. ST6, 1976, pp. 1229–1254.

18.4. D. F. Meinheit and J. O. Jirsa, "Shear Strength of Reinforced Concrete Beam-Column Connections," *J. Struct. Div.*, ASCE, vol. 107, no. ST11, 1981, pp. 2227–2244.

18.5. J. G. L. Marques and J. O. Jirsa, "A Study of Hooked Bar Anchorages in Beam-Column Joints," *J. ACI*, vol. 72, no. 5, 1975, pp. 198–209.

18.6. ACI-ASCE Committee 352, Recommendations for Design of Beam-Column Joints in Monolithic Reinforced Concrete Structures (ACI 352R-11), American Concrete Institute, Farmington Hills, MI, 2011.

18.7. N. S. Anderson and J. A. Ramirez, "Detailing of Stirrup Reinforcement," *ACI Struct. J.*, vol. 86, no. 5, 1989, pp. 507–515.

18.8. A. H. Mattock and J. F. Shen, "Joints Between Reinforced Concrete Members of Similar Depth," *ACI Struct. J.*, vol. 89, no. 3, 1992, pp. 290–295.

18.9. R. Park and T. Pauley, *Reinforced Concrete Structures*, John Wiley, New York, 1975.

18.10. L. B. Kriz and C. H. Raths, "Connections in Precast Concrete Structures—Strength of Corbels," *J. Prestressed Concr. Inst.*, vol. 10, no. 1, 1965, pp. 16–47.

18.11. A. H. Mattock, K. C. Chen, and K. Soongswang, "The Behavior of Reinforced Concrete Corbels," *J. Prestressed Concr. Inst.*, vol. 21, no. 2, 1976, pp. 52–77.

18.12. A. H. Mattock, "Design Proposals for Reinforced Concrete Corbels," *J. Prestressed Concr. Inst.*, vol. 21, no. 3, 1976, pp. 18–24.

PROBLEMS

18.1. An interior Type 1 joint, which is to be considered a part of the primary lateral-force-resisting system, is to be designed. The 16 in. square column, with main reinforcement consisting of four No. 11 (No. 36) bars, is intersected by two 12 × 18 in. beams in the *X* direction, reinforced with three No. 10 (No. 32) top bars and three No. 8 (No. 25) bottom bars. In the *Y* direction, there are two 12 × 22 in. girders, reinforced with three No. 11 (No. 36) top bars and three No. 9 (No. 29) bottom bars. Concrete cover is 2.5 in. to the center of the bars, except for the top reinforcement in the girders, which is carried just under the top reinforcement of the beams. Design and detail the joint, using $f_c' = 4000$ psi and $f_y = 60,000$ psi. Specify placement of all bars and cutoff points.

18.2. Figure P18.2 shows a typical exterior joint for a building with 24 in. square columns. A 16 × 18 in. normal beam terminates at the column; an 18 × 22 in. spandrel girder also frames into the column. Reinforcement is as shown in Fig. P18.2, and the column height is 12 ft. Design and detail the joint, specifying bar placement, cutoff points, and details, such as bar hook dimension.

FIGURE P18.2

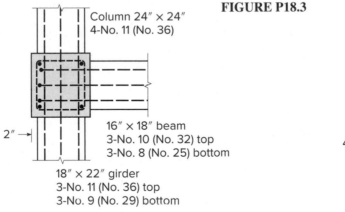

Column 24″ × 24″
4-No. 11 (No. 36)

2″ →

16″ × 18″ beam
3-No. 10 (No. 32) top
3-No. 8 (No. 25) bottom

18″ × 22″ girder
3-No. 11 (No. 36) top
3-No. 9 (No. 29) bottom

FIGURE P18.3

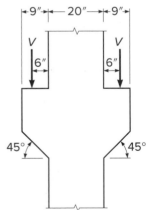

18.3. The precast columns of a proposed parking garage will incorporate symmetrical brackets to carry the end reactions of short girders that, in turn, carry long-span precast, prestressed double T floor units. The girder reactions will be applied 6 in. from the column face, as shown in Fig. P18.3, and a total width of bracket of 9 in. must be provided for proper bearing. Column width in the perpendicular direction is 20 in. Service load reactions applied at the top face of the brackets are 45 kips dead load and 36 kips live load. Select all unspecified concrete dimensions and design and detail the reinforcement. A corner angle is suggested at the outer top edge of the bracket. Column material strengths are $f_c' = 6000$ psi and $f_y = 60,000$ psi.

18.4. The stem of a 60 ft long, 8 ft wide simply supported single T beam rests on the ledge of the inverted T beam shown in Fig. P18.4. The T beam has a bearing area 6 in. thick and 4 in. parallel to the axis of the T. The applied service load is 85 psf dead load, including self-weight, and 50 psf live load. Design the connection detail under the stem using $f_c' = 5000$ psi and $f_y = 60,000$ psi.

FIGURE P18.4

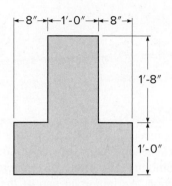

19 Concrete Building Systems

19.1 INTRODUCTION

Most of the material in the preceding chapters has pertained to the design of reinforced concrete *structural elements*, such as slabs, columns, beams, footings, and walls. These elements are combined in various ways to create *structural systems* for buildings and other construction. An important part of the total responsibility of the structural engineer is to select, from many alternatives, the best structural system for the given conditions. The wise choice of structural system is far more important, in its effect on overall economy and serviceability, than refinements in proportioning the individual members. Close cooperation with the architect in the early stages of a project is essential in developing a structure that not only meets functional and esthetic requirements but exploits to the fullest the special advantages of reinforced concrete, which include the following:

Versatility of form. Usually placed in the structure in the fluid state, the material is readily adaptable to a wide variety of architectural and functional requirements.

Durability. With proper concrete protection of the steel reinforcement, the structure will have long life, even under highly adverse climatic or environmental conditions.

Fire resistance. With proper protection for the reinforcement, a reinforced concrete structure provides the maximum in fire protection.[†]

Speed of construction. In terms of the entire period, from the date of approval of the contract drawings to the date of completion, a concrete building can often be completed in less time than a steel structure. Although the field erection of a steel building is more rapid, this phase must necessarily be preceded by prefabrication of all parts in the shop.

Cost. In many cases, the first cost of a concrete structure is less than that of a comparable steel structure. In almost every case, maintenance costs are less.

Availability of labor and material. It is always possible to make use of local sources of labor, and in many inaccessible areas, a nearby source of good aggregate can be found, so that only the cement and reinforcement need to be brought in from a remote source.

Two record-setting examples of good building design in concrete are shown in Figs. 19.1 and 19.2.

[†] Code requirements for fire protection are presented in Ref. 19.1.

FIGURE 19.1
View of 311 South Wacker
Drive under construction.
When completed, it was the
world's tallest concrete
building, with total height of
946 ft. (*Courtesy of Portland
Cement Association.*)

19.2 FLOOR AND ROOF SYSTEMS

The types of concrete floor and roof systems are so numerous as to defy concise classification. In steel construction, the designer usually is limited to using structural shapes that have been standardized in form and size by the relatively few producers in the field. In reinforced concrete, the engineer has almost complete control over the form of the structural parts of a building. In addition, many small producers of reinforced concrete structural elements and accessories can compete profitably in this field, since plant and equipment requirements are not excessive. This has resulted in the development of a wide variety of concrete systems. Only the more common types are mentioned in this text.

In general, reinforced concrete floor and roof systems can be classified as one-way systems, in which the main reinforcement in each structural element runs in one direction only, and two-way systems, in which the main reinforcement in at least one of the structural elements runs in perpendicular directions. Systems of each type can be identified in the following list:

(a) One-way slab supported by monolithic concrete beams
(b) One-way slab supported by steel beams (shear connectors are used for composite action in the direction of the beam span)
(c) One-way slab with cold-formed steel decking as form and reinforcement

FIGURE 19.2
The Burj Khalifa is the
current record holder not
only as the tallest reinforced
concrete building but also as
the tallest structure of any
type in the world, with a total
height of 2717 ft. (*Laborant/
Shutterstock*)

(d) One-way joist floor (also known as ribbed slab)
(e) Two-way slab supported by edge beams for each panel
(f) Flat slabs, with column capitals, drop panels, or both, but without beams
(g) Flat plates, without beams and with no drop panels or column capitals
(h) Two-way joist floors, with or without beams on the column lines

Each of these types is described briefly in the following sections. Additional information is found in Refs. 19.2 to 19.4. In addition to the cast-in-place floor and roof systems described in this section, a great variety of precast concrete systems have been devised. Some of these are described in Section 19.3.

a.   Monolithic Beam-and-Girder Floors

A beam-and-girder floor consists of a series of parallel beams supported at their extremities by girders, which in turn frame into concrete columns placed at more or less regular intervals over the entire floor area, as shown in Fig. 19.3. This framework is covered by a one-way reinforced concrete slab, the load from which is transmitted first to the beams and then to the girders and columns. The beams are usually spaced so that they come at the midpoints, at the third points, or at the quarter points of the girders. The arrangement of beams and spacing of columns should be determined by economical and practical considerations. These are affected by the planned use of the building, the size and shape of the ground area, and the load that must be carried. A comparison of a number of trial designs and estimates should be made if the size of the building warrants, and the most satisfactory arrangement selected. If the spans in one direction are not long, say 20 ft

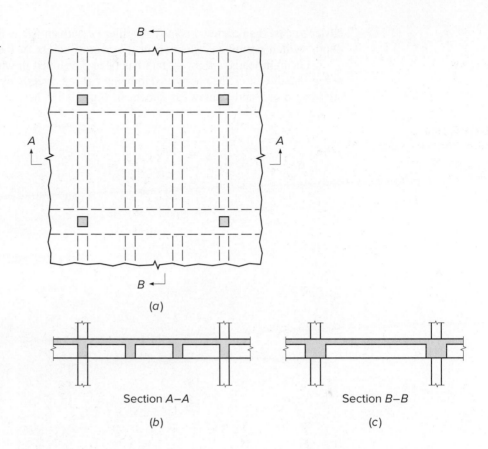

or less, the beams may be omitted altogether, and the slab, spanning in one direction, can be carried directly by girders spanning in the perpendicular direction on the column lines. Since the slabs, beams, and girders are built monolithically, the beams and girders are designed as T beams and advantage is taken of continuity.

Beam-and-girder floors are adaptable to any loads and to any spans that might be encountered in ordinary building construction. The normal maximum spread in live load values is from 40 to 400 psf, and the normal range in column spacings is from 16 to 32 ft.

The design and detailing of the joints where beams or girders frame into building columns should be given careful consideration, particularly for designs in which substantial horizontal loads are to be resisted by frame action of the building. In this case, the column region, within the depth of the beams framing into it, is subjected to significant horizontal shears as well as to axial and flexural loads. Special horizontal column ties must be included to avoid uncontrolled diagonal cracking and disintegration of the concrete, particularly if the joint is subjected to load reversals. Specific recommendations for the design of beam-column joints are found in Chapter 18 and Ref. 19.5. Joint design for buildings that resist seismic forces is subject to special ACI Code provisions (see Chapter 20).

In normal beam-and-girder construction, the depth of the beams may be as much as 3 times the web width. Improved economy, however, is achieved by using beams with webs that are generally wider and shallower, coupled with girders that have the same depth as the beams. The resulting girders, more often than not, have webs that are wider than their effective depths. Although the flexural steel in the members is increased because of the reduced effective depth compared with deeper members, the increases in material costs are more than paid for by savings in forming costs (one depth for all members) and easier construction (wider beams are easier to cast than narrow beams). Another key advantage is the reduced construction depth, which permits a reduction in the overall height of the building.

For light loads, a floor system has been developed in which the beams are omitted in one direction, the one-way slab being carried directly by column-line beams that are very broad and shallow, as shown in Fig. 19.4. These beams, supported directly

(*a*) Interior slab band

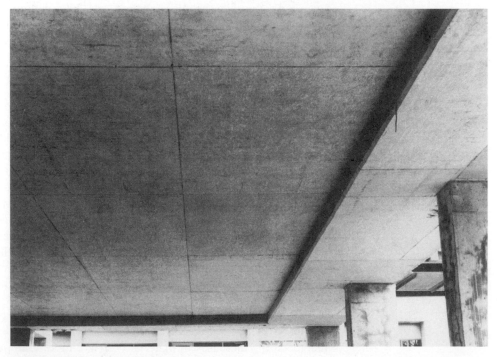

(*b*) Edge band at exterior column

by the columns, become little more than a thickened portion of the slab. This type of construction, in fact, is known as *banded slab construction*, and there are a number of advantages associated with its use, over and above those associated with shallow beam-and-girder construction. In the direction of the slab span, a haunched member is present, in effect, with the maximum effective depth at the location of greatest negative moment, across the support lines. Negative moments are small at the edge of the haunch, where the depth becomes less, and positive slab moments are reduced as well. The increased flexural steel in the beam (slab-band) resulting from the reduced effective depth is often outweighed by savings in the slab steel. Along with reduced construction depth, banded slab construction allows greater flexibility in locating columns, which may be displaced some distance from the centerline of the slab-band without significantly changing the structural action of the floor. Formwork is simplified because of the reduction in the number of framing members. For such systems, special attention should be given to design details at the beam-column joint. Transverse top steel may be required to distribute the column reaction over the width of the slab-band. In addition, punching shear failure is possible; this may be investigated using the same methods presented earlier for flat plates (see Section 13.7).

b. Composite Construction with Steel Beams

One-way reinforced concrete slabs are also frequently used in buildings for which the columns, beams, and girders consist of structural steel. The slab is normally designed for full continuity over the supporting beams, and the usual methods are followed. The spacing of the beams is usually 6 to 8 ft.

    To provide composite action, shear connectors are welded to the top of the steel beam and are embedded in the concrete slab, as shown in Fig. 19.5a. By preventing longitudinal slip between the slab and steel beam in the direction of the beam axis, the

FIGURE 19.5
Composite beam-and-slab floor.

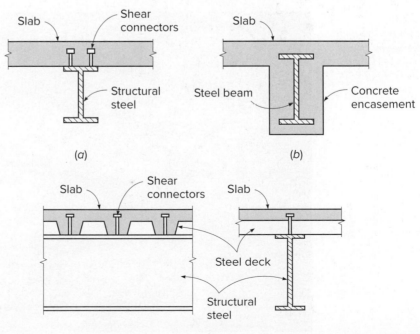

combined member is both stronger and stiffer than if composite action were not developed. Thus, for given loads and deflection limits, smaller and lighter steel beams can be used.

Composite floors may also use encased beams, as shown in Fig. 19.5*b*, offering the advantage of full fireproofing of the steel, but at the cost of more complicated formwork and possible difficulty in placing the concrete around and under the steel member. Such fully encased beams do not require shear connectors as a rule.

c. Steel Deck Reinforced Composite Slabs

It is nearly standard practice to use stay-in-place light-gage cold-formed steel deck panels in composite floor construction. As shown in Fig. 19.5*c*, the steel deck serves as a stay-in-place form and, with suitable detailing, the slab becomes composite with the steel deck, serving as the main tensile flexural steel. Suitable for relatively light floor loading and short spans, composite steel deck reinforced slabs are found in office buildings and apartment buildings, with column-line girders and beams in the perpendicular direction subdividing panels into spans up to about 12 ft. Temporary shoring may be used at the midspan or third point of the panels to avoid excessive stresses and deflections while the concrete is placed, when the steel deck panel alone must carry the load.

d. One-Way Joist Floors

A one-way joist floor consists of a series of small, closely spaced reinforced concrete T beams, framing into monolithically cast concrete girders, which are in turn carried by the building columns. The T beams, called *joists*, are formed by creating void spaces in what otherwise would be a solid slab. Usually these voids are formed using special steel pans, as shown in Fig. 19.6. Concrete is cast between the forms to create ribs and placed to a depth over the top of the forms so as to create a thin monolithic slab that becomes the T beam flange.

Since the strength of concrete in tension is small and is commonly neglected in design, elimination of much of the tension concrete in a slab by the use of pan

FIGURE 19.6
Steel forms for one-way joist floor. (*Photograph by Arthur H. Nilson.*)

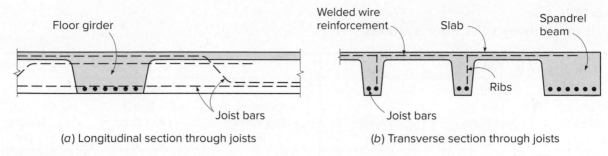

FIGURE 19.7
One-way joist floor cross sections: (*a*) cross section through supporting girder showing ends of joists and (*b*) cross section through typical joists.

forms results in a saving of weight with little change in the structural characteristics of the slab. Ribbed floors are economical for buildings, such as apartment houses, hotels, and hospitals, where the live loads are fairly small and the spans comparatively long. They are not suitable for heavy construction such as in warehouses, printing plants, and heavy manufacturing buildings.

Standard forms for the void spaces between ribs are either 20 or 30 in. wide and 8, 10, 12, 14, 16, or 20 in. deep. They are tapered in cross section, as shown in Fig. 19.7, generally at a slope of 1 to 12, to facilitate removal. Any joist width can be obtained by varying the spacing between pans. Tapered end pans are used where it is desired to obtain a wider joist near the end supports, such as may be required for high shear or negative bending moment. After the concrete has hardened, the steel pans are removed for reuse.

According to ACI Code 9.8.1, ribs must not be less than 4 in. wide and may not have a depth greater than 3.5 times the minimum web width. (For easier bar placement and placement of concrete, a minimum web width of 5 in. is desirable.) The clear spacing between ribs (determined by the pan width) must not exceed 30 in. The slab thickness over the top of the pans must not be less than one-twelfth of the clear distance between ribs, nor less than 2 in., according to ACI Code 9.8.3. Table 19.1 gives unit weights, in terms of psf of floor surface, for common combinations of joist width and depth, slab thickness, and form width.

Reinforcement for the joists usually consists of two bars in the positive bending region, with one bar discontinued where no longer needed or bent up to provide a part of the negative steel requirement over the supporting girders. Straight top bars are added over the support to provide for the negative bending moment. According to ACI Code 9.8.1, at least one bottom bar must be continuous over the support, or at noncontinuous supports, terminated with a standard hook or headed bar, as a measure to improve structural integrity in the event of major structural damage.

ACI Code 20.5.1.3 permits a reduced concrete cover of $\frac{3}{4}$ in. to be used for joist construction, just as for slabs. The thin slab (top flange) is reinforced mainly for temperature and shrinkage stresses, using welded wire reinforcement or small bars placed at right angles to the joists. The area of this reinforcement is usually 0.18 percent of the gross cross section of the concrete slab.

One-way joists are generally proportioned with the concrete providing all of the shear strength, with no stirrups used. A 10 percent increase in V_c above the values given by Eqs. (5.12*a*), (5.12*c*), and (5.12*d*) is permitted for joist construction, according to ACI Code 9.8.1, based on the possibility of redistribution of local overloads to adjacent joists. Tests have shown that while local redistribution does occur, the shear strength of the full system (all joists acting together) is enhanced by less than 10 percent (Ref. 19.6).

TABLE 19.1

Weight of one-way joist floor systems

3 in. Top Slab			$4\frac{1}{2}$ in. Top Slab		
Depth of Pan Form, in.	Width of Joist + Pan Form, in.	Weight, psf	Depth of Pan Form, in.	Width of Joist + Pan Form, in.	Weight, psf
8	5 + 20	60	8	5 + 20	79
8	5 + 30	54	8	5 + 30	72
10	5 + 20	67	10	5 + 20	85
10	5 + 30	58	10	5 + 30	77
12	5 + 20	74	12	5 + 20	92
12	5 + 30	63	12	5 + 30	82
14	5 + 30	68	14	5 + 30	87
14	6 + 30	72	14	6 + 30	91
16	6 + 30	78	16	6 + 30	97
16	7 + 30	83	16	7 + 30	101
20	6 + 30	91	20	6 + 30	109
20	7 + 30	96	20	7 + 30	115

Data Source: Adapted from Ref. 19.3.

The joists and the supporting girders are placed monolithically. Like the joists, the girders are designed as T beams. The shape of the girder cross section depends on the shape of the end pans that form the joists, as shown in Fig. 19.7*a*. If the girders are deeper than the joists, the thin concrete slab directly over the top of the pans is often neglected in the girder design. The girder width can be adjusted, as needed, by varying the placement of the end pans. The width of the web below the bottom of the joists must be at least 3 in. narrower than the flange (on either side) to allow for pan removal.

A type of one-way joist floor system has evolved known as a *joist-band system* in which the joists are supported by broad girders having the same total depth as the joists, as illustrated in Fig. 19.7. Separate beam forms are eliminated, and the same deck forms the soffit of both the joists and the girders. The simplified formwork, faster construction, level ceiling with no obstructing beams, and reduced overall height of walls, columns, and vertical utilities combine to achieve an overall reduction in cost in most cases.

In one-way joist floors, the thickness of the slab is often controlled by fire resistance requirements. For a rating of 2 hours, for example, the slab must be about $4\frac{1}{2}$ in. thick (Ref. 19.1). If 20 or 30 in. pan forms are used, slab span is small and slab strength is underutilized. This has led to what is known as the *wide module joist system*, or *skip joist system* (Ref. 19.7), as shown in Fig. 19.8. Such floors generally have 6 to 8 in. wide ribs that are 5 to 6 ft on centers, with a $4\frac{1}{2}$ in. top slab. These floors not only provide more efficient use of concrete in the slab but also require less formwork labor. By ACI Code 9.8.1.8, wide module joist ribs must be designed as ordinary T beams, because the clear spacing between ribs exceeds the 30 in. maximum for joist construction, and the special ACI Code provisions for joists do not apply. Concrete cover for reinforcement is as required for beams, not joists, and the

FIGURE 19.8
Skip joist system showing
wide spacing between ribs.
(*Photograph by David Darwin.*)

10 percent increase in V_c does not apply. Often the joists in wide module systems are carried by wide beams on the column lines, the depth of which is the same as that of the joists, to form a joist-band system equivalent to that described earlier.

Useful design information pertaining to one-way joist floors, including extensive load tables, are found in the *CRSI Design Guides* (Ref. 19.3). Suggested bar details and typical design drawings are found in the *ACI Detailing Manual* (Ref. 19.4).

e.   Two-Way Edge-Supported Slabs

Two-way solid slabs supported by beams on the column lines on all sides of each slab panel have been discussed in detail in Chapter 13. The perimeter beams are usually concrete cast monolithically with the slab, although they may also be structural steel, often encased in concrete for composite action and for improved fire resistance. For monolithic concrete, both the beams and the slabs are designed using the equivalent frame method described in Chapter 13.

Two-way solid slab systems are suitable for intermediate to heavy loads on spans up to about 30 ft. This range corresponds closely to that for beamless slabs with drop panels and column capitals, described in the following section. The latter are often preferred because of the complete elimination of obstructing beams below the slab.

For lighter loads and shorter spans, a two-way solid slab system has evolved in which the column-line beams are wide and shallow, such that a cross section through the floor in either direction resembles the slab-band shown earlier in Fig. 19.4. The result is a two-way slab-band floor that, from below, appears as a paneled ceiling. Advantages are similar to those given earlier for one-way slab-band floors and for joist-band systems.

f.   Beamless Flat Slabs with Drop Panels or Column Capitals

By suitably proportioning and reinforcing the slab, it is possible to eliminate supporting beams altogether. The slab is supported directly on the columns. In a rectangular or square region centered on the columns, the slab may be thickened and the column tops flared, as shown in Fig. 19.9. The thickened slab is termed a *drop panel*, and the column flare is referred to as a *column capital*. Both of these serve a double purpose: They increase the shear strength of the floor system in the critical region around the column, and they provide increased effective depth for the flexural steel in the region of high negative bending moment over the support. Beamless systems with drop panels or column capitals or both are termed *flat slab systems* (although almost all slabs in structural engineering practice are "flat" in the usual sense of the word), and are differentiated from flat plate systems, with absolutely no projections below the slab, which are described in the following section.

In general, flat slab construction is economical for live loads of 100 psf or more and for spans up to about 30 ft. It is widely used for storage warehouses, parking garages, and below-grade structures carrying heavy earth-fill loads, for example. For lighter loads such as in apartment houses, hotels, and office buildings, flat plates (Section 19.2g) or some form of joist construction (Sections 19.2d and h) will usually prove less expensive. For spans longer than about 30 ft, beams and girders are used because of the greater stiffness of that form of construction.

Flat slabs may be designed by the equivalent frame method described in detail in Chapter 13 or the strip method described in Chapter 24.

FIGURE 19.9
Flat slab garage floor with both drop panels and column capitals. (*Courtesy of Portland Cement Association.*)

FIGURE 19.10
Flat plate floor construction.
(*Courtesy of Portland Cement
Association.*)

g. Flat Plate Slabs

A flat plate floor is essentially a flat slab floor with the drop panels and column capitals omitted, so that a floor of uniform thickness is carried directly by prismatic columns. Flat plate floors have been found to be economical and otherwise advantageous for such uses as apartment buildings, as shown in Fig. 19.10, where the spans are moderate (up to about 30 ft) and loads relatively light. Prestressed concrete (Chapter 22) flat plate construction for residential and light commercial buildings has spans up to 40 ft. The construction depth for each floor is held to the absolute minimum, with resultant savings in the overall height of the building. The smooth underside of the slab can be painted directly and left exposed for ceiling, or plaster can be applied to the concrete. Minimum construction time and low labor costs result from the very simple formwork.

Certain problems associated with flat plate construction require special attention. Shear stresses near the columns may be very high, requiring the use of special types of slab reinforcement there. The transfer of moments from slab to columns may further increase these shear stresses and requires concentration of negative flexural steel in the region close to the columns. Both of these problems are treated in detail in Chapter 13. At the exterior columns, where such shear and moment transfer may cause particular difficulty, the design is much improved by extending the slab past the column in a short cantilever.

Some flat plate buildings are constructed by the lift slab method, shown in Fig. 19.11. A casting bed (often doubling as the ground-floor slab) is placed, steel columns are erected and braced, and at ground level successive slabs, which will later become the upper floors, are cast. A membrane or sprayed parting agent is laid down between successive pours so that each slab can be lifted in its turn, starting with the top.

FIGURE 19.11
Lift slab construction used
with flat plate floors; student
dormitory at Clemson
University, South Carolina.
(*Photograph by Arthur H.
Nilson.*)

FIGURE 19.12
Lifting collar embedded in
concrete slab.

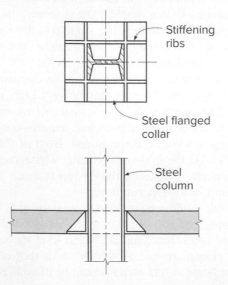

FIGURE 19.13
Two-way joist floor under
construction with steel dome
forms. (*Courtesy of Ceco
Construction Group.*)

Jacks placed atop the columns are connected to threaded rods extending down the faces of the columns and connecting, in turn, to lifting collars embedded in the slabs, as shown in Fig. 19.12. When a slab is in its final position, shear plates are welded to the column below the lifting collar, or other devices are used to transfer the vertical slab reaction. Lifting collars such as those shown in Fig. 19.12, in addition to providing anchorage for the lifting rods, serve to increase the effective size of the support for the slab and consequently improve the shear strength of the slab. The successful erection of structures using the lift slab method requires precise control of the lifting operation at all times, because even slight differences in level of the support collars may drastically change moments and shears in the slab, possibly leading to reversal of loading. Catastrophic accidents have resulted from failure to observe proper care in lifting or to provide adequate lateral bracing for the columns (Ref. 19.8). As a result of these accidents, this method of construction is used only by specialized contractors.

h. **Two-Way Joist Floors**

As in one-way floor systems, the dead weight of two-way slabs can be reduced considerably by creating void spaces in what would otherwise be a solid slab. For the most part, the concrete removed is in tension and ineffective, so the lighter floor has virtually the same structural characteristics as the corresponding solid floor. Voids are usually formed using dome-shaped steel pans that are removed for reuse after the slab has hardened. Forms are placed on a plywood platform as shown in Fig. 19.13. Note in the figure that domes have been omitted near the columns to obtain a solid slab in the region of negative bending moment and high shear. The lower flange of each dome contacts that of the adjacent dome, so that the concrete is cast entirely against a metal surface, resulting in an excellent finished appearance of the slab. A wafflelike appearance (these slabs are sometimes called waffle slabs) is imparted to the underside of the slab, which can be featured to architectural advantage, as shown in Fig. 19.14.

FIGURE 19.14
Regency House Apartments, San Antonio, with cantilevered two-way joist slab plus integral beams on column lines. (*Photograph by Arthur H. Nilson.*)

TABLE 19.2
Equivalent slab thickness and weight of two-way joist floor systems

Depth of Pan Form, in.	3 in. Top Slab		$4\frac{1}{2}$ in. Top Slab	
	Equivalent Uniform Thickness, in.	Weight, psf	Equivalent Uniform Thickness, in.	Weight, psf
36 in. Module (30 in. pans plus 6 in. ribs)				
8	5.7	71	7.2	90
10	6.4	80	7.9	99
12	7.2	90	8.7	109
14	8.0	100	9.5	119
16	8.9	111	10.3	129
20	10.6	132	12.1	151
24 in. Module (19 in. pans plus 5 in. ribs)				
8	6.3	79	7.8	98
10	7.3	91	8.8	110
12	8.2	103	9.8	122
14	9.3	116	10.7	134
16	10.3	129	11.8	148

Adapted from: CRSI Design Guides. Schaumburg, IL: Concrete Reinforcing Steel Institute, 2019.

Two-way joist floors are designed following the usual procedures for two-way solid slab systems, as presented in Chapter 13, with the solid regions at the columns considered as drop panels. Joists in each direction are divided into column strip joists and middle strip joists, the former including all joists that frame into the solid head. Each joist rib usually includes two bars for positive-moment resistance, and one may be discontinued where no longer required. Negative steel is provided by separate straight bars running in each direction over the columns.

In design calculations, the self-weight of two-way joist floors is considered to be uniformly distributed, based on an equivalent slab of uniform thickness having the same volume of concrete as the actual ribbed slab. Equivalent thicknesses and weights are given in Table 19.2 for standard 30 and 19 in. pans of various depths and for either a 3 in. top slab or $4\frac{1}{2}$ in. top slab, based on normalweight concrete (150 lb/ft^3).

19.3   PRECAST CONCRETE FOR BUILDINGS

Section 19.2 emphasized cast-in-place reinforced concrete structures. Construction of these structures requires a significant amount of skilled on-site labor. There is, however, another class of concrete construction for which the members are manufactured off site in precasting yards, under factory conditions, and subsequently assembled on site, a process that provides significant advantages in terms of economy and speed of construction.

Precast concrete construction involves the mass production of repetitive and often standardized units: columns, beams, floor and roof elements, and wall panels. On large jobs, precasting yards are sometimes constructed on or adjacent to the site. More frequently, these yards are stationary regional enterprises that supply precast members to sizable areas within reasonable shipping distances, on the order of 200 mi. Advantages of precast construction include less labor per unit because of mechanized series production; use of unskilled local labor, in contrast to skilled mobile construction labor; shorter construction time because site labor primarily involves only foundation construction and connecting the precast units; better quality control and higher concrete strength that are achievable under factory conditions; and greater independence of construction from weather and season. Disadvantages are the greater cost of transporting precast units, as compared with transporting materials, and the additional technical problems and costs of site connections of precast elements.

Precast construction is used in all major types of structures: industrial buildings, residential and office buildings, halls of sizable span, bridges, stadiums, and prisons. Precast members frequently are prestressed in the casting yard. In the context of the present chapter, it is irrelevant whether a precast member is also prestressed. Discussion is focused on types of precast members and precast structures and on methods of connection; these are essentially independent of whether the desired strength of the member was achieved with ordinary reinforcement or by prestressing. A broader discussion of precast construction, which includes planning, design, materials, manufacturing, handling, construction, and inspection, are found in Refs. 19.9 and 19.10.

a.   Types of Precast Members

A number of types of precast units are in common use. Though most are not formally standardized, they are widely available, with minor local variations. At the same time, the precasting process is sufficiently adaptable for special shapes developed for particular projects to be produced economically, provided that the number of repetitive units is sufficiently large. This is particularly important for exterior wall panels, which permit a wide variety of architectural treatments.

Wall panels are made in a considerable variety of shapes, depending on architectural requirements. The most frequent four shapes are shown in Fig. 19.15. These units are produced in one to four-story-high sections and up to 8 ft in width. They are used either as curtain walls attached to columns and beams or as bearing walls. To improve thermal insulation, sandwich panels are used that consist of an insulation core (such as foam glass, glass fiber, or expanded plastic) between two layers of normalweight or lightweight concrete. The two layers must be adequately interconnected through the core to act as one unit. A variety of surface finishes can be produced through the use of special exposed aggregates or of colored cement, sometimes employed in combination. The special design problems that arise in load-bearing wall panels, such as tilt-up construction, are discussed in Ref. 19.11.

Stresses in wall panels are frequently more severe in handling and during erection than in the finished structure, and the design must provide for these temporary conditions. Also, control of cracking is of greater importance in wall panels than in other precast units, for appearance more than for safety. To control cracking, the maximum tensile stress in the concrete, calculated by straight-line theory, should not exceed the modulus of rupture of the particular concrete with an adequate margin of safety. ACI Committee 533 (Ref. 19.12) recommends that tensile stresses for normalweight concrete be limited to $5\sqrt{f_c'}$ under the effects of form removal, handling, transportation, impact, and live load. Maximum tensile stresses equal to 75 and 85 percent of this value are recommended for all-lightweight and sand-lightweight concrete, respectively. A wealth of information on precast wall panels is found in Refs. 19.10 and 19.12.

Roof and floor elements are made in a wide variety of shapes adapted to specific conditions, such as span lengths, magnitude of loads, desired fire ratings, and

FIGURE 19.15
Precast concrete wall panels.

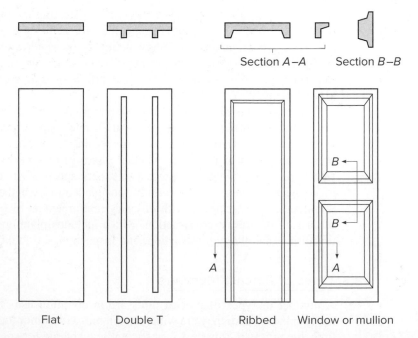

Section *A–A*   Section *B–B*

Flat   Double T   Ribbed   Window or mullion

FIGURE 19.16
Precast floor and roof elements.

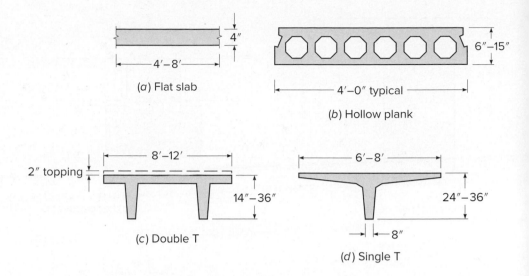

(*a*) Flat slab

(*b*) Hollow plank

(*c*) Double T

(*d*) Single T

appearance. Figure 19.16 shows typical examples of the most common shapes, arranged in approximate order of increasing span length, even though the spans covered by the various configurations overlap widely.

Flat slabs (Fig. 19.16*a*) are usually 4 in. thick, although they are used as thin as $2\frac{1}{2}$ in. when continuous over several spans, and are produced in widths of 4 to 8 ft and in lengths up to 36 ft. Depending on the magnitude of loads and on deflection limitations, they are used over roof and floor spans ranging from 8 to about 22 ft. For lower weight and better insulation and to cover longer spans, *hollow-core planks* (Fig. 19.16*b*) with a variety of shapes are used. Some of these are made by extrusion in special machines. Depths range from 6 to 15 in., with widths of 3 or 4 ft. Again depending on load and deflection requirements, they are used on roof spans from about 16 to 34 ft and on floor spans from 12 to 26 ft, which can be augmented to about 30 ft if a 2 in. topping is applied to act monolithically with the hollow plank.

For longer spans, *double T* members (Fig. 19.16*c*) are the most widely used shapes. Usual depths are from 14 to 36 in. They are used on roof spans up to 120 ft. When used as floor members, a concrete topping of at least 2 in. is usually applied to act monolithically with the precast members for spans up to about 50 ft, depending on load and deflection requirements. Finally, *single T* members are available in dimensions shown in Fig. 19.16*d*, mostly used for roof spans up to 100 ft and more.

For all of these units, the member itself or its flange constitutes the roof or floor slab. If the floor or roof proper is made of other material (for example, plywood, gypsum, and plank), it can be supported on *precast joists* in a variety of shapes for spans from about 15 to 60 ft. Reference 19.10 addresses the design of both reinforced and prestressed concrete floor and roof units.

The shape of *precast beams* depends chiefly on the manner of framing. If floor and roof members are supported on top of the beams, these are mostly rectangular in shape (Fig. 19.17*a*). To reduce total depth of floor and roof construction, the tops of beams are often made flush with the top surface of the floor elements. To provide bearing, the beams are then constructed as ledger beams (Fig. 19.17*b*) or L beams (Fig. 19.17*c*). Although these shapes pertain to building construction, precast beams or girders are also frequently used in highway bridges.

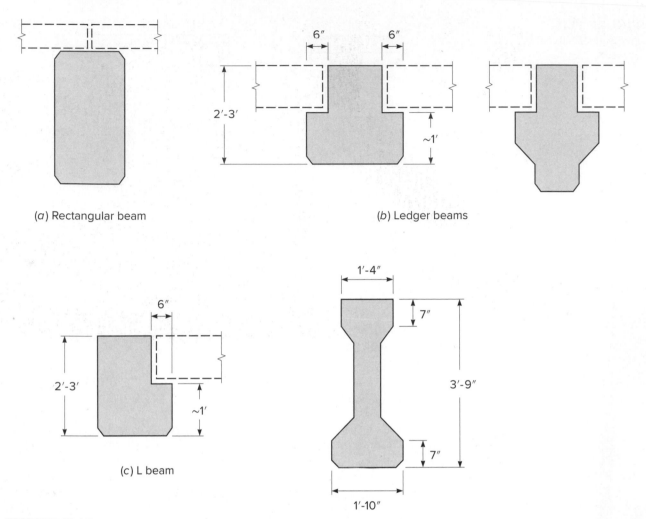

(*a*) Rectangular beam

(*b*) Ledger beams

(*c*) L beam

(*d*) AASHTO bridge girder

FIGURE 19.17
Precast beams and girders.

As an example, Fig. 19.17*d* shows one of the various AASHTO bridge girders, so named because they were developed by the American Association of State Highway and Transportation Officials.

If *precast columns* of single-story height are used so that the beams rest on top of the columns, simple prismatic columns are employed, which are available in sizes from about 12 × 12 to 24 × 24 in. (Fig. 19.18*a*). In this case, the beams are usually made continuous over the columns. Alternatively, in multistory construction, the columns can be made continuous for up to about six stories. In this case, integral brackets are frequently used to provide a bearing for the beams, as shown in Fig. 19.18*b* (see also Section 19.3b). Occasionally, T columns are used for direct support of double T floor members without the use of intermediate beams (Fig. 19.18*c*).

Figures 19.19 to 19.27 illustrate some of the many ways in which precast members have been used. Figure 19.19 shows a floor slab element being placed on precast columns with integral column capitals. The entire building, including elevator

FIGURE 19.18
Precast concrete columns.

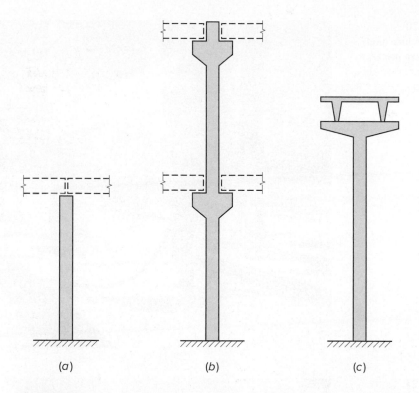

(a)   (b)   (c)

FIGURE 19.19
Precast slab element with precast columns, beams, and lateral framing. (*Photograph by Charles W. Dolan.*)

and stair shafts, is precast concrete. The photograph in Fig. 19.20 was taken in a precasting yard producing a variety of L, T, and rectangular shapes. Figure 19.21 shows symmetrical precast I beams, such as are used both for buildings and bridges. The projecting stirrup bars along the top flange provide secure interlock between the precast beams and a cast-in-place slab added later, ensuring composite action.

FIGURE 19.20
Sand-blasted architectural
finish applied to a precast
L beam. (*Photograph by
Charles W. Dolan.*)

FIGURE 19.21
Precast I beams designed for
composite action with a deck
slab to be cast in place. (*Photograph by Charles W. Dolan.*)

Figure 19.22 shows a multistory parking garage in which three-story precast columns support L-section and inverted T-section girders. The girders, in turn, carry 60 ft span prestressed single T beams, which provide the deck surface.

Figure 19.23 demonstrates that unusual architectural designs can be realized in precast concrete, as in this all-precast administration building. Wall panels are used to produce a curved facade. Wedge-shaped repetitive floor units span freely from the exterior facade to the interior curved beam and column framework. In the insurance

FIGURE 19.22
Precast parking garage at
Cornell University. (*Photo-
graph by Arthur H. Nilson.*)

FIGURE 19.23
All precast administration
building. (*Courtesy of Portland
Cement Association.*)

FIGURE 19.24
Precast girders with 99 ft
span and 44 in. depth for
a column-free interior.
(*Courtesy of Portland Cement
Association.*)

FIGURE 19.25
Precast roof and wall panels
combined with cast-in-place
frames and floor slabs.
(*Courtesy of Portland Cement
Association.*)

building shown in Fig. 19.24, 44 in. deep precast girders span 99 ft between exterior
walls supported on four points each and provide six floors of office space entirely
free of interior supports. The convention headquarters of Fig. 19.25 combines
cast-in-place frames and floor slabs with precast double T roof beams and precast

FIGURE 19.26
Precast room-sized modules
for a 21-story hotel. (*Courtesy
of Portland Cement Association.*)

wall panels of special design. Figure 19.26 shows a 21-story hotel under construction, which, except for the service units, consists entirely of box-shaped, room-sized modules completely prefabricated and stacked on top of each other. Abroad, such precast modules, with plumbing, wiring, and heating preinstalled, are widely used for multistory apartment buildings as an alternative to making similar apartment structures in precast wall, roof, and floor panels, which are more easily shipped but less easily erected than box-shaped modules.

Finally, Fig. 19.27 shows an example of the frequent combined use of structural steel with precast concrete. In this case, the framing of an eight-story hotel was done using bolted structural steel, while precast concrete floor and roof planks and precast wall panels were used for all other main structural components. This type of construction is economical for 6 to 12-story buildings, where it provides savings in both cost and construction time. It is one example of the increasingly important combined use of various structural materials and methods.

b. Connections

Cast-in-place reinforced concrete structures, by their very nature, tend to be monolithic and continuous. Connections, in the sense of joining two hitherto separate pieces, rarely occur in that type of construction. Precast structures, on the other

FIGURE 19.27
Steel framing combined with
precast concrete floor planks
for an eight-story hotel.
(*Courtesy of Arcelor Mittal.*)

hand, resemble steel construction in that the final structure consists of large numbers
of prefabricated elements that are connected on site to form the finished structure.
In both types of construction, such connections can be detailed to transmit gravity
forces only, gravity and horizontal forces, or moments in addition to these forces. In
the last case, a continuous structure is obtained much as in cast-in-place construction,
and connections that achieve such continuity by appropriate use of special hardware,
reinforcing steel, and concrete to transmit all tension, compression, and shear stresses
are sometimes called *hard* connections. In contrast, connections that transmit reactions in one direction only, analogous to rockers or rollers in steel structures, but
permit a limited amount of motion to relieve other forces, such as horizontal reaction components, are sometimes known as *soft* connections (Ref. 19.13). In almost
all precast connections, bearing plates or pads are used to ensure distribution and
reasonable uniformity of bearing pressures. Bearing plates are made of steel, while
bearing pads are made of materials such as chloroprene, fiber-reinforced polymers,
and Teflon. If bearing plates are used, and the plates on two members are suitably
joined by welding or other means, a hard connection is obtained in the sense that
horizontal, as well as vertical, forces are transmitted. On the other hand, bearing
pads transmit gravity loads but can permit sizable horizontal deformations and, thus,
relieve horizontal forces.

Precast concrete structures are subject to dimensional changes from creep, shrinkage, and relaxation of prestress in addition to temperature, while in steel structures only temperature changes produce dimensional variations. In the early development of precast construction, there was a tendency to use soft connections extensively to permit these dimensional changes to occur without causing restraint forces in the members, and particularly in the connections. Subsequent experience, however, has shown that the resulting structures possess insufficient stability against lateral forces, such as high wind and, particularly, earthquake effects. Therefore, current practice emphasizes the use of hard connections that provide a high degree of continuity (Refs. 19.10 and 19.14). When designing hard connections, provisions must be made to resist the restraint forces that are caused by the previously described volume changes (Ref. 19.10). Considerable information concerning this and other matters relating to connections is found in Refs. 19.10 and 19.14.

Bearing stresses on plain concrete are limited by ACI Code 22.8.3.2 to $0.85\phi f_c'$, except when the supporting area is wider on all sides than the loaded area A_1. In such a case this value of the permissible bearing stress may be multiplied by $\sqrt{A_2/A_1}$ but not more than 2.0, where A_2 is the maximum portion of the supporting surface that is geometrically similar to and concentric with the loading area (see Section 15.6b).

In the design of connections, it is prudent to use load factors that exceed those required for the connected members. This is so because connections are generally subject to high stress concentrations that preclude the development of much ductility. In contrast, the members connected are likely to possess considerable ductility if designed by usual ACI Code procedures and will give warning of impending collapse if overloading should take place. In addition, imperfections in connection geometry may cause large changes in the magnitude of stresses compared with those assumed in the design.

In designing members according to the ACI Code, load factors of 1.2 and 1.6 are applied to dead and live loads, D and L respectively, to determine the required strength. When volume change effects T are considered, they are often treated as dead load, and the factored load U is calculated from the equation $U = 1.2(D + T) + 1.6L$.

A wide variety of connection details for precast concrete building components have evolved, only a few of which are shown here as more or less representative connections. Many additional possibilities are described fully in Refs. 19.10 and 19.14.

Column base connections are generally accomplished using steel base plates that are anchored into the precast column. Figure 19.28a shows a column base detail with projecting base plate. Four anchor bolts are used, with double nuts facilitating erection and leveling of the column. Typically a minimum of 2 in. of nonshrink grout is used between the top of the pier, footing, or wall and the bottom of the steel base plate. Column reinforcement is welded to the top face of the base plate. Tests have confirmed that such column connections can transmit the full moment for which the column is designed, if properly detailed.

An alternative base detail is shown in Fig. 19.28b, with the dimensions of the base plate the same as, or slightly smaller than, the outside column dimensions. Anchor bolt pockets are provided, either centered on the column faces as shown or located at the corners. Bolt pockets are grouted after the nuts are tightened. Column bars, not shown here, would be welded to the top face of the base plate as before. Figure 19.29 shows the base plate detail, similar to Fig. 19.28b, that was used for the precast three-story columns in the parking garage shown in Fig. 19.22.

In Fig. 19.28c, the main column bars project from the ends of the precast member a sufficient distance to develop their strength by bond. The projecting bars are inserted into grout-filled holes cast in the foundation when it is placed.

FIGURE 19.28
Column base connections.
See Chapter 21 for bolt
anchorage details.

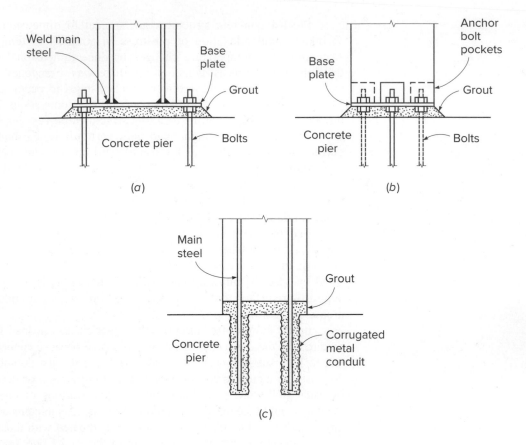

(a)   (b)

(c)

FIGURE 19.29
Detail at base of precast
column of Cornell University
parking garage shown in
Fig. 19.22. (*Photograph by
Arthur H. Nilson.*)

FIGURE 19.30
Beam-to-column
connections.

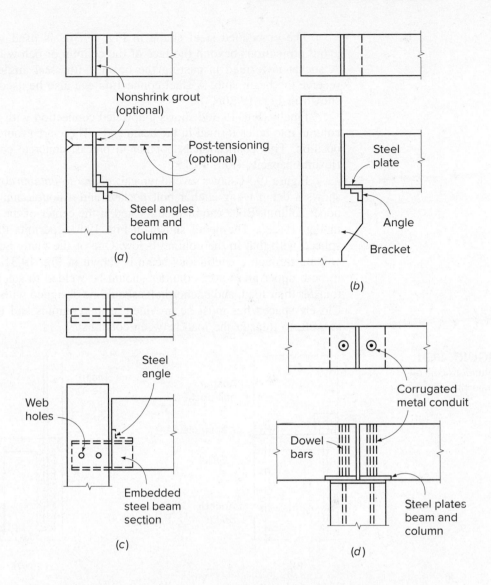

In all of the cases shown, confining steel should be provided around the anchor bolts in the form of closed ties. A minimum of four No. 3 (No. 10) ties is recommended, placed on 3 in. centers near the top surface of the pier or wall. Tie reinforcement in the columns should be provided as usual.

Figure 19.30 shows several *beam-to-column* connections. In all cases, rectangular beams are shown, but similar details apply to I or T beams. The figure shows only the basic geometry, and auxiliary reinforcement, anchors, and ties are omitted for the sake of clarity.

Figure 19.30a shows a joint detail with a concealed haunch. Well-anchored bearing angles are provided at the column seat and beam end. This type of connection may be used to provide vertical and horizontal reaction components, and with the addition of post-tensioned prestressing, provides moment resistance as well.

Figure 19.30b shows a typical bracket, common for industrial construction where the projecting bracket is not objectionable. The seat angle is welded to reinforcing bars anchored in the column. A steel bearing plate is used at the bottom of the beam and anchored into the concrete.

The embedded steel shape in Fig. 19.30*c* is used when it is necessary to avoid projections beyond the face of the column or below the bottom of the beam. A socket is formed in casting the beam, with steel angle or plate at its top, to receive the beam stub. A steel connection can also be used in place of the bracket shown in Fig. 19.30*b*.

Finally, Fig. 19.30*d* shows a doweled connection with bars projecting from the column into holes formed in the beam ends. These are grouted after the beams are in position. This connection is popular in precast concrete construction but has little flexural capacity (Ref. 19.15).

Figure 19.31 shows several typical *column-to-column* connections. Figure 19.31*a* shows a detail using anchor bolt pockets and a double-nut system for leveling the upper column. Bolts can also be located at the center of the column faces, as shown in Fig. 19.28*b*. The detail shown in Fig. 19.31*b* permits the main steel to be lap-spliced with that in the column below. One of the many possibilities for splicing a column through a continuous beam is shown in Fig. 19.31*c*. Main reinforcing bars in both upper and lower columns should be welded to steel cap and base plates to transfer their load, and anchor bolts should be designed with the same consideration. Closely spaced ties must be provided in the columns and in this case in the beam as well, to transfer the load between columns.

FIGURE 19.31
Column-to-column connections.

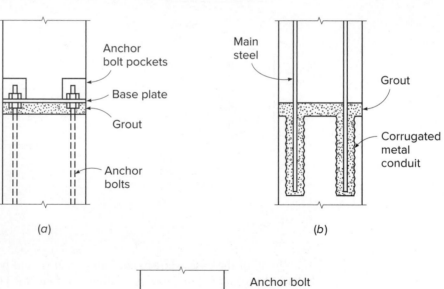

(a)   (b)

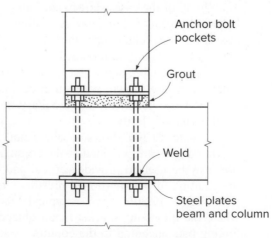

(c)

FIGURE 19.32
Slab-to-beam connections.

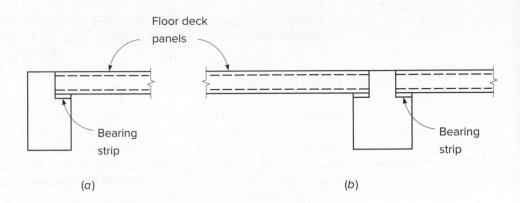

Slab-to-beam connections generally use some variation of the detail shown in Fig. 19.32. Support is provided by an L beam (Fig. 19.32*a*) or an inverted T beam (Fig. 19.32*b*) that is flush with the top of the precast floor planks. The detail shown is sufficient if no mechanical tie is required between the precast parts. Where a positive connection is required, steel plates are set into the top of the members, suitably anchored, and short connecting plates are welded so as to attach the built-in plates.

Basic tools for the design of precast concrete connections are the *shear friction design method* described in detail in Chapter 5 and the *strut-and-tie model* described in Chapter 17. Example 5.6 (Section 5.9) demonstrated the use of the shear-friction approach to determining the reinforcement for the end-bearing region of a precast concrete girder. The use of both the shear-friction method and a strut-and-tie model for joint behavior was shown in Section 18.7, and Example 18.5 presented the detailed design of a bracket for a precast concrete column. Additional design information pertaining to precast concrete connection design is found in Refs. 19.10, 19.13, and 19.14.

c. **Structural Integrity**

Precast concrete structures normally lack the joint continuity and high degree of redundancy characteristic of monolithic, cast-in-place reinforced concrete construction. Progressive collapse in the event of abnormal loading, in which the failure of one element leads to the collapse of another, and then another, can produce catastrophic results. For this reason, the structural integrity of precast concrete structures is specifically addressed in ACI Code 16.2.4 and 16.2.5. ACI Code 16.2.4 does not permit the use of "soft" connections that rely solely on friction caused by gravity forces. Full moment-resisting connections are unusual, but some positive means of connecting members to their supports, with due regard to the need to accommodate dimensional changes associated with creep, shrinkage, and temperature effects, is strongly recommended.

In addition, experience with precast structures has shown that the introduction of special reinforcement in the form of tension ties, though adding little to the cost of construction, can contribute greatly to maintaining structural integrity in the event of extraordinary loading, such as loads caused by extreme winds, earthquake, or explosion. This tension reinforcement is best arranged in a three-dimensional grid, usually on the column lines, tying the floors together vertically and in both horizontal directions. For precast concrete construction, ACI Code 16.2.1.8, 16.2.4, and 16.2.5 require that tension ties be provided in the transverse, longitudinal, and vertical directions of the structure and around its perimeter. Specific details vary widely. Although no specific guidance is offered in either the

ACI Code or Commentary regarding steel placement or design forces, valuable suggestions are found in Refs. 19.9, 19.10, and 19.14.

19.4 DIAPHRAGMS

Diaphragms are primarily horizontal elements, such as floors and roofs, that in addition to carrying out-of-plane loads also transfer lateral loads, such as from wind and earthquake, to vertical elements of the lateral-force-resisting system, typically consisting of walls and columns. Walls can also serve as diaphragms. Diaphragms are an integral part of most building systems and may consist of cast-in-place and precast concrete elements.

For many years, the role of roofs and floors in transferring loads to and from the vertical elements of the lateral-force-resisting system was taken for granted, but poor performance, primarily under seismic loading, has demonstrated the need to develop and codify effective design procedures (Ref. 19.16).

a. Load Transfer by Diaphragms

Figure 19.33 illustrates a number of the roles played by diaphragms in buildings. These roles include carrying out-of-plane gravity load, transferring loads to and distributing loads between the vertical elements of the lateral-force-resisting system, tying the vertical elements of the lateral-force-resisting system together, carrying thrust from inclined columns, and resisting soil pressure. The forces within diaphragms include those due to the application of lateral loads, those due to the distribution of loads from the diaphragms to the vertical elements of the lateral-force-resisting system, and those transferred from

FIGURE 19.33
Diaphragm actions.
(*Adapted from Ref. 19.17.*)

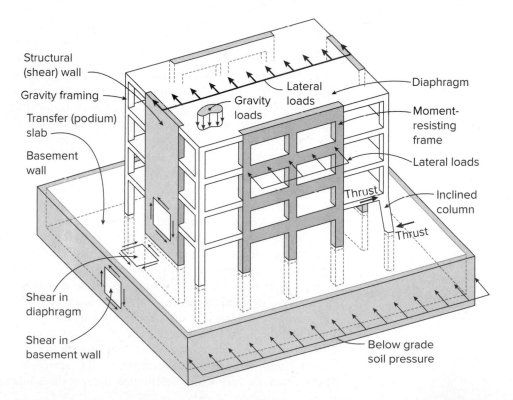

FIGURE 19.34
Idealization of diaphragm.
(*Adapted from Ref. 19.17.*)

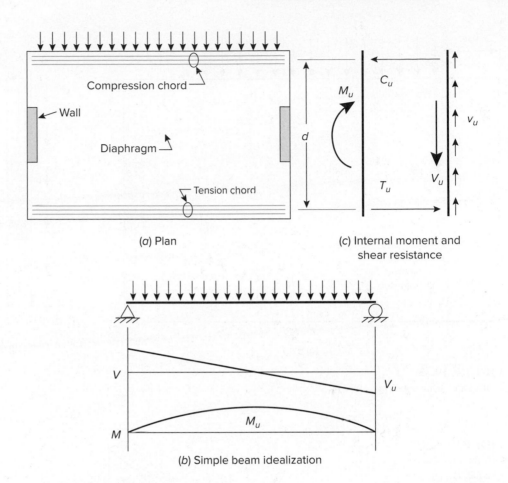

(*a*) Plan

(*c*) Internal moment and
shear resistance

(*b*) Simple beam idealization

the vertical elements of the lateral-force-resisting system to the diaphragms. Due to the latter two roles, regions of diaphragms must be designed as *distributor* or *collector* elements, the latter term being the one used by the ACI Code (Ref. 19.18).

The components of a diaphragm are illustrated in Figs. 19.34–19.36. The diaphragm shown in Fig. 19.34*a* is a floor slab that must carry a lateral load, supported by walls on each end. As shown in Fig. 19.34*b*, the diaphragm may be considered to perform as a beam with the bending moment carried primarily by *tension* and *compression chord*s. Shear is considered to be distributed uniformly through the depth of the diagram (Fig. 19.34*c*).

The lateral forces in a diaphragm must be transferred to the vertical elements of the lateral-force-resisting system. This transfer is done through regions in the diaphragm referred to as collectors. As shown in Fig. 19.35*a*, collectors may have the same width as the vertical elements of the lateral-force-resisting system, in this case walls, or may have a width that is greater than the width of the wall. Figure 19.35*b* illustrates the nature of the forces in the collector, which range from compression on one side of the wall to tension on the other. The forces increase from zero at the extremes of the collector to a maximum where they are transferred to the vertical element.

In the case shown in Fig. 19.36, forces from the wall are transferred to a floor slab that, in turn, transfers the forces to a basement wall. In this case, the collector serves as a "distributor," transferring compression forces on one side and tension forces on the other side, which are ultimately carried by the surrounding diaphragm away from the plane of the wall.

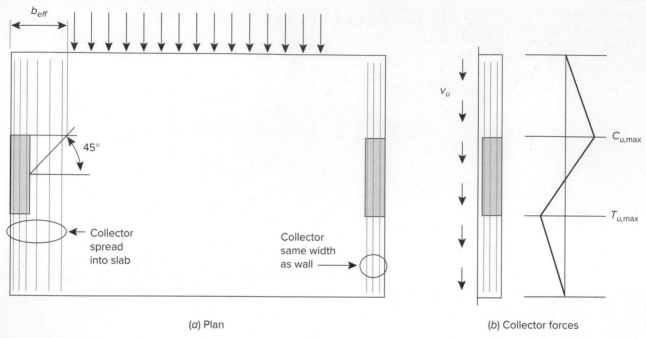

(a) Plan

(b) Collector forces

FIGURE 19.35
Collectors. (*Adapted from Ref. 19.17.*)

FIGURE 19.36
Distributor. (*Adapted from Ref. 19.17.*)

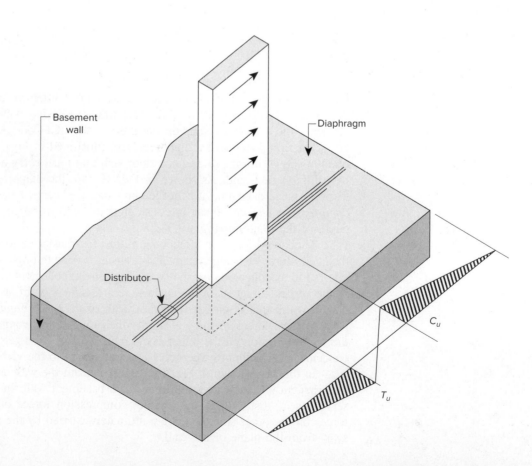

b. ACI Provisions for Diaphragms

Chapter 12 of the ACI Code deals with the design of diaphragms, including both cast-in-place and precast elements. Diaphragms in structures in seismic design categories D, E, and F are covered in ACI Code 18.12 and Chapter 20.

The ACI Code requires that the design process for diaphragms consider (1) in-plane forces due to lateral loads, (2) forces caused by differences in lateral deformation properties between the vertical elements connected by the diaphragm, (3) forces at connections between the diaphragm and vertical framing or nonstructural elements, (4) bracing forces, such as provided by sloped columns, and (5) out-of-plane forces due to gravity or other loads applied to the surface of the diaphragm, as illustrated in Fig. 19.33. The factored load combinations listed in Table 1.2 apply to diaphragms, as they do to other structural members.

A number of different modeling and analysis procedures are allowed by the ACI Code, which permits the use of "any set of reasonable and consistent assumptions" for diaphragm stiffness to calculate in-plane factored moments, shears, and axial loads. ACI 12.4.2.4 permits a range of options, including modeling the diaphragm as rigid or flexible, using an analysis that establishes upper and lower bounds for in-plane stiffness, using finite element models, and using strut-and-tie models, as discussed in Chapter 17. ASCE/SEI 7 (Ref. 19.19) includes modeling criteria for structures subjected to wind and earthquake loading. Strength reduction factors ϕ are shown in Table 1.3.

When designed for moment and axial load, diaphragms can be treated as beams or columns, as appropriate. Reinforcement resisting tension due to bending in the plane of the diaphragm must be located within $h/4$ of the tension edge of the diaphragm, where h is the diaphragm depth at that location. If the depth changes along the span, the reinforcement can be developed in adjacent segments of the diaphragm that are not within the $h/4$ limit.

For diaphragms that are cast-in-place, the nominal shear strength is given by

$$V_n = A_{cv} \left(2\lambda \sqrt{f_c'} + \rho_t f_y \right) \tag{19.1}$$

where A_{cv} is the gross area of the concrete bounded by the diaphragm thickness and depth, reduced to account for any void areas. The value of $\sqrt{f_c'}$ used in the calculation may not exceed 100 psi, and the reinforcement ratio ρ_t is for the distributed reinforcement parallel to the in-plane shear. For shear design, $\phi = 0.75$ or a lower value appropriate for seismic design, as discussed in Section 20.7.

The dimensions of cast-in-place diaphragms must be selected so that

$$V_u \le \phi 8 A_{cv} \sqrt{f_c'} \tag{19.2}$$

where $\sqrt{f_c'}$ used in the calculation may not exceed 100 psi.

The ACI Code also includes shear provisions for diaphragms with cast-in-place concrete toppings on precast elements and interconnected precast elements.

Where shear is transferred from a diaphragm to a collector or from a diaphragm or collector to a vertical element of the lateral-force-resisting system, the force transfer must be designed either in accordance with the shear-friction provisions described in Section 5.9 or, where mechanical connectors or dowels are used, consideration must be given to the effects uplift and rotation of the vertical elements of the lateral-force-resisting system.

ACI Code 12.5.4 requires that collectors extend from vertical elements, as needed, to transfer the shear force from the diaphragm to the vertical element

(Fig. 19.35). A collector may be discontinued along lengths of vertical elements where the transfer of forces is not required. Collectors are designed as tension members, compression members, or both. The length of a collector must be the greater of the length required to develop the reinforcement in tension and the length required to transfer the design forces to the vertical element through shear-friction, mechanical connectors, or other force transfer mechanisms.

The reinforcement used in diaphragms must be at least that required for temperature and shrinkage stresses for slabs, as described in Section 12.3a, and floors and roofs that serve as diaphragms must meet the reinforcement limits for one-way or two-way slabs, as appropriate. Reinforcement required to resist in-plane forces is in addition to that designed to resist other load effects, with the exception that shrinkage and temperature reinforcement may be used to resist diaphragm in-plane forces.

19.5 ENGINEERING DRAWINGS FOR BUILDINGS

Design information is conveyed to the builder mainly by engineering drawings. Their preparation is therefore a matter of utmost importance, and they should be carefully checked by the design engineer to ensure that concrete dimensions and reinforcement agree with the calculations.

Engineering drawings for buildings usually consist of a plan view of each floor showing overall dimensions and locating the main structural elements, cross-sectional views through typical members, and beam and slab schedules that give detailed information on the concrete dimensions and reinforcement in tabular form. Sectional views are usually drawn to a larger scale than the plan and serve to locate the steel and establish cutoff and bend points as well as to define the shape of the member. Usually a separate drawing is included that gives, in the form of schedules and cross sections, the details of columns and footings.

The construction documents, including the plans, specifications, and cost estimates, provide detailed descriptions of the material strengths. Additionally, many building officials, particularly in active seismic regions, require a description of the structural framing system, lateral-force-resisting system, and design live loads to be included on the structural drawings. Typical concrete design drawings and details are in Ref. 19.4. ACI Code Chapter 26 identifies the information needed in construction documents and the conformance requirements for acceptance of work.

REFERENCES

19.1. ACI Committee 216, *Code Requirements for Determining Fire Resistance of Concrete and Masonry Construction Assemblies*, ACI 216.1-14, American Concrete Institute, Farmington Hills, MI, 2014.

19.2. E. S. Hoffman, D. P. Gustafson, and A. J. Gouwens, *Structural Design Guide for the ACI Building Code*, Kluwer Academic Publishers, Boston, 1998.

19.3. *CRSI Design Guides*, Concrete Reinforcing Steel Institute, Schaumburg, IL, 2019.

19.4. *ACI Detailing Manual*, ACI Special Publication SP-66(04), American Concrete Institute, Farmington Hills, MI, 2004.

19.5. ACI Committee 352, *Recommendations for Design of Beam-Column Joints in Monolithic Reinforced Concrete Structures*, ACI 352R-02, American Concrete Institute, Farmington Hills, MI, 2002.

19.6. S. Ravikumar, D. Darwin, S. L. McCabe, and G. P. Pasley, "Shear Strength of Continuous Lightly Reinforced Concrete Joist Systems," *SM Report* No. 37, University of Kansas Center for Research, Lawrence, March 1994.

19.7. L. E. Svab and R. E. Jurewicz, "Wide Module Concrete Joist Construction," *Concr. Intl.*, vol. 12, no. 11, 1990, pp. 39–42.

19.8. C. F. Scribner and C. G. Culver, "Investigation of the Collapse of L'Ambiance Plaza," *J. Perform. Construct. Facil.* ASCE, vol. 2, no. 2, 1988, pp. 58–79.

19.9. ACI Committee 550, *Design Recommendations for Precast Concrete Structures*, ACI 550R-96, American Concrete Institute, Farmington Hills, MI, 1996.

19.10. *PCI Design Handbook*, 8th ed., Precast/Prestressed Concrete Institute, Chicago, 2017.

19.11. K. M. Kripanarayanan and M. Fintel, "Analysis and Design of Slender Tilt-Up Reinforced Concrete Wall Panels," *J. ACI*, vol. 71, no. 1, 1974, pp. 20–28.

19.12. ACI Committee 533, *Guide for Precast Concrete Wall Panels*, ACI 533R-11, American Concrete Institute, Farmington Hills, MI, 2011.

19.13. P. W. Birkeland and H. W. Birkeland, "Connections in Precast Concrete Construction," *J. ACI*, vol. 63, no. 3, 1966, pp. 345–368.

19.14. PCI Committee on Connection Details, *Design and Typical Details of Connections for Precast and Prestressed Concrete*, MNL-123-88, Precast/Prestressed Concrete Institute, Chicago, 1988.

19.15. C. W. Dolan, J. F. Stanton, and R. G. Anderson, "Moment Resistant Connections and Simple Connections," *J. Precast/Prestressed Concr. Inst.*, vol. 32, no. 2, 1987, pp. 62–74.

19.16. W. G. Corley, L. Cluff, S. Hilmy, W. Holmes, and J. Wight. "Concrete Parking Structures," Northridge Report vol. 2, *Earthquake Spectra*, vol. 11, 1996, Supplement C, pp. 75–98.

19.17. J. P. Moehle, J. D. Hooper, D. J. Kelly, and T. R. Meyer, "Seismic Design of Cast-in-Place Concrete Diaphragms, Chords, and Collectors: A Guide for Practicing Engineers," *NEHRP Seismic Design Technical Brief* No. 3, National Institute of Standards and Technology, Gaithersburg, MD, NIST GCR 10-917-4, 2010.

19.18. *Building Code Requirements for Structural Concrete and Commentary* (ACI 318-19), American Concrete Institute, Farmington Hills, MI, 2019.

19.19. *Minimum Design Loads for Buildings and Other Structures*, ASCE/SEI 7-16, American Society of Civil Engineers, Reston, VA, 2016.

20

Seismic Design

20.1 INTRODUCTION

Earthquakes result from the sudden movement of tectonic plates in the earth's crust. The movement takes place at fault lines, and the energy released is transmitted through the earth in the form of waves that cause ground motion many miles from the epicenter. Regions adjacent to active fault lines are the most prone to experience earthquakes. The map in Fig. 20.1 shows the *maximum considered ground motion* for the contiguous 48 states. The mapped values, expressed as a percent of gravity, represent the expected peak acceleration of a single-degree-of-freedom system with a 0.2 sec period and 5 percent of critical damping. Known as the *0.2 sec spectral response acceleration* S_S (subscript S for *short* period), it is used, along with the 1.0 sec spectral response acceleration S_1 (mapped in a similar manner), to establish the loading criteria for seismic design. Accelerations S_S and S_1 are based on historical records and local geology. For most of the country, they represent earthquake ground motion with a "likelihood of exceedance of 2 percent in 50 years," a value that is equivalent to a return period of about 2500 years (Ref. 20.1).

As experienced by structures, earthquakes consist of random horizontal and vertical movements of the earth's surface. As the ground moves, inertia tends to keep structures in place (Fig. 20.2), resulting in the imposition of displacements and forces that can have catastrophic results. The purpose of seismic design is to proportion structures so that they can withstand the displacements and the forces induced by the ground motion.

Historically in North America, seismic design has emphasized the effects of horizontal ground motion because the horizontal components of an earthquake usually exceed the vertical component and because structures are usually much stiffer and stronger in response to vertical loads than they are in response to horizontal loads. Experience has shown that the horizontal components are the most destructive. For structural design, the intensity of an earthquake is usually described in terms of the peak ground acceleration as a fraction of the acceleration of gravity, that is, $0.1g$, $0.2g$, or $0.3g$. Although peak acceleration is an important design parameter, the frequency characteristics and duration of an earthquake are also important; the closer the frequency of the earthquake motion is to the natural frequency of a structure and the longer the duration of the earthquake, the greater the potential for damage.

Based on elastic behavior, structures subjected to a major earthquake would be required to undergo large displacements. However, North American practice (Ref. 20.2) requires that structures be designed for only a fraction of the forces associated with those displacements. The relatively low design forces are justified by the observations that buildings designed for low forces have behaved satisfactorily and that structures

FIGURE 20.1
Map showing
maximum considered
earthquake ground
motion, 0.2 sec
spectral response
acceleration
(5 percent of critical
damping), for the
contiguous United
States. (*United States
Geological Survey*)

FIGURE 20.2
Structure subjected to ground
motion.

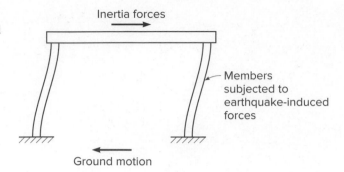

dissipate significant energy as the materials yield and behave inelastically. This non-linear behavior, however, usually translates into increased displacements, which may require significant ductility and result in major nonstructural damage. Displacements may also be of such a magnitude that the strength of the structure is affected by stability considerations, such as discussed for slender columns in Chapter 10.

Designers of structures that may be subjected to earthquakes, therefore, are faced with a choice: (1) providing adequate stiffness and strength to limit the response of structures to the elastic range or (2) providing lower-strength structures, with presumably lower initial costs, that have the ability to withstand large inelastic deformations while maintaining their load-carrying capability.

20.2   STRUCTURAL RESPONSE

The safety of a structure subjected to seismic loading rests on the designer's understanding of the response of the structure to ground motion. For many years, the goal of earthquake design in North America has been to construct buildings that will withstand *moderate earthquakes without damage and severe earthquakes without collapse*, although with the passage of time, the advantages of minimizing damage due to severe earthquakes have become apparent. Building codes have undergone regular modification as major earthquakes have exposed weaknesses in existing design criteria.

Design for earthquakes differs from design for gravity and wind loads in the relatively greater sensitivity of earthquake-induced forces to the geometry of the structure. Without careful design, forces and displacements can be concentrated in portions of a structure that are not capable of providing adequate strength or ductility. Steps to strengthen a member for one type of loading may actually increase the forces in the member and change the mode of failure from ductile to brittle.

a.   Structural Considerations

The closer the frequency of the ground motion is to one of the natural frequencies of a structure, the greater the likelihood of the structure experiencing resonance, resulting in an increase in both displacement and damage. Therefore, earthquake response depends strongly on the geometric properties of a structure, especially height. Tall buildings respond more strongly to long-period (low-frequency) ground motion, while short buildings respond more strongly to short-period (high-frequency) ground motion. Figure 20.3 shows the shapes for the principal modes of vibration of a three-story frame structure. The relative contribution of each mode to the lateral displacement of the structure depends on the frequency characteristics of the ground motion. The first mode (Fig. 20.3*a*) usually provides the greatest contribution to lateral displacement.

FIGURE 20.3
Modal shapes for a three-story building: (*a*) first mode; (*b*) second mode; (*c*) third mode. (*Adapted from Ref. 20.3.*)

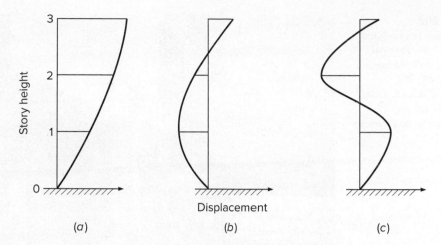

Displacement

(*a*)   (*b*)   (*c*)

FIGURE 20.4
Soft first story supporting a stiff upper structure.

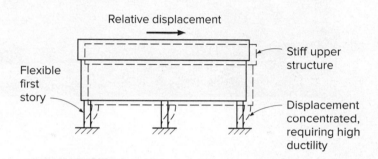

Relative displacement

Stiff upper structure

Flexible first story

Displacement concentrated, requiring high ductility

The taller a structure, the more susceptible it is to the effects of higher modes of vibration, which are generally additive to the effects of the lower modes and tend to have the greatest influence on the upper stories. Under any circumstances, the longer the duration of an earthquake, the greater the potential for damage.

The configuration of a structure also has a major effect on its response to an earthquake. Structures with a discontinuity in stiffness or geometry can be subjected to undesirably high displacements or forces. For example, the discontinuance of shear walls, infill walls, or even cladding at a particular story level, such as shown in Fig. 20.4, will have the result of concentrating the displacement in the open, or "soft," story. The high displacement will, in turn, require a large amount of ductility if the structure is not to fail. Such a design is not recommended, and the stiffening members should be continued to the foundation. The problems associated with a soft story are illustrated in Fig. 20.5, which shows the Olive View Hospital following the 1971 San Fernando earthquake. The high ductility "demand" could not be satisfied by the column at the right, with low amounts of transverse reinforcement. Even the columns at center, with significant transverse reinforcement, performed poorly because the transverse reinforcement was not continued into the joint, resulting in the formation of hinges at the column ends. Figure 20.6 illustrates structures with vertical geometric and plan irregularities, which result in torsion induced by ground motion.

Within a structure, stiffer members tend to pick up a greater portion of the load. When a frame is combined with a shear wall, this can have the positive effect of reducing the displacements of the structure and decreasing both structural and nonstructural damage. However, when the effects of higher stiffness members, such as masonry infill walls, are not considered in the design, unexpected and often undesirable results can occur.

FIGURE 20.5
Damage to soft story columns
in the Olive View Hospital
as a result of the 1971 San
Fernando earthquake.
(*Photograph by James L. Stratta.
Courtesy of the Federal
Emergency Management Agency.*)

FIGURE 20.6
Structures with (*a*) vertical
geometric and (*b*) plan
irregularities. (*Adapted
from Ref. 20.3.*)

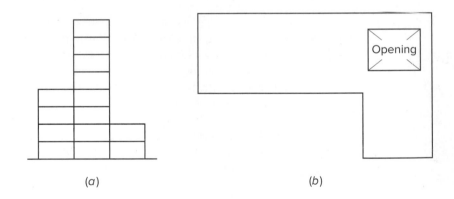

(*a*)   (*b*)

Finally, any discussion of structural considerations would be incomplete without emphasizing the need to provide adequate separation between structures. Lateral displacements can result in structures coming in contact during an earthquake, resulting in major damage due to hammering, as shown in Fig. 20.7. Spacing requirements to ensure that adjacent structures do not come into contact as the result of earthquake-induced motion are specified in Ref. 20.2.

b.   Member Considerations

Members designed for seismic loading must perform in a ductile fashion and dissipate energy in a manner that does not compromise the strength of the structure. Both the overall design and the structural details must be considered to meet this goal.

FIGURE 20.7
Damage caused by
hammering for buildings
with inadequate separation
in the 1985 Mexico City
earthquake. (*Photograph
courtesy of Jack Moehle*)

The principal method of ensuring ductility in members subject to shear and bending is to provide confinement for the concrete. This is accomplished through the use of closed *hoops* or spiral reinforcement, which enclose the core of beams and columns. Specific criteria are discussed in Sections 20.4, 20.5, and 20.6. When confinement is provided, beams and columns can undergo nonlinear cyclic bending while maintaining their flexural strength and without deteriorating due to diagonal tension cracking. The formation of *ductile hinges* allows reinforced concrete frames to dissipate energy.

Successful seismic design of frames requires that the structures be proportioned so that hinges occur at locations that least compromise strength. For a frame undergoing lateral displacement, such as shown in Fig. 20.8*a*, the flexural capacity of the members at a joint (Fig. 20.8*b*) should be such that the columns are stronger than the beams. In this way, hinges will form in the beams rather than the columns, minimizing the portion of the structure affected by nonlinear behavior and maintaining the overall vertical load capacity. For these reasons, the "weak beam–strong column" approach is used to design reinforced concrete frames subject to seismic loading.

When hinges form in a beam, or in extreme cases within a column, the moments at the end of the member, which are governed by flexural strength, determine the

FIGURE 20.8
Frame subjected to lateral loading: (*a*) deflected shape; (*b*) moments acting on beam-column joint; (*c*) deflected shape and forces acting on a beam; and (*d*) forces acting on faces of a joint due to lateral load.

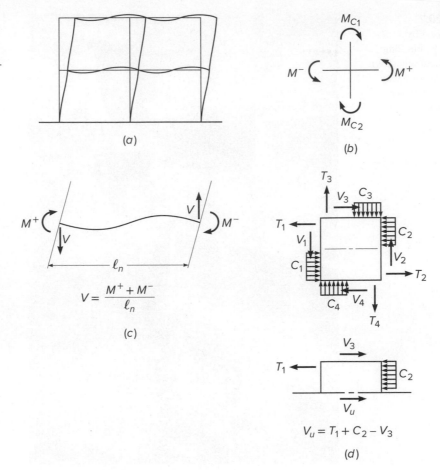

shear that must be carried, as illustrated in Fig. 20.8*c*. The shear V corresponding to a flexural failure at both ends of a beam or column is

$$V = \frac{M^+ + M^-}{\ell_n} \tag{20.1}$$

where M^+ and M^- = flexural capacities at the ends of the member
ℓ_n = clear span between supports

The member must be checked for adequacy under the shear V in addition to shear resulting from dead and live gravity loads. Transverse reinforcement is added, as required. For members with inadequate shear capacity, the response will be dominated by the formation of diagonal cracks, rather than ductile hinges, resulting in a substantial reduction in the energy dissipation capacity of the member.

If short members are used in a frame, the members may be strong in flexure compared to their shear capacity. An example would be columns in a structure with deep spandrel beams or with "nonstructural" walls with openings that expose a portion of the columns to the full lateral load. As a result, the exposed region, called a *captive column*, responds by undergoing a shear failure, as shown in Fig. 20.9.

The lateral displacement of a frame places beam-column joints under high shear stresses because of the change from positive to negative bending in the flexural members from one side of the joint to the other, as shown in Fig. 20.8*d*. The joint must be

FIGURE 20.9
Shear failure in a captive
column without adequate
transverse reinforcement.
(*Photograph courtesy of Jack
Moehle.*)

able to withstand the high shear stresses and allow for a change in bar stress from
tension to compression between the faces of the joint. Such a transfer of shear and
bond is often made difficult by congestion of reinforcement through the joint. Thus,
designers must ensure that joints not only have adequate strength but are also con-
structable. Two-way systems without beams are especially vulnerable because of low
ductility at the slab-column intersection.

Additional discussion of seismic design can be found in Refs. 20.3 to 20.8.

20.3   SEISMIC LOADING CRITERIA

In the United States, the design criteria for earthquake loading are based on design
procedures developed by the Building Seismic Safety Council (Ref. 20.1) and incorporated in *Minimum Design Loads for Buildings and Other Structures* (ASCE/SEI 7)
(Ref. 20.2). The values of the spectral response accelerations S_S and S_1 are obtained
from detailed maps produced by the United States Geological Survey[†] (such as,
Fig. 20.1) and included in ASCE/SEI 7. The values of S_S and S_1 are used to determine
the spectral response accelerations S_{DS} and S_{D1} that are used in design.

$$S_{DS} = \frac{2}{3} F_a S_S \tag{20.2}$$

$$S_{D1} = \frac{2}{3} F_v S_1 \tag{20.3}$$

where F_a and F_v are site coefficients that range from 0.8 to 2.4 and from 0.8 to 4.2,
respectively, as a function of the geotechnical properties of the building site and the

---

[†] A full set of maps is available at the United States Geological Survey website.

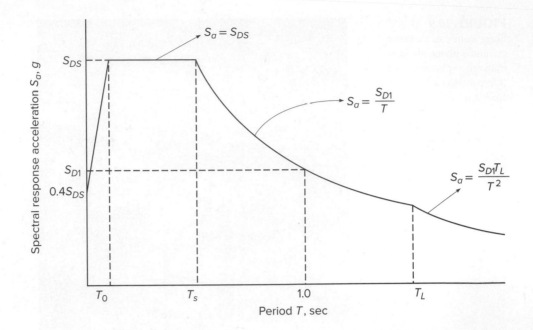

values of S_S and S_1, respectively. Higher values of F_a and F_v are possible for some sites. The coefficients F_a and F_v increase in magnitude as site conditions change from hard rock to thick, soft clays and (for softer foundations) as the values of S_S and S_1 decrease.

Both S_{DS} and S_{D1} are used to construct the design response spectrum shown in Fig. 20.10, which relates the spectral response acceleration S_a, used to calculate the earthquake force, to the fundamental period of the structure T. In the spectrum, $T_0 = 0.2S_{D1}/S_{DS}$, $T_S = S_{D1}/S_{DS}$, and T_L is the site-specific *long-period transition period*, which, like S_S and S_1, is obtained from maps provided by the U.S. Geological Survey.

Structures are assigned to one of six *Seismic Design Categories* (SDC) A through F as a function of (1) structure occupancy and use and (2) the values of S_{DS} and S_{D1}. Requirements for seismic design and detailing are minimal for SDCs A and B but become progressively more rigorous for SDCs C through F.

As presented in Table 1.2, earthquake loading is included in two combinations of factored load.

$$U = 1.2D + 1.0E + 1.0L + 0.2S \qquad (20.4)$$

$$U = 0.9D + 1.0E \qquad (20.5)$$

where D = dead load
  E = earthquake load
  L = live load
  S = snow load

The load factor for live load L may be reduced to 0.5, except for garages, areas occupied as places of public assembly, and areas where L is greater than 100 psf.

For SDC A, the earthquake load E is a horizontal load equal to 1 percent of the dead load D assigned to each floor. For SDC B through F, the values of the earthquake load E used in Eqs. (20.4) and (20.5) are, respectively,

$$E = E_h + E_v = \rho Q_E + 0.2S_{DS}D \qquad (20.6a)$$

$$E = E_h - E_v = \rho Q_E - 0.2S_{DS}D \qquad (20.6b)$$

where E_h = horizontal seismic load effect
E_v = vertical seismic load effect
Q_E = earthquake-induced horizontal seismic forces
ρ = redundancy factor

The factor ρ is taken as 1.0 for structures assigned to SDC B and C and as 1.3 for structures assigned to SDC D, E, and F, except for structures meeting specific criteria described in Ref. 20.2, in which case ρ may be taken as 1.0.

Combining Eq. (20.4) with Eq. (20.6a) and Eq. (20.5) with Eq. (20.6b) gives

$$U = (1.2 + 0.2S_{DS})D + \rho Q_E + 1.0L + 0.2S \qquad (20.7)$$

$$U = (0.9 - 0.2S_{DS})D + \rho Q_E \qquad (20.8)$$

Equations (20.4) and (20.6a) are used when dead load adds to the effects of horizontal ground motion, while Eqs. (20.5) and (20.6b) are used when dead load counteracts the effects of horizontal ground motion. Thus, the total load factor for dead load is greater than 1.2 in Eq. (20.7) and less than 0.9 in Eq. (20.8).

ASCE/SEI 7 specifies four procedures (if SDC A is included) for determining the earthquake load E_h. These procedures include three progressively more detailed methods that represent earthquake loading through the use of *equivalent static lateral loads, modal response spectrum analysis, and nonlinear seismic response-history analysis*. The method selected depends on the seismic design category. These procedures are discussed next.

a. Equivalent Lateral Force Procedure

According to ASCE/SEI 7 (Ref. 20.2), equivalent lateral force analysis may be applied to all structures with S_{DS} less than $0.33g$ and S_{D1} less than $0.133g$ (SDC B and C), as well as structures subjected to much higher design spectral response accelerations (SDC D, E, and F), if the structures meet certain requirements. More sophisticated dynamic analysis procedures must be used otherwise.

The equivalent lateral force procedure provides for the calculation of the total lateral force, defined as the design base shear V, which is then distributed over the height of the building. The design base shear V is calculated for a given direction of loading according to the equation

$$V = C_s W \qquad (20.9)$$

where W is the total dead load plus applicable portions of other loads and

$$C_s = \frac{S_{DS}}{R/I_e} \qquad (20.10)$$

which need not be greater than

$$C_s = \frac{S_{D1}}{T(R/I_e)} \qquad \text{for } T \leq T_L \qquad (20.11)$$

or

$$C_s = \frac{S_{D1}T_L}{T^2(R/I_e)} \qquad \text{for } T > T_L \qquad (20.12)$$

but may not be less than

$$C_s = 0.44IS_{DS} \geq 0.01 \qquad (20.13)$$

or where $S_1 \geq 0.6g$,

$$C_s = \frac{0.5S_1}{R/I_e} \tag{20.14}$$

where R = response modification factor (depends on structural system); values of R for most reinforced concrete structures range from 4 to 8, based on ability of structural system to sustain earthquake loading and to dissipate energy

I_e = occupancy important factor = 1.00, 1.25, or 1.50, depending upon the occupancy and use of structure

T = fundamental period of structure

According to ASCE/SEI 7, the period T can be calculated based on an analysis that accounts for the structural properties and deformational characteristics of the elements within the structure. Approximate methods may also be used in which the fundamental period of the structure may be approximated as

$$T_a = C_t h_n^x \tag{20.15}$$

where h_n = height above the base to the highest level of structure, ft

C_t = 0.016 for reinforced concrete moment-resisting frames in which frames resist 100 percent of required seismic force and are not enclosed or adjoined by more rigid components that will prevent frame from deflecting when subjected to seismic forces, and 0.020 for all other reinforced concrete buildings

x = 0.90 for C_t = 0.016 and 0.75 for C_t = 0.020

Alternately, for structures not exceeding 12 stories in height, in which the lateral-force-resisting system consists of a moment-resisting frame and the story height is at least 10 ft,

$$T_a = 0.1N \tag{20.16}$$

where N = number of stories.

For shear wall structures, ASCE/SEI 7 permits T to be approximated as

$$T_a = \frac{0.0019}{\sqrt{C_w}} h_n \tag{20.17}$$

where
$$C_w = \frac{100}{A_B} \sum_{i=1}^{x} \left(\frac{h_n}{h_i}\right)^2 \frac{A_i}{1 + 0.83\,(h_i/D_i)^2} \tag{20.18}$$

where A_B = base area of structure, ft^2

A_i = area of shear wall, ft^2

D_i = length of shear wall i, ft

x = number of shear walls in building that are effective in resisting lateral forces in direction under consideration

The lateral seismic force F_x at each level over the height of a structure is determined using Eqs. (20.19a) and (20.19b).

$$F_x = C_{vx}V \tag{20.19a}$$

$$C_{vx} = \frac{w_x h_x^k}{\sum_{i=1}^{n} w_i h_i^k} \tag{20.19b}$$

where C_{vx} = vertical distribution factor

V = total design lateral force or shear at the base of the structure from Eq. (20.9)

FIGURE 20.11

Forces based on ASCE/SEI 7 (Ref. 20.2) equivalent lateral force procedure: (*a*) structure; (*b*) distribution of lateral forces over height; and (*c*) story shears.

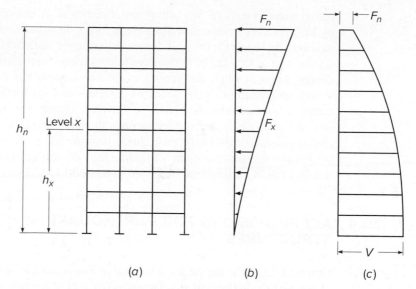

(*a*)   (*b*)   (*c*)

F_x = lateral seismic force induced at level x

w_x, w_i = portion of W at level x and level i, respectively

h_x, h_i = height to level x and level i, respectively

k = exponent related to structural period = 1 for $T \leq 0.5$ sec and = 2 for $T \geq 2.5$ sec. For $0.5 < T < 2.5$, k is determined by linear interpolation or set to a value of 2

The design shear at any story V_x equals the sum of the forces F_x at and above that story. For a 10-story building with a uniform mass distribution over the height and $T = 1.0$ sec, the lateral forces and story shears are distributed as shown in Fig. 20.11.

At each level, V_x is distributed in proportion to the stiffness of the elements in the vertical lateral-force-resisting system. To account for unintentional building irregularities that may cause a horizontal torsional moment, a minimum 5 percent eccentricity must be applied if the vertical lateral-force-resisting systems are connected by a floor system that is rigid in its own plane.

In addition to the criteria just described, ASCE/SEI 7 includes criteria to account for overturning effects and provides limits on story drift. $P-\Delta$ effects must be considered (as discussed in Chapter 10), and the effects of upward loads must be accounted for in the design of horizontal cantilever components and prestressed members.

b. Dynamic Lateral Force Procedures

ASCE/SEI 7 includes dynamic lateral force procedures that involve the use of (1) modal response spectra, which provide the earthquake-induced forces as a function of the natural periods of the structure, or (2) a nonlinear time-history analysis of the structural response based on a series of ground motion acceleration histories that are representative of ground motion expected at the site. Both procedures require the development of a mathematical model of the structure to represent the spatial distribution of mass and stiffness. Response spectra, such as shown in Fig. 20.10, are used to calculate peak forces for a "sufficient number of modes to obtain a combined modal mass participation of 100 percent of the structure's mass" (Ref. 20.2). All modes with periods less than 0.05 sec are permitted to be represented by a single rigid-body mode that has a period of 0.05 sec. Alternatively, an analysis may be used that includes a "sufficient number of modes to obtain the combined modal mass participation of at least 90 percent of the actual mass in each of two orthogonal directions." A site-specific analysis is required by Ref. 20.2 in certain cases. Since these forces do not always act in the same direction, as shown in Fig. 20.3, the peak forces are averaged statistically, in most cases using the square root of the sum of the squares to obtain equivalent static lateral forces for use in design. In cases where the periods in the translational and torsional modes are closely spaced and result in significant cross correlation of the modes, the *complete quadratic combination* method is used (Ref. 20.9). When a nonlinear time-history analysis is used, design forces are obtained directly from the analyses. Both modal response spectrum and nonlinear time-history procedures provide more realistic representations of the seismically induced forces in a structure than do equivalent lateral force analyses. The details of these methods are presented in Refs. 20.1, 20.2, and 20.8.

20.4 ACI PROVISIONS FOR EARTHQUAKE-RESISTANT STRUCTURES

Criteria for seismic design are contained in Earthquake-Resistant Structures, Chapter 18 of the ACI Code (Ref. 20.10). The principal goal of the provisions is to ensure adequate toughness under inelastic displacement reversals brought on by earthquake loading. The provisions accomplish this goal by requiring the designer to provide for concrete confinement and inelastic rotation capacity. The provisions apply to frames, walls, coupling beams, diaphragms, trusses, and foundations in structures assigned to Seismic Design Categories D, E, and F and to frames, including two-way slab systems, and precast walls in structures assigned to Seismic Design Category C. Structural systems in SDC D, E, and F are referred to as *special*, while systems in SDC C are referred to as *intermediate*.

The requirements in ACI Code 18.3.1 for frame structures assigned to SDC B, described as *ordinary moment frames*, are limited. Beams must have at least two longitudinal bars that are continuous along both the top and bottom faces of the beam, the continuous bottom bars must have an area equal to at least one-fourth the maximum area of the bottom bars along the span, and the bars must be developed at the face of the supports; and columns with a clear height ℓ_u less than or equal to 5 times the column dimension c_1 in the direction of bending must be designed for shear, as required for intermediate frames (described in Section 20.8) in accordance with ACI Code 18.3.3. Beam-column joints must meet the requirements in ACI Code Chapter 15 for joint shear V_u at midheight of the joint based on tensile and compressive beam forces and column shear consistent with beam nominal moment strengths M_n, as will be demonstrated in Example 20.2. There are no special requirements in ACI Code Chapter 18 for structures assigned to SDC A.

The ACI provisions are based on many of the observations made earlier in this chapter. The effect of nonstructural elements on overall structural response must be considered, as must the response of the nonstructural elements themselves. Structural elements that are not specifically proportioned to carry earthquake loads must also be considered.

The load factors used for earthquake loads are given in Eqs. (20.4) and (20.5). The strength reduction factors used for seismic design are the same as those used for nonseismic design (Table 1.3), with the additional requirements that $\phi = 0.60$ for shear, if the nominal shear capacity of a member is less than the shear based on the nominal flexural strength [see Eq. (20.1)], and $\phi = 0.85$ for shear in joints and diagonally reinforced coupling beams.

To ensure adequate ductility and toughness under inelastic rotation, ACI Code 19.2.1.1 sets a minimum concrete strength of 3000 psi for special moment frames,

FIGURE 20.12
Example of transverse
reinforcement in columns;
consecutive crossties
engaging the same
longitudinal bars must have
90° hooks on opposite sides
of columns. (*Adapted from
Ref. 20.10.*)

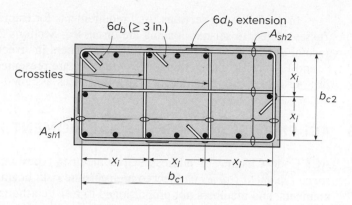

$x_i \leq 14$ in.; $x_i \leq 8$ in. for columns
with $P_u > 0.3 f_c' A_g$ or $f_c' > 10,000$ psi

as well as for special structural walls with Grade 60 or 80 reinforcement. The minimum concrete strength increases to 5000 psi for special structural walls with Grade 100 reinforcement. For lightweight aggregate concrete, an *upper limit* of 5000 psi is placed on concrete strength; this limit is based on a lack of experimental evidence for higher-strength lightweight concretes.

Under ACI Code 20.2.2.5, reinforcing steel must meet the requirements of Grades 60, 80, and 100 ASTM A706 bars (see Table 2.4) (although Grade 100 was not available at the time of publication of ACI 318-19) for special structural walls and Grades 60 and 80 ASTM A706 bars for special structural frames. ASTM A706 specifies *maximum* yield strengths 18 ksi above the specified minimum yield strength and minimum tensile strengths 20 ksi above the specified minimum yield strengths. The actual tensile strength must be at least 1.25 times the actual yield strength. In addition to reinforcement manufactured in accordance with ASTM A706, the Code allows the use of Grade 60 reinforcement meeting the requirements of ASTM A615, provided that the actual yield strength does not exceed the specified yield by more than 18 ksi, the actual tensile strength exceeds the actual yield strength by at least 25 percent, and the bars satisfy the specific elongation requirements specified in ACI Code 20.2.2.5. The upper limits on yield strength are used to limit the maximum moment capacity of the section because of the dependency of the earthquake-induced shear on the moment capacity [Eq. (20.1)]. The minimum ratio of tensile strength to yield strength helps provide adequate inelastic rotation capacity. Evidence reported in Ref. 20.11 indicates that an increase in the ratio of the ultimate moment to the yield moment results in an increase in the nonlinear deformation capacity of flexural members.

Confinement for concrete is provided by transverse reinforcement consisting of stirrups, hoops, and crossties. To ensure adequate anchorage, a *seismic hook* [with a bend not less than 135° and a 6 bar diameter (but not less than 3 in.) extension that engages the longitudinal reinforcement and projects into the interior of the stirrup or hoop][†] is used on stirrups, hoops, and crossties. *Hoops*, shown in Figs. 8.12a, c–e and 20.12, are closed ties that can be made up of several reinforcing elements, each having seismic hooks at both ends, or continuously wound ties with seismic hooks at both ends. A *crosstie* (see Fig. 20.12) is a continuous reinforcing bar with a seismic hook at one end and a hook with not less than a 90° bend and at least a 6 bar diameter extension at the other end. The hooks on crossties must engage peripheral longitudinal reinforcing bars.

[†] The term *seismic hook* is used in the ACI Code and this text. Seismic hooks also include 90° hooks on circular hoops. There are, however, no differences between the requirements for standard hooks for stirrups, ties, and hoops with 90° and 135° bends and those for seismic hooks.

In the following sections, ACI requirements for frames, walls, diaphragms, and trusses subject to seismic loading are discussed. Sections 20.5 and 20.6 describe the general design and detailing criteria for members in structures assigned to SDC D, E, and F. Specific shear strength requirements are presented in Section 20.7. Section 20.8 describes requirements for frame structures assigned to SDC C.

20.5   ACI PROVISIONS FOR SPECIAL MOMENT FRAMES

ACI Code Chapter 18 addresses five member types in frame structures, termed *special moment frames*, subject to high seismic risk: beams, columns, joints, precast members, and members not proportioned to resist earthquake forces. Two-way slabs without beams are prohibited as lateral-force-resisting systems in structures assigned to SDC D, E, and F.

a.   Beams

The requirements for beams in special moment frames are covered in ACI Code 18.6. The members must have a clear span ℓ_n equal to at least $4d$ and a width b_w not less than $0.3h$ or 10 in. The projection of the width b_w beyond the width of the supporting column c_2 on each side may not exceed the smaller of the width of the supporting member c_2 or 0.75 times the dimension of the supporting member in the direction of the span c_1, as shown in Fig. 20.13. The minimum clear span-to-depth ratio helps ensure that flexural rather than shear strength dominates member behavior under inelastic load reversals. Minimum web dimensions help provide adequate confinement for the concrete, whereas the width relative to the support (typically a column) is limited to provide adequate moment transfer between beams and columns.

In accordance with ACI Code 18.6.3, both top and bottom minimum flexural steel is required. $A_{s,\min}$ should not be less than that given by Eq. (4.37a) but need not be greater than four-thirds of that required by analysis, with a minimum of two reinforcing bars, top and bottom, throughout the member. In addition, the positive moment capacity at the face of columns must be at least one-half of the negative moment strength at the same location, and neither positive nor negative moment strength at any section in a member may be less than one-fourth of the maximum moment strength at either end of the member. These criteria are designed to provide for ductile behavior throughout the member, although the minimum of two reinforcing bars on the top and bottom is based principally on construction requirements. Maximum reinforcement ratios of 0.025 for Grade 60 reinforcement and 0.020 for Grade 80 reinforcement are set to limit problems with steel congestion and to ensure adequate member size for carrying shear that is governed by the flexural capacity of the member [Eq. (20.1)].

To obtain ductile performance, the location of lap splices is limited. They may not be used within joints, within twice the member depth from the face of a joint, or within a distance $2h$ of critical sections where flexural yielding is likely to occur as a result of lateral displacements of the frame beyond the elastic range. Lap splices must be enclosed by hoops or spirals with a maximum spacing of one-fourth of the effective depth or 4 in. Welded and mechanical connections may be used, provided that they are not used within a distance equal to twice the member depth from the face of a column or beam or sections where yielding of the reinforcement is likely to occur due to inelastic displacements under lateral load, in accordance with ACI Code 18.2.7 and 18.2.8.

FIGURE 20.13
Maximum effective width of wide beam and required transverse reinforcement. (*Adapted from Ref. 20.10.*)

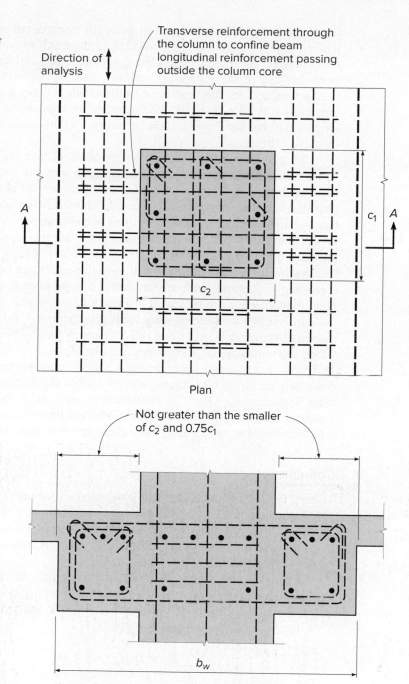

Direction of analysis

Transverse reinforcement through the column to confine beam longitudinal reinforcement passing outside the column core

c_1

c_2

Plan

Not greater than the smaller of c_2 and $0.75c_1$

b_w

Note:
Transverse reinforcement in column above and below the joint not shown for clarity

Section *A–A*

Transverse reinforcement is required throughout beams in frames resisting earthquake-induced forces. According to ACI Code 18.6.4, transverse reinforcement in the form of hoops must be used over a length equal to twice the beam depth measured from the face of the supporting column toward midspan, at both ends of the beam, and over lengths equal to twice the beam depth on both sides of a section where flexural yielding is likely to occur as the result of lateral displacements of the frame beyond the elastic range. The first hoop must be located not more than 2 in. from the face of the supporting member, and the maximum spacing of the hoops must not exceed one-fourth of the effective depth, 6 times the diameter of the smallest longitudinal bar for Grade 60 reinforcement or 5 times the diameter of the smallest longitudinal bar for Grade 80 reinforcement, in both cases excluding skin reinforcement, or 6 in.

To provide adequate support for longitudinal bars on the perimeter of a beam when the bars are placed in compression due to inelastic rotation, ACI Code 18.6.4.2 requires that hoops be arranged so that every corner and alternate longitudinal bar is provided lateral support by ties, in accordance with ACI Code 25.7.2.3. Arrangements meeting these criteria (although without seismic hooks) are illustrated in Fig. 9.3. Where hoops are not required, stirrups with seismic hooks at both ends must be provided throughout the member, with a maximum spacing of one-half of the effective depth. Hoops can be made up of a single reinforcing bar or two reinforcing bars consisting of a stirrup with seismic hooks at both ends and a crosstie. Examples of hoop reinforcement are presented in Figs. 8.12a, c–e and 20.12.

In beams with factored axial loads greater than $A_g f_c'/10$, the hoops must be as required for columns by ACI Code 18.7.5.2 through 18.7.5.4 (see following Section 20.5b). Along the remaining length, stirrups must be replaced by hoops satisfying ACI Code 18.7.5.2 with a spacing s not exceeding 6 times the diameter of the smallest longitudinal beam bars for Grade 60 reinforcement or 5 times the diameter of the smallest longitudinal beam bars for Grade 80 reinforcement, or 6 in. If the concrete cover over the transverse reinforcement exceeds 4 in., additional transverse reinforcement with a cover not exceeding 4 in. and spacing not exceeding 12 in. must be added.

b.   Columns

To help ensure constructability and adequate confinement of the concrete, ACI Code 18.7.2 requires that columns in special moment frames have (1) a minimum cross-sectional dimension of at least 12 in. when measured on a straight line passing through the geometric centroid and (2) a ratio of the shortest cross-sectional dimension to the perpendicular dimension of at least 0.4.

To obtain a weak beam–strong column design, ACI Code 18.7.3 requires that the nominal flexural strengths of the columns framing into a joint exceed the nominal flexural strengths of the girders framing into the joint by at least 20 percent. This requirement is expressed as

$$\Sigma M_{nc} \geq \frac{6}{5} \Sigma M_{nb} \qquad (20.20)$$

where ΣM_{nc} = sum of nominal flexural strengths of columns framing into joint, evaluated at the faces of the joint; values of M_{nc} are based on the factored axial load, consistent with the direction of the lateral forces, resulting in the lowest flexural strength

ΣM_{nb} = sum of nominal flexural strengths of beams framing into joint, evaluated at the faces of the joint; in T beam construction, where the slab is in tension under moment at the face of the joint, slab reinforcement within the effective flange width (see Section 4.7) is assumed to contribute to flexural strength if the slab reinforcement is developed at the critical section for flexure

As shown in Fig. 20.8b, the flexural strengths are summed so that the column moments oppose the beam moments. Equation (20.20) must be satisfied for beam moments acting both clockwise and counterclockwise on the joint. According to ACI Code 18.7.3.1, it need not be satisfied where the column is discontinuous above the joint and the column factored axial compressive force under load combinations, including earthquake effect E, is less than $A_g f_c'/10$.

If Eq. (20.20) is not satisfied for beam moments acting in both directions, the lateral strength and stiffness of the columns framing into the joint must be ignored when determining the strength and stiffness of the structures, and the columns must be designed under the provisions of ACI Code 18.14 for members that are not designated as part of the seismic-force-resisting system, as described in Section 20.5d. If the stiffness of the columns increases the design base shear or the effects of torsion, they must be included in the analysis, but still may not be considered as contributing to structural capacity.

In accordance with ACI Code 18.7.4, the column reinforcement ratio based on the gross section ρ_g must meet the requirement: $0.01 \leq \rho_g \leq 0.06$. Columns with circular hoops must have at least six longitudinal bars. To limit the potential for a bond failure of the longitudinal column bars (Refs. 20.12 and 20.13), the bars must be selected such that 1.25 times the development length is less than one-half of the column clear height. Welded splices and mechanical connections in columns must satisfy the same requirements specified for flexural members. Lap splices must be designed for tension, are permitted only in the center half of columns, and must be enclosed by transverse reinforcement satisfying ACI Code 18.7.5.2 and 18.7.5.3.

ACI Code 18.7.5 specifies the use of minimum transverse reinforcement over length ℓ_o from each joint face and on both sides of any section where flexural yielding is likely because of inelastic lateral displacement of the frame. This section is referenced by other sections of the code. The length ℓ_o may not be less than (1) the depth of the column at the joint face or at the section where flexural yielding is likely to occur, (2) one-sixth of the clear span of the member, or (3) 18 in.

Following ACI Code 18.7.5.2, the transverse reinforcement may consist of single or overlapping spirals satisfying Eq. (9.5) and the provisions of ACI Code 25.7.3 (see Section 9.2), circular hoops, or rectilinear hoops (Fig. 20.12) with or without crossties. The crossties may be the same size or smaller than the bars used for the hoops but must be at least No. 3 (No. 10) bars for No. 10 (No. 32) and smaller longitudinal reinforcement and No. 4 (No. 13) bars for No. 11 (No. 36) and larger longitudinal reinforcement or bundled longitudinal reinforcement. The bars so supported may not be spaced more than 14 in., as shown in Fig. 20.12. Where $P_u > 0.3 f_c' A_g$ or $f_c' > 10,000$ psi in columns with rectilinear hoops, every longitudinal bar or bundle of bars around the perimeter of the column core must have lateral support provided by the corner of a hoop or by a seismic hook, and the value of h_x may not exceed 8 in. P_u is taken as the highest axial compression consistent with factored load combinations including E.

In accordance with ACI Code 18.7.5.3, the spacing of transverse reinforcement within ℓ_o may not exceed one-quarter of the minimum member dimension, 6 times the diameter of the longitudinal bar for Grade 60 reinforcement, or 5 times the diameter of the longitudinal bar for Grade 80 reinforcement, or

$$s_o = 4 + \frac{14 - h_x}{3} \tag{20.21a}$$

$$4 \text{ in.} \leq s_o \leq 6 \text{ in.} \tag{20.21b}$$

where h_x is the maximum horizontal spacing of hoop or crosstie legs on all faces of the column (largest value of x_i in Fig. 20.12).

In accordance with ACI Code 18.7.5.4, minimum transverse reinforcement is specified in terms of the ratio of the volume of the transverse reinforcement to the volume of the core confined by the reinforcement (measured out-to-out of the confining steel) ρ_s for spirals or circular hoop reinforcement as

$$\rho_s = 0.12 \frac{f_c'}{f_{yt}} \tag{20.22}$$

but not less than specified in Eq. (9.5), where f_{yt} is the specified yield strength of transverse reinforcement. In addition, for $P_u > 0.3 f_c' A_g$ or $f_c' > 10,000$ psi, the minimum transverse reinforcement must be at least

$$\rho_s = 0.35 k_f \frac{P_u}{f_{yt} A_{ch}} \tag{20.23}$$

where the concrete strength factor $k_f = (f_c'/25,000) + 0.6 \geq 1.0$ and A_{ch} = cross-sectional area of column core, measured out-to-out of transverse reinforcement.

To provide similar confinement using rectangular hoop reinforcement, ACI Code 18.7.5.4 requires a minimum total cross-sectional area of transverse reinforcement A_{sh} along the length of the longitudinal reinforcement that may not be less than given in Eqs. (20.24) and (20.25).

$$\frac{A_{sh}}{s b_c} = 0.3 \left(\frac{A_g}{A_{ch}} - 1 \right) \frac{f_c'}{f_{yt}} \tag{20.24}$$

$$\frac{A_{sh}}{s b_c} = 0.09 \frac{f_c'}{f_{yt}} \tag{20.25}$$

where A_{ch} is defined following Eq. (20.23), s = spacing of transverse reinforcement, and b_c = cross-sectional dimension of column core, measured to outside edges of transverse reinforcement composing A_{sh}. For $P_u > 0.3 f_c' A_g$ or $f_c' > 10,000$ psi, the minimum transverse reinforcement may not be less than

$$\frac{A_{sh}}{s b_c} = 0.2 k_f \, k_n \frac{P_u}{f_{yt} A_{ch}} \tag{20.26}$$

where the concrete strength factor k_f is defined after Eq. (20.23) and confinement effectiveness factor $k_n = n_l/(n_l - 2)$, with n_l = number of longitudinal bars or bar bundles around the perimeter of a column core that are laterally supported by the corner of hoops or by seismic hooks.

Equations (20.24) through (20.26) must be evaluated for A_{sh} in both the 1 and 2 directions, as indicated in Fig. 20.12.

For regions outside of ℓ_o, when the minimum transverse reinforcement defined above is not provided, the spacing of spiral or hoop reinforcement may not exceed 6 or 5 times the diameter of the longitudinal column bars, respectively, for Grades 60 and 80 reinforcement, or 6 in.

To account for the major ductility demands that are placed on columns that support rigid members (see Figs. 20.4 and 20.5), ACI Code 18.7.5.6 specifies that, for such columns, the minimum transverse reinforcement requirements must be satisfied throughout the *full column height* if the factored axial compressive force related to earthquake effect exceeds $f_c' A_g/10$, and that the transverse reinforcement must extend into the discontinued stiff member for at least the development length of the largest longitudinal reinforcement for walls and at least 12 in. into foundations. ACI Code 18.7.5.6 requires an increase in the limit of $f_c' A_g/10$ to $f_c' A_g/4$ when design forces have been magnified to account for the overstrength of the vertical elements of the seismic-force-resisting system.

If the concrete cover outside the confining transverse reinforcement is greater than 4 in., the Code requires the addition of transverse reinforcement with a cover of 4 in. or less to limit the potential hazard caused by spalling of the concrete shell away from the column.

EXAMPLE 20.1   **Relative flexural strengths of members at a joint and minimum transverse column reinforcement.**   The exterior joint shown in Fig. 20.14 is part of a reinforced concrete frame designed to resist earthquake loads. A 6 in. slab, not shown, is reinforced with No. 5 (No. 16) bars spaced 10 in. center to center at the same level as the flexural steel in the beams. The member section dimensions and reinforcement are as shown. The frame story height is 12 ft. Concrete is normalweight. Material strengths are $f'_c = 4000$ psi and $f_y = 60,000$ psi. The maximum factored axial load on the upper column framing into the joint is 2210 kips, and the maximum factored axial load on the lower column is 2306 kips. Determine if the nominal flexural strengths of the columns exceed those of the beams by at least 20 percent, as required by Eq. (20.20), and determine the minimum transverse reinforcement required over the length ℓ_o in the columns.

FIGURE 20.14

Exterior beam-column joint for Examples 20.1 and 20.2: (*a*) plan view; (*b*) cross section through spandrel beam; and (*c*) cross section through normal beam. Note that confining reinforcement is not shown, except for column hoops and crossties in (*a*).

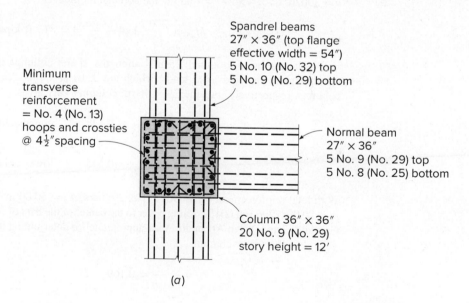

Spandrel beams
27″ × 36″ (top flange
effective width = 54″)
5 No. 10 (No. 32) top
5 No. 9 (No. 29) bottom

Minimum transverse reinforcement = No. 4 (No. 13) hoops and crossties @ 4½″ spacing

Normal beam
27″ × 36″
5 No. 9 (No. 29) top
5 No. 8 (No. 25) bottom

Column 36″ × 36″
20 No. 9 (No. 29)
story height = 12′

(*a*)

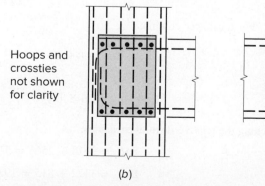

Hoops and crossties not shown for clarity

(*b*)   (*c*)

SOLUTION.   Checking the relative flexural strengths in the frame of the spandrel beams will be sufficient, since this is clearly the controlling case for the joint. In addition, because the beam reinforcement is the same on both sides of the joint, a single comparison will suffice for both clockwise and counterclockwise beam moments.

The negative nominal flexural strength of the beam at the joint is governed by the top steel, which consists of five No. 10 (No. 32) bars in the beams plus four No. 5 (No. 16) bars in the slab within the effective width of the top flange, $A_s = 6.35 + 1.24 = 7.59$ in^2. The yield force in the steel is

$$A_s f_y = 7.59 \times 60 = 455 \text{ kips}$$

The effective depth is $d = 36.0 - 1.5 - 0.5 - 1.27/2 = 33.4$ in., and with stress block depth $a = 455/(0.85 \times 4 \times 27) = 4.96$ in., the nominal moment is

$$M_{nb} = \frac{455}{12}\left(33.4 - \frac{4.96}{2}\right) = 1172 \text{ ft-kips}$$

The positive nominal flexural strength of the beam at the joint is determined by the bottom steel, five No. 9 (No. 29) bars, $A_s = 5.00$ in^2. The yield force in the steel is

$$A_s f_y = 5.00 \times 60 = 300 \text{ kips}$$

The effective depth is $d = 36.0 - 1.5 - 0.5 - 1.128/2 = 33.4$ in., and with stress block depth $a = 300/(0.85 \times 4 \times 54) = 1.63$ in., the nominal moment is

$$M_{nb} = \frac{300}{12}\left(33.4 - \frac{1.63}{2}\right) = 815 \text{ ft-kips}$$

The minimum nominal flexural strengths of the columns in this example depend on the maximum factored axial loads, which are 2210 and 2306 kips for the upper and lower columns, respectively. For the 36 × 36 in. columns, this gives

$$\frac{P_u}{f_c' A_g} = \frac{2210}{4 \times 1296} = 0.426 \qquad \text{upper column}$$

$$\frac{P_u}{f_c' A_g} = \frac{2306}{4 \times 1296} = 0.445 \qquad \text{lower column}$$

With total reinforcement of 20 No. 9 (No. 29) bars, $A_{st} = 20.00$ in^2 and the reinforcement ratio $\rho_g = 20.00/1296 = 0.00154$. Using cover to the center of the bars of 3 in., $\gamma = (36 - 6)/36 = 0.83$, Graphs A.7 and A.8 in Appendix A are appropriate for determining the flexural capacity.

For the upper column,

$$R_n = \frac{M_{nc}}{f_c' A_g h} = 0.169$$

$$M_{nc} = 0.169 \times 4 \times 1296 \times \frac{36}{12} = 2628 \text{ ft-kips}$$

For the lower column,

$$R_n = \frac{M_{nc}}{f_c' A_g h} = 0.165$$

$$M_{nc} = 0.165 \times 4 \times 1296 \times \frac{36}{12} = 2566 \text{ ft-kips}$$

Checking the relative flexural capacities,

$$\Sigma M_{nc} = 2628 + 2566 = 5194 \text{ ft-kips}$$

$$\Sigma M_{nb} = 1172 + 815 = 1987 \text{ ft-kips}$$

By inspection, $\Sigma M_{nc} \geq \frac{6}{5} \Sigma M_{nb}$.

Minimum transverse reinforcement is required over a length ℓ_o on either side of the joint. According to ACI Code 18.7.5, ℓ_o is the greater of (1) the depth $h = 36$ in., (2) one-sixth of the clear span $= (12 \times 12 - 36)/6 = 18$ in., or (3) 18 in. Every corner and alternate longitudinal bar must have lateral support, and because the spacing of crossties and legs of hoops is limited to a maximum of 8 in. within the plane of the transverse reinforcement because $P_u > 0.3 f_c' A_g$, the scheme shown in Fig. 20.14a will be used, giving a maximum spacing just over 7.5 in. The maximum spacing of transverse reinforcement s is limited to the smallest of one-quarter of the minimum member dimension $= 36/4 = 9$ in., 6 times the diameter of the longitudinal bar $6 \times 1.128 = 6.77$ in., and using $h_x = 8$ in. in Eq. (20.21a),

$$s_o = 4 + \frac{14 - h_x}{3} = 4 + \frac{14 - 8}{3} = 6 \text{ in.}$$

with 4 in. $\leq s_o \leq 6$ in. Although, s_o may be 6 in., a value less than 6 in. may be selected to limit the size of the ties and crossties.

Because the column is square and has equal reinforcement on all sides, A_{sh} will be the same in both directions. Using No. 4 (No. 13) bars, the cross-sectional dimension of the column core, measured to the outside edges of the confining steel, is $b_c = 33$ in., and the cross-sectional area of column core, also measured to the outside edges of the confining steel, is $A_{ch} = 33 \times 33 = 1089$ in^2. For $f_{yt} = 60$ ksi, the total area of transverse reinforcement A_{sh} is based on the greatest value calculated using Eqs. (20.24) through (20.26), with Eq. (20.26) required because $P_u > 0.3 f_c' A_g$.

$$\frac{A_{sh}}{sb_c} = 0.3 \left(\frac{A_g}{A_{ch}} - 1 \right) \frac{f_c'}{f_{yt}} = 0.3 \left(\frac{1296}{1089} - 1 \right) \frac{4}{60} = 0.0038$$

$$\frac{A_{sh}}{sb_c} = 0.09 \frac{f_c'}{f_{yt}} = 0.09 \frac{4}{60} = 0.0060$$

For use in Eq. (20.26), $k_f = (f_c'/25,000) + 0.6 \geq 1.0$ equals 1.0 and $k_n = n_l/(n_l - 2) = 20/(20 - 2) = 1.11$, where $n_l =$ number of longitudinal bars or bar bundles around the perimeter of a column core that are laterally supported by the corner of hoops or by seismic hooks.

$$\frac{A_{sh}}{sb_c} = 0.2 k_f k_n \frac{P_u}{f_{yt} A_{ch}} = 0.2 \times 1.0 \times 1.11 \frac{2306}{60 \times 1089} = 0.0078$$

Eq. (20.26) controls. Using six No. 4 bars in each direction, as shown in Fig. 20.14a gives $A_{sh} = 1.20$ in. and spacing $s = A_{sh}/(0.0093 \times b_c) = 1.20/(0.0078 \times 33) = 4.66$ in. A tie spacing s of 4.5 in. will be used.

c. Joints and Development of Reinforcement

The design of beam-column joints is discussed in Section 18.2. The forces acting on a joint subjected to lateral loads are illustrated in Figs. 18.4 and 20.8d. The factored shear acting on a joint is

$$V_u = T_1 + C_2 - V_{col}$$
$$= T_1 + T_2 - V_{col} \tag{20.27}$$

where $T_1 =$ tensile force in negative moment beam steel on one side of a joint
$T_2 =$ tensile force in positive moment beam steel on one side of a joint
$C_2 =$ compressive force counteracting T_2
$V_{col} =$ shear in the column at top and bottom faces of the joint corresponding to the net moment in the joint and points of inflection at midheight of columns (see Fig. 18.5). $V_{col} = V_3 = V_4$ in Figs. 18.4 and 20.8. Calculation of V_{col} is demonstrated in Example 20.2.

TABLE 20.1
Nominal joint shear strength V_n (from Ref. 20.10)

Column	Beam in Direction of V_n	Confinement by Transverse Beams According to ACI Code 15.2.8	V_n
Continuous or meets ACI Code 15.2.6	Continuous or meets ACI Code 15.2.7	Confined	$20\lambda\sqrt{f_c'}A_j$
		Not confined	$15\lambda\sqrt{f_c'}A_j$
	Other	Confined	$15\lambda\sqrt{f_c'}A_j$
		Not confined	$12\lambda\sqrt{f_c'}A_j$
Other	Continuous or meets 15.2.7	Confined	$15\lambda\sqrt{f_c'}A_j$
		Not confined	$12\lambda\sqrt{f_c'}A_j$
	Other	Confined	$12\lambda\sqrt{f_c'}A_j$
		Not confined	$8\lambda\sqrt{f_c'}A_j$

For seismic design, the forces T_1 and $T_2 (= C_2)$ must be based on a stress in the flexural tension reinforcement of $1.25f_y$. In accordance with ACI Code 18.8.4, the nominal shear capacity of a joint depends on the degree of confinement provided by members framing into the joint, as shown in Table 20.1.

When using Table 20.1, $\lambda = 0.75$ for lightweight concrete and 1.0 for normalweight concrete and A_j is the effective cross-sectional area of the joint in a plane parallel to the plane of reinforcement generating shear in the joint, illustrated in Fig. 20.15. The joint depth is the overall depth of the column. For beams framing into a support of larger width, the effective width of the joint is the smaller of (1) beam width plus joint depth or (2) twice the smaller perpendicular distance from the longitudinal axis of the beam to the column side.

FIGURE 20.15
Effective area of joint A_j, which must be considered separately for forces in each direction of framing. Note that the joint illustrated does not meet conditions necessary to be considered as confined because the forming members do not cover at least $\frac{3}{4}$ of each joint face. (*Adapted from Ref. 20.10.*)

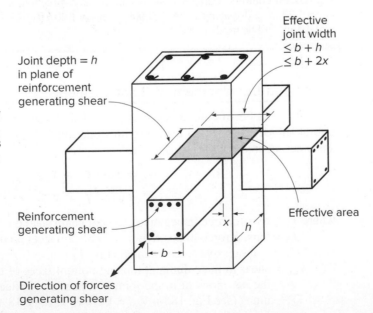

Table 20.1 cites three sections of the ACI Code: ACI Code 15.2.6 requires that (1) the column extend above joint at least one column depth h measured in the direction of joint shear, and (2) longitudinal and transverse reinforcement from the column below joint be continued through the extension. ACI Code 15.2.7 requires that (1) the beam extend at least one beam depth h beyond joint face, and (2) longitudinal and transverse reinforcement from beam on opposite side of the joint be continued through the extension. ACI Code 15.2.8 requires that two transverse beams satisfying the following criteria be provided: (1) The width of each transverse beam must be at least three-quarters of the width of the column face into which the beam frames; (2) the transverse beams must extend at least one beam depth h beyond the joint faces; and (3) the transverse beams must contain at least two continuous top and bottom bars providing the minimum flexural reinforcement given in Eq. (4.37a) and No. 3 (No. 10) or larger stirrups providing the minimum shear reinforcement given in Eq. (5.13a) with maximum spacing not exceeding that shown in Fig. 5.15 for $V_c \leq 4\sqrt{f_c'}\,b_w d$ and half the values shown in Fig. 5.15 for $V_c > 4\sqrt{f_c'}\,b_w d$.

To provide adequate confinement within a joint, ACI Code 18.8.3 requires that the transverse reinforcement used in columns must be continued through the joint, in accordance with ACI Code 18.7.5, as described in Section 20.5b. This reinforcement may be reduced by one-half within the depth h of the shallowest framing member, and the spacing of spirals or hoops may be increased to 6 in., if beams or girders frame into all four sides of the joint and the flexural members cover at least three-fourths of the column width.

For joints where the beam is wider than the column, transverse reinforcement, as required for columns (ACI Code 18.6.4), must be provided to confine the flexural steel in the beam, as shown in Fig. 20.13, unless confinement is provided by a transverse flexural member. When beam negative moment reinforcement is provided by headed deformed bars that terminate in the joint, the column must extend above the top of the joint a distance at least equal to the depth of the joint h in the direction of the span or the headed bars must be enclosed by additional vertical joint reinforcement providing equivalent confinement to the top face of the joint.

To provide adequate development of beam reinforcement passing through a joint, ACI Code 18.8.2 requires that the column dimension parallel to the beam reinforcement be at least $(20/\lambda)$ times the diameter of the largest longitudinal bar for Grade 60 reinforcement and 26 times the bar diameter for Grade 80 reinforcement. In addition, the depth h of the joint may not be less than one-half of depth h of any beam framing into the joint that causes joint shear as part of the seismic-force-resisting system. Beam longitudinal reinforcement that is terminated within a column must be extended to the far face of the column core. The reinforcement must be anchored in compression as described in Section 6.8 (ACI Code 25.4.9). The reinforcement may be anchored in tension by a hook, a straight bar, or a head. Bars anchored by hooks must, in addition to satisfying the requirements described in Section 6.4 (ACI Code 25.4.3), also satisfy ACI Code 18.8.5, which requires that the development length of bars with standard hooks ℓ_{dh} not be less than $8d_b$, 6 in., or

$$\ell_{dh} = \frac{f_y d_b}{65\lambda\sqrt{f_c'}} \tag{20.28}$$

where $\lambda = 0.75$ for lightweight concrete and 1.0 for normalweight concrete. The development length ℓ_{dh} for lightweight concrete may not be less than $10d_b$ or 7.5 in. The hook must be located within the confined core of the column or the boundary element of a wall or diaphragm (discussed in Section 20.6a) and bent into the joint.

For straight bars anchored within a column core, the development length ℓ_d of bottom bars must be at least 2.5 times the value required for hooks; ℓ_d for top bars must be at least 3.25 times the length required for hooks.

According to ACI Code 18.8.5, straight bars that are terminated at a joint must pass through the confined core of a column or a boundary element. Because of the lower degree of confinement provided outside of the confined region, the Code requires that any portion of the straight embedment length that is not within the core be increased by a factor of 1.6. Thus, the required development length ℓ_{dm} of a bar that is not entirely embedded in confined concrete is

$$\ell_{dm} = 1.6 \, (\ell_d - \ell_{dc}) + \ell_{dc} \qquad (20.29a)$$

$$\ell_{dm} = 1.6\ell_d - 0.6\ell_{dc} \qquad (20.29b)$$

where ℓ_d = required development length for a straight bar embedded in confined concrete

ℓ_{dc} = length embedded in confined concrete

Bars anchored by heads must satisfy ACI Code 25.4.4, as described in Section 6.5.

EXAMPLE 20.2   **Design of exterior joint.**   Design the joint shown in Fig. 20.14.

SOLUTION.   As discussed in Chapter 18, a joint must be detailed so that the beam and column bars do not interfere with each other and so that placement and consolidation of the concrete are practical. Bar placement is shown in Fig. 20.14.

Development of the spandrel beam flexural steel within the joint is checked based on the requirement that the column dimension be at least $(20/\lambda) = 20$ times the bar diameter of the largest bars. This requirement is met for the No. 10 (No. 32) bars used as top reinforcement.

$$20 \times 1.27 = 25.4 \text{ in.} < 36 \text{ in.}$$

The flexural steel in the normal beam must be anchored within the core of the column based on Eqs. (6.8) and (20.28), but not less than $8d_b$ or 6 in. For the No. 9 top bars, the transverse reinforcement in the column (Fig. 20.14a) continued through the joint gives $\psi_r = 1.0$. Solving Eq. (6.8) gives

$$\ell_{dh} = \left(\frac{f_y\psi_e\psi_r\psi_o\psi_c}{55\lambda\sqrt{f_c'}}\right)d_b^{1.5} = \left(\frac{60{,}000 \times 1.0 \times 1.0 \times 1.0 \times 0.867}{55 \times 1.0\sqrt{4000}}\right)1.128^{1.5} = 17.9 \text{ in.}$$

Checking Eq. (20.28) gives

$$\ell_{dh} = \frac{60{,}000 \times 1.128}{65 \times 1.0\sqrt{4000}} = 16.5 \text{ in.}$$

Finally, checking development in compression using Table 6.5 from Section 6.8 with $\psi_r = 0.75$,[†]

$$\ell_{dc} \geq \left(\frac{f_y\psi_r}{50\lambda\sqrt{f_c'}}\right)d_b = \left(\frac{60{,}000 \times 0.75}{50 \times 1.0\sqrt{4000}}\right)1.128 = 16.1 \text{ in.}$$

$$\ell_{dc} \geq 0.0003\,f_y\psi_r d_b = 0.0003 \times 60{,}000 \times 1.0 \times 1.128 = 20.3 \text{ in.}$$

[†] Although spirals, ties, or hoops in a plane perpendicular to the bars being developed in compression are not provided, as required by ACI Code 25.4.9 to use $\psi_r = 0.75$, by the ties within the joint, a value of $\psi_r = 0.75$ is typically used by designers because the confinement provided by the longitudinal column reinforcement passing through the joint is considered to provide equivalent confinement in the plane perpendicular to the bars.

FIGURE 20.16

Free-body diagrams in plane of spandrel beam for Example 20.2: (a) column and joint region and (b) forces acting on joint due to lateral load.

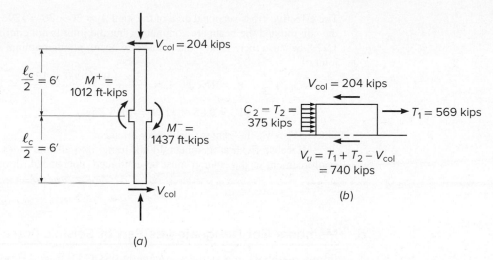

The second equation controls. In accordance with guidance in ACI Commentary 18.8.2.2, ℓ_{dc} is "measured from the critical section to the onset of the bend for hooked bars," giving a total required length of $20.3 + 4 \times 1.128 + 1.128 = 25.9$ in., which controls but can be accommodated within the column.

The same requirements hold true for the No. 8 (No. 25) bottom bars, which must also be anchored in both tension and compression (ACI Code 18.8.2) because lateral loading will subject the beam to both positive and negative bending moments at the exterior joint. By inspection, development will be adequate for the No. 8 bars. All flexural steel from the normal beam must be anchored using hooks extended to the far face of the column core, as shown in Fig. 20.14b.

To check the shear strength of the joint, the shear forces acting on the joint must be calculated based on a stress of $1.25f_y$ in the flexural reinforcement. By inspection, shear in the plane of the spandrel beam will control.

The tensile force in the negative reinforcement, including the bars within the effective width of the flange, is

$$T_1 = 1.25 \times 7.59 \times 60 = 569 \text{ kips}$$

For an effective depth of 33.4 in. (Example 20.1) and a depth of stress block $a = 569/(0.85 \times 4 \times 27) = 6.20$ in., the moment due to negative bending is

$$M^- = \frac{569}{12}\left(33.4 - \frac{6.20}{2}\right) = 1437 \text{ ft-kips}$$

For positive bending on the other side of the column,

$$T_2 = 1.25 \times 5.00 \times 60 = 375 \text{ kips}$$

$$a = \frac{375}{0.85 \times 4 \times 54} = 2.04 \text{ in.}$$

$$M^+ = \frac{375}{12}\left(33.4 - \frac{2.04}{2}\right) = 1012 \text{ ft-kips}$$

The column shear corresponding to the sum of the moments M^+ and M^- and based on the free body of the column between assumed midheight inflection points, as shown in Fig. 20.16a, is $V_{col} = (1437 + 1012)/12 = 204$ kips. The shear forces acting on the joint are shown in Fig. 20.16b, and the factored joint shear is

$$V_u = T_1 + T_2 - V_{col} = 569 + 375 - 204 = 740 \text{ kips}$$

The effective cross-sectional area of the joint $A_j = 36 \times 36 = 1296$ in^2. Using Table 20.1, both the column and the beam are continuous, but the joint is not confined according to ACI Code 15.2.8 because there is only one transverse beam, giving nominal and design capacities of the joint of

$$V_n = 15\lambda\sqrt{f_c'}\,A_j = \frac{15 \times 1.0\sqrt{4000} \times 1296}{1000} = 1229 \text{ kips}$$

$$\phi V_n = 0.85 \times 1229 = 1045 \text{ kips}$$

Since $\phi V_n > V_u$, the joint is satisfactory for shear.

Because flexural members do not frame into all four sides of the joint, the transverse reinforcement in the column must be continued, unchanged, through the joint.

d. Members Not Designated as Part of Seismic-Force-Resisting System

Frame members in structures assigned to SDC D, E, and F that are assumed not to contribute to the structure's ability to carry earthquake forces must still be able to support the factored gravity loads [see Eqs. (20.4) and (20.5)] for which they are designed as the structures undergo lateral displacement. To provide adequate strength and ductility, ACI Code 18.14.2 requires that these members be designed based on moments corresponding to the design displacement, which ACI Commentary 18.14 suggests should be based on models that will provide a conservatively large estimate of displacement. In this case, ACI Code 5.3 permits the load factor for live load L to be reduced to 0.5, except for garages, places of public assembly, and areas where $L > 100$ psf.

When the induced moments and shears, combined with the factored gravity moments and shears (see Table 1.2), do not exceed the design capacity of a frame member, ACI Code 18.14.3 requires that beams contain minimum longitudinal top and bottom reinforcement as provided in Eq. (4.37a), a reinforcement ratio not greater than 0.025, and at least two continuous bars top and bottom. In addition, stirrups are required with a maximum spacing of $d/2$ throughout.

For members with factored axial forces exceeding $A_g f_c'/10$, the longitudinal reinforcement must meet the requirements for columns proportioned for earthquake loads, and the transverse reinforcement must consist of hoops and crossties, as used in columns designed for seismic loading (as required by ACI Code 18.7.5.2) with a maximum spacing of 6 times the diameter of the smallest longitudinal bar or 6 in. For columns, the maximum longitudinal spacing of the ties also may not exceed 6 times the diameter of the smallest longitudinal bar or 6 in. throughout the full column height. Hoops and crossties are required over the length ℓ_o from each joint face (Section 20.5b). In addition, the transverse reinforcement must carry shear induced by inelastic rotation at the ends of the member, as required by ACI Code 18.6.5 (discussed in Section 20.7). Members with factored gravity axial forces exceeding 35 percent of the axial capacity without eccentricity $0.35P_o$ must be designed with transverse reinforcement equal to at least one-half of that required by ACI Code 18.7.5.4 [see Eqs. (20.21) through (20.26)] over the length ℓ_o from each joint face.

If the induced moments or shears under the design lateral displacements exceed the design moment or shear strengths, or if such a calculation is not made, ACI Code 18.14.3.3 requires that the members meet the material criteria for concrete and steel in ACI Code 18.2.5 and 18.2.6 (see Section 20.4), along with criteria for mechanical and welded splices (ACI Code 18.2.7 and 18.2.8, respectively). Beams must meet the same criteria as beams in which the induced moments and shears do not exceed the design strengths, including the requirement that stirrups may not be spaced at greater than $d/2$ throughout the length of the member, along with the requirement that the shear capacity of the member must be adequate to carry forces induced by flexural yielding in accordance with ACI Code 18.6.5 [see Fig. 20.19 and Eq. (20.32) in Section 20.7]. Columns must meet the requirements for the longitudinal reinforcement, transverse reinforcement, and shear capacity specified for columns designed for earthquake-induced lateral loading (ACI Code 18.7.4, 18.7.5, and 18.7.6). Joints must be detailed in accordance with the requirements for intermediate moment frames, as discussed in Section 20.8.

To reduce the potential for a punching shear failure for slab-column connections in two-way slabs without beams, ACI Code 18.14.5 requires that stirrups or headed studs satisfying the requirements of ACI Code 8.7.6 or 8.7.7, respectively (see Section 13.7), and providing v_s of at least $3.5\sqrt{f_c'}$ extend at least 4 times the slab thickness away from the face of the support, unless the design story drift ratio Δ_x/h_{sx} (relative lateral displacement under design load from the top to the bottom of a story Δ_x divided by the height of the story h_{sx}) does not exceed the larger of 0.005 and $[0.035 - 0.05(v_{uv}/\phi v_c)]$. The design story drift ratio is equal to the larger of the design story drift ratios of the stories above and below the slab-column connection. The shear strength of the concrete v_c is defined by Eqs. (13.7a) through (13.7c), and v_{uv} is the factored shear force on the slab critical section for two-way action without moment transfer.

Wall piers not designated as part of the seismic-force-resisting system must satisfy the requirements for wall piers that are designated as part of the seismic-force-resisting system, in accordance with ACI Code 18.10.8 (see Section 20.6c). If the general building code includes provisions to account for overstrength of the seismic-force-resisting system, the design shear force can be calculated as the overstrength factor Ω_o, which accounts for components that cannot provide reliable inelastic response or energy dissipation (Ref. 20.1), times the shear induced under design displacements δ_u.

20.6 ACI PROVISIONS FOR SPECIAL STRUCTURAL WALLS, COUPLING BEAMS, DIAPHRAGMS, AND TRUSSES

ACI Code Chapter 18 includes requirements for stiff structural systems and members that carry earthquake forces or distribute earthquake forces between portions of structures that carry earthquake forces. Structural walls, coupling beams, wall piers, diaphragms, trusses, struts, ties, chords, and collector elements are in this category. The general requirements for these members are presented in this section. The requirements for shear design are presented in Section 20.7c.

a. Structural Walls

To ensure adequate ductility, ACI Code 18.10.2 requires that structural walls have minimum shear reinforcement ratios in both the longitudinal and transverse directions ρ_ℓ and ρ_t of 0.0025 and a maximum reinforcement spacing of 18 in. If the factored shear force assigned to a wall exceeds $2\lambda\sqrt{f_c'}A_{cv}$, where A_{cv} is the net area of the concrete section bounded by the web thickness and the length of the section in the direction of the factored shear force, at least two curtains of reinforcement must be used. If, however, the factored shear is not greater than $\lambda\sqrt{f_c'}A_{cv}$, the minimum reinforcement criteria of ACI Code 11.6 govern (see Section 14.5).

To ensure that it will reach its yield strength, ACI Code 18.10.2.3 requires that longitudinal reinforcement extend 12 ft above the point where it is no longer needed to resist flexure, but not more than ℓ_d above the *next* floor level. Where yielding is likely to occur due to lateral displacements, ℓ_d should be based on $1.25f_y$. Lap splices of longitudinal reinforcement within boundary regions (described in the following paragraph) are not permitted over a height equal to the height of story x, h_{sx} above (but not more than 20 ft) and ℓ_d below critical sections where yielding of the longitudinal reinforcement is likely to occur due to lateral displacements, as shown in Fig. 20.17.

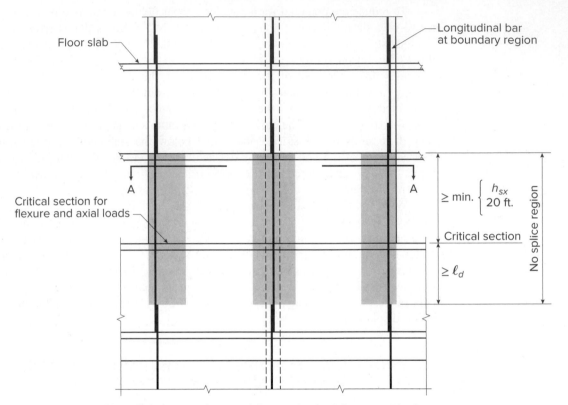

Note: For clarity, only part of the required reinforcement is shown.

(*a*) Elevation

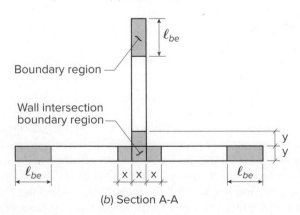

(*b*) Section A-A

FIGURE 20.17
Heights within wall boundary regions where lap splices are not permitted (ℓ_{be} = length of boundary element from compression face of member; x and y = thicknesses of walls). (*Adapted from Ref. 20.10.*)

FIGURE 20.18

Cross sections of structural walls with boundary elements.

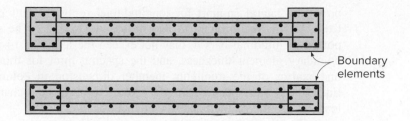

Boundary elements

Boundary elements are added along the edges of structural walls and diaphragms to increase strength and ductility. The elements include added longitudinal and transverse reinforcement and may lie entirely within the thickness of the wall or may require a larger cross section, as shown in Fig. 20.18. Under certain conditions, openings must be bordered by boundary elements. Under the provisions of ACI Code 18.10.6.2 for walls with $h_{wcs}/\ell_w \geq 2.0$, where h_{wcs} is the height of the entire structural wall above the critical section for flexural and axial loads and ℓ_w is the length of the wall, that are continuous from the base of the structure to the top of a wall, compression zones must be reinforced with special boundary elements when the depth to the neutral axis c exceeds the value shown in Eq. (20.30).

$$c \geq \frac{\ell_w}{600(1.5\ \delta_u/h_w)} \tag{20.30}$$

where c is the largest neutral axis depth calculated for the factored axial force and nominal moment strength consistent with the direction of the design displacement δ_u. In Eq. (20.30), δ_u/h_w is not taken less than 0.005. The requirement in Eq. (20.30) is expressed as $(1.5\delta_u/h_{wcs}) \geq (\ell_w/600c)$ in the ACI Code. When special boundary elements are required based on Eq. (20.30), the reinforcement in the boundary element must be extended vertically above and below the critical section a distance equal to the greater of ℓ_w or $M_u/4V_u$. In addition, one of two criteria must be satified: Either $b \geq \sqrt{0.025\ell_w c}$ or $\delta_c/h_{wcs} \geq 1.5\delta_u h_{wcs}$, where b is the width of the compression face, δ_c is that wall deformation capacity, and

$$\frac{\delta_c}{h_{wcs}} = \frac{1}{100}\left(4 - \frac{1}{50}\left(\frac{\ell_w}{b}\right)\left(\frac{c}{b}\right) - \frac{V_e}{8\sqrt{f_c'}A_{cv}}\right) \geq 0.015 \tag{20.31}$$

As an alternative to applying Eq. (20.30), ACI Code 18.10.6.3 may be used to establish the location of boundary elements at boundaries and around openings in walls where the maximum extreme fiber compressive stress under factored loads exceeds $0.2f_c'$. Stresses are calculated based on a linear elastic model using the gross cross section $[\sigma = (P/A) \pm (My/I)]$. The boundary elements may be discontinued once the calculated compressive stress drops below $0.15f_c'$. The confinement provided by the boundary element increases both the ductility of the wall and its ability to carry repeated cycles of loading.

Where required, the boundary element must extend horizontally from the extreme compressive fiber a distance not less than $c - 0.1\ell_w$ or $c/2$, whichever is greater, where c is defined following Eq. (20.30). When flanged sections are used, the boundary element is defined based on the effective flange width and extends at least 12 in. into the web. Transverse reinforcement within the boundary element must meet many of the detailing and spacing requirements for columns discussed in Section 20.5b, including the type of reinforcement, and the need for bends to engage peripheral longitudinal steel, crossties alternating end for end, and hoops and crossties providing lateral support for longituduinal reinforcement in accordance with ACI Code 18.7.5.2(a) through (d) and ACI Code 18.7.5.3. The spacing of laterally supported longitudinal bars h_x may not exceed the lesser of 14 in. and two-thirds of the boundary element thickness, and the spacing limit for transverse reinforcement of one-quarter of the minimum member dimension in columns [described prior to Eq. (20.21) and specified in ACI Code 18.7.5.3(a)] is changed to one-third of the least dimension of the boundary element.

Concrete within the thickness of the floor system must have a compressive strength of at least $0.70 f_c'$ of the wall; and for a distance above and below the critical section equal to at least the greater of ℓ_w and $M_u/4V_u$, web vertical reinforcement must be laterally supported by the corner of a hoop or by a crosstie with seismic hooks at each end. The transverse reinforcement must be spaced vertically not more than 12 in. and meet the size requirements of column ties. The transverse reinforcement within a boundary element must extend into the support a distance equal to at least the development length of the largest longitudinal reinforcement, except where the boundary element terminates at a footing or mat, in which case the transverse reinforcement must extend at least 12 in. into the foundation.

Horizontal reinforcement in the wall web must extend to within 6 in. of the end of the wall and be anchored to develop f_y within the confined core of the boundary element using standard hooks or heads. In cases where the confined boundary element is wide enough to develop the horizontal web reinforcement and $A_s f_y/s$ of the horizontal web reinforcement does not exceed $A_s f_y/s$ of the boundary element transverse reinforcement parallel to the horizontal web reinforcement, the horizontal web reinforcement may be terminated without a standard hook or head.

When boundary elements are not required and when the longitudinal reinforcement ratio in the wall boundary is greater than $400/f_y$, the transverse reinforcement at the boundary must consist of hoops or spirals at the wall boundary with crossties or legs, in accordance with ACI Code 18.7.5.2, that are not spaced more than 14 in. on center extending into the wall a distance of $c - 0.1\ell_w$ or $c/2$, whichever is greater. The transverse reinforcement may not be spaced at more than the values given in Table 20.2. The transverse reinforcement in such cases must be anchored with a standard hook around the edge reinforcement, or the edge reinforcement must be enclosed in U stirrups of the same size and spacing as the transverse reinforcement. This requirement need not be met if the factored shear force V_u is less than $\lambda \sqrt{f_c'} A_{cv}$.

b. Coupling Beams

Coupling beams connect structural walls, as shown in Fig. 20.19a. Under lateral loading, they can increase the stiffness of the structure and dissipate energy. Deeper coupling beams can be subjected to significant shear, which is carried effectively by diagonal reinforcement. According to ACI Code 18.10.7, coupling beams with clear span-to-total-depth ratios ℓ_n/h of 4 or greater may be designed using the criteria for flexural members described in Section 20.5a. In this case, however, the limitations on width-to-depth ratio and total width for flexural members need not be applied if it can be shown by analysis that the beam has adequate lateral stability. Coupling beams with ℓ_n/h less than 2 and a factored shear $V_u > 4\lambda \sqrt{f_c'} A_{cw}$, where A_{cw} is the concrete area resisting shear $= b_w h$, must be reinforced with two intersecting groups of diagonal reinforcement, as shown in Fig. 20.19b and c, unless it can be shown that the loss of stiffness and strength in the beams will not impair the vertical load-carrying capacity of the structure, egress from the structure, or the integrity of nonstructural components and their connections to the structure. Coupling beams with $2 \le \ell_n/h < 4$ may

TABLE 20.2

Maximum vertical spacing of transverse reinforcement at wall boundary in members not requiring boundary elements

Grade of Primary Flexural Reinforcing Bar	Transverse Reinforcement Required	Maximum Vertical Spacing of Transverse Reinforcement[a]	
60	Within the greater of ℓ_w and $M_u/4V_u$ above and below critical sections[b]	Lesser of	$6d_b$
			6 in.
	Other locations	Lesser of	$8d_b$
			8 in.
80	Within the greater of ℓ_w and $M_u/4V_u$ above and below critical sections[b]	Lesser of	$5d_b$
			6 in.
	Other locations	Lesser of	$6d_b$
			6 in.
100	Within the greater of ℓ_w and $M_u/4V_u$ above and below critical sections[b]	Lesser of	$4d_b$
			6 in.
	Other locations	Lesser of	$6d_b$
			6 in.

[a] d_b is the diameter of the smallest primary flexural reinforcing bar.
[b] Critical sections defined as locations where yielding of longitudinal reinforcement is likely to occur as a result of lateral displacements.
(*Source: ACI 318-19. Building Code Requirements for Structural Concrete and Commentary. American Concrete Institute, 2019.*)

be designed using the criteria for flexural members or may be reinforced using two intersecting groups of diagonally placed bars that are symmetrical about the midspan. Such reinforcement is not effective unless it is placed at a steep angle (Refs. 20.14 and 20.15) and thus is not permitted for coupling beams with $\ell_n/h \geq 4$. The criteria for shear reinforcement in coupling beams are discussed in Section 20.7c. In accordance with ACI Code 18.10.2.5, the development length of both longitudinal and diagonal reinforcement extending from a coupling beam into a wall or wall pier is 1.25 times the value calculated for f_y in tension. As shown in Fig. 20.19, there is no requirement that the horizontal reinforcement in a coupling beam be developed at the wall.

c. Wall Piers

Wall piers are narrow vertical wall segments, usually around windows and doors, with dimensions and reinforcement that are intended to result in the lateral shear demand being limited by flexural yielding of the vertical reinforcement in the pier. The desirability of the shear force being governed by the flexural strength is demonstrated by the brittle shear failure of the captive column shown in Fig. 20.9. Pier width ℓ_w and height h_w are defined in Fig. 20.20.

According to ACI Commentary 18.10.8, wall piers with a ratio of width to wall thickness $\ell_w/b_w \leq 2.5$ behave as columns, and according to ACI Code 18.10.8.1, must meet the requirements for columns in special moment frames in ACI Code 18.7.4 through 18.7.6 (see Section 20.5b).

FIGURE 20.19
Coupled shear walls and
coupling beam. (*Adapted from:
ACI 318-19. Building Code
Requirements for Structural
Concrete and Commentary.
American Concrete Institute,
2019.*)

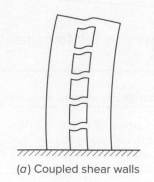

(*a*) Coupled shear walls

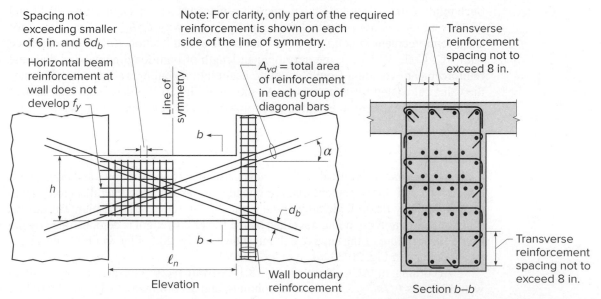

(*b*) Coupling beam with confinement of individual diagonals

(*c*) Coupling beam with full confinement of diagonally reinforced concrete beam section

FIGURE 20.20

Required horizontal reinforcement in wall segments above and below wall piers at the edge of a wall. (*Adapted from Ref. 20.10.*)

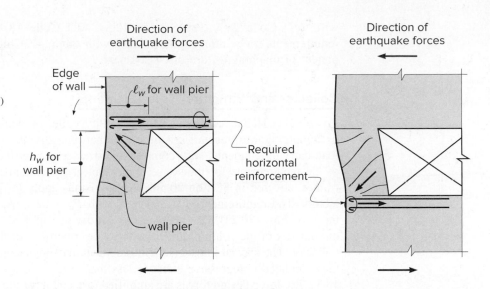

For wall piers with $\ell_w/b_w > 2.5$, ACI Code 18.10.8.1 requires that the shear force be determined as it is for columns, with joint faces taken as the top and bottom of the clear height of the wall pier (see Section 20.7b). When the general building code includes overstrength provisions, the design shear force does not need to exceed the overstrength factor Ω_o times the factored shear calculated by analysis of the structure for the earthquake load effects. The shear force and shear reinforcement must meet the requirements for special structural walls, the transverse reinforcement must consist of hoops with a spacing not greater than 6 in., except single-leg horizontal reinforcement parallel to ℓ_w may be used where the shear reinforcement is placed in a single layer. The single-leg horizontal reinforcement must have 180° bends at each end that engage wall pier boundary longitudinal reinforcement. The transverse reinforcement must extend at least 12 in. above and below the clear height of the wall pier. Special boundary elements must be provided if required by 18.10.6.3, as discussed in Section 20.6a.

In accordance with ACI Code 18.10.8.2, for wall piers at the edge of a wall, horizontal reinforcement must be provided in adjacent wall segments above and below the wall pier and designed to transfer the design shear force from the wall pier into the adjacent wall segments, as shown in Fig. 20.20.

d. Structural Diaphragms

As discussed in Section 19.4, floors and roofs serve as structural diaphragms in buildings. In addition to supporting vertical dead, live, and snow loads, they connect and transfer lateral forces between the members in the vertical lateral-force-resisting system and support other building elements, such as partitions, that may resist horizontal forces but do not act as part of the vertical lateral-force-resisting system. Floor and roof slabs that act as diaphragms may be monolithic with the other horizontal elements in the structures or may include a topping slab. ACI Code 18.12.6 requires that concrete slabs and composite topping slabs designed as structural diaphragms to transmit earthquake forces be at least 2 in. thick. Topping slabs placed over precast floor or roof elements that do not rely on composite action must be at least $2\frac{1}{2}$ in. thick.

Design procedures for diaphragms are discussed in Section 19.4 and covered by ACI Code Chapter 12 and 18.12. As described in Section 19.4, reinforcement must meet the requirements for shrinkage and temperature steel in slabs in accordance with ACI Code 24.4 (see Section 12.3). ACI Code 18.12.7 includes additional requirements for reinforcement in slabs with toppings. Collectors, an integral component of diaphragms, are discussed next.

e. Collector and Structural Truss Elements

To provide adequate confinement and ductility, collector elements, which act in tension or compression to transmit seismic forces between structural diaphragms and vertical load-carrying elements, with compressive stresses greater than $0.2f_c'$ must meet most of the requirements for transverse reinforcement for columns in seismic-load-resisting frames, covered in Section 20.5b. The provisions apply to the full length of the collectors. The requirements for members with $P_u > 0.3f_c' A_g$ or $f_c' > 10,000$ psi and those of Eqs. (20.23), (20.24), and (20.26) are not applied, and the spacing limit of one-quarter of the column dimension becomes one-third of the least dimension of the collector. The special transverse reinforcement may be discontinued at a section where the calculated compressive stress is less than $0.15f_c'$, in accordance with ACI Code 18.12.7.6. If the design forces are amplified to account for the overstrength of the vertical elements of the seismic-force-resisting system, the limit of $0.2f_c'$ is increased to $0.5 f_c'$ and the limit of $0.15f_c'$ is increased to $0.4f_c'$. To limit the buckling of reinforcement and provide adequate bar development near for splices and anchorage zones, ACI Code 18.12.7.7 requires that longitudinal reinforcement have a center-to-center spacing of at least 3 longitudinal bar diameters and not less than $1\frac{1}{2}$ in., and a concrete clear cover of at least $2\frac{1}{2}$ longitudinal bar diameters, not less than 2 in. In addition to meeting the requirements for transverse reinforcement in ACI Code 18.12.7.6, the area of transverse shear reinforcement A_v must equal the greater of $0.75\sqrt{f_c'}\, b_w\, s/f_{yt}$ and $50b_w s/f_{yt}$.

Transverse reinforcement is also required for truss elements with compressive stresses greater than $0.2f_c'$, but in contrast with collector elements, must be continued over the full length of the element. As required by ACI Code 18.12.12 for trusses, the requirements for columns in ACI Code 18.7.5.2, 18.7.5.3, and 18.7.5.7 are used, as described in Section 20.5b

Compressive stresses in collector and truss elements are calculated for the factored forces using a linear elastic model and the gross section properties of the elements. Continuous reinforcement in stiff structural systems must be anchored and spliced to develop f_y in tension.

20.7 ACI PROVISIONS FOR SHEAR STRENGTH

a. Beams

A prime concern in the design of seismically loaded structures is the shear induced in members due to nonlinear behavior in flexure [Eq. (20.1)]. As discussed in Section 20.2, increasing the flexural strength of beams and columns may increase the shear in these members if the structure is subjected to severe lateral loading. As a result, the ACI Code requires that beams and columns in frames that are part of the seismic-force-resisting system (including some members that are not designed to carry lateral loads) be designed for the combined effects of factored gravity load and shear induced by the formation of plastic hinges at the ends of the members.

For beams, ACI Code 18.6.5 requires that the design shear force V_e be based on the shear induced by moments of opposite sign corresponding to the *probable flexural strength M_{pr}* plus the factored tributary gravity load and vertical earthquake loads along

FIGURE 20.21
Forces considered in the shear design of flexural members subjected to seismic loading.

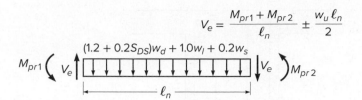

$$V_e = \frac{M_{pr1} + M_{pr2}}{\ell_n} \pm \frac{w_u \ell_n}{2}$$

the span. Loading corresponding to this case is shown in Fig. 20.21. The probable flexural strength M_{pr} is based on the reinforcing steel achieving a stress of $1.25f_y$.

$$M_{pr} = 1.25 A_s f_y \left(d - \frac{a}{2} \right) \tag{20.32a}$$

where

$$a = \frac{1.25 f_y A_s}{0.85 f'_c b} \tag{20.32b}$$

The shear V_e is given by

$$V_e = \frac{M_{pr1} + M_{pr2}}{\ell_n} \pm \frac{w_u \ell_n}{2} \tag{20.33}$$

where M_{pr1} and M_{pr2} = probable moment strengths at two ends of member when moments are acting in the same sense

ℓ_n = length of member between faces of supports

w_u = factored uniform load based on $(1.2 + 0.2 S_{DS})D + 1.0L + 0.2S$ [see Eq. (20.7)]

Equation (20.33) should be evaluated separately for moments at both ends acting in the clockwise and then counterclockwise directions.

To provide adequate ductility and concrete confinement, ACI Code 18.6.5.1 requires that the transverse reinforcement within a length equal to twice the member depth from the face of the support (length defined in ACI Code 18.6.4.1), at both ends of the flexural member, must be designed based on a concrete shear capacity $V_c = 0$, when the earthquake-induced shear force in Eq. (20.33) $(M_{pr1} + M_{pr2})/\ell_n + 0.2 S_{DS} w_u \ell_n$ is one-half or more of the maximum required shear strength within that length and the factored axial compressive force in the member, including earthquake effects, is below $A_g f'_c / 20$.

EXAMPLE 20.3   **Beam shear design.**   An 18 in. wide by 24 in. deep reinforced concrete beam spans between two interior columns in a building frame designed for a region of high seismic risk. The clear span is 24 ft, and the reinforcement at the face of the support consists of four No. 10 (No. 32) top bars and four No. 8 (No. 25) bottom bars. The effective depth is 21.4 in. for both top and bottom steel. $S_{DS} = 0.20$. The maximum factored shear $w_u \ell_n / 2 = [(1.2 + 0.2 S_{DS})w_d + 1.0 w_l] \ell_n / 2 = 32$ kips at each end of the beam. Materials strengths are $f'_c = 5000$ psi and $f_y = 60,000$ psi. Design the shear reinforcement for the regions adjacent to the column faces.

SOLUTION.   The probable moment strengths M_{pr} are based on a steel stress of $1.25f_y$. For negative bending, the area of steel is $A_s = 5.08$ in^2 at both ends of the beam, the stress block depth is $a = 1.25 \times 5.08 \times 60/(0.85 \times 5 \times 18) = 4.98$ in., and the probable strength is

$$M_{pr1} = 1.25 A_s f_y \left(d - \frac{a}{2} \right) = \frac{1.25 \times 5.08 \times 60}{12} \left(21.4 - \frac{4.98}{2} \right) = 600 \text{ ft-kips}$$

FIGURE 20.22
Configuration of hoop
reinforcement for beam in
Example 20.3.

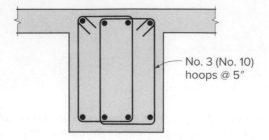

No. 3 (No. 10)
hoops @ 5″

For positive bending, the area of steel is $A_s = 3.16$ in^2, the effective width is 90 in., the stress block depth $a = 1.25 \times 3.16 \times 60/(0.85 \times 5 \times 90) = 0.62$ in., and the probable strength is

$$M_{pr2} = \frac{1.25 \times 3.16 \times 60}{12} \left(21.4 - \frac{0.62}{2}\right) = 417 \text{ ft-kips}$$

As given in the problem statement, the effect of factored uniform loads $w_u \ell_n/2 = [(1.2 + 0.2S_{DS})w_d + 1.0w_l]\ell_n/2 = 32$ kips giving a design shear force at each end of the beam, according to Eq. (20.32), of

$$V_e = \frac{M_{pr1} + M_{pr2}}{\ell_n} \pm \frac{w_u \ell_n}{2} = \frac{600 + 417}{24} + 32 = 42 + 32 = 74 \text{ kips}$$

Since the earthquake-induced force not counting the addition effect of $0.2S_{DS}w_u\ell_n$, 42 kips, is greater than one-half of the maximum required shear strength, the transverse hoop reinforcement must be designed to resist the full value of V_e (that is, $\phi V_s \geq V_e$) over a length $2h = 48$ in. from the face of the column, in accordance with ACI Code 18.6.5.1. The maximum spacing of the hoops s is based on the smaller of $d/4 = 5.4$ in., $6d_b$ for the smallest longitudinal bars = 6 in., or 6 in. for the hoop bars [assumed to be No. 3 (No. 10) bars] = 9 in., or 12 in. A spacing $s = 5$ in. will be used.

The area of shear reinforcement within a distance s is

$$A_v = \frac{(V_e/\phi)s}{f_{yt}d} = \frac{(74/0.75)5}{60 \times 21.4} = 0.38 \text{ in}^2$$

Providing support for corner and alternate longitudinal bars, in accordance with ACI Code 18.6.4 (and ACI Code 25.7.2.3), leads to the use of overlapping hoop reinforcement, shown in Fig. 20.22, and a total area of transverse steel $A_v = 0.44$ in^2.

The first hoop is placed 2 in. from the face of the column. The other hoops are spaced at 5 in. within 48 in. from each column face. Transverse reinforcement for the balance of the beam is calculated based on the value of V_e at that location and a nonzero concrete contribution V_c. The stirrups must have seismic hooks and a maximum spacing of $d/2$.

b. Columns

In accordance with ACI Code 18.7.6.1, shear provisions similar to those used for beams to account for the formation of inelastic hinges must also be applied to members with axial loads greater than $A_g f_c'/10$. In this case, the loading is illustrated in Fig. 20.23a, and the factored shear is

$$V_e = \frac{M_{pr1} + M_{pr2}}{\ell_u} \tag{20.34}$$

where ℓ_u is the clear distance between beams and M_{pr1} and M_{pr2} are based on a steel tensile strength of $1.25f_y$.

FIGURE 20.23
(a) Forces considered in the shear design of columns subjected to seismic loading. (b) Column interaction diagram used to determine maximum probable moment strengths. Note that M_{pr} for columns is usually governed by M_{pr} of the girders framing into a joint, rather than M_{max}.

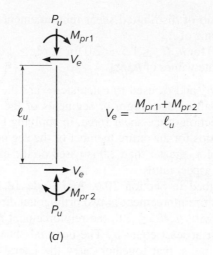

$$V_e = \frac{M_{pr1} + M_{pr2}}{\ell_u}$$

(a)

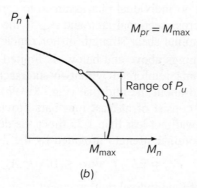

(b)

In Eq. (20.34), M_{pr1} and M_{pr2} are the maximum probable moment strengths for the range of factored axial loads to which the column will be subjected, as shown in Fig. 20.23b; V_e, however, need not be greater than a value based on M_{pr} for the transverse members framing into the joint. For most frames, the latter will control. Of course, V_e may not be less than that obtained from the analysis of the structure under factored loads.

The ACI Code requires that the transverse reinforcement in a column over a length ℓ_o (the greater of the depth of the member at the joint face, one-sixth of the clear span, or 18 in.) from each joint face be proportioned to resist shear based on a concrete shear capacity $V_c = 0$ when (1) the earthquake-induced shear force is one-half or more of the maximum required shear strength within those lengths and (2) the factored axial compressive force, including earthquake effects, is less than $A_g f_c' / 20$.

c. Walls, Coupling Beams, Wall Piers, Diaphragms, and Trusses

The factored shear force V_u for walls, coupling beams, diaphragms, and trusses is obtained from analysis based on the factored (including earthquake) loads.

In accordance with ACI Code 18.10.4.1, the nominal shear strength V_n of structural walls and wall segments is taken as

$$V_n = \left(\alpha_c \lambda \sqrt{f_c'} + \rho_t f_y \right) A_{cv} \tag{20.35}$$

where A_{cv} = gross area of concrete section bounded by the web thickness and length of the section in the direction of shear force

ρ_t = ratio of distributed shear reinforcement on a plane perpendicular to the plane of A_{cv}

α_c = 3.0 for $h_w/\ell_w \leq 1.5$, = 2.0 for $h_w/\ell_w \geq 2.0$, and varies linearly for intermediate values of h_w/ℓ_w

The values of h_w and ℓ_w used to calculate α_c are the height and length, respectively, of the entire wall or diaphragm or segments of the wall or diaphragm. ℓ_w is measured in the direction of the shear force. In applying Eq. (20.35), the ratio h_w/ℓ_w is the larger of the ratios for the entire member or the segment of the member being considered. The use of α_c greater than 2.0 is based on the higher shear strength observed for walls with low aspect ratios.

As described in Section 20.6, ACI Code 18.10.2 requires that walls contain distributed shear reinforcement in two orthogonal directions in the plane of the member. For walls with $h_w/\ell_w \leq 2.0$, the reinforcement ratio for steel crossing the plane of A_{cv}, ρ_ℓ, must at least equal ρ_t. The nominal shear strength of all wall segments, including wall piers, that together carry the lateral force is limited to a maximum value of $8\sqrt{f_c'}A_{cv}$, with no individual pier assumed to carry greater than $10\sqrt{f_c'}A_{cw}$, where A_{cv} is the total cross-sectional area and A_{cw} is the cross-sectional area of an individual pier. The nominal shear strength of horizontal wall segments (regions of a wall bounded by openings above and below) is limited to $10\sqrt{f_c'}A_{cw}$.

For coupling beams reinforced with two intersecting groups of diagonally placed bars symmetrical about the midspan (Fig. 20.19b), each group of the diagonally placed bars must consist of at least four bars provided in two or more layers and embedded into the wall not less than 1.25 times the development length required for f_y in tension. The nominal strength provided by the diagonal bars is given by

$$V_n = 2A_{vd}f_y\sin\alpha \leq 10\sqrt{f_c'}\,A_{cw} \tag{20.36}$$

where A_{vd} = total area of longitudinal reinforcement in an individual diagonal
A_{cw} = area of concrete section resisting shear
α = angle between diagonal reinforcement and longitudinal axis of coupling beam

The upper limit in Eq. (20.36) is a safe upper bound based on the experimental observation that coupling beams remain ductile at shear forces exceeding this value (Ref. 20.15).

ACI Code 18.10.7 allows two options for providing confinement for coupling beams. In the first, shown in Fig. 20.19b, each group of diagonal bars must be enclosed by transverse reinforcement having out-to-out dimensions not smaller than $b_w/2$ in the direction parallel to b_w and $b_w/5$ along the other sides (that is, perpendicular to b_w). The transverse reinforcement must consist of hoops satisfying Eqs. (20.24) and (20.25), with a spacing measured parallel to the diagonal bars satisfying Eq. (20.21), but not exceeding 6 times the diameter of the diagonal bars, and have spacing of crossties or legs of hoops measured perpendicular to the diagonal bars not exceeding 14 in. When computing A_g for use in Eqs. (9.5) and (20.24), the concrete cover required by ACI Code 20.5.1 (see Section 4.5a) is assumed on all four sides of each group of diagonal bars. The transverse reinforcement must be continued through the intersection of the diagonal bars. Additional longitudinal and transverse reinforcement must be distributed around the beam perimeter with a total area in each direction not less than $0.002b_ws$ and spacing not greater than 12 in.

For the second option, shown in Fig. 20.19c, ACI Code 18.10.7 allows hoops to be provided for the entire beam cross section satisfying Eqs. (20.24) and (20.25), with longitudinal spacing not exceeding the smaller of 6 in. and 6 times the diameter of the diagonal bars, and with spacing of crossties or legs of hoops both vertically and horizontally in the plane of the beam cross section not exceeding 8 in. Each crosstie and each hoop leg must engage a longitudinal bar of equal or larger diameter. Because construction is easier, the second option is used far more often in practice than the first.

According to ACI Code 12.5.3, the maximum nominal shear strength of diaphragms is

$$V_n = \left(2\lambda\sqrt{f_c'} + \rho_t f_y\right)A_{cv} \le 8\sqrt{f_c'}A_{cv} \tag{20.37}$$

where A_{cv} is the gross area of the concrete bounded by the diaphragm thickness and length of the diaphragm in the direction of the shear force, and $\sqrt{f_c'}$ used to calculate V_n may not exceed 100 psi. Web reinforcement in the diaphragm is distributed uniformly in both directions.

EXAMPLE 20.4   **Coupling beam design.**   A coupling beam with a clear span $\ell_n = 72$ in., a depth $h = 72$ in., a thickness $b_w = 18$ in. will be subjected to a factored shear force $V_u = 340$ kips. Design the coupling beam using $f_c' = 4000$ psi and $f_y = 60,000$ psi.

SOLUTION.   The design approach depends on the span-to-depth ratio ℓ_n/h and the factored shear V_u relative to $4\lambda\sqrt{f_c'}A_{cw}$. $\ell_n/h = 72/72 = 1.0$ and $4\lambda\sqrt{f_c'}A_{cw} = 4 \times 1.0\sqrt{4000} \times 18 \times 72/1000 = 328$ kips. As described in Section 20.6b, because ℓ_n/h is less than 2, and V_u is greater than $4\lambda\sqrt{f_c'}A_{cw}$, the coupling beam must be reinforced with two intersecting groups of diagonal reinforcement.

To calculate the angle α between the diagonal and the longitudinal axis of the coupling beam (Fig. 20.19), assume an out-to-out spacing of the outer bars in the diagonal reinforcement of 8 in. and a clear spacing between the outer bars and the top of the beam as the bars in the diagonal exit the coupling beam of 3 in., giving $\alpha = \tan^{-1}[(72 - 8 - 2 \times 3)/72] = 38.8$ deg.

Solving Eq. (20.36) for A_{vd} and substituting V_u/ϕ for V_n gives

$$A_{vd} = V_u/(2\phi f_y \sin\alpha) = 340/(2 \times 0.75 \times 60 \times \sin 38.8\text{ deg}) = 6.02\text{ in}^2$$

Four No. 11 (No. 36) bars, with a total area of 6.24 in^2, are selected. By inspection, $V_n < 10\sqrt{f_c'}A_{cw}$, satisfying the upper limit shown in Eq. (20.36). The No. 11 (No. 36) bars must extend 1.25 times ℓ_d for Grade 60 reinforcement into the wall.

As shown in Fig 20.19 and explained in Section 20.7c, there are two options for providing confinement for coupling beams, (1) ties around each set of diagonal bars, as shown in Fig. 20.19b, and (2) transverse reinforcement around the full coupling beam, as shown in Fig. 20.19c.

Option 1: The tentative dimensions of the diagonal bars are shown in Fig. 20.24. Outside dimensions of the transverse reinforcement of 9 in. both parallel and perpendicular to the side face of the beam will satisfy the requirements that out-to-out dimensions must be at least $b_w/2$ in the direction parallel to b_w and $b_w/5$ along the other sides. For the four No. 11 (No. 36) bars, h_x is approximately 6.6 in. Under Option 1, the maximum spacing of ties measured parallel to the diagonal bars s is given by s_o in Eq. (20.21a) but must have values between 4 and 6 in., as given in Eq. (20.21b) and not greater than $6d_b = 6 \times 1.41 = 8.46$ in.

$$s = s_o = 4 + \left(\frac{14 - h_x}{3}\right) = 4 + \left(\frac{14 - 6.6}{3}\right) = 6.47\text{ in.}$$

The maximum spacing controls, giving the spacing between transverse reinforcement $s = 6$ in. The area of the ties must also satisfy Eqs. (20.24) and (20.25) in both directions

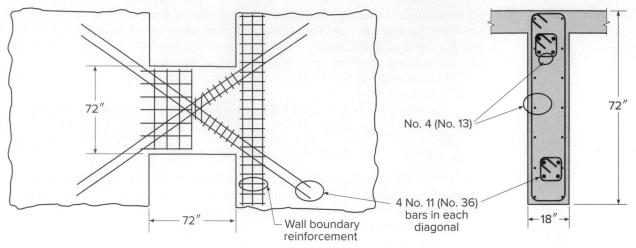

No. 4 (No. 13)

4 No. 11 (No. 36) bars in each diagonal

Wall boundary reinforcement

(a) Coupling beam with confinement of individual diagonals

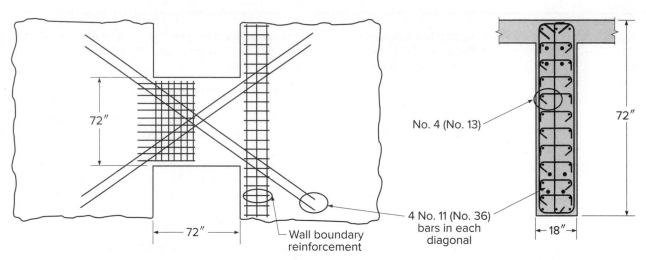

No. 4 (No. 13)

4 No. 11 (No. 36) bars in each diagonal

Wall boundary reinforcement

(b) Coupling beam with full confinement of diagonally reinforced beam section

FIGURE 20.24
Designs of coupling beams in Example 20.4.

(parallel and perpendicular to b_w). Solving Eqs. (20.24) and (20.25) for the area of the ties A_{sh} gives, respectively,

$$A_{sh} = 0.3\left(s b_c f_c'/f_{yt}\right)\left[(A_g/A_{ch}) - 1\right]$$

$$A_{sh} = 0.09\, s b_c f_c'/f_{yt}$$

A_g is calculated based on the concrete cover required by ACI Code 20.5.1, which is $\frac{3}{4}$ in. for walls. Thus, $A_g = (9.0 + 2 \times 0.75) \times (9.0 + 2 \times 0.75) = 110$ in^2, and $A_{ch} = 9.0 \times 9.0 = 81$ in^2. Because $b_c = 9.0$ in. both parallel and perpendicular to b_w, the required area of the ties is the same in the two directions. The values of A_{sh} based on Eqs. (20.24) and (20.25) are

$$A_{sh} = 0.3(6 \times 9 \times 4/60)[(110/81) - 1] = 0.39 \text{ in}^2$$

$$A_{sh} = 0.09 \times 6 \times 9 \times 4/60 = 0.32 \text{ in}^2$$

The required area can be provided by a single No. 4 (No. 13) bar tie, which has two legs in each direction, giving $A_{sh} = 0.40$ in^2. The spacing between crossties or legs of hoops perpendicular to the diagonal bars cannot exceed 14 in., a requirement that is satisfied by the center-to-center spacing of the No. 11 (No. 36) bars, which is 6.6 in. The transverse reinforcement must be continued through the intersection of the diagonals.

As a final step, additional longitudinal and transverse reinforcement is required around the perimeter of the beam with a total area of $0.002b_w s$ and a spacing not greater than 12 in. With two layers of No. 4 (No. 13) bars and $b_w = 18$ in., $s = 0.40/(0.002 \times 18) = 11.1$ in. Starting 3 in. in from each boundary, seven No. 4 (No. 13) bars, spaced at 11 in., will be used on each face of the beam for a total of fourteen No. 4 (No. 13) bars each way. As described in Section 20.6b, the horizontal bars need not be developed within the wall and, based on experience, will project just 6 in. into the wall. The final design for Option 1 is shown in Fig. 20.24a.

Option 2: Option 2 requires transverse reinforcement around the full coupling beam that satisfies Eqs. (20.24) and (20.25) in both directions. The spacing of the vertical reinforcement may not exceed 6 in. or 6 times the diameter of the diagonal bars. For No. 11 (No. 36) diagonal bars, the maximum spacing is the smaller of 6 in. and $6d_b = 6 \times 1.41 = 8.5$ in, with 6 in. controlling. In addition, the spacing of the crossties or legs of hoops may not exceed 8 in. in either the horizontal or vertical direction. To provide adequate reinforcement in the steps that follow, spacings s of 5 and 6 in. will be used, respectively, providing 14 vertical and 12 horizontal bars on each face of the beam.

To solve for the area of confining reinforcement, $A_g = 18.0 \times 72.0 = 1296$ in^2, and $A_{ch} = 16 \times 70 = 1120$ in^2.

The area of reinforcement A_{sh} in the vertical direction based on the larger value obtained from Eqs. (20.24) and (20.25) is calculated using $s = 5$ in. and $b_c = 16$ in.

$$A_{sh} = 0.3(5 \times 16 \times 4/60)[(1296/1120) - 1] = 0.27 \text{ in}^2$$

$$A_{sh} = 0.09 \times 5 \times 16 \times 4/60 = 0.48 \text{ in}^2$$

The area of reinforcement A_{sh} in the vertical direction based on the larger value obtained from Eqs. (20.24) and (20.25) is calculated using $s = 6$ in. and $b_c = 70$ in.

$$A_{sh} = 0.3(6 \times 70 \times 4/60)[(1296/1120) - 1] = 1.32 \text{ in}^2$$

$$A_{sh} = 0.09 \times 6 \times 70 \times 4/60 = 2.52 \text{ in}^2$$

In the vertical direction, a single No. 4 (No. 13) tie with two legs plus the additional required crosstie, also No. 4 (No. 13), to limit the spacing of the crossties or legs of hoops to a maximum of 8 in. provides an area of 0.60 in^2, above the requirement of 0.48 in^2 based on Eq. (20.25). In the horizontal direction, two No. 4 (No. 13) tie legs plus 12 crossties spaced at 5 in. provides an area of $14 \times 0.20 = 2.80$ in^2, above the requirement of 2.52 in^2 based on Eq. (20.25). As with Option 1, the horizontal bars will project 6 in. into the wall. The final design for Option 2 is shown in Fig. 20.24b.

20.8 ACI PROVISIONS FOR INTERMEDIATE MOMENT FRAMES

ACI Code 18.4 governs the design of frames in structures assigned to SDC C. The requirements include specified loading and detailing requirements. Unlike structures assigned to SDC D, E, and F, two-way slab systems without beams are allowed to serve as lateral-force-resisting systems. Walls, diaphragms, and trusses in regions of moderate seismic risk are designed using the main part of the Code.

ACI Code 18.4.2.3 and 18.4.3.1, addressing the shear design of beams and columns, respectively, offer two options for the shear design of frame members. The first option is similar to that illustrated in Figs. 20.21 and 20.23 and Eqs. (20.33) and (20.34), with the exception that the probable strengths M_{pr} are replaced by the nominal strengths M_n. For beams, f_y is substituted for $1.25f_y$ in Eq. (20.30). For columns, the moments used at the top and bottom of the column [Fig. 20.23 and Eq. (20.34)] are based on the capacity of the column alone (not considering the moment capacity of the beams framing into the joints) under the factored axial load P_u that results in the maximum nominal moment capacity.

As an alternative to designing for shear induced by the formation of hinges at the ends of beams, ACI Code 18.4.2.3 allows shear design to be based on load combinations that include an earthquake effect that is twice that required by the governing building code. Thus, Eq. (20.4) becomes

$$U = 1.2D + 2.0E + 1.0L + 0.2S \qquad (20.38)$$

where E is given in Eq. (20.6a).

For columns, the second option in ACI Code 18.4.3.1 allows the maximum shear to be obtained from factored load combinations that include E, with E multiplied by the overstrength factor Ω_o, giving

$$U = 1.2D + \Omega_o E + 1.0L + 0.2S \qquad (20.39)$$

where $\Omega_o = 3.0$ for intermediate moment frames (Ref. 20.10).

For beams and columns, the Code prescribes detailing requirements that are not as stringent as those used in regions of high seismic risk, but that provide greater confinement and increased ductility compared to those used in structures not designed for earthquake loading. For beams, the positive-moment strength at the face of a joint must be at least one-third of the negative-moment strength at the joint, in accordance with ACI Code 18.4.2.2. Both the positive and negative-moment strength along the full length of a beam must be at least one-fifth of the maximum moment strength at the face of either joint. Hoops are required at both ends of beams over a length equal to twice the member depth; the first hoop must be placed within 2 in. of the face of the support, and the maximum spacing in this region may not exceed one-fourth of the effective depth, 8 times the diameter of the smallest longitudinal bar, 24 times the stirrup diameter, or 12 in. The maximum stirrup spacing elsewhere in beams is one-half of the effective depth.

For columns, within length ℓ_o from the joint face, the hoop spacing s_o may not exceed 8 times the diameter of the smallest longitudinal bar or 8 in. for Grade 60 reinforcement, 6 times the diameter of the smallest longitudinal bar or 6 in. for Grade 80 reinforcement, or one-half of the smallest cross-sectional dimension of the column. The length ℓ_o must be greater than one-sixth of the column clear span, the maximum cross-sectional dimension of the member, or 18 in. The first hoop must be located not more than $s_o/2$ from the joint face. Outside of ℓ_o, the spacing of the transverse reinforcement must satisfy the requirements for shear reinforcement described in Section 5.5d in accordance with ACI Code 10.7.6.5.2.

Like columns in special moment frames, columns in intermediate moment frames must be designed to provide for ductile behavior when supporting discontinuous stiff members, such as walls. Columns in intermediate moment frames must contain transverse reinforcement with spacing s_o over the full height beneath the level at which the discontinuity occurs if the portion of factored axial compressive force in the columns related to earthquake effects exceeds $A_g f_c'/10$. If design forces have been magnified to account for the overstrength of the vertical elements of the seismic-force-resisting system, the limit of $A_g f_c'/10$ be increased to $A_g f_c'/4$. Transverse reinforcement shall extend above and below the column in accordance with ACI Code 18.7.5.6(b). This transverse reinforcement must extend above and below the columns into the discontinued stiff member for at least the development length of the largest longitudinal reinforcement for walls and at least 12 in. into foundations.

As required by ACI Code 18.4.4, joints in intermediate moment frames must be confined with ties, spirals, or hoops but must have not less than two layers of horizontal transverse reinforcement within the depth of the shallowest beam framing into the joint. Spacing requirements for transverse reinforcement through the deepest beam framing into the joint are the same as for columns in intermediate moment frames. The strut-and-tie method must be used to design the joint if the depth of a beam framing into the joint exceeds twice the depth of the column. Longitudinal reinforcement terminated in a joint must extend to the far face of the joint core and be developed in tension as required for special moment frames in accordance with ACI Code 18.8.5, as described in Section 20.5c, and in compression in accordance with ACI Code 25.4.9, as described in Section 6.8. If transverse reinforcement is required, slab-column joints must contain at least one layer of reinforcement between the top and bottom slab reinforcement. In accordance with ACI Code 18.4.4.7, the factored shear force with a joint V_u is calculated based on the nominal moment capacity of the beams framing into the joint, as described in Section 18.2d. The nominal shear capacity V_n is calculated in accordance with the provisions for special moment frames described in ACI Code 18.8.4.3 and shown in Table 20.1.

For two-way slabs without beams, ACI Code 18.4 requires design for earthquake effects using Eqs. (20.4) and (20.5). Under these loading conditions, the reinforcement provided to resist the unbalanced moment transferred between the slab and the column M_{sc} (see Section 13.8) must be placed within the column strip. Reinforcement to resist the fraction of the unbalanced moment M_{sc} defined by Eq. (13.12a), $M_{ub} = \gamma_f M_{sc}$, but not less than one-half of the reinforcement in the column strip at the support must be concentrated near the column. This reinforcement is placed within an effective slab width located between lines $1.5h$ on either side of the column or column capital, where h is the total thickness of the slab or drop panel.

To ensure ductile behavior throughout two-way slabs without beams, at least one-quarter of the top reinforcement at the support in column strips must be continuous throughout the span, as must bottom reinforcement equal to at least one-third of the top reinforcement at the support in column strips. A minimum of one-half of all bottom middle strip reinforcement and all bottom column strip reinforcement at midspan in both column and middle strips must be continuous and develop its yield strength at the face of columns, capitals, brackets, and walls. For discontinuous edges of the slab, both the top and bottom reinforcement must be developed at the face of columns, capitals, brackets, and walls. Finally, at critical sections for two-way shear at columns (Section 13.7a), V_u may not exceed $0.4\phi V_c$. The latter provision may be waived if the requirements of ACI Code 18.14.5 for slab-column connections in members not designated as part of the seismic-force-resisting system are met (see Section 20.5b). If transverse reinforcement is required in slab-column joints, at least one layer of reinforcement is required between the top and bottom slab reinforcement.

REFERENCES

20.1. *2015 NEHRP Recommended Provisions for Seismic Regulations for New Buildings and Other Structures*, (FEMA P-1050), Part 1—Provisions, Part 2—Commentary, Building Seismic Safety Council, Washington, DC, 2015.

20.2. *Minimum Design Loads for Buildings and Other Structures*, ASCE/SEI 7-16, American Society of Civil Engineers, Reston, VA, 2016.

20.3. S. K. Ghosh, A. W. Domel Jr., and D. A. Fanella, *Design of Concrete Buildings for Earthquake and Wind Forces*, 2nd ed., Portland Cement Association, Skokie, IL, and International Conference of Building Officials, Whittier, CA, 1995.

20.4. W.-F. Chen and C. Scrawthorn, eds., *Earthquake Engineering Handbook*, CRC Press, Boca Raton, FL, 2003.

20.5. T. Paulay and M. J. N. Priestly, *Seismic Design of Reinforced Concrete and Masonry Buildings*, John Wiley & Sons, Inc., New York, 1992.

20.6. G. G. Penelis and A. J. Kappos, *Earthquake-Resistant Concrete Structures*, CRC Press, New York, 1997.

20.7. *Concrete Structures in Earthquake Regions*: *Design & Analysis*, E. Booth, ed., Longman Scientific & Technical, Harlow, England, 1994.

20.8. J. P. Moehle, *Seismic Design of Reinforced Concrete Buildings*, McGraw-Hill Education, New York, NY, 2014.

20.9. *Seismic Analyisis of Safety-Related Nuclear Structures*, ASCE/SEI 4-98, American Society of Civil Engineers, Reston, VA, 2000.

20.10. *Building Code Requirements for Structural Concrete* (ACI 318-19) *and Commentary*, American Concrete Institute, Farmington Hills, MI, 2019.

20.11. ACI Committee 352, *Recommendations for Design of Beam-Column Joints in Monolithic Reinforced Concrete Structures*, ACI 352R-02, American Concrete Institute, Farmington Hills, MI, 2008.

20.12. T. Ichinose, "Splitting Bond Failure of Columns under Seismic Action," *ACI Struct. J.,* vol. 92, no. 5, 1995, pp. 535–541.

20.13. D. Sokoli and W. M. Ghannoum, "High-Strength Reinforcement in Columns under High Shear Stresses," *ACI Struct. J,* vol. 113, no. 3, 2016, pp. 605–614.

20.14. T. Paulay and J. R. Binney, "Diagonally Reinforced Coupling Beams of Shear Walls," in *Shear in Reinforced Concrete*, SP-42, American Concrete Institute, Detroit, MI, 1974, pp. 579–798.

20.15. G. B. Barney, K. N. Shiu, B. G. Rabbat, A. E. Fiorato, H. G. Russell, and W. G. Corley, *Behavior of Coupling Beams under Load Reversals* (RD068.01), Portland Cement Association, Skokie, IL, 1980.

PROBLEMS

20.1. An interior column joint in a reinforced concrete frame structure assigned to SDC D consists of 28 in. wide by 20 in. deep beams and 36 in. wide by 20 in. deep girders framing into a 28×28 in. column. The slab thickness is 5 in., and the effective overhanging flange width on either side of the web of the flexural members is 40 in. Girder reinforcement at the joint consists of five No. 10 (No. 32) top bars and five No. 8 (No. 25) bottom bars. Beam reinforcement consists of four No. 10 (No. 32) top bars and four No. 8 (No. 25) bottom bars. As the flexural steel crosses the joint, the top and bottom girder bars rest on the respective top and bottom beam bars. Column reinforcement consists of 20 No. 7 (No. 22) bars evenly spaced around the perimeter of the column, similar to the placement shown in Fig. 20.14. Clear cover on the outermost main flexural and column longitudinal reinforcement is 2 in. Assume No. 4 (No. 13) stirrups and ties. For earthquake loading, the maximum factored axial load on the upper column framing into the joint is 1098 kips, and the maximum factored axial load on the lower column is 1160 kips. For a frame story height of 13 ft, determine if the nominal flexural strengths of the columns exceed those of the beams and girders by at least 20 percent and determine the minimum transverse reinforcement required in the columns adjacent to the beams. Use $f'_c = 4000$ psi and $f_y = 60,000$ psi.

20.2. Design the joint and the transverse column reinforcement for the members described in Problem 20.1. The factored shears due to earthquake load are 29 kips in the upper column and 31 kips in the lower column. Minimum factored axial loads are 21 and 25 kips below the forces specified in Problem 20.1 for the upper and lower columns, respectively.

20.3. In Example 20.1, the columns are spaced 28 ft on center in the direction of the spandrel beams. The total dead load on the spandrel beam (including self-weight) is 2 kips/ft and the total live load is 0.93 kips/ft. Design the spandrel beam transverse reinforcement for a building subject to high seismic risk.

20.4. Repeat Problem 20.3 for an intermediate frame.

20.5. An interior column joint in a reinforced concrete frame structure assigned to SDC E consists of 20 in. wide by 28 in. deep beams and girders framing into 26 × 26 in. columns. The slab thickness is 4 in., and the effective overhanging flange width on either side of the web of the flexural members is 32 in. Girder and beam reinforcement at the joint consists of four No. 10 (No. 32) top bars and four No. 9 (No. 29) bottom bars. As the flexural steel crosses the joint, the top and bottom girder bars are outside the respective top and bottom beam bars. Column reinforcement consists of eight No. 10 (No. 32) bars at the corners and midsides of the column and eight No. 9 (No. 29) bars at the alternate sites around the perimeter of the column. Clear cover on the outermost main flexural and column longitudinal reinforcement is 2 in. Assume No. 4 (No. 13) stirrups and ties. For earthquake loading, the maximum factored axial load on the upper column framing into the joint is 1100 kips, and the maximum factored axial load on the lower column is 1230 kips. The story height is 12 ft, and the columns are spaced 24 ft on center in the direction of the girders. Use $f'_c = 4000$ psi and $f_y = 60,000$ psi.

(*a*) Determine if the nominal flexural strengths of the columns exceed those of the beams and girders by at least 20 percent, and determine the minimum transverse reinforcement required in the columns adjacent to the beams.

(*b*) The total dead load on the girder (including self-weight) is 2.8 kips/ft, and the total live load is 1.3 kips/ft. Design the girder transverse reinforcement.

21

Anchoring to Concrete

21.1 INTRODUCTION

The majority of connections in concrete structures are monolithic, as described in Chapter 18. Nonetheless, most buildings require some members to be connected and some loads to be transferred through mechanical *attachments*. These attachments consist of steel plates, steel weldments, or precast components that are fastened to the supporting concrete using *anchors*. An attachment is either cast into the concrete or post-installed using mechanical or adhesive anchors. Examples of attachments include column base plates, column brackets, wall corbels, stair supports, and mechanical equipment supports. Figure 21.1 shows a precast column bracket with the anchor

(*a*)

(*b*)

FIGURE 21.1
(*a*) Precast column bracket and (*b*) final placement in column.
(*Photographs courtesy of Rocky Mountain Prestress, LLC.*)

(a)

FIGURE 21.3
Mechanical equipment post-installed in a concrete slab.
(*Photograph by Charles W. Dolan.*)

(b)

FIGURE 21.2
(*a*) Steel corbel post-installed to a cast-in-place wall (*Photograph courtesy of Hilti Inc.*) and (*b*) cast-in steel bracket. (*Photograph courtesy of Rocky Mountain Prestress, LLC.*)

FIGURE 21.4
Tension anchor failure of a Cazaly hanger.
(*Photograph by Charles W. Dolan.*)

reinforcement extended and the bracket cast into a column. Figure 21.2 shows post-installed and cast-in steel corbels, and Fig. 21.3 shows mechanical equipment fastened to the soffit of a concrete slab. Figure 21.4 illustrates a tension breakout failure of a Cazaly hanger attached using anchors at the end of a precast concrete floor panel.

21.2   BEHAVIOR OF ANCHORS

Anchors are steel elements that are cast into the concrete, or adhesively bonded or mechanically post-installed in hardened concrete. The *effective embedded depth* h_{ef} is shown in Fig. 21.5.

The variety of possible anchors and their varied failure modes require the engineer to determine the strength associated with each possible failure mode and base the design on the lowest strength to assure overall structural safety (Refs. 21.1 and 21.2). Anchor failures include those based on:

Steel strength in shear or tension
Concrete breakout in shear or tension
Concrete pullout in tension
Side face blowout
Pryout in shear
Combined tension and shear
Bond failure of adhesive

The failure modes result from tension, shear, or combined loading. Figure 21.6 summarizes the failure modes. Except for the adhesive anchor (Fig. 21.6*f*), headed studs are illustrated. Screw anchors are discussed in Section 21.16. The failure modes, however, are common to all anchor types. In addition to failure mode, the design must also consider any eccentric or unequal loading of individual anchors.

This chapter deals with the anchors shown in Fig. 21.5. Anchorage of reinforcement is addressed in Chapter 6. Groups of closely spaced straight reinforcing bars or bars anchored by hooks or heads should also be checked for the possibility of concrete breakout.

FIGURE 21.5

Anchor types. (*a*) Cast-in anchors: headed studs and bolts. Post-installed anchors: (*b*) adhesive anchor; (*c*) drop-in type displacement-controlled expansion anchor or undercut anchor; and (*d*) torque-controlled expansion anchors.

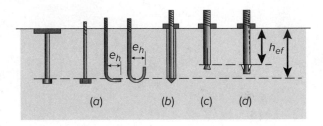

FIGURE 21.6

Anchor failure modes: (*a*) concrete breakout in shear; (*b*) concrete breakout in tension; (*c*) steel failure in shear; (*d*) steel failure in tension; (*e*) concrete pullout in tension; (*f*) bond failure of adhesive; (*g*) combined tension and shear; and (*h*) side-face blowout.

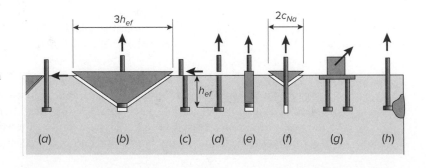

a. Steel Strength in Shear or Tension

When used in design, the shear or tensile strength of the anchor itself (Fig. 21.6c and d) is based on the specified tensile strength of the steel. Strength should be based on the net area of the anchor when threads are present. Tensile strength rather than yield strength is used to calculate anchor capacity to allow for direct comparison with the concrete breakout capacity and to allow individual anchors to yield and distribute load to adjacent anchors.

b. Concrete Breakout in Shear or Tension

Concrete breakout occurs when an anchor in tension generates tensile stresses on the surface of a prism of concrete radiating out from the head of the anchor (Fig. 21.6b). Shear prisms form between the anchor and the edge of the concrete (Fig. 21.6a). When the stresses are high enough for the concrete to fracture, the prism of concrete surrounding the anchor separates from the surrounding concrete. Attachments with more than one anchor, as shown in Fig. 21.6g, may generate overlapping breakout regions. In design, concrete breakout capacity must be modified to account for this overlap.

c. Concrete Pullout in Tension

As the head size on cast-in anchors diminishes, the anchor can pull out by creating a cylinder of concrete directly above the anchor head (Fig. 21.6e). The pullout strength is a function of the area of the head less the area of the shaft. Normally proportioned headed studs and bolts have sufficient head area that concrete breakout usually occurs before concrete pullout can develop.

d. Side-Face Blowout

Anchors with deep embedment and thin side cover may fail by concrete spalling on the side face around the embedded head with no major breakout occurring at the top concrete surface, as shown in Fig. 21.6h.

e. Pryout in Shear

If an attachment similar to that shown in Fig. 21.6g has short anchors, is located away from an edge, and is subjected to high shear load, the plate may bend and the anchors on the back side of the attachment may rotate upward, leading to a pryout failure. Longer anchors are less prone to pryout failures.

f. Combined Tension and Shear

Anchors on some attachments, such as shown in Fig. 21.6g, are subjected to both shear and tensile loading. Determination of the strength of these attachments requires examining the interaction between the effects of the shear and the tensile loads.

g. Bond Failure of Adhesive

Adhesive anchors develop a bond between the anchor, the epoxy, and the concrete. A typical bond failure results in the anchor pulling out due to a concrete breakout around the upper portion combined with an adhesive failure along the lower portion of the anchor, as shown in Fig. 21.6f.

21.3   CONCRETE BREAKOUT CAPACITY

For many years, the design of concrete anchors in tension was based on the assumption that a 45° concrete cone formed at failure (Refs. 21.3 to 21.5). The strength associated with a 45° failure cone correlates with the square of the embedment depth of the anchor h_{ef}^2. Individual anchors were assumed to develop their full tensile capacity if embedded at least 8 times their diameter (Ref. 21.6). The 45° cone failure model, however, was found to both over and underpredict test results, especially for shallow anchors. The variation between predicted behavior and test results led to the development of the *concrete capacity design* (*CCD*) method for determination of concrete breakout capacity (Refs. 21.2 and 21.7). The CCD method evolved after extensive testing and statistical analysis that allowed correlation between strength and key material and geometric parameters. Fracture mechanics models and the experimental research resulted in the observation that a prism with a failure angle of approximately 35°, such as shown in Fig. 21.7, provides a statistically more reliable strength prediction than does the 45° model. The strength of an anchor with a 35° failure cone correlates to the embedment depth of the anchor raised to the 1.5 power $h_{ef}^{1.5}$. Underlying research indicates that the predicted strength becomes unconservative for higher-strength concrete. Consequently, the concrete strength used for calculations is limited to 10,000 psi for cast-in anchors and 8,000 psi for post-installed anchors due to the lack of test data.

The initial research examined anchor behavior in uncracked concrete and concluded with a caution that anchor behavior in cracked concrete may differ considerably from that in uncracked concrete. Cracking can result from early form removal, handling, temperature, and shrinkage. Consequently, all concrete should be assumed cracked unless it is shown by analysis that the concrete is uncracked.

Behavior in cracked concrete is critical to anchors placed in negative moment regions and in structures subjected to earthquake loading (Refs. 21.8 and 21.9). Cracked concrete specimens exhibit a loss of capacity ranging up to 30 percent for cast-in-place bolts and headed studs and 40 to 60 percent for post-installed anchors compared to uncracked concrete specimens. The greater loss in post-installed anchors is attributed to a variation in the ability of the mechanical expansion devices on the

FIGURE 21.7
Single stud breakout prism
indicating a failure angle
of approximately 35°.
(*Photograph by Charles W. Dolan.*)

anchor to fully engage the concrete. Tests of anchors expanded to only 50 percent of their specified limits confirmed this loss in evaluation tests. The observed anchor strength reduction in cracked concrete led to the conclusion that cracked concrete should be the basis for design unless analysis indicates that the concrete will remain uncracked throughout its service life.

Precast members are often assumed to remain uncracked during removal from the forms, lifting and handling, and throughout their design life. An industry study on headed stud anchors in uncracked concrete confirmed the applicability of the concrete capacity design approach and provided an alternative set of equations for the *PCI Design Handbook* (Refs. 21.10 to 21.12). The equations in the *PCI Design Handbook* must be modified when used for anchors in cracked concrete. Research sponsored by PCI concluded that the strength prediction equations were valid for concrete strengths as low as $f_c' = 1000$ psi (Ref. 21.13).

ACI 355.2 *Qualification of Post-Installed Mechanical Anchors in Concrete and Commentary* (Ref. 21.14) provides test acceptance criteria for the certification of post-installed anchors in cracked concrete and ACI 355.4 *Qualification of Post-Installed Adhesive Anchors in Concrete* (Ref. 21.15) provides acceptance criteria for adhesive anchors. Manufacturers use these criteria to establish the anchor capacity and installation conditions associated with their anchors.

Analyses of test results led to the conclusion that the statistically significant variables for predicting anchor capacity in tension are the tensile strength of the concrete and the anchor embedment depth. The basic equations resulting from the analyses are used to establish the mean tensile and shear strengths for individual anchors. The nominal breakout strength of a single anchor is established at the 5 percent fractile.[†] Using the 5 percent fractile, rather than the mean strength, provides an appropriate level of structural reliability (see Section 1.4).

21.4 ANCHOR DESIGN

Attachments usually include multiple anchors arranged in *groups*. The design of attachment anchors involves the calculation of the strength for each anchor in the group based on the possible failure modes and uses the lowest individual strength as the basis for the nominal strength of the *anchor group*. The strength reduction factor corresponding to the failure mode associated with the lowest strength anchor is then used to calculate the design strength for comparison with the factored loads. Tensile and shear failure modes must be checked and, in the event that both tension and shear are present, verification of the capacity due to the interaction of the two loads is required.

Attachment capacity is based on the strength on an individual anchor and then adjusted for anchor groups. The overlap of the concrete breakout prisms for anchors in tension that are placed less than $3h_{ef}$ apart on an attachment or anchors with an edge distance less than $1.5h_{ef}$ results in a concrete breakout capacity that is less than the sum of the individual anchor breakout capacities. The calculated strength of anchor groups placed near an edge requires further modification because concrete may split near individual anchors and the total failure breakout prism may not be mobilized. Adhesive anchors are addressed separately in Section 21.13. Post-installed screw anchors are addressed in Section 21.14 and shear lug attachments in Section 21.17.

[†] Nominal strength for which there is 90 percent confidence that there is a 95 percent probability that the predicted anchor strength will be exceeded by the actual strength.

21.5 ACI CODE PROVISIONS FOR CONCRETE BREAKOUT CAPACITY

Chapter 17 of the ACI Code (Ref. 21.16) presents equations for each possible failure mode. Temporary lifting attachments, specialty attachments, and anchors subject to high cycle fatigue or impact loading are outside the scope of the Code. The Code addresses cast-in headed studs, bolts and hooked rods, post-installed bolts, screw anchors, adhesively bonded anchors, and shear lugs. The Code places particular attention to the concrete breakout capacity of the cast-in and adhesive anchors. The capacity of post-installed mechanical, undercut, and expansion anchors require manufacturers' certification and are only addressed in the Code through the selection of strength reduction factors. The ACI Code requires that anchor and anchor group design be based on design models in substantial agreement with test data. The equations that follow are from the ACI Code and are deemed to satisfy this requirement.

Reinforcement placed within the anchor breakout prism affects anchor performance. *Supplementary reinforcement* is added to reduce spalling and control crack width. Supplementary reinforcement restrains the potential concrete breakout but is

TABLE 21.1
Summary of strength reduction factors for anchors

Type of Anchor	Anchor Category	Anchor Category[a]	Strength Reduction Factor, ϕ	
			Tension	Shear
Strength of steel	Ductile		0.75	0.65
	Brittle[b]		0.65	0.60
Cast-in anchor governed by concrete breakout, bond or side-face blowout	Supplemental steel present		0.75	0.75
	Supplemental steel not present		0.70	0.70
Post-installed or screw anchor governed by concrete breakout, bond, or side-face blowout	Supplemental steel present	Category 1[c]	0.75	0.75
		Category 2[d]	0.65	
		Category 3[e]	0.55	
	Supplemental steel not present	Category 1[c]	0.65	0.70
		Category 2[d]	0.55	
		Category 3[e]	0.45	
Cast-in or screw anchor governed by concrete pullout or pryout strength			0.70	0.70
Post-installed or screw anchor governed by concrete pullout or pryout strength		Category 1[c]	0.65	0.70
		Category 2[d]	0.55	
		Category 3[e]	0.45	

[a] Strength and sensitivity categories for post-installed anchors are established by the ACI 355.2. The effects of variability in anchor torque during installation, tolerance on drilled hole size, and energy level used in setting anchors are considered; for expansion and undercut anchors approved for use in cracked concrete, increased crack widths are considered. ACI 355.4 tests for sensitivity to installation procedures determine the category for a particular adhesive anchor system considering the influence of adhesive mixing and the influence of hole cleaning in dry, saturated, and water-filled/underwater bore holes.

[b] Brittle steel elements have an elongation less than 14 percent when tested by their appropriate ASTM method.

[c] Category 1, low sensitivity to installation and high reliability.

[d] Category 2, medium sensitivity to installation and medium reliability.

[e] Category 3, high sensitivity to installation and low reliability.

TABLE 21.2
Minimum and critical edge distance and anchor spacing[a]

Anchor Type	Minimum Edge Distance[b]	Critical Edge Distance,[c] c_{ac}	Minimum Anchor Spacing
Cast-in	Minimum cover[d]		$4d_a$
Adhesive	$6d_a$	$2h_{ef}$	$6d_a$
Undercut or expansion[e]	$6d_a$	$2.5h_{ef}$	$6d_a$
Torque controlled	$8d_a$	$4h_{ef}$	$6d_a$
Displacement controlled	$10d_a$	$4h_{ef}$	$6d_a$
Screw anchor	$6d_a$	$4h_{ef}$	Greater of $6d_a$ or $0.6h_{ef}$

[a] Minimum and critical edge distances can be reduced if validated by tests following ACI 355.2 or ACI 355.4 protocols.
[b] Minimum edge distance is the greater of the required reinforcement cover for a bar of equivalent size or the value in the table.
[c] The critical edge distance, c_{ac}, is required to develop the basic strength as controlled by concrete breakout or bond of a post-installed anchor in tension in uncracked concrete without supplementary reinforcement.
[d] Cast-in anchors to be torqued require $6d_a$ edge distance and not less that the required cover for a comparable reinforcing bar.
[e] h_{ef} is the lesser of (a) the actual h_{ef}, (b) 2/3 of the slab thickness, h_a, and (c) the slab thickness less 4 in.

not designed to transfer the full design load from the anchors into the structural member. *Anchor reinforcement* is added to transfer the full anchor load into the structural member and is discussed in Section 21.12. When anchor reinforcement is provided, the calculation of concrete breakout capacity is not required.

As for other members in reinforced concrete structures, anchor assemblies are designed so that the design strength, the product of the nominal strength S_n and the strength reduction factor ϕ, exceeds the factored load U or $\phi S_n \geq U$ (see Section 1.4). The strength reduction factor for anchors is a function of the type of anchor, type of loading, the presence of supplementary reinforcement, and the installation conditions. The strength reduction factors listed in Table 21.1 are derived from an analysis of test results. In general, cast-in-place anchors have higher strength reduction factors than post-installed or adhesive anchors. The strength reduction factors for post-installed anchors reflect both the difficulty of installation and the ability of the post-installed anchor to expand properly. Post-installed anchor details are often proprietary and vary between anchor suppliers.

To reduce local cracking and splitting, anchor placement must satisfy the minimum and critical edge distances given in Table 21.2. These distances are based on both the embedment depth h_{ef} and the anchor diameter d_a. Use of an edge distance below that given in Table 21.2 is allowed if the lower distance is validated by tests according to ACI 355.2 or ACI 355.4 and supplementary reinforcement is present. Embedment is slabs requires h_{ef} to be less than or equal to $\frac{2}{3}$ the thickness of the slab for post-installed expansion, screw, or undercut anchors. In all cases, attachments with anchors loaded in shear require a reinforcing bar parallel to the edge between the attachment and the edge.

21.6   STEEL STRENGTH

Nominal steel tensile and shear strengths, N_{sa} and V_{sa}, respectively, are based on the assumption that the anchors will yield and thus distribute the load to all anchors in the group. The calculated nominal strength of each anchor is then based on the specified tensile strength of the steel f_{uta} and the net cross-sectional area of the anchor A_{se}. A subscript extension on the area denotes whether the area is for tension $A_{se,N}$ or shear $A_{se,V}$.

TABLE 21.3

Nominal steel anchor capacity

Anchor Type	Tension Capacity	Shear Capacity
Headed studs	$N_{sa} = nA_{se,N} f_{uta}$	$V_{sa} = nA_{se,V} f_{uta}$
Headed bolts or hooked bolt anchors	$N_{sa} = nA_{se,N} f_{uta}$	$V_{sa} = n0.6A_{se,V} f_{uta}$
Post-installed	Per manufacturer's specifications	Per manufacturer's specifications

ANSI/ASME B1.1 (Ref. 21.18) defines $A_{se,N}$ and $A_{se,V}$ for threaded bolts and hooked bolts as $\frac{\pi}{4}\left(d_a - \frac{0.9743}{n_t}\right)^2$ where d_a is the diameter of the anchor and n_t is the number of threads per inch.

The specified tensile strength is used because the steel used in many anchors does not have a well-defined yield point. Because tensile strength is higher than yield strength, the value most often used in reinforced concrete design, the strength reduction factors for anchors in tension given in Table 21.1 are lower than those associated with steel yielding to be compatible with the higher steel strength and with the load factors given in ASCE 7 (Ref. 21.17). According to ACI Code 17.6.1.2, the specified tensile strength f_{uta} may not exceed the lesser of 1.9 times the specified yield strength of an anchor f_{ya} and 125,000 psi. The later two conditions are to limit displacement of inserts with large post-yield behavior and should not exceed the specified tensile strength. The nominal capacities for steel tension and shear strengths are summarized in Table 21.3 for n anchors in an anchor group.

EXAMPLE 21.1   **Steel tensile strength.**   Determine the nominal steel tensile capacity of the anchor group for the attachment shown in Fig. 21.8. The attachment is anchored with six $\frac{1}{2}$ in. diameter headed studs with $f_{uta} = 65$ ksi.

FIGURE 21.8
Attachment details for
Examples 21.1 through 21.9.

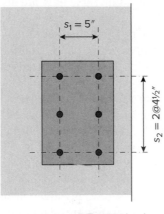

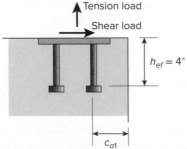

SOLUTION. The area of a $\frac{1}{2}$ in. diameter headed stud is 0.20 in². Based on the equations in Table 21.3, the six studs shown in Fig. 21.8 then have a nominal tensile strength of

$$N_{sa} = nA_{se,N}f_{uta} = 6 \times 0.20 \times 65 = 78.0 \text{ kips}$$

This example uses the ASTM specification for headed studs that specifies the tensile strength as $f_{uta} = 65$ ksi.

EXAMPLE 21.2   **Steel shear strength.**   Determine the nominal steel shear capacity of the anchor group for the attachment shown in Fig. 21.8 if the attachment is anchored with six $\frac{3}{4}$ in. diameter A307 threaded bolts, each with 10 threads per inch and specified tensile strength $f_{uta} = 60$ ksi. There is no grout pad under the plate.

SOLUTION.   From Table 21.3, the effective area of a $\frac{3}{4}$ in. diameter headed bolt is

$$A_{se,V} = \frac{\pi}{4}\left(d_a - \frac{0.9743}{n_t}\right)^2 = \frac{\pi}{4}\left(\frac{3}{4} - \frac{0.9743}{10}\right)^2 = 0.334 \text{ in}^{2\dagger}$$

An A307 bolt has an elongation of 18 percent, thus qualifying as a ductile member. For the six bolts shown in Fig. 21.8, $n = 6$ bolts, giving the nominal shear steel strength as

$$V_{sa} = n0.6A_{se,V}f_{uta} = 6 \times 0.6 \times 0.334 \times 60 = 72.1 \text{ kips}$$

21.7 CONCRETE BREAKOUT CAPACITY OF SINGLE CAST-IN AND POST-INSTALLED, UNDERCUT, AND SCREW ANCHORS

The concrete breakout capacity of an anchor group is based on the breakout capacity of a single anchor in cracked concrete and then adjusted for the effect of anchors in a group. The individual anchor capacity is further modified to account for concrete cracking, the distance to the edge of the member from the closest anchors, eccentricity of the load on the attachment, anchor pullout, and anchor pryout effects.

a.   Tensile Breakout Strength of a Single Anchor

The basic tensile breakout strength of a single anchor N_b in cracked concrete is

$$N_b = k_c\lambda_a\sqrt{f_c'}\,h_{ef}^{1.5} \tag{21.1}$$

where k_c is 24 for cast-in anchors, 17 for post-installed anchors, and λ_a is the modification factor for lightweight concrete in the anchor zone. The value of λ_a is 1.0λ for cast-in and undercut anchors and 0.8λ for expansion, undercut, and screw anchors. Equation (21.1) is limited to anchors with diameters of 4 in. or less due to the lack of test data on larger bolts. ACI Code 17.6.2.2 allows the value of k_c to increase above 17 for post-installed anchors if supported by tests based on an ACI 355.2 or ACI 355.4 evaluation for post-installed anchors that justifies the higher value. In no case is k_c allowed to exceed 24.

---

† The area of a bolt is usually determined from tables such as the AISC *Design Guides*. The calculation given is for illustration purposes.

FIGURE 21.9
Single anchor breakout prism
projections shear and tension.

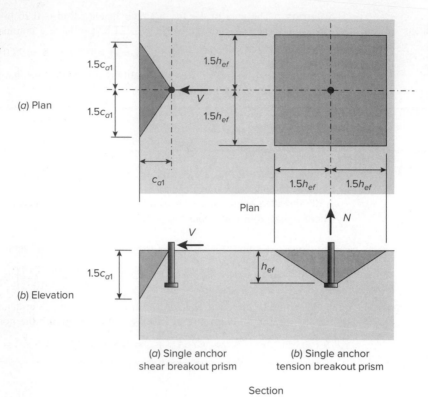

(a) Plan

(b) Elevation

Plan

Section

(a) Single anchor
shear breakout prism

(b) Single anchor
tension breakout prism

Test data for bolts with deep effective embedment lengths indicate a greater strength than predicted by Eq. (21.1). Thus, the basic breakout strength of headed studs or threaded bolts with h_{ef} between 11 and 25 in. may be calculated as

$$N_b = 16\lambda_a \sqrt{f_c'}\, h_{ef}^{5/3} \tag{21.2}$$

b. Shear Breakout Strength of a Single Anchor

The calculated concrete breakout strength in shear of a single anchor in concrete is the lesser of Eqs. (21.3) and (21.4).

$$V_b = \left(7\left(\frac{\ell_e}{d_a}\right)^{0.2}\sqrt{d_a}\right)\lambda_a \sqrt{f_c'}\,(c_{a1})^{1.5} \tag{21.3}$$

where d_a is the diameter of the anchor, c_{a1} is the distance from the edge of the concrete to the center of the shaft of the first anchor (see Fig. 21.9a), and λ_a is defined in Section 21.7a. The value for ℓ_e, the load-bearing length of the anchor for shear, is equal to h_{ef} for anchors with constant stiffness over their full length of embedment or $2d_a$ for torque-controlled expansion bolts separated from the expansion sleeve. In all cases, ℓ_e must be $\leq 8d_a$. For headed studs, headed bolts, or hooked bolts continuously welded to steel plates with a thickness of at least $\frac{3}{8}$ in., the value of 7 in Eq. (21.3) may be increased to 8, provided the capacity does not exceed V_b calculated as

$$V_b = 9\lambda_a \sqrt{f_c'}\,(c_{a1})^{1.5} \tag{21.4}$$

c.   Concrete Breakout Strength of Anchor Groups

The tension breakout strength of a single anchor is based on a 35° breakout prism. This results in a prism of concrete extending $1.5h_{ef}$ from the center of the anchor, as shown in Fig. 21.9b. The tension breakout prism for a single anchor has a single anchor projected area $A_{Nco} = (3h_{ef})^2 = 9h_{ef}^2$. The shear breakout prism extends from the anchor downward toward the free edge, giving a shear breakout area for a single anchor A_{Vco} on the front face of the concrete. For example, $A_{Vco} = 2(1.5c_{a1}$ wide$) \times (1.5c_{a1}$ deep$) = 4.5\,c_{a1}^2$ for the case shown in Fig. 21.9a. ACI Code 17.6.2.1 allows the projected area to be increased if a plate or washer is added to the head of the anchor. The extension of the base of the prism cannot exceed the thickness of the plate or washer.

Anchors spaced less than $3h_{ef}$ apart have overlapping breakout prisms, as shown in Fig. 21.10. Experimental evaluation of these groups indicates that the strength of groups in tension can be accounted for by multiplying the breakout strength of a single anchor N_b by the ratio of the projected breakout area of the anchor group A_{Nc} to the projected breakout area of a single anchor A_{Nco} (Ref. 21.19). Calculation of the anchor group breakout prism projected area is affected by both the spacing of the anchors and the distance to the edge of the concrete. Calculation of tensile strength is a function of the anchor embedment depth h_{ef}, the distances to the edges c_{a1} and c_{a2}, where c_{a2} is the distance from center of an anchor shaft to the edge of the concrete in the direction perpendicular to c_{a1}, and the anchor spacing s_1. Typical conditions for anchors in tension are illustrated in Fig. 21.10a through c.

Similar to the approach taken for tension breakout, the shear capacity of individual anchors in a group is obtained by multiplying the shear capacity of a single anchor by the ratio of the projected area of the shear anchor group A_{Vc} to the projected area for a single anchor A_{Vco}. Anchors placed perpendicular to the edge of the concrete offer two possible failure modes. The first mode is based on the anchor nearest the edge carrying half the load and failing first (Fig. 21.10f). The second mode is based on the anchor closest to the edge "riding with the failure breakout" with the entire load carried by the anchor farthest from the edge (Fig. 21.10g). Where the load is placed symmetrically on two or more anchors, the projected area is a function of the slab thickness h_a, the anchor spacing s_1, and the distance from the edge c_{a1} (Fig. 21.10h). For slabs with a thickness greater than c_{a1}, h_a is replaced with c_{a1}, as shown in Fig. 21.10g.

Anchor strength is further modified for the condition of the concrete immediately around the individual anchor. Modification factors for anchor concrete breakout strength (presented next in Sections 21.7d and e) are designated as ψ, followed by subscripts defining the particular condition and direction of loading. The subscript c indicates modification for cracked or uncracked concrete, the subscript ed for edge distance, the subscript ec for load eccentricity, the subscript cp for anchor pullout, and the subscript h for the slab thickness. Combining these effects, the concrete breakout capacity for a single anchor in tension is

$$N_{cb} = \frac{A_{Nc}}{A_{Nco}}\, \psi_{ed,N}\, \psi_{c,N}\, \psi_{cp,N}\, N_b \qquad (21.5)$$

and for an anchor group in tension,

$$N_{cbg} = \frac{A_{Nc}}{A_{Nco}}\, \psi_{ec,N}\, \psi_{ed,N}\, \psi_{c,N}\, \psi_{cp,N}\, N_b \qquad (21.6)$$

Tension Breakout Areas A_{Nc}
Plan

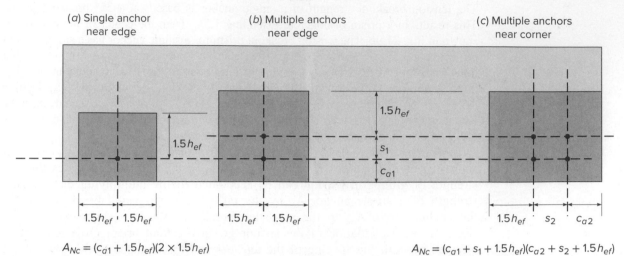

(a) Single anchor near edge

$A_{Nc} = (c_{a1} + 1.5h_{ef})(2 \times 1.5h_{ef})$
If $c_{a1} < 1.5h_{ef}$

(b) Multiple anchors near edge

$A_{Nc} = (c_{a1} + s_1 + 1.5h_{ef})(2 \times 1.5h_{ef})$
If c_{a1} and $s_1 < 1.5h_{ef}$

(c) Multiple anchors near corner

$A_{Nc} = (c_{a1} + s_1 + 1.5h_{ef})(c_{a2} + s_2 + 1.5h_{ef})$
If c_{a1}, c_{a2}, s_1 and $s_2 < 1.5h_{ef}$

Shear Breakout Areas A_{Vc}
Plan

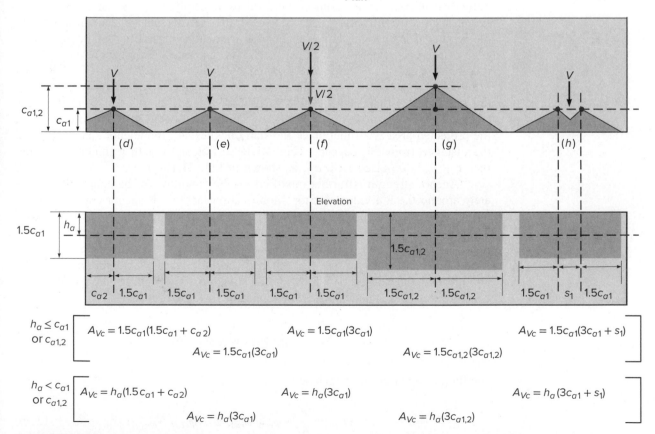

Elevation

$h_a \leq c_{a1}$ or $c_{a1,2}$

$A_{Vc} = 1.5c_{a1}(1.5c_{a1} + c_{a2})$

$A_{Vc} = 1.5c_{a1}(3c_{a1})$

$A_{Vc} = 1.5c_{a1}(3c_{a1})$

$A_{Vc} = 1.5c_{a1,2}(3c_{a1,2})$

$A_{Vc} = 1.5c_{a1}(3c_{a1} + s_1)$

$h_a < c_{a1}$ or $c_{a1,2}$

$A_{Vc} = h_a(1.5c_{a1} + c_{a2})$

$A_{Vc} = h_a(3c_{a1})$

$A_{Vc} = h_a(3c_{a1})$

$A_{Vc} = h_a(3c_{a1,2})$

$A_{Vc} = h_a(3c_{a1} + s_1)$

FIGURE 21.10
Calculation of anchor group breakout projected areas for tension and shear near edges.

The concrete breakout capacity in shear for a single anchor loaded *perpendicular* to the edge is

$$V_{cb} = \frac{A_{Vc}}{A_{Vco}} \, \psi_{ed,V} \, \psi_{c,V} \, \psi_{h,V} \, V_b \tag{21.7}$$

and for a group of anchors loaded perpendicular to the edge,

$$V_{cbg} = \frac{A_{Vc}}{A_{Vco}} \, \psi_{ec,V} \, \psi_{ed,V} \, \psi_{c,V} \, \psi_{h,V} \, V_b \tag{21.8}$$

ACI Code 17.7.2.1 permits shear forces *parallel* to an edge to be computed as twice the value determined in Eqs. (21.7) and (21.8).

In some instances, such as anchors subjected to tension near the edges of walls or other locations where there is less than $1.5h_{ef}$ from three surfaces or anchors subjected to shear in narrow sections of limited thickness with both edge distances c_{a2} and the member thickness less than $1.5c_{a1}$, using the ratios of projected areas results in an overestimation of strength. In these situations, an adjusted effective embedment depth or edge distance is used in addition to the ratio of breakout areas to bring the calculated nominal strengths in line with test results. ACI Code 17.6.2.1.2 and 17.7.2.1.2 provide guidance for these conditions.

d. Modification Factors for Concrete Cracking, Edge Distance, and Slab Thickness

Research cited in Refs. 21.2 and 21.19 indicates that anchor capacity will be reduced if the concrete is cracked at service load. For this reason, Eqs. (21.1) through (21.4) were developed assuming cracked concrete. If analysis indicates that the concrete is uncracked under service load, then both the tensile and shear capacity may be increased. Modification factors for anchors placed in uncracked concrete are summarized in Table 21.4.

TABLE 21.4
Breakout modification factors for concrete cracking near an anchor

Concrete Condition	Tension $\psi_{c,N}$ Factor	Shear[a] $\psi_{c,V}$ Factor
Cracked	1.0	1.0
Cracked		1.2^b
Cracked		1.4^c
Uncracked cast-in	1.25	1.4
Uncracked post-installed and k_c is 17	1.4	
Uncracked and capacity determined by ACI 355.2	1.0	

[a] ACI Code does not differentiate between cast-in and post-installed anchors for shear.

[b] ACI Code requires a No. 4 (No. 13) or larger bar between the anchors and the edge of concrete.

[c] ACI Code requires a No. 4 (No. 13) or larger bar between the anchors and the edge and enclosed in stirrups not more than 4 in. on centers.

TABLE 21.5
Breakout modification factors for edge distance

Condition	Tension		Shear
	$\psi_{ed,N}$ [a]	$\psi_{ed,Na}$ [b]	$\psi_{ed,V}$
$c_{a,min}$ [c] $\geq 1.5\,h_{ef}$ $c_{a,min} \geq 1.5\,c_{Na}$	1.0	1.0	
$c_{a,min} < 1.5\,h_{ef}$ $c_{a,min} < 1.5\,c_{Na}$	$0.7 + 0.3\,\dfrac{c_{a,min}}{1.5h_{ef}}$	$0.7 + 0.3\,\dfrac{c_{a,min}}{c_{Na}}$	
$c_{a2} \geq 1.5\,c_{a1}$			1.0
$c_{a2} \leq 1.5\,c_{a1}$			$0.7 + 0.3\,\dfrac{c_{a2}}{1.5c_{a1}}$

[a] For cast-in or post-installed anchors.
[b] For adhesive anchors.
[c] Minimum distance from center of anchor shaft to edge of concrete.

The strength of an individual anchor near one or more edges must be further adjusted due to localized cracking. Modification factors for anchors near free edges are summarized in Table 21.5 using the notation illustrated in Fig. 21.10. In situations where the slab thickness h_a is less than $1.5c_{a1}$, a further shear edge distance modification of $\psi_{h,V} = \sqrt{1.5c_{a1}/h_a} \leq 1.0$ is required.

EXAMPLE 21.3   **Concrete breakout strength.**   Determine the concrete tensile breakout capacity of the anchor group in Fig. 21.8, given that the load is concentrically applied and the attachment is in concrete that analysis indicates is cracked during service load. The anchors are cast in 5000 psi normalweight concrete with six $\frac{1}{2}$ in. diameter headed studs and $c_{a1} = 8$ in.

Solution.   From Fig. 21.8, $s_1 = 5$ in., $s_2 = 4.5$ in., and $h_{ef} = 4$ in. For normalweight concrete $\lambda = \lambda_a = 1.0$. From Table 21.4 for cracked concrete, $\psi_{c,N} = 1.0$, the pullout modification factor is $\psi_{cp,N} = 1.0$, and because the value of c_{a1} is greater than $1.5h_{ef}$, from Table 21.4 $\psi_{ed,N} = 1.0$. The load is concentric, so $\psi_{ec,N} = 1.0$. The attachment has cast-in headed anchors resulting in a value of $k_c = 24$.

The projected area of a single anchor is

$$A_{Nco} = 9\,h_{ef}^2 = 9 \times 4^2 = 144 \text{ in}^2$$

The projected area of the anchor group is

$$A_{Nc} = (3h_{ef} + s_1) \times (3h_{ef} + 2s_2) = (3 \times 4 + 5) \times (3 \times 4 + 2 \times 4.5) = 357 \text{ in}^2$$

From Eq. (21.1), the tensile concrete breakout strength of a single anchor is

$$N_b = k_c\lambda_a \sqrt{f_c'}\, h_{ef}^{1.5} = 24 \times 1.0 \sqrt{5000} \times 4^{1.5} = 13,580 \text{ lb}$$

Lastly, from Eq. (21.5)

$$N_{cbg} = \frac{A_{Nc}}{A_{Nco}}\, \psi_{ec,N}\, \psi_{ed,N}\, \psi_{c,N}\, \psi_{cp,N}\, N_b = \frac{357}{144} \times 1.0 \times 1.0 \times 1.0 \times 1.0 \times 13.58 = 33.7 \text{ kips}$$

EXAMPLE 21.4 **Concrete breakout strength.** Determine the required length of six $\frac{1}{2}$ in. diameter headed studs shown in Fig. 21.8 for a factored tensile load of 35 kips and the conditions given in Example 21.3 are present.

SOLUTION. From Table 21.1, the ϕ-factor for a cast-in headed ductile anchor is 0.75. Thus, the required nominal strength of the anchor group is $N_u/\phi = 35/0.75 = 46.7$ kips. Using the information from Example 21.3, $A_{Nco} = 9h_{ef}^2$, and after combining terms, $A_{Nc} = 9h_{ef}^2 + (6s_1 + 3s_2)h_{ef} + 2s_1s_2$. The tensile capacity of a single anchor is $N_b = k_c\lambda_a \sqrt{f_c'} \, h_{ef}^{1.5} = 24 \times 1.0 \sqrt{5000} \times h_{ef}^{1.5} = 1697 \, h_{ef}^{1.5}$. Combining the terms and inserting the values for $s_1 = 5$ in. and $s_2 = 4.5$ in. in Eq. (21.6) gives

$$N_{cbg} = N_u/\phi = (A_{Nc}/A_{Nco}) \, \psi_{ec,N} \, \psi_{ed,N} \, \psi_{c,N} \, \psi_{cp,N} \, N_b$$

$$46{,}700 = [(9h_{ef}^2 + (6 \times 5 + 3 \times 4.5)h_{ef} + 2 \times 5 \times 4.5)/9h_{ef}^2] \times 1.0 \times 1.0 \times 1.0 \times 1.0 \times 1697 h_{ef}^{1.5}$$

Using an equation solver or by trial, h_{ef} must be at least 5.85 in. Use $h_{ef} = 6$ in. Headed studs are available in $\frac{1}{2}$ in. and 1 in. increments. Thus, an iterative solution is often easier than finding a closed-form solution; otherwise an Excel or Mathcad calculation sheet with h_{ef} as a variable provides an effective solution tool.

EXAMPLE 21.5 **Shear breakout strength.** Determine the shear breakout capacity of the anchor group shown in Fig. 21.8. The load is concentrically applied, the attachment has six $\frac{3}{4}$ in. diameter headed bolts cast in 5000 psi normalweight concrete that analysis indicates is uncracked during service load, $h_{ef} = 4.5$ in., and $c_{a1} = 8$ in.

SOLUTION. From Fig. 21.8, $s_1 = 5.0$ in., $s_2 = 4.5$ in., and $h_{ef} = 4$ in. For normalweight concrete, λ and $\lambda_a = 1.0$. From Table 21.4 for uncracked concrete and cast-in anchors, $\psi_{c,V} = 1.2$, and because the value of c_{a2} is greater than c_{a1} by inspection, from Table 21.5 $\psi_{ed,V} = 1.0$. The load is concentric, so $\psi_{ec,V} = 1.0$, and the concrete thickness is greater than $1.5 \, c_{a1}$ so $\psi_{h,V} = 1.0$.

Assume that one-half of the load is carried by the front row of anchors. The projected area of a single anchor is

$$A_{Vco} = 4.5c_{a1}^2 = 4.5 \times 8^2 = 288 \text{ in}^2$$

The projected area of the anchor group is

$$A_{Nc} = 1.5c_{a1}(3c_{a1} + 2s_2) = 1.5 \times 8 \times (3 \times 8 + 2 \times 4.5) = 396 \text{ in}^2$$

For a single anchor, $d_a = 0.75$ in. and $\ell_e = h_{ef}$. Then from Eq. (21.3), the shear strength of a single anchor is

$$V_b = \left(7\left(\frac{\ell_e}{d_a}\right)^{0.2} \sqrt{d_a}\right) \lambda_a \sqrt{f_c'} \, (c_{a1})^{1.5} = \left(7 \left(\frac{4}{0.75}\right)^{0.2} \sqrt{0.75}\right) 1.0 \sqrt{5000}(8)^{1.5} = 13{,}560 \text{ lb}$$

The anchor group capacity is

$$\frac{V_{cbg}}{2} = \frac{A_{Vc}}{A_{Vco}} \, \psi_{ec,V} \, \psi_{ed,V} \, \psi_{c,V} \, \psi_{h,V} \, V_b = \frac{396}{288} \, 1.0 \times 1.0 \times 1.2 \times 1.0 \times 13.56 = 22.4 \text{ kips}$$

Thus,

$$V_{cbg} = 44.8 \text{ kips}$$

The strength of the anchor must also be checked for the case in which the back row of anchors carries the entire load. For this condition the edge distance c_{a1} increases to $8 + 5$ in. $= 13$ in. With this change,

$$A_{Vco} = 4.5 \, c_{a1}^2 = 4.5 \times 13^2 = 760.5 \text{ in}^2$$

$$A_{Vc} = 1.5 \, c_{a1} \, (3c_{a1} + 2s_2) = 1.5 \times 13 \times (3 \times 13 + 2 \times 4.5) = 936 \text{ in}^2$$

$$V_b = \left(7\left(\frac{\ell_e}{d_a}\right)^{0.2} \sqrt{d_a}\right)\lambda_a \sqrt{f_c'} \, (c_{a1})^{1.5} = \left(7\left(\frac{4}{0.75}\right)^{0.2} \sqrt{0.75}\right)1.0 \sqrt{5000}(13)^{1.5} = 28{,}080 \text{ lb}$$

$$V_{cbg} = \frac{A_{Vc}}{A_{Vco}} \, \psi_{ec,V} \, \psi_{ed,V} \, \psi_{c,V} \, \psi_{h,V} \, V_b = \frac{936}{760.5} \times 1.0 \times 1.0 \times 1.2 \times 1.0 \times 28.08 = 41.5 \text{ kips}$$

Thus, the second condition limits the nominal strength of the anchor group to $V_{cbg} = 41.5$ kips. The anchor group must be detailed with a No. 4 (No. 13) bar or larger between the edge and the anchor to be consistent with the assumed value of $\psi_{c,v}$.

The solution to Example 21.5 is consistent with the code requirements. At the same time, the holes in the plate are designed to fit over the cast-in bolts and there are tolerances on the hole size. Consequently, some designers recommend only using the bolts closest to the edge to establish the anchor group capacity under the assumption that the hole tolerance may not allow the back holes to completely engage the bolts. Under this restriction, the capacity of the attachment in Example 21.5 would be 22.4 kips.

e.   Modification for Eccentrically Applied Loads

The strength of an anchor group is limited by the strength of the most severely loaded anchor. Eccentrically loaded attachments result in loads being redistributed such that some anchors are more severely loaded, as shown in Fig. 21.11. Depending on the eccentricity of the normal load or the magnitude of the moment, anchors may remain

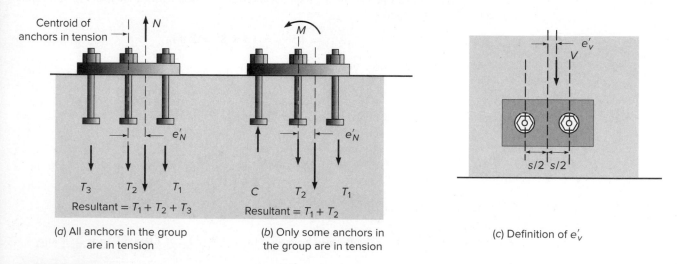

(a) All anchors in the group are in tension

(b) Only some anchors in the group are in tension

(c) Definition of e_V'

FIGURE 21.11
Definition of loading eccentricity: (a) e_N' when all anchors are in tension and (b) e_N' for when one or more anchors are in compression and (c) e_V' for shear loadings.

TABLE 21.6
Breakout modification factors for eccentricity

Condition	Tension		Shear
	$\psi_{ec,N}{}^a$	$\psi_{ec,Na}{}^b$	$\psi_{ec,V}$
Anchor group loaded eccentrically in tension	$\dfrac{1}{\left(1 + \dfrac{2e'_N}{3h_{ef}}\right)} \le 1.0$	$\dfrac{1}{\left(1 + \dfrac{e'_N}{e_{Na}}\right)} \le 1.0$	
Anchor group loaded eccentrically in shear			$\dfrac{1}{\left(1 + \dfrac{2e'_V}{3h_{ef}}\right)} \le 1.0$

^{a} For cast-in, post-installed, undercut, and screw anchors
^{b} For adhesive anchors

in tension (Fig. 21.11a) or be placed in compression (Fig. 21.11b). In either case, the anchor to the right in the figure carries a higher load than the remaining anchors and will fail first. The capacity of the anchor or the anchor group in concrete breakout in tension or in shear must be modified to correct for this load redistribution. Table 21.6 summarizes the eccentricity modification factors. When adjusting for eccentricity, the values of e'_N and e'_V, respectively, are calculated only for those anchors in tension or those anchors loaded toward the edge in shear. In cases where the eccentricity occurs in two orthogonal directions, the modification factor is calculated for each direction and the product of the two factors used in Eqs. (21.6) and (21.8).

21.8 PULLOUT STRENGTH OF ANCHORS

The pullout strength of a single cast-in, post-installed, undercut, or screw anchor in tension is based on the head of the anchor pulling or slipping through the concrete, creating a cylindrical failure (Fig. 21.6e). For cast-in, post-installed, undercut, and screw anchors, the nominal pullout strength in tension N_{pn} is

$$N_{pn} = \psi_{c,p}\, N_p \tag{21.9}$$

where N_p is the strength of an individual anchor given in Eq. (21.10) or (21.11) (Refs. 21.20 and 21.21) and in concrete when analysis indicates no cracking $\psi_{c,p} = 1.4$, otherwise $\psi_{c,p} = 1.0$. The individual anchor strength is based on the net area of the anchor head directly bearing in the concrete A_{brg} or on the geometry of the anchor. A_{brg} is typically calculated as the area of the head of the bolt or stud less the area of the shaft. The anchor capacity for a single headed stud or bolt adjusted to the 5 percent fractile is

$$N_p = 8A_{brg}\, f'_c \tag{21.10}$$

and for a hooked bolt is

$$N_p = 0.9f'_c e_h d_a \tag{21.11}$$

where e_h is the distance from the inner surface of the shaft of a J- or L-bolt to the outer tip of the J- or L-bolt and d_a is the diameter of the hooked bolt. The ACI Code requires that e_h, illustrated in Fig. 21.5a, meet the requirement $3d_a \leq e_h < 4.5d_a$. The pullout strength of a post-installed anchor is supplied by the manufacturer based on an evaluation performed in accordance with ACI 355.2 and reduced to the 5 percent fractile.

The pullout strength is further modified depending on whether the anchor is in cracked or uncracked concrete. For anchors located in areas where the analysis indicates the concrete is cracked, $\psi_{c,p} = 1.0$, and where the analysis indicates the concrete is uncracked, $\psi_{c,p} = 1.4$.

EXAMPLE 21.6   **Pullout strength.**   Calculate the pullout strength of an anchor in Example 21.3, where A_{brg} is 0.589 in^2.

SOLUTION. $\psi_{cp} = 1.0$ for cracked concrete, so the pullout strength of one anchor is

$$N_{pn} = N_p = 8A_{brg} f_c' = 8 \times 0.589 \times 5000 = 23.6 \text{ kips}$$

21.9   SIDE-FACE BLOWOUT

Anchors with deeper embedment but thinner side cover may fail by concrete spalling on the side face around the embedded head while no major breakout occurs at the top concrete surface, as illustrated in Fig. 21.6h. For a single headed anchor with deep embedment close to an edge ($h_{ef} > 2.5c_{a1}$), the nominal side-face blowout strength N_{sb} is given in ACI Code 17.4.4 as

$$N_{sb} = 160 \, c_{a1} \sqrt{A_{brg}} \, \lambda_a \, \sqrt{f_c'} \tag{21.12}$$

If c_{a2} for the single headed anchor is less than $3c_{a1}$, the value of N_{sb} is modified by the factor $(1 + c_{a2}/c_{a1})/4$ where the ratio c_{a2}/c_{a1} must be greater than 1 and is limited to be less than or equal to 3. For headed anchor groups with deep embedment close to an edge ($h_{ef} > 2.5c_{a1}$) and with anchor spacing less than $6c_{a1}$, the nominal strength of those anchors susceptible to a side-face blowout failure N_{sbg} is

$$N_{sbg} = 1 + \left(\frac{s}{6c_{a1}} \right) N_{sb} \tag{21.13}$$

where s is the distance between the anchors nearest the edge and N_{sb} is given in Eq. (21.12) without modification for a perpendicular edge distance.

EXAMPLE 21.7   **Side-face blowout strength.**   Calculate the side blowout strength of the anchor in Example 21.4 where $h_{ef} = 6$ in., $s = 4.5$ in., $c_{a1} = 8$ in., and $A_{brg} = 0.589$ in^2.

SOLUTION.   The spacing modification factor for the anchors in bearing from Eq. (21.13) is

$$\left(1 + \frac{s_1}{6c_{a1}} \right) = \left(1 + \frac{4.5}{6 \times 8} \right) = 1.09$$

The value for N_{sb} from Eq. (21.12) is

$$N_{sb} = 160 c_{a1} \sqrt{A_{brg}} \, \lambda_a \, \sqrt{f_c'} = 160 \times 8 \sqrt{0.589} \times 1.0 \times \sqrt{5000} = 69.5 \text{ kips}$$

The side blowout strength of a single anchor exceeds the steel tensile strength of a single anchor calculated in Example 21.1, so side blowout is not a limiting condition for this anchor group. The spacing modification factor is greater than 1.0, so no further calculations are necessary.

21.10   PRYOUT OF ANCHORS

Pryout is a phenomenon that occurs with short anchors for attachments loaded in shear (Ref. 21.13). As an anchor group moves laterally, the individual anchors can fail due to a shear failure of the steel, concrete breakout, or rotation of short anchors, in which the end of the anchor displaces in the direction opposite to the applied shear force, a phenomenon known as pryout. The nominal pryout strength for a single anchor V_{cp} is given in ACI Code 17.7.3 as

$$V_{cp} = k_{cp}N_{cp} \tag{21.14}$$

The pryout strength for a group of anchors V_{cpg} is

$$V_{cpg} = k_{cp}N_{cpg} \tag{21.15}$$

For cast-in, expansion, and undercut anchors, N_{cp} and N_{cpg} may be taken as N_{cb} or N_{cbg} from Eqs. (21.5) and (21.6), respectively. For both single anchors and anchor groups, $k_{cp} = 1.0$ for h_{ef} less than 2.5 in. and $k_{cp} = 2.0$ for h_{ef} greater than or equal to 2.5 in.

EXAMPLE 21.8   **Pryout strength.**   Calculate the pryout strength of the anchor group in Example 21.5.

SOLUTION. h_{ef} is greater than 2.5 in., so $k_{cp} = 2.0$. The results from Example 21.3 provide an anchor group capacity of $N_{cbg} = 33.7$ kips. The pryout capacity from Eq. (21.15) is then

$$V_{cpg} = k_{cp}N_{cbg} = 2 \times 33.7 = 67.4 \text{ kips}$$

EXAMPLE 21.9   **Summary of anchor group capacity.**   Determine the nominal and design capacity for the anchor group in Fig. 21.8.

SOLUTION.   The nominal failure loads for each possible condition are summarized in Table 21.7 using the results from Examples 21.1, 21.3, 21.5, 21.6, 21.7, and 21.8. From Table 21.7, the lowest nominal capacity in tension is 33.7 kips and results from tension concrete breakout, as indicated in the shaded cells. The corresponding design capacity is 21.5 kips when the associated strength reduction factor is applied. The nominal and design capacities in shear are 41.5 kips and 29.1 kips, respectively.

TABLE 21.7
Summary of failure modes

Failure Mode	Example	Nominal Tension (kips)	Strength Reduction Factor, ϕ	Design Tension (kips)	Example	Nominal Shear (kips)	Strength Reduction Factor, ϕ	Design Shear (kips)
Steel	21.1	78.0	0.75	58.5	—	46.8	0.65	30.4
Concrete breakout	21.3	33.7	0.70	21.5	21.5	41.5	0.70	29.1
Pullout	21.5	141	0.70	98.0				
Side-face blowout	21.7	69.5	0.70	48.6				
Pryout					21.8	67.4	0.70	47.2

The possibility exists, due to variation in strength reduction factors, that the governing load could be based on nominal strength in one failure mode and design strength in another failure mode. The ACI Code requires the design strength be less than or equal to the factored load; therefore, the lowest *design* load, regardless of lowest nominal strength failure mode, governs the design.

21.11   COMBINED SHEAR AND NORMAL FORCE

Attachments such as the corbels shown in Figs. 21.1 and 21.2 generate both tensile and shear forces in the anchors. Experimental studies indicate that the interaction can be represented using a curvilinear relationship, such as shown in Fig. 21.12 (Ref. 21.22). ACI Code 17.8 simplifies this curvilinear relationship by using a trilinear approximation of the interaction behavior. The trilinear relationship allows the full tensile strength of the anchor to be used if $V_{ua}/\phi V_n$ is less than 0.2 and the full shear strength to be used if $N_{ua}/\phi N_n$ is less than 0.2. Between these two limits the load combination must satisfy Eq. (21.16).

$$\frac{N_{ua}}{\phi N_u} + \frac{V_{ua}}{\phi V_u} \le 1.2 \tag{21.16}$$

FIGURE 21.12
Shear and tensile load interaction.

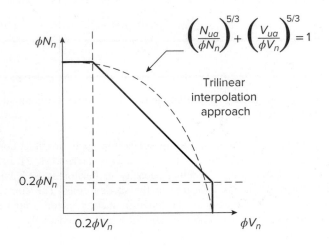

$$\left(\frac{N_{ua}}{\phi N_n}\right)^{5/3} + \left(\frac{V_{ua}}{\phi V_n}\right)^{5/3} = 1$$

Trilinear interpolation approach

EXAMPLE 21.10   **Anchor design example.**   Design the anchor group for the attachment shown in Fig. 21.13 using $\frac{5}{8}$ in. diameter headed studs with $A_{brg} = 0.59$ in.^{2} and $f_{uta} = 65$ ksi. The normalweight concrete has a compressive strength of 3500 psi and analysis indicates that concrete will remain uncracked during the service live. The corbel carries a factored load of 35 kips. The attachment is fabricated from $\frac{3}{4}$ in. thick plate, is located at far enough from any edge to preclude edge effects, and no supplementary reinforcement is present.

SOLUTION.   The load is resisted by four studs in shear, and the tension generated by the load eccentricity is resisted by the top two studs. With no supplementary reinforcement, $\phi = 0.70$ from Table 21.1. With no edge distance issues, eccentricity, or pullout restrictions, $\psi_{ed,N}$ and $\psi_{cp,N}$ equal 1.0. The concrete is uncracked, so $\psi_{c,N}$ is 1.25 from Table 21.4.

FIGURE 21.13
Steel attachment for
Example 21.10.

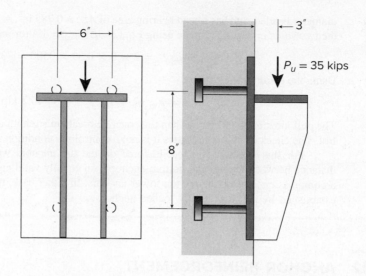

The shear capacity is controlled by the steel strength of the anchors. Thus, the design shear capacity for four anchors is

$$V_n = A_{se,V} f_{uta} = 4 \times 0.31 \times 65 = 80.6 \text{ kips}$$

which is greater than the 35 kip applied load. Since c_{a1} is at least 10 in., pryout need not be checked.

The tensile load on the top two anchors is assumed to be resolved from a couple between the studs. Thus, for a vertical stud spacing of $s = 8$ in.,

$$N_u = P_u \times \text{eccentricity}/s = (35 \times 3)/8 = 13.1 \text{ kips}$$

for which, $N_N = N_u/\phi = 13.1/0.70 = 18.75$ kips.

If the tensile strength capacity is controlled by tensile concrete breakout of the two top anchors, then for a spacing between the studs of $s = 6$ in.,

$$A_{Nco} = 9h_{ef}^2 = 9 \times 8^2 = 576 \text{ in}^2.$$

and

$$A_{Nc} = (3h_{ef} + s)\,3h_{ef} = 9h_{ef}^2 + 18h_{ef}$$

Combining Eqs. (21.1) and (21.5) gives

$$N_{cbg} = \frac{A_{Nc}}{A_{Nco}}\,\psi_{ed,N}\,\psi_{c,N}\,\psi_{cp,N}\,k_c\lambda_a\,\sqrt{f_c'}\,h_{ef}^{1.5}$$

$$18.75 = \frac{9h_{ef}^2 + 18h_{ef}^2}{9h_{ef}^2}\,1.0 \times 1.25 \times 1.0 \times 24 \times 1.0\,\sqrt{3500} \times h_{ef}^{1.5}$$

Solving by trial for the required embedment depth gives $h_{ef} = 4$ in. and $N_{cbg} = N_N = 21.3$ kips. Because both shear and tensile forces are present, the anchors must be checked for combined effects. Then from Eq. (21.16),

$$\frac{N_u}{\phi N_N} + \frac{V_u}{\phi V_N} = \frac{13.1}{0.70 \times 21.3} + \frac{35}{0.70 \times 80.6} = 1.50$$

This exceeds the code requirement of 1.2, so the embedment length must be increased. Using a 5 in. embedment length and the plate thickness of $\frac{3}{4}$ in. gives $h_{ef} = 5.75$ in., $N_N = 33.0$ kips, and a combined ratio from Eq. (21.16) of 1.19, less than the maximum values of 1.2. A $\frac{5}{8}$ in.

diameter headed stud has a head bearing area of $A_{brg} = 0.785$ in². A check of pullout capacity of the two studs in tension is made using Eq. (21.10). $\psi_{c,p} = 1.4$ for uncracked concrete.

$$N_p = n8A_{brg} f_c' = 2 \times 8 \times 0.59 \times 3500 = 33.0 \text{ kips}$$

Using the result in Eq. (21.9),

$$N_{pn} = \psi_{c,p} N_p = 1.4 \times 33.0 = 46.2 \text{ kips}$$

The pullout strength is greater than the concrete breakout strength and, therefore, does not control. The check of pryout indicates it is not a controlling condition.

For this solution, the tensile force on the top anchors was calculated based on the distance between the top and bottom anchors. An equally valid approach would be to assume a compression centroid below the lower anchors. In either case, the anchor group capacity is established by the most highly loaded anchor.

21.12 ANCHOR REINFORCEMENT

The ACI Code identifies two types of reinforcement for use with anchors. *Supplementary reinforcement* assists in controlling crack width and preventing spalling. *Anchor reinforcement* transfers the total factored load to the supporting structure. To be effective, the anchor reinforcement must be aligned with the direction of the applied load and be developed in both the concrete breakout zone associated with the anchor and the underlying concrete, as shown in Fig. 21.14. The detailing shown in Fig. 21.14 requires that the anchor reinforcement be placed close to the surface for shear loads, commensurate with the cover requirements. Hairpin

FIGURE 21.14
Anchor reinforcement for (*a*) tension and (*b*) shear.

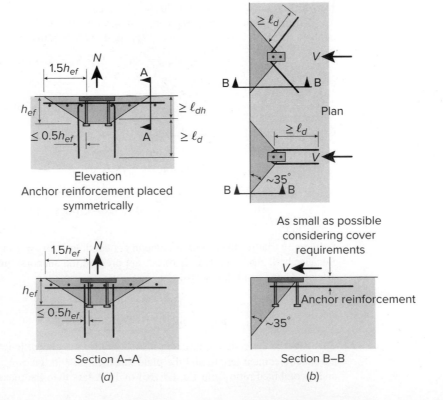

bars are often used for this purpose. Attachments in plastic hinge regions, or areas where analysis indicates substantial cracking may be present, must be detailed to include anchor reinforcement.

ACI Code 17.5.2.1 states that when anchor reinforcement is used, a calculation of the concrete breakout capacity is not required. In many instances, however, the addition of anchorage reinforcement is not practical. For example, examining Fig. 21.14, an anchor group in a thin slab loaded normal to the surface may not have sufficient length available below the breakout prism for the development of the anchor reinforcement. Similarly, the development length of a hook for a No. 5 (No. 16) or larger bar may exceed the embedment length of the anchor. Anchor reinforcement is useful in shear where the studs may be captured by a hairpin, as shown in Fig. 21.14b, in lieu of having to develop a hook. A strength reduction factor of $\phi = 0.75$ is used when determining the area of the anchor reinforcement.

21.13 ADHESIVE ANCHORS

Following the collapse of the ceiling panels in the "Big Dig" in Boston, the National Transportation Safety Board requested ACI to develop criteria for adhesive anchors (Ref. 21.23). Adhesive anchors are sensitive to a number of factors, including installation temperature, moisture, and sustained loading. To provide uniformity of installation and use, anchor systems must be qualified in accordance with procedures described in ACI 355.4 *Acceptance Criteria for Qualification of Post-Installed Adhesive Anchors in Concrete* (Ref. 21.15) and must be installed by qualified technicians. The ACI Code requires anchors installed horizontally or in an upward sloping orientation to be subject to continuous inspection during construction.

Adhesive anchors have failure modes similar to cast-in or post-installed anchors in addition to the possibility of a bond failure (Refs. 21.24 and 21.25). Bond failures arise when the adhesive undergoes a shear failure between the hardened adhesive and the concrete. This results in a bond pullout failure accompanied by a concrete breakout near the surface, as shown in Fig. 21.6f. ACI Code 17.5.2 places a number of restrictions on the installation of the anchor so that the performance of adhesive anchors correlates with the qualification tests of ACI 355.4. These restrictions include:

> *The minimum concrete age is 21 days.* This provision allows moisture to be used in the hydraulic cement reaction and not be available to disrupt the adhesive cure.

> *A concrete strength that is equal to or greater than 2500 psi.* Qualification tests are conducted using concrete with compressive strengths of at least 2500 psi and data on lower strength is very limited.

> *Rotary impact or rock drills must be used to drill the holes for adhesive anchors.* These tools create a rough irregular surface to improve mechanical interlock between the adhesive and the concrete. Holes made with coring bits are much smoother and have less interlock.

> *Installation temperature is at least 50°F.* This minimum temperature is required for the adhesive to cure properly.

Adhesive anchors require bond to prevent pullout and to mobilize a concrete breakout failure mode. The *characteristic bond stresses* in uncracked concrete τ_{uncr} and in cracked concrete τ_{cr} are provided by the manufacturer based on the 5 percent

TABLE 21.8

Minimum characteristic bond stress

Installation and Service Environment[a,b]	Moisture Content of Concrete at Time of Installation	Peak In-Service Temperature of Concrete, °F	τ_{cr} psi	τ_{uncr} psi
Outdoor	Dry to fully saturated	175	200	650
Indoor	Dry	110	300	1000

[a] Where anchor design includes sustained loads, the values of τ_{cr} and τ_{uncr} should be multiplied by 0.4.

[b] Where the anchor design includes earthquake loads for structures assigned to Seismic Design Category C, D, E, or F, the value of τ_{cr} should be multiplied by 0.8 and the value of τ_{uncr} should be multiplied by 0.4.

fractile results derived from the tests specified in ACI 355.4. Table 21.8 provides conservative values for the characteristic bond stresses given in the ACI Code. The characteristic bond stress is multiplied by 0.55 for sustained tension load and 0.40 if the anchor is subjected to sustained shear load. These values or higher values used for design must be included in the construction documents to transmit the performance requirements to the contractor. The values in the table are selected to be compatible with both the installation and service environments. For example, if adhesive anchors are installed before a building is enclosed, as shown in Fig. 1.16, the environment would be "outdoor."

a.   Basic Bond Strength

The bond stress is not uniform over the entire embedded length, and consequently, the projected area of concrete breakout strength is limited to a width of $2c_{Na}$, where c_{Na} is defined by Eq. (21.21). The formulation of the bond capacity of adhesive anchors is a function of both c_{Na} and h_{ef}. The basic bond strength of an adhesive anchor is

$$N_{ba} = \lambda_a \tau_{cr} \pi d_a\, h_{ef} \tag{21.17}$$

where τ_{cr} is the characteristic bond stress for cracked concrete, d_a is the anchor diameter, and h_{ef} is the effective embedment depth. For the calculation of adhesive anchor bond strength, the value of λ_a is 0.6λ. The tensile breakout capacity of a single adhesive anchor is

$$N_a = \frac{A_{Na}}{A_{Nao}} \psi_{ed,Na}\, \psi_{cp,Na}\, N_{ba} \tag{21.18}$$

and for an anchor group in tension

$$N_{ag} = \frac{A_{Na}}{A_{Nao}} \psi_{ec,Na}\, \psi_{ed,Na}\, \psi_{cp,Na}\, N_{ba} \tag{21.19}$$

Modification factors $\psi_{ed,Na}$ and $\psi_{ec,Na}$ are given in Tables 21.5 and 21.6, respectively. The modification factor $\psi_{cp,Na}$ equals 1.0 for $c_{a,min} \geq c_{ac}$ and equals $c_{a,min}/c_{ac}$ for $c_{a,min} \leq c_{ac}$. Critical edge distances are given in Table 21.2. The projected area of a single anchor A_{Nao} is

$$A_{Nao} = (2c_{Na})^2 \tag{21.20}$$

where

$$c_{Na} = 10d_a \sqrt{\frac{\tau_{uncr}}{1100}} \tag{21.21}$$

and the constant 1100 carries the units of lb/in^2.

EXAMPLE 21.11   **Adhesive anchor design.**   Determine if the anchor group for the attachment shown in Fig. 21.15 is adequate to carry the 5 kip factored sustained load using the characteristic bond stresses from Table 21.8. The A36 steel anchors are $\frac{3}{4}$ in. in diameter, have a net area of 0.334 in^2, $f_{uta} = 60$ ksi, and are embedded 8 in. into an exterior concrete wall with $f'_c = 4000$ psi. The anchor group is well away from any edges, supplementary reinforcement is present, and the anchors are considered Category 2—medium sensitivity and reliability.

SOLUTION.   From Table 21.1, the strength reduction factor for a Category 2 anchor with supplementary reinforcement is 0.65. All ψ values equal 1.0, as there is no eccentricity or edge distance constraint. From Table 21.1, $\phi = 0.65$.

The shear on the anchor group is equal to the applied load and is 5 kips. The design shear capacity of a single anchor based on the steel strength is

$$\phi V_{sa} = \phi n A_{se,V}\, 0.6\, f_{uta} = 0.65 \times 4 \times 0.334 \times 0.6 \times 60.0 = 31.3 \text{ kips}$$

The tensile breakout capacity of the anchor is as follows, noting that the ACI Code requires that τ_{uncr} to be multiplied by 0.55 for the sustained tensile loads. From Table 21.8, $\tau_{uncr} = 650$ and $\tau_{cr} = 200$.

$$c_{Na} = 10 d_a \sqrt{\frac{0.55 \tau_{uncr}}{1100}} = 10 \times 0.75 \sqrt{\frac{0.55 \times 650}{1100}} = 4.28 \text{ in.}$$

The basic projected area is $A_{Nao} = (2 c_{Na})^2 = (2 \times 4.28)^2 = 73.3$ in^2, and the projected area of the two top anchors is $A_{Na} = 2 c_{Na}(2 c_{Na} + s) = 2 \times 4.28 \times (2 \times 4.28 + 6) = 124.6$ in^2.

The basic bond strength of one adhesive anchor with a factor of 0.55 applied to τ_{cr} for sustained loads is

$$N_{ba} = 0.55 \tau_{cr} \pi d_u\, h_{ef} = 0.55 \times 200 \times \pi \times 0.75 \times 8 = 2.07 \text{ kips}$$

The two top anchors then provide a capacity of

$$N_a = \frac{A_{Na}}{A_{Nao}} \psi_{ec,Na} \psi_{ed,Na} \psi_{pc,Na} N_{ba} = \frac{73.3}{124.6} \times 1 \times 1 \times 1 \times 2.07 = 3.52 \text{ kips}$$

The design load on the anchors is the applied load P_u times the eccentricity divided by the distance between the anchors, and thus, $N_u = 5 \times 3/8 = 1.875$ kips, which is more than 20 percent of the nominal capacity, so the combined loading must be checked, giving

$$\frac{N_u}{\phi N_a} + \frac{V_u}{\phi V_{sa}} = \frac{1.875}{0.65 \times 3.52} + \frac{5}{0.65 \times 31.3} = 1.07 < 1.2$$

FIGURE 21.15
Adhesive anchor attachment for Example 21.11.

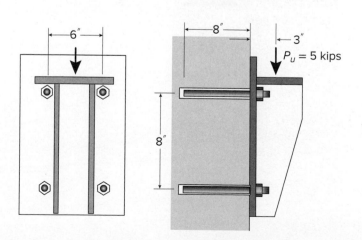

The interaction meets the code requirements, and thus, the attachment can carry the applied load. Comparing the anchor group in Example 21.10, the capacity of the attachment with adhesive anchors is approximately one-seventh the capacity of the attachment with cast-in anchors. This comparison clearly shows that cast-in anchors are structurally more efficient than adhesive anchors when the latter are designed based on the minimum characteristic bond stresses. In addition, a comparison of the characteristic bond stress values in the ACI Code with those obtained for commercial adhesive anchors derived through the ACI 355.4 qualification process (see, for example, values shown in Problem 21.8 at the end of this chapter) demonstrates that the Code values are conservative.

21.14   SCREW ANCHORS

Screw anchors consist of a steel shank and an upset thread that engages into the concrete as the anchor is installed (Fig. 21.16). Screw anchors installed between $5\,d_a$ and $10\,d_a$ develop concrete breakout capacity using the same procedures as cast-in anchors (Ref. 21.26). The hole for the screw anchor is drilled with the same diameter drill as the nominal diameter of the anchor, not the upset dimension of the threads. The effective embedment length is adjusted based on the qualifying test data and depends on the geometric configuration of the screw threads. This adjustment is provided by the manufacturer. The edge distance and minimum spacing for screw anchors considers the lateral pressure and disruption to the concrete from the installation process. This edge distance is important in corner installations where premature failures of the anchors have been observed.

EXAMPLE 21.12   **Screw anchor strength.**   Calculate the concrete breakout capacity for a $\frac{5}{8}$ in. diameter screw anchor with a cross-sectional area of 0.31 in^2 and an effective embedment length $h_{ef} = 4.125$ in. (Fig. 21.16). The concrete $f_c' = 5000$ psi and $f_{uta} = 60,000$ psi and is considered ductile. The installation is sufficiently away from edges that edge effects need not be

FIGURE 21.16
Screw anchor installation.
(*Courtesy of Simpson Strong-Tie Co., Inc.*)

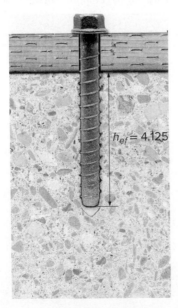

$h_{ef} = 4.125$

considered, the concrete is considered to crack during its lifetime, and no supplemental reinforcement is present.

SOLUTION. The strength reduction factor for steel in a screw anchor is $\phi = 0.70$. Based on Table 21.3, the tensile steel strength is

$$\phi N_{sa} = \phi A_{se,N} f_{uta} = 0.70 \times 0.31 \times 60 = 13.0 \text{ kips}$$

The effective depth of the anchor is 4.125 in. based on the length of the original screw less the washer and the plate. From Fig. 21.10,

$$A_{Nc} = 9h_{ef}^2 = 9 \times 4.125^2 = 153 \text{ in}^2$$

which for this case is both the individual and group area. For post-installed anchors in normalweight concrete, $k_c = 17$, $\lambda_a = 1.0$. Assuming the concrete is likely to crack and no edge effects, all ψ factors are equal to 1.0. Then, from Eq. (21.1) and the group capacity identical to the individual capacity and for $\phi = 0.70$ for a post-installed anchor with no supplemental reinforcement, the design strength is

$$\phi N_b = \phi k_c \lambda_a \sqrt{f_c'}\, h_{ef}^{1.5} = 0.70 \times 17 \times 1.0 \times \sqrt{5000} \times 4.125^{1.5} = 7050 \text{ lb}$$

For this installation, the breakout strength is less than the steel strength and is the controlling condition. The design strength of the anchor is 7.05 kips.

21.15 EARTHQUAKE DESIGN

The ACI Code requires that earthquake provisions be applied to all attachments when earthquake loading E is a load case, regardless of whether E controls the design. For seismic installations, all post-installed anchors must be certified under the seismic protocols in ACI 355.2 or ACI 355.4. Ideally, the anchors should be designed so that steel yielding is the controlling condition. Provisions requiring ductility assist in leading to yielding as the preferred failure mode (Refs. 21.27 to 21.29). The ACI Code offers four approaches to the design of anchors subjected to earthquake loads, one of which must be met to satisfy the code intent.

a. Concrete Breakout Strength Must Exceed the Steel Strength

This requirement provides ductility by having the anchor steel yield before any possible concrete failure due to concrete breakout, side blowout, or pullout strength of the anchor. The provision requires consideration of strain hardening of the steel to preclude concrete failure as the steel stress increases due to strain hardening. The provision also requires that anchor loads be transmitted through a ductile steel element. Ductility is attained by incorporation of a *stretch length*. The stretch length is an unbonded portion of the anchor at least $8d_a$ long over which the anchor is allowed to yield. The stretch length ensures there is a region where deformation can occur.

b. Load Transfer through a Ductile Yield Mechanism

This option allows the selection of materials to provide a ductile load path to the attachment. With this option, the anchor need not be the ductile element. The selection of the ductile mechanism should avoid concrete failure as the load increases due to strain hardening.

c. **Transfer the Load Elastically without Yielding**

The ACI Code allows the attachment to be designed to carry the entire load without yielding. When following this option, the load on the anchor must remain below the yield strength of the steel and the nominal strengths for all concrete breakout failure modes.

d. **Design for the Maximum Earthquake Load**

The attachment and anchors may be designed for the maximum design load combination that include E, with E increased by the overstrength factor Ω_o (see Chapter 20).

Additional cracking due to earthquake motions reduces the anchor capacity. To address this concern, the design capacity of anchors is reduced to 0.75ϕ times the nominal strength in breakout, pullout, or side-face blowout. The ACI Code does not allow anchors in plastic hinge regions due to the extensive cracking and a lack of test data. Anchors that must be placed in these regions should include anchor reinforcement detailed to mobilize both the attachment and the supporting structure. Minimum edge distances for anchors in seismic applications are the same as those given in Table 21.2.

If the earthquake contribution to V_u does not exceed 20 percent of the total factored shear load, shear capacity may be calculated using the procedures described in Section 21.7c; otherwise, the requirements 21.15a through 21.15d must be met. Anchors should be protected against buckling when load reversals are present. Detailed information on these options can be found in ACI Code 17.10 and Ref. 21.30.

21.16 SHEAR LUG ATTACHMENTS

A shear lug attachment consists of a base plate, a single lug plate, multiple lug plates, or orthogonal lug plates, plus anchors. Shear lugs provide additional shear resistance for attachments. They are constructed from rectangular steel plates or steel shapes composed of plate-like elements welded to a base plate and recessed into the concrete (Fig. 21.17). The shear lugs react against the concrete in bearing. The shear lug location below the base plate creates an eccentricity of the shear forces resulting in uplift on the base plate. This moment requires the shear lug attachment be secured with at least four anchors to restrain any uplift. Attachments fail by fracture of the concrete in front of the shear lug, indicated in light shading in Fig. 21.18, or by shear breakout failure at an edge, indicated by the lower dotted line in Fig. 21.18.

Shear lug attachments may be cast-in or post-installed. All elements of the attachment are assumed to move as an integral member, and failure occurs at small displacements. The strength of the attachment drops off markedly once concrete fracture occurs, which is often followed by a secondary fracture of the anchors. The ACI Code addresses shear lugs consisting of welded plates or plate-like elements. Shear lugs based on other formats, such as welded pipe sections, are allowed if the strength can be demonstrated by test or analysis.

The design approach in ACI Code 17.11 addresses the failure of the concrete and does not address the strength of the weldment or welding used to fabricate the attachment. The experimentally derived strength equations assume a uniform bearing stress of $1.7 f_c'$ on the effective area of the plate. Underlying experimental work indicates that only a portion of the plate area, that acting over a depth from the

FIGURE 21.17
Cast-in-place and post-installed shear lug attachments.

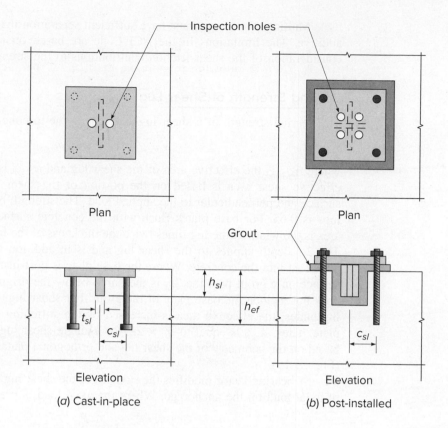

Plan

Plan

Inspection holes

Grout

h_{sl}

h_{ef}

t_{sl}

c_{sl}

c_{sl}

Elevation

Elevation

(a) Cast-in-place

(b) Post-installed

FIGURE 21.18
Shear lug attachment failure planes.

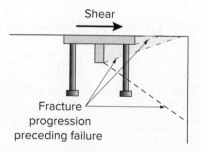

Shear

Fracture progression preceding failure

concrete surface equal to twice the plate thickness, is available to provide resistance (Ref. 21.31). The strength calculation only considers the shear lugs and is somewhat more conservative than if the embedded plate and anchor effects are included. Base plates are typically thick enough to provide moment resistance so that pryout of the anchor does not occur. If shear lug stiffeners are used, the length of the plate perpendicular to the applied shear must be at least half the length of the primary bearing plate.

The loading of the shear lug base plate can affect performance. If the attachment is in tension, the anchors must be far enough from the shear lug to reduce interaction between the two elements. Two conditions must be met to accomplish this separation. First, h_{ef}/h_{sl} must be greater than or equal to 2.5. Second, h_{ef}/c_{sl} must also be greater than or equal to 2.5. The dimensions associated with these variables are illustrated in Fig. 21.17. Inspection holes seen in Fig. 21.17 provide for visual confirmation that concrete is consolidated beneath the shear lug.

Multiple shear lugs must have sufficient separation that the fracture planes are additive. The limitations in the ACI Code are based on results of two tests and consideration of the shear-friction contributions to the strength (Ref. 21.32).

a.   Bearing Strength of Shear Lug Anchorages

The bearing strength of a shear lug attachment due to concrete failure is given by

$$V_{brg,sl} = 1.7 f_c' A_{ef,sl} \psi_{brg,sl} \tag{21.22}$$

where $A_{ef,sl}$ is the effective area of the shear lug and $\psi_{brg,sl}$ is the bearing factor. The effective shear area is based on the portion of the shear lug below the concrete surface and perpendicular to the applied load. The strength reduction factor for shear lugs is 0.65. For base plates flush with the concrete surface, the effective bearing area is the width of the lug times twice the thickness of the lug plate $2t_{sl}$ (Fig. 21.19). The $2t_{sl}$ depth applies to the shear lug and is in addition to the thickness of base plate cast into the concrete. Where the base plate is post-installed above the concrete surface on a grout pad, the $2t_{sl}$ is measured below the original concrete surface and does not include the base plate or the grout. For shear lugs with orthogonal stiffening plates, the effective area is increased by the stiffening effects of the transverse plate. Thus, $A_{ef,sl}$ is equal to $b_{sl} \times 2t_{sl}$ for a single shear lug and must be calculated based on the geometry of the shear lug and orthogonal plates for stiffened shear lugs (Fig. 21.19).

A bearing factor modifies the strength of the shear lug anchorage for the effects of axial load on the anchorage. Where the axial load is present,

$$\psi_{brg,sl} = 1 + \frac{P_u}{nN_{sa}} \leq 1.0 \tag{21.23a}$$

where P_u is the axial load and is negative for tension and n is the number of anchors in tension. When no axial load is present,

$$\psi_{brg,sl} = 1.0 \tag{21.23b}$$

and where the applied load is compression, P_u is positive

$$\psi_{brg,sl} = 1 + 4 \frac{P_u}{A_{bp} f_c'} \leq 2.0 \tag{21.23c}$$

and A_{bp} is the area of the base plate.

b.   Concrete Shear Breakout Strength Near Edges

The shear strength of a shear lug attachment near an edge of the concrete follows the procedures laid out in Section 21.7b for shear breakout strength. A breakout

FIGURE 21.19
Effective area of shear lugs.

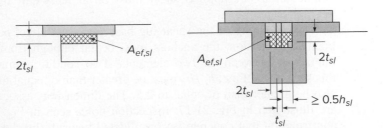

FIGURE 21.20
Concrete breakout of shear
lug attachment.

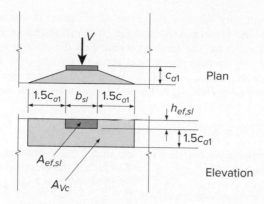

prism extends out and down from the shear lug as seen in Fig. 21.20. The concrete
shear breakout strength is given by Eq. (21.4). The nominal strength of the shear lug
attachment is given by Eq. (21.7) with the width of the breakout prism increased by
the width of the shear lug b_{sl}.

EXAMPLE 21.13   **Shear lug design.**   Design a shear lug attachment to carry a factored column compressive
axial load of 25 kips and a factored shear of 30 kips on the 12 in. × 12 in. base plate shown
in Fig. 21.21. Use screw anchors with a $f_{uta} = 50$ ksi, $f_c' = 5000$ psi, and a 1 in. thick shear
lug. The attachment is sufficiently far from an edge that edge effects may be ignored and the
eccentricity between the shear force and shear resistance is 2.5 in.

SOLUTION.   The depth of the shear lug effective bearing area is $2t_{sl} = 2$ in. From Eq. (21.23c)
the bearing factor is

$$\psi_{brg,sl} = 1 + 4\,\frac{P_u}{A_{bp}\,f_c'} = 1 + 4\,\frac{25{,}000}{12 \times 12 \times 5000} = 1.14$$

Then, rearranging Eq. (21.22), noting that $A_{ef,sl} = b_{sl} \times 2t_{sb}$, and for a shear lug $\phi = 0.65$

$$b_{sl} = \frac{V_{brg,sl}}{\phi 1.7 f_c' 2 t_{sl} \psi_{brg,sl}} = \frac{30}{0.65 \times 1.7 \times 5000 \times 2 \times 1 \times 1.14} = 2.38 \text{ in.}$$

In anticipation of interaction with the tensile strength requirements of the anchors, use
a 4 in. long shear lug, giving an effective bearing area of $A_{brg,sl} = 2 \times 4 = 8$ in.[2] The shear
strength of the attachment based on Eq. (21.22) is

$$\phi V_{brg,sl} = \phi 1.7 f_c' A_{ef,sl} \psi_{brg,sl} = 0.65 \times 1.7 \times 5000 \times 8 \times 1.14 = 50{,}390 \text{ lb}$$

FIGURE 21.21
Shear lug design example
configuration.

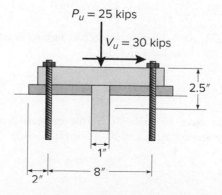

The eccentricity between the center of the shear resistance and the top of the plate consists of the base plate thickness and the grout pad thickness. For this example, the eccentricity is given as 2.5 in. The resulting moment on the plate is $V_u \times e = 30$ kips $\times 2.5$ in. $= 75$ in-kips. Assume the base plate rotates about its leading edge; then the tensile force on the anchors is $T_u = M_u/10$ in. $= 7.5$ kips. Try a $\frac{5}{8}$ in. diameter screw anchor with $h_{ef} = 6$ in. For a post-installed anchor, $k_c = 17$, and for normalweight concrete, $\lambda_a = 1.0$. The basic breakout strength from Eq. (21.1) is

$$N_b = k_c \lambda_a \sqrt{f_c'} h_{ef}^{1.5} = 17 \times 1.0 \times \sqrt{5000} \times 6^{1.5} = 17.7 \text{ kips}$$

The spacing between anchors is 8 in., less than the $3h_{ef}$ distance, so the breakout area of the anchor group is $A_{Nc} = 9h_{ef}^{1.5} = 9 \times 6^{1.5} = 324$ in^2 and the group area is

$$A_{Nco} = 3h_{ef} \times (3h_{ef} + s) = 3 \times 5 \times (3 \times 5 + 8) = 468 \text{ in}^2$$

The design concrete breakout strength in tension is then

$$\phi N_{cbg} = \phi \frac{A_{Nco}}{A_{Nc}} \psi_{ed,N} \psi_{c,N} \psi_{ec,N} N_b = 0.65 \times \frac{468}{324} \times 1.0 \times 1.0 \times 1.0 \times 17.7 = 16.6 \text{ kips}$$

The steel tensile strength of the steel screw anchor is $A_{sa} f_{uta} = 0.31 \times 50 = 15.5$ kips, and for the two anchors $2\phi A_{sa} f_{uta} = 2 \times 0.65 \times 0.31 \times 50 = 20.2$ kips, indicating that the concrete breakout strength controls. The factored load is greater than 20 percent of the nominal capacity; therefore, the interaction must be checked using Eq. (21.16)

$$\frac{V_u}{\phi V_n} + \frac{T_u}{\phi T_n} = \frac{30}{50.3} + \frac{7.5}{16.6} = 1.05 < 1.2$$

which is adequate. Therefore, the design requires a 1 in. thick by 4 in. long shear lug and four $\frac{5}{8}$ in. diameter screw anchors with an effective embedment of 6 in. If the moment for the tensile force were taken about the anchor centers, the tensile force would be 9.4 kips and the interaction ratio would increase to 1.16, which is again satisfactory.

EXAMPLE 21.14   **Shear lug strength.**   Using the shear lug attachment from Example 21.13, determine if the attachment is adequate when the face of the shear lug is 20 in. from an edge of the concrete. The slab is considered cracked, has no supplemental reinforcement, and is thick enough that $\psi_{h,V} = 1$.

SOLUTION.   The solution is identical to Example 21.13, except that the concrete breakout due to proximity to the edge must be checked. The edge distance is $c_{a1} = 20$ in. From Eq. (21.4), the basic shear breakout is

$$V_b = 9\lambda_a \sqrt{f_c'} c_{a1}^{1.5} = 9 \times 1.0 \times \sqrt{5000} \times 20^{1.5} = 56,920 \text{ lb}$$

With only one shear lug, $A_{Vco} = A_{vo}$ and the distance to the edge is greater than h_{ef}; thus $\psi_{ed,V} = 1.0$. The nominal shear strength from Eq. (21.8) is

$$\phi V_{cbg} = \phi \frac{A_{Vco}}{A_{Vc}} \psi_{ed,V} \psi_{c,V} \psi_{ec,V} \psi_{h,V} V_b = 0.65 \times \frac{1}{1} \times 1.0 \times 1.0 \times 1.0 \times 1.0 \times 56.9 = 37.0 \text{ kips}$$

Maintaining $h_{ef} = 6$ in. from Example 21.13, the concrete breakout strength is 16.6 kips. Checking the force interaction for the anchors using Eq. (21.16)

$$\frac{V_u}{\phi V_n} + \frac{T_u}{\phi T_n} = \frac{30}{37.0} + \frac{7.5}{16.6} = 1.26 > 1.2$$

concluding that the attachment is not adequate. Increasing h_{ef} to 7 in. results in the nominal tension breakout force increasing to 20.0 kips and ϕV_{cbg} remains 37.0 kips. The interaction resultant is

$$\frac{V_u}{\phi V_n} + \frac{T_u}{\phi T_n} = \frac{30}{37.0} + \frac{7.5}{20.0} = 1.18 < 1.2$$

concluding that the placement near the edge requires that the effective embedment depth of the anchor be increased to 7 in.

REFERENCES

21.1. *Design of Fastenings in Concrete*, Comite Euro-International du Beton (CEB), Thomas Telford Services Ltd., London, 1997.

21.2. W. Fuchs, R. Eligehausen, and J. Breen, "Concrete Capacity Design (CCD) Approach for Fastening to Concrete," *ACI Struct. J.*, vol. 92, no. 1, 1995, pp. 73–93. Also discussion, *ACI Struct. J.*, vol. 92, no. 6, 1995, pp. 787–802.

21.3. P. D. Courtois, "Industrial Research on Connections for Precast and In-situ Concrete," *Mechanical Fasteners for Concrete*, ACI SP-22, 1969, pp. 43–52.

21.4. A. F. Shaikh and W. Yi, "In-Place Strength of Welded Studs," *PCI J.*, vol. 30, no. 2, 1985, pp. 56–81.

21.5. PCI Industry Handbook Committee, *PCI Design Handbook*: *Precast and Prestressed Concrete*, 7th ed., MNL-120-10, 2010.

21.6. N. M. Hawkins, "Strength in Shear and Tension of Cast-in-Place Anchor Bolts," *Anchorage to Concrete*, ACI SP-103, 1987, pp. 233–255.

21.7. ACI Committee 349, *Code Requirements for Nuclear Safety Related Concrete Structures*, American Concrete Institute, Farmington Hills, MI, 2001.

21.8. R. Eligehausen and T. Balogh, "Behavior of Fasteners Loaded in Tension in Cracked Reinforced Concrete," *ACI Struct. J.*, vol. 92, no. 3, 1995, pp. 365–379.

21.9. R. A. Cook and R. E. Klingner, "Behavior of Ductile Multiple-Anchor Steel-to-Concrete Connections with Surface-Mounted Baseplates," *Anchors in Concrete*: *Design and Behavior*, ACI SP-130, American Concrete Institute, Farmington Hills, MI, 1992, pp. 61–122.

21.10. N. S. Anderson, and D. F. Meinheit, "A Review of Headed Stud Design Criteria," *PCI J.*, vol. 52, no. 1, 2007, pp. 82–100.

21.11. N. S. Anderson and D. F. Meinheit, "Pryout Capacity of Cast-In Headed Stud Anchors," *PCI J.*, vol. 50, no. 2, 2005, pp. 90–112.

21.12. PCI Industry Handbook Committee, *PCI Design Handbook*: *Precast and Prestressed Concrete*, 7th ed., MNL-120, 2010.

21.13. J. B Winter and C. W. Dolan, "Concrete Breakout Capacity of Cast-in-Place Concrete Anchors in Early Age Concrete," *PCI J.*, vol. 59, no. 2, pp. 114–131.

21.14. ACI Committee 355.2-07, *Qualification of Post-Installed Mechanical Anchors in Concrete and Commentary*, American Concrete Institute, Farmington Hills, MI, 2007.

21.15. ACI Committee 355.4-10, *Acceptance Criteria for Qualification of Post-Installed Adhesive Anchors in Concrete*, American Concrete Institute, Farmington Hills, MI, 2010.

21.16. *Building Code for Structural Concrete and Commentary* (ACI 318-14), American Concrete Institute, Farmington Hills, MI, 2014.

21.17. *Minimum Design Loads for Buildings and Other Structures*, ASCE 7-10, American Society of Civil Engineers, Reston, VA 2010.

21.18. ANSI/ASME B1.1, "Unified Inch Screw Threads (UN and UNR Thread Form)," American Society of Mechanical Engineers, New York, 2003.

21.19. R. A. Cook and R. E. Klingner, "Ductile Multiple-Anchor Steel-to-Concrete Connections," *J. Struct. Eng.*, ASCE, vol. 118, no. 6, 1992, pp. 1645–1665.

21.20. D. Kuhn and F. Shaikh, "Slip-Pullout Strength of Hooked Anchors," *Research Report*, University of Wisconsin-Milwaukee, submitted to the National Codes and Standards Council, 1996.

21.21. E. J. Primavera, J.-P. Pinelli, and E. H. Kalajian, "Tensile Behavior of Cast-in-Place and Undercut Anchors in High-Strength Concrete," *ACI Struct. J.*, vol. 94, no. 5, 1997, pp. 583–594.

21.22. D. Lotze, R. E. Klingner, and H. L. Graves III, "Static Behavior of Anchors under Combinations of Tension and Shear Loading," *ACI Struct. J.*, vol. 98, no. 4, 2001, pp. 525–536.

21.23. B. Hansen, "Investigators Fault Epoxy 'Creep' in Big Dig Collapse," *Civil Eng.*, vol. 77, no. 9, 2009, pp 20–21.

21.24. R. A. Cook, J. Kunz, W. Fuchs, and R. C. Konz, "Behavior and Design of Single Adhesive Anchors under Tensile Load in Uncracked Concrete," *ACI Struct. J.*, vol. 95, no. 1, 1998, pp. 9–26.

21.25. R. Eligehausen, R. A. Cook, and J. Appl, "Behavior and Design of Adhesive Bonded Anchors," *ACI Struct. J.*, vol. 103, no. 6, 2006, pp. 822–831.

21.26. J. Olsen, T. Pregartner, and A. J. Lamanna, "Basis for Design of Screw Anchors in Concrete," *ACI Struct. J.*, vol. 109, no. 4, 2012, pp. 559–568.

21.27. M. Hoehler and R. Eligehausen, "Behavior and Testing of Anchors in Simulated Seismic Cracks," *ACI Struct. J.*, vol. 105, no. 3, 2008, pp. 348–357.

21.28. Y. Zhang, R. E. Klingner, and H. L. Graves III, "Seismic Response of Multiple-Anchor Connections to Concrete," *ACI Struct. J.*, vol. 98, no. 6, 2001, pp. 811–822.

21.29. E. Vintzeleou and R. Eligehausen, "Behavior of Fasteners under Monotonic or Cyclic Shear Displacements," *Anchors in Concrete*: *Design and Behavior*, ACI SP-130, American Concrete Institute, Farmington Hills, MI, 1992, pp. 181–204.

21.30. FEMA P-750, "NEHRP Recommended Seismic Provisions for New Buildings and Other Structures," *FEMA-BSS Council*, Washington, 2009.

21.31. R. A. Cook and H. Michler, "Behavior and Design of Anchorages with Shear Lugs," 3rd International Symposium on Connections between Steel and Concrete, Stuttgart, Germany, 2017, pp. 560–570.

21.32. J. V. Rotz and M. Reifschneider, "Experimental Evaluation of Steel Embedments Subjected to Combined Axial and Shear," Bechtel Power Corp., Ann Arbor, MI, 1984, 100 pages.

PROBLEMS

21.1. Determine the required embedment h_{ef} for the $\frac{5}{8}$ in. diameter hex head bolt shown in Fig. P21.1. The bolt is located away from edges in 4000 psi cracked concrete with no supplementary reinforcement and carrying a factored sustained dead load of 8400 lb. Given: $f_y = 36$ ksi, $f_{uta} = 58$ ksi, $A_{se,N} = 0.226$, and $A_{brg} = 0.454$ in^2 ($h_{ef} = 4$ in.).

FIGURE P21.1

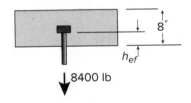

21.2. An insert with four $\frac{1}{2}$ in. diameter headed studs shown in Fig. P21.2 is embedded 4 in. into an uncracked concrete panel well away from any edges. $f'_c = 5000$ psi and stud yield strength $f_{uta} = 65,000$ psi. Determine the concentric tensile design load for the insert ($N_u = 21.7$ kips).

FIGURE P21.2

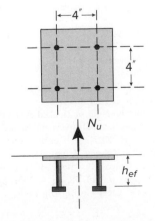

21.3. The insert with four $\frac{1}{2}$ in. diameter headed studs shown in Fig. P21.3 is embedded 4 in. into an uncracked concrete panel with the center of the insert 6 in. from any edges. $f'_c = 5000$ psi and stud tensile strength $f_{uta} = 65$ ksi. Determine the shear design load for the insert.

FIGURE P21.3

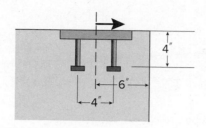

21.4. The insert with four $\frac{1}{2}$ in. diameter headed studs shown in Fig. P21.4 is embedded 4 in. into an uncracked concrete panel near a corner. $f'_c = 5000$ psi and stud tensile strength $f_{uta} = 65$ ksi. Determine the tensile design load for the insert.

FIGURE P21.4

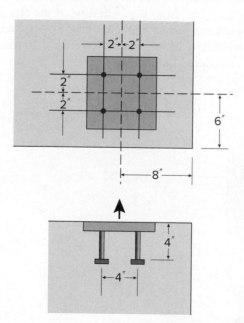

21.5.   The attachment shown in Fig. P21.5 carries a concentrically placed factored load of 18 kips. Given that the four headed studs are $\frac{1}{2}$ in. in diameter, have a specified tensile strength of 65 ksi, and are welded to a $\frac{3}{4}$ in. plate, determine h_{ef} for the anchor group. The concrete may be considered cracked, $f_c' = 4000$ psi, and the attachment is located away from any edge.

FIGURE P21.5

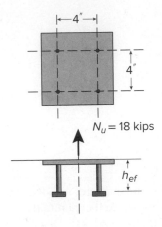

21.6.   The attachment shown in Fig. P21.6 carries a factored load of 21.5 kips. The load is placed $\frac{1}{2}$ in. from the centerline of the attachment. Given that the four headed studs are $\frac{5}{8}$ in. in diameter, have a specified tensile strength of 65 ksi, and are welded to a $\frac{3}{4}$ in. plate, determine h_{ef} for the attachment. The concrete may be considered uncracked, $f_c' = 4000$ psi, and the attachment located is away from any edge.

FIGURE P21.6

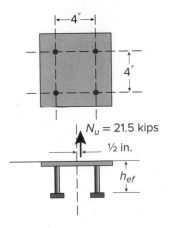

21.7. The insert with four $\frac{1}{2}$ in. diameter headed studs shown in Fig. P21.7 is embedded 4 in. into an uncracked concrete panel near a corner. $f_c' = 5000$ psi, stud tensile strength $f_{uta} = 65$ ksi, and supplemental reinforcement is present. Determine the design shear load for the insert.

FIGURE P21.7

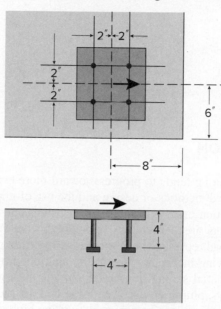

21.8. Determine the capacity of the attachment shown in Fig. 21.15 so that it is adequate to carry the sustained applied load using characteristic bond stresses of $\tau_{cr} = 1800$ psi and $\tau_{uncr} = 2600$ psi. The anchors are $\frac{3}{4}$ in. in diameter, have a cross-sectional area of 0.334 in^2, and are embedded 8 in. into an exterior concrete wall with a compressive strength of 4000 psi. $f_{uta} = 65$ ksi. The attachment is well away from any edges, supplementary reinforcement is present, and the anchors are considered Category 2—medium sensitivity and reliability. Comment on your solution.

21.9. Design a shear lug attachment for the column loads indicated in Fig. P21.9. Use screw anchors to secure the $12 \times 16 \times 1$ in. plate on a $\frac{3}{4}$ in. thick grout pad and a shear lug to provide shear resistance. The factored loads are 45 kips axial compression and 25 kips shear. The face of the 1 in. thick shear lug is located 20 in. from the edge of the concrete. Material properties are $f_c' = 4000$ psi and $f_{uta} = 50$ ksi.

FIGURE P21.9

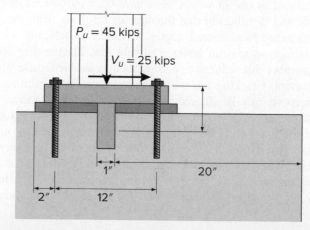

22 *Prestressed Concrete*

22.1 INTRODUCTION

Structural engineering tends to progress toward more economical structures through gradually improved methods of design and the use of higher-strength materials. This results in a reduction of cross-sectional dimensions and consequent weight savings. Such developments are particularly important in the field of reinforced concrete, where the dead load represents a substantial part of the total load. Also, in multistory buildings, any saving in depth of members, multiplied by the number of stories, can represent a substantial saving in total height, load on foundations, length of heating and electrical ducts, plumbing risers, and wall and partition surfaces.

Significant savings can be achieved by using high-strength concrete and steel in conjunction with present-day design methods, which permit an accurate appraisal of member strength. However, there are limitations to this development, due mainly to the interrelated problems of cracking and deflection at service loads. The efficient use of high-strength steel is limited by the fact that the amount of cracking (width and number of cracks) is proportional to the strain, and therefore the stress, in the steel. Although a moderate amount of cracking is normally not objectionable in structural concrete, excessive cracking is undesirable in that it exposes the reinforcement to corrosion, it may be visually offensive, and it may trigger a premature failure by diagonal tension. The use of high-strength materials is further limited by deflection considerations, particularly when refined analysis is used. The slender members that result may permit deflections that are functionally or visually unacceptable. This is further aggravated by cracking, which reduces the flexural stiffness of members.

These limiting features of ordinary reinforced concrete have been largely overcome by the development of *prestressed concrete*. A prestressed concrete member can be defined as one in which there have been introduced internal stresses of such magnitude and distribution that the stresses resulting from the given external loading are counteracted to a desired degree. Concrete is basically a compressive material, with its strength in tension being relatively low. Prestressing applies a precompression to the member that reduces or eliminates undesirable tensile stresses that would otherwise be present. Cracking under service loads can be minimized or even avoided entirely. Deflections may be limited to an acceptable value, and in some cases, members can be designed to have zero deflection under the combined effects of service load and prestress force. Deflection and crack control, achieved through prestressing, permit the engineer to make use of efficient and economical high-strength steels in the form of strands, wires, or bars, in conjunction with high-strength concretes. Thus, prestressing results in the overall improvement in performance of structural concrete used for ordinary loads and spans and extends the range of application far beyond the limits for ordinary reinforced concrete, not only leading to much longer spans, but also permitting innovative new structural forms to be employed.

22.2 EFFECTS OF PRESTRESSING

There are at least three ways to look at the prestressing of concrete: (a) as a method of achieving *concrete stress control*, by which the concrete is precompressed so that tension normally resulting from the applied loads is reduced or eliminated, (b) as a means for introducing *equivalent loads* on the concrete member so that the effects of the applied loads are counteracted to the desired degree, and (c) as a *special variation of reinforced concrete* in which prestrained high-strength steel is used, usually in conjunction with high-strength concrete. Each of these viewpoints is useful in the analysis and design of prestressed concrete structures, and they are illustrated in the following paragraphs.

a. Concrete Stress Control by Prestressing

Many important features of prestressed concrete can be demonstrated by simple examples. Consider first the plain, unreinforced concrete beam with a rectangular cross section shown in Fig. 22.1*a*. It carries a single concentrated load at the center of its span. (The self-weight of the member is neglected here.) As the load W is gradually applied, longitudinal flexural stresses are induced. If the concrete is stressed only within its elastic range, the flexural stress distribution at midspan is linear.

At a relatively low load, the tensile stress in the concrete at the bottom of the beam reaches the tensile strength of the concrete f_r, and a crack forms. Because no restraint is provided against upward extension of the crack, the beam will collapse without further increase of load.

Now consider an otherwise identical beam, shown in Fig. 22.1*b*, in which a longitudinal axial force P is introduced prior to the vertical loading. The longitudinal prestressing force produces a uniform axial compression $f_c = P/A_c$, where A_c is the cross-sectional area of the concrete. The force can be adjusted in magnitude so that when the transverse load Q is applied, the superposition of stresses due to P and Q results in zero tensile stress at the bottom of the beam as shown. Tensile stress in the concrete may be eliminated in this way or reduced to a specified amount.

It would be more logical to apply the prestressing force near the bottom of the beam, to compensate more effectively for the load-induced tension. A possible design specification, for example, might be to introduce the maximum compression at the bottom of the beam without causing tension at the top, when only the prestressing force acts. It is easily shown that, for a beam with a rectangular cross section, the point of application of the prestressing force should be at the lower third point of the section depth to achieve this. The force P, with the same value as before, but applied with eccentricity $e = h/6$ relative to the concrete centroid, produces a longitudinal compressive stress distribution varying linearly from zero at the top surface to a maximum of $2f_c = P/A_c + Pec_2/I_c$ at the bottom, where f_c is the concrete stress at the concrete centroid, c_2 is the distance from the concrete centroid to the bottom of the beam, and I_c is the moment of inertia of the cross section. This is shown in Fig. 22.1*c*. The stress at the bottom is exactly twice the value produced before by axial prestressing.

Consequently, the transverse load can now be twice as great as before, or $2Q$, and still cause no tensile stress. In fact, the final stress distribution resulting from

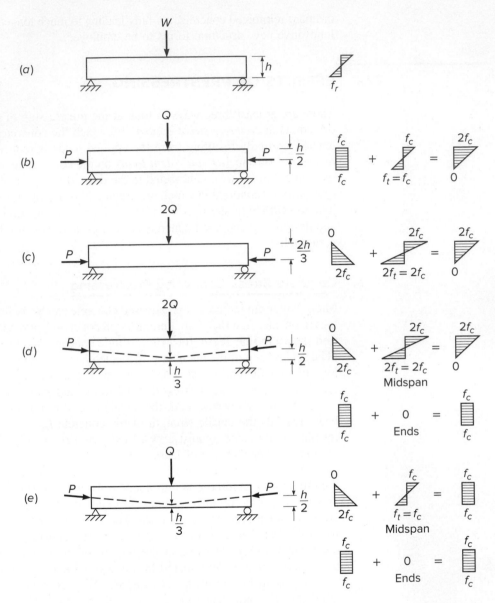

the superposition of load and prestressing force in Fig. 22.1*c* is identical to that of Fig. 22.1*b*, with the same prestressing force, although the load is twice as great. The advantage of eccentric prestressing is obvious.

The methods by which concrete members are prestressed is discussed in Section 22.3. For present purposes, it is sufficient to know that one practical method of prestressing uses high-strength steel tendons passing through a conduit embedded in the concrete beam. The tendon is anchored, under high tension, at both ends of the beam, thereby causing a longitudinal compressive stress in the concrete. The prestress force of Fig. 22.1*b* and *c* could easily have been applied in this way.

A significant improvement can be made, however, by using a prestressing tendon with variable eccentricity with respect to the concrete centroid, as shown in Fig. 22.1*d*. The load 2*Q* produces a bending moment that varies linearly along the span, from zero at the supports to maximum at midspan. Intuitively, one suspects that the best arrangement of prestressing would produce a *countermoment* that acts in the opposite sense to the load-induced moment and that would vary in the same way. This would be achieved by giving the tendon an eccentricity that varies linearly, from zero at the supports to maximum at midspan. This is shown in Fig. 22.1*d*. The stresses at midspan are the same as those in Fig. 22.1*c*, both when the load 2*Q* acts and when it does not. At the supports, where only the prestress force with zero eccentricity acts, a uniform compression stress f_c is obtained as shown.

For each characteristic load distribution, there is a *best tendon profile* that produces a prestress moment diagram that corresponds to that of the applied load. If the prestress countermoment is made exactly equal and opposite to the load-induced moment, the result is a beam that is subject only to uniform axial compressive stress in the concrete all along the span. Such a beam would be free of flexural cracking, and theoretically it would not be deflected up or down when that particular load is in place, compared to its position as originally cast. Such a result would be obtained for a load of $\frac{1}{2} \times 2Q = Q$, as shown in Fig. 22.1*e*, for example.

Some important conclusions can be drawn from these simple examples as follows:

1. Prestressing can control or even eliminate concrete tensile stress for specified loads.
2. Eccentric prestress is usually much more efficient than concentric prestress.
3. Variable eccentricity is usually preferable to constant eccentricity, from the viewpoints of both stress control and deflection control.

b.   Equivalent Loads

The effect of a change in the vertical alignment of a prestressing tendon is to produce a vertical force on the concrete beam. That force, together with the prestressing force acting at the ends of the beam through the tendon anchorages, can be looked upon as a system of external loads.

In Fig. 22.2*a*, for example, a tendon that applies force *P* at the centroid of the concrete section at the ends of a beam and that has a uniform slope at angle θ between the ends and midspan introduces a transverse force $2P \sin \theta$ at the point of change of slope at midspan. At the anchorages, the vertical component of the prestressing force is $P \sin \theta$ and the horizontal component is $P \cos \theta$. The horizontal component is very nearly equal to *P* for the usual flat slope angles. The moment diagram for the beam of Fig. 22.2*a* is seen to have the same form as that for any center-loaded simple span.

The beam of Fig. 22.2*b*, with a curved tendon, is subject to a vertical upward load from the tendon as well as the forces *P* at each end. The exact distribution of the load depends on the profile of the tendon. A tendon with a parabolic profile, for example, produces a uniformly distributed load. In this case, the moment diagram is parabolic, as it is for a uniformly loaded simple span.

If a straight tendon is used with constant eccentricity, as shown in Fig. 22.2*c*, there are no vertical forces on the concrete, but the beam is subject to a moment *Pe* at each end, as well as the axial force *P*, and a diagram of constant moment results.

The end moment must also be accounted for in the beam shown in Fig. 22.2*d*, in which a parabolic tendon is used that does not pass through the concrete centroid at the ends of the span. In this case, a uniformly distributed upward load plus end anchorage forces are produced, as shown in Fig. 22.2*b*, but in addition, the end moments $M = Pe \cos \theta$ must be accounted for.

It may be evident that for any arrangement of applied loads, a tendon profile can be selected so that the equivalent loads acting on the beam from the tendon are just equal and opposite to the applied loads. The result would be a state of pure compressive stress in the concrete, as discussed in somewhat different terms in reference to

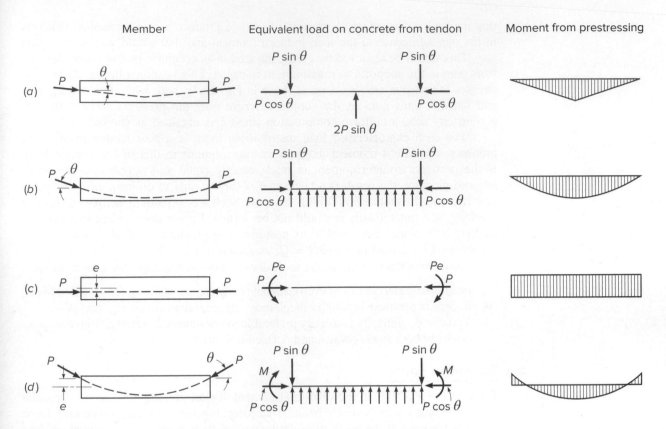

FIGURE 22.2
Equivalent loads and moments produced by prestressing tendons.

stress control and Fig. 22.1*e*. An advantage of the equivalent load concept is that it leads the designer to select what is probably the best tendon profile for a particular loading.

c. Prestressed Concrete as a Variation of Reinforced Concrete

In the descriptions of the effects of prestressing in Sections 22.2a and b, it was implied that the prestress force remained constant as the vertical load was introduced, that the concrete responded elastically, and that no concrete cracking occurred. These conditions may prevail up to about the service load level, but if the loads should be increased much beyond that, flexural tensile stresses eventually exceed the modulus of rupture and cracks will form. Loads, however, can usually be increased much beyond the cracking load in well-designed prestressed beams, and depending on the level of prestress, the beam response at service load may vary from uncracked, to minor cracking, to fully cracked, as occurs for an ordinary reinforced concrete beam.

Eventually both the steel and concrete at the cracked section are stressed into the inelastic range. The condition at incipient failure is shown in Fig. 22.3, which shows a beam carrying a *factored load* equal to some multiple of the expected service load. The beam undoubtedly would be in a partially cracked state; a possible pattern of flexural cracking is shown in Fig. 22.3*a*.

At the maximum moment section, only the concrete in compression is effective, and all of the tension is taken by the steel. The external moment from the applied loads is resisted by the internal force couple $Cz = Tz$. The behavior at this stage is almost

FIGURE 22.3
Prestressed concrete beam at load near flexural failure: (*a*) beam with factored load applied and (*b*) equilibrium of forces on left half of beam.

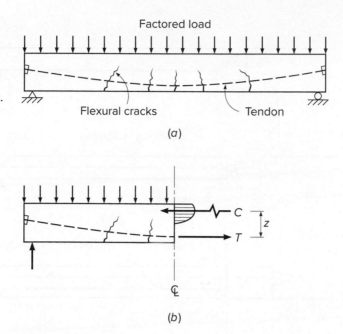

identical to that of an ordinary reinforced concrete beam at overload. The main difference is that the very high-strength steel used must be *prestrained* before loads are applied to the beam; otherwise, the high steel stresses would produce excessive concrete cracking and large beam deflections.

Each of the three viewpoints described—concrete stress control, equivalent loads, and reinforced concrete using prestrained steel—is useful in the analysis and design of prestressed concrete beams, and none of the three is sufficient in itself. Neither an elastic stress analysis nor an equivalent load analysis provides information about strength or safety margin. However, the stress analysis is helpful in predicting the extent of cracking, and the equivalent load analysis is often the best way to calculate deflections. Strength analysis is essential to evaluate safety against collapse, but it tells nothing about cracking or deflections of the beam under service conditions.

22.3 SOURCES OF PRESTRESS FORCE

Prestress can be applied to a concrete member in many ways. Perhaps the most obvious method of precompressing is to use jacks reacting against abutments, as shown in Fig. 22.4*a*. Such a scheme has been employed for large projects. Many variations are possible, including replacing the jacks with compression struts after the desired stress in the concrete is obtained or using inexpensive jacks that remain in place in the structure, in some cases with a cement grout used as the hydraulic fluid. The principal difficulty associated with such a system is that even a slight movement of the abutments drastically reduces the prestress force.

In most cases, the same result is more conveniently obtained by tying the jack bases together with wires or strands, as shown in Fig. 22.4*b*. These wires or strands may be external, located on each side of the beam; more commonly, they are passed through a hollow conduit embedded in the concrete beam. Usually, one end of the prestressing tendon is anchored, and all of the force is applied at the other end. After

FIGURE 22.4

Prestressing methods:
(*a*) post-tensioning by
jacking against abutments;
(*b*) post-tensioning with
jacks reacting against beam;
and (*c*) pretensioning with
tendon stressed between
fixed external anchorages.

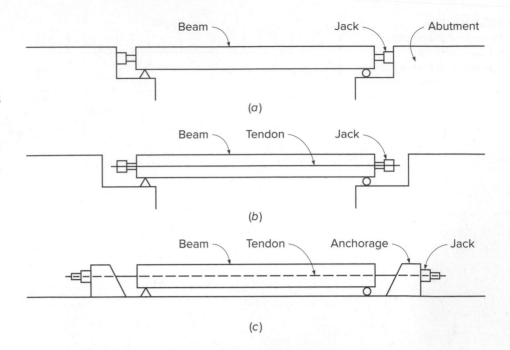

reaching the desired prestress force, the tendon is wedged against the concrete and the jacking equipment is removed for reuse. In this type of prestressing, the entire system is self-contained and is independent of relative displacement of the supports.

Another method of prestressing that is widely used is illustrated by Fig. 22.4*c*. The prestressing strands are tensioned between massive abutments in a casting yard prior to placing the concrete in the beam forms. The concrete is placed around the tensioned strands, and after the concrete has attained sufficient strength, the jacking pressure is released. This transfers the prestressing force to the concrete by bond and friction along the strands, chiefly at the outer ends.

It is essential, in all three cases shown in Fig. 22.4, that the beam be supported in such a way as to permit the member to shorten axially without restraint so that the prestressing force can be transferred to the concrete.

Other means for introducing the desired prestressing force have been attempted on an experimental basis. Thermal prestressing can be achieved by preheating the steel by electrical or other means. Anchored against the ends of the concrete beam while in the extended state, the steel cools and tends to contract. The prestress force is developed through the restrained contraction. The use of expanding cement in concrete members has been tried with varying success. The volumetric expansion, restrained by steel strands or by fixed abutments, produces the prestress force.

Most of the patented systems for applying prestress in current use are variations of those shown in Fig. 22.4*b* and *c*. Such systems can generally be classified as *pretensioning* or *post-tensioning* systems. In the case of pretensioning, the tendons are stressed before the concrete is placed, as in Fig. 22.4*c*. This system is well suited for mass production, since casting beds can be made several hundred feet long, the entire length cast at once, and individual beams fabricated to the desired length in a single casting. Figure 22.5 shows workers using a hydraulic jack to tension strands at the anchorage of a long pretensioning bed. Although each tendon is individually stressed in this case, large capacity jacks are often used to tension all strands simultaneously.

FIGURE 22.5
Massive strand jacking abutment at the end of a long pretensioning bed. (*Photograph by Charles W. Dolan.*)

In post-tensioned construction, shown in Fig. 22.4*b*, the tendons are tensioned after the concrete is placed and has gained its strength. Usually, a hollow conduit or sleeve is provided in the beam, through which the tendon is passed. In some cases, tendons are placed in the interior of hollow box-section beams. The jacking force is usually applied against the ends of the hardened concrete, eliminating the need for massive abutments. In Fig. 22.6, a tendon consisting of many individual strands is being post-tensioned using a portable hydraulic jack.

A large number of particular systems, steel elements, jacks, and anchorage fittings have been developed in this country and abroad, many of which differ from each other only in minor details (Refs. 22.1 to 22.8). As far as the designer of prestressed concrete structures is concerned, it is unnecessary and perhaps even undesirable to specify in detail the technique that is to be followed and the equipment to be used. It is common practice to specify only the magnitude and line of action of the prestress force. The contractor is then free, in bidding the work, to receive quotations from several different prestressing subcontractors, with resultant cost savings. It is evident, however, that the designer must have some knowledge of the details of the various systems contemplated for use, so that in selecting cross-sectional dimensions, any one of several systems can be accommodated. Once a system is selected, it is reviewed by the designer for conformance with the structural requirements.

FIGURE 22.6
Post-tensioning a 12-strand
tendon using a multistrand
jack. (*Photograph by
Charles W. Dolan.*)

22.4 PRESTRESSING STEELS

Early attempts at prestressing concrete were unsuccessful because steel with ordinary structural strength was used. The low prestress obtainable in such rods was quickly lost due to shrinkage and creep in the concrete.

Such changes in length of concrete have much less effect on prestress force if that force is obtained using highly stressed steel wires or strands. In Fig. 22.7a, a concrete member of length ℓ is prestressed using steel bars with ordinary strength stressed to 24,000 psi. With $E_s = 29 \times 10^6$ psi, the unit strain ε_s required to produce the desired stress in the steel of 24,000 psi is

$$\varepsilon_s = \frac{\Delta\ell}{\ell} = \frac{f_s}{E_s} = \frac{24,000}{29 \times 10^6} = 8.0 \times 10^{-4}$$

However, the long-term strain in the concrete due to shrinkage and creep alone, if the prestress force were maintained over a long period, would be on the order of 8.0×10^{-4} and would be sufficient to completely relieve the steel of all stress.

Alternatively, suppose that the beam is prestressed using high-strength steel stressed to 150,000 psi. The elastic modulus of steel does not vary greatly, and the same value of 29×10^6 psi is assumed here. Then in this case, the unit strain required to produce the desired stress in the steel is

$$\varepsilon_s = \frac{150,000}{29 \times 10^6} = 51.7 \times 10^{-4}$$

If shrinkage and creep strain are the same as before, the net strain in the steel after these losses is

$$\varepsilon_{s,\text{net}} = (51.7 - 8.0) \times 10^{-4} = 43.7 \times 10^{-4}$$

FIGURE 22.7
Loss of prestress due to
concrete shrinkage and creep.

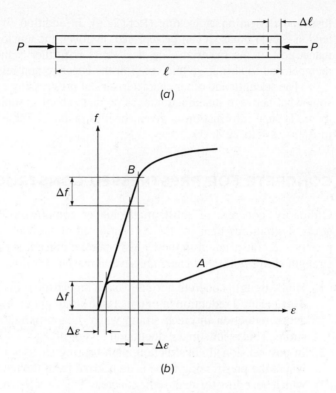

(a)

(b)

and the corresponding stress after losses is

$$f_s = \varepsilon_{s,net} E_s = (43.7 \times 10^{-4})(29 \times 10^6) = 127,000 \text{ psi}$$

This represents a stress loss of about 15 percent, compared with 100 percent loss in the beam using ordinary steel. It is apparent that the amount of stress lost because of shrinkage and creep is independent of the original stress in the steel. Therefore, the higher the original stress, the lower the percentage loss. This is illustrated graphically by the stress-strain curves of Fig. 22.7b. Curve A is representative of ordinary reinforcing bars, with a yield stress of 60,000 psi, while curve B represents high tensile steel, with a tensile strength of 270,000 psi. The stress change Δf resulting from a certain change in strain $\Delta\varepsilon$ is seen to have much less effect when high steel stress levels are attained. Prestressing of concrete is therefore practical only when steels of very high strength are used.

Prestressing steel is most commonly used in the form of individual wires, strands made up of seven wires, and alloy-steel bars. The physical properties of these have been discussed in Section 2.16, and typical stress-strain curves appear in Fig. 2.19. Virtually all strands in use are low-relaxation (Section 2.16c).

The tensile stress permitted by ACI Code 20.3.2.5 in prestressing wires, strands, or bars is dependent upon the type of loading. When the jacking force is first applied, a maximum stress of $0.80f_{pu}$, $0.94f_{py}$, or the manufacturer's maximum recommended value is allowed, whichever is smaller, where f_{pu} is the tensile strength of the steel and f_{py} is the yield strength. Immediately after transfer of prestress force at post-tensioning anchorages, the stress is limited to $0.70f_{pu}$. The ACI Code does not specify a stress at transfer for pretensioned members. Industry practice, based on earlier ACI Code requirements, is for the stress after transfer to be $0.74f_{pu}$. The justification for a higher allowable stress during the stretching operation is that the steel stress is known quite precisely at this stage. Hydraulic jacking pressure and total steel elongation are quantities that are easily measured, and quality control specifications require correlation of load and deflection at jacking (Ref. 22.9). In addition, if an accidentally deficient tendon should break, it can be replaced; in effect, the tensioning operation is a performance test of the material. The steel stress is further reduced during the life of the member due to shrinkage and creep in the concrete and relaxation in the steel.

The strength and other characteristics of prestressing wire, strands, and bars vary somewhat between manufacturers, as do methods of grouping tendons and anchoring them. Typical information is given for illustration in Table A.15 of Appendix A and in Refs. 22.1 to 22.8.

22.5 CONCRETE FOR PRESTRESSED CONSTRUCTION

Ordinarily, concrete of substantially higher compressive strength is used for prestressed structures than for those constructed of ordinary reinforced concrete. Most prestressed construction in the United States at present is designed for a compressive strength above 5000 psi. There are several reasons for this:

1. High-strength concrete normally has a higher modulus of elasticity (see Fig. 2.3). This means a reduction in initial elastic strain under application of prestress force and a reduction in creep strain, which is approximately proportional to elastic strain. This results in a reduction in loss of prestress.
2. In post-tensioned construction, high bearing stresses result at the ends of beams where the prestressing force is transferred from the tendons to anchorage fittings, which bear directly against the concrete. This problem can be met by increasing the size of the anchorage fitting or by increasing the bearing capacity of the concrete by increasing its compressive strength. The latter is usually more economical.
3. In pretensioned construction, where transfer by bond is customary, the use of high-strength concrete permits shorter transfer and development lengths.
4. A substantial part of the prestressed construction in the United States is precast, with the concrete mixed, placed, and cured under carefully controlled conditions that facilitate obtaining higher strengths.

The strain characteristics of concrete under short-term and sustained loads assume an even greater importance in prestressed structures than in reinforced concrete structures because of the influence of strain on loss of prestress force. Strains due to stress, together with volume changes due to shrinkage and temperature changes, may have considerable influence on prestressed structures. In this connection, it is suggested that the reader review Sections 2.8 to 2.11, which discuss in some detail the compressive and tensile strengths of concrete under short-term and sustained loads and the changes in concrete volume that occur due to shrinkage and temperature change.

The allowable stresses in the concrete, according to ACI Code 24.5.3, depend upon the stage of loading and the behavior expected of the member. ACI Code 24.5.2 defines three classifications of behavior, depending on the extreme fiber stress f_t at service load in the precompressed tensile zone. The three classifications are U, T, and C. Class U flexural members are assumed to behave as uncracked members. Class T members represent a transition between uncracked and cracked flexural members, while Class C members are assumed to behave as cracked flexural members. Permissible stresses for these three classifications are given in Table 22.1.

In Table 22.1, f'_{ci} is the compressive strength of the concrete at the time of initial prestress and f'_c the specified compressive strength of the concrete. In parts e and f of Table 22.1, *sustained load* is any part of the service load that is sustained for a sufficient period of time to cause significant time-dependent deflections, whereas *total load* refers to the total service load, a part of which may be transient

TABLE 22.1
Permissible calculated stresses in concrete in prestressed flexural members

Condition	Class		
	U	T	C*
a. Extreme fiber stress in compression immediately after transfer (except as in b)	$0.60f'_{ci}$	$0.60f'_{ci}$	$0.60f'_{ci}$
b. Extreme fiber stress in compression at ends of simply supported members	$0.70f'_{ci}$	$0.70f'_{ci}$	$0.70f'_{ci}$
c. Extreme fiber stress in tension immediately after transfer (except as in d)	$3\sqrt{f'_{ci}}$	$3\sqrt{f'_{ci}}$	$3\sqrt{f'_{ci}}$
d. Extreme fiber stress in tension immediately after transfer at the end of simply supported members[†]	$6\sqrt{f'_{ci}}$	$6\sqrt{f'_{ci}}$	$6\sqrt{f'_{ci}}$
e. Extreme fiber stress in compression due to prestress plus sustained load	$0.45f'_c$	$0.45f'_c$	—
f. Extreme fiber stress in compression due to prestress plus total load	$0.60f'_c$	$0.60f'_c$	—
g. Extreme fiber stress in tension f_t in precompressed tensile zone under service load	$\leq 7.5\sqrt{f'_c}$	$>7.5\sqrt{f'_c}$ and $\leq 12\sqrt{f'_c}$	$>12\sqrt{f'_c}$
h. Prestressed two-way slabs	$\leq 6\sqrt{f'_c}$		

* There are no service stress requirements for Class C.

† When calculated tensile stresses exceed these values, bonded auxiliary prestressed or nonprestressed reinforcement must be provided in the tensile zone to resist the total tensile force in the concrete calculated with the assumption of an uncracked section.

or temporary live load. Thus, sustained load would include dead load and may or may not include service live load, depending on its duration. If the live load duration is short or intermittent, the higher limit of part f is permitted.

Two-way slabs are designated as Class U flexural members with f_t limited to values $\leq 6\sqrt{f'_c}$. Class C flexural members have no service level stress requirements but must satisfy strength and serviceability requirements. Service load stress calculations are based on uncracked section properties for Class U and T flexural members and on the cracked section properties for Class C members.

22.6 ELASTIC FLEXURAL ANALYSIS

It has been noted earlier in this text that the design of concrete structures may be based either on providing sufficient strength, which would be used fully only if the expected loads were increased by an overload factor, or on keeping material stresses within permissible limits when actual service loads act. In the case of ordinary reinforced concrete members, strength design is used. Members are proportioned on the basis of strength requirements and then checked for satisfactory service load behavior, notably with respect to deflection and cracking. The design is then modified if necessary.

Class C members are principally designed based on strength. Class U and T members, however, are proportioned so that stresses in the concrete and steel at actual service loads are within permissible limits. These limits are a fractional part of the actual capacities of the materials. There is some logic to this approach, since an important objective of prestressing is to improve the performance of members at service loads. Consequently, service load requirements often control the amount of prestress force used in Class U and T members. Design based on service loads may usually be carried out assuming elastic behavior of both the concrete and the steel, since stresses are relatively low in each.

Regardless of the starting point chosen for the design, a structural member must be satisfactory at all stages of its loading history. Accordingly, prestressed members proportioned on the basis of permissible stresses must be checked to ensure that sufficient strength is provided should overloads occur, and deflection and cracking under service loads should be investigated. Consistent with most U.S. practice, the design of prestressed concrete beams starts with a consideration of stress limits, after which strength and other properties are checked.

It is convenient to think of prestressing forces as a system of external forces acting on a concrete member, which must be in equilibrium under the action of those forces. Figure 22.8a shows a simple-span prestressed beam with curved tendons, typical of many post-tensioned members. The portion of the beam to the left of a vertical cutting plane x-x is taken as a free body, with forces acting as shown in Fig. 22.8b. The force P at the left end is exerted on the concrete through the tendon anchorage, while the force P at the cutting plane x-x results from combined shear and normal stresses acting at the concrete surface at that location. The direction of P is tangent to the curve of the tendon at each location. Note the presence of the force N, acting on the concrete from the tendon, due to tendon curvature. This force is distributed in some manner along the length of the tendon, the exact distribution depending upon the tendon profile. Its resultant and the direction in which the resultant acts can be found from the force diagram of Fig. 22.8c.

It is convenient when working with the prestressing force P to divide it into its components in the horizontal and vertical directions. The horizontal component (Fig. 22.8d) is $H = P \cos \theta$, and the vertical component is $V = H \tan \theta = P \sin \theta$, where θ is the angle of inclination of the tendon centroid at the particular section. Since the slope angle is normally quite small, the cosine of θ is very close to unity and it is sufficient for most calculations to take $H = P$.

The magnitude of the prestress force is not constant. The *jacking force P_j* is immediately reduced to what is termed the *initial prestress force P_i* because of elastic shortening of the concrete upon transfer, slip of the tendon as the force is transferred from the jacks to the beam ends, and loss due to friction between the tendon and the concrete (post-tensioning) or between the tendon and the strand alignment devices (pretensioning). There is a further reduction of force from P_i to the *effective prestress P_e*, occurring over a long period of time at a gradually decreasing rate, because of concrete creep under the sustained prestress force, concrete shrinkage, and relaxation of stress in the steel. Methods for predicting losses are discussed in

FIGURE 22.8
Prestressing forces acting on concrete.

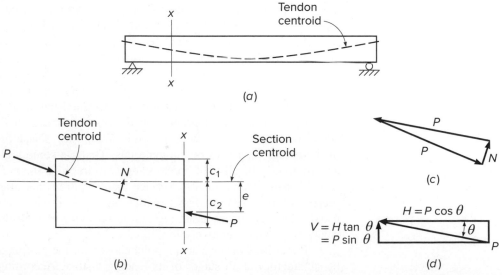

FIGURE 22.9
Concrete stress distributions in beams: (*a*) effect of prestress; (*b*) effect of prestress plus self-weight of beam; and (*c*) effect of prestress, self-weight, and external dead and live service loads.

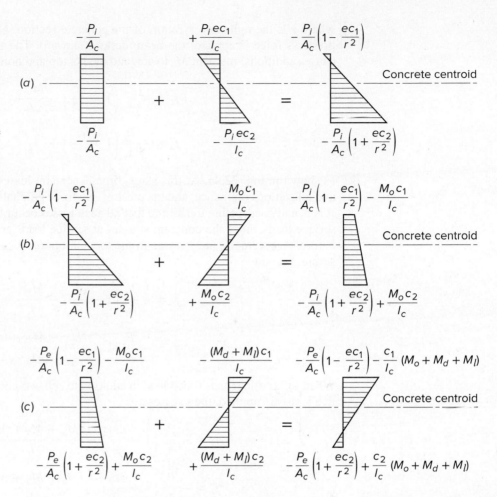

Section 22.13. Of primary interest to the designer are the initial prestress P_i immediately after transfer and the final or effective prestress P_e after all losses.

In developing elastic equations for flexural stress, the effects of prestress force, self-weight moment, and dead and live load moments are calculated separately, and the separate stresses are superimposed. When the initial prestress force P_i is applied with an eccentricity e below the centroid of the cross section with area A_c and top and bottom fiber distances c_1 and c_2, respectively, it causes the compressive stress $-P_i/A_c$ and the bending stresses $+P_i e c_1/I_c$ and $-P_i e c_2/I_c$ in the top and bottom fibers, respectively (compressive stresses are designated as negative, tensile stresses as positive[†]), as shown in Fig. 22.9*a*. Then, at the top fiber, the stress is

$$f_1 = -\frac{P_i}{A_c} + \frac{P_i e c_1}{I_c} = -\frac{P_i}{A_c}\left(1 - \frac{e c_1}{r^2}\right) \tag{22.1a}$$

and at the bottom fiber

$$f_2 = -\frac{P_i}{A_c} - \frac{P_i e c_2}{I_c} = -\frac{P_i}{A_c}\left(1 + \frac{e c_2}{r^2}\right) \tag{22.1b}$$

[†] Designating tension positive is consistent throughout this text. Many practicing engineers prefer to use tension as negative in recognition that compression is the preferred stress state in concrete.

where r is the radius of gyration of the concrete section. Normally, as the eccentric prestress force is applied, the beam deflects upward. The beam self-weight w_o then causes additional moment M_o to act, and the net top and bottom fiber stresses become

$$f_1 = -\frac{P_i}{A_c}\left(1 - \frac{ec_1}{r^2}\right) - \frac{M_o c_1}{I_c} \tag{22.2a}$$

$$f_2 = -\frac{P_i}{A_c}\left(1 + \frac{ec_2}{r^2}\right) + \frac{M_o c_2}{I_c} \tag{22.2b}$$

as shown in Fig. 22.9b. At this stage, time-dependent losses due to shrinkage, creep, and relaxation commence, and the prestressing force gradually decreases from P_i to P_e. It is usually acceptable to assume that all such losses occur prior to the application of service loads, since the concrete stresses at service loads are critical after losses, not before. Accordingly, the stresses in the top and bottom fiber, with P_e and beam load acting, become

$$f_1 = -\frac{P_e}{A_c}\left(1 - \frac{ec_1}{r^2}\right) - \frac{M_o c_1}{I_c} \tag{22.3a}$$

$$f_2 = -\frac{P_e}{A_c}\left(1 + \frac{ec_2}{r^2}\right) + \frac{M_o c_2}{I_c} \tag{22.3b}$$

When full service loads (dead load in addition to self-weight of the beam, plus service live load) are applied, the stresses are

$$f_1 = -\frac{P_e}{A_c}\left(1 - \frac{ec_1}{r^2}\right) - \frac{(M_o + M_d + M_l)c_1}{I_c} \tag{22.4a}$$

$$f_2 = -\frac{P_e}{A_c}\left(1 + \frac{ec_2}{r^2}\right) + \frac{(M_o + M_d + M_l)c_2}{I_c} \tag{22.4b}$$

as shown in Fig. 22.9c.

It is necessary, in reviewing the adequacy of a beam (or in designing a beam on the basis of permissible stresses), that the stresses in the extreme fibers remain within specified limits under any combination of loadings that can occur. Normally, the stresses at the section of maximum moment, in a properly designed beam, must stay within the limit states defined by the distributions shown in Fig. 22.10 as the beam passes from the unloaded stage (P_i plus self-weight) to the loaded stage (P_e plus full service loads). In the figure, f_{ci} and f_{ti} are the permissible compressive and tensile stresses, respectively, in the concrete immediately after transfer, and f_{cs} and f_{ts} are the permissible compressive and tensile stresses at service loads given in Table 22.1.

FIGURE 22.10

Stress limits: (a) unloaded beam, with initial prestress plus self-weight, and (b) loaded beam, with effective prestress, self-weight, and full service load.

f_{ti}

f_{ci}

(a) Unloaded

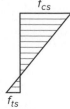

f_{cs}

f_{ts}

(b) Loaded

FIGURE 22.11
Location of kern points.

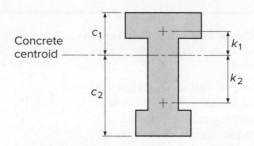

In calculating the section properties A_c, I_c, etc., to be used in the above equations, it is relevant that, in post-tensioned construction, the tendons are usually grouted in the conduits after tensioning. Before grouting, stresses should be based on the net section with holes deducted. After grouting, the transformed section should be used with holes considered filled with concrete and with the steel replaced with an equivalent area of concrete. However, it is satisfactory, unless the holes are quite large, to compute section properties on the basis of the gross concrete section. Similarly, while in pretensioned beams the properties of the transformed section should be used, it makes little difference if calculations are based on properties of the gross concrete section.[†]

It is useful to establish the location of the upper and lower *kern points* of a cross section. These are defined as the limiting points inside which the prestress force resultant may be applied without causing tension anywhere in the cross section. Their locations are obtained by writing the expression for the tensile fiber stress due to application of an eccentric prestress force acting alone and setting this expression equal to zero to solve for the required eccentricity. In Fig. 22.11, to locate the upper kern-point distance k_1 from the neutral axis, let the prestress force resultant P act at that point. Then the bottom fiber stress is

$$f_2 = -\frac{P}{A_c}\left(1 + \frac{ec_2}{r^2}\right) = 0$$

Thus, with

$$1 + \frac{ec_2}{r^2} = 0$$

one obtains the corresponding eccentricity

$$e = k_1 = -\frac{r^2}{c_2} \tag{22.5a}$$

Similarly, the lower kern-point distance k_2 is

$$k_2 = \frac{r^2}{c_1} \tag{22.5b}$$

The region between these two limiting points is known as the *kern*, or in some cases the *core*, of the section.

[†] ACI Code 4.12.2.4 contains the following provision: "Effect of loss of area due to open ducts shall be considered in calculating section properties before grout in post-tensioning ducts has attained design strength." Earlier versions of the ACI Code Commentary stated: "If the effect of the open duct area on design is deemed negligible, section properties may be based on total area. In post-tensioned members after grouting and in pretensioned members, section properties may be based on effective sections using transformed areas of bonded prestressing steel and nonprestressed gross sections, or net sections."

EXAMPLE 22.1   **Pretensioned I beam with constant eccentricity.**   A simply supported symmetrical I beam shown in cross section in Fig. 22.12a is used on a 40 ft simple span. It has the following section properties:

Moment of inertia: $I_c = 12{,}000 \text{ in}^4$
Concrete area: $A_c = 176 \text{ in}^2$
Radius of gyration: $r^2 = 68.2 \text{ in}^2$
Section modulus: $S = 1000 \text{ in}^3$
Self-weight: $w_o = 0.183 \text{ kips/ft}$

and is to carry a superimposed dead plus live load (considered "sustained," not short-term) of 0.750 kips/ft in addition to its own weight. The beam will be pretensioned with multiple seven-wire strands with the centroid at a constant eccentricity of 7.91 in. The prestress force P_i immediately after transfer will be 158 kips; after time-dependent losses, the force reduces to $P_e = 134$ kips. The specified compressive strength of the concrete $f_c' = 5000$ psi, and at the time of prestressing the strength will be $f_{ci}' = 3750$ psi. Calculate the concrete flexural stresses at the midspan section of the beam at the time of transfer, and after all losses with full service load in place. Compare with ACI allowable stresses for a Class U member.

SOLUTION.   Stresses in the concrete along the length of the beam resulting from the initial prestress force of 158 kips may be found by Eqs. (22.1a) and (22.1b):

$$f_1 = -\frac{158{,}000}{176}\left(1 - \frac{7.91 \times 12}{68.2}\right) = +352 \text{ psi}$$

$$f_2 = -\frac{158{,}000}{176}\left(1 + \frac{7.91 \times 12}{68.2}\right) = -2147 \text{ psi}$$

The self-weight of the beam causes the immediate superposition of a moment at midspan of

$$M_o = 0.183 \times \frac{40^2}{8} = 36.6 \text{ ft-kips}$$

and corresponding stresses of $M_o/S = 36{,}600 \times 12/1000 = 439$ psi, so that the net stresses at the top and bottom of the concrete section at midspan due to initial prestress and self-weight, from Eqs. (22.2a) and (22.2b), are

$$f_1 = +352 - 439 = -87 \text{ psi}$$

$$f_2 = -2147 + 439 = -1708 \text{ psi}$$

After losses, the prestress force is reduced to 134 kips, and the concrete stresses at midspan due to that force plus self-weight are

$$f_1 = +352 \times \frac{134}{158} - 439 = -140 \text{ psi}$$

$$f_2 = -2147 \times \frac{134}{158} + 439 = -1382 \text{ psi}$$

and stresses at the end of the beam are

$$f_1 = +352\left(\frac{134}{158}\right) = 299$$

$$f_2 = -2147\left(\frac{134}{158}\right) = -1821$$

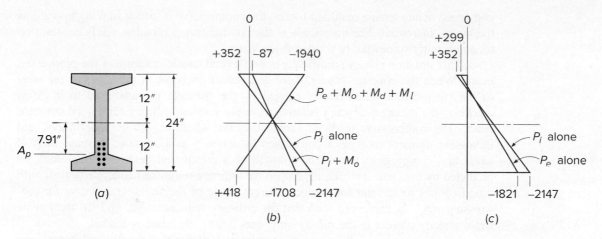

FIGURE 22.12
Pretensioned I beam. Design example: (*a*) cross section; (*b*) stresses at midspan (psi); and (*c*) stresses at ends (psi).

The superimposed load of 0.750 kip/ft produces a midspan moment of $M_d + M_l = 0.750 \times 40^2/8 = 150$ ft-kips and the corresponding stresses of $150,000 \times 12/1000 = 1800$ psi in compression and tension at the top and bottom of the beam, respectively. Thus, the service load stresses at the top and bottom faces at midspan are

$$f_1 = -140 - 1800 = -1940 \text{ psi}$$

$$f_2 = -1382 + 1800 = +418 \text{ psi}$$

Concrete stresses at midspan are shown in Fig. 22.12*b* and at the beam end in Fig. 22.12*c*. According to the ACI Code (see Table 22.1), the stresses permitted in the concrete are

$$\text{Tension at transfer: } f_{ti} = 3\sqrt{3750} = +184 \text{ psi}$$

$$\text{Compression at transfer: } f_{ci} = 0.70 \times 3750 = -2625 \text{ psi}$$

$$\text{Tension at service load: } f_{ts} = 7.5\sqrt{5000} = +530 \text{ psi}$$

$$\text{Compression at service load: } f_{cs} = 0.45 \times 5000 = -2250 \text{ psi}$$

At the initial stage, with prestress plus self-weight in place, the actual compressive stress of 1708 psi is well below the limit of 2250 psi, and no tension acts at the top, although 184 psi is allowed. While more prestress force or more eccentricity might be suggested to more fully utilize the section, to attempt to do so in this beam, with constant eccentricity, would violate limits at the support, where self-weight moment is zero. It is apparent that at the supports, the initial prestress force acting alone produces tension of 352 psi at the top of the beam (Fig. 22.12*c*), barely below the value of $6\sqrt{3750} = 367$ permitted at the beam end, so very little improvement can be made. The compressive stress at the supports is -2147 psi, well below the magnitude of the permitted value of $0.70f'_{ci} = -2625$ psi. Finally, at full service load, the tension of 418 psi is under the allowed 530 psi, and compression of 1940 psi is well below the permitted 2250 psi.

22.7 FLEXURAL STRENGTH

In an ordinary reinforced concrete beam, the stress in the tensile steel and the compressive force in the concrete increase in proportion to the applied moment up to and somewhat beyond service load, with the distance between the two internal stress resultants remaining essentially constant. In contrast to this behavior, in a prestressed beam, increased moment is resisted by a proportionate increase in the distance between the compressive and tensile resultant forces, the compressive resultant moving upward as the load is increased. The magnitude of the internal forces remains nearly constant up to, and usually somewhat beyond, service loads.

This situation changes drastically upon flexural tensile cracking of the prestressed beam. When the concrete cracks, there is a sudden increase in the stress in the steel as the tension that was formerly carried by the concrete is transferred to it. After cracking, the prestressed beam behaves essentially as an ordinary reinforced concrete beam. The compressive resultant cannot continue to move upward indefinitely, and increasing moment must be accompanied by a nearly proportionate increase in steel stress and compressive force. The strength of a prestressed beam can, therefore, be predicted by the same methods developed for ordinary reinforced concrete beams, with modifications to account for (a) the different shape of the stress-strain curve for prestressing steel, as compared with that for ordinary reinforcement, and (b) the tensile strain already present in the prestressing steel before the beam is loaded.

Highly accurate predictions of the flexural strength of prestressed beams can be made based on a *strain compatibility analysis* that accounts for these factors in a rational and explicit way (Ref. 22.1). For ordinary design purposes, certain approximate relationships have been derived. ACI Code 20.3.2.3 and the accompanying ACI Commentary 20.3.2.3 include approximate equations for flexural strength that are summarized in the following paragraphs.

a.   ## Stress in the Prestressed Steel at Flexural Failure

When a prestressed concrete beam fails in flexure, the prestressing steel is at a stress f_{ps} that is higher than the effective prestress f_{pe} but below the tensile strength f_{pu}. If the effective prestress $f_{pe} = P_e/A_{ps}$ is not less than $0.50f_{pu}$, ACI Code 20.3.2.3 permits use of certain approximate equations for f_{ps}. These equations appear quite complex as they are presented in the ACI Code, mainly because they are written in general form to account for differences in type of prestressing steel and to apply to beams in which nonprestressed bar reinforcement may be included in the flexural tension zone, the compression region, or both. Separate equations are given for members with bonded tendons and unbonded tendons because, in the latter case, the increase in steel stress at the maximum moment section as the beam is overloaded is much less than if the steel were bonded throughout its length.

For the basic case, in which the prestressed steel provides all of the flexural reinforcement, the ACI Code equations can be stated in simplified form as follows:

1. For members with bonded tendons:

$$f_{ps} = f_{pu}\left(1 - \frac{\gamma_p}{\beta_1}\frac{\rho_p f_{pu}}{f_c'}\right) \tag{22.6}$$

where $\rho_p = A_{ps}/bd_p$, d_p = effective depth to the prestressing steel centroid, b = width of compression face, β_1 = the familiar relations between stress block depth and depth to the neutral axis [Eq. (4.22)], and γ_p is a factor that depends on the type of prestressing steel used, as follows:

$$\gamma_p = \begin{cases} 0.55 & \text{for } f_{py}/f_{pu} \geq 0.80 \text{ (typical high-strength bars)} \\ 0.40 & \text{for } f_{py}/f_{pu} \geq 0.85 \text{ (typical ordinary strand)} \\ 0.28 & \text{for } f_{py}/f_{pu} \geq 0.90 \text{ (typical low-relaxation strand)} \end{cases}$$

Virtually all strand in current use satisfies the criteria for low relaxation.

2. For members with unbonded tendons and with a span-depth ratio of 35 or less (this includes most beams),

$$f_{ps} = f_{pe} + 10,000 + \frac{f_c'}{100\rho_p} \tag{22.7}$$

but not greater than f_{py} and not greater than $f_{pe} + 60,000$ psi.

3. For members with unbonded tendons and with span-depth ratio greater than 35 (applying to many slabs),

$$f_{ps} = f_{pe} + 10,000 + \frac{f_c'}{300\rho_p} \tag{22.8}$$

but not greater than f_{py} and not greater than $f_{pe} + 30,000$ psi.

b.   Nominal Flexural Strength and Design Strength

With the stress in the prestressed tensile steel when the member fails in flexure established by Eq. (22.6), (22.7), or (22.8), the nominal flexural strength can be calculated by methods and equations that correspond directly with those used for ordinary reinforced concrete beams. For rectangular cross sections, or flanged sections such as I or T beams in which the stress block depth is equal to or less than the average flange thickness, the nominal flexural strength is

$$M_n = A_{ps} f_{ps} \left(d_p - \frac{a}{2} \right) \tag{22.9}$$

where

$$a = \frac{A_{ps} f_{ps}}{0.85 f_c' b} \tag{22.10}$$

Equations (22.9) and (22.10) can be combined as follows:

$$M_n = \rho_p f_{ps} b d_p^2 \left(1 - 0.588 \frac{\rho_p f_{ps}}{f_c'} \right) \tag{22.11}$$

where $\rho_p = A_{ps}/b_f d_p$, which assumes the compression block is in the flange. Values of ρ_p are typically much lower than reinforcement ratios for reinforced concrete. In all cases, the *flexural design strength* is taken equal to ϕM_n, where ϕ is the strength reduction factor for flexure (see Section 22.7c).

If the stress block depth exceeds the average flange thickness, the method for calculating flexural strength is exactly analogous to that used for ordinary reinforced concrete I and T beams. The total prestressed tensile steel area is divided into two parts for computational purposes. The first part A_{pf}, acting at the stress f_{ps}, provides a tensile force to balance the compression in the overhanging parts of the flange. Thus,

$$A_{pf} = 0.85 \frac{f_c'}{f_{ps}} (b - b_w) h_f \tag{22.12}$$

The remaining prestressed steel area

$$A_{pw} = A_{ps} - A_{pf} \tag{22.13}$$

provides tension to balance the compression in the web. The total resisting moment is the sum of the contributions of the two force couples:

$$M_n = A_{pw} f_{ps} \left(d_p - \frac{a}{2} \right) + A_{pf} f_{ps} \left(d_p - \frac{h_f}{2} \right) \tag{22.14a}$$

or

$$M_n = A_{pw} f_{ps} \left(d_p - \frac{a}{2} \right) + 0.85 f_c' \, (b - b_w) h_f \left(d_p - \frac{h_f}{2} \right) \qquad (22.14b)$$

where

$$a = \frac{A_{pw} f_{ps}}{0.85 f_c' b_w} \qquad (22.15)$$

As before, the design strength is taken as ϕM_n, where ϕ is typically 0.90, as discussed in Section 22.7c.

If, after a prestressed beam is designed by elastic methods at service loads, it has inadequate strength to provide the required safety margin under factored load, nonprestressed reinforcement can be added on the tension side and works in combination with the prestressing steel to provide the needed strength. Such nonprestressed steel, with area A_s, can be assumed to act at its yield stress f_y, to contribute a tension force at the nominal moment of $A_s f_y$. The reader should consult ACI Code and Commentary 20.3.2.3 for equations for prestressed steel stress at failure and for flexural strength, which are direct extensions of those given above.

c.   Limits for Reinforcement

The ACI Code classifies prestressed concrete flexural members as tension-controlled or compression-controlled based on the net tensile strain ε_t in the same manner as done for ordinary reinforced concrete beams. Section 4.3d describes the strain distributions and the variation of strength reduction factors associated with limitations on the net tensile strain. Recall that the net tensile strain excludes strains due to creep, shrinkage, temperature, and effective prestress. To maintain a strength reduction factor ϕ of 0.90 and ensure that if flexural failure were to occur, it would be a ductile failure, a net tensile strain of at least 0.005 is required. Due to the complexity of computing net tensile strain in prestressed members, it is easier to perform the check using the c/d_t ratio. From Fig. 4.9a, this simplifies to

$$\frac{c}{d_t} \le 0.375 \qquad (22.16)$$

where d_t is the distance from the extreme compressive fiber to the extreme tensile steel. In many cases, d_t is the same as d_p, the distance from the extreme compressive fiber to the centroid of the prestressed reinforcement. However, when supplemental nonprestressed steel is used or the prestressing strands are distributed through the depth of the section, d_t is greater than d_p. Prestressed members with $P_u \le 0.1 f_c' A_g$ must meet the c/d_t limit. For prestressed members with $P_u \ge 0.1 f_c' A_g$, the strength reduction factor ϕ must be determined as shown in Fig. 4.8. (see Ref. 22.1).

A *minimum tensile reinforcement ratio* is required for ordinary reinforced concrete beams, so that the beams are safe from sudden failure upon the formation of flexural cracks. Because of the same concern, ACI Code 7.6.2, 8.6.2, and 9.6.2 require that the total tensile reinforcement in members with bonded prestressed reinforcement be adequate to support a factored load of at least 1.2 times the cracking load of the beam, calculated on the basis of a modulus of rupture of $7.5\sqrt{f_c'}$. A similar requirement is not placed on members with unbonded prestressed reinforcement. Unlike members with bonded reinforcement, which are subject to tendon failure when the concrete cracks and the tensile force in the concrete is suddenly transferred to the bonded steel, abrupt failure does not occur in beams with unbonded tendons because the reinforcement can undergo slip, which distributes the increased strain along the length of the tendon, lowering the magnitude of the increased stress in the tendon.

d. Minimum Bonded Reinforcement

To control cracking in beams and one-way prestressed slabs with *unbonded tendons*, some bonded reinforcement must be added in the form of nonprestressed reinforcing bars, uniformly distributed over the tension zone as close as permissible to the extreme tension fiber. According to ACI Code 7.6.2 and 9.6.2, the minimum amount of such reinforcement is

$$A_s = 0.004A_{ct} \tag{22.17}$$

where A_{ct} is the area of that part of the cross section between the flexural tension face and the centroid of the gross concrete cross section. ACI Code 8.6.2 provides requirements for two-way slabs.

EXAMPLE 22.2   **Flexural strength of pretensioned I beam.**   The prestressed I beam shown in cross section in Fig. 22.13 is pretensioned using five low relaxation stress-relieved Grade 270 $\frac{1}{2}$ in. diameter strands, carrying effective prestress $f_{pe} = 160$ ksi. Concrete strength is $f_c' = 4000$ psi. Calculate the design strength of the beam.

SOLUTION.   The effective prestress in the strands of 160 ksi is well above $0.50 \times 270 = 135$ ksi, confirming that the approximate ACI equations are applicable. $A_{ps} = 5 \times 0.153 = 0.765$ in^2. The tensile reinforcement ratio is

$$\rho_p = \frac{0.765}{12 \times 17.19} = 0.0037$$

and the steel stress f_{ps} when the beam fails in flexure is found from Eq. (22.6) to be

$$f_{ps} = f_{pu}\left(1 - \frac{\gamma_p}{\beta_1}\frac{\rho_p f_{pu}}{f_c'}\right) = 270\left(1 - \frac{0.28}{0.85}\frac{0.0037 \times 270}{4}\right) = 248 \text{ ksi}$$

Next, it is necessary to check whether the stress block depth is greater or less than the average flange thickness of 4.5 in. On the assumption that it is not greater than the flange thickness, Eq. (22.10) is used:

$$a = \frac{A_p f_{ps}}{0.85f_c'b} = \frac{0.765 \times 248}{0.85 \times 4 \times 12} = 4.65 \text{ in.}$$

FIGURE 22.13
Post-tensioned beam of
Example 22.2.

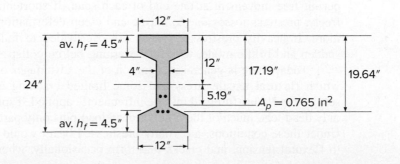

It is concluded from this trial calculation that a actually exceeds h_f, so the trial calculation is not valid and equations for flanged members must be used. The steel that acts with the overhanging flanges is found from Eq. (22.12) to be

$$A_{pf} = \frac{0.85 \times 4(12 - 4)4.5}{248} = 0.494 \text{ in}^2$$

and from Eq. (22.13), the steel acting with the web is

$$A_{pw} = 0.765 - 0.494 = 0.271 \text{ in}^2$$

The actual stress block depth is now found from Eq. (22.15):

$$a = \frac{0.271 \times 248}{0.85 \times 4 \times 4} = 4.94 \text{ in.}$$

$$c = \frac{a}{\beta_1} = \frac{4.94}{0.85} = 5.81$$

A check should now be made to determine if the beam can be considered tension-controlled. As shown in Fig. 22.13, $d_t = 19.64$ in. From Eq. (22.16),

$$\frac{c}{d_t} = \frac{5.81}{19.64} = 0.296$$

This is less than 0.375 for $\varepsilon_t \geq 0.005$, confirming that this can be considered to be a tension-controlled prestressed beam, and $\phi = 0.90$. The nominal flexural strength, from Eq. (22.14b), is

$$M_n = 0.271 \times 248(17.19 - 2.47) + 0.85 \times 4(12 - 4)4.5(17.19 - 2.25)$$

$$= 2818 \text{ in-kips} = 235 \text{ ft-kips}$$

and, finally, the design strength is $\phi M_n = 211$ ft-kips.

22.8 PARTIAL PRESTRESSING

Early in the development of prestressed concrete, the goal of prestressing was the complete elimination of concrete tensile stress at service load. This kind of design, in which the service load tensile stress limit $f_{ts} = 0$, is often referred to as *full prestressing*.

While full prestressing offers many advantages over nonprestressed construction, some problems can arise. Heavily prestressed beams, particularly those for which full live load is seldom in place, may have excessively large upward deflection, or camber, which increases with time because of concrete creep under the eccentric prestress force. Fully prestressed beams may also have a tendency for severe longitudinal shortening, causing large restraint forces unless special provision is made to permit free movement at one end of each span. If shortening is permitted to occur freely, prestress losses due to elastic and creep deformation may be large. Furthermore, if heavily prestressed beams are overloaded to failure, they may fail in a sudden and brittle mode, with little warning before collapse.

Today there is general recognition of the advantages of *partial prestressing*, in which flexural tensile stress and some limited cracking are permitted under full service load. That full load may be infrequently applied. Typically, many beams carry only dead load much of the time, or dead load plus only part of the service live load. Under these conditions, a partially prestressed beam would normally not be subject to flexural tension, and cracks that form occasionally, when the full live load is in place, would close completely when that live load is removed. Controlled cracks prove no more objectionable in prestressed concrete structures than in reinforced concrete structures. With partial prestressing, excessive camber and troublesome axial shortening are avoided. Should overloading occur, there will be ample warning of distress, with extensive cracking and large deflections (Refs. 22.10 to 22.13).

Although the amount of prestressing steel may be reduced in partially prestressed beams compared with fully prestressed beams, a proper safety margin must still be maintained, and to achieve the necessary flexural strength, partially prestressed beams may require additional tensile reinforcement. In fact, partially prestressed beams are often defined as beams in which (1) flexural cracking is permitted at full service load and (2) the main flexural tension reinforcement includes both prestressed and nonprestressed steel. Analysis indicates, and tests confirm, that such nonprestressed steel is fully stressed to f_y at flexural failure.

The ACI Code does not specifically mention partial prestressing but does include the concept explicitly in the classification of flexural members. Class U members generally fit the historical category of fully prestressed members. Class T flexural members require service level stress checks and have maximum allowable tensile stresses above the modulus of rupture. Class C flexural members do not require stress checks at service load but do require crack control checks (Section 22.18). The designations of Class T and C flexural members bring the ACI Code into closer agreement with European practice (Refs. 22.13 to 22.15).

The three classes of prestressed flexural members, U, T, and C (see Table 22.1), provide the designer with considerable flexibility in achieving economical designs. To attain the required strength, supplemental reinforcement in the form of nonprestressed ordinary steel or unstressed prestressing strand may be required. Reinforcing bars are less expensive than high-strength prestressing steel. Strand, however, at twice the cost of ordinary reinforcement, provides 3 times the strength. Labor costs for bar placement are generally similar to those for placing unstressed strand. Similarly, the addition of a small number of strands in a plant prestressing bed is often more economical than adding reinforcing bars; however, a strain compatibility analysis is required for flexural strength. The designer may select the service level performance strategy best suited for the project. A criterion that includes no tensile stress under dead load and a tensile stress less than the modulus of rupture at the service live load is possible with Class U and T flexural members, while Class C members use prestressing primarily for deflection control.

The choice of a suitable degree of prestress is governed by a number of factors. These include the nature of the loading (for example, highway or railroad bridges, and storage warehouses), the ratio of live to dead load, the frequency of occurrence of the full service load, and the presence of a corrosive environment.

22.9 FLEXURAL DESIGN BASED ON CONCRETE STRESS LIMITS

As in reinforced concrete, problems in prestressed concrete can be separated generally as analysis problems or design problems. For the former, with the applied loads, the concrete cross section, steel area, and the amount and point of application of the prestress force known, Eqs. (22.1) to (22.4) permit the direct calculation of the resulting concrete stresses. The equations in Section 22.7 predict the flexural strength. However, if the dimensions of a concrete section, the steel area and centroid location, and the amount of prestress are to be found—given the loads, limiting stresses, and required strength–the problem is complicated by the many interrelated variables.

There are at least three practical approaches to the flexural design of a prestressed concrete member. Some engineers prefer to assume a concrete section, calculate the required prestress force and eccentricities for what is probably the controlling load stage, then check the stresses at all stages using the preceding equations, and finally check the flexural strength. The trial section is then revised if necessary. If a beam is to be chosen from a limited number of standard shapes, as is often the case for shorter spans and ordinary loads, this procedure is probably best. For longer spans or when customized shapes are used, a more efficient member may result by designing the cross section so that the specified concrete stress limits of Table 22.1 are closely matched. This cross section, close to "ideal" from the limit stress viewpoint, may then be modified to meet functional requirements (for example, providing a broad top flange for a bridge deck) or to meet strength requirements, if necessary. Equations facilitating this approach are developed in this section. A third method of design is based on load balancing, using the concept of equivalent loads (see Section 22.2b). A trial section is chosen, after which the prestress force and tendon profile are selected to provide uplift forces as to just balance a specified load. Modifications may then be made, if needed, to satisfy stress limits or strength requirements. This third approach is developed in Section 22.12.

Notation is established pertaining to the allowable concrete stresses at limiting stages as follows:

f_{ci} = allowable compressive stress immediately after transfer

f_{ti} = allowable tensile stress immediately after transfer

f_{cs} = allowable compressive stress at service load, after all losses

f_{ts} = allowable tensile stress at service load, after all losses

The values of these limit stresses are normally set by specification (see Table 22.1).

a. Beams with Variable Eccentricity

For a typical Class U or T beam in which the tendon eccentricity is permitted to vary along the span, flexural stress distributions in the concrete at the maximum moment section are shown in Fig. 22.14a. The eccentric prestress force, having an initial value of P_i, produces the linear stress distribution (1). Because of the upward camber of the beam as that force is applied, the self-weight of the member is immediately introduced, the flexural stresses resulting from the moment M_o are superimposed, and the distribution (2) is the first that is actually attained. At this stage, the tension at the top surface is not to exceed f_{ti}, and the compression at the bottom surface is not to exceed f_{ci}, as shown in Fig. 22.14a.

It is assumed that all the losses occur at this stage, and that the stress distribution changes to distribution (3). The losses produce a reduction of tension in the amount Δf_1 at the top surface and a reduction of compression in the amount Δf_2 at the bottom surface.

As the superimposed dead load moment M_d and the service live load moment M_l are introduced, the associated flexural stresses, when superimposed on stresses already present, produce distribution (4). At this stage, the tension at the bottom surface must not be greater than f_{ts}, and the compression at the top of the section must not exceed f_{cs}.

The requirements for the sections moduli S_1 and S_2 with respect to the top and bottom surfaces, respectively, are

$$S_1 \geq \frac{M_d + M_l}{f_{1r}} \qquad (a)$$

FIGURE 22.14
Flexural stress distributions
for beams with variable
eccentricity: (*a*) maximum
moment section and
(*b*) support section.

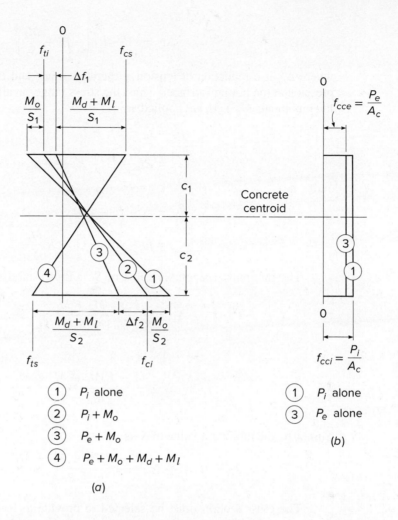

$$S_2 \geq \frac{M_d + M_l}{f_{2r}} \tag{b}$$

where the available stress ranges f_{1r} and f_{2r} at the top and bottom face can be calculated from the specified stress limits f_{ti}, f_{cs}, f_{ts}, and f_{ci}, once the stress changes Δf_1 and Δf_2, associated with prestress loss are known.

The effectiveness ratio R accounts for the loss of prestress and is defined as

$$R = \frac{P_e}{P_i} \tag{22.18}$$

Thus, the loss in prestress force is

$$P_i - P_e = (1 - R)P_i \tag{22.19}$$

The changes in stress at the top and bottom faces, Δf_1 and Δf_2, as losses occur, are equal to $(1 - R)$ times the corresponding stresses due to the initial prestress force P_i *acting alone*:

$$\Delta f_1 = (1 - R)\left(f_{ti} + \frac{M_o}{S_1}\right) \tag{c}$$

$$\Delta f_2 = (1 - R)\left(-f_{ci} + \frac{M_o}{S_2}\right) \tag{d}$$

where Δf_1 is a reduction of tension at the top surface and Δf_2 is a reduction of compression at the bottom surface.[†] Thus, the stress ranges available as the superimposed load moments $M_d + M_l$ are applied are

$$f_{1r} = f_{ti} - \Delta f_1 - f_{cs}$$

$$= Rf_{ti} - (1 - R)\frac{M_o}{S_1} - f_{cs} \tag{e}$$

and

$$f_{2r} = f_{ts} - f_{ci} - \Delta f_2$$

$$= f_{ts} - Rf_{ci} - (1 - R)\frac{M_o}{S_2} \tag{f}$$

The minimum acceptable value of S_1 is thus established:

$$S_1 \geq \frac{M_d + M_l}{Rf_{ti} - (1 - R)M_o/S_1 - f_{cs}}$$

or

$$S_1 \geq \frac{(1 - R)M_o + M_d + M_l}{Rf_{ti} - f_{cs}} \tag{22.20}$$

Similarly, the minimum value of S_2 is

$$S_2 \geq \frac{(1 - R)M_o + M_d + M_l}{f_{ts} - Rf_{ci}} \tag{22.21}$$

The cross section must be selected to provide at least these values of S_1 and S_2. Furthermore, since $I_c = S_1c_1 = S_2c_2$, the centroidal axis must be located such that

$$\frac{c_1}{c_2} = \frac{S_2}{S_1} \tag{g}$$

or in terms of the total section depth $h = c_1 + c_2$

$$\frac{c_1}{h} = \frac{S_2}{S_1 + S_2} \tag{22.22}$$

From Fig. 22.14a, the concrete centroidal stress under initial conditions f_{cci} is given by

$$f_{cci} = f_{ti} - \frac{c_1}{h}(f_{ti} - f_{ci}) \tag{22.23}$$

The initial prestress force is easily obtained by multiplying the value of the concrete centroidal stress by the concrete cross-sectional area A_c.

$$P_i = A_c f_{cci} \tag{22.24}$$

[†] Note that the stress limits such as f_{ti} and other specific points along the stress axis are considered signed quantities, whereas stress changes such as M_o/S_1 and Δf_2 are taken as absolute values.

The eccentricity of the prestress force may be found by considering the flexural stresses that must be imparted by the bending moment $P_i e$. With reference to Fig. 22.14, the flexural stress at the top surface of the beam resulting from the eccentric prestress force alone is

$$\frac{P_i e}{S_1} = (f_{ti} - f_{cci}) + \frac{M_o}{S_1} \tag{h}$$

from which the required eccentricity is

$$e = (f_{ti} - f_{cci})\frac{S_1}{P_i} + \frac{M_o}{P_i} \tag{22.25}$$

Summarizing the design process to determine the best cross section and the required prestress force and eccentricity based on stress limitations: the required section moduli with respect to the top and bottom surfaces of the member are found from Eqs. (22.20) and (22.21) with the centroidal axis located using Eq. (22.22). Concrete dimensions are chosen to satisfy these requirements as nearly as possible. The concrete centroidal stress for this ideal section is given by Eq. (22.23), the desired initial prestress force by Eq. (22.24), and its eccentricity by Eq. (22.25).

In practical situations, very seldom will the concrete section chosen have exactly the required values of S_1 and S_2 as found by this method, nor will the concrete centroid be exactly at the theoretically ideal level. Rounding concrete dimensions upward, providing broad flanges for functional reasons, or using standardized cross-sectional shapes results in a member whose section properties exceed the minimum requirements. In such a case, the stresses in the concrete as the member passes from the unloaded stage to the full service load stage stays within the allowable limits, but the limit stresses are not obtained exactly. An infinite number of combinations of prestress force and eccentricity satisfy the requirements. Usually, the design requiring the lowest value of prestress force, and the largest practical eccentricity, is the most economical.

The total eccentricity in Eq. (22.25) includes the term M_o/P_i. As long as the beam is deep enough to allow this full eccentricity, the girder dead load moment is carried with no additional penalty in terms of prestress force, section, or stress range. This ability to carry the beam dead load "free" is a major contribution of variable eccentricity.

The stress distributions shown in Fig. 22.14a, on which the design equations are based, apply at the maximum moment section of the member. Elsewhere, M_o is less, and, consequently, the prestress eccentricity or the force must be reduced if the stress limits f_{ti} and f_{ci} are not to be exceeded. In many cases, tendon eccentricity is reduced to zero at the support sections, where all moments due to transverse load are zero. In this case, the stress distributions of Fig. 22.14b are obtained. The stress in the concrete is uniformly equal to the centroidal value f_{cci} under conditions of initial prestress and f_{cce} after losses.

EXAMPLE 22.3   **Design of beam with variable eccentricity tendons.**   A post-tensioned prestressed concrete beam is to carry an intermittent live load of 1000 lb/ft and superimposed dead load of 500 lb/ft, in addition to its own weight, on a 40 ft simple span. Normalweight concrete is used with compressive strength $f'_c = 6000$ psi. It is estimated that, at the time of transfer, the concrete will have attained 70 percent of f'_c, or 4200 psi. Time-dependent losses may be assumed to be 15 percent of the initial prestress, giving an effectiveness ratio of 0.85. Determine the required concrete dimensions, magnitude of prestress force, and eccentricity of the steel centroid based on ACI stress limitations for a Class U beam, as given in Sections 22.4 and 22.5.

SOLUTION.   Referring to Table 22.1, the stress limits are

$$f_{ci} = -0.60 \times 4200 = -2520 \text{ psi}$$

$$f_{ti} = 3\sqrt{4200} = +194 \text{ psi}$$

$$f_{cs} = -0.60 \times 6000 = -3600 \text{ psi}$$

$$f_{ts} = 7.5\sqrt{6000} = +581 \text{ psi}$$

The self-weight of the girder is estimated at 250 lb/ft. The service moments due to transverse loading are

$$M_o = \frac{1}{8} \times 0.250 \times 40^2 = 50 \text{ ft-kips}$$

$$M_d + M_l = \frac{1}{8} \times 1.500 \times 40^2 = 300 \text{ ft-kips}$$

The required section moduli with respect to the top and bottom surfaces of the concrete beam are found from Eqs. (22.20) and (22.21).

$$S_1 \geq \frac{(1-R)M_o + M_d + M_l}{Rf_{ti} - f_{cs}} = \frac{(0.15 \times 50 + 300)12,000}{0.85 \times 194 + 3600} = 980 \text{ in}^3$$

$$S_2 \geq \frac{(1-R)M_o + M_d + M_l}{f_{ts} - Rf_{ci}} = \frac{(0.15 \times 50 + 300)12,000}{581 + 0.85 \times 2520} = 1355 \text{ in}^3$$

The values obtained for S_1 and S_2 suggest that an asymmetrical section is most appropriate. However, a symmetrical section is selected for simplicity and to ensure sufficient compression area for flexural strength. The 28 in. deep I section shown in Fig. 22.15a meets the requirements and has the following properties:

$$I_c = 19,904 \text{ in}^4$$

$$S = 1422 \text{ in}^3$$

$$A_c = 240 \text{ in}^2$$

$$r^2 = 82.9 \text{ in}^2$$

$$w_o = 250 \text{ lb/ft (as assumed)}$$

Next, the concrete centroidal stress is found from Eq. (22.23):

$$f_{cci} = f_{ti} - \frac{c_1}{h}(f_{ti} - f_{ci}) = 194 - \frac{1}{2}(195 + 2520) = -1163 \text{ psi}$$

and from Eq. (22.24) the initial prestress force is

$$P_i = A_c f_{cci} = 240 \times 1.163 = 279 \text{ kips}$$

From Eq. (22.25), the required tendon eccentricity at the maximum moment section of the beam is

$$e = (f_{ti} - f_{cci})\frac{S_1}{P_i} + \frac{M_o}{P_i} = (195 + 1163)\frac{1422}{279,000} + \frac{50 \times 12,000}{279,000}$$

$$= 9.07 \text{ in.}$$

Elsewhere along the span, the eccentricity is reduced so that the concrete stress limits are not violated.

The required initial prestress force of 279 kips is provided using tendons consisting of $\frac{1}{2}$ in. diameter Grade 270 low-relaxation strands (see Section 2.16). The minimum tensile

FIGURE 22.15
Design example of beam with variable eccentricity of tendons: (*a*) cross section dimensions and (*b*) concrete stresses at midspan (psi).

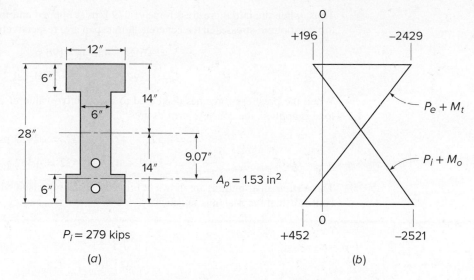

(*a*)

(*b*)

strength is $f_{pu} = 270$ ksi, and the yield strength may be taken as $f_{py} = 0.90 \times 270 = 243$ ksi. According to industry practice (see Section 22.4), the stress in the strand immediately after transfer should not exceed $0.74\,f_{pu} = 200$ ksi. The required area of prestressing steel is

$$A_{ps} = \frac{279}{200} = 1.40 \text{ in}^2$$

The cross-sectional area of one $\frac{1}{2}$ in. diameter strand is 0.153 in^2; hence, the number of strands required is

$$\text{Number of strands} = \frac{1.40}{0.153} = 9.2$$

Two five-strand tendons are selected, as shown in Fig. 22.15*a*; each five-strand tendon is stressed to 139.5 kips immediately following transfer, somewhat less than the maximum of $0.74\,f_{pu}$.

It is good practice to check the calculations by confirming that stress limits are not exceeded at critical load stages. The top and bottom surface concrete stresses produced, in this case, by the separate loadings are

$$P_i: \quad f_1 = -\frac{279{,}000}{240}\left(1 - \frac{9.07 \times 14}{82.9}\right) = +618 \text{ psi}$$

$$f_2 = -\frac{279{,}000}{240}\left(1 + \frac{9.07 \times 14}{82.9}\right) = -2943 \text{ psi}$$

$$P_e: \quad f_1 = 0.85 \times 618 = +525 \text{ psi}$$

$$f_2 = 0.85(-2943) = -2502 \text{ psi}$$

$$M_o: \quad f_1 = -\frac{50 \times 12{,}000}{1422} = -422 \text{ psi}$$

$$f_2 = +422 \text{ psi}$$

$$M_d + M_l: \quad f_1 = -\frac{300 \times 12{,}000}{1422} = -2532 \text{ psi}$$

$$f_2 = +2532 \text{ psi}$$

Thus, when the initial prestress force of 279 kips is applied and the beam self-weight acts, the top and bottom stresses in the concrete at midspan are, respectively,

$$f_1 = +618 - 422 = +196 \text{ psi}$$

$$f_2 = -2943 + 422 = -2521 \text{ psi}$$

When the prestress force has decreased to its effective value of 237 kips and the full service load is applied, the concrete stresses are

$$f_1 = +525 - 422 - 2532 = -2429 \text{ psi}$$

$$f_2 = -2502 + 422 + 2532 = +452 \text{ psi}$$

These stress distributions are shown in Fig. 22.15b. Comparison with the specified limit stresses confirms that the design is satisfactory.

b. Beams with Constant Eccentricity

The design method presented in the previous section was based on stress conditions at the maximum moment section of a beam, with the maximum value of moment M_o resulting from the self-weight immediately being superimposed. If P_i and e were to be held constant along the span, as is often convenient in pretensioned prestressed construction, then the stress limits f_{ti} and f_{ci} would be exceeded elsewhere along the span, where M_o is less than its maximum value. To avoid this condition, the constant eccentricity must be less than that given by Eq. (22.25). Its maximum value is given by conditions at the support of a simple span, where M_o is zero.

Figure 22.16 shows the flexural stress distributions at the support and midspan sections for a beam with constant eccentricity. In this case, the stress limits f_{ti} and f_{ci} are not to be violated when the eccentric prestress moment acts alone, as at the supports. The stress changes Δf_1 and Δf_2 as losses occurring are equal to $(1 - R)$ times the top and bottom surface stresses, respectively, due to initial prestress alone:

$$\Delta f_1 = (1 - R) \, (f_{ti}) \tag{a}$$

$$\Delta f_2 = (1 - R) \, (-f_{ci}) \tag{b}$$

In this case, the available stress ranges between limit stresses must provide for the effect of M_o as well as M_d and M_l, as seen from Fig. 22.16a, and are

$$f_{1r} = f_{ti} - \Delta f_1 - f_{cs}$$

$$= R f_{ti} - f_{cs} \tag{c}$$

$$f_{2r} = f_{ts} - f_{ci} - \Delta f_2$$

$$= f_{ts} - R f_{ci} \tag{d}$$

and the requirements on the section moduli are that

$$S_1 \geq \frac{M_o + M_d + M_l}{R f_{ti} - f_{cs}} \tag{22.26}$$

$$S_2 \geq \frac{M_o + M_d + M_l}{f_{ts} - R f_{ci}} \tag{22.27}$$

FIGURE 22.16
Flexural stress distributions
for beam with constant
eccentricity of tendons:
(*a*) maximum moment
section and (*b*) support
section.

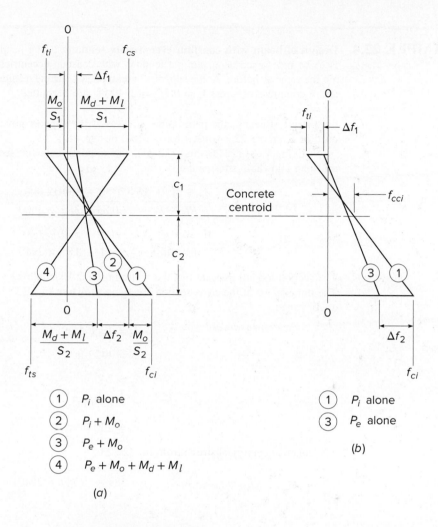

(1) P_i alone

(2) $P_i + M_o$

(3) $P_e + M_o$

(4) $P_e + M_o + M_d + M_l$

(*a*)

(1) P_i alone

(3) P_e alone

(*b*)

The concrete centroidal stress may be found by Eq. (22.23) and the initial prestress force by Eq. (22.24) as before. However, the expression for required eccentricity differs. In this case, referring to Fig. 22.16*b*,

$$\frac{P_i e}{S_1} = f_{ti} - f_{cci} \qquad (e)$$

from which the required eccentricity is

$$e = (f_{ti} - f_{cci}) \frac{S_1}{P_i} \qquad (22.28)$$

A significant difference between beams with variable eccentricity and those with constant eccentricity is noted by comparing Eqs. (22.20) and (22.21) with the corresponding Eqs. (22.26) and (22.27). In the first case, the section modulus requirement is governed mainly by the superimposed load moments M_d and M_l. Almost all of the self-weight is carried "free," that is, without increasing section modulus or prestress force, by the simple expedient of increasing the eccentricity along the span by the amount M_o/P_i. In the second case, the eccentricity is controlled by conditions at the supports, where M_o is zero, and the full moment M_o due to self-weight must be included in determining section moduli. Nevertheless, beams with constant eccentricity are often used for practical reasons.

EXAMPLE 22.4   **Design of beam with constant eccentricity tendons.**   The beam in the preceding example is to be redesigned using straight tendons with constant eccentricity. All other design criteria are the same as before. At the supports, a temporary concrete tensile stress $f_{ti} = 6\sqrt{f'_{ci}} = 389$ psi and a compressive stress $f_{ci} = 0.7f'_{ci} = 2940$ psi are permitted.

SOLUTION.   Because the permissible stresses at the support govern in this case, the section used in Example 22.3 is used for the trial design.

Using Eqs. (22.26) and (22.27), the requirements for section moduli based on the midspan allowable stresses are

$$S_1 \geq \frac{M_o + M_d + M_l}{Rf_{ti} - f_{cs}} = \frac{(50 + 300)12,000}{0.85 \times 389 + 3600} = 1069 \text{ in}^3$$

$$S_2 \geq \frac{M_o + M_d + M_l}{f_{ts} - Rf_{ci}} = \frac{(50 + 300)12,000}{581 + 0.85 \times 2940} = 1363 \text{ in}^3$$

The section moduli indicate that the section used in Example 22.3 works in this case, as well. The dimensions of the cross section are again shown in Fig. 22.17a. The following properties are obtained:

$$l_c = 19,904 \text{ in}^4$$

$$S = 1422 \text{ in}^3$$

$$A_c = 240 \text{ in}^2$$

$$r^2 = 82.9 \text{ in}^2$$

$$w_o = 250 \text{ lb/ft}$$

The concrete centroidal stress, from Eq. (22.23), is

$$f_{cci} = f_{ti} - \frac{c_1}{h}(f_{ti} - f_{ci}) = 389 - \frac{1}{2}(389 + 2940) = -1276 \text{ psi}$$

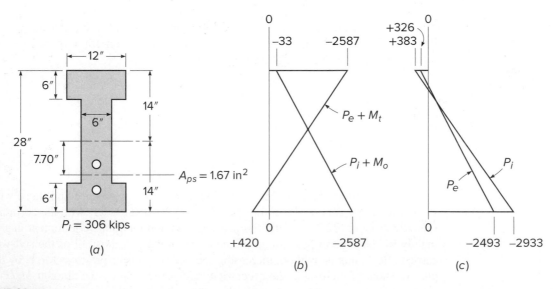

(a)   *(b)*   *(c)*

FIGURE 22.17
Design example of beam with constant eccentricity of tendons: (*a*) cross section dimensions; (*b*) stresses at midspan (psi); and (*c*) stresses at supports (psi).

and from Eq. (22.24), the initial prestress force is

$$P_i = A_c f_{cci} = 240 \times 1.276 = 306 \text{ kips}$$

From Eq. (22.28), the required constant eccentricity is

$$e = (f_{ti} - f_{cci})\frac{S_1}{P_i} = (389 + 1276)\frac{1422}{306,000} = 7.74 \text{ in.}$$

An eccentricity $e = 7.70$ in. is selected.

Again, two tendons are used to provide the required prestress force, each composed of multiple Grade 270 low-relaxation strands. With the maximum permissible stress in the tendon just after transfer of 199 ksi, the total required steel area is

$$A_{ps} = \frac{306}{200} = 1.53 \text{ in}^2$$

This is just over the area provided by the ten $\frac{1}{2}$ in. strands used in Example 22.3. In their place, ten 0.60 in. diameter strands are selected (Table A.15) providing a total area of 1.74 in. Two identical five-strand tendons are used as before, in this case being stressed to a total of 306 kips.

The calculations are checked by verifying the concrete stresses at the top and bottom of the beam for the critical load stages. The component stress contributions are

$$P_i: \qquad f_1 = -\frac{306,000}{240}\left(1 - \frac{7.70 \times 14.0}{82.9}\right) = +383 \text{ psi}$$

$$f_2 = -\frac{306,000}{240}\left(1 + \frac{7.70 \times 14.0}{82.9}\right) = -2933 \text{ psi}$$

$$P_e: \qquad f_1 = 0.85 \times 383 = +326 \text{ psi}$$

$$f_2 = 0.85(-2933) = -2493 \text{ psi}$$

$$M_o: \qquad f_1 = -\frac{50 \times 12,000}{1422} = -422 \text{ psi}$$

$$f_2 = +422 \text{ psi}$$

$$M_d + M_l: \qquad f_1 = -\frac{300 \times 12,000}{1422} = -2532 \text{ psi}$$

$$f_2 = +2532 \text{ psi}$$

Superimposing the appropriate stress contributions, the stress distributions in the concrete at midspan and at the supports are obtained, as shown in Fig. 22.17b and c, respectively. When the initial prestress force of 306 kips acts alone, as at the supports, the stresses at the top and bottom surfaces are

$$f_1 = +383 \text{ psi}$$

$$f_2 = -2933 \text{ psi}$$

After losses, the prestress force is reduced to 260 kips and the support stresses are reduced accordingly. At midspan, the beam weight is immediately superimposed, and stresses resulting from P_i plus M_o are

$$f_1 = +383 - 422 = -39 \text{ psi}$$

$$f_2 = -2933 + 422 = -2411 \text{ psi}$$

When the full service load acts, together with P_e, the midspan stresses are

$$f_1 = +326 - 422 - 2532 = -2628 \text{ psi}$$

$$f_2 = -2493 + 422 + 2532 = +461 \text{ psi}$$

If we check against the specified limiting stresses, it is evident that the design is satisfactory in this respect at the critical load stages and locations, including the allowable stress at the end of the simply supported beam of $0.70f'_{ci}$ and, with the addition of the beam self-weight, the criterion of $0.60f'_{ci}$ at midspan.

22.10 SHAPE SELECTION

One of the special features of prestressed concrete design is the freedom to select cross-section proportions and dimensions to suit the special requirements of the job at hand. The member depth can be changed, the web thickness modified, and the flange widths and thicknesses varied independently to produce a beam with nearly ideal proportions for a given case.

Several common precast shapes are shown in Fig. 22.18. Some of these are standardized and mass-produced, employing reusable steel or fiberglass forms. Others are individually proportioned for large and important works. The double T (Fig. 22.18a) is probably the most widely used cross section in U.S. prestressed construction. A flat surface is provided, 4 to 15 ft wide. Slab thicknesses and web depths vary, depending upon requirements. Spans to 60 ft are not unusual. The single T (Fig. 22.18b) is more appropriate for longer spans, to 120 ft, and heavier loads. The I and bulb T sections (Fig. 22.18c and d) are widely used for bridge spans and roof girders up to about 140 ft, while the channel slab (Fig. 22.18e) is suitable for floors in the intermediate span range. The box girder (Fig. 22.18f) and custom shapes are advantageous for bridges of intermediate and major span. The inverted T section (Fig. 22.18g) provides a bearing ledge to carry the ends of precast deck members spanning in the perpendicular direction. Local precasting plants can provide catalogs of available shapes. This information is also available in the *PCI Design Handbook* (Ref. 22.8).

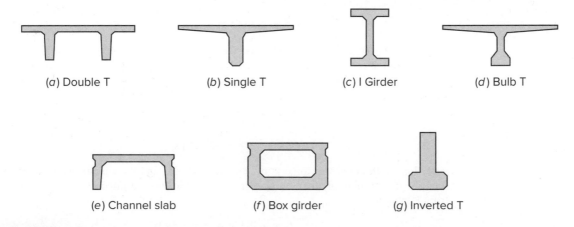

(a) Double T   (b) Single T   (c) I Girder   (d) Bulb T

(e) Channel slab   (f) Box girder   (g) Inverted T

FIGURE 22.18
Typical beam cross sections.

As indicated, the cross section may be symmetrical or unsymmetrical. An unsymmetrical section is a good choice (1) if the available stress ranges f_{1r} and f_{2r} at the top and bottom surfaces are not the same; (2) if the beam must provide a flat, useful surface as well as offering load-carrying capacity; (3) if the beam is to become a part of composite construction, with a cast-in-place slab acting together with a precast web; or (4) if the beam must provide support surfaces, as shown in Fig. 22.18*g*. In addition, T sections provide increased flexural strength, since the internal arm of the resisting couple at maximum design load is greater than that for rectangular sections.

Generally speaking, I, T, and box sections with relatively thin webs and flanges are more efficient than members with thicker parts. However, several factors limit the gain in efficiency that may be obtained in this way. These include the instability of very thin overhanging compression parts, the vulnerability of thin parts to breakage in handling (in the case of precast construction), and the practical difficulty of placing concrete in very thin elements. The designer must also recognize the need to provide adequate spacing and concrete protection for tendons and anchorages, the importance of construction depth limitations, and the need for lateral stability if the beam is not braced by other members against buckling (Ref. 22.16).

22.11 TENDON PROFILES

The equations developed in Section 22.9a for members with variable tendon eccentricity establish the requirements for section modulus, prestress force, and eccentricity at the maximum moment section of the member. Elsewhere along the span, the eccentricity of the steel must be reduced if the concrete stress limits for the unloaded stage are not to be exceeded. (Alternatively, the section must be increased, as demonstrated in Section 22.9b.) Conversely, there is a minimum eccentricity, or upper limit for the steel centroid, such that the limiting concrete stresses are not exceeded when the beam is in the full service load stage.

Limiting locations for the prestressing steel centroid at any point along the span can be established using Eqs. (22.2) and (22.4), which give the values of concrete stress at the top and bottom of the beam in the unloaded and service load stages, respectively. The stresses produced for those load stages should be compared with the limiting stresses applicable in a particular case, such as the ACI stress limits of Table 22.1. This permits a solution for tendon eccentricity e as a function of distance x along the span.

To indicate that both eccentricity e and moments M_o or M_t are functions of distance x from the support, they are written as $e(x)$ and $M_o(x)$ or $M_t(x)$, respectively. In writing statements of inequality, it is convenient to designate tensile stress as larger than zero and compressive stress as smaller than zero. Thus, $+450 > -1350$, and $-600 > -1140$, for example.

Considering first the unloaded stage, the tensile stress at the top of the beam must not exceed f_{ti}. From Eq. (22.2b),

$$f_{ti} \geq -\frac{P_i}{A_c}\left[1 - \frac{e(x)c_1}{r^2}\right] - \frac{M_o(x)}{S_1} \tag{a}$$

Solving for the maximum eccentricity gives

$$e(x) \leq \frac{f_{ti}\,S_1}{P_i} + \frac{S_1}{A_c} + \frac{M_o(x)}{P_i} \tag{22.29}$$

At the bottom of the unloaded beam, the stress must not exceed the limiting initial compression. From Eq. (22.2*b*),

$$f_{ci} \leq -\frac{P_i}{A_c}\left[1 + \frac{e(x)c_2}{r^2}\right] + \frac{M_o(x)}{S_2} \tag{b}$$

Hence, the second lower limit for the steel centroid is

$$e(x) \leq -\frac{f_{ci}\,S_2}{P_i} - \frac{S_2}{A_c} + \frac{M_o(x)}{P_i} \tag{22.30}$$

Now considering the member in the fully loaded stage, the upper limit values for the eccentricity may be found. From Eq. (22.4*a*),

$$f_{cs} \leq -\frac{P_e}{A_c}\left[1 - \frac{e(x)c_1}{r^2}\right] - \frac{M_t(x)}{S_1} \tag{c}$$

from which

$$e(x) \geq \frac{f_{cs}\,S_1}{P_e} + \frac{S_1}{A_c} + \frac{M_t(x)}{P_e} \tag{22.31}$$

and using Eq. (22.4*b*)

$$f_{ts} \geq -\frac{P_e}{A_c}\left[1 + \frac{e(x)c_2}{r^2}\right] + \frac{M_t(x)}{S_2} \tag{d}$$

from which

$$e(x) \geq -\frac{f_{ts}\,S_2}{P_e} - \frac{S_2}{A_c} + \frac{M_t(x)}{P_e} \tag{22.32}$$

Using Eqs. (22.29) and (22.30), the lower limit of tendon eccentricity is established at successive points along the span. Then, using Eqs. (22.31) and (22.32), the corresponding upper limit is established. This upper limit may well be negative, indicating that the tendon centroid may be above the concrete centroid at that location.

It is often convenient to plot the envelope of acceptable tendon profiles, as done in Fig. 22.19, for a typical case in which both dead and live loads are uniformly distributed. Any tendon centroid falling completely within the shaded zone would be satisfactory from the point of view of concrete stress limits. It should be emphasized that it is only the tendon centroid that must be within the shaded zone; individual strands are often outside of it.

The tendon profile actually used is often a parabolic curve or a catenary in the case of post-tensioned beams. The duct containing the prestressing steel is draped to the desired shape and held in that position by wiring it to the transverse web reinforcement, after which the concrete may be placed. In pretensioned beams, *deflected* or *harped tendons* are often used. The strands are held down at midspan, at the third points, or at the quarter points of the span and held up at the ends, so that a smooth curve is approximated to a greater or lesser degree.

In practical cases, it is often not necessary to make a centroid zone diagram, as is shown in Fig. 22.19. By placing the centroid at its known location at midspan, at or close to the concrete centroid at the supports, and with a near-parabolic shape

FIGURE 22.19
Typical limiting zone for
centroid of prestressing steel.

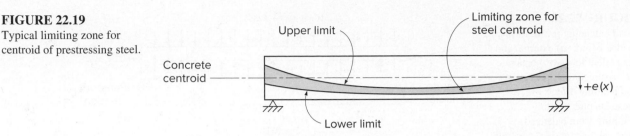

between those control points, satisfaction of the limiting stress requirements is
ensured. With nonprismatic beams, beams in which a curved concrete centroidal axis
is employed, or with continuous beams, diagrams such as Fig. 22.19 are a great aid.

22.12   FLEXURAL DESIGN BASED ON LOAD BALANCING

It was pointed out in Section 22.2b that the effect of a change in the alignment of a
prestressing tendon in a beam is to produce a vertical force on the beam at that location. Prestressing a member with curved or deflected tendons thus has the effect of
introducing a set of equivalent loads, and these may be treated just as any other loads
in finding moments or deflections. Each particular tendon profile produces its own
unique set of equivalent forces. Typical tendon profiles, with corresponding equivalent loads and moment diagrams, were illustrated in Fig. 22.2. Both Fig. 22.2 and
Section 22.2b should be reviewed carefully.

The equivalent load concept offers an alternative approach to the determination
of required prestress force and eccentricity. The prestress force and tendon profile
can be established so that external loads that act are exactly counteracted by the
vertical forces resulting from prestressing. The net result, for that particular set of
external loads, is that the beam is subjected only to axial compression and no bending moment. The selection of the load to be balanced is left to the judgment of the
designer. Often the balanced load chosen is the sum of the self-weight and superimposed dead load.

The design approach described in this section was introduced in the
United States by T. Y. Lin in 1963 and is known as the *load-balancing method*.
The fundamentals are illustrated in the context of the simply supported, uniformly
loaded beam shown in Fig. 22.20a. The beam is to be designed for a balanced load
consisting of its own weight w_o, the superimposed dead load w_d, and some fractional
part of the live load, denoted by $k_b w_l$. Since the external load is uniformly distributed, it is reasonable to adopt a tendon having a parabolic shape. It is easily shown
that a parabolic tendon produces a uniformly distributed upward load equal to

$$w_p = \frac{8Py}{\ell^2} \tag{22.33}$$

where P = magnitude of prestress force
  y = maximum sag of tendon measured with respect to the chord between its
    endpoints
  ℓ = span length

If the downward load exactly equals the upward load from the tendon, these
two loads cancel and no bending stress is produced, as shown in Fig. 22.20b. The
bending stresses due to prestress eccentricity are equal and opposite to the bending

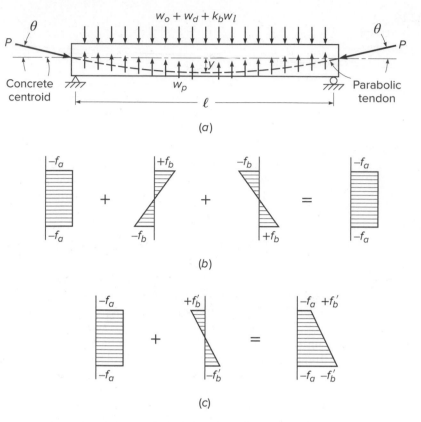

stresses resulting from the external load. The net resulting stress is uniform compression f_a equal to that produced by the axial force $P \cos \theta$. Excluding consideration of time-dependent effects, the beam would show no vertical deflection.

If the live load is removed or increased, then bending stresses and deflections results because of the *unbalanced* portion of the load. Stresses resulting from this differential loading must be calculated and superimposed on the axial compression to obtain the net stresses for the unbalanced state. Referring to Fig. 22.20*c*, the bending stresses f_b' resulting from removal of the partial live loading are superimposed on the uniform compressive stress f_a, resulting from the combination of eccentric prestress force and full balanced load to produce the final stress distribution shown.

Loads other than uniformly distributed would lead naturally to the selection of other tendon configurations. For example, if the external load consisted of a single concentration at midspan, a deflected tendon such as that of Fig. 22.2*a* would be chosen, with maximum eccentricity at midspan, varying linearly to zero eccentricity at the supports. A third-point loading would lead the designer to select a tendon deflected at the third points. A uniformly loaded cantilever beam would best be stressed using a tendon in which the eccentricity varied parabolically, from zero at the free end to y at the fixed support, in which case the upward reaction of the tendon would be

$$w_p = \frac{2Py}{\ell^2} \tag{22.34}$$

It should be clear that, for simple spans designed by the load-balancing concept, it is necessary for the tendon to have zero eccentricity at the supports because the moment due to superimposed loads is zero there. Any tendon eccentricity would produce an unbalanced moment (in itself an equivalent load) equal to the horizontal component of the prestress force times its eccentricity. At the simply supported ends, the requirement of zero eccentricity must be retained.

In practice, the load-balancing method of design starts with selection of a trial beam cross section, based on experience and judgment. An appropriate span-depth ratio is often applied. The tendon profile is selected using the maximum available eccentricity, and the prestress force is calculated. The trial design is then checked to ensure that concrete stresses are within the allowable limits should the live load be totally absent or fully in place, when bending stresses are superimposed on the axial compressive stresses. There is no assurance that the section will be adequate for these load stages or that adequate strength will be provided should the member be overloaded. Revision to correct any deficiency may be necessary.

It should further be observed that obtaining a uniform compressive concrete stress at the balanced load stage does not ensure that the member has zero deflection at this stage. The reason is that the uniform stress distribution is made up of two parts: that from the eccentric prestress force and that from the external loads. The prestress force varies with time because of shrinkage, creep, and relaxation, changing the vertical deflection associated with the prestress force. Concurrently, the beam experiences creep deflection under the combined effects of the diminishing prestress force and the external loads, a part of which may be sustained and a part of which may be short-term. However, if load balancing is carried out based on the effective prestress force P_e plus self-weight and external dead load only, the result may be near-zero deflection for that combination.

The load-balancing method provides the engineer with a useful tool. For simple spans, it leads the designer to choose a sensible tendon profile and focuses attention very early on the matter of deflection. But the most important advantages become evident in the design of indeterminate prestressed members, including both continuous beams and two-way slabs. For such cases, only the unbalanced load need be considered in conjunction with the axial compression, greatly simplifying the analysis effort.

EXAMPLE 22.5   **Beam design initiating with load balancing.**   A post-tensioned beam is to be designed to carry a uniformly distributed load over a 30 ft span, as shown in Fig. 22.21. In addition to its own weight, it must carry a dead load of 150 lb/ft and a service live load of 600 lb/ft. Concrete strength of 4000 psi is attained at 28 days; at the time of transfer of the prestress force, the strength is 3000 psi. Prestress loss may be assumed at 20 percent of P_i. On the basis that about one-quarter of the live load will be sustained over a substantial time period, k_b of 0.25 is used in determining the balanced load.

SOLUTION.   On the basis of an arbitrarily chosen span-depth ratio of 18, a 20 in. deep, 10 in. wide trial section is selected. The calculated self-weight of the beam is 208 lb/ft, and the selected load to be balanced is

$$w_{bal} = w_o + w_d + k_b w_l = 208 + 150 + 0.25 \times 600 = 508 \text{ lb/ft}$$

Based on a minimum concrete cover from the steel centroid to the bottom face of the beam of 4 in., the maximum eccentricity that can be used for the 20 in. trial section is 6 in. A parabolic tendon is selected to produce a uniformly distributed upward tendon load. To equilibrate the sustained downward loading, the prestress force P_e after losses, from Eq. (22.33), should be

$$P_e = \frac{w_{bal}\ell^2}{8y} = \frac{508 \times 900}{8 \times 0.5} = 114{,}000 \text{ lb}$$

FIGURE 22.21

Example of design by load balancing: (*a*) beam profile and cross section and (*b*) flexural stresses at maximum moment section (psi).

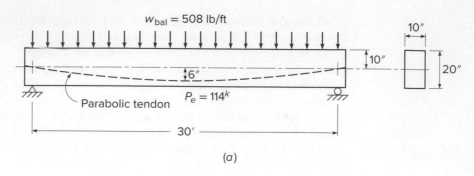

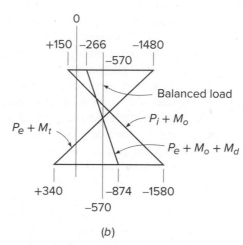

and the corresponding initial prestress force is

$$P_i = \frac{P_e}{R} = \frac{114,000}{0.8} = 143,000 \text{ lb}$$

For the balanced load stage, the concrete is subjected to a uniform compressive stress of

$$f_{bal} = \frac{114,000}{200} = -570 \text{ psi}$$

as shown in Fig. 22.21*b*. Should the partial live load of 150 lb/ft be removed, the stresses to be superimposed on f_{bal} result from a net *upward* load of 150 lb/ft. The section modulus for the trial beam is 667 in³ and

$$M_{unbal} = 150 \times \frac{900}{8} = 16,900 \text{ ft-lb}$$

Hence, the unbalanced bending stresses at the top and bottom faces are

$$f_{unbal} = 16,900 \times \frac{12}{667} = 304 \text{ psi}$$

Thus, the net stresses are

$$f_1 = -570 + 304 = -266 \text{ psi}$$
$$f_2 = -570 - 304 = -874 \text{ psi}$$

Similarly, if the *full* live load should act, the stresses to be superimposed are those resulting from a net *downward* load of 450 lb/ft. The resulting stresses in the concrete at full service load are

$$f_1 = -570 - 910 = -1480 \text{ psi}$$
$$f_2 = -570 + 910 = +340 \text{ psi}$$

Stresses in the concrete with live load absent and live load fully in place are shown in Fig. 22.21b.

It is also necessary to investigate the stresses in the initial unloaded stage, when the member is subjected to P_i plus moment due to its own weight.

$$M_o = 208 \times \frac{900}{8} = 23,400 \text{ ft-lb}$$

Hence, in the initial stage:

$$f_1 = -\frac{143,000}{200}\left(1 - \frac{6 \times 10}{33.35}\right) - \frac{23,400 \times 12}{667} = +150 \text{ psi}$$

$$f_2 = -\frac{143,000}{200}\left(1 + \frac{6 \times 10}{33.35}\right) + \frac{23,400 \times 12}{667} = -1580 \text{ psi}$$

The stresses in the unloaded and full service load stages must be checked against these permitted by the ACI Code. With $f_c' = 4000$ psi and $f_{ci}' = 3000$ psi, the stresses permitted for a Class U member are

$$f_{ti} = +165 \text{ psi} \qquad f_{ts} = +474 \text{ psi}$$

$$f_{ci} = -1800 \text{ psi} \qquad f_{cs} = -2400 \text{ psi}$$

The actual stresses, shown in Fig. 22.21b, are within these limits and acceptably close to the allowable limits, and no revision is needed in the trial 10×20 in. cross section on the basis of stress limits.

The flexural strength of the members must now be checked, to ensure that an adequate margin of safety against collapse has been provided. The required P_i of 143,000 lb is provided using Grade 270 strand, with $f_{pu} = 270,000$ psi and $f_{py} = 243,000$ psi. Referring to Section 22.4, the initial stress immediately after transfer must not exceed $0.74 \times 270,000 = 200,000$ psi. Accordingly, the required area of tendon steel is

$$A_{ps} = 143,000/200,000 = 0.72 \text{ in}^2$$

This is provided using five $\frac{1}{2}$ in. strands, giving an actual area of 0.765 in^2 (Table A.15). The resulting stresses at the initial and final stages are

$$f_{pi} = \frac{143,000}{0.765} = 187,000 \text{ psi}$$

$$f_{pe} = \frac{114,000}{0.765} = 149,000 \text{ psi}$$

Using the ACI approximate equation for steel stress at failure [see Eq. (22.6)], with $\rho_p = 0.765/160 = 0.0048$ and $\gamma_p = 0.40$ for the ordinary Grade 270 tendons, the stress f_{ps} is given by

$$f_{ps} = f_{pu}\left(1 - \frac{\gamma_p}{\beta_1}\frac{\rho_p f_{pu}}{f_c'}\right)$$

$$= 270\left(1 - \frac{0.40}{0.85}\frac{0.0048 \times 270}{4}\right)$$

$$= 229 \text{ ksi}$$

Then

$$a = \frac{A_{ps}f_{ps}}{0.85f_c'b}$$

$$= \frac{0.765 \times 229}{0.85 \times 4 \times 10} = 5.15 \text{ in.}$$

Placing the two five-strand tendons in two layers results in $d_t = 16.6$ in. Then

$$c = \frac{5.15}{0.85} = 6.06$$

$$\frac{c}{d_t} = \frac{6.06}{16.6} = 0.365$$

This is less than the limit $c/d_t = 0.375$, thus $\phi = 0.90$. The nominal flexural strength is

$$M_n = A_{ps}\,f_{ps}\left(d - \frac{a}{2}\right) = 0.765 \times 229{,}000\left(16 - \frac{5.15}{2}\right)\frac{1}{12}$$

$$= 196{,}000 \text{ ft-lb}$$

and the design strength with $\phi = 0.90$ is

$$\phi M_n = 0.90 \times 196{,}000 = 176{,}000 \text{ ft-lb}$$

The ACI load factors with respect to dead and live loads are, respectively, 1.2 and 1.6. Calculating the factored load,

$$w_u = 1.2(208 + 150) + 1.6(600) = 1390 \text{ lb/ft}$$

$$M_u = \frac{1390(900)}{8} = 156{,}000 \text{ ft-lb}$$

Thus, $\phi M_n > M_u$, and the design is judged satisfactory.

22.13 LOSS OF PRESTRESS

As discussed in Section 22.6, the initial prestress force P_i immediately after transfer is less than the jacking force P_j because of elastic shortening of the concrete, slip at the anchorages, and frictional losses along the tendons. The force is reduced further, after a period of many months or even years, due to length changes resulting from shrinkage and creep of the concrete and relaxation of the highly stressed steel; eventually it attains its effective value P_e. In the preceding sections of this chapter, losses were accounted for, making use of an assumed effectiveness ratio $R = P_e/P_i$. Losses have no effect on the nominal strength of a member with bonded tendons, but overestimation or underestimation of losses may have a pronounced effect on service conditions including camber, deflection, and cracking.

ACI Committee 423 prepared a comprehensive guide to estimating prestress losses (Ref. 22.17). The report indicates that the average prestress loss, based on data reported in the literature for all beam types, is 41 ksi with a coefficient of variation of 38 percent.

ACI Code 20.3.2.6.1 requires that prestress loss be considered in the calculation of final stresses and provides a list of categories to be considered. The estimation of losses can be made on several different levels. Lump-sum losses are used initially for some designs. For cases where greater accuracy is required or for confirmation of lump sum losses, it is necessary to estimate the separate losses, taking account of the conditions of member geometry, material properties, and construction methods that apply. Accuracy of loss estimation can be improved still further by accounting for the interdependence of time-dependent losses, using the summation of losses in a sequence of discrete time steps. These methods are discussed briefly in the following paragraphs.

TABLE 22.2
Estimate of prestress losses

Type of Beam Section	Level	Wires or Strands with f_{pu} = 235, 250, or 270 ksi[a]
Rectangular beams, solid slabs	Upper bound	33.0 ksi
	Average	30.0 ksi
Box girder	Upper bound	25.0 ksi
	Average	23.0 ksi
I girder	Average	$33.0[1 - 0.15(f_c' - 6.0)/6.0] + 6.0$
Single T, double T, hollow core and voided slab	Upper bound	$39.0[1 - 0.15(f_c' - 6.0)/6.0] + 6.0$
	Average	$33.0[1 - 0.15(f_c' - 6.0)/6.0] + 6.0$

[a] Values are for fully prestressed beams; reductions are allowed for partial prestress. Losses due to friction are excluded. Friction losses should be calculated according to Section 22.13b. For low-relaxation strands, the values specified may be reduced by 4.0 ksi for box girders; 6.0 ksi for rectangular beams, solid slabs, and I girders; and 8.0 ksi for single T's, double T's, and hollow core and voided slabs.
Table adapted from Ref. 22.18.

a. Lump-Sum Estimates of Losses

It was recognized very early in the development of prestressed concrete that there was a need for approximate expressions to be used to estimate prestress losses in design. Many thousands of successful prestressed structures have been built based on such estimates, and where member sizes, spans, materials, construction procedures, prestress forces, and environmental conditions are not out of the ordinary, this approach is satisfactory. For such conditions, the American Association of State Highway and Transportation Officials (AASHTO, Ref. 22.18) has recommended the values in Table 22.2 for preliminary design or for certain controlled precasting conditions. Losses due to friction must be added to these values for post-tensioned members. These may be calculated separately by the equations of Section 22.13b.

The AASHTO recommended losses of Table 22.2 include losses due to elastic shortening, creep, shrinkage, and relaxation (see Section 22.13b). Thus, for comparison with R values for estimating losses, such as were employed for the preceding examples, which included only the time-dependent losses due to shrinkage, creep, and relaxation, elastic shortening losses should be estimated by the methods discussed in Section 22.13b and deducted from the total.

b. Estimate of Separate Losses

A separate estimate of individual losses is made for most designs and specifically required when using the ACI Code. Such an analysis is complicated by the interdependence of time-dependent losses. For example, the relaxation of stress in the tendons is affected by length changes due to creep of concrete. Rate of creep, in turn, is altered by change in tendon stress. In the following six sections, losses are treated as if they occurred independently, although certain arbitrary adjustments are included to account for the interdependence of time-dependent losses. If greater refinement is necessary, a step-by-step approach like that mentioned in Section 22.13c may be used (see also Refs. 22.8 and 22.17).

(1) SLIP AT THE ANCHORAGES   As the load is transferred to the anchorage device in post-tensioned construction, a slight inward movement of the tendon occurs as the wedges seat themselves and as the anchorage itself deforms under stress. The amount of movement varies greatly, depending on the type of anchorage and on construction techniques. The amount of movement due to seating and stress deformation associated with any particular type of anchorage is best established by test. Once this amount $\Delta\ell$ is determined, the stress loss is easily calculated from

$$\Delta f_{s,\text{slip}} = \frac{\Delta\ell}{\ell} E_s \qquad (22.35)$$

It is significant to note that the amount of slip is nearly independent of the tendon length. For this reason, the stress loss is large for short tendons and relatively small for long tendons. The practical consequence of this is that it is most difficult to post-tension short tendons with any degree of accuracy. Prestressing plants with long-line stressing beds stress the tendons to compensate for anchor seating losses.

(2) ELASTIC SHORTENING OF THE CONCRETE   In pretensioned members, as the tendon force is transferred from the fixed abutments to the concrete beam, elastic instantaneous compressive strain takes place in the concrete, tending to reduce the stress in the bonded prestressing steel. The steel stress loss is

$$\Delta f_{s,\text{elastic}} = E_s \frac{f_c}{E_c} = nf_c \qquad (22.36)$$

where f_c is the concrete stress at the level of the steel centroid immediately after prestress is applied:

$$f_c = -\frac{P_i}{A_c}\left(1 + \frac{e^2}{r^2}\right) + \frac{M_o e}{I_c} \qquad (22.37)$$

If the tendons are placed with significantly different effective depths, the stress loss in each can be calculated separately.

In computing f_c by Eq. (22.37), the prestress force used should be that after the losses being calculated have occurred. It is usually adequate to estimate this as about 10 percent less than P_j.

In post-tensioned members, if all of the strands are tensioned at one time, there is no loss due to elastic shortening, because this shortening occurs as the jacking force is applied and before the prestressing force is measured. On the other hand, if various strands are tensioned sequentially, the stress loss in each strand varies, being a maximum in the first strand tensioned and zero in the last strand. In most cases, it is sufficiently accurate to calculate the loss in the first strand and to apply one-half that value to all strands.

(3) FRICTIONAL LOSSES   Losses due to friction, as the tendon is stressed in post-tensioned members, are usually separated for convenience into two parts: curvature friction and wobble friction. The first is due to intentional bends in the specified tendon profile and the second to the unintentional variation of the tendon from its intended profile. It is apparent that even a "straight" tendon duct will have some unintentional misalignment so that wobble friction must always be considered in post-tensioned work. Usually, curvature friction must be considered as well. The force at the jacking end of the tendon P_o, required to produce the force P_x at any point x along the tendon, can be found from the expression

$$P_o = P_x e^{K\ell_x + \mu\alpha} \tag{22.38a}$$

where e = base of natural logarithms

$\quad \ell_x$ = tendon length from jacking end to point x

$\quad \alpha$ = angular change of tendon from jacking end to point x, rad

$\quad K$ = wobble friction coefficient, lb/lb per ft

$\quad \mu$ = curvature friction coefficient

There has been much research on frictional losses in prestressed construction, particularly with regard to the values of K and μ. These vary appreciably, depending on construction methods and materials used. The values in Table 22.3 may be used as a guide. A report by ACI Committee 423 (Ref. 22.17) provides detailed information on the selection of wobble and friction coefficients.

If one accepts the approximation that the normal pressure on the duct causing the frictional force results from the undiminished initial tension all the way around the curve, the following simplified expression for loss in tension is obtained:

$$P_o = P_x(1 + K\ell_x + \mu\alpha) \tag{22.38b}$$

where α is the angle between the tangents at the ends. The simplified Eq. (22.38b) is valid for values of $K\ell_x + \mu\alpha$ not greater than 0.30.

The loss of prestress for the entire tendon length can be calculated by segments, with each segment assumed to consist of either a circular arc or a length of tangent.

(4) Creep of Concrete Shortening of concrete under sustained load has been discussed in Section 2.8. It can be expressed in terms of the creep coefficient C_c. Creep shortening may be several times the initial elastic shortening, and it is evident that it results in loss of prestress force. The stress loss can be calculated from

$$\Delta f_{s,\text{creep}} = C_c n f_c \tag{22.39}$$

Ultimate values of C_c for different concrete strengths for average conditions of humidity C_{cu} are given in Table 2.2.

TABLE 22.3
Friction coefficients for post-tensioned tendons

Type of Tendon	Wobble Coefficient K, per ft	Curvature Coefficient μ
Grouted tendons in metal sheathing		
Wire tendons	0.0010–0.0015	0.15–0.25
High-strength bars	0.0001–0.0007	0.08–0.30
Seven-wire strand	0.0002–0.0010	0.15–0.25
Unbonded tendons		
Extruded wire tendons	0.0002–0.0010	0.01–0.05
Extruded seven-wire strand	0.0002–0.0010	0.01–0.05
Lubricated seven-wire strand	0.0002–0.0010	0.12–0.18

Table adapted from Ref. 22.17.

In Eq. (22.39), the concrete stress f_c to be used is that at the level of the steel centroid, when the eccentric prestress force plus all sustained loads are acting. Equation (22.37) can be used, except that the moment M_o should be replaced by the moment due to *all* dead loads plus that due to any portion of the live load that may be considered sustained.

It should be noted that the prestress force causing creep is not constant but diminishes with the passage of time due to relaxation of the steel, shrinkage of the concrete, and length changes associated with creep itself. To account for this, it is recommended that the prestress force causing creep be assumed at 10 percent less than the initial value P_i.

(5) SHRINKAGE OF CONCRETE   It is apparent that a decrease in the length of a member due to shrinkage of the concrete is just as detrimental as length changes due to stress, creep, or other causes. As discussed in Section 2.11, the shrinkage strain ε_{sh} may vary between about 0.0004 and 0.0008. A typical value of 0.0006 may be used in lieu of specific data. The steel stress loss resulting from shrinkage is

$$\Delta f_{s,\text{shrink}} = \varepsilon_{sh} E_s \tag{22.40}$$

Only that part of the shrinkage that occurs after transfer of prestress force to the concrete need be considered. For pretensioned members, transfer commonly takes place just 18 hours after placing the concrete, and nearly all the shrinkage takes place after that time. However, post-tensioned members are seldom stressed at an age earlier than 7 days. About 15 percent of ultimate shrinkage may occur within 7 days, under typical conditions, and about 40 percent by the age of 28 days.

(6) RELAXATION OF STEEL   The phenomenon of relaxation, similar to creep, was discussed in Section 2.16c. Loss of stress due to relaxation varies depending upon the stress in the steel, and may be estimated using Eqs. (2.11) and (2.12). To allow for the gradual reduction of steel stress resulting from the combined effects of creep, shrinkage, and relaxation, the relaxation calculation can be based on a prestress force 10 percent less than P_i.

It is interesting to observe that the largest part of the relaxation loss occurs shortly after the steel is stretched. For stresses of $0.80 f_{pu}$ and higher, even a very short period of loading produces substantial relaxation, and this in turn reduces the relaxation that occurs later at a lower stress level. The relaxation rate can thus be artificially accelerated by temporary overtensioning. This technique is the basis for producing low-relaxation steel.

c. Loss Estimation by the Time-Step Method

The loss calculations of the preceding paragraphs recognized the interdependence of creep, shrinkage, and relaxation losses in an approximate way, by an arbitrary reduction of 10 percent of the initial prestress force P_i to obtain the force for which creep and relaxation losses were calculated. For cases requiring greater accuracy, losses can be calculated for discrete time steps over the period of interest. The prestress force causing losses during any time step is taken equal to the value at the end of the preceding time step, accounting for losses due to all causes up to that time. Accuracy can be improved to any desired degree by reducing the length and increasing the number of time steps.

A step-by-step method developed by the Committee on Prestress Losses of the Prestressed Concrete Institute uses only a small number of time steps and is adequate for ordinary cases (Ref. 22.17).

22.14 SHEAR, DIAGONAL TENSION, AND WEB REINFORCEMENT

In prestressed concrete beams at service load, there are two factors that greatly reduce the intensity of diagonal tensile stresses, compared with stresses that would exist if no prestress force were present. The first of these results from the combination of longitudinal compressive stress and shearing stress. An ordinary tensile reinforced concrete beam under load is shown in Fig. 22.22a. The stresses acting on a small element of the beam taken near the neutral axis and near the support are shown in (b). It is found by means of Mohr's circle of stress (c) that the principal stresses act at 45° to the axis of the beam (d) and are numerically equal to the shear stress intensity; thus,

$$t_1 = t_2 = v \qquad (a)$$

Now suppose that the same beam, with the same loads, is subjected to a precompression stress in the amount c, as shown in Fig. 22.23a and b. From Mohr's circle (Fig. 22.23c), the principal tensile stress is

$$t_1 = -\frac{c}{2} + \sqrt{v^2 + \left(\frac{c}{2}\right)^2} \qquad (b)$$

and the direction of the principal tension with respect to the beam axis is

$$\tan 2\alpha = \frac{2v}{c} \qquad (c)$$

as shown in Fig. 22.23d.

Comparison of Eq. (a) with Eq. (b) and Fig. 22.22c with Fig. 22.23c shows that, with the same shear stress intensity, the principal tension in the prestressed beam is much reduced.

The second factor working to reduce the intensity of the diagonal tension at service loads results from the slope of the tendons. Normally, this slope is such as to produce a shear due to the prestress force that is opposite in direction to the

FIGURE 22.22

Principal stress analysis for an ordinary reinforced concrete beam.

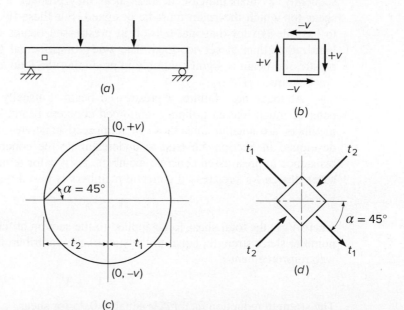

FIGURE 22.23

Principal stress analysis for a prestressed concrete beam.

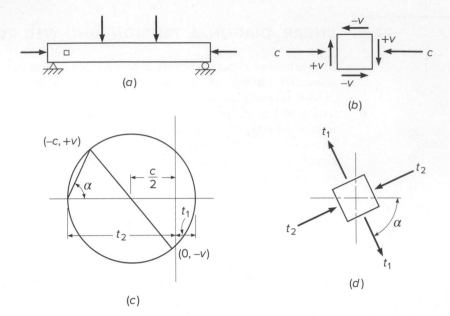

load-imposed shear. The magnitude of this *countershear* is $V_p = P_e \sin \theta$, where θ is the slope of the tendon at the section considered (see Fig. 22.8).

It is important to note, however, that in spite of these characteristics of prestressed beams at service loads, an investigation of diagonal tensile stresses at service loads does not ensure an adequate margin of safety against failure. In Fig. 22.23c, it is evident that a relatively small decrease in compressive stress and increase in shear stress, which may occur when the beam is overloaded, produces a disproportionately large increase in the resulting principal tension. In addition to this effect, if the countershear of inclined tendons is used to reduce design shear, its contribution does not increase directly with load, but much more slowly (see Section 22.7). Consequently, a small increase in total shear may produce a large increase in the net shear for which the beam must be designed. For these two reasons, it is necessary to base design for diagonal tension in prestressed beams on conditions at factored load rather than at service load. The study of principal stresses in the uncracked prestressed beam is significant only in predicting the load at which the first diagonal crack forms.

At loads near failure, a prestressed beam is usually extensively cracked and behaves much like an ordinary reinforced concrete beam. The cracks in prestressed members are smaller than those in reinforced concrete members. The procedures developed in Section 5.5 tend to underestimate the concrete contribution to shear resistance in prestressed concrete members. As it is for reinforced concrete members, shear design for prestressed concrete members is based on the relation

$$V_u \leq \phi V_n \qquad (22.41)$$

where V_u is the total shear force applied to the section at factored loads and V_n is the nominal shear strength, equal to the sum of the contributions of the concrete V_c and web reinforcement V_s:

$$V_n = V_c + V_s \qquad (22.42)$$

The strength reduction factor ϕ is equal to 0.75 for shear.

In calculating the factored load shear V_u, the first critical section is assumed to be at a distance $h/2$ from the face of a support, and sections located a distance less than $h/2$ are designed for the shear calculated at $h/2$.

The shear force V_c resisted by the concrete after cracking has occurred is taken equal to the shear that causes the first diagonal crack. Two types of diagonal cracks have been observed in tests of prestressed concrete beams:

1. **Flexure-shear cracks**, occurring at nominal shear V_{ci}, start as nearly vertical flexural cracks at the tension face of the beam, then spread diagonally upward (under the influence of diagonal tension) toward the compression face. These are common in beams with a low value of prestress force.
2. **Web-shear cracks**, occurring at nominal shear V_{cw}, start in the web due to high diagonal tension, then spread diagonally both upward and downward. These are often found in beams with thin webs with high prestress force.

On the basis of extensive tests, it was established that the shear causing flexure shear cracking can be found using the expression

$$V_{ci} = 0.6\sqrt{f_c'}\,b_w d_p + V_{cr,o+d+l} \qquad (a)$$

where $V_{cr,o+d+l}$ is the shear force, due to total load, at which the flexural crack forms at the section considered, and $0.6\sqrt{f_c'}\,b_w d_p$ represents an additional shear force required to transform the flexural crack into an inclined crack.

While self-weight is generally uniformly distributed, the superimposed dead and live loads may have any distribution. Consequently, it is convenient to separate the total shear into V_o caused by the beam self-weight (without load factor) and V_{cr}, the additional shear force, due to superimposed dead and live loads, corresponding to flexural cracking. Thus,

$$V_{ci} = 0.6\sqrt{f_c'}\,b_w d_p + V_o + V_{cr} \qquad (b)$$

The shear V_{cr} due to superimposed loads can then be found conveniently from

$$V_{cr} = \frac{V_{d+l}}{M_{d+l}}\,M_{cre} \qquad (c)$$

where V_{d+l}/M_{d+l}, the ratio of superimposed dead and live load shear to moment, remains constant as the load increases to the cracking load, and the cracking moment is

$$M_{cre} = \frac{I_c}{y_t}\left(6\lambda\sqrt{f_c'} + f_{pe} - f_o\right) \qquad (22.43)$$

where y_t = distance from concrete centroid to tension face
λ = lightweight concrete modification factor (see Section 5.5)
f_{pe} = compressive stress at tension face resulting from effective prestress force alone
f_o = stress due to beam self-weight (unfactored) at extreme fiber of section where tensile stress is caused by externally applied dead and live loads[†]

The first term inside the parentheses is a conservative estimate of the modulus of rupture. The bottom-fiber stress due to self-weight is subtracted here because self-weight is considered separately in Eq. (b). Thus, Eq. (b) becomes

$$V_{ci} = 0.6\lambda\sqrt{f_c'}\,b_w d_p + V_o + \frac{V_{d+l}}{M_{d+l}}\,M_{cre} \qquad (22.44)$$

[†] All stresses are used with absolute value here, consistent with ACI convention.

Tests indicate that V_{ci} need not be taken less than $1.7\lambda\sqrt{f_c'}b_w d$. The values of d and d_p need not be taken less than $0.80h$ for this and all other equations relating to shear, according to the ACI Code, unless specifically noted otherwise. Additionally, the values V_{d+l} and M_{d+l} should be calculated for the load combination causing the maximum moment in the section. Because V_{d+l} is the incremental load above the beam self-weight, the ACI Code uses the notation $V_i M_{cre}/M_{max}$, noting that M_{cre} comes from Eq. (22.43).

The shear force causing web-shear cracking can be found from an exact principal stress calculation, in which the principal tensile stress is set equal to the direct tensile capacity of the concrete (conservatively taken equal to $4\lambda\sqrt{f_c'}$ according to the ACI Code). Alternatively, the ACI Code permits use of the approximate expression

$$V_{cw} = (3.5\lambda\sqrt{f_c'} + 0.3f_{pc})b_w d_p + V_p \tag{22.45}$$

in which f_{pc} is the compressive stress in the concrete, after losses, at the centroid of the concrete section (or at the junction of the web and the flange when the centroid lies in the flange) and V_p is the vertical component of the effective prestress force. In a pretensioned beam, the $0.3f_{pc}$ contribution to V_{cw} should be adjusted from zero at the beam end to its full value one transfer length (see Section 22.15b) in from the end of the beam.

After V_{ci} and V_{cw} have been calculated, then V_c, the shear resistance provided by the concrete, is taken equal to the smaller of the two values.

Calculating M_{cre}, V_{ci}, and V_{cw} for a prestressed beam is a tedious matter because many of the parameters vary along the member axis. For hand calculations, the required quantities may be found at discrete intervals along the span, such as at $\ell/2$, $\ell/3$, and $\ell/6$, and at $h/2$ from the support face, and stirrups spaced accordingly, or computer spreadsheets may be used.

To shorten the calculation required, the ACI Code includes, as a conservative alternative to the above procedure, an equation for finding the concrete shear resistance V_c directly:

$$V_c = \left(0.6\lambda\sqrt{f_c'} + 700\,\frac{V_u d_p}{M_u}\right) b_w d \tag{22.46}$$

in which M_u is the bending moment occurring simultaneously with shear force V_u, but $V_u d_p/M_u$ is not to be taken greater than 1.0, and d is the effective depth including prestressed and nonprestressed reinforcement. When this equation is used, V_c need not be taken less than $2\lambda\sqrt{f_c'}b_w d_p$ and must not be taken greater than $5\lambda\sqrt{f_c'}b_w d_p$. While Eq. (22.46) is temptingly easy to use and may be adequate for uniformly loaded members of minor importance, its use is apt to result in highly uneconomical designs for I beams with medium and long spans and for composite construction (Ref. 22.19).

When shear reinforcement perpendicular to the axis of the beam is used, its contribution to shear strength of a prestressed beam is

$$V_s = \frac{A_v f_{yt} d}{s} \tag{22.47}$$

the same as for a nonprestressed member. According to the ACI Code, the value of V_s must not be taken greater than $8\sqrt{f_c'}b_w d$.

The total nominal shear strength V_n is found by summing the contributions of the concrete and steel, as indicated by Eq. (22.42):

$$V_n = V_c + \frac{A_v f_{yt} d}{s} \tag{22.48}$$

Then, from Eq. (22.41),

$$V_u = \phi V_n = \phi(V_c + V_s)$$

from which

$$V_u = \phi\left(V_c + \frac{A_v f_{yt} d}{s}\right) \tag{22.49}$$

The required cross-sectional area of one stirrup A_v can be calculated by suitable transposition of Eq. (22.49).

$$A_v = \frac{(V_u - \phi V_c)s}{\phi f_{yt} d} \tag{22.50}$$

Normally, in practical design, the engineer selects a trial stirrup size, for which the required spacing is found. Thus, a more convenient form of the last equation is

$$s = \frac{\phi A_v f_{yt} d}{V_u - \phi V_c} \tag{22.51}$$

A minimum area of shear reinforcement is required in all prestressed concrete members where the total factored shear force is greater than one-half the design shear strength provided by the concrete ϕV_c. Exceptions are made for slabs and footings, concrete-joist floor construction, and certain very shallow beams, according to the ACI Code. The ACI Code exempts members from minimum shear reinforcement if tests demonstrate the member has sufficient strength. Many precast T and double-T beam producers invoke this exemption. The minimum area of shear reinforcement to be provided in all other cases is equal to the smaller of

$$A_{v,min} = 0.75\sqrt{f_c'}\,\frac{b_w s}{f_{yt}} \le 50\,\frac{b_w s}{f_{yt}} \tag{22.52}$$

and

$$A_{v,min} = \frac{A_{ps} f_{pu}}{80 f_{yt}}\frac{s}{d}\sqrt{\frac{d}{b_w}} \tag{22.53}$$

in which A_{ps} is the cross-sectional area of the prestressing steel, f_{pu} is the tensile strength of the prestressing steel, and all other terms are as defined above.

The ACI Code contains certain restrictions on the maximum spacing of web reinforcement to ensure that any potential diagonal crack is crossed by at least a minimum amount of web steel. For prestressed members, this maximum spacing is not to exceed the smaller of $0.75h$ or 24 in. If the value V_s exceeds $4\sqrt{f_c'}b_w d_p$, these limits are reduced by one-half.

EXAMPLE 22.6   **Design of shear reinforcement.**   The unsymmetrical I beam shown in Fig. 22.24 carries an effective prestress force of 288 kips and supports a superimposed dead load of 345 lb/ft and service live load of 900 lb/ft, in addition to its own weight of 255 lb/ft, on a 50 ft simple span. At the maximum moment section, the effective depth to the main steel is 24.5 in. (eccentricity 11.4 in.). The strands are deflected upward starting 15 ft from the support, and eccentricity is reduced linearly to zero at the support. If concrete with $f_c' = 5000$ psi and stirrups with $f_{yt} = 60,000$ psi are used, and if the prestressed strands have strength $f_{pu} = 270$ ksi, what is the required stirrup spacing at a point 10 ft from the support?

FIGURE 22.24
Post-tensioned beam in
Example 22.6.

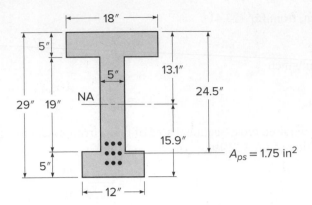

SOLUTION.   For a cross section with the given dimensions, it is easily confirmed that $I_c = 24{,}200$ in^4, $A_c = 245$ in^2, and $r^2 = I_c/A_c = 99$ in^2. At a distance 10 ft from the support centerline, the tendon eccentricity is

$$e = 11.4 \times \frac{10}{15} = 7.6 \text{ in.}$$

corresponding to an effective depth d from the compression face of 20.7 in. According to the ACI Code, the larger value of $d = 0.80 \times 29 = 23.2$ in. is used. Calculation of V_{ci} is based on Eqs. (22.43) and (22.44). The bottom-fiber stress due to effective prestress acting alone is

$$f_{pe} = f_{2pe} = -\frac{P_e}{A_c}\left(1 + \frac{ec_2}{r^2}\right) = -\frac{288{,}000}{245}\left(1 + \frac{7.6 \times 15.9}{99}\right) = -2600 \text{ psi}$$

The moment and shear at the section due to beam load alone are, respectively,

$$M_{o,10} = \frac{w_o\, x}{2}\,(\ell - x) = 0.255 \times 5 \times 40 = 51 \text{ ft-kips}$$

$$V_{o,10} = w_o\left(\frac{\ell}{2} - x\right) = 0.255 \times 15 = 3.8 \text{ kips}$$

and the bottom-fiber stress due to this load is

$$f_{2o} = \frac{51 \times 12{,}000 \times 15.9}{24{,}200} = 402 \text{ psi}$$

Then, from Eq. (22.43),

$$M_{cre} = \frac{24{,}200(425 + 2600 - 402)}{15.9 \times 12} = 333{,}000 \text{ ft-lb}$$

The ratio of superimposed load shear to moment at the section is

$$\frac{V_{d+l}}{M_{d+l}} = \frac{\ell - 2x}{x(\ell - x)} = \frac{30}{400} = 0.075 \text{ ft}^{-1}$$

Equation (22.44) is then used to determine the shear force at which flexure-shear cracks can be expected to form.

$$V_{ci} = \left[0.6 \times 1\sqrt{5000}(5 \times 23.2) + 3800 + 0.075 \times 330{,}000\right] \times \frac{1}{1000} = 33.5 \text{ kips}$$

The lower limit of $1.7 \times 1\sqrt{5000}(5 \times 23.2)/1000 = 13.9$ kips does not control.

Calculation of V_{cw} is based on Eq. (22.45). The slope θ of the tendons at the section under consideration is such that $\sin\theta \approx \tan\theta = 11.4/(15 \times 12) = 0.063$. Consequently, the vertical component of the effective prestress force is $V_p = 0.063 \times 288 = 18.1$ kips. The concrete compressive stress at the section centroid is

$$f_{pc} = \frac{P_e}{A_c} = \frac{288{,}000}{245} = 1180 \text{ psi}$$

Equation (22.45) can now be used to find the shear at which web-shear cracks should occur.

$$V_{cw} = \left[(3.5 \times 1\sqrt{5000} + 0.3 \times 1180)5 \times 23.2 + 18,100\right] \times \frac{1}{1000} = 87.9 \text{ kips}$$

Thus, in the present case,

$$V_c = V_{ci} = 33.5 \text{ kips}$$

At the section considered, the total shear force at factored loads is

$$V_u = 1.2 \times 0.600 \times 15 + 1.6 \times 0.900 \times 15 = 32.4 \text{ kips}$$

When No. 3 (No. 10) U stirrups are used, for which $A_v = 2 \times 0.11 = 0.22 \text{ in}^2$, the required spacing is found from Eq. (22.51) to be

$$s = \frac{\phi A_v f_{yt} d}{V_u - \phi V_c} = \frac{0.75 \times 0.22 \times 60,000 \times 23.2}{32,400 - 0.75 \times 33,500} = 32 \text{ in.}$$

Equation (22.53) is then applied to establish a maximum spacing criterion.

$$0.22 = \frac{1.75}{80} \times \frac{270}{60} \times \frac{s}{23.2}\sqrt{\frac{23.2}{5}} = 0.0091s$$

$$s = 24.1 \text{ in.}$$

The other criteria for maximum spacing, $\frac{3}{4} \times 29 = 22$ in. and 24 in., however, control here. Open U stirrups are used, at a spacing of 22 in.

For comparison, the concrete shear is calculated on the basis of Eq. (22.46). The ratio V_u/M_u is 0.075, and

$$V_c = \left(0.6 \times 1\sqrt{5000} + 700 \times \frac{0.075}{12} \times 23.2\right)(5 \times 23.2) \times \frac{1}{1000} = 16.7 \text{ kips}$$

The lower and upper limits, $2 \times 1\sqrt{5000}(5 \times 23.2)/1000 = 16.4$ kips and $5 \times 1\sqrt{5000}(5 \times 23.2)/1000 = 41.0$ kips, do not control. Thus, on the basis of V_c obtained from Eq. (22.46), the required spacing of No. 3 (No. 10) U stirrups is

$$s = \frac{0.75 \times 0.22 \times 60,000 \times 23.2}{32,400 - 0.75 \times 16,700} = 11.6 \text{ in.}$$

For the present case, an I-section beam of intermediate span, nearly 2 times the web steel is required at the location investigated if the alternative expression giving V_c directly is used.

22.15 TRANSFER LENGTH AND DEVELOPMENT LENGTH

There are two separate sources of bond stress in prestressed concrete beams: (1) flexural bond, which exists in pretensioned construction between the tendons and the concrete and in grouted post-tensioned construction between the tendons and the grout, and between the conduit (if any) and concrete, and (2) prestress transfer bond, generally applicable to pretensioned members only.

a. Flexural Bond

Flexural bond stresses arise due to the change in tension along the tendon resulting from differences in bending moment at adjacent sections. They are proportional to the rate of change of bending moment, hence to the shear force, at a given location along the span. Provided the concrete member is uncracked, flexural bond stress is very low. After cracking, it is higher by an order of magnitude. However, flexural bond stress need not be considered in designing prestressed concrete beams, provided that adequate end anchorage is furnished for the tendon, in the form of either mechanical anchorage (post-tensioning) or strand embedment (pretensioning).

b.   Transfer Length and Development Length

For pretensioned beams, when the external jacking force is released, the prestressing force is transferred from the steel to the concrete near the ends of the member by bond, over a distance which is known as the *transfer length*. The transfer length depends upon a number of factors, including the steel stress, the configuration of the steel cross section (such as strands vs. wires), the condition of the surface of the steel, and the suddenness with which the jacking force is released. Based on tests of seven-wire prestressing strand (Ref. 22.20), the effective prestress f_{pe} in the steel may be assumed to act at a transfer length from the end of the member equal to

$$\ell_t = \frac{f_{pe}}{3000} d_b \qquad (a)$$

where ℓ_t = transfer length, in.
$\quad d_b$ = nominal strand diameter, in.
$\quad f_{pe}$ = effective prestress, psi

The same tests indicate that the additional distance past the original transfer length necessary to develop the failure strength of the steel is closely represented by the expression

$$\ell_t' = \left(\frac{f_{ps} - f_{pe}}{1000}\right) d_b \qquad (b)$$

where the quantity in parentheses is the stress increment above the effective prestress level, in psi units, to reach the calculated steel stress at failure f_{ps}. Thus, the total development length at failure is

$$\ell_d = \ell_t + \ell_t' \qquad (c)$$

or

$$\ell_d = \left(\frac{f_{ps} - \frac{2}{3} f_{pe}}{1000}\right) d_b \qquad (22.54)$$

The ACI Code does not require that flexural bond stress be checked in either pretensioned or post-tensioned members, but for pretensioned strand it is required that the full development length, given by Eq. (22.54), be provided beyond the critical bending section. Investigation may be limited to those cross sections nearest each end of the member that are required to develop their full flexural strength under the specified factored load. Strands not extending to the end of the member require double the development length.

The development length of prestressing strand affects both shear and flexural strength at the end of pretensioned beams. The prestress component of the concrete shear contribution in Eq. (22.45) is usually considered to vary linearly from zero at the beam end to its full value of $0.3f_{pc}$ at the end of the transfer length ℓ_t, according to ACI Commentary 21.2.3; and the flexural strength reduction factor $\phi = 0.75$ from the end of the member to the end of the transfer length and then varies linearly from 0.75 to 0.9 from the end of the transfer length to the end of the development length ℓ_d, according to ACI Code 21.2.3. These reductions are especially relevant if concentrated loads are applied between the beam end and the end of the development length.

22.16   ANCHORAGE ZONE DESIGN

In prestressed concrete beams, the prestressing force is introduced as a load concentration acting over a relatively small fraction of the total member depth. For post-tensioned beams with mechanical anchorage, the load is applied at the end face, while for pretensioned beams it is introduced somewhat more gradually over the transfer length. In either case, the compressive stress distribution in the concrete becomes linear, conforming to that dictated by the overall eccentricity of the applied forces, only after a distance from the end roughly equal to the depth of the beam.

This transition of longitudinal compressive stress, from concentrated to linearly distributed, produces transverse (vertical) tensile stresses that may lead to longitudinal cracking of the member. The pattern and magnitude of the concrete stresses depend on the location and distribution of the concentrated forces applied by the tendons. Numerous studies have been made using the methods of classical elasticity, photoelasticity, and finite element analysis, and typical results are given in Fig. 22.25. Here the beam is loaded uniformly over a height equal to $h/8$ at an eccentricity of $3h/8$. Contour lines are drawn through points of equal vertical tension, with coefficients expressing the ratio of vertical stress to average longitudinal compression. Typically, there are high *bursting stresses* along the axis of the load a short distance inside the end zone and high *spalling stresses* at the loaded face.

FIGURE 22.25
Contours of equal vertical stress. (*Adapted from Ref. 22.16.*)

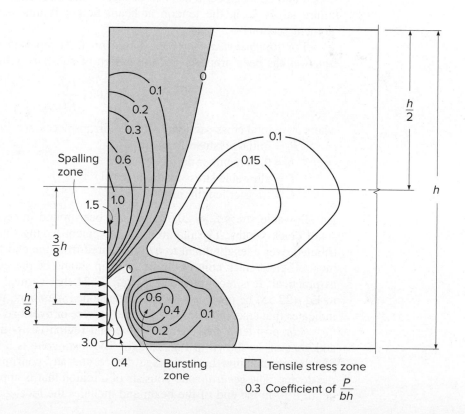

FIGURE 22.26
Post-tensioned I beam with
rectangular end block.

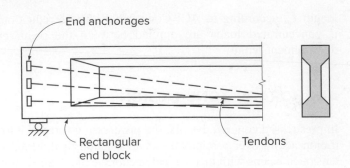

In many post-tensioned prestressed I beams, solid end blocks are provided, as shown in Fig. 22.26. While these are often necessary to accommodate end-anchorage hardware and supplemental reinforcement, they are of little use in reducing transverse tension or avoiding cracking.

Steel reinforcement for end-zone stresses may be in the form of vertical bars of relatively small diameter and close spacing and should be well anchored at the top and bottom of the member. Closed stirrups are commonly used, with auxiliary horizontal bars inside the 90° bends.

Rational design of the reinforcement for end zones must recognize that horizontal cracking is likely. If adequate reinforcement is provided, so that the cracks are restricted to a few inches in length and to 0.01 in. or less in width, these cracks are not detrimental to the performance of the beam either at service load or at the factored load stage. Information on control of cracking of thin webs in pretensioned girders is found in Ref. 22.21. It should be noted that end-zone stresses in pretensioned and bonded post-tensioned beams do not increase in proportion to loads. The failure stress f_{ps} in the tendon at beam failure is attained only at the maximum moment section.

For *pretensioned members*, based on tests reported in Ref. 22.21, a simple equation has been proposed for the design of end-zone reinforcement:

$$A_t = 0.021 \frac{P_i h}{f_s \ell_t} \tag{22.55}$$

where A_t = total cross-sectional area of stirrups necessary, in^2
$\qquad P_i$ = initial prestress force, lb
$\qquad h$ = total member depth, in.
$\qquad f_s$ = allowable stress in stirrups, psi
$\qquad \ell_t$ = transfer length, in.

An allowable stress $f_s = 20{,}000$ psi has been found in tests to produce acceptably small crack widths. The required reinforcement having total area A_t should be distributed over a length equal to $h/5$ measured from the end face of the beam, and for most efficient crack control the first stirrup should be placed as close to the end face as practical. It is recommended in Ref. 22.21 that vertical reinforcement according to Eq. (22.55) be provided for *all* pretensioned members, unless tests or experience indicates that cracking does not occur at service or overload stages.

For post-tensioned members, the end region is divided into two zones, local and general, as shown in Fig. 22.27a. The *local zone* is a rectangular prism immediately surrounding the anchorage device and any confining reinforcement around the device. The *general zone* consists of a region that is approximately one structural depth h from the end of the beam and includes the local zone. For internal anchors,

FIGURE 22.27
Post-tensioned end
block: (*a*) local and
general zone and
(*b*) strut-and-tie model.

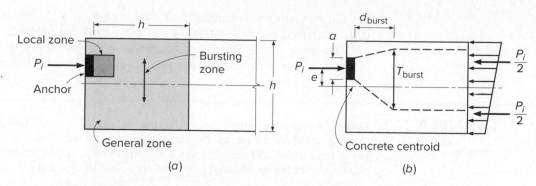

(a) (b)

such as used in slabs, the general zone extends a distance h ahead of and behind the anchorage hardware. Stresses in the local zone are determined based on tests. The post-tensioning supplier specifies the reinforcement details for the local zone.

Stress variations in the general zone are nonlinear and are characterized by a transition from the local zone to an assumed uniform stress gradient a distance h from the anchor. Reinforcement in the general zone may be designed by one of three methods. These methods include equilibrium-based plasticity models, such as the strut-and-tie model, linear stress analysis such as finite element analysis, and simplified elasticity solutions similar to the photoelastic model shown graphically in Fig. 22.25 or elasticity analyses described in Ref. 22.22. Simplified equations are not permitted for nonrectangular cross sections, where multiple anchorages are used (unless closely spaced), or where discontinuities disrupt the force flow path.

Strut-and-tie design approaches for highway girder anchorages are detailed in the *AASHTO LRFD Bridge Design Specifications* (Refs. 22.18 and 22.22). An abbreviated version of the AASHTO Specifications is incorporated in ACI Commentary 25.9.4. ACI Code 25.9.3 requires that complex, multistrand anchorage systems conform to the full AASHTO Specifications.

For the common case of a rectangular end block and simple anchorage (Fig. 22.27b), ACI Commentary 25.9.4 offers simplified equations based on test results and strut-and-tie modeling. The magnitude of the bursting force T_{burst} and the location of its centroid distance from the front of the anchor d_{burst} may be calculated as

$$T_{burst} = 0.25 \Sigma P_{pu} \left(1 - \frac{h_{anc}}{h} \right) \tag{22.56}$$

and

$$d_{burst} = 0.5(h - 2e_{anc}) \tag{22.57}$$

where ΣP_{pu} = sum of total factored post-tensioning force
   e_{anc} = absolute value of eccentricity of anchorage device to centroid of concrete section
   h = depth of cross section
   h_{anc} = depth of anchorage device

The use of the factored post-tensioning force P_{pu} recognizes that the tendon force is acting as a load. Hence, the maximum jacking stress $0.80f_{pu}$ is multiplied by a load factor of 1.2 to calculate P_{pu}.

$$P_{pu} = 1.2(0.80f_{pu})A_{ps} = 0.96f_{pu}A_{ps} \tag{22.58}$$

Transverse reinforcement with total area $A_s = T_{burst}/\phi f_y$ is added in a region that is centered on the location d_{burst} to carry the bursting force.

In cases where the simplified equations do not apply, a strut-and-tie model (Chapter 17) or finite element analysis may be required to design the bursting zone.

EXAMPLE 22.7   **Design of end-zone reinforcement for post-tensioned beam.**   End-zone reinforcement is to be designed for the rectangular post-tensioned beam shown in Fig. 22.28. The initial pre-stress force P_i of 250 kips is applied by two closely spaced tendons having a combined eccentricity of 8.0 in. Material properties are $f'_{ci} = 4250$ psi and $f_y = 60,000$ psi.

SOLUTION.   The rectangular section and the closely spaced anchorage devices allow the use of the simplified ACI equations.

$$d_{burst} = 0.5(h - 2e_{anc}) = 0.5(30 - 2 \times 8) = 7 \text{ in.}$$

The initial prestressing force is 250 kips, which corresponds to a tendon stress level of $0.82f_{py}$. The maximum jacking stress level in the tendons is $0.94f_{py}$, or $0.80f_{pu}$. In this example, only the initial prestress is provided. Hence, the factored tendon force is calculated as

$$P_{pu} = 1.2 \left(\frac{0.94}{0.82}\right) 250 = 344 \text{ kips}$$

for which

$$T_{burst} = 0.25\Sigma P_{pu} \left(1 - \frac{h_{anc}}{h}\right) = 0.25 \times 344 \left(1 - \frac{6}{30}\right) = 68.8 \text{ kips}$$

The area of steel needed to resist T_{burst} is

$$A_s = \frac{T_{burst}}{\phi f_y} = \frac{68.8}{0.85 \times 60} = 1.35 \text{ in}^2$$

Using No. 4 (No. 13) closed stirrups with an area of $2 \times 0.20 \text{ in}^2$ gives

$$n = \frac{1.35}{2 \times 0.20} = 3.4 \text{ stirrups}$$

Four No. 4 (No. 13) closed stirrups are selected. The first stirrup is placed $2\frac{1}{2}$ in. from the anchor plate, and the other three stirrups are placed 3 in. on center, as shown in Fig. 22.28b, centering the stirrups a distance d_{burst} from the anchor plate. The closed stirrups ensure that anchorage requirements are satisfied. Details of the reinforcement in the local zone are not shown.

FIGURE 22.28
Design of post-tensioned anchor zone: (a) section at end anchors and (b) end zone reinforcement.

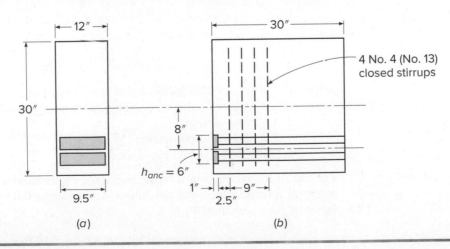

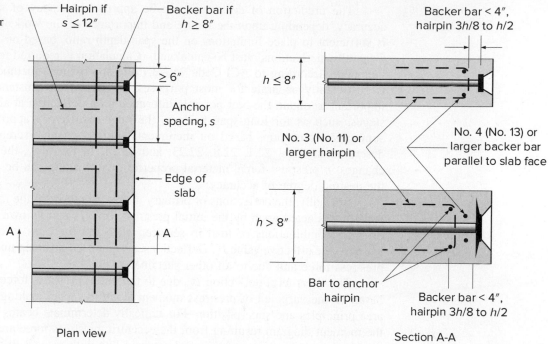

FIGURE 22.29
Monostrand anchor reinforcement.

Plan view

Section A-A

Anchors for monostrand tendons are designed to transfer the jacking force directly to the concrete. Stresses under the anchor plates are acceptable providing the ACI Code stressing limits are observed. Cracking is controlled by the placement of reinforcement *backer bars* within 4 in. of the anchors. Hairpin bars are placed between anchors when the anchor spacing is 12 in. or less and a longitudinal bar is used to secure the hairpin bars. The ACI Code differentiates details between members 8 in. or less deep and those greater than 8 in deep (Fig. 22.29).

22.17 DEFLECTION

Deflection of the slender, relatively flexible beams that are made possible by prestressing must be predicted with care. Many members, satisfactory in all other respects, have proved to be unserviceable because of excessive deformation. In some cases, the absolute amount of deflection is excessive. Often, it is the differential deformation between adjacent members (for example, precast roof-deck units) that causes problems. More often than not, any difficulties that occur are associated with upward deflection due to the sustained prestress load. Such difficulties are easily avoided by proper consideration in design.

When the prestress force is first applied, a beam will normally camber upward. With the passage of time, concrete shrinkage and creep will cause a gradual reduction of prestress force. In spite of this, the upward deflection usually increases, due to the differential creep, affecting the highly stressed bottom fibers more than the top. With the application of superimposed dead and live loads, this upward deflection is partially or completely overcome, and zero or downward deflection obtained. Clearly, in computing deformation, careful attention must be paid to both the age of the concrete at the time of load application and the duration of the loading.

The prediction of deflection can be approached at any of several levels of accuracy, depending upon the nature and importance of the work. In some cases, it is sufficient to place limitations on the span-depth ratio, based on past experience. Generally, deflections must be calculated. (Calculation is required for *all* prestressed members, according to ACI Code 24.2.3.) The approximate method described here is sufficiently accurate for most purposes. In special circumstances, where it is important to obtain the best possible information on deflection at all important load stages, such as for long-span bridges, the only satisfactory approach is to use a summation procedure based on incremental deflection at discrete time steps, as described in Refs. 22.1, 22.8, 22.23, and 22.24. In this way, the time-dependent changes in prestress force, material properties, and loading can be accounted for to the desired degree of accuracy.

Normally, the deflections of primary interest are those at the initial stage, when the beam is acted upon by the initial prestress force P_i and its own weight, and one or more combinations of load in service, when the prestress force is reduced by losses to the effective value P_e. Deflections are modified by creep under the sustained prestress force and due to all other sustained loads.

The short-term deflection Δ_{pi} due to the initial prestress force P_i can be found based on the variation of prestress moment along the span, making use of moment-area principles and superposition. For statically determinate beams, the ordinates of the moment diagram resulting from the eccentric prestress force are directly proportional to the eccentricity of the steel centroid line with respect to the concrete centroid. For indeterminate beams, eccentricity should be measured to the thrust line rather than to the steel centroid (see Ref. 22.1). In either case, the effect of prestress can also be regarded in terms of equivalent loads and deflections found using familiar deflection equations.

The downward deflection Δ_o due to girder self-weight, which is usually uniformly distributed, is easily found by conventional means. Thus, the net deflection obtained immediately upon prestressing is

$$\Delta = -\Delta_{pi} + \Delta_o \tag{22.59}$$

where the negative sign indicates upward displacement.

Long-term deflections due to prestress occur as that force is gradually reducing from P_i to P_e. This can be accounted for in an approximate way by assuming that creep occurs under a constant prestress force equal to the average of the initial and final values. Corresponding to this assumption, the total deflection resulting from prestress alone is

$$\Delta = -\Delta_{pe} - \frac{\Delta_{pi} + \Delta_{pe}}{2} C_c \tag{22.60}$$

where

$$\Delta_{pe} = \Delta_{pi} \frac{P_e}{P_i}$$

and C_c is set equal to the ultimate creep coefficient C_u for the concrete (see Table 2.2).

The long-term deflection due to self-weight is also increased by creep and can be obtained by applying the creep coefficient directly to the instantaneous value. Thus, the total member deflection, after losses and creep deflections, when effective prestress and self-weight act, is

$$\Delta = -\Delta_{pe} - \frac{\Delta_{pi} + \Delta_{pe}}{2} C_c + \Delta_o (1 + C_c) \tag{22.61}$$

TABLE 22.4
Deflection and crack width requirements for prestressed concrete members

	Class		
Condition	U	T	C
Assumed behavior	Uncracked	Transition between cracked and uncracked	Cracked
Deflection calculation basis	Gross section	Cracked section— bilinear behavior	Cracked section— bilinear behavior

The deflection due to superimposed loads can now be added, with the creep coefficient introduced to account for the long-term effect of the sustained loads, to obtain the net deflection at full service loading:

$$\Delta = -\Delta_{pe} - \frac{\Delta_{pi} + \Delta_{pe}}{2} C_c + (\Delta_o + \Delta_d)(1 + C_c) + \Delta_l \qquad (22.62)$$

where Δ_d and Δ_l are the immediate deflections due to superimposed dead and live loads, respectively.

The selection of section properties for the calculation of deflections is dependent upon the cracking in the section. Table 22.4 defines the appropriate section properties and deflection calculation methodology for Class U, T, and C members (Refs. 22.1, 22.22, and 22.23). Bilinear behavior in Table 22.4 implies that deflections based on loads up to the cracking moment are based on the gross section, and deflections on loads greater than the cracking load are based on the effective cracked section properties rather than the modified properties used for reinforced concrete (Ref. 22.8).

EXAMPLE 22.8   **Long-term deflections.** The 40-ft simply supported T beam shown in Fig. 22.30 is prestressed with a force of 314 kips, using a parabolic tendon with an eccentricity of 3 in. above the concrete centroid at the supports and 7.9 in. below the centroid at midspan. After time-dependent losses have occurred, this prestress is reduced to 267 kips. In addition to its own weight of 330 lb/ft, the girder must carry a short-term superimposed live load of 900 lb/ft. Estimate the deflection at all critical stages of loading. The creep coefficient $C_c = 2.0$, $E_c = 4 \times 10^6$ psi, and modulus of rupture = 530 psi.

SOLUTION. It is easily confirmed that the stress in the bottom fiber when the beam carries the maximum load to be considered is 80 psi compression, meeting the requirements for a Class U member. All deflection calculations can, therefore, be based on the moment of inertia of the gross concrete section $I_c = 15,800$ in^4. It is convenient to calculate the deflection due to prestress and that due to girder load separately, superimposing the results later. For the eccentricities of the tendon profile shown in Fig. 22.30b, the application of $P_i = 314$ kips causes the moments shown in Fig. 22.30c. Applying the second moment-area theorem by taking moments of the M/EI diagram between midspan and the support, about the support, produces the vertical displacement between those two points as follows:

$$\Delta_{pi} = \frac{-\left(3.42 \times 10^6 \times 240 \times \frac{2}{3} \times 240 \times \frac{5}{8}\right) + (0.942 \times 10^6 \times 240 \times 120)}{4 \times 10^6 \times 15,800} = -0.87 \text{ in.}$$

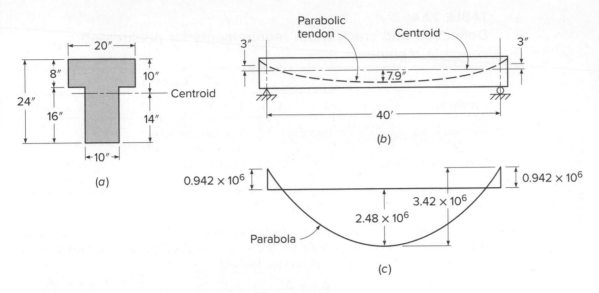

FIGURE 22.30

T beam of Example 22.8: (a) cross section; (b) tendon profile; and (c) moment from initial prestressing force (in-lb).

the minus sign indicating upward deflection, or camber, due to initial prestress alone. The downward deflection due to the self-weight of the girder is calculated by the well-known equation

$$\Delta_o = \frac{5w\ell^4}{384EI} = \frac{5 \times 330 \times 40^4 \times 12^4}{384 \times 12 \times 4 \times 10^6 \times 15,800} = +0.30 \text{ in.}$$

When these two results are superimposed, the net upward deflection when initial prestress and girder load act together is

$$-\Delta_{pi} + \Delta_o = -0.87 + 0.30 = -0.57 \text{ in.}$$

Shrinkage and creep of the concrete cause a gradual reduction of prestress force from $P_i = 314$ kips to $P_e = 267$ kips and reduce the bending moment due to prestress proportionately. Concrete creep, however, acts to increase both the upward deflection component due to the prestress force and the downward deflection component due to the girder load. The net deflection after these changes take place is found using Eq. (22.60), with $\Delta_{pe} = -0.87 \times 267/314 = -0.74$ in.:

$$\Delta = -0.74 - \frac{0.87 + 0.74}{2} \times 2.0 + 0.30(1 + 2.0)$$
$$= -0.74 - 1.61 + 0.90 = -1.45 \text{ in.}$$

In spite of prestress loss, the upward deflection is considerably larger than before. Finally, as the 900 lb/ft short-term superimposed load is applied, the net deflection is

$$\Delta = -1.45 + 0.30\left(\frac{900}{330}\right) = -0.63 \text{ in.}$$

Thus, a net upward deflection of about 1/750 times the span is obtained when the member carries its full superimposed load.

22.18 CRACK CONTROL FOR CLASS C FLEXURAL MEMBERS

The service level stress limitations for Class U and T flexural members are sufficient to control cracking at service loads. Class C flexural members must satisfy the crack control provisions for ordinary reinforced concrete members, modified by ACI Code 24.3. These requirements take the form of limitations on tendon spacing and on the change in stress in the prestressing tendon under service load.

For Class C prestressed flexural members not subjected to fatigue or aggressive exposure, the spacing of bonded reinforcement nearest the extreme tension face may not exceed that given for nonprestressed concrete in Section 7.3. Aggressive conditions occur where the tendons may be exposed to chemical attack and include seawater and corrosive industrial environments. In these situations, the designer should increase the concrete cover or reduce the tensile stresses, based on professional judgment, commensurate with the exposure risk.

The spacing requirements for reinforcement in Class C members may be satisfied by using nonprestressed bonded tendons. The spacing between bonded tendons, however, may not exceed two-thirds of the maximum spacing for nonprestressed reinforcement given in Eq. (7.3). When both conventional reinforcement and bonded tendons are used to meet the spacing requirements, the spacing between a tendon and a bar may not exceed five-sixths of that permitted in Eq. (7.3). When applying Eq. (7.3), Δf_{ps} is substituted for f_s, where Δf_{ps} is the difference between the tendon stress at service loads based on a cracked section and the decompression stress f_{dc}, which is equal to the stress in the tendon when concrete stress at the level of the tendon is zero. ACI Code 24.3.2.2 permits f_{dc} to be taken as the effective prestress f_{pe}. The magnitude of Δf_{ps} is limited to a maximum of 36 ksi. When Δf_{ps} is less than 20 ksi, the reduced spacing requirements need not be applied. If the effective depth of the member exceeds 36 in., additional skin reinforcement along the sides of the member web, as described in Section 7.3, is required to prevent excessive surface crack widths above the main flexural reinforcement.

REFERENCES

22.1. C. W. Dolan and H. R. Hamilton, *Prestressed Concrete: Building, Design and Construction*, Springer Nature, Switzerland, 2019.

22.2. A. E. Naaman, *Prestressed Concrete Analysis and Design: Fundamentals*, 2nd ed., TechnoPress 3000, Ann Arbor, Michigan, 2004.

22.3. T. Y. Lin and N. H. Burns, *Design of Prestressed Concrete Structures*, 3rd ed., John Wiley, New York, 1981.

22.4. J. R. Libby, *Modern Prestressed Concrete*, 3rd ed., Van Nostrand Reinhold, New York, 1984.

22.5. E. G. Nawy, *Prestressed Concrete*, 5th ed., Prentice-Hall, Englewood Cliffs, NJ, 2005.

22.6. M. P. Collins and D. Mitchell, *Prestressed Concrete Structures*, Prentice-Hall, Englewood Cliffs, NJ, 1991.

22.7. *Post-Tensioning Manual*, 6th ed., Post-Tensioning Institute, Phoenix, AZ, 2006.

22.8. *PCI Design Handbook*, 8th ed., Precast/Prestressed Concrete Institute, Chicago, IL, 2018.

22.9. *Quality Control for Plants and Production of Structural Precast and Prestressed Concrete Products*, MNL 116-99, Precast/Prestressed Concrete Institute, Chicago, IL, 1999.

22.10. P. W. Abeles, "Design of Partially Prestressed Concrete Beams," *J. ACI*, vol. 64, no. 10, 1967, pp. 669–677.

22.11. A. H. Nilson, "Discussion of 'Design of Partially Prestressed Concrete Beams' by P. W. Abeles" (Ref. 22.10), *J. ACI*, vol. 65, no. 4, 1968, pp. 345–347.

22.12. P. W. Abeles and B. K. Bardhan-Roy, *Prestressed Concrete Designer's Handbook*, 3rd ed., Cement and Concrete Association, London, 1981.

22.13. ACI Committee 423, *State-of-the-Art Report on Partially Prestressed Concrete* (ACI 423.5R-99), American Concrete Institute, Farmington Hills, MI, 1999.

22.14. CEB-FIP Joint Committee, "International Recommendations for the Design and Construction of Prestressed Concrete Structures," Cement and Concrete Association, London, 1970.

22.15. *Code of Practice for the Structural Use of Concrete*, CP110, British Standards Institution, London, 1972.

22.16. Y. Guyon, *Limit State Design of Prestressed Concrete*, vols. 1 and 2, John Wiley, New York, 1972.

22.17. ACI Committee 423, *Guide to Estimating Prestress Loss* (ACI 423.10R-14), American Concrete Institute, Farmington Hills, MI, 2014.

22.18. *AASHTO LRFD Bridge Design Specifications*, 6th ed., American Association of State Highway and Bridge Officials, Washington, DC, 2012.

22.19. J. G. MacGregor and J. M. Hanson, "Proposed Changes in Shear Provisions for Reinforced and Prestressed Concrete Beams," *J. ACI*, vol. 66, no. 4, 1969, pp. 276–288.

22.20. N. W. Hanson and P. W. Kaar, "Flexural Bond Tests of Pretensioned Prestressed Beams," *J. ACI*, vol. 30, no. 7, 1959, pp. 783–802.

22.21. W. T. Marshall and A. H. Mattock, "Control of Horizontal Cracking in the Ends of Pretensioned Prestressed Concrete Girders," *J. Prestressed Concr. Inst.*, vol. 5, no. 5, 1962, pp. 56–74.

22.22. J. E. Breen, O. Burdet, C. Roberts, D. Sanders, G. Wollman, and B. Falconer, "Anchorage Zone Reinforcement for Post-Tension Concrete Girders," NCHRP Report 356, Transportation Research Board, National Academy Press, Washington, DC, 1994.

22.23. ACI Committee 435, "Deflection of Prestressed Concrete Members," *J. ACI*, vol. 60, no. 12, 1963, pp. 1697–1728.

22.24. D. E. Branson, *Deformation of Concrete Structures*, McGraw-Hill, New York, 1977.

PROBLEMS

22.1. A rectangular concrete beam with width $b = 11$ in. and total depth $h = 28$ in. is post-tensioned using a single parabolic tendon with an eccentricity of 7.8 in. at midspan and 0 in. at the simple supports. The initial prestress force $P_i = 336$ kips, and the effectiveness ratio $R = 0.84$. The member is to carry superimposed dead and live loads of 300 and 1000 lb/ft, respectively, uniformly distributed over the 40 ft span. Specified concrete strength $f_c' = 5000$ psi, and at the time of transfer $f_{ci}' = 4000$ psi. Determine the flexural stress distributions in the concrete at midspan (*a*) for initial conditions before application of superimposed load and (*b*) at full service load. Compare with the ACI limit stresses for Class U members.

22.2. A pretensioned prestressed beam has a rectangular cross section of 6 in. width and 20 in. total depth. It is built using normal-density concrete with a specified strength $f_c' = 4000$ psi and a strength at transfer of $f_{ci}' = 3000$ psi. Stress limits are as follows: $f_{ti} = 164$ psi, $f_{ci} = -1800$ psi, $f_{ts} = 474$ psi, and $f_{cs} = -1800$ psi. The effectiveness ratio R may be assumed equal to 0.80. For these conditions, find the initial prestress force P_i and eccentricity e to maximize the superimposed load moment $M_d + M_l$ that can be carried without exceeding the stress limits. What uniformly distributed load can be carried on a 30 ft simple span? What tendon profile would you recommend?

22.3. A 24 in. deep, 8 ft wide, 70 ft long double T beam carries a girder weight of 420 plf, a sustained dead load of 10 psf, and superimposed live load of 30 psf. It is prestressed with twelve $\frac{1}{2}$ in. diameter 270 ksi low-relaxation strands with a midspan eccentricity of 12.3 in. and an end eccentricity of 5.0 in. Section properties are $A_c = 401$ in^2, $I_g = 20{,}985$ in^4, $y_b = 17.15$ in., $y_t = 6.85$ in., and $h_f = 2$ in. Concrete strength is 5000 psi, $f_{ci}' = 3500$ psi, and $R = 0.80$. Determine the class of the beam.

22.4. Determine if the strength of the double T beam in Problem 22.3 is adequate.

22.5. The hollow core section shown in Fig. P22.5 is prestressed with four $\frac{1}{2}$ in. diameter, 270 ksi low-relaxation strands and is simply supported on masonry walls with a span length of 20 ft, center to center of the supports. In addition to its self-weight, the section carries a superimposed live load of 225 psf. Material properties are $f_c' = 5000$ psi and $f_{ci}' = 3500$ psi. Determine (a) if service load stresses in the section are suitable for a Class U flexural member using $R = 0.82$ and (b) if the section has sufficient capacity for the specified loads.

FIGURE P22.5

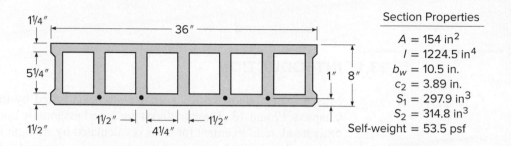

Section Properties

$A = 154$ in^2
$I = 1224.5$ in^4
$b_w = 10.5$ in.
$c_2 = 3.89$ in.
$S_1 = 297.9$ in^3
$S_2 = 314.8$ in^3
Self-weight = 53.5 psf

22.6. For the beam in Problem 22.5, make a detailed computation of the losses in the prestressing force. Compare your results to the assumed value of $R = 0.82$.

22.7. Establish the required spacing of No. 3 (No. 10) stirrups at a beam cross section subject to factored load shear V_u of 35.55 kips and moment M_u of 474 ft-kips. Web width $b_w = 5$ in., effective depth $d = 24$ in., and total depth $h = 30$ in. The concrete shear contribution may be based on the approximate relationship of Eq. (22.46). Use $f_y = 60,000$ psi for stirrup steel, and take $f_c' = 5000$ psi.

22.8. The concrete T beam shown in Fig. P22.8 is post-tensioned at an initial force $P_i = 229$ kips, which reduces after 1 year to an effective value $P_e = 183$ kips. In addition to its own weight, the beam carries a superimposed short-term live load of 21.5 kips at midspan. Using the approximate method described in Section 22.17, find (a) the initial deflection of the unloaded girder and (b) the deflection at the age of 1 year of the loaded girder. The following data are given: $A_c = 450$ in^2, $c_1 = 8$ in., $I_c = 24,600$ in^4, $E_c = 3,500,000$ psi, $C_c = 2.5$.

FIGURE P22.8

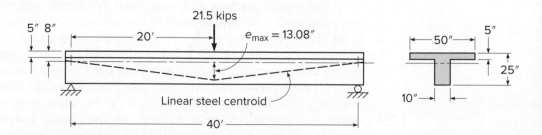

23

Yield Line Analysis for Slabs

23.1 INTRODUCTION

Most concrete slabs are designed for moments found by the methods described in Chapters 12 and 13. These methods are based essentially upon elastic theory. On the other hand, reinforcement for slabs is calculated by strength methods that account for the actual inelastic behavior of members at the factored load stage. A corresponding contradiction exists in the process by which beams and frames are analyzed and designed, as was discussed in Chapter 11, and the concept of limit, or plastic, analysis of reinforced concrete was introduced. Limit analysis not only eliminates the inconsistency of combining elastic analysis with inelastic design but also accounts for the reserve strength characteristic of most reinforced concrete structures and permits, within limits, an arbitrary readjustment of moments found by elastic analysis to arrive at design moments that permit more practical reinforcing arrangements.

For slabs, there is still another good reason for interest in limit analysis. The direct design and equivalent frame methods are restricted in important ways. Slab panels must be square or rectangular. They must be supported along two opposite sides (one-way slabs), two pairs of opposite sides (two-way edge-supported slabs), or by a fairly regular array of columns (flat plates and related forms). Loads must be uniformly distributed, at least within the bounds of any single panel. There can be no large openings. But in practice, many slabs do not meet these restrictions. Solutions are needed, for example, for round or triangular slabs, slabs with large openings, slabs supported on two or three edges only, and slabs carrying concentrated loads. Limit analysis provides a powerful and versatile tool for treating such problems.

Full plastic analysis of a continuous reinforced concrete beam or frame is computationally intensive because of the need to calculate the rotation requirement at all plastic hinges and to check rotation capacity at each hinge to ensure that it is adequate. Consequently, for beams and frames, the very simplified approach to plastic moment redistribution of ACI Code 6.6.5 is used. However, for slabs, which typically have tensile reinforcement ratios much below the balanced value and consequently have large rotation capacity, it can be safely assumed that the necessary ductility is present. Practical methods for the plastic analysis of slabs are thus possible and have been developed. *Yield line theory*, presented in this chapter, is one of these. Although the ACI Code contains no specific provisions for limit or plastic analysis of slabs, ACI Code 1.10 permits use of "any system of design or construction," the adequacy of which has been shown by successful use, analysis, or tests, and ACI Commentary 8.2.1 refers specifically to yield line analysis as an acceptable approach.

Yield line analysis for slabs was first proposed by Ingerslev (Ref. 23.1) and was greatly extended by Johansen (Refs. 23.2 and 23.3). Early publications were mainly in Danish, and it was not until Hognestad's English language summary (Ref. 23.4) of Johansen's work that the method received wide attention. Since that time, a number of important publications on the method have appeared (Refs. 23.5 through 23.15). A particularly useful and comprehensive treatment is found in Ref. 23.15.

The *plastic hinge* was introduced in Section 11.9 as a location along a member in a continuous beam or frame at which, upon overloading, there would be large inelastic rotation at essentially a constant resisting moment. For slabs, the corresponding mechanism is the *yield line*. For the overloaded slab, the resisting moment per unit length measured along a yield line is constant as inelastic rotation occurs; the yield line serves as an axis of rotation for the slab segment.

Figure 23.1a shows a simply supported, uniformly loaded reinforced concrete slab. It is assumed to be underreinforced (as are almost all slabs), with $\rho < \rho_{max}$. The elastic moment diagram is shown in Fig. 23.1b. As the load is increased, when the applied moment becomes equal to the flexural capacity of the slab cross section, the tensile steel starts to yield along the transverse line of maximum moment.

Upon yielding, the curvature of the slab at the yielding section increases sharply, and deflection increases disproportionately. The elastic curvatures along the slab span are small compared with the curvature resulting from plastic deformation at the yield line, and it is acceptable to consider that the slab segments between the yield line and supports remain rigid, with all the curvature occurring at the yield line, as shown in Fig. 23.1c. The "hinge" that forms at the yield line rotates with essentially constant resistance, according to the relation shown earlier in Fig. 11.13a. The resistance per unit width of slab is the nominal flexural strength of the slab; that is, $m_p = m_n$, where m_n is calculated by the usual equations. For design purposes, m_p would be taken equal to ϕm_n, with ϕ typically equal to 0.90, since ρ is well below ρ_{max} for most slabs.

FIGURE 23.1
Simply supported, uniformly loaded one-way slab.

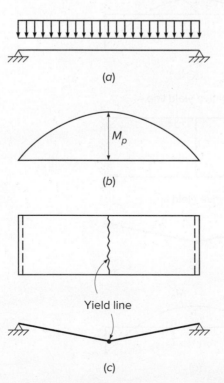

(a)

M_p

(b)

Yield line

(c)

For a statically determinate slab like that in Fig. 23.1, the formation of one yield line results in collapse. A "mechanism" forms; that is, the segments of the slab between the hinge and the supports are able to move without an increase in load. Indeterminate structures, however, can usually sustain their loads without collapse even after the formation of one or more yield lines. When it is loaded uniformly, the fixed-fixed slab in Fig. 23.2a, assumed here to be equally reinforced for positive and negative moments, has an elastic distribution of moments, as shown in Fig. 23.2b. As the load is gradually increased, the more highly stressed sections at the support start yielding. Rotations occur at the support line hinges, but restraining moments of constant value m_p continue to act. The load can be increased further, until the moment at midspan becomes equal to the moment capacity there, and a third yield line forms, as shown in Fig. 23.2c. The slab is now a mechanism, large deflections occur, and collapse takes place.

The moment diagram just before failure is shown in Fig. 23.2d. Note that the ratio of elastic positive to negative moments of 1:2 no longer holds. Due to inelastic deformation, the ratio of these moments just before collapse is 1:1 for this particular structure. Redistribution of moments was discussed earlier in Section 11.9, and it was pointed out that the moment ratios at the collapse stage depend upon the reinforcement provided, not upon the results of elastic analysis.

FIGURE 23.2
Fixed-end, uniformly loaded one-way slab.

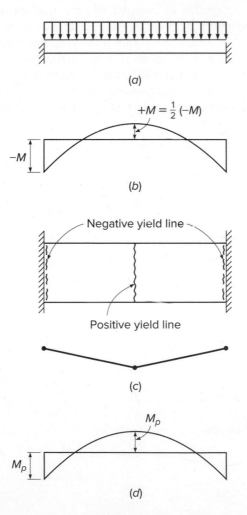

$+M = \frac{1}{2}(-M)$

$-M$

(a)

(b)

Negative yield line

Positive yield line

(c)

M_p

M_p

(d)

23.2   UPPER AND LOWER BOUND THEOREMS

Plastic analysis methods such as the yield line theory derive from the general theory of structural plasticity, which states that the collapse load of a structure lies between two limits, an upper bound and a lower bound of the true collapse load. These limits can be found by well-established methods. A full solution by the theory of plasticity would attempt to make the lower and upper bounds converge to a single correct solution.

The lower bound theorem and the upper bound theorem, when applied to slabs, can be stated as follows:

Lower bound theorem: If, for a given external load, it is possible to find a distribution of moments that satisfies equilibrium requirements, with the moment not exceeding the yield moment at any location, and if the boundary conditions are satisfied, then the given load is a lower bound of the true carrying capacity.

Upper bound theorem: If, for a small increment of displacement, the internal work done by the slab, assuming that the moment at every plastic hinge is equal to the yield moment and that boundary conditions are satisfied, is equal to the external work done by the given load for that same small increment of displacement, then that load is an upper bound of the true carrying capacity.

If the lower bound conditions are satisfied, the slab can certainly carry the given load, although a higher load may be carried if internal redistributions of moment occur. If the upper bound conditions are satisfied, a load greater than the given load will certainly cause failure, although a lower load may produce collapse if the selected failure mechanism is incorrect in any sense. When the upper and lower bound solutions contain the same hinges, forces, and displacements and converge on a single solution, that solution is *unique* and represents the true strength of the structure.

In practice, in the plastic analysis of structures, one works with either the lower bound theorem or the upper bound theorem, not both, and precautions are taken to ensure that the predicted failure load at least closely approaches the correct value.

The yield line method of analysis for slabs is an upper bound method, and consequently, the failure load calculated for a slab with known flexural resistances may be higher than the true value. This is certainly a concern, as the designer would naturally prefer to be correct, or at least on the safe side. However, procedures can be incorporated in yield line analysis to help ensure that the calculated capacity is correct. Such procedures are illustrated by the examples in Sections 23.4 and 23.5.

23.3   RULES FOR YIELD LINES

The location and orientation of the yield line were evident for the simple slab in Fig. 23.1. Similarly, the yield lines were easily established for the one-way indeterminate slab in Fig. 23.2. For other cases, it is helpful to have a set of guidelines for drawing yield lines and locating axes of rotation. When a slab is on the verge of collapse because of the existence of a sufficient number of real or plastic hinges to form a mechanism, axes of rotation are located along the lines of support or over point supports such as columns. The slab segments can be considered to rotate as rigid bodies in space about these axes of rotation. The yield line between any two adjacent slab segments is a straight line, being the intersection of two essentially plane surfaces. Because the yield line (as a line of intersection of two planes) contains all points common to these two planes, it must contain the point of intersection (if any) of the two axes of rotation, which is also common to the two planes. That is, the yield line (or yield line extended) must pass through the point of intersection of the axes of rotation of the two adjacent slab segments.

The terms *positive yield line* and *negative yield line* are used to distinguish between those associated with tension at the bottom and tension at the top of the slab, respectively.

Guidelines for establishing axes of rotation and yield lines are summarized as follows:

1. Yield lines are straight lines because they represent the intersection of two planes.
2. Yield lines represent axes of rotation.
3. The supported edges of the slab also establish axes of rotation. If the edge is fixed, a negative yield line may form, providing constant resistance to rotation. If the edge is simply supported, the axis of rotation provides zero restraint.
4. An axis of rotation passes over any column support. Its orientation depends on other considerations.
5. Yield lines form under concentrated loads, radiating outward from the point of application.
6. A yield line between two slab segments must pass through the point of intersection of the axes of rotation of the adjacent slab segments.

In Fig. 23.3, which shows a slab simply supported along its four sides, rotation of slab segments A and B is about ab and cd, respectively. The yield line ef between these two segments is a straight line passing through f, the point of intersection of the axes of rotation.

Illustrations are given in Fig. 23.4 of the application of the guidelines to the establishment of yield line locations and failure mechanisms for a number of slabs with various support conditions. Figure 23.4a shows a slab continuous over parallel supports. Axes of rotation are situated along the supports (negative yield lines indicated as dashed wavy lines) and near midspan, parallel to the supports (positive yield line indicated as solid wavy lines). The particular location of the positive yield line in this case and the other cases in Fig. 23.4 depends upon the distribution of loading and the reinforcement of the slab. Methods for determining its location are discussed later.

For the continuous slab on nonparallel supports, shown in Fig. 23.4b, the midspan yield line (extended) must pass through the intersection of the axes of rotation over the supports. In Fig. 23.4c there are axes of rotation over all four simple supports. Positive yield lines form along the lines of intersection of the rotating segments of the slab. A rectangular two-way slab on simple supports is shown in Fig. 23.4d. The diagonal yield lines must pass through the corners, while the central yield line

FIGURE 23.3
Two-way slab with simply supported edges.

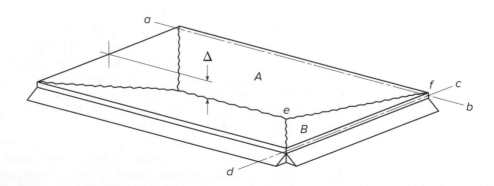

FIGURE 23.4
Typical yield line patterns.

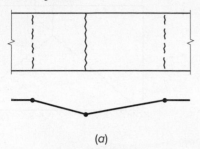

(a)

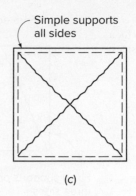

Simple supports
all sides

(c)

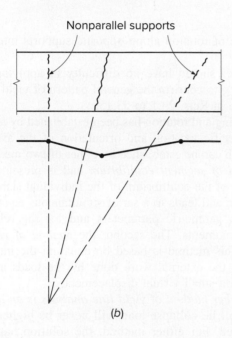

Nonparallel supports

(b)

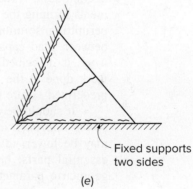

Simple supports
all sides

(d)

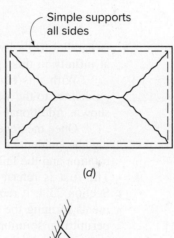

Fixed supports
two sides

(e)

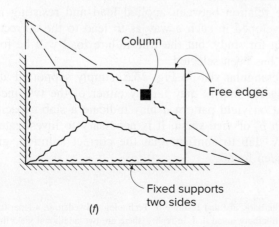

Column

Free edges

Fixed supports
two sides

(f)

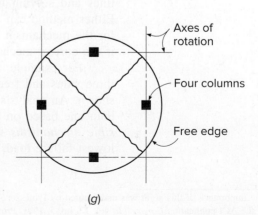

Axes of
rotation

Four columns

Free edge

(g)

FIGURE 23.5
Alternative mechanisms for a
slab supported on three sides.

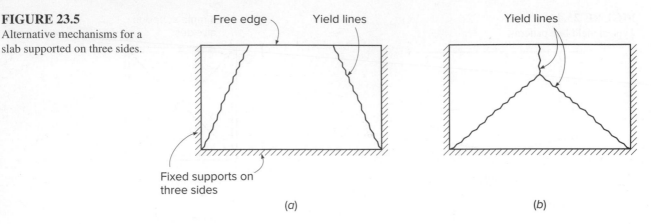

is parallel to the two long sides (axes of rotation along opposite supports intersect at infinity in this case).

With this background, the reader should have no difficulty in applying the guidelines to the slabs in Fig. 23.4e to g to confirm the general pattern of yield lines shown. Additional examples are found in Refs. 23.1 to 23.15.

Once the general pattern of yielding and rotation has been established by applying the guidelines just stated, the specific location and orientation of the axes of rotation and the failure load for the slab can be established by either of two methods. The first is referred to as the *method of segment equilibrium* and is presented in Section 23.4. It requires consideration of the equilibrium of the individual slab segments forming the collapse mechanism and leads to a set of simultaneous equations permitting solution for the unknown geometric parameters and for the relation between load capacity and resisting moments. The second, the *method of virtual work*, is described in Section 23.5. This method is based on equating the internal work done at the plastic hinges with the external work done by the loads as the predefined failure mechanism is given a small virtual displacement.

It should be emphasized that *either method of yield line analysis is an upper bound approach* in the sense that the true collapse load will never be higher, but may be lower, than the load predicted. For either method, the solution has two essential parts: (1) establishing the correct failure pattern and (2) finding the geometric parameters that define the exact location and orientation of the yield lines and solving for the relation between applied load and resisting moments. Either method can be developed in such a way as to lead to the correct solution for the mechanism chosen for study, but the true failure load will be found only if the correct mechanism has been selected.

For example, the rectangular slab in Fig. 23.5, simply supported along only three sides and free along the fourth, may fail by either of the two mechanisms shown. An analysis based on yield pattern *a* may indicate a slab capacity higher than one based on pattern *b*, or vice versa. It is necessary to investigate *all possible mechanisms* for any slab to confirm that the correct solution, giving the lowest failure load, has been found.[†]

[†] The importance of this point was underscored by Professor Arne Hillerborg, of Lund Institute of Technology, Sweden, in a letter to the editor of the ACI publication *Concr. Intl*, vol. 13, no. 5, 1991. Professor Hillerborg noted that, in reality, there are two additional yield line patterns for a slab such as shown in Fig. 23.5. For a particular set of dimensions and reinforcement, both of these gave a lower failure load than did the mechanism shown in Fig. 23.5a.

The method of segment equilibrium should not be confused with a true equilibrium method such as the strip method described in Chapter 24. A true equilibrium method is a lower bound method of analysis; that is, it will always give a *lower bound* of the true capacity of the slab.

23.4 ANALYSIS BY SEGMENT EQUILIBRIUM

Once the general pattern of yielding and rotation has been established by applying the guidelines of Section 23.3, the location and orientation of axes of rotation and the failure load for the slab can be established based on the equilibrium of the various segments of the slab. Each segment, studied as a free body, must be in equilibrium under the action of the applied loads, the moments along the yield lines, and the reactions or shear along the support lines. Because the yield moments are principal moments, twisting moments are zero along the yield lines, and in most cases the shearing forces are also zero. Only the unit moment m generally is considered in writing equilibrium equations.

EXAMPLE 23.1   **Segment equilibrium analysis of one-way slab.**   The method is demonstrated first with respect to the one-way, uniformly loaded, continuous slab of Fig. 23.6a. The slab has a 10 ft span and is reinforced to provide a resistance to positive bending $\phi m_n = 5.0$ ft-kips/ft through the span. In addition, negative steel over the supports provides moment capacities of 5.0 ft-kips/ft at A and 7.5 ft-kips/ft at C. Determine the load capacity of the slab.

SOLUTION.   The number of equilibrium equations required depends upon the number of unknowns. One unknown is always the relation between the resisting moments of the slab and the load. Other unknowns are needed to define the locations of yield lines. In the present instance, one additional equation suffices to define the distance of the yield line from the

FIGURE 23.6
Analysis of a one-way slab by segment equilibrium equations.

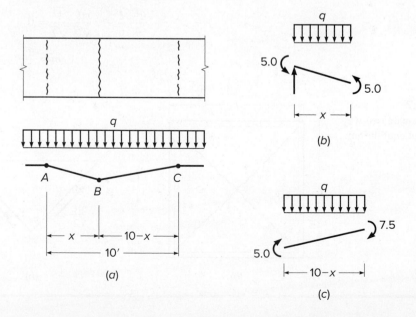

supports. Taking the left segment of the slab as a free body and writing the equation for moment equilibrium about the left support line (see Fig. 23.6b) lead to

$$\frac{qx^2}{2} - 10.0 = 0 \qquad (a)$$

Similarly, for the right slab segment,

$$\frac{q}{2}(10 - x)^2 - 12.5 = 0 \qquad (b)$$

Solving Eqs. (a) and (b) simultaneously for q and x results in

$$q = 0.89 \text{ kip/ft}^2 \qquad x = 4.75 \text{ ft}$$

If a slab is reinforced in orthogonal directions so that the resisting moment is the same in these two directions, the moment capacity of the slab is the same along any other line, regardless of direction. Such a slab is said to be *isotropically* reinforced. If, however, the strengths are different in two perpendicular directions, the slab is called *orthogonally anisotropic*, or simply *orthotropic*. Only isotropic slabs are discussed in this section. Orthotropic reinforcement, which is very common in practice, is discussed in Section 23.6.

It is convenient in yield line analysis to represent moments with vectors. The standard convention, in which the moment acts in a clockwise direction when viewed along the vector arrow, is followed. Treatment of moments as vector quantities is illustrated by the following example.

EXAMPLE 23.2   **Segment equilibrium analysis of square slab.**   A square slab is simply supported along all sides and is to be isotropically reinforced. Determine the resisting moment $m = \phi m_n$ per linear foot required just to sustain a uniformly distributed factored load of q psf.

Solution.   Conditions of symmetry indicate the yield line pattern shown in Fig. 23.7a. Considering the moment equilibrium of any one of the identical slab segments about its support (see Fig. 23.7b), one obtains

$$\frac{q\ell^2}{4} \frac{\ell}{6} - 2\frac{m\ell}{\sqrt{2}} \frac{1}{\sqrt{2}} = 0$$

$$m = \frac{q\ell^2}{24}$$

FIGURE 23.7
Analysis of a square two-way slab by segment equilibrium equations.

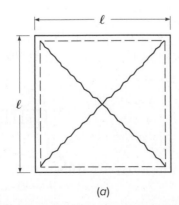

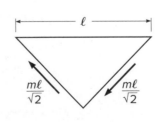

(a)   (b)

In both examples just given, the resisting moment was constant along any particular yield line; that is, the reinforcing bars were of constant diameter and equally spaced along a given yield line. On the other hand, it is recalled that, by the elastic methods of slab analysis presented in Chapter 13, reinforcing bars generally have a different spacing and may be of different diameter in middle strips compared with column or edge strips. A slab designed by elastic methods, leading to such variations, can easily be analyzed for strength by the yield line method. It is merely necessary to subdivide a yield line into its component parts, within any one of which the resisting moment per unit length of hinge is constant. Either the equilibrium equations of this section or the work equations of Section 23.5 can be modified in this way.

23.5 ANALYSIS BY VIRTUAL WORK

Alternative to the method of Section 23.4 is a method of analysis using the principle of virtual work. Since the moments and loads are in equilibrium when the yield line pattern has formed, an infinitesimal increase in load will cause the structure to deflect further. The external work done by the loads to cause a small arbitrary virtual deflection must equal the internal work done as the slab rotates at the yield lines to accommodate this deflection. The slab is therefore given a virtual displacement, and the corresponding rotations at the various yield lines can be calculated. By equating internal and external work, the relation between the applied loads and the resisting moments of the slab is obtained. Elastic rotations and deflections within the slab element are not considered when writing the work equations, as they are very small compared with the plastic deformations.

a. External Work Done by Loads

An external load acting on a slab segment, as a small virtual displacement is imposed, does work equal to the product of its constant magnitude and the distance through which the point of application of the load moves. If the load is distributed over a length or an area, rather than concentrated, the work can be calculated as the product of the total load and the displacement of the point of application of its resultant.

Figure 23.8 illustrates the basis for external work calculation for several types of loads. If a square slab carrying a single concentrated load at its center (Fig. 23.8a) is given a virtual displacement defined by a unit value under the load, the external work is

$$W_e = P \times 1 \tag{a}$$

If the slab shown in Fig. 23.8b, supported along three sides and free along the fourth, is loaded with a line load w per unit length along the free edge, and if that edge is given a virtual displacement having unit value along the central part, the external work is

$$W_e = (2wa) \times \frac{1}{2} + wb = w(a + b) \tag{b}$$

When a distributed load q per unit area acts on a triangular segment defined by a hinge and yield lines, such as Fig. 23.8c,

$$W_e = \frac{qab}{2} \times \frac{1}{3} = \frac{qab}{6} \tag{c}$$

FIGURE 23.8
External work basis for
various types of loads.

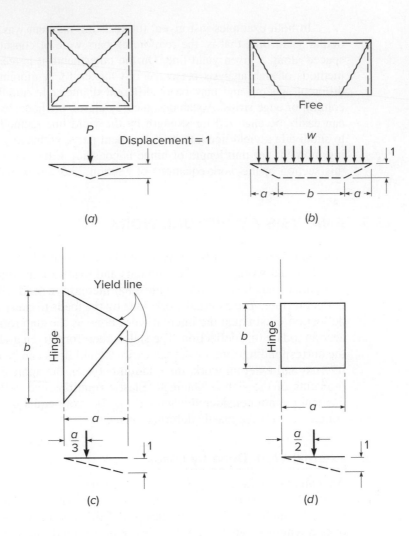

(a) (b)

(c) (d)

while for the rectangular slab segment shown in Fig. 23.8d, carrying a distributed load
q per unit area, the external work is

$$W_e = \frac{qab}{2} \qquad (d)$$

More complicated trapezoidal shapes may always be subdivided into component triangles and rectangles. The total external work is then calculated by summing the work
done by loads on the individual parts of the failure mechanism, with all displacements
keyed to a unit value assigned somewhere in the system. There is no difficulty in combining the work done by concentrated loads, line loads, and distributed loads, if these
act in combination.

b. Internal Work Done by Resisting Moments

The internal work done during the assigned virtual displacement is found by summing
the products of yield moment m per unit length of hinge times the plastic rotation θ
at the respective yield lines, consistent with the virtual displacement. If the resisting

moment m is constant along a yield line of length ℓ, and if a rotation θ is experienced, the internal work is

$$W_i = m\ell\theta \qquad (e)$$

If the resisting moment varies, as would be the case if bar size or spacing were not constant along the yield line, the yield line is divided into n segments, within each one of which the moment is constant. The internal work is then

$$W_i = (m_1\ell_1 + m_2\ell_2 + \cdots + m_n\ell_n)\theta \qquad (f)$$

For the entire system, the total internal work done is the sum of the contributions from all yield lines. In all cases, the internal work contributed is positive, regardless of the sign of m, because the rotation is in the same direction as the moment. External work, on the other hand, may be either positive or negative, depending on the direction of the displacement of the point of application of the force resultant.

EXAMPLE 23.3 **Virtual work analysis of one-way slab.**   Determine the load capacity of the one-way uniformly loaded continuous slab shown in Fig. 23.9, using the method of virtual work. The resisting moments of the slab are 5.0, 5.0, and 7.5 ft-kips/ft at A, B, and C, respectively.

SOLUTION.   A unit deflection is given to the slab at B. Then the external work done by the load is the sum of the loads times their displacements and is equal to

$$\frac{qx}{2} + \frac{q}{2}(10 - x)$$

The rotations at the hinges are calculated in terms of the unit deflection (Fig. 23.9) and are

$$\theta_A = \theta_{B1} = \frac{1}{x} \qquad \theta_{B2} = \theta_C = \frac{1}{10 - x}$$

The internal work is the sum of the moments times their corresponding rotation angles:

$$5 \times \frac{1}{x} \times 2 + 5 \times \frac{1}{10 - x} + 7.5 \times \frac{1}{10 - x}$$

FIGURE 23.9
Virtual work analysis of
one-way slab.

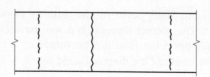

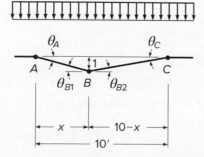

Equating the external and internal work gives

$$\frac{qx}{2} + 5q - \frac{qx}{2} = \frac{10}{x} + \frac{5}{10 - x} + \frac{7.5}{10 - x}$$

$$5q = \frac{10}{x} + \frac{25}{2(10 - x)}$$

$$q = \frac{2}{x} + \frac{5}{2(10 - x)}$$

To determine the minimum value of q, this expression is differentiated with respect to x and set equal to zero:

$$\frac{dq}{dx} = -\frac{2}{x^2} + \frac{5}{2(10 - x)^2} = 0$$

from which

$$x = 4.75 \text{ ft}$$

Substituting this value in the preceding expression for q, one obtains

$$q = 0.89 \text{ kips/ft}^2$$

as before.

In many cases, particularly those with yield lines established by several unknown dimensions (such as Fig. 23.4f), direct solution by virtual work would become quite tedious. The ordinary derivatives in Example 23.3 would be replaced by several partial derivatives, producing a set of equations to be solved simultaneously. In such cases it is often more convenient to select an arbitrary succession of possible yield line locations, solve the resulting mechanisms for the unknown load (or unknown moment), and determine the correct minimum load (or maximum moment) by trial.

EXAMPLE 23.4   **Virtual work analysis of rectangular slab.**   The two-way slab shown in Fig. 23.10 is simply supported on all four sides and carries a uniformly distributed load of q psf. Determine the required moment resistance for the slab, which is to be isotropically reinforced.

SOLUTION.   Positive yield lines form in the pattern shown in Fig. 23.10a, with the dimension a unknown. The correct dimension a will be such as to maximize the moment resistance required to support the load q. The values of a and m are found by trial.

In Fig. 23.10a the length of the diagonal yield line is $\sqrt{25 + a^2}$. From similar triangles,

$$b = 5\,\frac{\sqrt{25 + a^2}}{a} \qquad c = a\,\frac{\sqrt{25 + a^2}}{5}$$

Then the rotation of the plastic hinge at the diagonal yield line corresponding to a unit deflection at the center of the slab (see Fig. 23.10b) is

$$\theta_1 = \frac{1}{b} + \frac{1}{c} = \frac{a}{5\sqrt{25 + a^2}} + \frac{5}{a\sqrt{25 + a^2}} = \frac{1}{\sqrt{25 + a^2}}\left(\frac{a}{5} + \frac{5}{a}\right)$$

The rotation of the yield line parallel to the long edges of the slab (see Fig. 23.10c) is

$$\theta_2 = \frac{1}{5} + \frac{1}{5} = 0.40$$

FIGURE 23.10
Virtual work analysis for rectangular two-way slab.

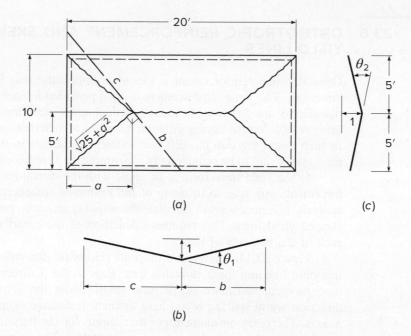

For a first trial, let $a = 6$ ft. Then the length of the diagonal yield line is

$$\sqrt{25 + 36} = 7.81 \text{ ft}$$

The rotation at the diagonal yield line is

$$\theta_1 = \frac{1}{7.81}\left(\frac{6}{5} + \frac{5}{6}\right) = 0.261$$

At the central yield line, it is $\theta_2 = 0.40$. The internal work done as the incremental deflection is applied is

$$W_i = (m \times 7.81 \times 0.261 \times 4) + (m \times 8 \times 0.40) = 11.36m$$

The external work done during the same deflection is

$$W_e = \left(10 \times 6 \times \tfrac{1}{2}q \times \tfrac{1}{3} \times 2\right) + \left(8 \times 5q \times \tfrac{1}{2} \times 2\right) + \left(12 \times 5 \times \tfrac{1}{2}q \times \tfrac{1}{3} \times 2\right) = 80q$$

Equating W_i and W_e, one obtains

$$m = \frac{80q}{11.36} = 7.05q$$

Successive trials for different values of a result in the following data:

a	W_i	W_e	m
6.0	11.36m	80.0q	7.05q
6.5	11.08m	78.4q	7.08q
7.0	10.87m	76.6q	7.04q
7.5	10.69m	75.0q	7.02q

It is evident that the yield line pattern defined by $a = 6.5$ ft is critical. The required resisting moment for the given slab is $7.08q$.

23.6   ORTHOTROPIC REINFORCEMENT AND SKEWED YIELD LINES

Generally slab reinforcement is placed orthogonally, that is, in two perpendicular directions. The same reinforcement is often provided in each direction, but the effective depths are different. In many practical cases, economical designs are obtained using reinforcement having different bar areas or different spacings in each direction. In such cases, the slab has different moment capacities in the two orthogonal directions and is said to be orthogonally anisotropic, or simply orthotropic.

Often yield lines form at an angle with the directions established by the reinforcement; this was so in many of the examples considered earlier. For yield line analysis, it is necessary to calculate the resisting moment, per unit length, along such skewed yield lines. This requires calculation of the contribution to resistance from each of the two sets of bars.

Figure 23.11a shows an orthogonal grid of reinforcement, with angle α between the yield line and the X direction bars. Bars in the X direction are at spacing v and have moment resistance m_y per unit length about the Y axis, while bars in the Y direction are at spacing u and have moment resistance m_x per unit length about the X axis. The resisting moment per unit length for the bars in the Y and X directions are determined separately, with reference to Fig. 23.11b and c, respectively.

For the Y direction bars, the resisting moment *per bar* about the X axis is $m_x u$, and the component of that resistance about the α axis is $m_x u \cos \alpha$. The resisting moment per unit length along the α axis provided by the Y direction bars is therefore

$$m_{\alpha y} = \frac{m_x u \cos \alpha}{u/\cos \alpha} = m_x \cos^2 \alpha \qquad (a)$$

FIGURE 23.11
Yield line skewed with orthotropic reinforcement: (a) orthogonal grid and yield line; (b) Y direction bars; and (c) X direction bars.

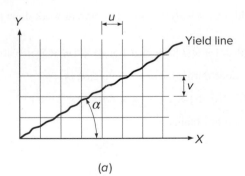

(a)

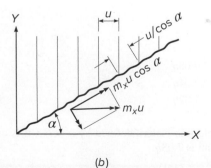

(b)

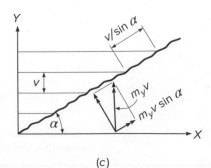

(c)

For the bars in the X direction, the resisting moment per bar about the Y axis is $m_y v$, and the component of that resistance about the α axis is $m_y v \sin \alpha$. Thus, the resisting moment per unit length along the α axis provided by the X direction bars is

$$m_{\alpha y} = \frac{m_y v \sin \alpha}{v/\sin \alpha} = m_y \sin^2 \alpha \qquad (b)$$

Thus, for the combined sets of bars, the resisting moment per unit length measured along the α axis is given by the sum of the resistances from Eqs. (a) and (b):

$$m_\alpha = m_x \cos^2 \alpha + m_y \sin^2 \alpha \qquad (23.1)$$

For the special case where $m_x = m_y = m$, with the same reinforcement provided in each direction,

$$m_\alpha = m(\cos^2 \alpha + \sin^2 \alpha) = m \qquad (23.2)$$

The slab is said to be *isotropically reinforced*, with the same resistance per unit length regardless of the orientation of the yield line.

The analysis just presented neglects any consideration of strain compatibility along the yield line and assumes that the displacements at the level of the steel during yielding, which are essentially perpendicular to the yield line, are sufficient to produce yielding in both sets of bars. This is reasonably in accordance with test data, except for values of α close to 0 to 90°. For such cases, it would be conservative to neglect the contribution of the bars nearly parallel to the yield line.

It has been shown that the analysis of an orthotropic slab can be simplified to that of a related isotropic slab, referred to as the *affine slab*, provided that the ratio of negative to positive reinforcement areas is the same in both directions. The horizontal dimensions and slab loads must be modified to permit this transformation. Details are found in Refs. 23.1 to 23.5.

EXAMPLE 23.5   **Resisting moment along a skewed yield line.**   The balcony slab in Fig. 23.12 has fixed supports along two adjacent sides and is unsupported along the third side. It is reinforced for positive bending with No. 5 (No. 16) Grade 60 bars at 10 in. spacing and 5.5 in. effective

FIGURE 23.12
Skewed yield line example.

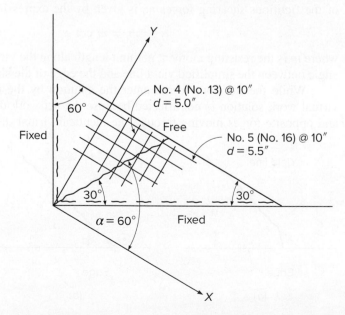

depth, parallel to the free edge, and No. 4 (No. 13) bars at 10 in. spacing and 5.0 in. effective depth perpendicular to that edge. Concrete strength and steel yield stress are 4000 and 60,000 psi, respectively. One possible failure mechanism includes a positive yield line at 30° with the long edge, as shown. Find the total resisting moment along the positive yield line provided by the two sets of bars.

SOLUTION. It is easily confirmed that the resisting moment about the X axis provided by the Y direction bars is $m_x = 5.21$ ft-kips/ft, and the resisting moment about the Y axis provided by the X direction bars is $m_y = 8.70$ ft-kips/ft (both with $\phi = 0.90$ included). The yield line makes an angle of 60° with the X axis bars. With cos $\alpha = 0.500$ and sin $\alpha = 0.866$, from Eq. (23.1) the resisting moment along the α axis is

$$m_\alpha = 5.21 \times 0.500^2 + 8.70 \times 0.866^2 = 7.83 \text{ ft-kips/ft}$$

23.7 SPECIAL CONDITIONS AT EDGES AND CORNERS

Certain simplifications were made in defining yield line patterns in some of the preceding examples in the vicinity of edges and corners. In some cases, such as Fig. 23.4b and f, positive yield lines were shown intersecting an edge at an angle. Actually, at a free or simply supported edge, both bending and twisting moments should theoretically be zero. The principal stress directions are parallel and perpendicular to the edge, and consequently the yield lines should enter an edge perpendicular to it. Tests confirm that this is the case, but the yield lines generally turn only quite close to the edge, the distance t in Fig. 23.13 being small compared to the dimensions of the slab (Ref. 23.4).

Referring to Fig. 23.13, the actual yield line of a can be simplified by extending the yield line in a straight line to the edge, as in b, if a pair of concentrated shearing forces m_t is introduced at the corners of the slab segments. The force m_t acting downward at the acute corner (circled cross) and the force m_t acting upward at the obtuse corner (circled dot) together are the static equivalent of twisting moments and shearing forces near the edge. It is shown in Ref. 23.4 that the magnitude of the fictitious shearing forces m_t is given by the expression

$$m_t = m \cot \alpha \tag{23.3}$$

where m is the resisting moment per unit length along the yield line and α is the acute angle between the simplified yield line and the edge of the slab.

While the fictitious forces enter the solution by the equilibrium method, the virtual work solution is not affected because the net work done by the pair of equal and opposite forces moving through the identical virtual displacement is zero.

FIGURE 23.13
Conditions at edge of slab:
(a) actual yield line and
(b) simplified yield line.

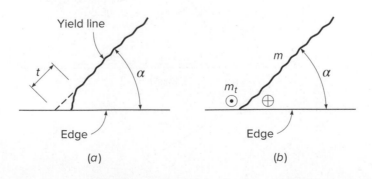

(a) (b)

FIGURE 23.14
Corner conditions.

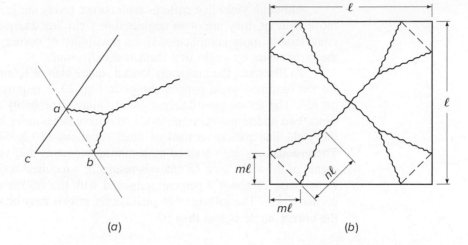

(a)   (b)

Also, in the preceding examples, it was assumed that yield lines enter the corners between the two intersecting sides. An alternative possibility is that the yield line forks before it reaches the corner, forming what is known as a *corner lever*, shown in Fig. 23.14a.

If the corner is not held down, the triangular element *abc* will pivot about the axis *ab* and lift off the supports. The development of such a corner lever is clearly shown in Fig. 23.15. The photograph shows a model reinforced concrete slab that was tested under uniformly distributed load. The edges were simply supported and were not restrained against upward movement. If the corner is held down, a similar situation occurs, except that the line *ab* becomes a yield line. If cracking at the corners of such a slab is to be controlled, top steel more or less perpendicular to the line *ab* must be provided. The direction taken by the positive yield lines near the corner indicates the desirability of supplementary bottom-slab reinforcement at the corners, placed approximately parallel to the line *ab* (see Section 13.1).

FIGURE 23.15
Development of corner levers in a simply supported, uniformly loaded slab.
(*Photograph by Arthur H. Nilson.*)

Although yield line patterns with corner levers are generally more critical than those without, they are often neglected in yield line analysis. The analysis becomes considerably more complicated if the possibility of corner levers is introduced, and the error made by neglecting them is usually small.

To illustrate, the uniformly loaded square slab of Example 23.2, when analyzed for the assumed yield pattern shown in Fig. 23.7, required a moment capacity of $q\ell^2/24$. The actual yield line pattern at failure is probably as shown in Fig. 23.14b. Since two additional parameters m and n have necessarily been introduced to define the yield line pattern, a total of three equations of equilibrium is now necessary. These equations are obtained by summing moments and vertical forces on the segments of the slab. Such an analysis results in a required resisting moment of $q\ell^2/22$, an increase of about 9 percent compared with the results of an analysis neglecting corner levers. The influence of such corner effects may be considerably larger when the corner angle is less than 90°.

23.8   FAN PATTERNS AT CONCENTRATED LOADS

If a concentrated load acts on a reinforced concrete slab at an interior location, away from any edge or corner, a negative yield line will form in a more or less circular pattern, as in Fig. 23.16a, with positive yield lines radiating outward from the load point. If the positive resisting moment per unit length is m and the negative resisting moment m', the moments per unit length acting along the edges of a single element of the fan, having a central angle β and radius r, are as shown in Fig. 23.16b. For small values of the angle β, the arc along the negative yield line can be represented as a straight line of length $r\beta$.

Figure 23.16c shows the moment resultant obtained by vector addition of the positive moments mr acting along the radial edges of the fan segment. The vector

FIGURE 23.16
Yield fan geometry at concentrated load: (a) yield fan; (b) moment vectors acting on fan segment; (c) resultant of positive-moment vectors; and (d) edge view of fan segment.

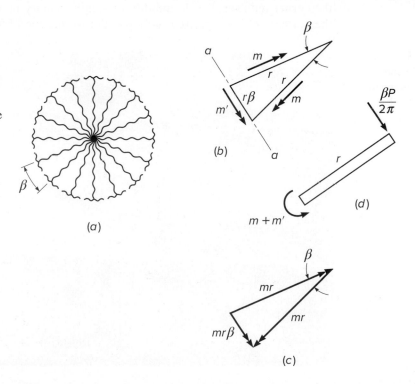

sum is equal to $mr\beta$, acting along the length $r\beta$, and the resultant positive moment, per unit length, is therefore m. This acts in the same direction as the negative moment m', as shown in Fig. 23.16d. Figure 23.16d also shows the fractional part of the total load P that acts on the fan segment.

Taking moments about the axis $a - a$ gives

$$(m + m')r\beta - \frac{\beta Pr}{2\pi} = 0$$

from which

$$P = 2\pi(m + m') \tag{23.4}$$

The collapse load P is seen to be independent of the fan radius r. Thus, with only a concentrated load acting, a complete fan of any radius could form with no change in collapse load.

It follows that Eq. (23.4) also gives the collapse load for a fixed-edge slab of any shape, carrying only a concentrated load P. The only necessary condition is that the boundary must be capable of a restraining moment equal to m' at all points. Finally, Eq. (23.4) is useful in establishing whether flexural failure will occur before a punching shear failure under a concentrated load.

Other load cases of practical interest, including a concentrated load near or at a free edge, and a concentrated corner load, are treated in Ref. 23.5. Loads distributed over small areas and load combinations are discussed in Ref. 23.12.

23.9 LIMITATIONS OF YIELD LINE THEORY

The usefulness of yield line theory should be apparent from the preceding sections. In general, elastic solutions are available only for restricted conditions, usually uniformly loaded rectangular slabs and slab systems. They do not account for the effects of inelastic action, except empirically. By yield line analysis, a rational determination of flexural strength may be had for slabs of any shape, supported in a variety of ways, with concentrated loads as well as distributed and partially distributed loads. The effects of holes of any size can be included. It is thus seen to be a powerful analytical tool for the structural engineer and has been suggested as an approach for the assessment of structures against progressive collapse (Ref. 23.16). By removing a column, yield line analyses provide an indication of the residual structural capacity.

On the other hand, as an upper bound method, it will predict a collapse load that may be greater than the true collapse load. The actual capacity will be less than predicted if the selected mechanism is not the controlling one or if the specific locations of yield lines are not exactly correct. Most engineers would prefer an approach that would be in error, if at all, on the safe side. In this respect, the strip method of Chapter 24 is distinctly superior.

Beyond this, it should be evident that yield line theory provides, in essence, a method for determining the capacity of trial designs, arrived at by some other means, rather than for determining the amount and spacing of reinforcement. It is not, strictly speaking, a design method. To illustrate, yield line theory provides no inducement for the designer to place steel at anything other than a uniform lateral spacing along a yield line. It is necessary to consider the results of elastic analysis of a flat plate, for example, to recognize that reinforcement in that case should be placed in strong bands across the columns.

In applying yield line analysis to slabs, it must be remembered that the analysis is predicated upon available rotation capacity at the yield lines. If the slab reinforcement happens to correspond closely to the elastic distribution of moments in the slab, little rotation is required. If, on the other hand, there is a marked difference, it is possible that the required rotation will exceed the available rotation capacity, in which case the slab will fail prematurely. However, in general, because slabs are typically rather lightly reinforced, they will have adequate rotation capacity to attain the collapse loads predicted by yield line analysis.

It should also be borne in mind that the yield line analysis focuses entirely on the flexural capacity of the slab. It is presumed that earlier failure will not occur due to shear or torsion and that cracking and deflections at service load will not be excessive. ACI Code 8.2.1 calls attention specifically to the need to meet "all serviceability conditions, including limits on deflections," and ACI Commentary 8.2.1 calls attention to the need for "evaluation of the stress conditions around the supports in relation to shear and torsion as well as flexure."

REFERENCES

23.1. A. Ingerslev, "The Strength of Rectangular Slabs," *J. Inst. Struct. Eng.*, London, vol. 1, no. 1, 1923, pp. 3–14.

23.2. K. W. Johansen, *Brutlinieteorier*, Jul. Gjellerups Forlag, Copenhagen, 1943 (see also *Yield Line Theory*, English translation, Cement and Concrete Association, London, 1962).

23.3. K. W. Johansen, *Pladeformler*, 2nd ed., Polyteknisk Forening, Copenhagen, 1949 (see also *Yield Line Formulae for Slabs*, English translation, Cement and Concrete Association, London, 1972).

23.4. E. Hognestad, "Yield Line Theory for the Ultimate Flexural Strength of Reinforced Concrete Slabs," *J. ACI*, vol. 24, no. 7, 1953, pp. 637–656.

23.5. L. L. Jones and R. H. Wood, *Yield Line Analysis of Slabs*, American Elsevier, New York, 1967.

23.6. R. Taylor, D. R. H. Maher, and B. Hayes, "Effect of the Arrangement of Reinforcement on the Behavior of Reinforced Concrete Slabs," *Mag. Concr. Res.*, vol. 18, no. 55, 1966, pp. 85–94.

23.7. R. Lenschow and M. A. Sozen, "A Yield Criterion for Reinforced Concrete Slabs," *J. ACI*, vol. 64, no. 5, 1967, pp. 266–273.

23.8. S. H. Simmonds and A. Ghali, "Yield Line Design of Slabs," *J. Struct. Div.*, ASCE, vol. 102, no. ST1, 1976, pp. 109–123.

23.9. K. H. Chu and R. B. Singh, "Yield Analysis of Balcony Floor Slabs," *J. ACI*, vol. 63, no. 5, 1966, pp. 571–586.

23.10. E. C. Demsky and D. S. Hatcher, "Yield Line Analysis of Slabs Supported on Three Sides," *J. ACI*, vol. 66, no. 9, 1969, pp. 741–744.

23.11. A. Zaslavsky and C. H. Avraham, "Yield Line Design of Rectangular Reinforced Concrete Balconies," *J. ACI*, vol. 67, no. 1, 1970, pp. 53–56.

23.12. H. Gesund, "Limit Design of Slabs for Concentrated Loads," *J. Struct. Div.*, ASCE, vol. 107, no. ST9, 1981, pp. 1839–1856.

23.13. A. Zaslavsky, "Yield Line Analysis of Rectangular Slabs with Central Openings," *J. ACI*, vol. 64, no. 12, 1967, pp. 838–844.

23.14. S. Islam and R. Park, "Yield Line Analysis of Two-Way Reinforced Concrete Slabs with Openings," *J. Inst. Struct. Eng.*, vol. 49, no. 6, 1971, pp. 269–276.

23.15. R. Park and W. L. Gamble, *Reinforced Concrete Slabs*, 2nd ed., John Wiley & Sons, New York, 1999.

23.16. K. V. Udilovich, I. B. Shleykov, and M. V. Banchuzhnyy, "Design of Flat-Plate Floors for Progressive Collapse Using Yield-line Analysis," *Concr. Intl.*, vol. 32, no. 7, 2010, pp. 37–44.

PROBLEMS

23.1. A square slab measuring 10 ft on each side is simply supported on three sides and unsupported along the fourth. It is reinforced for positive bending with an isotropic mat of steel providing resistance ϕm_n of 7000 ft-lb/ft in each of the two principal directions. Determine the uniformly distributed load that would cause flexural failure, using the method of virtual work.

23.2. The triangular slab shown in Fig. P23.2 has fixed supports along the two perpendicular edges and is free of any support along the diagonal edge. Negative reinforcement perpendicular to the supported edges provides design strength $\phi m_n = 4$ ft-kips/ft. The slab is reinforced for positive bending by an orthogonal grid providing resistance $\phi m_n = 2.67$ ft-kips/ft in all directions. Find the total factored load w_u that produces flexural failure. A virtual work solution is suggested.

FIGURE P23.2

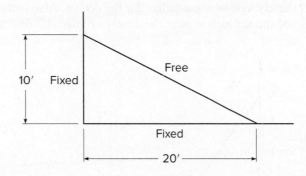

23.3. The one-way reinforced concrete slab shown in Fig. P23.3 spans 20 ft. It is simply supported at its left edge, fully fixed at its right edge, and free of support along the two long sides. Reinforcement provides design strength $\phi m_n = 5$ ft-kips/ft in positive bending and $\phi m_n = 7.5$ ft-kips/ft in negative bending at the right edge. Using the equilibrium method, find the factored load q_u uniformly distributed over the surface that would cause flexural failure.

FIGURE P23.3

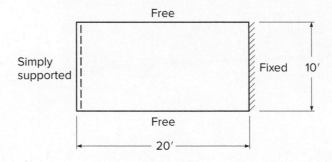

23.4. Solve Problem 23.3 using the method of virtual work.

23.5.   The triangular slab shown in Fig. P23.5 is to serve as weather protection over a loading dock. Support conditions are essentially fixed along *AB* and *BC*, and *AC* is a free edge. In addition to self weight, a superimposed dead load of 15 psf and service live load of 40 psf must be provided for. Material strengths are $f_c' = 4000$ psi and $f_y = 60,000$ psi. Using yield line analysis, find the required slab thickness h and find the reinforcement required at critical sections. Neglect corner pivots. Use a maximum reinforcement ratio of 0.005. Select bar sizes and spacings, and provide a sketch summarizing important aspects of the design. Make an approximate, conservative check of safety against shear failure for the design. Also include a conservative estimate of the deflection near the center of edge *AC* due to a full live load.

FIGURE P23.5

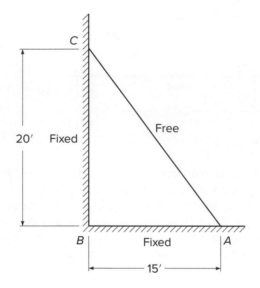

23.6. The square concrete slab shown in Fig. P23.6 is supported by monolithic concrete walls providing full vertical and rotational restraint along two adjacent edges, and by a 6 in. diameter steel pipe column, near the outer corner, that offers negligible rotational restraint. It is reinforced for positive bending by an orthogonal grid of bars parallel to the walls, providing design moment capacity $\phi m_n = 6.5$ ft-kips/ft in all directions. Negative reinforcement perpendicular to the walls and negative bars at the outer corner parallel to the slab diagonal provide $\phi m_n = 8.9$ ft-kips/ft. Neglecting corner pivots, find the total factored uniformly distributed load q_u that initiates flexural failure. Solution by the method of virtual work is recommended, with collapse geometry established by successive trials. Yield line lengths and perpendicular distances are most easily found graphically. Include a check of the shear capacity of the slab, using approximate methods. The steel column is capped with a 12 × 12 in. plate providing bearing.

FIGURE P23.6

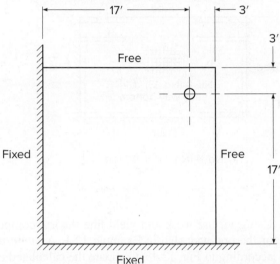

23.7. The square slab shown in Fig. P23.7 is supported by, and is monolithic with, a reinforced concrete wall along the edge CD that provides full fixity, and also is supported by a masonry wall along AB that provides a simply supported line. It is to carry a factored load $q_u = 300$ psf including its self weight. Assuming a uniform 6 in. slab thickness, find the required reinforcement. Include a sketch summarizing details of your design, indicating placement and length of all reinforcing bars. Also check the shear capacity of the structure, making whatever assumptions appear reasonable and necessary. Use $f'_c = 4000$ psi and $f_y = 60,000$ psi.

FIGURE P23.7

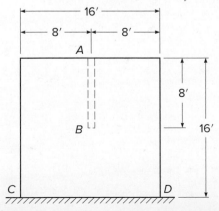

23.8. The slab of Fig. P23.8 is supported by three fixed edges but has no support along one long side. It has a uniform thickness of 7 in., resulting in effective depths in the long direction of 6.0 in. and in the short direction of 5.5 in. Bottom reinforcement consists of No. 4 (No. 13) bars at 14 in. centers in each direction, continued to the supports and the free edge. Top negative steel along the supported edges consists of No. 4 (No. 13) bars at 12 in. on centers, except that in a 2 ft wide "strong band" parallel and adjacent to the free edge, four No. 5 (No. 16) bars are used. All negative bars extend past the points of inflection, as required by ACI Code. Material strengths are $f_c' = 4000$ psi and $f_y = 60,000$ psi. Using the yield line method, determine the factored load q_u that can be carried.

FIGURE P23.8

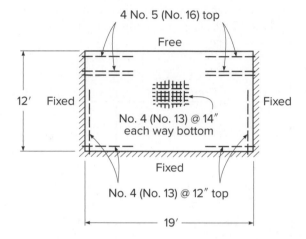

23.9. Using virtual work and yield line theory, compute the flexural collapse load of the one-way slab in Example 12.1. Assume that all straight bars are used, according to Fig. 12.4b. Compare the calculated collapse load with the original factored design load, and comment on differences.

23.10. Using virtual work and yield line theory, compute the flexural collapse load of the two-way column-supported flat plate of Example 13.2. To simplify the calculations, assume that all positive moment bars are carried to the edges of the panels, not cut off in the span. Consider all possible failure mechanisms, including a circular fan around the column. Neglect corner effects. Compare the calculated collapse load with the original factored design load and comment on differences.

24

Strip Method for Slabs

24.1 INTRODUCTION

In Section 23.2, the upper and lower bound theorems of the theory of plasticity were presented, and it was pointed out that the yield line method of slab analysis was an *upper bound approach* to determining the flexural strength of slabs. An upper bound analysis, if in error, will be so on the unsafe side. The actual carrying capacity will be less than, or at best equal to, the capacity predicted, which is certainly a cause for concern in design. Also, when applying the yield line method, it is necessary to assume that the distribution of reinforcement is known over the whole slab. It follows that the yield line approach is a tool to *analyze* the capacity of a given slab and can be used for *design* only in an iterative sense, for calculating the capacities of trial designs with varying reinforcement until a satisfactory arrangement is found.

These circumstances motivated Hillerborg to develop what is known as the *strip method* for slab design, his first results being published in Swedish in 1956 (Ref. 24.1). In contrast to yield line analysis, the strip method is a *lower bound approach*, based on satisfaction of equilibrium requirements everywhere in the slab. By the strip method (sometimes referred to as the *equilibrium theory*), a moment field is first determined that fulfills equilibrium requirements, after which the reinforcement in the slab at each point is designed for this moment field. If a distribution of moments can be found that satisfies both equilibrium and boundary conditions for a given external loading, and if the yield moment capacity of the slab is nowhere exceeded, then the given external loading will represent a lower bound of the true carrying capacity.

The strip method gives results on the safe side, which is certainly preferable in practice, and differences from the true carrying capacity never impair safety. The strip method is a *design* method, by which the needed reinforcement can be calculated. It encourages the designer to vary the reinforcement in a logical way, leading to an economical arrangement of steel, as well as a safe design. It is generally simple to use, even for slabs with holes or irregular boundaries.

In his original work in 1956, Hillerborg set forth the basic principles for edge-supported slabs and introduced the expression *strip method* (Ref. 24.1). He later expanded the method to include the practical design of slabs on columns and L-shaped slabs (Refs. 24.2 and 24.3). The first treatment of the subject in English was by Crawford (Ref. 24.4). In 1964, Blakey translated the earlier Hillerborg work into English (Ref. 24.5). Important contributions, particularly regarding continuity conditions, have been made by Kemp (Refs. 24.6 and 24.7) and Wood and Armer (Refs. 24.8, 24.9, and 24.10). Load tests of slabs designed by the strip method were carried out by Armer (Ref. 24.11) and confirmed that the method produces safe and satisfactory designs. In 1975, Hillerborg produced Ref. 24.12 "for the practical designer, helping him in the simplest possible way to produce safe designs for most of the slabs that he will meet in practice, including slabs that are irregular in plan or that carry unevenly distributed loads." Subsequently, he published a paper in which he summarized what has become known as the *advanced strip method*, pertaining to the design of slabs supported on columns, reentrant corners, or interior walls (Ref. 24.13). Useful summaries of both the simple and advanced strip methods can be found in Refs. 24.14 and 24.15.

The strip method is appealing not only because it is safe, economical, and versatile over a broad range of applications but also because it formalizes procedures followed instinctively by competent designers in placing reinforcement in the best possible position. In contrast with the yield line method, which provides no inducement to vary bar spacing, the strip method encourages the use of strong bands of steel where needed, such as around openings or over columns, improving economy and reducing the likelihood of excessive cracking or large deflections under service loading.

24.2 BASIC PRINCIPLES

The governing equilibrium equation for a small slab element having sides dx and dy is

$$\frac{\partial^2 m_x}{\partial x^2} + \frac{\partial^2 m_y}{\partial y^2} - 2\frac{\partial^2 m_{xy}}{\partial x \partial y} = -q \tag{24.1}$$

where q = external load per unit area
m_x, m_y = bending moments per unit width in X and Y directions, respectively
m_{xy} = twisting moment (Ref. 24.16)

According to the lower bound theorem, any combination of m_x, m_y, and m_{xy} that satisfies the equilibrium equation at all points in the slab and that meets boundary conditions is a valid solution, provided that the reinforcement is placed to carry these moments.

The basis for the simple strip method is that the torsional moment is chosen equal to zero; that is, no load is assumed to be resisted by the twisting strength of the slab. Therefore, if the reinforcement is parallel to the axes in a rectilinear coordinate system,

$$m_{xy} = 0$$

The equilibrium equation then reduces to

$$\frac{\partial^2 m_x}{\partial x^2} + \frac{\partial^2 m_y}{\partial y^2} = -q \tag{24.2}$$

This equation can be split conveniently into two parts, representing twistless beam strip action

$$\frac{\partial^2 m_x}{\partial x^2} = -kq \tag{24.3a}$$

and

$$\frac{\partial^2 m_y}{\partial y^2} = -(1 - k)q \tag{24.3b}$$

where the proportion of load taken by the strips is k in the X direction and $1 - k$ in the Y direction. In many regions in slabs, the value of k will be either 0 or 1. With

$k = 0$, all of the load is dispersed by strips in the Y direction; with $k = 1$, all of the load is carried in the X direction. In other regions, it may be reasonable to assume that the load is divided equally in the two directions (that is, $k = 0.5$).

24.3 CHOICE OF LOAD DISTRIBUTION

Theoretically, the load q can be divided arbitrarily between the X and Y directions. Different divisions, of course, lead to different patterns of reinforcement, and not all are equally appropriate. The desired goal is to arrive at an arrangement of steel that is safe and economical and that avoids problems at the service load level associated with excessive cracking or deflections. In general, the designer may be guided by knowledge of the general distribution of elastic moments.

To see an example of the strip method and to illustrate the choices open to the designer, consider the square, simply supported slab shown in Fig. 24.1, with side length a and a uniformly distributed factored load q per unit area.

The simplest load distribution is obtained by setting $k = 0.5$ over the entire slab, as shown in Fig. 24.1. The load on all strips in each direction is then $q/2$, as illustrated by the load dispersion arrows of Fig. 24.1a. This gives maximum moments

$$m_x = m_y = \frac{qa^2}{16} \tag{24.4}$$

over the whole slab, as shown in Fig. 24.1c, with a uniform lateral distribution across the width of the critical section, as in Fig. 24.1d.

FIGURE 24.1
Square slab with load shared equally in two directions.

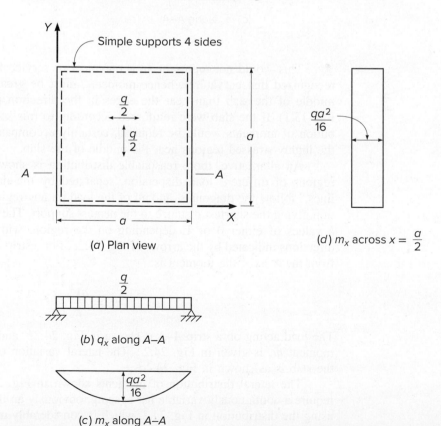

(a) Plan view

(b) q_x along A–A

(c) m_x along A–A

(d) m_x across $x = \dfrac{a}{2}$

FIGURE 24.2
Square slab with load
dispersion lines following
diagonals.

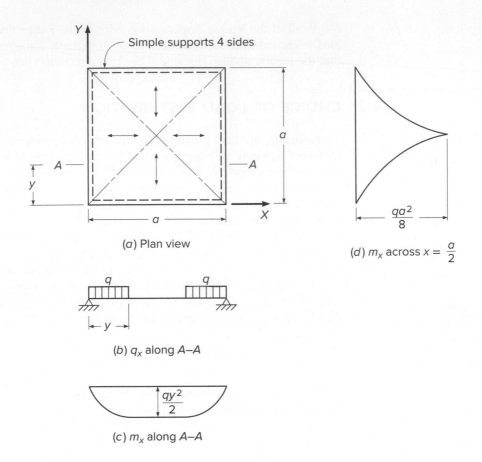

(a) Plan view

(b) q_x along A–A

(c) m_x along A–A

(d) m_x across $x = \dfrac{a}{2}$

This would not represent an economical or serviceable solution because it is recognized that curvatures, hence moments, must be greater in the strips near the middle of the slab than near the edges in the direction parallel to the edge (see Fig. 13.1). If the slab were reinforced according to this solution, extensive redistribution of moments would be required, certainly accompanied by much cracking in the highly stressed regions near the middle of the slab.

An alternative, more reasonable distribution is shown in Fig. 24.2. Here the regions of different load dispersion, separated by the dash-dotted "discontinuity lines," follow the diagonals, and all of the load on any region is carried in the direction giving the shortest distance to the nearest support. The solution proceeds, giving k values of either 0 or 1, depending on the region, with load transmitted in the directions indicated by the arrows of Fig. 24.2a. For a strip A–A at a distance $y \leq a/2$ from the X axis, the moment is

$$m_x = \frac{qy^2}{2} \tag{24.5}$$

The load acting on a strip A–A is shown in Fig. 24.2b, and the resulting diagram of moment m_x is given in Fig. 24.2c. The lateral variation of m_x across the width of the slab is as shown in Fig. 24.2d.

The lateral distribution of moments shown in Fig. 24.2d would theoretically require a continuously variable bar spacing, obviously an impracticality. One way of using the distribution in Fig. 24.2, which is considerably more economical than that in Fig. 24.1, would be to reinforce for the *average* moment over a certain width, approximating the actual lateral variation shown in Fig. 24.2*d* in a stepwise manner. Hillerborg notes that this is not strictly in accordance with the equilibrium theory and that the design is no longer certainly on the safe side, but other conservative assumptions, for example, neglect of membrane strength in the slab and neglect of strain hardening of the reinforcement, would surely compensate for the slight reduction in safety margin.

A third alternative distribution is shown in Fig. 24.3. Here the division is made so that the load is carried to the nearest support, as before, but load near the diagonals has been divided, with one-half taken in each direction. Thus, *k* is given values

FIGURE 24.3
Square slab with load near diagonals shared equally in two directions.

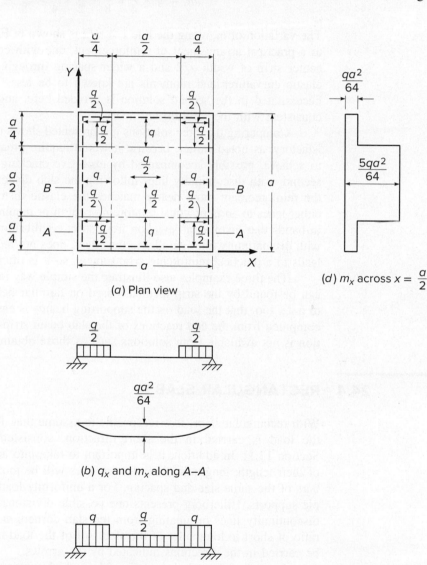

(a) Plan view

(b) q_x and m_x along *A–A*

(c) q_x and m_x along *B–B*

(d) m_x across $x = \dfrac{a}{2}$

of 0 or 1 along the middle edges and a value of 0.5 in the corners and center of the slab, with load dispersion in the directions indicated by the arrows shown in Fig. 24.3a. Two different strip loadings are now identified. For an X direction strip along section A–A, the maximum moment is

$$m_x = \frac{q}{2} \times \frac{a}{4} \times \frac{a}{8} = \frac{qa^2}{64} \tag{24.6a}$$

and for a strip along section B–B, the maximum moment is

$$m_x = q \times \frac{a}{4} \times \frac{a}{8} + \frac{q}{2} \times \frac{a}{4} \times \frac{3a}{8} = \frac{5qa^2}{64} \tag{24.6b}$$

The variation of m_x along the line $x = a/2$ is shown in Fig. 24.3d. This design leads to a practical arrangement of reinforcement, one with constant spacing through the center strip of width $a/2$ and a wider spacing through the outer strips, where the elastic curvatures and moments are known to be less. The averaging of moments necessitated in the second solution is avoided here, and the third solution is fully consistent with the equilibrium theory.

Comparing the three solutions just presented shows that the first would be unsatisfactory, as noted earlier, because it would require great redistribution of moments to achieve, possibly accompanied by excessive cracking and large deflections. The second, with discontinuity lines following the slab diagonals, has the advantage that the reinforcement more nearly matches the elastic distribution of moments, but it either leads to an impractical reinforcing pattern or requires an averaging of moments in bands that involves a deviation from strict equilibrium theory. The third solution, with discontinuity lines parallel to the edges, does not require moment averaging and leads to a practical reinforcing arrangement, so it is often preferred.

The three examples also illustrate the simple way in which moments in the slab can be found by the strip method, based on familiar beam analysis. It is important to note, too, that the load on the supporting beams is easily found because it can be computed from the end reactions of the slab beam strips in all cases. This information is not available from solutions such as those obtained by the yield line theory.

24.4   RECTANGULAR SLABS

With rectangular slabs, it is reasonable to assume that, throughout most of the area, the load is carried in the short direction, consistent with elastic theory (see Section 13.1). In addition, it is important to take into account the fact that because of their length, longitudinal reinforcing bars will be more expensive than transverse bars of the same size and spacing. For a uniformly loaded rectangular slab on simple supports, Hillerborg presents one possible division, as shown in Fig. 24.4, with discontinuity lines originating from the slab corners at an angle depending on the ratio of short to long sides of the slab. All of the load in each region is assumed to be carried in the directions indicated by the arrows.

Instead of the solution of Fig. 24.4, which requires continuously varying reinforcement to be strictly correct, Hillerborg suggests that the load can be distributed as shown in Fig. 24.5, with discontinuity lines parallel to the sides of the slab. For such cases, it is reasonable to take edge bands of width equal to one-fourth the short span dimension. Here the load in the corners is divided equally in the X and Y directions as shown, while elsewhere all of the load is carried in the direction indicated by the arrows.

FIGURE 24.4
Rectangular slab with
discontinuity lines
originating at the corners.

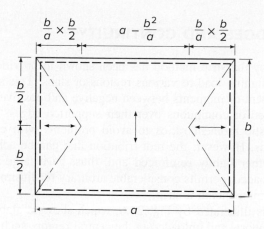

FIGURE 24.5
Discontinuity lines parallel to
the sides for a rectangular
slab.

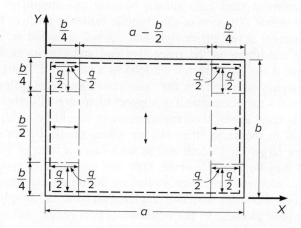

The second, preferred arrangement, shown in Fig. 24.5, gives slab moments as follows:

In the X direction:

Side strips $b/4$ wide: $m_x = \dfrac{q}{2} \times \dfrac{b}{4} \times \dfrac{b}{8} = \dfrac{qb^2}{64}$ (24.7a)

Middle strips $b/2$ wide: $m_x = q \times \dfrac{b}{4} \times \dfrac{b}{8} = \dfrac{qb^2}{32}$ (24.7b)

In the Y direction:

Side strips $b/4$ wide: $m_y = \dfrac{qb^2}{64}$ (24.8a)

Middle strips a-$b/2$ wide: $m_y = \dfrac{qb^2}{8}$ (24.8b)

This distribution, requiring no averaging of moments across band widths, is always on the safe side and is both simple and economical.

24.5   FIXED EDGES AND CONTINUITY

Designing by the strip method has been shown to provide a large amount of flexibility in assigning load to various regions of slabs. This same flexibility extends to the assignment of moments between negative and positive bending sections of slabs that are fixed or continuous over their supported edges. Some attention should be paid to elastic moment ratios to avoid problems with cracking and deflection at service loads. However, the redistribution that can be achieved in slabs, which are typically rather lightly reinforced and, thus, have large plastic rotation capacities when overloaded, permits considerable arbitrary readjustment of the ratio of negative to positive moments in a strip.

This is illustrated by Fig. 24.6, which shows a slab strip carrying loads only near the supports and unloaded in the central region, such as often occurs in designing by the strip method. It is convenient if the unloaded region is subject to a constant moment (and zero shear), because this simplifies the selection of positive reinforcement. The sum of the absolute values of positive span moment and negative end moment at the left or right end, shown as m_l and m_r in Fig. 24.6, depends only on the conditions at the respective end and is numerically equal to the negative moment if the strip carries the load as a cantilever. Thus, in determining moments for design, one calculates the "cantilever" moments, selects the span moment, and determines the corresponding support moments. Hillerborg notes that, as a general rule for fixed edges, the support moment should be about 1.5 to 2.5 times the span moment in the same strip. Higher values should be chosen for longitudinal strips that are largely unloaded, and in such cases a ratio of support to span moment of 3 to 4 may be used. However, little will be gained by using such a high ratio if the positive moment steel is controlled by minimum requirements of the ACI Code.

For slab strips with one end fixed and one end simply supported, the dual goals of constant moment in the unloaded central region and a suitable ratio of negative to positive moments govern the location to be chosen for the discontinuity lines. Figure 24.7a shows a uniformly loaded rectangular slab having two adjacent edges fixed and the other two edges simply supported. Note that although the middle strips have the same width as those of Fig. 24.5, the discontinuity lines are shifted to account for the greater stiffness of the strips with fixed ends. Their location is defined by a coefficient α, with a value clearly less than 0.5 for the slab shown, its exact value yet to be determined. It will be seen that the selection of α relates directly to the ratio of negative to positive moments in the strips.

The moment curve of Fig. 24.7b is chosen so that moment is constant over the unloaded part, that is, shearing force is zero. With constant moment, the positive steel can be fully stressed over most of the strip. The maximum positive moment in the X direction middle strip is then

$$m_{xf} = \frac{\alpha q b}{2} \times \frac{\alpha b}{4} = \alpha^2 \frac{q b^2}{8} \tag{24.9}$$

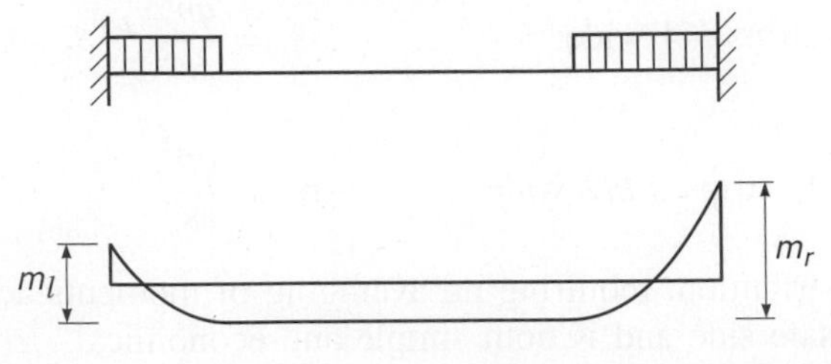

FIGURE 24.6
Slab strip with central region unloaded.

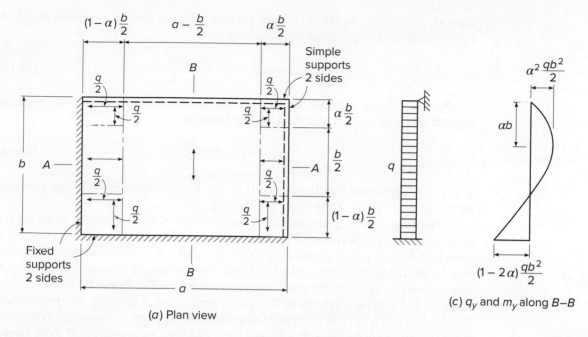

(a) Plan view

(c) q_y and m_y along B–B

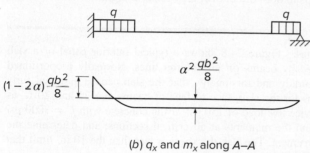

(b) q_x and m_x along A–A

FIGURE 24.7
Rectangular slab with two edges fixed and two edges simply supported.

The cantilever moment at the left support is

$$m_x = (1 - \alpha)\frac{qb}{2}(1 - \alpha)\frac{b}{4} = (1 - \alpha)^2\frac{qb^2}{8} \tag{24.10}$$

and so the negative moment at the left support is

$$m_{xs} = (1 - \alpha)^2\frac{qb^2}{8} - \alpha^2\frac{qb^2}{8} = (1 - 2\alpha)\frac{qb^2}{8} \tag{24.11}$$

For reference, the ratio of negative to positive moments in the X direction middle strip is

$$\frac{m_{xs}}{m_{xf}} = \frac{1 - 2\alpha}{\alpha^2} \tag{24.12}$$

The moments in the X direction edge strips are one-half of those in the middle strips because the load is one-half as great.

It is reasonable to choose the same ratio between support and span moments in the Y direction as in the X direction. Accordingly, the distance from the right support, Fig. 24.7c, to the maximum positive moment section is chosen as ab. It follows that the maximum positive moment is

$$m_{yf} = \alpha q b \times \frac{ab}{2} = \alpha^2 \frac{qb^2}{2} \qquad (24.13)$$

Applying the same methods as used for the X direction shows that the negative support moment in the Y direction middle strips is

$$m_{ys} = (1 - 2\alpha)\frac{qb^2}{2} \qquad (24.14)$$

It is easily confirmed that the moments in the Y direction edge strips are just one-eighth of those in the Y direction middle strip.

With the above expressions, all of the moments in the slab can be found once a suitable value for α is chosen. From Eq. (24.12), it can be confirmed that values of α from 0.35 to 0.39 give corresponding ratios of negative to positive moments from 2.45 to 1.45, the range recommended by Hillerborg. For example, if it is decided that support moments are to be twice the span moments, the value of α should be 0.366, and the negative and positive moments in the central strip in the Y direction are, respectively, $0.134qb^2$ and $0.067qb^2$. In the middle strip in the X direction, moments are one-fourth those values; and in the edge strips in both directions, they are one-eighth of those values.

EXAMPLE 24.1   **Rectangular slab with fixed edges.**   Figure 24.8 shows a typical interior panel of a slab floor in which support is provided by beams on all column lines. Normally proportioned beams are stiff enough, both flexurally and torsionally, that the slab can be assumed fully restrained on all sides. Clear spans for the slab, face to face of beams, are 25 and 20 ft, as shown. The floor must carry a service live load of 150 psf, using concrete with $f_c' = 3000$ psi and steel with $f_y = 60,000$ psi. Find the moments at all critical sections, and determine the required slab thickness and reinforcement. The slab thickness is less than the 10 in. limit that would require inclusion of size effects.

SOLUTION.   The minimum slab thickness required by the ACI Code can be found from Eq. (13.4b), with $\ell_n = 25$ ft and $\beta = 1.25$:

$$h = \frac{25 \times 12(0.8 + 60/200)}{36 + 9 \times 1.25} = 6.98 \text{ in.}$$

A total thickness of 7 in. is selected, for which $q_d = 150 \times 7/12 = 87.5$ psf. Applying the load factors of 1.2 and 1.6 to dead load and live load, respectively, determines that the total factored load for design is 340 psf. For strip analysis, discontinuity lines are selected as shown in Fig. 24.8, with edge strips of width $b/4 = 20/4 = 5$ ft. In the corners, the load is divided equally in the two directions; elsewhere, 100 percent of the load is assigned to the direction indicated by the arrows. A ratio of support moment to span moment of 2.0 is used. Calculation of moments then proceeds as follows:

X direction middle strip:

$$\text{Cantilever:} \quad m_x = \frac{qb^2}{32} = 340 \times \frac{400}{32} = 4250 \text{ ft-lb/ft}$$

$$\text{Negative:} \quad m_{xs} = 4250 \times \frac{2}{3} = 2833$$

$$\text{Positive:} \quad m_{xf} = 4250 \times \frac{1}{3} = 1417$$

FIGURE 24.8
Design example: two-way
slab with fixed edges.

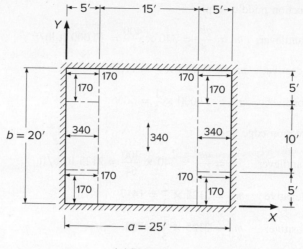

(a) Plan view

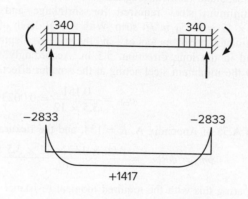

(b) X direction middle strip

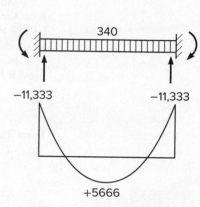

(d) Y direction middle strip

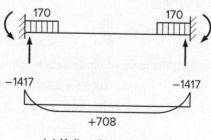

(c) X direction edge strip

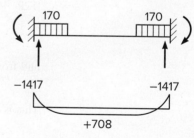

(e) Y direction edge strip

X direction edge strips:

Cantilever: $m_x = \dfrac{qb^2}{64} = 340 \times \dfrac{400}{64} = 2125$ ft-lb/ft

Negative: $m_{xs} = 2125 \times \dfrac{2}{3} = 1417$

Positive: $m_{xf} = 2125 \times \dfrac{1}{3} = 708$

Y direction middle strip:

Cantilever: $m_y = \dfrac{qb^2}{8} = 340 \times \dfrac{400}{8} = 17{,}000$ ft-lb/ft

Negative: $m_{ys} = 17{,}000 \times \dfrac{2}{3} = 11{,}333$

Positive: $m_{yf} = 17{,}000 \times \dfrac{1}{3} = 5666$

Y direction edge strips:

Cantilever: $m_y = \dfrac{qb^2}{64} = 340 \times \dfrac{400}{64} = 2125$ ft-lb/ft

Negative: $m_{ys} = 2125 \times \dfrac{2}{3} = 1417$

Positive: $m_{yf} = 2125 \times \dfrac{1}{3} = 708$

Strip loads and moment diagrams are as shown in Fig. 24.8. According to ACI Code 8.6, the minimum steel required for shrinkage and temperature crack control is $0.0018 \times 7 \times 12 = 0.151$ in^2/ft strip. With a total depth of 7 in., with $\frac{3}{4}$ in. concrete cover, and with estimated bar diameters of $\frac{1}{2}$ in., the effective depth of the slab in the short direction is 6 in., and in the long direction, 5.5 in. Accordingly, the flexural reinforcement ratio provided by the minimum steel acting at the smaller effective depth is

$$\rho_{\min} = \frac{0.151}{5.5 \times 12} = 0.0023$$

From Table A.5a of Appendix A, $R = 134$, and the flexural design strength is

$$\phi m_n = \phi R b d^2 = \frac{0.90 \times 134 \times 12 \times 5.5^2}{12} = 3648 \text{ ft-lb/ft}$$

Comparing this with the required moment resistance shows that the minimum steel is adequate in the *X* direction in both middle and edge strips and in the *Y* direction edge strips. No. 4 (No. 13) bars at 14 in. spacing provides the needed area. In the *Y* direction middle strip, for negative bending,

$$R = \frac{m_u}{\phi b d^2} = \frac{11{,}333 \times 12}{0.90 \times 12 \times 6^2} = 350$$

and from Table A.5a, the required reinforcement ratio is 0.0063. The required steel is then

$$A_s = 0.0063 \times 12 \times 6 = 0.45 \text{ in}^2/\text{ft}$$

This is provided using No. 5 (No. 16) bars at 8 in. on centers. For positive bending,

$$R = \frac{5666 \times 12}{0.90 \times 12 \times 6^2} = 175$$

for which $\rho = 0.0030$, and the required positive steel area per strip is

$$A_s = 0.0030 \times 12 \times 6 = 0.22 \text{ in}^2/\text{ft}$$

to be provided by No. 4 (No. 13) bars on 10 in. centers. Note that all bar spacings are less than $2h = 2 \times 7 = 14$ in., as required by the Code, and that the reinforcement ratios are well below the value for a tension-controlled section of 0.0135.

Negative bar cutoff points can easily be calculated from the moment diagrams. For the *X* direction middle strip, the point of inflection a distance *x* from the left edge is found as follows:

$$1700x - 2833 - 340\left(\frac{x^2}{2}\right) = 0$$

$$x = 2.11 \text{ ft}$$

According to the Code, the negative bars must be continued at least d or $12d_b$ beyond that point, requiring a 6 in. extension in this case. Thus, the negative bars are cut off $2.11 + 0.50 = 2.61$ ft, say 2 ft 8 in., from the face of support. The same result is obtained for the X direction edge strips and the Y direction edge strips. For the Y direction middle strip, the distance $y = 4.23$ ft from face of support to inflection point is found in a similar manner. In this case, with No. 5 (No. 10) bars used, the required extension is 7.5 in., giving a total length past the face of supports of $4.23 + 0.63 = 4.86$ ft or 4 ft 11 in. All positive bars are carried 6 in. into the face of the supporting beams. If the slab was part of a moment resisting frame, the cutoff locations would have to be in accordance with Fig. 13.10.

24.6 UNSUPPORTED EDGES

The slabs considered in the preceding sections, together with the supporting beams, could also have been designed by the methods of Chapter 13. The real power of the strip method becomes evident when dealing with nonstandard problems, such as slabs with an unsupported edge, slabs with holes, or slabs with reentrant corners (L-shaped slabs).

For a slab with one edge unsupported, for example, a reasonable basis for analysis by the simple strip method is that a strip along the unsupported edge takes a greater load per unit area than the actual unit load acting, that is, the strip along the unsupported edge acts as a support for the strips at right angles. Such strips have been referred to by Wood and Armer as *strong bands* (Ref. 24.8). A strong band is, in effect, an integral beam, usually having the same total depth as the remainder of the slab but containing a concentration of reinforcement. The strip may be made deeper than the rest of the slab to increase its carrying capacity, but this is not usually necessary.

Figure 24.9a shows a rectangular slab carrying a uniformly distributed factored load q per unit area, with fixed edges along three sides and no support along one short side. Discontinuity lines are chosen as shown. The load on a unit middle strip in the X direction, shown in Fig. 24.9b, includes the downward load q in the region adjacent to the fixed left edge and the upward reaction kq in the region adjacent to the free edge. Summing moments about the left end, with moments positive clockwise and with the unknown support moment denoted m_{xs}, gives

$$m_{xs} + \frac{qb^2}{32} - \frac{kqb}{4}\left(a - \frac{b}{8}\right) = 0$$

from which

$$k = \frac{1 + 32m_{xs}/qb^2}{8(a/b) - 1} \qquad (24.15)$$

Thus, k can be calculated after the support moment is selected.

The appropriate value of m_{xs} to be used in Eq. (24.15) depends on the shape of the slab. If a is large relative to b, the strong band in the Y direction at the edge will be relatively stiff, and the moment at the left support in the X direction strips will approach the elastic value for a propped cantilever. If the slab is nearly square, the deflection of the strong band will tend to increase the support moment; a value about one-half the free cantilever moment might be selected (Ref. 24.14).

Once m_{xs} is selected and k is known, it is easily shown that the maximum span moment occurs when

$$x = (1 - k)\frac{b}{4}$$

FIGURE 24.9
Slab with free edge along
short side.

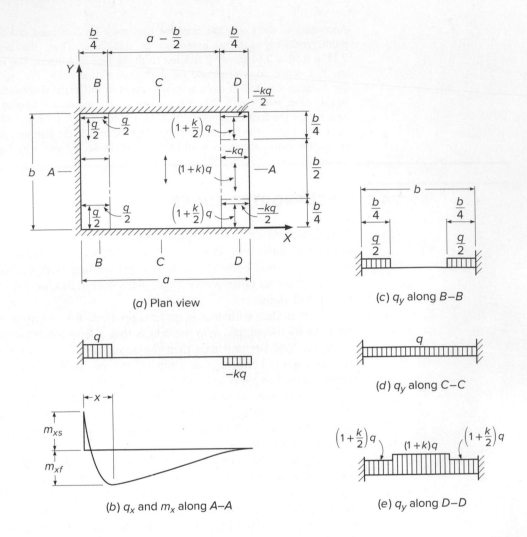

(a) Plan view

(c) q_y along B–B

(d) q_y along C–C

(b) q_x and m_x along A–A

(e) q_y along D–D

It has a value

$$m_{xf} = \frac{kqb^2}{32}\left(\frac{8a}{b} - 3 + k\right) \qquad (24.16)$$

The moments in the X direction edge strips are one-half of those in the middle strip. In the Y direction middle strip, Fig. 24.9d, the cantilever moment is $qb^2/8$. Adopting a ratio of support to span moment of 2 results in support and span moments, respectively, of

$$m_{ys} = \frac{qb^2}{12} \qquad (24.17a)$$

$$m_{yf} = \frac{qb^2}{24} \qquad (24.17b)$$

Moments in the Y direction strip adjacent to the fixed edge, Fig. 24.9c, are one-eighth of those values. In the Y direction strip along the free edge, Fig. 24.9e, moments can, with slight conservatism, be made equal to $(1 + k)$ times those in the Y direction middle strip.

If the unsupported edge is in the long-span direction, then a significant fraction of the load in the slab central region is carried in the direction perpendicular to the long edges, and the simple distribution shown in Fig. 24.10a is more suitable. A strong band along the free edge serves as an integral edge beam, with width βb normally chosen as low as possible considering limitations on tensile reinforcement ratio in the strong band.

For a Y direction strip, with moments positive clockwise,

$$m_{ys} + \frac{1}{2} k_1 q (1 - \beta)^2 b^2 - k_2 q \beta b^2 (1 - \beta/2) = 0$$

from which

$$k_2 = \frac{k_1 (1 - \beta)^2 + 2 m_{ys}/q b^2}{\beta(2 - \beta)} \tag{24.18}$$

The value of k_1 may be selected to make use of the minimum steel in the X direction required by ACI Code 8.6. In choosing m_{ys} to be used in Eq. (24.18) for calculating k_2, one should again recognize that the deflection of the strong band along the free edge will tend to increase the Y direction moment at the supported edge above the propped cantilever value based on zero deflection. A value for m_{ys} of about one-half the free cantilever moment may be appropriate in typical cases. A high ratio of a/b permits greater deflection of the free edge through the central region, tending to increase the support moment, and a low ratio restricts deflection, reducing the support moment.

FIGURE 24.10
Slab with free edge in long-span direction.

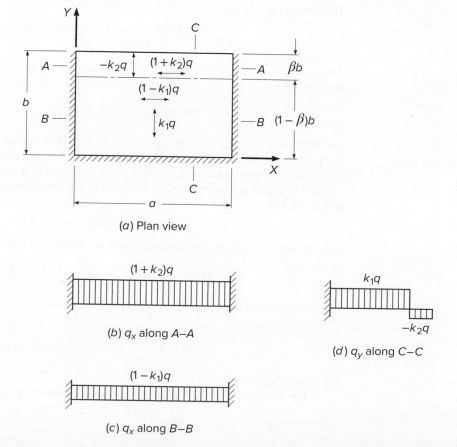

(a) Plan view

(b) q_x along A–A

(c) q_x along B–B

(d) q_y along C–C

EXAMPLE 24.2 **Rectangular slab with long edge unsupported.** The 12×19 ft slab shown in Fig. 24.11a, with three fixed edges and one long edge unsupported, must carry a uniformly distributed service live load of 125 psf; $f_c' = 4000$ psi and $f_y = 60,000$ psi. Select an appropriate slab thickness, determine all factored moments in the slab, and select reinforcing bars and spacings for the slab.

SOLUTION. The minimum thickness requirements of the ACI Code do not really apply to the type of slab considered here. However, Table 13.2, which controls for beamless flat plates, can be applied conservatively because although the present slab is beamless along the free edge, it has infinitely stiff supports on the other three edges. From that table, with $\ell_n = 19$ ft,

$$h = \frac{19 \times 12}{33} = 6.91 \text{ in.}$$

A total thickness of 7 in. will be selected. The slab dead load is $150 \times \frac{7}{12} = 88$ psf, and the total factored design load is $1.2 \times 88 + 1.6 \times 125 = 306$ psf.

A strong band 2 ft wide will be provided for support along the free edge. In the main slab, a value $k_1 = 0.45$ is selected, resulting in a slab load in the Y direction of $0.45 \times 306 = 138$ psf and in the X direction of $0.55 \times 306 = 168$ psf.

First, with regard to the Y direction slab strips, the negative moment at the supported edge is chosen as one-half the free cantilever value, which in turn is approximated based on 138 psf over an 11 ft distance from the support face to the center of the strong band. The restraining moment is thus

$$m_{ys} = \frac{1}{2} \times \frac{138 \times 11^2}{2} = 4175 \text{ ft-lb/ft}$$

Then, from Eq. (24.18)

$$k_2 = \frac{0.45(5/6)^2 - 2 \times 4175/(306 \times 144)}{(1/6)(2 - 1/6)} = 0.403$$

Thus, an uplift of $0.403 \times 306 = 123$ psf is provided for the Y direction strips by the strong band, as shown in Fig. 24.11d. For this loading, the negative moment at the left support is

$$m_{ys} = 138 \times \frac{10^2}{2} - 123 \times 2 \times 11 = 4194 \text{ ft-lb/ft}$$

The difference from the original value of 4175 ft-lb/ft is caused by numerical rounding of the load terms. The statically consistent value of 4194 ft-lb/ft is used for design. The maximum positive moment in the Y direction strips is located at the point of zero shear. With y_1 as the distance of that point from the free edge to the zero shear location, and with reference to Fig. 24.11d,

$$123 \times 2 - 138(y_1 - 2) = 0$$

from which $y_1 = 3.78$ ft. The maximum positive moment, found at that location, is

$$m_{yf} = 123 \times 2(3.78 - 1) - 138 \times \frac{1.78^2}{2} = 465 \text{ ft-lb/ft}$$

For later reference in cutting off bars, the point of inflection is located a distance y_2 from the free edge:

$$123 \times 2(y_2 - 1) - \frac{138}{2}(y_2 - 2)^2 = 0$$

resulting in $y_2 = 6.38$ ft.

For the X direction slab strips, the cantilever moment is

Cantilever: $$m_x = \frac{168 \times 19^2}{8} = 7581 \text{ ft-lb/ft}$$

FIGURE 24.11
Design example: slab with
long edge unsupported.

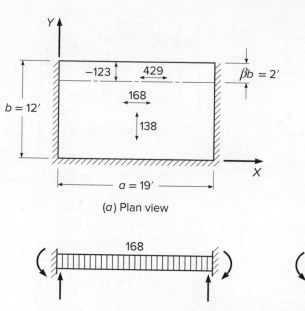

(a) Plan view

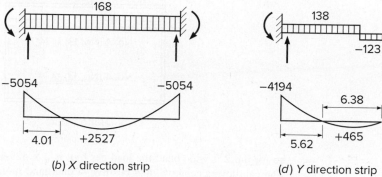

(b) X direction strip

(d) Y direction strip

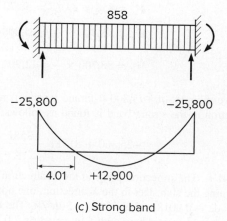

(c) Strong band

A ratio of negative to positive moments of 2.0 is chosen here, resulting in negative and positive moments, respectively, of

Negative:
$$m_{xs} = 7581 \times \frac{2}{3} = 5054 \text{ ft-lb/ft}$$

Positive:
$$m_{xf} = 7581 \times \frac{1}{3} = 2527 \text{ ft-lb/ft}$$

as shown in Fig. 24.11b.

The unit load on the strong band in the X direction is

$$(1 + k_2)q = (1 + 0.403) \times 306 = 429 \text{ psf}$$

FIGURE 24.11
(*Continued*)

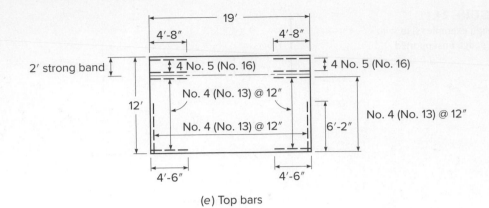

(e) Top bars

(f) Bottom bars

so for the 2 ft wide band the load per foot is $2 \times 429 = 858$ psf, as indicated in Fig. 24.11c. The cantilever, negative, and positive strong band moments are, respectively,

Cantilever: $\qquad\qquad M_x = 858 \times 19^2/8 = 38{,}700$ ft-lb

Negative: $\qquad\qquad M_{xs} = 38{,}700 \times \dfrac{2}{3} = 25{,}800$ ft-lb

Positive: $\qquad\qquad M_{xf} = 38{,}700 \times \dfrac{1}{3} = 12{,}900$ ft-lb

With a negative moment of $-25{,}800$ ft-lb and a support reaction of $858 \times \frac{19}{2} = 8151$ lb, the point of inflection in the strong band is found as follows:

$$-25{,}800 + 8151x - \frac{858x^2}{2} = 0$$

giving $x = 4.01$ ft. The inflection point in the X direction slab strips is at the same location.

In designing the slab steel in the X direction, one notes that the minimum steel required by the ACI Code is $0.0018 \times 7 \times 12 = 0.15$ in²/ft. The effective slab depth in the X direction, assuming $\frac{1}{2}$ in. diameter bars with $\frac{3}{4}$ in. cover, is $7.0 - 1.0 = 6.0$ in. The corresponding flexural reinforcement ratio in the X direction is $\rho = 0.15/(12 \times 6) = 0.0021$. From Table A.5a, $R = 124$, and the design strength is

$$\phi m_n = \phi R b d^2 = \frac{0.90 \times 124 \times 12 \times 6^2}{12} = 4018 \text{ ft-lb/ft}$$

It is seen that the minimum slab steel required by the Code provides for the positive bending moment of 2527 ft-lb/ft. The requirement of 0.15 in²/ft is met by No. 4 (No. 13) bars at the maximum permitted spacing of $2h = 14$ in., providing 0.17 in²/ft. The X direction negative moment of 5054 ft-lb/ft requires

$$R = \frac{m_u}{\phi b d^2} = \frac{5054 \times 12}{0.90 \times 12 \times 6^2} = 156$$

and Table A.5a indicates that the required $\rho = 0.0027$. Thus, the negative bar requirement is $A_s = 0.0027 \times 12 \times 6 = 0.19$ in²/ft. This is provided by No. 4 (No. 13) bars at 12 in. spacing, continued $4.01 \times 12 + 6 = 54$ in., or 4 ft 6 in., from the support face.

In the Y direction, the effective depth is one bar diameter less than in the X direction, or 5.5 in. Thus, the flexural reinforcement ratio provided by the shrinkage and temperature steel is $\rho = 0.15/(12 \times 5.5) = 0.0023$. This results in $R = 135$, so the design strength is

$$\phi m_n = \frac{0.90 \times 135 \times 12 \times 5.5^2}{12} = 3675 \text{ ft-lb/ft}$$

well above the requirement for positive bending of 465 ft-lb/ft. No. 4 (No. 13) bars at 14 in. are satisfactory for positive steel in this direction also. For the negative moment of 4194 ft-lb/ft,

$$R = \frac{4194 \times 12}{0.90 \times 12 \times 5.5^2} = 154$$

and from Table A.5a, the required $\rho = 0.0027$. The corresponding steel requirement is $0.0027 \times 12 \times 5.5 = 0.18$ in²/ft. No. 4 (No. 13) bars at 12 in. are used, and they are extended $5.62 \times 12 + 6 = 74$ in., or 6 ft 2 in., past the support face.

In the strong band, the positive moment of 12,900 ft-lb requires

$$R = \frac{12,900 \times 12}{0.90 \times 24 \times 6^2} = 199$$

The corresponding reinforcement ratio is 0.0034, and the required bar area is $0.0034 \times 24 \times 6 = 0.49$ in². This can be provided by two No. 5 (No. 16) bars. For the negative moment of 25,800 ft-lb,

$$R = \frac{25,800 \times 12}{0.90 \times 24 \times 6^2} = 398$$

resulting in $\rho = 0.0070$, and required steel equal to $0.0070 \times 24 \times 6 = 1.01$ in². Four No. 5 (No. 16) bars, providing an area of 1.23 in², are used, and they are cut off $4.01 \times 12 + 7.5 = 56$ in., or 4 ft 8 in., from the support face.

The final arrangement of bar reinforcement is shown in Fig. 24.11e and f. Negative bar cutoff locations are as indicated, and development by embedded lengths into the supports is provided. All positive bars in the slab and strong band are carried 6 in. into the support faces.

A design problem commonly met in practice is that of a slab supported along three edges and unsupported along the fourth, with a distributed load that increases linearly from zero along the free edge to a maximum at the opposite supported edge. Examples include the wall of a rectangular tank subjected to liquid pressure and earth-retaining walls with buttresses or counterforts (see Section 16.1).

Figure 24.12 shows such a slab, with load of intensity q_0 at the long, supported edge, reducing to zero at the free edge. In the main part of the slab, a constant load $k_2 q_0$ is carried in the X direction, as shown in Fig. 24.12c; thus, a constant load $k_2 q_0$ is deducted from the linear varying load in the Y direction, as shown in Fig. 24.12d. Along the free edge, a strong band of width βb is provided, carrying a load $k_1 q_0$, as in Fig. 24.12a, and so providing an uplift load equal to that amount at the end of the Y direction strip in Fig. 24.12d. The choice of k_1 and k_2 depends on the ratio of a/b. If this ratio is high, k_2 should be chosen with regard to the minimum slab reinforcement required by the ACI Code. The value of k_1 is then calculated by statics, based on a selected value of the restraining moment at the fixed edge, say one-half of the free cantilever value. In many cases it is convenient to let k_1 equal k_2. Then it is the support moment that follows from statics. The value of β is selected

FIGURE 24.12
Slab with one free edge and
linearly varying load.

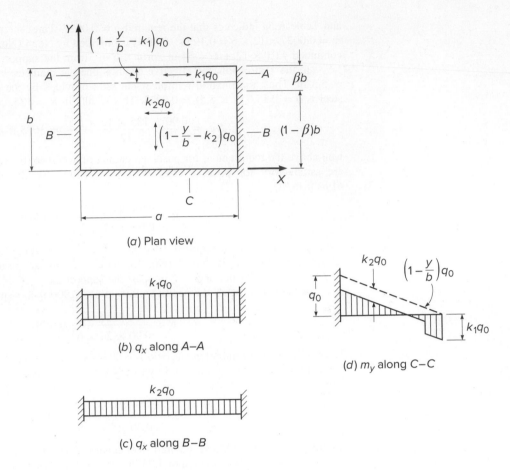

(a) Plan view

(b) q_x along A–A

(c) q_x along B–B

(d) m_y along C–C

as low as possible considering the upper limit on tensile reinforcement ratio in the
strong band imposed by the Code for beams. The strong band is designed for a load
of intensity $k_1 q_0$ distributed uniformly over its width βb.

24.7   SLABS WITH HOLES

Slabs with small openings can usually be designed as if there were no openings,
replacing the interrupted steel with bands of reinforcing bars of equivalent area on
either side of the opening in each direction (see Section 13.10). Slabs with larger
openings must be treated more rigorously. The strip method offers a rational and
safe basis for design in such cases. Integral load-carrying beams are provided along
the edges of the opening, usually having the same depth as the remainder of the slab
but with extra reinforcement, to pick up the load from the affected regions and
transmit it to the supports. In general, these integral beams should be chosen so as
to carry the loads most directly to the supported edges of the slab. The width of the
strong bands should be selected so that the reinforcement ratios ρ are at or below
the value required to produce a tension-controlled member (that is, $\epsilon_t \geq 0.005$ and
$\phi = 0.90$). Doing so ensures ductile behavior of the slab.

Use of the strip method for analysis and design of a slab with a large central
hole is illustrated by the following example.

EXAMPLE 24.3   **Rectangular slab with central opening.**   Figure 24.13*a* shows a 16 × 28 ft slab with fixed supports along all four sides. A central opening 4 × 8 ft must be accommodated. Estimated slab thickness, from Eq. (13.4*b*), is 7 in. The slab is to carry a uniformly distributed factored load of 300 psf, including self weight. Devise an appropriate system of strong bands to reinforce the opening, and determine moments to be resisted at all critical sections of the slab.

SOLUTION.   The basic pattern of discontinuity lines and load dispersion are selected according to Fig. 24.5. Edge strips are defined having width $\frac{16}{4} = 4$ ft. In the corners, the load is equally divided in the two directions. In the central region, 100 percent of the load is assigned to the Y direction, while along the central part of the short edges, 100 percent of the load is carried in the X direction. Moments for this "basic case" without the hole are calculated and later used as a guide in selecting moments for the actual slab with hole. A ratio of support to span moments of 2.0 is used generally, as for the previous examples. Moments for the slab, neglecting the hole, would then be as follows:

X direction middle strips:

Cantilever: $$m_x = \frac{qb^2}{32} = 300 \times \frac{16^2}{32} = 2400 \text{ ft-lb/ft}$$

Negative: $$m_{xs} = 2400 \times \frac{2}{3} = 1600$$

Positive: $$m_{xf} = 2400 \times \frac{1}{3} = 800$$

X direction edge-strip moments are one-half of the middle-strip moments.

Y direction middle strips:

Cantilever: $$m_y = \frac{qb^2}{8} = 300 \times \frac{16^2}{8} = 9600 \text{ ft-lb/ft}$$

Negative: $$m_{ys} = 9600 \times \frac{2}{3} = 6400$$

Positive: $$m_{yf} = 9600 \times \frac{1}{3} = 3200$$

Y direction edge-strip moments are one-half of the middle-strip moments.

Because of the hole, certain strips lack support at one end. To support them, 1 ft wide strong bands are provided in the X direction at the long edges of the hole and 2 ft wide strong bands in the Y direction on each side of the hole. The Y direction bands provide for the reactions of the X direction bands. With the distribution of loads shown in Fig. 24.13*a*, strip reactions and moments are found as follows:

Strip A–A
It may at first be assumed that propped cantilever action is obtained, with the restraint moment along the slab edge taken as 6400 ft-lb/ft, the same as for the basic case. Summing moments about the left end of the loaded strip then results in

$$q_1 = \frac{300 \times 6 \times 3 - 6400}{1 \times 5.5} = -182 \text{ psf}$$

The negative value indicates that the cantilever strips are serving as supports for strip D–D, and in turn for the strong bands in the Y direction, which is hardly a reasonable assumption. Instead, a discontinuity line is assumed 5 ft from the support, as shown in Fig. 24.13*b*, terminating the cantilever and leaving the 1 ft strip D–D along the edge of the opening in the X direction to carry its own load. It follows that the support moment in the cantilever strip is

Negative: $$m_{ys} = 300 \times 5 \times 2.5 = 3750 \text{ ft-lb/ft}$$

FIGURE 24.13
Design example: slab with
central hole.

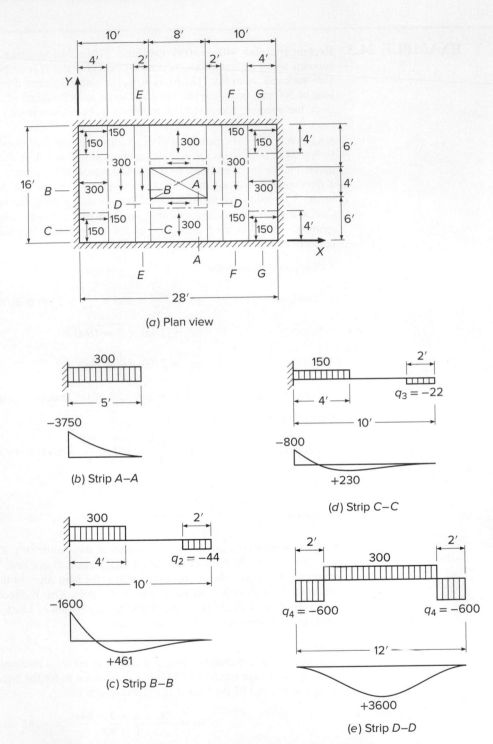

(a) Plan view

(b) Strip A–A

(c) Strip B–B

(d) Strip C–C

(e) Strip D–D

Strip B–B
The restraint moment at the supported edge is taken to be the same as the basic case, that is,
1600 ft-lb/ft. Summing moments about the left end of the strip of Fig. 24.13c then results
in an uplift reaction at the right end, to be provided by strip E–E, of

$$q_2 = \frac{300 \times 4 \times 2 - 1600}{2 \times 9} = 44 \text{ psf}$$

FIGURE 24.13
(*Continued*)

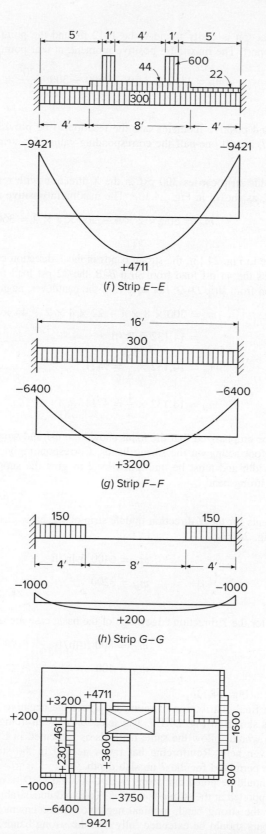

(*f*) Strip *E–E*

(*g*) Strip *F–F*

(*h*) Strip *G–G*

(*i*) Factored moments—symmetrical about both axes

The left reaction is easily found to be 1112 lb, and the point of zero shear is 3.70 ft from the left support. The maximum positive moment, at that point, is

Positive:
$$m_{xf} = 1112 \times 3.70 - 1600 - 300\,\frac{3.70^2}{2} = 461 \text{ ft-lb/ft}$$

Strip C–C

Negative and positive moments and the reaction to be provided by strip E–E, as shown in Fig. 24.13d, are all one-half the corresponding values for strip B–B.

Strip D–D

The 1 ft wide strip carries 300 psf in the X direction with reactions provided by the strong bands E–E, as shown in Fig. 24.13e. The maximum positive moment is

$$m_{xf} = 600 \times 2 \times 5 - 300 \times 4 \times 2 = 3600 \text{ ft-lb/ft}$$

Strip E–E

In reference to Fig. 24.13f, the strong bands in the Y direction carry the directly applied load of 300 psf plus the 44 psf load from strip B–B, the 22 psf load from strip C–C, and the 600 psf end reaction from strip D–D. For strip E–E the cantilever, negative, and positive moments are

Cantilever:
$$m_y = 300 \times 8 \times 4 + 22 \times 4 \times 2 + 44 \times 4 \times 6 + 600 \times 1 \times 5.5$$

$$= 14{,}132 \text{ ft-lb/ft}$$

Negative:
$$m_{ys} = 14{,}132 \times \frac{2}{3} = 9421$$

Positive:
$$m_{yf} = 14{,}132 \times \frac{1}{3} = 4711$$

It should be emphasized that the loads shown are psf and would be multiplied by 2 to obtain loads per foot acting on the strong bands. Correspondingly, the moments just obtained are per foot width and must be multiplied by 2 to give the support and span moments for the 2 ft wide strong band.

Strip F–F

The moments for the Y direction middle strip of the basic case may be used without change; thus, in Fig. 24.13g,

Negative:
$$m_{ys} = 6400 \text{ ft-lb/ft}$$

Positive:
$$m_{yf} = 3200$$

Strip G–G

Moments for the Y direction edge strips of the basic case are used without change, resulting in

Negative:
$$m_{ys} = 800 \text{ ft-lb/ft}$$

Positive:
$$m_{yf} = 400$$

as shown in Fig. 24.13h.

The final distribution of moments across the negative and positive critical sections of the slab is shown in Fig. 24.13i. The selection of reinforcing bars and determination of cut-off points would follow the same methods as presented in Examples 24.1 and 24.2 and will not be given here. Reinforcing bar ratios needed in the strong bands are well below the maximum permitted for the 7 in. slab depth.

It should be noted that strips B–B, C–C, and D–D have been designed as if they were simply supported at the strong band E–E. To avoid undesirably wide cracks where these strips pass over the strong band, nominal negative reinforcement should be added in this region. Positive bars should be extended fully into the strong bands.

24.8   ADVANCED STRIP METHOD

The simple strip method described in the earlier sections of this chapter is not directly suitable for the design of slabs supported by columns (for example, flat plates) or slabs supported at reentrant corners.[†] For such cases, Hillerborg introduced the advanced strip method (Rcfs. 24.2, 24.5, 24.12, and 24.13).

Fundamental to the advanced strip method is the corner-supported element, such as that shown shaded in Fig. 24.14a. The corner-supported element is a rectangular region of the slab with the following properties:

1. The edges are parallel to the reinforcement directions.
2. It carries a uniform load q per unit area.

FIGURE 24.14
Slab with central supporting column.

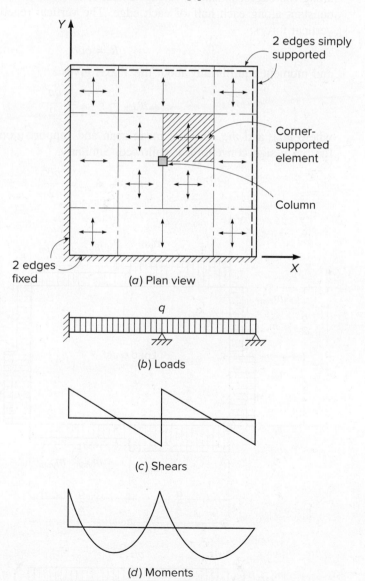

(a) Plan view

(b) Loads

(c) Shears

(d) Moments

[†] However, Wood and Armer, in Ref. 24.8, suggest that beamless slabs with column supports can be solved by the simple strip method through the use of strong bands between columns or between columns and exterior walls.

3. It is supported at only one corner.
4. No shear forces act along the edges.
5. No twisting moments act along the edges.
6. All bending moments acting along an edge have the same sign or are zero.
7. The bending moments along the edges are the factored moments used to design the reinforcing bars.

A uniformly loaded strip in the X direction, shown in Fig. 24.14b, thus has shear and moment diagrams as shown in Fig. 24.14c and d, respectively. Maximum moments are located at the lines of zero shear. The outer edges of the corner-supported element are defined at the lines of zero shear in both the X and Y directions.

A typical corner-supported element, with an assumed distribution of moments along the edges, is shown in Fig. 24.15. It is assumed that the bending moment is constant along each half of each edge. The vertical reaction is found by summing vertical forces:

$$R = qab \tag{24.19}$$

and moment equilibrium about the Y axis gives

$$m_{xfm} - m_{xsm} = \frac{qa^2}{2} \tag{24.20}$$

where m_{xfm} and m_{xsm} are the mean span and support moments per unit width, and the beam sign convention is followed. Similarly,

$$m_{yfm} - m_{ysm} = \frac{qb^2}{2} \tag{24.21}$$

FIGURE 24.15
Corner-supported element.

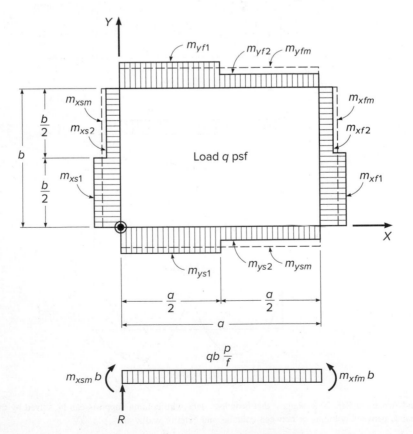

The last two equations are identical with the condition for a corresponding part of a simple strip—Eq. (24.20) spanning in the X direction and Eq. (24.21) in the Y direction—supported at the axis and carrying the load qb or qa per foot. So if the corner-supported element forms a part of a strip, that part should carry 100 percent of the load q in each direction. (This requirement was discussed earlier in Chapter 13 and is simply a requirement of static equilibrium.)

The distribution of moments within the boundaries of a corner-supported element is complex. With the load on the element carried by a single vertical reaction at one corner, strong twisting moments must be present within the element; this contrasts with the assumptions of the simple strip method used previously.

The moment field within a corner-supported element and its edge moments have been explored in great detail in Ref. 24.12. It is essential that the edge moments, given in Fig. 24.15, are used to design the reinforcing bars (that is, nowhere within the element is a bar subjected to a greater moment than at the edges). To meet this requirement, a limitation must be put on the moment distribution along the edges. Based on his studies (Ref. 24.12), Hillerborg has recommended the following restriction on edge moments:

$$m_{xf2} - m_{xs2} = \alpha \frac{qa^2}{2} \tag{24.22a}$$

with

$$0.25 \leq \alpha \leq 0.7 \tag{24.22b}$$

where m_{xf2} and m_{xs2} are the positive and negative X direction moments, respectively, in the outer half of the element, as shown in Fig. 24.15. The corresponding restriction applies in the Y direction. He notes further that for most practical applications, the edge moment distribution shown in Fig. 24.16 is appropriate, with

$$m_{xf1} = m_{xf2} = m_{xfm} \tag{24.23}$$

$$m_{xs2} = 0 \tag{24.24a}$$

$$m_{xs1} = 2m_{xsm} \tag{24.24b}$$

FIGURE 24.16

Recommended distribution of moments for typical corner-supported element.

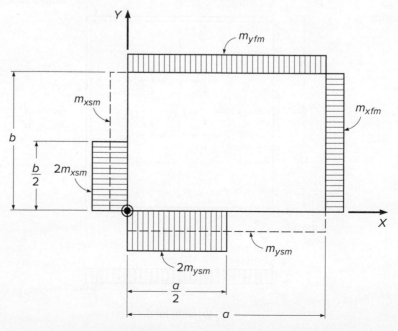

(Alternatively, it is suggested in Ref. 24.14 that negative support moments across the column line be taken at $1.5m_{xsm}$ in the half-element width by the column and at $0.5m_{xsm}$ in the remaining outside half-element width.) Positive reinforcement in the span should be carried through the whole corner-supported element. The negative reinforcement corresponding to $m_{xs1} - m_{xs2}$ in Fig. 24.15 must be extended at least $0.6a$ from the support. The remaining negative steel, if any, should be carried through the whole corner-supported element. The corresponding restrictions apply in the Y direction.

In practical applications, corner-supported elements are combined with each other and with parts of one-way strips, as shown in Fig. 24.14, to form a system of strips. In this system, each strip carries the total load q, as discussed earlier. In laying out the elements and strips, the concentrated corner support for the element may be assumed to be at the center of the supporting column, as shown in Fig. 24.14, unless supports are of significant size. In that case, the corner support may be taken at the corner of the column, and an ordinary simple strip may be included that spans between the column faces, along the edge of the corner-supported elements. Note in the figure that the corner regions of the slab are not included in the main strips that include the corner-supported elements. These may safely be designed for one-third of the corresponding moments in the main strips (Ref. 24.13).

EXAMPLE 24.4   **Edge-supported flat plate with central column.**   Figure 24.17a illustrates a flat plate with overall dimensions 34×34 ft, with fixed supports along the left and lower edges in the sketch, hinged supports at the right and upper edges, and a single central column 16 in. square. It must carry a service live load of 40 psf over its entire surface plus its own weight and an additional superimposed dead load of 7 psf. Find the moments at all critical sections, and determine the required slab thickness and reinforcement. Material strengths are specified at $f_y = 60,000$ psi and $f'_c = 4000$ psi.

FIGURE 24.17

Design example: edge-supported flat plate with central column.

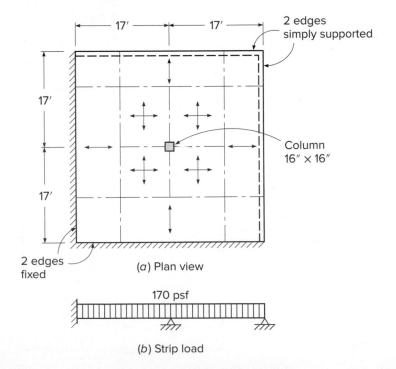

(a) Plan view

(b) Strip load

FIGURE 24.17
(*Continued*)

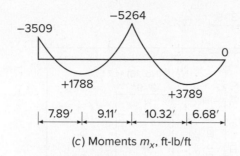

(*c*) Moments m_x, ft-lb/ft

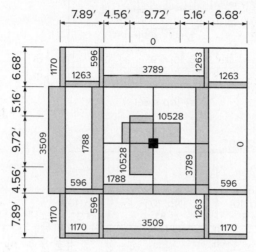

(*d*) Factored moments per 1 ft strip

SOLUTION.   A trial slab depth is chosen based on Table 13.2, which governs for flat plates. It is conservative for the present case, where continuous support is provided along the outer edges.

$$h = \frac{17 \times 12}{33} = 6.18 \text{ in.}$$

A thickness of 6.5 in. is tentatively selected, for which the self weight is $150 \times 6.5/12 = 81$ psf. The total factored load to be carried is thus

$$q_u = 1.2\,(81 + 7) + 1.6 \times 40 = 170 \text{ psf}$$

The average strip moments in the X direction in the central region caused by the load of 170 psf are found by elastic theory and are shown in Fig. 15.17*c*. The analysis in the Y direction is identical. The points of zero shear (and maximum moments) are located 9.11 ft to the left of the column and 10.32 ft to the right, as indicated. These dimensions determine the size of the four corner-supported elements.

Moments in the slab are then determined according to the preceding recommendations. At the fixed edge along the left side of the main strips, the moment m_{xs} is simply the moment per foot strip from the elastic analysis, 3509 ft-lb/ft. At the left edge of the corner-supported element in the left span,

$$m_{xf1} = m_{xf2} = m_{xfm} = 1788 \text{ ft-lb/ft}$$

Along the centerline of the slab, over the column, following the recommendations shown in Fig. 24.16,

$$m_{xs2} = 0$$

$$m_{xs1} = 2m_{xsm} = 10{,}528 \text{ ft-lb/ft}$$

FIGURE 24.17
(*Continued*)

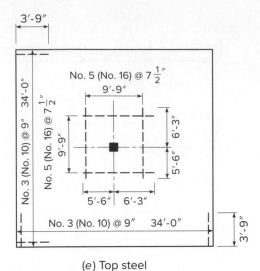

(*e*) Top steel

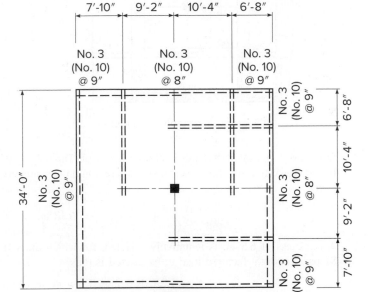

(*f*) Bottom steel

At the right edge of the corner-supported element in the right span,

$$m_{xf1} = m_{xf2} = m_{xfm} = 3789 \text{ ft-lb/ft}$$

At the outer, hinge-supported edge, all moments are zero. Make a check of the α values, using Eq. (24.22*b*), and note from Eq. (24.20) that $qa^2/2 = m_{xfm} - m_{xsm}$. Thus, in the left span,

$$\alpha = \frac{m_{xf2} - m_{xs2}}{qa^2/2} = \frac{1788 - 0}{1788 + 5264} = 0.25$$

and in the right span,

$$\alpha = \frac{3789 - 0}{3789 + 5264} = 0.42$$

Because both values are within the range of 0.25 to 0.75, the proposed distribution of moments is satisfactory. If the first value had been below the lower limit of 0.25, the negative moment in the column half-strip might have been reduced from 10,528 ft-lb/ft, and the negative moment in the adjacent half-strip might have been increased above the 0 value used. Alternatively, the total negative moment over the column might have been somewhat decreased, with a corresponding increase in span moments.

Moments in the Y direction correspond throughout, and all results are summarized in Fig. 24.17d. Moments in the strips adjacent to the supported edges are set equal to one-third of those in the adjacent main strips.

With moments per ft strip known at all critical sections, the required reinforcement is easily found. With a $\frac{3}{4}$ in. concrete cover and $\frac{1}{2}$ in. bar diameter, in general the effective depth of the slab is 5.5 in. Where bar stacking occurs—that is, over the central column and near the intersection of the two fixed edges—an average effective depth equal to 5.25 in. is used. This results in reinforcement identical in the two directions and simplifies construction.

For the 6.5 in. thick slab, minimum steel for shrinkage and temperature crack control is $0.0018 \times 6.5 \times 12 = 0.140$ in^2/ft strip, which is provided by No. 3 (No. 10) bars at 9 in. spacing. The corresponding flexural reinforcement ratio is

$$\rho_{\min} = \frac{0.140}{5.5 \times 12} = 0.0021$$

Interpolating from Table A.5a of Appendix A makes $R = 124$, and the design strength is

$$\phi m_n = \phi R b d^2 = 0.90 \times 124 \times 12 \times 5.5^2 / 12 = 3376 \text{ ft-lb/ft}$$

In comparison with the required strengths summarized in Fig. 24.17d, this is adequate everywhere except for particular regions as follows:

Negative steel over column:

$$R = \frac{m_u}{\phi b d^2} = \frac{10,528 \times 12}{0.90 \times 12 \times 5.25^2} = 424$$

for which $\rho = 0.0076$ (from Table A.5a), and $A_s = 0.0076 \times 12 \times 5.25 = 0.48$ in^2/ft. This is provided using No. 5 (No. 16) bars at 7.5 in. spacing. They are continued a distance $0.6 \times 9.11 = 5.47$ ft, say 5 ft 6 in., to the left of the column centerline, and $0.6 \times 10.32 = 6.19$ ft, say 6 ft 3 in., to the right.

Negative steel along fixed edges:

$$R = \frac{3509 \times 12}{0.90 \times 12 \times 5.50^2} = 129$$

for which $\rho = 0.0022$ and $A_s = 0.0022 \times 12 \times 5.5 = 0.15$ in^2/ft. No. 3 (No. 10) bars at 9 in. spacing is adequate. The point of inflection for the slab in this region is easily found to be 3.30 ft from the fixed edge. The negative bars are extended 5.5 in. beyond that point, resulting in a cutoff 45 in., or 3 ft 9 in., from the support face.

Positive steel in outer spans:

$$R = \frac{3789 \times 12}{0.90 \times 12 \times 5.50^2} = 139$$

resulting in $\rho = 0.0024$ and $A_s = 0.0024 \times 12 \times 5.5 = 0.16$ in^2/ft. No. 3 (No. 10) bars at 8 in. spacing are used. In all cases, the maximum spacing of $2h = 13$ in. is satisfied. Alternatively, No. 4 (No. 13) bars at 13 in. in lieu of the No. 3 (No. 10) bars provide $R = 163$ and will simplify placement costs.

Bar size and spacing and cutoff points for the top and bottom steel are summarized in Fig. 24.17e and f, respectively.

Finally, the load carried by the central column is

$$P = 170 \times 19.43 \times 19.43 = 64{,}200 \text{ lb}$$

Investigating punching shear at a critical section taken $d/2$ from the face of the 16 in. column, with reference to Eq. (13.7a) and with $b_o = 4 \times (16.00 + 5.25) = 85$ in., gives

$$\phi V_c = 4\phi \sqrt{f_c'}\, b_o d = 4 \times 0.75 \sqrt{4000} \times 85 \times 5.25 = 84{,}700 \text{ lb}$$

This is well above the applied shear of 64,200 lb, confirming that the slab thickness is adequate and that no shear reinforcement is required.

24.9 COMPARISONS OF YIELD LINE AND STRIP METHODS FOR SLAB ANALYSIS AND DESIGN

The conventional methods of slab analysis and design, as described in Chapter 13, are limited to applications in which slab panels are supported on opposite sides or on all four sides by beams or walls or to the case of flat plates and related forms supported by a relatively regular array of columns. In all cases, slab panels must be square or rectangular, loads must be uniformly distributed within each panel, and slabs must be free of significant holes.

Both the yield line theory and the strip method offer the designer rational methods for slab analysis and design over a much broader range, including the following:

1. Boundaries of any shape, including rectangular, triangular, circular, and L-shaped boundaries with reentrant corners
2. Supported or unsupported edges, skewed supports, column supports, or various combinations of these conditions
3. Uniformly distributed loads, loads distributed over partial panel areas, linear varying distributed loads, line loads, and concentrated loads
4. Slabs with significant holes

The most important difference between the strip method and the yield line method is the fact that the strip method produces results that are always on the safe side, but yield line analysis may result in unsafe designs. A slab designed by the strip method may possibly carry a higher load than estimated, through internal force redistributions, before collapse; a slab analyzed by yield line procedures may fail at a lower load than anticipated if an incorrect mechanism has been selected as the basis or if the defining dimensions are incorrect.

Beyond this, it should be realized that the strip method is a tool for *design*, by which the slab thickness and reinforcing bar size and distribution may be selected to resist the specified loads. In contrast, the yield line theory offers only a means for *analyzing the capacity of a given slab*, with known reinforcement. According to the yield line approach, the design process is actually a matter of reviewing the capacities of a number of trial designs and alternative reinforcing patterns. All possible yield line patterns must be investigated and specific dimensions varied to be sure that the correct solution has been found. Except for simple cases, this is likely to be a time-consuming process.

Neither the strip method nor the yield line approach provides any information regarding cracking or deflections at service load. Both focus attention strictly on flexural strength. However, by the strip method, if care is taken at least to approximate the elastic distribution of moments, little difficulty should be experienced with excessive cracking. The methods for deflection prediction presented in Section 13.11 can, without difficulty, be adapted for use with the strip method, because the concepts are fully compatible.

With regard to economy of reinforcement, it might be supposed that use of the strip method, which always leads to designs on the safe side, might result in more expensive structures than the yield line theory. Comparisons, however, indicate that in most cases this is not so (Refs. 24.8 and 24.12). Through proper use of the strip method, reinforcing bars are placed in a nonuniform way in the slab (for example, in strong bands around openings) where they are used to best effect; yield line methods, on the other hand, often lead to uniform bar spacings, which may mean that individual bars are used inefficiently.

Many tests have been conducted on slabs designed by the strip method (Ref. 24.11; also, see the summary in Ref. 24.12). These tests included square slabs, rectangular slabs, slabs with both fixed and simply supported edges, slabs supported directly by columns, and slabs with large openings. The conclusions drawn determine that the strip method provides for safe design with respect to nominal strength and that at service load, behavior with respect to cracking and deflections is generally satisfactory. The method has been widely and successfully used in the Scandinavian countries since the 1960s.

REFERENCES

24.1. A. Hillerborg, "Equilibrium Theory for Reinforced Concrete Slabs" (in Swedish), *Betong*, vol. 41, no. 4, 1956, pp. 171–182.

24.2. A. Hillerborg, *Strip Method for Slabs on Columns, L-Shaped Plates, Etc.* (in Swedish), Svenska Riksbyggen, Stockholm, 1959.

24.3. A. Hillerborg, "A Plastic Theory for the Design of Reinforced Concrete Slabs," *Proc. Sixth Congr.*, International Association for Bridge and Structural Engineering, Stockholm, 1960.

24.4. R. E. Crawford, *Limit Design of Reinforced Concrete Slabs*, Thesis submitted to University of Illinois for the degree of Ph.D., Urbana, IL, 1962.

24.5. F. A. Blakey, *Strip Method for Slabs on Columns, L-Shaped Plates, Etc.* (translation of Ref. 24.2), Commonwealth Scientific and Industrial Research Organization, Melbourne, 1964.

24.6. K. O. Kemp, "A Lower Bound Solution to the Collapse of an Orthotropically Reinforced Slab on Simple Supports," *Mag. Concr. Res.*, vol. 14, no. 41, 1962, pp. 79–84.

24.7. K. O. Kemp, "Continuity Conditions in the Strip Method of Slab Design," *Proc. Inst. Civ. Eng.*, vol. 45, 1970, p. 283 (supplement paper 7268s).

24.8. R. H. Wood and G. S. T. Armer, "The Theory of the Strip Method for the Design of Slabs," *Proc. Inst. Civ. Eng.*, vol. 41, 1968, pp. 285–311.

24.9. R. H. Wood, "The Reinforcement of Slabs in Accordance with a Predetermined Field of Moments," *Concrete*, vol. 2, no. 2, 1968, pp. 69–76.

24.10. G. S. T. Armer, "The Strip Method: A New Approach to the Design of Slabs," *Concrete*, vol. 2, no. 9, 1968, pp. 358–363.

24.11. G. S. T. Armer, "Ultimate Load Tests of Slabs Designed by the Strip Method," *Proc. Inst. Civ. Eng.*, vol. 41, 1968, pp. 313–331.

24.12. A. Hillerborg, *Strip Method of Design*, Viewpoint Publications, Cement and Concrete Association, Wexham Springs, Slough, England, 1975.

24.13. A. Hillerborg, "The Advanced Strip Method—A Simple Design Tool," *Mag. Concr. Res.*, vol. 34, no. 121, 1982, pp. 175–181.

24.14. R. Park and W. L. Gamble, *Reinforced Concrete Slabs*, 2nd ed. (Chapter 6), John Wiley, New York, 2000, pp. 232–302.

24.15. A. Hillerborg, *Strip Method Design Handbook*, E & FN Spon/Chapman & Hill, London, 1996.

24.16. S. Timoshenko and S. Woinowsky-Krieger, *Theory of Plates and Shells*, 2nd ed., McGraw-Hill, New York, 1959.

PROBLEMS

Note: For all the following problems, use material strengths $f_y = 60,000$ psi and $f_c' = 4000$ psi. All ACI Code requirements for minimum steel, maximum spacings, bar cutoff, and special corner reinforcement are applicable.

24.1. The square slab of Fig. P24.1 is simply supported by masonry walls along all four sides. It is to carry a service live load of 100 psf in addition to its self weight. Specify a suitable load distribution; determine moments at all controlling sections; and select the slab thickness, reinforcing bars, and spacing.

FIGURE P24.1

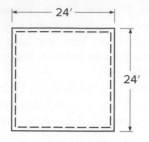

24.2   The rectangular slab shown in Fig. P24.2 is a typical interior panel of a large floor system having beams on all column lines. Columns and beams are sufficiently stiff that the slab can be considered fully restrained along all sides. A live load of 100 psf and a superimposed dead load of 30 psf must be carried in addition to the slab self weight. Determine the required slab thickness, and specify all reinforcing bars and spacings. Cutoff points for negative bars should be specified; all positive steel may be carried into the supporting beams. Take support moments to be 2 times the span moments in the strips.

FIGURE P24.2

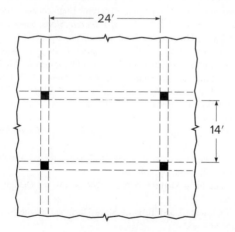

24.3.   The slab of Fig. P24.3 may be considered fully fixed along three edges, but it is without support along the fourth, long side. It must carry a uniformly distributed live load of 80 psf plus an external dead load of 40 psf. Specify a suitable slab depth, and determine reinforcement and cutoff points.

FIGURE P24.3

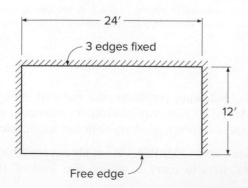

24.4. Figure P24.4 shows a counterfort retaining wall (see Section 16.9) consisting of a base slab and a main vertical wall of constant thickness retaining the earth. Counterfort walls spaced at 19 ft on centers along the wall provide additional support for the main slab. Each section of the main wall, which is 16 ft high and 18 ft long, may be considered fully fixed at its base and also along its two vertical sides (because of full continuity and identical loadings on all such panels). The top of the main wall is without support. The horizontal earth pressure varies from 0 at the top of the wall to 587 psf at the top of the base slab. Determine a suitable thickness for the main wall, and select reinforcing bars and spacing.

FIGURE P24.4

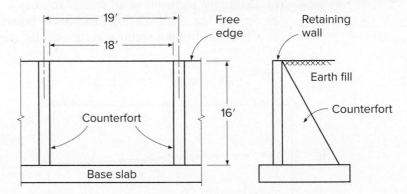

24.5. The triangular slab shown in Fig. P24.5, providing cover over a loading dock, is fully fixed along two adjacent sides and free of support along the diagonal edge. A uniform snow load of 60 psf is anticipated. Dead load of 10 psf acts, in addition to self weight. Determine the required slab depth and specify all reinforcement. (*Hint*: The main bottom reinforcement should be parallel to the free edge, and the negative reinforcement should be perpendicular to the supported edges.)

FIGURE P24.5

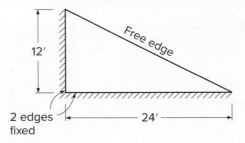

24.6. Figure P24.6 shows a rectangular slab with a large opening near one corner. It is simply supported along one long side and the adjacent short side, and

FIGURE P24.6

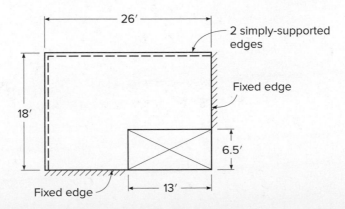

the two edges adjacent to the opening are fully fixed. A factored load of 250 psf must be carried. Find the required slab thickness, and specify all reinforcement.

24.7. The roof deck slab of Fig. P24.7 is intended to carry a total factored load, including self-weight, of 165 psf. It has fixed supports along the two long sides and one short side, but the fourth edge must be free of any support. Two 16 in. square columns are located as shown.

(*a*) Determine an acceptable slab thickness.

(*b*) Select appropriate load dispersion lines.

(*c*) Determine moments at all critical sections.

(*d*) Specify bar sizes, spacings, and cutoff points.

(*e*) Check controlling sections in the slab for shear strength.

FIGURE P24.7

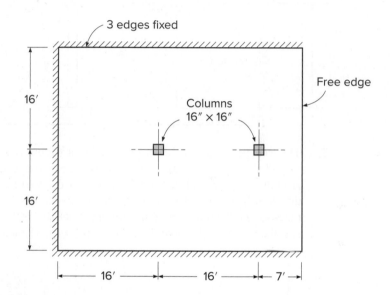

APPENDIX

Design Aids

TABLE A.1
Designations, diameters, areas, and weights of standard bars

Bar No. Inch-Pound[a]	Bar No. SI[b]	Diameter, in.	Cross-Sectional Area, in²	Nominal Weight, lb/ft
3	10	$\frac{3}{8} = 0.375$	0.11	0.376
4	13	$\frac{1}{2} = 0.500$	0.20	0.668
5	16	$\frac{5}{8} = 0.625$	0.31	1.043
6	19	$\frac{3}{4} = 0.750$	0.44	1.502
7	22	$\frac{7}{8} = 0.875$	0.60	2.044
8	25	$1 = 1.000$	0.79	2.670
9	29	$1\frac{1}{8} = 1.128^{c}$	1.00	3.400
10	32	$1\frac{1}{4} = 1.270^{c}$	1.27	4.303
11	36	$1\frac{3}{8} = 1.410^{c}$	1.56	5.313
14	43	$1\frac{3}{4} = 1.693^{c}$	2.25	7.650
18	57	$2\frac{1}{4} = 2.257^{c}$	4.00	13.600

[a]Based on the number of eighths of an inch included in the nominal diameter of the bars. The nominal diameter of a deformed bar is equivalent to the diameter of a plain bar having the same weight per foot as the deformed bar.

[b]Bar number approximates the number of millimeters included in the nominal diameter of the bar. Bars are marked with this designation.

[c]Approximate to nearest $\frac{1}{8}$ in.

TABLE A.2
Areas of groups of standard bars, in^2

Bar No.							Number of Bars						
Inch-Pound	SI	1	2	3	4	5	6	7	8	9	10	11	12
4	13	0.20	0.40	0.60	0.80	1.00	1.20	1.40	1.60	1.80	2.00	2.20	2.40
5	16	0.31	0.62	0.93	1.24	1.55	1.86	2.17	2.48	2.79	3.10	3.41	3.72
6	19	0.44	0.88	1.32	1.76	2.20	2.64	3.08	3.52	3.96	4.40	4.84	5.28
7	22	0.60	1.20	1.80	2.40	3.00	3.60	4.20	4.80	5.40	6.00	6.60	7.20
8	25	0.79	1.58	2.37	3.16	3.95	4.74	5.53	6.32	7.11	7.90	8.69	9.48
9	29	1.00	2.00	3.00	4.00	5.00	6.00	7.00	8.00	9.00	10.00	11.00	12.00
10	32	1.27	2.54	3.81	5.08	6.35	7.62	8.89	10.16	11.43	12.70	13.97	15.24
11	36	1.56	3.12	4.68	6.24	7.80	9.36	10.92	12.48	14.04	15.60	17.16	18.72
14	43	2.25	4.50	6.75	9.00	11.25	13.50	15.75	18.00	20.25	22.50	24.75	27.00
18	57	4.00	8.00	12.00	16.00	20.00	24.00	28.00	32.00	36.00	40.00	44.00	48.00

TABLE A.3
Areas of bars in slabs and walls, in^2/ft

Spacing, in.	Inch-Pound: SI:	Bar No.								
		3 10	4 13	5 16	6 19	7 22	8 25	9 29	10 32	11 36
3		0.44	0.78	1.23	1.77	2.40	3.14	4.00	5.06	6.25
$3\frac{1}{2}$		0.38	0.67	1.05	1.51	2.06	2.69	3.43	4.34	5.36
4		0.33	0.59	0.92	1.32	1.80	2.36	3.00	3.80	4.68
$4\frac{1}{2}$		0.29	0.52	0.82	1.18	1.60	2.09	2.67	3.37	4.17
5		0.26	0.47	0.74	1.06	1.44	1.88	2.40	3.04	3.75
$5\frac{1}{2}$		0.24	0.43	0.67	0.96	1.31	1.71	2.18	2.76	3.41
6		0.22	0.39	0.61	0.88	1.20	1.57	2.00	2.53	3.12
$6\frac{1}{2}$		0.20	0.36	0.57	0.82	1.11	1.45	1.85	2.34	2.89
7		0.19	0.34	0.53	0.76	1.03	1.35	1.71	2.17	2.68
$7\frac{1}{2}$		0.18	0.31	0.49	0.71	0.96	1.26	1.60	2.02	2.50
8		0.17	0.29	0.46	0.66	0.90	1.18	1.50	1.89	2.34
9		0.15	0.26	0.41	0.59	0.80	1.05	1.33	1.69	2.08
10		0.13	0.24	0.37	0.53	0.72	0.94	1.20	1.52	1.87
12		0.11	0.20	0.31	0.44	0.60	0.78	1.00	1.27	1.56

TABLE A.4
Limiting steel reinforcement ratios for tension-controlled members

f_y, psi	f_c', psi	β_1	$\rho_{max}{}^a$	$\rho_{min} = \dfrac{200}{f_y}$	$\rho_{min} = \dfrac{3\sqrt{f_c'}}{f_y}$
40,000	3000	0.85	0.0220	0.0050	0.0041
	4000	0.85	0.0294	0.0050	0.0047
	5000	0.80	0.0346	0.0050	0.0053
	6000	0.75	0.0389	0.0050	0.0058
	7000	0.70	0.0423	0.0050	0.0063
	8000	0.65	0.0449	0.0050	0.0067
	9000	0.65	0.0505	0.0050	0.0071
	10000	0.65	0.0562	0.0050	0.0075
60,000[b]	3000	0.85	0.0135	0.0033	0.0027
	4000	0.85	0.0181	0.0033	0.0032
	5000	0.80	0.0213	0.0033	0.0035
	6000	0.75	0.0239	0.0033	0.0039
	7000	0.70	0.0260	0.0033	0.0042
	8000	0.65	0.0276	0.0033	0.0045
	9000	0.65	0.0311	0.0033	0.0047
	10000	0.65	0.0345	0.0033	0.0050
80,000	3000	0.85	0.0093	0.0025	0.0021
	4000	0.85	0.0124	0.0025	0.0024
	5000	0.80	0.0146	0.0025	0.0027
	6000	0.75	0.0164	0.0025	0.0029
	7000	0.70	0.0178	0.0025	0.0031
	8000	0.65	0.0189	0.0025	0.0034
	9000	0.65	0.0213	0.0025	0.0036
	10000	0.65	0.0237	0.0025	0.0038
100,000	3000	0.85	0.0069	0.0020	0.0016
	4000	0.85	0.0092	0.0020	0.0019
	5000	0.80	0.0108	0.0020	0.0021
	6000	0.75	0.0121	0.0020	0.0023
	7000	0.70	0.0132	0.0020	0.0025
	8000	0.65	0.0140	0.0020	0.0027
	9000	0.65	0.0158	0.0020	0.0028
	10000	0.65	0.0175	0.0020	0.0030

$^a\rho = 0.85\,\beta_1\dfrac{f_c'}{f_y}\dfrac{\varepsilon_u}{\varepsilon_u + \varepsilon_y + 0.003}$

^{b} Calculated using $\varepsilon_y = 0.002$

TABLE A.5a

Flexural resistance factor: $R = \rho f_y\left(1 - 0.588\dfrac{\rho f_y}{f_c'}\right) = \dfrac{M_n}{bd^2} = \dfrac{M_u}{\phi bd^2}$, psi

	$f_y = 40{,}000$ psi				$f_y = 60{,}000$ psi				$f_y = 80{,}000$ psi			
	f_c', psi				f_c', psi				f_c', psi			
ρ	3000	4000	5000	6000	3000	4000	5000	6000	3000	4000	5000	6000
0.0005	20	20	20	20	30	30	30	30	40	40	40	40
0.0010	40	40	40	40	59	59	60	60	79	79	79	79
0.0015	59	59	60	60	88	89	89	89	117	118	118	119
0.0020	79	79	79	79	117	118	118	119	155	156	157	157
0.0025	98	99	99	99	146	147	147	148	192	194	195	196
0.0030	117	118	118	119	174	175	176	177	229	232	233	234
0.0035	136	137	138	138	201	204	205	206	265	268	271	272
0.0040	155	156	157	157	229	232	233	234	300	305	308	310
0.0045	174	175	176	177	256	259	261	263	335	341	345	347
0.0050	192	194	195	196	282	287	289	291	369	376	381	384
0.0055	211	213	214	215	309	314	317	319	402	412	417	421
0.0060	229	232	233	234	335	341	345	347	435	446	453	457
0.0065	247	250	252	253	360	368	372	375	467	480	488	494
0.0070	265	268	271	272	385	394	399	403	499	514	523	529
0.0075	282	287	289	291	410	420	426	430	529	547	558	565
0.0080	300	305	308	310	435	446	453	457	560	580	592	600
0.0085	317	323	326	329	459	472	479	485	589	612	626	635
0.0090	335	341	345	347	483	497	506	511	618	644	659	669
0.0095	352	359	363	366	506	522	532	538	647	675	692	703
0.0100	369	376	381	384	529	547	558	565	675	706	725	737
0.0105	385	394	399	403	552	572	583	591	702	736	757	771
0.0110	402	412	417	421	575	596	609	617	728	766	789	804
0.0115	419	429	435	439	597	620	634	643	754	796	820	837
0.0120	435	446	453	457	618	644	659	669		825	852	870
0.0125	451	463	471	476	640	667	684	695		853	882	902
0.0130	467	480	488	494	661	691	708	720		881	913	934
0.0135	483	497	506	511	681	714	733	746		909	943	966
0.0140	499	514	523	529	702	736	757	771		936	972	997
0.0145	514	531	540	547	722	759	781	796		962	1002	1028
0.0150	529	547	558	565	741	781	805	821		988	1031	1059
0.0155	545	563	575	582	760	803	828	845		1014	1059	1089
0.0160	560	580	592	600		825	852	870			1087	1119
0.0165	575	596	609	617		846	875	894			1115	1149
0.0170	589	612	626	635		867	898	918			1142	1179
0.0175	604	628	642	652		888	920	942			1170	1208
0.0180	618	644	659	669		909	943	966			1196	1237
0.0185	633	660	676	686		929	965	989				1265
0.0190	647	675	692	703		949	987	1013				1294
0.0195	661	691	708	720		969	1009	1036				1322
0.0200	675	706	725	737		988	1031	1059				1349

TABLE A.5b

Flexural resistance factor: $R = \rho f_y \left(1 - 0.588 \dfrac{\rho f_y}{f_c'}\right) = \dfrac{M_n}{bd^2} = \dfrac{M_u}{\phi bd^2}$, psi

	$f_y = 40,000$ psi				$f_y = 60,000$ psi				$f_y = 80,000$ psi			
	f_c', psi				f_c', psi				f_c', psi			
ρ	3000	4000	5000	6000	3000	4000	5000	6000	3000	4000	5000	6000
0.003	117	118	118	119	174	175	176	177	229	232	233	234
0.004	155	156	157	157	229	232	233	234	300	305	308	310
0.005	192	194	195	196	282	287	289	291	369	376	381	384
0.006	229	232	233	234	335	341	345	347	435	446	453	457
0.007	265	268	271	272	385	394	399	403	499	514	523	529
0.008	300	305	308	310	435	446	453	457	560	580	592	600
0.009	335	341	345	347	483	497	506	511	618	644	659	669
0.010	369	376	381	384	529	547	558	565	675	706	725	737
0.011	402	412	417	421	575	596	609	617	728	766	789	804
0.012	435	446	453	457	618	644	659	669		825	852	870
0.013	467	480	488	494	661	691	708	720		881	913	934
0.014	499	514	523	529	702	736	757	771		936	972	997
0.015	529	547	558	565	741	781	805	821		988	1031	1059
0.016	560	580	592	600		825	852	870			1087	1119
0.017	589	612	626	635		867	898	918			1142	1179
0.018	618	644	659	669		909	943	966			1196	1237
0.019	647	675	692	703		949	987	1013				1294
0.020	675	706	725	737		988	1031	1059				1349
0.021	702	736	757	771			1073	1104				
0.022	728	766	789	804			1115	1149				
0.023	754	796	820	837			1156	1193				
0.024		825	852	870			1196	1237				
0.025		853	882	902				1280				
0.026		881	913	934				1322				
0.027		909	943	966				1363				
0.028		936	972	997								
0.029		962	1002	1028								
0.030		988	1031	1059								
0.031		1014	1059	1089								
0.032			1087	1119								
0.033			1115	1149								
0.034			1142	1179								
0.035			1170	1208								
0.036			1196	1237								
0.037				1265								
0.038				1294								
0.039				1322								
0.040				1349								
0.041				1376								

TABLE A.6
Parameters k and j for elastic, cracked section beam analysis, where
$k = \sqrt{2\rho n + (\rho n)^2} - \rho n; \; j = 1 - \frac{1}{3}k$

	$n = 7$		$n = 8$		$n = 9$		$n = 10$	
ρ	k	j	k	j	k	j	k	j
0.0010	0.112	0.963	0.119	0.960	0.125	0.958	0.132	0.956
0.0020	0.154	0.949	0.164	0.945	0.173	0.942	0.180	0.940
0.0030	0.185	0.938	0.196	0.935	0.207	0.931	0.217	0.928
0.0040	0.210	0.930	0.223	0.926	0.235	0.922	0.246	0.918
0.0050	0.232	0.923	0.246	0.918	0.258	0.914	0.270	0.910
0.0054	0.240	0.920	0.254	0.915	0.267	0.911	0.279	0.907
0.0058	0.247	0.918	0.262	0.913	0.275	0.908	0.287	0.904
0.0062	0.254	0.915	0.269	0.910	0.283	0.906	0.296	0.901
0.0066	0.261	0.913	0.276	0.908	0.290	0.903	0.303	0.899
0.0070	0.268	0.911	0.283	0.906	0.298	0.901	0.311	0.896
0.0072	0.271	0.910	0.287	0.904	0.301	0.900	0.314	0.895
0.0074	0.274	0.909	0.290	0.903	0.304	0.899	0.318	0.894
0.0076	0.277	0.908	0.293	0.902	0.308	0.897	0.321	0.893
0.0078	0.280	0.907	0.296	0.901	0.311	0.896	0.325	0.892
0.0080	0.283	0.906	0.299	0.900	0.314	0.895	0.328	0.891
0.0082	0.286	0.905	0.303	0.899	0.317	0.894	0.331	0.890
0.0084	0.289	0.904	0.306	0.898	0.321	0.893	0.334	0.889
0.0086	0.292	0.903	0.308	0.897	0.324	0.892	0.338	0.887
0.0088	0.295	0.902	0.311	0.896	0.327	0.891	0.341	0.886
0.0090	0.298	0.901	0.314	0.895	0.330	0.890	0.344	0.885
0.0092	0.300	0.900	0.317	0.894	0.332	0.889	0.347	0.884
0.0094	0.303	0.899	0.320	0.893	0.335	0.888	0.350	0.883
0.0096	0.306	0.898	0.323	0.892	0.338	0.887	0.353	0.882
0.0098	0.308	0.897	0.325	0.892	0.341	0.886	0.355	0.882
0.0100	0.311	0.896	0.328	0.891	0.344	0.885	0.358	0.881
0.0104	0.316	0.895	0.333	0.889	0.349	0.884	0.364	0.879
0.0108	0.321	0.893	0.338	0.887	0.354	0.882	0.369	0.877
0.0112	0.325	0.892	0.343	0.886	0.359	0.880	0.374	0.875
0.0116	0.330	0.890	0.348	0.884	0.364	0.879	0.379	0.874
0.0120	0.334	0.889	0.353	0.882	0.369	0.877	0.384	0.872
0.0124	0.339	0.887	0.357	0.881	0.374	0.875	0.389	0.870
0.0128	0.343	0.886	0.362	0.879	0.378	0.874	0.394	0.867
0.0132	0.347	0.884	0.366	0.878	0.383	0.872	0.398	0.867
0.0136	0.351	0.883	0.370	0.877	0.387	0.871	0.403	0.866
0.0140	0.355	0.882	0.374	0.875	0.392	0.869	0.407	0.864
0.0144	0.359	0.880	0.378	0.874	0.396	0.868	0.412	0.863
0.0148	0.363	0.879	0.382	0.873	0.400	0.867	0.416	0.861
0.0152	0.367	0.878	0.386	0.871	0.404	0.865	0.420	0.860
0.0156	0.371	0.876	0.390	0.870	0.408	0.864	0.424	0.859
0.0160	0.374	0.875	0.394	0.869	0.412	0.863	0.428	0.857
0.0170	0.383	0.872	0.403	0.867	0.421	0.860	0.437	0.854
0.0180	0.392	0.869	0.412	0.863	0.430	0.857	0.446	0.851
0.0190	0.400	0.867	0.420	0.860	0.438	0.854	0.455	0.848
0.0200	0.407	0.864	0.428	0.857	0.446	0.851	0.463	0.846

TABLE A.7
Maximum number of bars as a single layer in beam stems

$\frac{3}{4}$ in. Maximum Size Aggregate, No. 4 (No. 13) Stirrups[a]

Bar No.		Beam Width b_w, in.										
Inch-Pound	SI	8	10	12	14	16	18	20	22	24	26	28	30
5	16	2	4	5	6	7	8	10	11	12	13	15	16
6	19	2	3	4	6	7	8	9	10	11	12	14	15
7	22	2	3	4	5	6	7	8	9	10	11	12	13
8	25	2	3	4	5	6	7	8	9	10	11	12	13
9	29	1	2	3	4	5	6	7	8	9	9	10	11
10	32	1	2	3	4	5	6	6	7	8	9	10	10
11	36	1	2	3	3	4	5	5	6	7	8	8	9
14	43	1	2	2	3	3	4	5	5	6	6	7	8
18	57	1	1	2	2	3	3	4	4	4	5	5	6

1 in. Maximum Size Aggregate, No. 4 (No. 13) Stirrups[a]

Bar No.		Beam Width b_w, in.										
Inch-Pound	SI	8	10	12	14	16	18	20	22	24	26	28	30
5	16	2	3	4	5	6	7	8	9	10	11	12	13
6	19	2	3	4	5	6	7	8	9	9	10	11	12
7	22	1	2	3	4	5	6	7	8	9	10	10	11
8	25	1	2	3	4	5	6	7	7	8	9	10	11
9	29	1	2	3	4	5	6	7	7	8	9	9	10
10	32	1	2	3	4	5	6	6	7	7	8	9	10

[a]Minimum concrete cover assumed to be $1\frac{1}{2}$ in. to the No. 4 (No. 13) stirrup.

Source: Adapted from Ref. 4.8.

TABLE A.8
Minimum number of Grade 60 bars as a single layer in beam stems governed by crack control requirements of the ACI Code

(a) 2 in. clear cover, sides and bottom

Minimum Number of Bars as a Single Layer of a Beam Stem

Bar No.		Beam Stem Width b_w, in.														
Inch-Pound	SI	8	10	12	14	16	18	20	22	24	26	28	30	32	34	36
3–14	10–43	1	1	2	2	3	3	3	3	3	4	4	4	4	4	5
18	57	1	1	2	2	2	3	3	3	3	3	4	4	4	4	4

(b) $1\frac{1}{2}$ in. clear cover, sides and bottom

Minimum Number of Bars as a Single Layer of a Beam Stem

Bar No.		Beam Stem Width b_w, in.														
Inch-Pound	SI	8	10	12	14	16	18	20	22	24	26	28	30	32	34	36
3–4	10–13	1	1	2	2	3	3	3	3	3	4	4	4	4	4	4
5–14	16–43	1	1	2	2	3	3	3	3	3	3	4	4	4	4	4
18	57	1	1	2	2	2	3	3	3	3	3	4	4	4	4	4

TABLE A.9

Design strength ϕM_n for slab sections 12 in. wide, ft-kips; $f_y = 60$ ksi;
$\phi M_n = \phi \rho f_y b d^2 (1 - 0.59 \rho f_y / f_c')$

f_c', psi	ρ	\multicolumn{13}{c}{Effective Depth d, in.}
		3.0	3.5	4.0	4.5	5.0	5.5	6.0	6.5	7.0	8.0	9.0	10.0	12.0
3000	0.002	0.9	1.3	1.7	2.1	2.6	3.2	3.8	4.5	5.2	6.7	8.5	10.5	15.2
	0.003	1.4	1.9	2.5	3.2	3.9	4.7	5.6	6.6	7.7	10.0	12.7	15.6	22.5
	0.004	1.9	2.5	3.3	4.2	5.1	6.2	7.4	8.7	10.1	13.2	16.7	20.6	29.6
	0.005	2.3	3.1	4.1	5.1	6.4	7.7	9.1	10.7	12.4	16.3	20.6	25.4	36.6
	0.006	2.7	3.7	4.8	6.1	7.5	9.1	10.8	12.7	14.8	19.3	24.4	30.1	43.4
	0.007	3.1	4.2	5.5	7.0	8.7	10.5	12.5	14.7	17.0	22.2	28.1	34.7	49.9
	0.008	3.5	4.8	6.3	7.9	9.8	11.8	14.1	16.5	19.2	25.0	31.7	39.1	56.3
	0.009	3.9	5.3	7.0	8.8	10.9	13.1	15.6	18.4	21.3	27.8	35.2	43.4	62.6
	0.010	4.3	5.8	7.6	9.6	11.9	14.4	17.1	20.1	23.3	30.5	38.6	47.6	68.6
	0.011	4.7	6.3	8.3	10.5	12.9	15.6	18.6	21.8	25.3	33.1	41.9	51.7	74.4
4000	0.002	1.0	1.3	1.7	2.1	2.7	3.2	3.8	4.5	5.2	6.8	8.6	10.6	15.3
	0.003	1.4	1.9	2.5	3.2	3.9	4.8	5.7	6.7	7.7	10.1	12.8	15.8	22.7
	0.004	1.9	2.6	3.3	4.2	5.2	6.3	7.5	8.8	10.2	13.3	16.9	20.8	30.0
	0.005	2.3	3.2	4.1	5.2	6.5	7.8	9.3	10.9	12.6	16.5	20.9	25.8	37.2
	0.006	2.8	3.8	4.9	6.2	7.7	9.3	11.0	13.0	15.0	19.6	24.9	30.7	44.2
	0.007	3.2	4.3	5.7	7.2	8.9	10.7	12.8	15.0	17.4	22.7	28.7	35.5	51.1
	0.008	3.6	4.9	6.4	8.1	10.0	12.1	14.5	17.0	19.7	25.7	32.5	40.1	57.8
	0.009	4.0	5.5	7.2	9.1	11.2	13.5	16.1	18.9	21.9	28.6	36.2	44.7	64.4
	0.010	4.4	6.0	7.9	10.0	12.3	14.9	17.7	20.8	24.1	31.5	39.9	49.2	70.9
	0.011	4.8	6.6	8.6	10.9	13.4	16.2	19.3	22.7	26.3	34.3	43.4	53.6	77.2
	0.012	5.2	7.1	9.3	11.7	14.5	17.5	20.9	24.5	28.4	37.1	46.9	57.9	83.4
	0.013	5.6	7.6	9.9	12.6	15.5	18.8	22.4	26.2	30.4	39.8	50.3	62.1	89.5
	0.014	6.0	8.1	10.6	13.4	16.6	20.0	23.8	28.0	32.5	42.4	53.6	66.2	95.4
	0.015	6.3	8.6	11.2	14.2	17.6	21.2	25.3	29.7	34.4	45.0	56.9	70.2	101.2
5000	0.002	1.0	1.3	1.7	2.2	2.7	3.2	3.8	4.5	5.2	6.8	8.6	10.6	15.3
	0.003	1.4	1.9	2.5	3.2	4.0	4.8	5.7	6.7	7.8	10.1	12.8	15.9	22.8
	0.004	1.9	2.6	3.4	4.3	5.2	6.3	7.6	8.9	10.3	13.4	17.0	21.0	30.2
	0.005	2.3	3.2	4.2	5.3	6.5	7.9	9.4	11.0	12.8	16.7	21.1	26.0	37.5
	0.006	2.8	3.8	5.0	6.3	7.8	9.4	11.2	13.1	15.2	19.9	25.1	31.0	44.7
	0.007	3.2	4.4	5.7	7.3	9.0	10.9	12.9	15.2	17.6	23.0	29.1	35.9	51.7
	0.008	3.7	5.0	6.5	8.3	10.2	12.3	14.7	17.2	20.0	26.1	33.0	40.8	58.7
	0.009	4.1	5.6	7.3	9.2	11.4	13.8	16.4	19.2	22.3	29.1	36.9	45.5	65.5
	0.010	4.5	6.1	8.0	10.2	12.5	15.2	18.1	21.2	24.6	32.1	40.6	50.2	72.3
	0.011	4.9	6.7	8.8	11.1	13.7	16.6	19.7	23.1	26.8	35.1	44.4	54.8	78.9
	0.012	5.3	7.3	9.5	12.0	14.8	17.9	21.3	25.1	29.1	37.9	48.0	59.3	85.4
	0.013	5.7	7.8	10.2	12.9	15.9	19.3	22.9	26.9	31.2	40.8	51.6	63.7	91.8
	0.014	6.1	8.3	10.9	13.8	17.0	20.6	24.5	28.8	33.4	43.6	55.2	68.1	98.1
	0.015	6.5	8.9	11.6	14.7	18.1	21.9	26.1	30.6	35.5	46.3	58.6	72.4	104.3
	0.016	6.9	9.4	12.3	15.5	19.2	23.2	27.6	32.4	37.5	49.0	62.1	76.6	110.3
	0.017	7.3	9.9	12.9	16.4	20.2	24.4	29.1	34.1	39.6	51.7	65.4	80.8	116.3

TABLE A.10

Simplified tension development length in bar diameters ℓ_d/d_b for uncoated bars and normalweight concrete

		No. 6 (No. 19) and Smaller[a]			No. 7 and Larger		
		f_c', psi			f_c', psi		
	f_y, ksi	4000	5000	6000	4000	5000	6000
(1) Bottom Bars							
Spacing, cover and ties as per Case *a* or *b*	40	26	23	21	32	29	26
	60	38	34	31	48	43	39
	80	59	53	48	73	66	60
	100	83	74	68	103	92	84
Other cases	40	38	34	31	48	43	39
	60	57	51	47	72	64	59
	80	88	79	72	110	98	90
	100	124	111	101	155	138	126
(2) Top Bars							
Spacing, cover and ties as per Case *a* or *b*	40	33	30	27	42	37	34
	60	50	45	41	62	56	51
	80	76	68	62	95	85	78
	100	107	96	88	134	120	110
Other cases	40	50	45	41	62	56	51
	60	74	67	61	93	83	76
	80	114	102	93	142	127	116
	100	161	144	131	201	180	164

Case *a:* Clear spacing of bars being developed or spliced $\geq d_b$, clear cover $\geq d_b$, and stirrups or ties throughout ℓ_d not less than the Code minimum.

Case *b:* Clear spacing of bars being developed or spliced $\geq 2d_b$, and clear cover not less than d_b.

[a]ACI Committee 408 recommends that the values indicated for bar sizes No. 7 (No. 22) and larger be used for all bar sizes.

TABLE A.11

Development length in compression, in., for normalweight concrete

ℓ_{dc} = greater of $(f_y/50\sqrt{f_c'})d_b$ or $0.0003d_b$ (Minimum length 8 in. in all cases.)

Bar No.		f_y ksi	f_c', psi							
			3000		4000		5000		6000	
Inch-Pound	SI		Basic ℓ_{dc}	Confined	Basic ℓ_{dc}	Confined	Basic ℓ_{dc}	Confined	Basic ℓ_{dc}	Confined
3	10	40	8	8	8	8	8	8	8	8
		60	9	8	8	8	8	8	8	8
		80	11	9	10	8	9	8	9	8
4	13	40	8	8	8	8	8	8	8	8
		60	11	9	10	8	9	8	9	8
		80	15	11	13	10	12	9	12	9
5	16	40	10	8	8	8	8	8	8	8
		60	14	11	12	9	12	9	12	9
		80	19	14	16	12	15	12	15	12
6	19	40	11	9	10	8	9	8	9	8
		60	17	13	15	11	14	11	14	11
		80	22	17	19	15	18	14	18	14
7	22	40	13	10	12	9	11	8	11	8
		60	20	15	17	13	16	12	16	12
		80	26	20	23	17	21	16	21	16
8	25	40	15	11	13	10	12	9	12	9
		60	22	17	19	15	18	14	18	14
		80	30	22	26	19	24	18	24	18
9	29	40	17	13	15	11	14	11	14	11
		60	25	19	22	17	21	16	21	16
		80	33	25	29	22	28	21	28	21
10	32	40	19	14	17	13	16	12	16	12
		60	28	21	25	19	23	18	23	18
		80	38	28	33	25	31	23	31	23
11	36	40	21	16	18	14	17	13	17	13
		60	31	24	27	21	26	20	26	20
		80	42	31	36	27	34	26	34	26
14	43	40	25	19	22	17	21	16	21	16
		60	38	28	33	25	31	23	31	23
		80	50	38	43	33	41	31	41	31
18	57	40	33	25	29	22	28	21	28	21
		60	50	38	43	33	41	31	41	31
		80	66	50	58	43	55	41	55	41

TABLE A.12

Common stock styles of welded wire reinforcement (WWR)

Steel Designation[a]	Steel Area, in^2/ft		Weight (Approximate), lb per 100 ft^2
	Longitudinal	Transverse	
Rolls			
6 × 6-W1.4 × W1.4	0.028	0.028	21
6 × 6-W2.0 × W2.0	0.040	0.040	30
6 × 6-W2.9 × W2.9	0.058	0.058	43
6 × 6-W4.0 × W4.0	0.080	0.080	60
4 × 4-W1.4 × W1.4	0.042	0.042	30
4 × 4-W2.0 × W2.0	0.060	0.060	44
4 × 4-W2.9 × W2.9	0.087	0.087	63
4 × 4-W4.0 × W4.0	0.120	0.120	87
Sheets			
6 × 6-W2.9 × W2.9	0.058	0.058	43
6 × 6-W4.0 × W4.0	0.080	0.080	60
6 × 6-W5.5 × W5.5	0.110	0.110	82
4 × 4-W4.0 × W4.0	0.120	0.120	87

[a]The designation W indicates plain wire; WWR is also available as deformed wire, designated with a D.

TABLE A.13a

Coefficients for slabs with variable moment of inertia[a]

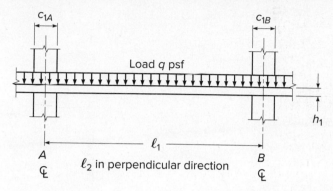

Load q psf

c_{1A}

c_{1B}

h_1

ℓ_1

ℓ_2 in perpendicular direction

A
$\mathbb{C}$

B
$\mathbb{C}$

Column Dimension		Uniform Load FEM = Coeff. $(q\ell_2\ell_1^2)$		Stiffness Factor[b]		Carryover Factor	
c_{1A}/ℓ_1	c_{1B}/ℓ_1	M_{AB}	M_{BA}	k_{AB}	k_{BA}	COF_{AB}	COF_{BA}
0.00	0.00	0.083	0.083	4.00	4.00	0.500	0.500
	0.05	0.083	0.084	4.01	4.04	0.504	0.500
	0.10	0.082	0.086	4.03	4.15	0.513	0.499
	0.15	0.081	0.089	4.07	4.32	0.528	0.498
	0.20	0.079	0.093	4.12	4.56	0.548	0.495
	0.25	0.077	0.097	4.18	4.88	0.573	0.491
0.05	0.05	0.084	0.084	4.05	4.05	0.503	0.503
	0.10	0.083	0.086	4.07	4.15	0.513	0.503
	0.15	0.081	0.089	4.11	4.33	0.528	0.501
	0.20	0.080	0.092	4.16	4.58	0.548	0.499
	0.25	0.078	0.096	4.22	4.89	0.573	0.494
0.10	0.10	0.085	0.085	4.18	4.18	0.513	0.513
	0.15	0.083	0.088	4.22	4.36	0.528	0.511
	0.20	0.082	0.091	4.27	4.61	0.548	0.508
	0.25	0.080	0.095	4.34	4.93	0.573	0.504
0.15	0.15	0.086	0.086	4.40	4.40	0.526	0.526
	0.20	0.084	0.090	4.46	4.65	0.546	0.523
	0.25	0.083	0.094	4.53	4.98	0.571	0.519
0.20	0.20	0.088	0.088	4.72	4.72	0.543	0.543
	0.25	0.086	0.092	4.79	5.05	0.568	0.539
0.25	0.25	0.090	0.090	5.14	5.14	0.563	0.563

[a]Applicable when $c_1/\ell_1 = c_2/\ell_2$. For other relationships between these ratios, the constants will be slightly in error.

[b]Stiffness is $K_{AB} = k_{AB}E(\ell_2 h_1^3/12\ell_1)$ and $K_{BA} = k_{BA}E(\ell_2 h_1^3/12\ell_1)$.

TABLE A.13b

Coefficients for slabs with variable moment of inertia[a]

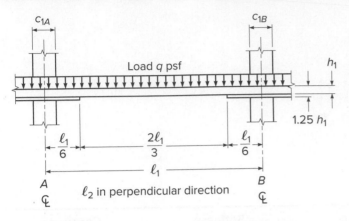

Column Dimension		Uniform Load FEM = Coeff. $(q\ell_2\ell_1^2)$		Stiffness Factor[b]		Carryover Factor	
c_{1A}/ℓ_1	c_{1B}/ℓ_1	M_{AB}	M_{BA}	k_{AB}	k_{BA}	COF_{AB}	COF_{BA}
0.00	0.00	0.088	0.088	4.78	4.78	0.541	0.541
	0.05	0.087	0.089	4.80	4.82	0.545	0.541
	0.10	0.087	0.090	4.83	4.94	0.553	0.541
	0.15	0.085	0.093	4.87	5.12	0.567	0.540
	0.20	0.084	0.096	4.93	5.36	0.585	0.537
	0.25	0.082	0.100	5.00	5.68	0.606	0.534
0.05	0.05	0.088	0.088	4.84	4.84	0.545	0.545
	0.10	0.087	0.090	4.87	4.95	0.553	0.544
	0.15	0.085	0.093	4.91	5.13	0.567	0.543
	0.20	0.084	0.096	4.97	5.38	0.584	0.541
	0.25	0.082	0.100	5.05	5.70	0.606	0.537
0.10	0.10	0.089	0.089	4.98	4.98	0.553	0.553
	0.15	0.088	0.092	5.03	5.16	0.566	0.551
	0.20	0.086	0.094	5.09	5.42	0.584	0.549
	0.25	0.084	0.099	5.17	5.74	0.606	0.546
0.15	0.15	0.090	0.090	5.22	5.22	0.565	0.565
	0.20	0.089	0.094	5.28	5.47	0.583	0.563
	0.25	0.087	0.097	5.37	5.80	0.604	0.559
0.20	0.20	0.092	0.092	5.55	5.55	0.580	0.580
	0.25	0.090	0.096	5.64	5.88	0.602	0.577
0.25	0.25	0.094	0.094	5.98	5.98	0.598	0.598

[a] Applicable when $c_1/\ell_1 = c_2/\ell_2$. For other relationships between these ratios, the constants will be slightly in error.

[b] Stiffness is $K_{AB} = k_{AB}E(\ell_2 h_1^3/12\ell_1)$ and $K_{BA} = k_{BA}E(\ell_2 h_1^3/12\ell_1)$.

TABLE A.13c
Stiffness factors for columns with variable moment of inertia[a]

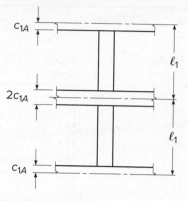

Slab Half-depth c_{1A}/ℓ_1	Stiffness Factor k_{AB}	Slab Half-depth c_{1A}/ℓ_1	Stiffness Factor k_{AB}
0.00	4.00	0.14	9.43
0.02	4.43	0.16	11.01
0.04	4.94	0.18	13.01
0.06	5.54	0.20	15.56
0.08	6.25	0.22	18.87
0.10	7.11	0.24	23.26
0.12	8.15		

[a]Simmonds, Sidney H., and Janko Misic. "Design Factors for the Equivalent Frame Method." *Journal Proceedings* 68, no. 11 (November 1, 1971): 825–31.

TABLE A.14
Size and pitch of spirals, ACI Code

Diameter of Column, in.	Out to Out of Spiral, in.	f_c', psi		
		3000	4000	5000
f_{yt} = 40,000 psi				
14, 15	11, 12	$\frac{3}{8}-1\frac{3}{4}$	$\frac{1}{2}-2\frac{1}{2}$	$\frac{1}{2}-1\frac{3}{4}$
16	13	$\frac{3}{8}-1\frac{3}{4}$	$\frac{1}{2}-2\frac{1}{2}$	$\frac{1}{2}-2$
17–19	14–16	$\frac{3}{8}-1\frac{3}{4}$	$\frac{1}{2}-2\frac{1}{2}$	$\frac{1}{2}-2$
20–23	17–20	$\frac{3}{8}-1\frac{3}{4}$	$\frac{1}{2}-2\frac{1}{2}$	$\frac{1}{2}-2$
24–30	21–27	$\frac{3}{8}-2$	$\frac{1}{2}-2\frac{1}{2}$	$\frac{1}{2}-2$
f_{yt} = 60,000 psi				
14, 15	11, 12	$\frac{3}{8}-2\frac{3}{4}$	$\frac{3}{8}-2$	$\frac{1}{2}-2\frac{3}{4}$
16–23	13–20	$\frac{3}{8}-2\frac{3}{4}$	$\frac{3}{8}-2$	$\frac{1}{2}-3$
24–29	21–26	$\frac{3}{8}-3$	$\frac{3}{8}-2\frac{1}{4}$	$\frac{1}{2}-3$
30	27	$\frac{3}{8}-3$	$\frac{3}{8}-2\frac{1}{4}$	$\frac{1}{2}-3\frac{1}{4}$

TABLE A.15

Properties of prestressing steels

Seven-Wire Strand, $f_{pu} = 270$ ksi						
Nominal Diameter (in.)	Area in^2	Weight plf	$0.7f_{pu}A_{ps}$ kips	$0.75f_{pu}A_{ps}$ kips	$0.8f_{pu}A_{ps}$ kips	$f_{pu}A_{ps}$ kips
0.375	0.085	0.29	16.1	17.3	18.4	23.0
0.438	0.115	0.39	21.7	23.3	24.8	31.0
0.500	0.153	0.52	28.9	31.0	33.0	41.3
0.520	0.167	0.57	31.5	33.8	36.0	45.0
0.563	0.192	0.65	36.2	38.8	41.4	51.7
0.600	0.217	0.74	41.0	44.0	46.9	58.6
0.700	0.294	1.00	55.6	59.6	63.5	79.4

Prestressing Wire						
Diameter	Area in^2	Weight plf	f_{pu} ksi	$0.7f_{pu}A_{ps}$ kips	$0.8f_{pu}A_{ps}$ kips	$f_{pu}A_{ps}$ kips
0.192	0.0290	0.098	250	5.07	5.80	7.25
0.196	0.0302	0.100	250	5.28	6.04	7.55
0.250	0.0491	0.170	240	8.25	9.43	11.78
0.276	0.0598	0.200	235	9.84	11.24	14.05

Deformed Prestressing Bars, $f_{pu} = 150$ ksi					
Nominal Diameter (in.)	Area in^2	Weight plf	$0.7f_{pu}A_{ps}$ kips	$0.8f_{pu}A_{ps}$ kips	$f_{pu}A_{ps}$ kips
$\frac{5}{8}$	0.28	0.98	29.4	33.6	42.0
$\frac{3}{4}$	0.42	1.49	44.1	50.4	63.0
1	0.85	3.01	89.3	102.0	127.5
$1\frac{1}{4}$	1.25	4.39	131.3	150.0	187.5
$1\frac{3}{8}$	1.58	5.56	165.9	159.6	237.0
$1\frac{3}{4}$	2.58	9.10	270.9	309.6	387.0
$2\frac{1}{2}$	5.16	18.20	541.8	619.2	774.0
3	6.85	24.09	719.3	822.0	1027.5

GRAPH A.1a
Moment capacity of
rectangular sections.
(Graph for educational
purposes only.)

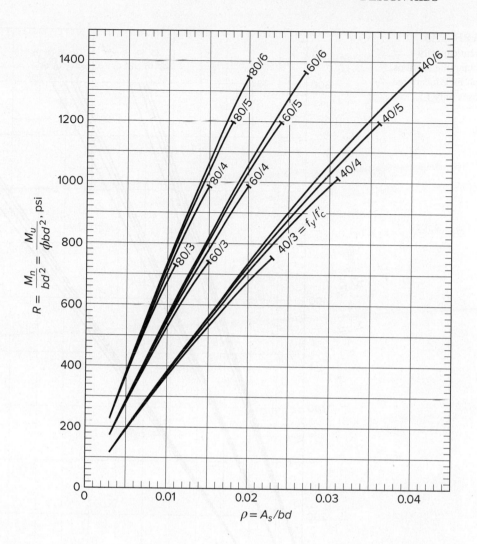

GRAPH A.1b
Moment capacity
of rectangular sections.
(Graph for educational
purposes only.)

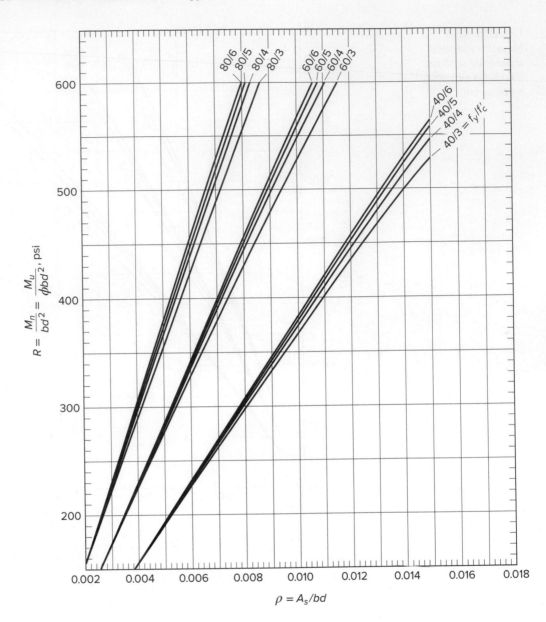

GRAPH A.2
Location of points where
bars can be bent up or cut off
for simply supported beams
uniformly loaded. (Graph for
educational purposes only.)

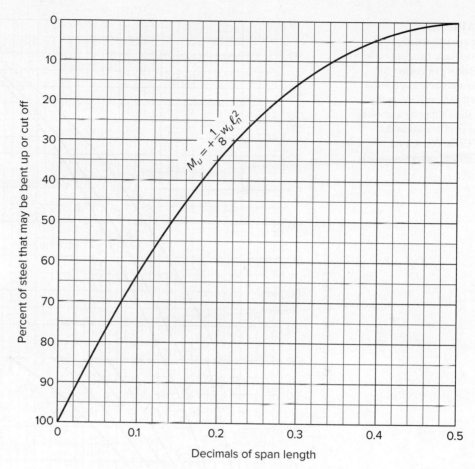

$$M_u = +\frac{1}{8} w_u \ell_n^2$$

Percent of steel that may be bent up or cut off

Decimals of span length

GRAPH A.3

Approximate locations of
points where bars can be bent
up or down or cut off for
continuous beams uniformly
loaded and built integrally
with their supports according
to the coefficients in the
ACI Code. (Graph for
educational purposes only.)

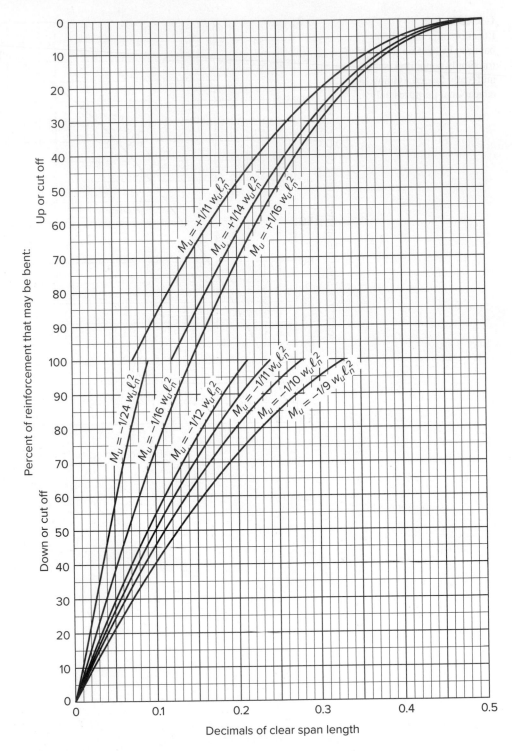

Percent of reinforcement that may be bent:

Up or cut off

Down or cut off

$M_u = +\frac{1}{11} w_u \ell_n^2$

$M_u = +\frac{1}{14} w_u \ell_n^2$

$M_u = +\frac{1}{16} w_u \ell_n^2$

$M_u = -\frac{1}{24} w_u \ell_n^2$

$M_u = -\frac{1}{16} w_u \ell_n^2$

$M_u = -\frac{1}{12} w_u \ell_n^2$

$M_u = -\frac{1}{11} w_u \ell_n^2$

$M_u = -\frac{1}{10} w_u \ell_n^2$

$M_u = -\frac{1}{9} w_u \ell_n^2$

Decimals of clear span length

GRAPH A.4
Interpolation charts for
lateral distribution of slab
moments. (Graph for
educational purposes only.)

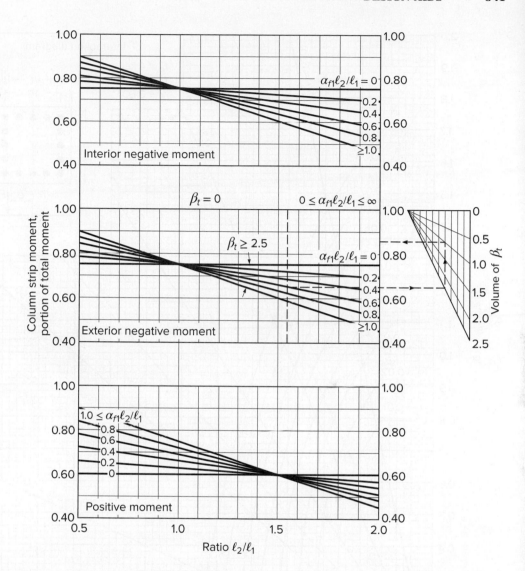

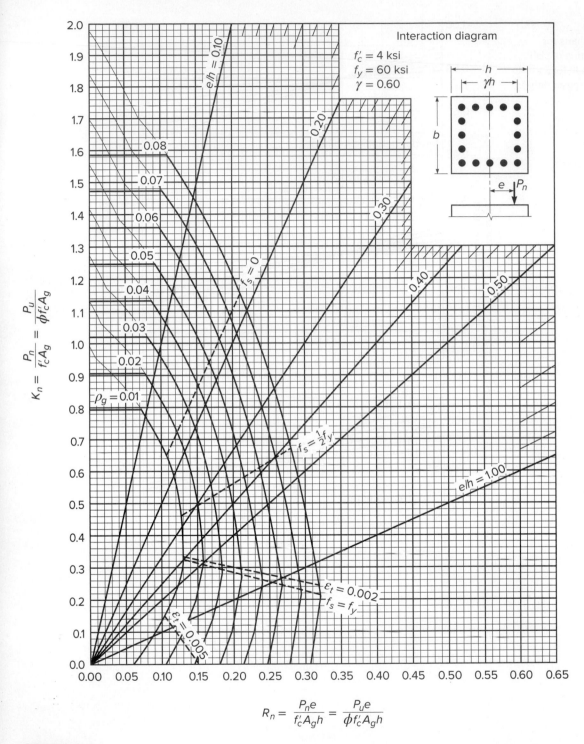

$$R_n = \frac{P_n e}{f'_c A_g h} = \frac{P_u e}{\phi f'_c A_g h}$$

GRAPH A.5

Column strength interaction diagram for rectangular section with bars on four faces and $\gamma = 0.60$. (Graph for educational purposes only.)

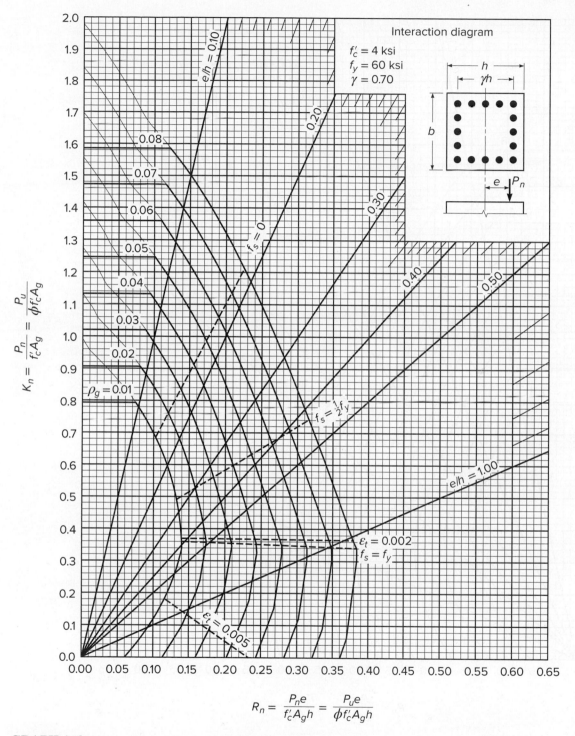

GRAPH A.6
Column strength interaction diagram for rectangular section with bars on four faces and $\gamma = 0.70$. (Graph for educational purposes only.)

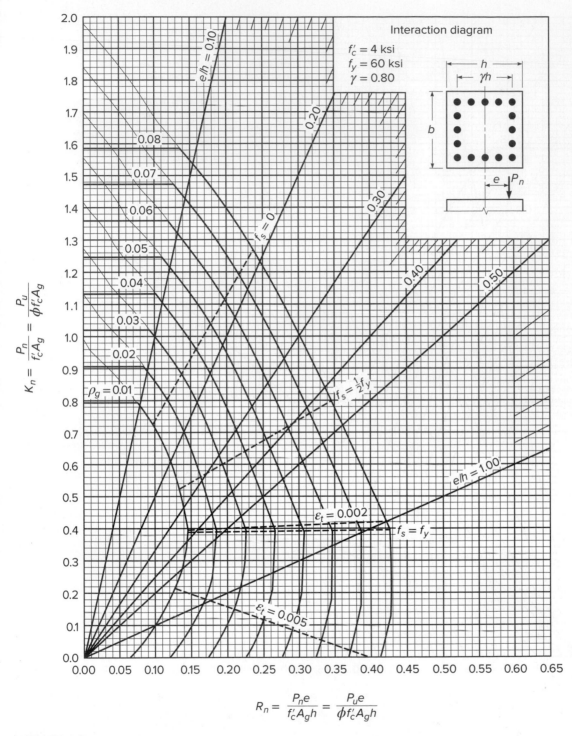

GRAPH A.7
Column strength interaction diagram for rectangular section with bars on four faces and $\gamma = 0.80$. (Graph for educational purposes only.)

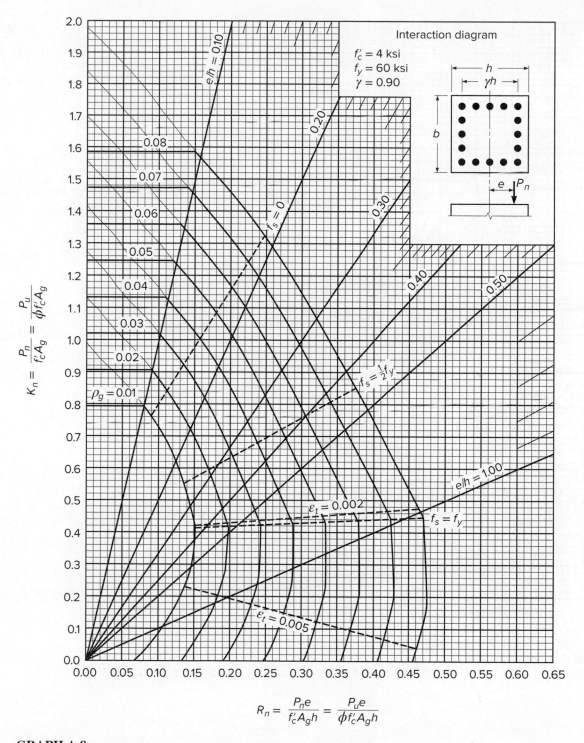

Interaction diagram

$f'_c = 4$ ksi
$f_y = 60$ ksi
$\gamma = 0.90$

$$K_n = \frac{P_n}{f'_c A_g} = \frac{P_u}{\phi f'_c A_g}$$

$$R_n = \frac{P_n e}{f'_c A_g h} = \frac{P_u e}{\phi f'_c A_g h}$$

GRAPH A.8

Column strength interaction diagram for rectangular section with bars on four faces and $\gamma = 0.90$. (Graph for educational purposes only.)

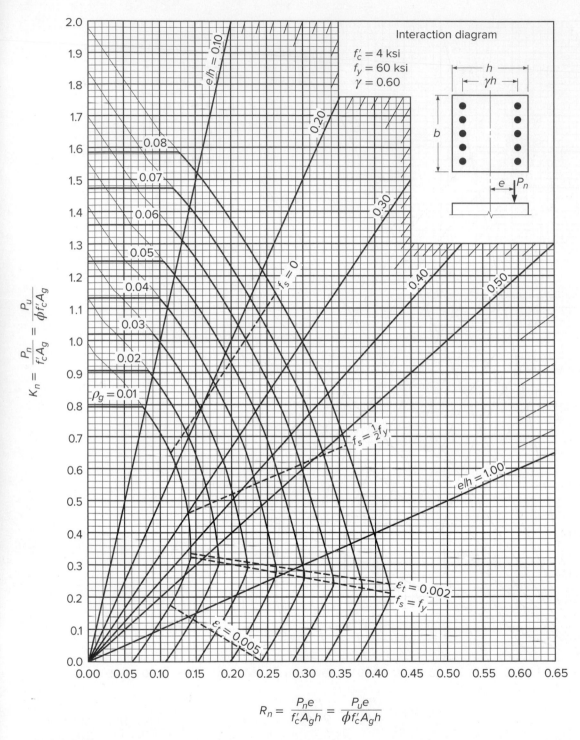

Interaction diagram
$f'_c = 4$ ksi
$f_y = 60$ ksi
$\gamma = 0.60$

$$K_n = \frac{P_n}{f'_c A_g} = \frac{P_u}{\phi f'_c A_g}$$

$$R_n = \frac{P_n e}{f'_c A_g h} = \frac{P_u e}{\phi f'_c A_g h}$$

GRAPH A.9

Column strength interaction diagram for rectangular section with bars on end faces and $\gamma = 0.60$. (Graph for educational purposes only.)

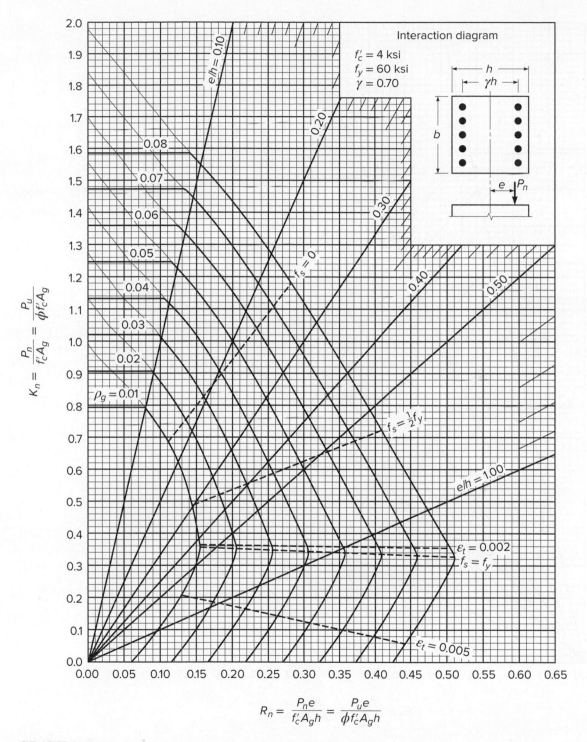

GRAPH A.10

Column strength interaction diagram for rectangular section with bars on end faces and $\gamma = 0.70$. (Graph for educational purposes only.)

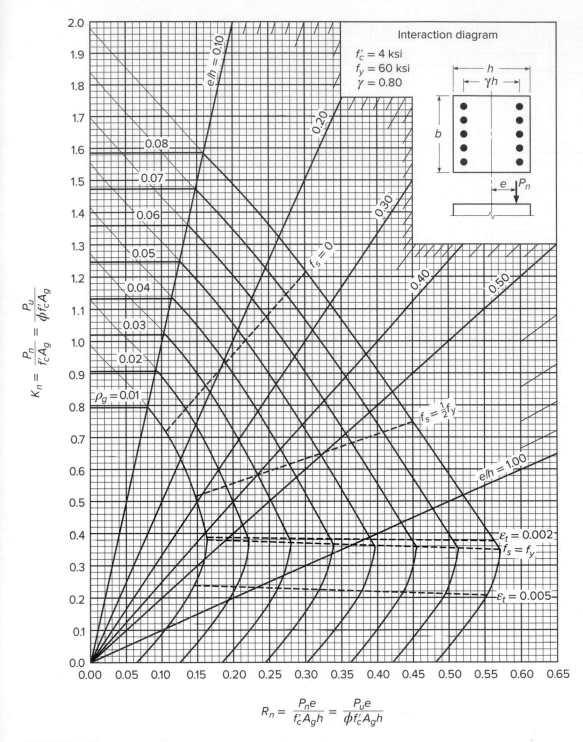

$$R_n = \frac{P_n e}{f'_c A_g h} = \frac{P_u e}{\phi f'_c A_g h}$$

GRAPH A.11

Column strength interaction diagram for rectangular section with bars on end faces and $\gamma = 0.80$. (Graph for educational purposes only.)

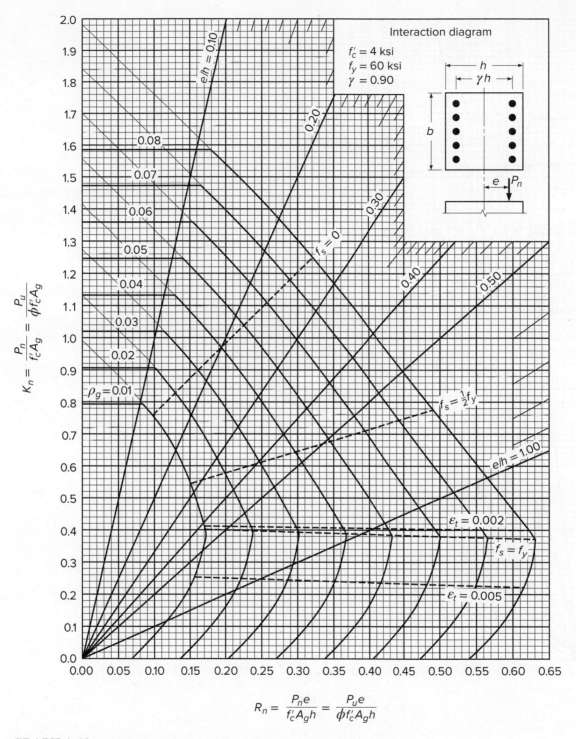

GRAPH A.12

Column strength interaction diagram for rectangular section with bars on end faces and $\gamma = 0.90$. (Graph for educational purposes only.)

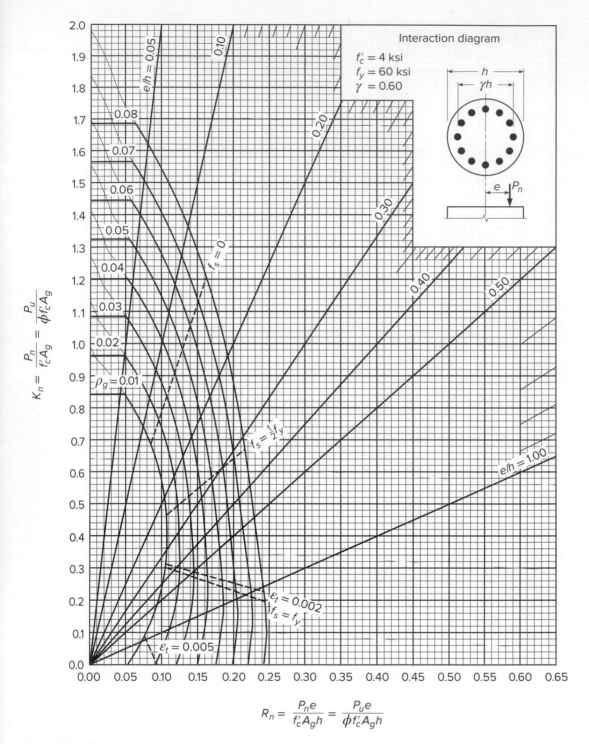

$$R_n = \frac{P_n e}{f_c' A_g h} = \frac{P_u e}{\phi f_c' A_g h}$$

GRAPH A.13
Column strength interaction diagram for circular section with $\gamma = 0.60$. (Graph for educational purposes only.)

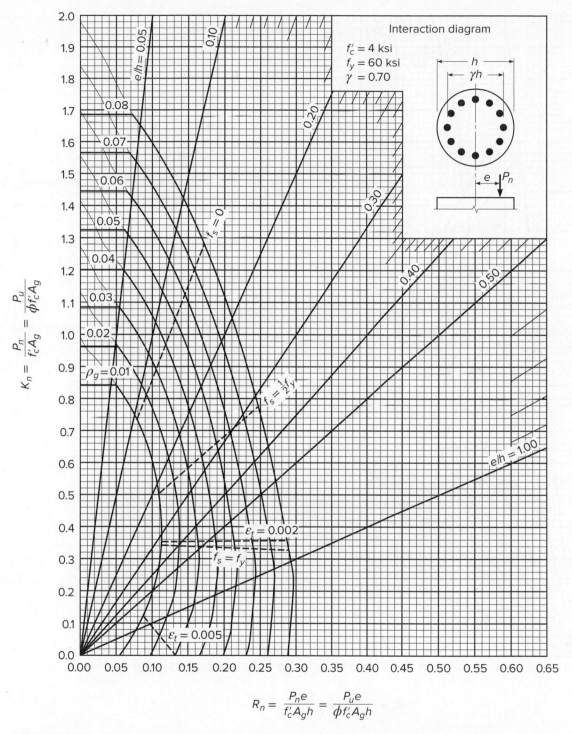

$$K_n = \frac{P_n}{f'_c A_g} = \frac{P_u}{\phi f'_c A_g}$$

Interaction diagram

f'_c = 4 ksi
f_y = 60 ksi
γ = 0.70

e/h = 0.05
0.10
0.08
0.07
0.06
0.05
0.04
0.03
0.02
ρ_g = 0.01
0.20
0.30
0.40
0.50
$f_s = 0$
$f_s = \frac{1}{2} f_y$
e/h = 1.00
ε_t = 0.002
$f_s = f_y$
ε_t = 0.005

$$R_n = \frac{P_n e}{f'_c A_g h} = \frac{P_u e}{\phi f'_c A_g h}$$

GRAPH A.14
Column strength interaction diagram for circular section with $\gamma = 0.70$. (Graph for educational purposes only.)

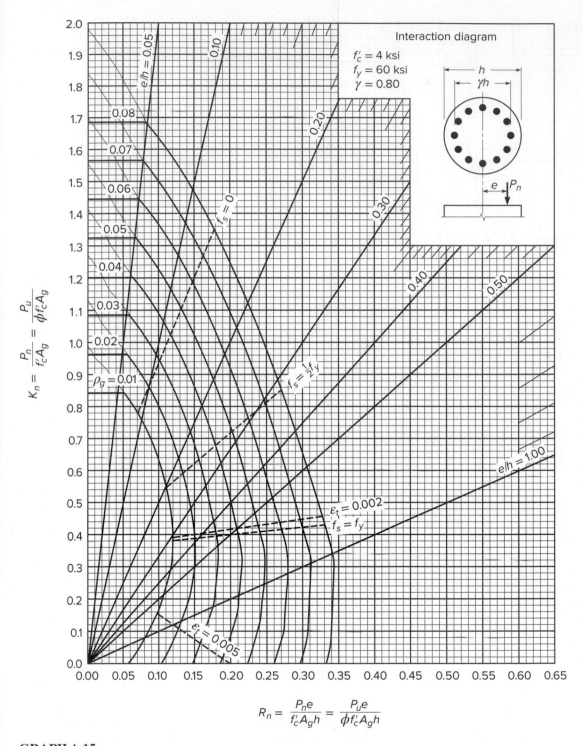

Interaction diagram

$f'_c = 4$ ksi
$f_y = 60$ ksi
$\gamma = 0.80$

$$K_n = \frac{P_n}{f'_c A_g} = \frac{P_u}{\phi f'_c A_g}$$

$$R_n = \frac{P_n e}{f'_c A_g h} = \frac{P_u e}{\phi f'_c A_g h}$$

GRAPH A.15

Column strength interaction diagram for circular section with $\gamma = 0.80$. (Graph for educational purposes only.)

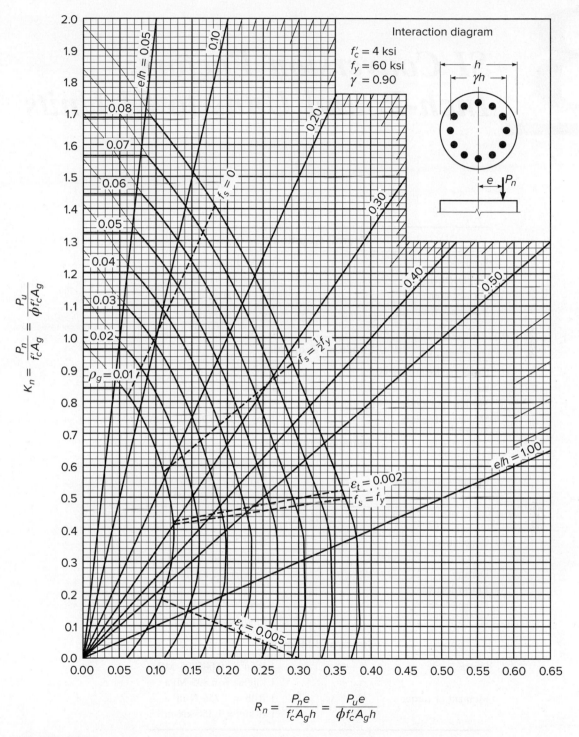

GRAPH A.16
Column strength interaction diagram for circular section with $\gamma = 0.90$. (Graph for educational purposes only.)

B

SI Conversion Factors: Inch-Pound Units to SI Units

Overall Geometry	
Spans	1 ft = 0.3048 m
Displacements	1 in. = 25.4 mm
Surface area	1 ft^2 = 0.0929 m^2
Volume	1 ft^3 = 0.0283 m^3
	1 yd^3 = 0.765 m^3

Structural Properties	
Cross-sectional dimensions	1 in. = 25.4 mm
Area	1 in^2 = 645.2 mm^2
Section modulus	1 in^3 = 16.39 × 10^3 mm^3
Moment of inertia	1 in^4 = 0.4162 × 10^6 mm^4

Material Properties	
Density	1 lb/ft^3 = 16.03 kg/m^3
Modulus and stress	1 lb/in^2 = 0.006895 MPa
	1 kip/in^2 = 6.895 MPa

Loadings	
Concentrated loads	1 lb = 4.448 N
	1 kip = 4.448 kN
Density	1 lb/ft^3 = 0.1571 kN/m^3
Linear loads	1 kip/ft = 14.59 kN/m
Surface loads	1 lb/ft^2 = 0.0479 kN/m^2
	1 kip/ft^2 = 47.9 kN/m^2

Stress and Moments	
Stress	1 lb/in^2 = 0.006895 MPa
	1 kip/in^2 = 6.895 MPa
Moment or torque	1 ft-lb = 1.356 N-m
	1 ft-kip = 1.356 kN-m

Author Index

Subject Index